HANDBOOK OF SUSTAINABLE POLYMERS

HANDBOOK OF SUSTAINABLE POLYMERS

PROCESSING AND APPLICATIONS

edited by
Vijay Kumar Thakur
Manju Kumari Thakur

CRC Press
Taylor & Francis Group
Boca Raton London New York

CRC Press is an imprint of the
Taylor & Francis Group, an **informa** business

Handbook of Sustainable Polymers: Processing and Applications

First published 2016 by Pan Stanford Publishing Pte. Ltd.

Published 2023 by CRC Press
Taylor & Francis Group
6000 Broken Sound Parkway NW, Suite 300
Boca Raton, FL 33487-2742

ISBN 13: 978-981-4613-53-8 (hbk)

Visit the Taylor & Francis Web site at
http://www.taylorandfrancis.com

and the CRC Press Web site at
http://www.crcpress.com

British Library Cataloguing-in-Publication Data
A catalogue record for this book is available from the British Library.

To my parents and teachers, who helped me become what I am today.

Vijay Kumar Thakur

Contents

Preface xvii

1. Sustainable Polymers and Applications **1**

Rita Mahanta and Ranadeep Mahanta

1.1 Introduction 1

1.2 Basics of Polymers 2

1.2.1 Polymer Families 3

1.2.1.1 Synthetic polymers 3

1.2.1.2 Natural polymers 5

1.2.2.1 Addition polymerization 6

1.2.2.2 Condensation polymerization 7

1.3 Classification 9

1.4 Introduction to Polysaccharides 11

1.4.1 Examples of Shorthand Notation for Polysaccharides 12

1.4.2 Polysaccharide Classification 12

1.4.3 Scheme of Polysaccharide Classification 13

1.4.3.1 General scheme for classification of polysaccharides 13

1.4.4 Sources of Polysaccharides and Related Classification 13

1.4.5 Structural Classification of Polysaccharides 14

1.4.6 Classification of Polysaccharides on Monomer Composition 15

1.4.7 Polysaccharide Classification by Charge 15

1.4.8 Characterization of Polysaccharides 15

1.4.9 Characterization of Starch 16

1.4.10 Characterization of Locust Bean Gum 18

1.4.11 Characterization of Tara Gum 18

1.4.12 Characterization of Konjac Flour 18
1.4.13 Characterization of Gum Arabic 19
1.4.14 Characterization of Gum Tragacanth 20
1.4.15 Characterization of Gum Karaya 20
1.4.16 Characterization of Carrageenan 21
1.4.17 Characterization of Agar 22
1.4.18 Characterization of Alginate 22
1.4.19 Characterization of Pectin 23
1.4.20 Characterization of Xanthan Gum 24
1.4.21 Characterization of Gellan Gum 24
1.4.22 Characterization of Cellulose 25
1.4.23 Characterization of Chitin 26
1.5 Proteins and Polypeptides 27
1.6 Polynucleotides 30
1.6.1 Polynucleotide Formation 32
1.6.2 Types and Character of Polynucleotides 34
1.7 Synthesis of DNA 37
1.8 Fundamentals of DNA Replication 38
1.9 RNA Synthesis 40
1.10 Applications of Polymers 42
1.11 Synthetic Water-Soluble Polymers 42
1.11.1 Polyethylene Glycol 42
1.11.2 Polyvinyl Pyrrolidone 43
1.11.3 Polyvinyl Alcohol 45
1.11.4 Polyacrylic Acid 45
1.11.5 Polyacrylamide 45
1.11.6 Divinyl Ether-Maleic Anhydride 46
1.11.7 Polyoxazoline 47
1.11.8 Polyphosphates 47
1.11.9 Polyphosphazenes 48
1.12 Natural Water-Soluble Polymers 48
1.12.1 Xanthan Gum 48
1.12.2 Pectins 49
1.12.3 Chitosan Derivatives 51

1.12.4 Dextran 51
1.12.5 Carrageenan 52
1.12.6 Guar Gum 53
1.12.7 Cellulose Ethers 53
1.12.8 Starch or Starch-Based Derivatives 55
1.12.9 Albumin 56

2. Tara Gum, the New Biopolymer for Various Applications: Hydrodynamic Properties 59

Martin Alberto Masuelli

2.1 Introduction 59
2.1.1 Botanical Aspect 60
2.1.2 Obtension Process 61
2.1.2.1 Food grade 62
2.1.2.2 Industrial grade 63
2.1.2.3 Pharmaceutical biopolymer 63
2.1.2.4 Food hydrocolloid 65
2.2 Mineral-Petroleum Applications 69
2.2.1 Mineral Application 69
2.2.2 Petroleum Applications 72
2.3 Experimental 73
2.3.1 Sample Preparation 73
2.3.2 Capillary Viscometry 74
2.3.3 Density Measurement 74
2.4 Physics Chemistry Properties 74
2.4.1 Viscosity 74
2.4.2 Density 75
2.4.3 Intrinsic Viscosity 78
2.4.4 Huggins' and Kraemer's Methods 78
2.4.5 Hydrodynamic Properties: Molecular Weight and Mark-Houwink Parameters 79
2.4.6 Hydration Values 81
2.4.6.1 Perrin number 83
2.4.6.2 Einstein viscosity increment 84

3. High-Value Utilization of Natural Cellulose: Cellulose-Based Biocomposite Materials 99

Yong Zhang and Juming Yao

3.1 Introduction 99
3.2 Methods of Cellulose Graft Copolymerization 100
3.2.1 Chemical Methods 101
3.2.2 Radiation-Induced Grafting 102
3.2.2.1 Low-energy radiation-induced grafting 102
3.2.2.2 High-energy radiation-induced grafting 102
3.2.2.3 Microwave radiation-induced grafting 103
3.2.3 Plasma-Initiated Grafting 103
3.3 Water Treatment Application 104
3.3.1 Modified Cellulose 104
3.3.1.1 Anionic cellulose 104
3.3.1.2 Cationic cellulose 105
3.3.2 Cellulose-Based Flocculant 107
3.3.2.1 Nonionic cellulose-based flocculant 107
3.3.2.2 Anionic cellulose-based flocculant 107
3.4 Biomedical Material Application 108
3.4.1 Wound Closure 108
3.4.2 Drug Delivery System 109
3.4.3 Vascular Grafts 110
3.4.4 Scaffold 112
3.5 Eco-Friendly Superabsorbent Polymer 113
3.5.1 Superabsorbent Polymer in Agricultural Application 113
3.5.2 Superabsorbent Polymer in Hygiene Application 118
3.6 Nanocrystalline Cellulose 119
3.6.1 Introduction to Nanocrystalline Cellulose 120

3.6.2 Nanocrystalline Cellulose as Reinforce Phase 123
3.7 Summary 126

4. Bioprocessing of Natural Plant Cellulosic Polymer Materials 141

C. Vigneswaran, P. Kandhavadivu, and M. Ananthasubramanian

4.1 Introduction 141
4.2 Biotechnology: Scope and Importance 142
4.3 Enzyme Structures and Mechanisms 143
4.3.1 Enzyme Kinetics and Their Reactions 144
4.3.2 Applications of Enzymes in Textile Wet Processing 145
4.4 Cotton Morphology and Chemistry 146
4.5 Textile Processing of Cotton Materials 148
4.6 Desizing of Cotton Fabric 149
4.6.1 Alpha Amylase Enzyme 150
4.6.2 Enzymatic Desizing of Cotton Fabric 151
4.6.3 Assessment of Enzyme Desizing 152
4.7 Scouring of Cotton Fabrics 152
4.7.1 Drawbacks Associated with Alkaline Scouring 153
4.7.2 Overview of Enzymatic Scouring 154
4.7.3 Pectinases in Bioscouring of Cotton 155
4.7.4 Pectin 156
4.7.5 Pectinase Treatments and Evaluation Methods 157
4.7.6 Lipases in Processing of Cotton Fabrics 157
4.7.7 Applications of Lipases in Cotton Processing 157
4.8 Proteases in Cotton Fabric Processing 158
4.8.1 Fabric Weight Loss and Strength Loss 159
4.8.2 Fabric Absorbency, Wettability, and Dyeability 159
4.8.3 Effects of Process Parameters on Bioscouring 160

4.9 Cellulases on Cotton Fabrics 160
4.10 Bleaching of Cotton Fabrics 161
4.10.1 Enzymatic Bleaching 162
4.11 Mixed Enzymatic Process in Scouring of Cotton Fabrics 162
4.11.1 Binary Combinations of Enzymes 162
4.11.2 Ternary and Quadruplet Combinations of Enzymes 164
4.12 Ultrasonic Treatment 165
4.12.1 Sonication: Basic Principle 165
4.12.2 Ultrasonic Application: Textile Wet Processing 166
4.13 Aerodynamic System 168
4.13.1 Effect of Air Pressure on Enzyme Activity 168
4.14 Summary 169

5. Biodegradable Polymers: Definition, Classification and Application as an Alternative to Plastic 181

Neha Patni and Shibu G. Pillai

5.1 Introduction 181
5.2 Plastic Spectrum 183
5.3 Biodegradable Polymer 185
5.4 Classification of Plastics 187
5.5 Classification of Polymers 188
5.5.1 On the Basis of Degradability 188
5.5.2 From BioMass 188
5.5.2.1 From polysaccharides 188
5.5.2.2 Cellulose 189
5.5.2.3 Starch 189
5.5.2.4 Chitin and chitosan 190
5.5.2.5 Pectin 190
5.5.2.6 Proteins 190
5.5.2.7 Soybean proteins 191
5.5.2.8 Corn proteins 191

5.5.2.9 Wheat gluten and casein 191
5.5.2.10 Gelatin 191
5.5.3 Polymers Obtained by Microbial Production 192
5.5.3.1 Polyhydroxyalkanoates 192
5.5.4 Conventionally and Chemically Synthesized Biodegradable Polymers with Monomers Obtained from Agro-Resource 192
5.5.4.1 Polylactic acid 192
5.5.4.2 Petroleum-based polyesters 193
5.5.4.3 Polycaprolactone 193
5.5.4.4 Aliphatic copolyesters 194
5.5.4.5 Aromatic copolyesters 194
5.5.4.6 Polyvinyl alcohol 194
5.6 Mechanism of Biodegradability of Polymers 194
5.6.1 Thermal Depolymerization 195
5.6.2 Photolytic Oxidation 195
5.6.3 Moisture Absorption 197
5.6.4 Microbial Action 197
5.6.4.1 Fungi 199
5.6.4.2 Bacteria 199
5.6.4.3 Enzymes 199
5.7 Tests for Biodegradable Polymers 200
5.7.1 Testing Methods 201
5.8 Current Trend of Polymer Usage 202
5.9 Biodegradable Polymers as an Alternative to Plastic 203
5.9.1 Biodegradable Polymer from Casein 203
5.9.1.1 Why casein aerogel composites? 203
5.9.1.2 Presence of casein in milk 204
5.9.1.3 Casein structure 204
5.9.1.4 Extraction of casein from milk 205
5.9.1.5 Biodegradability of casein polymer composite 205

5.9.1.6 Applications of biodegradable polymer 205
5.9.2 Polymer from Wheat Gluten 206
5.9.2.1 Conversion of WG into Composites 207
5.9.2.2 Biodegradability of WG 208
5.10 Summary and Future Perspectives 208

6. Potential Applications of Sustainable Polymers in Functionalization of Cellulosic Textile Materials 215

Nabil A. Ibrahim and Basma M. Eid

6.1 Introduction 215
6.2 Cellulose 216
6.2.1 Cellulose Fibers 216
6.2.1.1 Pretreatment processes 217
6.2.1.2 Environmental concerns 218
6.2.2 Coloration Process 220
6.2.2.1 Environmental Concerns 220
6.2.3 Finishing Process 221
6.3 Biopolymers Classifications and Applications for Functional Cellulose Textiles 222
6.3.1 Sustainable Biopolymers 222
6.3.2 Applications of Sustainable Biopolymers in Functionalization of Cellulosic Textiles 223
6.3.2.1 Options for enhancing functionalization of textiles 225
6.3.3 Antimicrobial Functional Finishes 225
6.3.3.1 Chitosan and chitosan derivatives 227
6.3.3.2 Application methods 227
6.3.3.3 Sericin protein 227
6.3.3.4 Cyclodextrins 236
6.3.3.5 Na-Alginate 239
6.3.3.6 Aloe vera gel 240
6.3.4 UV Protection 240
6.3.5 Flame Retardancy 244

6.3.6 Aroma Finishing 247
6.3.7 Other Functional Properties 248
6.4 Evaluation of Functional Finishes 249
6.5 Future Scope 252

7. Research Progress in Polymer Complexes and Their Applications 265

Altaf H. Basta and Houssni El Saied

7.1 Introduction 265
7.2 Structures of Cellulose Derivative–Metal Complexes 266
7.3 Functional Properties of Cellulose Derivative–Metal Complexes 272
7.3.1 Thermal Stability 272
7.3.2 Electrical Properties 278
7.3.3 Electrical Conductivity 281
7.4 Applications of Cellulose Derivative–Metal Complexes for Paper Production 285
7.4.1 Application of Cellulose Ether–Fe(III) Complexes 287
7.4.2 Application of Carboxymethyl Cellulose–Cu(II) Complexes 293
7.4.3 Evaluation of CMC–Cu(II) Complexes in the Production of Durable Paper Sheets 299
7.5 Conclusions 301

8. Hemicellulose: Isolation and Its Application in Pharmacy 305

Muchlisyam, Jansen Silalahi, and Urip Harahap

8.1 Introduction of Hemicellulose 305
8.2 Sources and Structure of Hemicellulose 308
8.2.1 Structure and Chemical Composition 311
8.2.2 Isolation of Hemicellulose 312
8.3 Characterization of Corn Cobs Hemicelluloses 315
8.3.1 Characterization of Hemicellulose by High-Performance Liquid Chromatography 315

8.3.2 Characterization of Corn Cobs Hemicellulose Using Infrared Spectrophotometry 316
8.4 Chemical Structure and Properties of Hemicelluloses 318
8.5 Application of Corn Cobs Hemicellulose as Chelating Agent for Toxic Metal 320
8.5.1 Test to Corn Cobs Hemicellulose Power Strap Respect for Cadmium Ion and Plumbum Ions with Titrimetric in the pH Bases 321
8.5.2 The Identification of Cadmium Corn Cobs Hemicellulose Bond and Plumbum Corn Cobs Hemicellulose Bond with Color Reaction and FTIR Vibration 323
8.5.2.1 Identification with the color reaction 323
8.5.2.2 Identification by FTIR 323
8.6 Cadmium and Plumbum Absorption 325
8.7 The in vivo Corn Cobs Hemicellulose Testing in Rabbits 329
8.7.1 Determination of Cadmium and Plumbum Levels in Blood of Rabbits 330
8.7.2 Binding Ability Corn Cobs Hemicellulose toward Cadmium in vivo 331
8.8 Conclusion 334

9. Optical Anisotropy of Cellulose Esters and Its Application to Optical Functional Films 341

Kultida Songsurang, Hikaru Shimada, Shogo Nobukawa, and Masayuki Yamaguchi

9.1 Optical Properties of Polymeric Materials 342
9.2 Wavelength Dispersion of Orientation Birefringence 347
9.3 Optical Anisotropy of Solution-Cast Cellulose Ester Films 350
9.3.1 In-Plane Birefringence of Cellulose Esters 350

9.3.2 Out-of-Plane Birefringence of Cellulose Esters 355

9.4 Optical Properties of Solution-Cast Cellulose Triacetate Films 357

9.4.1 Effect of Film Thickness 357

9.4.2 Effect of Evaporation Rate 363

9.4.3 Effect of Solvent Type 364

9.4.4 Effect of Plasticizer 366

9.4.5 Effect of Hot-Stretching 368

9.4.6 Effect of Crystallization 371

9.5 Advanced Method to Control Orientation Anisotropy 371

9.6 Conclusions 378

10. Mechanical Performance of *Eulaliopsis binata* Biofiber–Based Green Composites 385

Manju Kumari Thakur and Vijay Kumar Thakur

10.1 Introduction 385

10.2 Materials and Methods 386

10.2.1 Materials 386

10.2.2 Preparation of Polymer Composites 387

10.2.3 Characterization of Green Polymer Composites 387

10.3 Results and Discussion 387

10.3.1 Morphological and Thermal Study of the Green Composites 391

10.4 Conclusions 392

11. Development and Applications of Sustainable Polylactic Acid Parts 397

Khosrow Khodabakhshi and Morteza Ehsani

11.1 Introduction 397

11.2 Lactide 398

11.2.1 Production 398

11.2.1.1 Prepolymerization 399

11.2.1.2 Lactide synthesis 399

11.2.1.3 Purification of lactide 400
11.2.2 Properties 401
11.3 PLA Processing and Polymerization 402
11.3.1 Properties Related to PLA Extrusion 402
11.3.2 Polymerization Methods 404
11.3.2.1 Direct polycondensation 404
11.3.2.2 Conventional production via ring opening polymerization 405
11.3.3 Reactive Extrusion 406
11.4 PLA Nanocomposites 415
11.4.1 General Modifications 415
11.4.2 Biodegradability 416
11.4.3 Melt Rheology 417
11.5 Polylactic Acid Properties 417
11.5.1 Chemical Structure 417
11.5.2 Thermal Properties 418
11.5.3 Optical Properties 422
11.5.4 Degradation 424
11.5.5 Rheology 425
11.5.6 Mechanical Properties 426
11.5.7 Permeability, Sorption, and Diffusion 433
11.6 Applications 434
11.6.1 Medical 434
11.6.2 Packaging 436
11.6.3 Fibers 437
11.7 Conclusions 438

12. Pectin and Its Roles in Transdermal Drug Delivery 453

Anuar Nor Khaizan and Tin Wui Wong

12.1 Introduction 453
12.2 Pectin in Drug Delivery 454
12.3 Pectin in Transdermal Drug Delivery 458
12.3.1 Permeation Enhancement Mechanism 463
12.4 Conclusion 465

13. Development in Tunicate Cellulose **473**

Guangzhe Piao and Dawei Zhang

13.1 Lyotropic N*-LCs of Tunicate Cellulose Nanocrystals 473
13.1.1 Introduction 473
13.1.2 Experimental 476
13.1.2.1 Preparation of T-CN suspension 476
13.1.2.2 Ultrasonic process 477
13.1.3 Results and Discussion 477
13.1.3.1 N*-LCs of T-CNs 477
13.1.3.2 Effects of sonication on T-CN suspension 479
13.1.3.3 N*-LC film of T-CNs 481
13.2 T-CN Nanocomposites 483
13.2.1 Introduction 483
13.2.2 Experimental 484
13.2.3 Results and Discussion 485
13.2.3.1 Mechanical property 485
13.2.3.2 FT-IR analysis 486
13.2.3.3 SEM analysis 487
13.2.3.4 TEM analysis 488
13.3 Summary 489

14. Synthesis and Characterization of Bacterial Cellulose-Based Composites and Their Applications **495**

Muenduen Phisalaphong and Suchata Kirdponpattara

14.1 Introduction 495
14.2 Synthesis of BC-Based Composites 497
14.2.1 Biosynthesis 498
14.2.2 Impregnation 499
14.2.3 Blending 500
14.3 BC-Based Composites 500
14.3.1 BC-Alginate 500
14.3.1.1 Alginate 500

14.3.1.2 Characterization 501
14.3.1.3 Application 504
14.3.2 BC–Gelatin 505
14.3.2.1 Gelatin 505
14.3.2.2 Characterization 506
14.3.2.3 Application 509
14.3.3 BC–Aloe vera 510
14.3.3.1 Aloe vera 510
14.3.3.2 Characterization 511
14.3.3.3 Application 513
14.3.4 BC-Chitosan 513
14.3.4.1 Chitosan 513
14.3.4.2 Characterization 515
14.3.4.3 Application 518
14.3.5 BC–PVA 520
14.3.5.1 PVA 520
14.3.5.2 Characterization 521
14.3.5.3 Application 523
14.4 Conclusion 523

15. Chitosan Polymer as Novel Drug Carrier 533

Aakanchha Jain, Sourabh Jain, Vikas Jain, and Dharmveer Kohli

15.1 Chitosan as Polymer 533
15.1.1 Grades of Chitosan and Their Solubilities 534
15.1.2 Low-Molecular-Weight Chitosan/Chito-Oligomers 535
15.1.2.1 Applications 537
15.1.3 Structural Unit of Chitosan: D-Glucosamine 537
15.2 Chitosan Conjugates 538
15.3 Derivatives of Chitosan 539
15.4 Chitosan-Based Novel Drug Delivery Systems 542

15.4.1 Chitosan as Novel Drug Carrier 542
15.4.2 Chitosan DNA Complex 544
15.4.3 Chitosan as Biomaterial 545
15.4.3.1 Beads of chitosan/ microspheres: 545
15.4.3.2 Chitosan sponge 545
15.4.3.3 Fibers 546
15.5 Toxicity of Chitosan Nanoparticles 546
15.6 Electrokinetic and Stability Behavior of Chitosan 546
15.7 Aspects of Chitosan Uncovered in This Chapter 547
15.8 Conclusion 547

16. Chitosan and Its Roles in Transdermal Drug Delivery 557

Asif Nawaz, Nauman Rahim Khan, and Tin Wui Wong

16.1 Introduction 557
16.2 Stratum Corneum 559
16.3 Chitosan 560
16.4 Transdermal Drug Delivery 564
16.4.1 Chitosan as Transdermal Permeation Enhancer 564
16.4.2 Chitosan-Based Transdermal Drug Delivery Systems 565
16.4.2.1 Hydrogel 565
16.4.2.2 Film/membrane 566
16.4.2.3 Nanoparticle 567
16.4.2.4 Microparticle 569
16.4.2.5 Nanocapsule/microcapsule 570
16.4.2.6 Liposome 571
16.4.2.7 Polymersome 571
16.4.3 Chitosan-Based Transdermal Gene Delivery 572
16.4.4 Chitosan Skin Toxicity 573
16.5 Conclusion 574

17. Chiral Electrochemical Sensors Based on Molecularly Imprinted Polymers with Pharmaceutical Applications 587

Bogdan-Cezar Iacob, Ede Bodoki, and Radu Oprean

17.1 Introduction 587
17.2 Molecular Imprinting Approach 588
17.3 Chiral Electrochemical Sensors 593
17.4 Chiral Electrochemical MIP-Based Sensors 594
17.4.1 MIP Anchoring Onto the Transducer 595
17.4.2 Chiral MIP Applications 601
17.4.2.1 Amino acids 601
17.4.2.2 Monosaccharides 606
17.5 Conclusions and Future Perspectives 608

18. Molecularly Imprinted Polymers: A Versatile Tool in Pharmaceutical Applications 615

Sushma Pardeshi, Rita Dhodapkar, and Anupama Kumar

18.1 Introduction 615
18.1.1 Molecular Imprinting and Molecularly Imprinted Polymers 616
18.1.1.1 Molecular imprinting strategies 617
18.1.1.2 Characteristic of MIPs 619
18.1.2 Progress in MIP Technology 619
18.2 Computational Modeling for Rational Design of MIPs for Pharmaceutical Templates 620
18.2.1 An Overview of Computational Methods Used in Rational Design of MIPs 620
18.2.1.1 Molecular mechanics and molecular dynamics simulations 621
18.2.1.2 Quantum chemical methods 621
18.2.2 Computational Modeling for the Rational MIP Design 622
18.2.2.1 Rational selection of monomers and porogen and characterization

of template–monomer prepolymerization complex 623
18.2.2.2 Optimization of MIP performance through molecular mechanics and molecular dynamics simulation 625
18.2.2.3 Optimization of MIP performance through quantum chemical methods 626
18.2.2.4 Computational modeling: A tool for study of thermodynamic and spectroscopic properties of MIP 628
18.3 Polymerization Methods for Synthesis of MIPs for Pharmaceutical Applications 632
18.3.1 Bulk Polymerization Method 632
18.3.2 Precipitation Polymerization 634
18.3.3 Suspension Polymerization 637
18.3.4 Reverse Micro-Emulsion Polymerization 638
18.4 Application of MIPs in Pharmaceutical Industries 640
18.4.1 Molecularly Imprinted Solid Phase Extraction of Pharmaceuticals 640
18.4.1.1 MISPE procedure 641
18.4.2 MIPs as the Stationary Phase for Analytical Racemic Separations 643
18.4.3 MIPs as Drug Delivery Devices 644
18.5 Conclusions 646

19. Chitosan and Nano Chitosan: Properties and Application to Textiles 659

D. P. Chattopadhyay and M. S. Inamdar

19.1 Introduction 659
19.2 Chemistry of Chitin and Chitosan 661
19.3 Production of Chitin and Chitosan 663

19.4 Physical Properties of Chitosan 666
19.4.1 Degree of Deacetylation 666
19.4.2 Solubility 667
19.4.3 Molecular Weight 668
19.4.4 Viscosity Behavior 673
19.4.5 Chitosan Hydrogel 678
19.4.6 Chitosan Membrane 679
19.5 Chemical Properties of Chitosan 680
19.6 Applications of Chitosan 680
19.6.1 Chitosan in Agriculture 681
19.6.2 Chitosan in Food Technology 681
19.6.3 Chitosan in Cosmetics 682
19.6.4 Chitosan in Biomedicine 682
19.6.5 Chitosan in Paper Industry 683
19.6.6 Chitosan in Chromatography 683
19.6.7 Chitosan in Solid-State Batteries 683
19.6.8 Chitosan in Biocatalysis 684
19.6.9 Chitosan in Textiles 684
19.6.9.1 Chitosan fibers 684
19.6.9.2 Application of chitosan on cotton fabric 686
19.7 Nano Chitosan Applications to Textiles 708
19.7.1 Nano Concept 708
19.7.2 Synthesis and Characterization Nano Chitosan 711
19.7.3 Effect of Nano Chitosan Treatment on Cotton Fabric 720
19.8 Chitosan in Water Processing Used in Textile Industry 726
19.9 Conclusions 726

20. Silver Nanoparticle–Incorporated Hydrogels: Synthesis and Antibacterial Applications 743

P. Souda and Lisa Sreejith

20.1 Introduction 743

20.2 Ex situ Approach for the Synthesis of Silver/Hydrogel Composite 744
20.3 In situ Approach for the Synthesis of Silver/Hydrogel Composite 745
20.4 Swelling Studies of the Gel 746
20.5 Action of Silver Nanoparticle/Hydrogelon Microbes 747
20.6 Identification of Silver Particles in the Hydrogel/Metal Nanocomposite 748

21. Biodegradable Polyesters: From Monomer to Application **753**

Geert-Jan Graulus, Thomas Billiet, Sandra Van Vlierberghe, Hugo Thienpont, Heidi Ottevaere, and Peter Dubruel

21.1 Introduction 753
21.2 Polyesters 755
21.2.1 Polycondensation 755
21.2.2 Ring-Opening Polymerisations of Lactones 757
21.2.2.1 Anionic ring-opening polymerisation 757
21.2.2.2 Metal-insertion ring-opening polymerisation 759
21.2.2.3 Cationic ring-opening polymerisation 760
21.2.2.4 Organo-catalytic polymerisation 762
21.2.3 Ring-Opening Polymerisation of Cyclic Ketene Acetals 764
21.2.4 Enzymatic Polyester Synthesis 765
21.2.5 Other Synthesis Routes 766
21.3 Biodegradation 767
21.4 Processing of Polyesters 772
21.4.1 Production of Polymer Fibres 773
21.4.2 Production of Films and Sheets 775
21.4.3 Production of Three-Dimensional Objects 776
21.4.4 Additive Manufacturing 778

21.5 Applications 778
21.5.1 The Role of Biodegradability in the Biomedical Field 779
21.5.2 Biodegradable Polymers and Photonics 782
21.5.2.1 Photonics in polymer chemistry 782
21.5.2.2 Degradable polymers in photonic applications 784
21.5.3 Biodegradation as Emerging Trend in Packaging 785
21.5.3.1 Commodity packaging 786
21.5.3.2 Food packaging 787
21.5.4 Agricultural and Veterinary Applications 787
21.5.5 Other Applications 788
21.6 General Conclusions and Future Prospects 789

22. Synthetic–Natural Hybrid Polymers Based on Polyurethane: Structures, Properties, and Applications **803**

Veronica Saucedo-Rivalcoba, Ana Laura Martínez-Hernández, and Carlos Velasco-Santos

22.1 Introduction 803
22.1.1 Polyurethane Synthesis 805
22.2 Polyols, Isocyanates, and Chain Extenders Based on Natural Sources 808
22.2.1 Oils 808
22.2.2 Polysaccharides 821
22.2.3 Amine and Amino-Acid Groups 832
22.3 Polyurethane Membranes 836
22.3.1 A New Tendency in the Environmental Industry: Polyurethane Hybrid Membranes 841
22.4 Conclusions 848

23. Biodegradable Poly(Lactic Acid) and Its Composites **857**

Tao Yu and Yan Li

23.1 Introduction 857
23.2 Biodegradable PLA 858
23.2.1 Synthesis of PLA 858
23.2.1.1 Direct condensation 858
23.2.1.2 Ring-opening polymerization 860
23.2.1.3 Chain extension 862
23.2.2 Modification of PLA 864
23.2.3 Processing of PLA 869
23.2.3.1 Extrusion 869
23.2.3.2 Injection molding 870
23.2.3.3 Cast film 871
23.2.3.4 Stretch blow molding 873
23.2.3.5 Thermoforming 873
23.2.3.6 Foaming 874
23.2.3.7 Spinning 876
23.3 PLA-Based Composite 877
23.3.1 Natural Fiber-Reinforced Composites 877
23.3.2 Nanoparticle-Reinforced Composites 880
23.3.3 Surface Treatment Method 882
23.3.3.1 For natural fiber–reinforced composites 882
23.3.3.2 For nanoparticle-reinforced composites 886
23.3.4 Processing of the Composites 888
23.3.4.1 Natural fiber–reinforced composite 888
23.3.4.2 Nanoparticle-reinforced composites 889
23.4 Applications of PLA and PLA-Based Composites 891

24. The Main Characteristics, Properties, Improvements, and Market Data of Polyhydroxyalkanoates **899**

Elodie Bugnicourt, Patrizia Cinelli, Andrea Lazzeri, and Vera Alvarez

24.1 Introduction 899
24.2 Polyhydroxybutyrate 901
24.2.1 General Properties of Polyhydroxybutyrate 902
24.2.2 Thermal and Mechanical Properties of Polyhydroxyalkanoates 903
24.2.3 Improvements of PHB: Blending of PHB with Other Polymers or Plasticizers 905
24.3 Comparison with Other Biodegradable and Traditional Polymers 907
24.4 PHB-Based Nanocomposites 908
24.4.1 PHAs/Clay Nanocomposites 908
24.4.2 PHAs/Nanocellulose Nanocomposites 911
24.5 Market Aspects of Polyhydroxyalkanoates 912
24.5.1 Industrial Production of PHAs: History and Current Landscape 912
24.5.2 Feedstocks Used in PHA Production 913
24.5.3 PHA Production Capacity and Economic Aspects 915
24.5.4 PHA Applications and Commercial Products 918
24.6 Conclusions 919

25. Polymers and Polymer Composites for Electrolyte Application **929**

Aleksandra Kiyanitsa, Sergey Gaidukov, and Gerda Gaidukova

25.1 The Basic Theory of Polymer Electrolytes 929
25.1.1 Definitions 930
25.1.1.1 Polymer 930
25.1.1.2 Composite 930
25.1.1.3 Cell 931

25.1.1.4 Electrolyte 931
25.1.1.5 Battery 932
25.2 Polymers and Polymer Composites as Material for Electrolyte Application 934
25.2.1 Polymer Electrolyte 934
25.2.2 Solid Polymer Electrolyte 934
25.2.2.1 Fillers, salts, and other additives for composites preparation 934
25.2.3 Requirements for Polymer Electrolyte 938
25.3 Applications of Polymer Electrolytes 939

Index 947

Preface

Synthetic polymers have been used in abundance for the past couples of decades for a number of applications. Indeed these polymers have become an essential part of almost all industries and human life. These polymers have versatile qualities of strength, lightness, durability, and resistance to degradation. They have become an important commodity to enhance the comfort and quality of life. However, due to the rising environmental awareness and energy concerns, their use has resulted in a number of problems, especially regarding their biodegradability and recycling. Owing to these serious concerns regarding the use of synthetic polymers, a great amount of awareness is being created in the society encouraging the green consumerism as an initial approach to curb these issues. As a result of waste accumulation at the end of the life cycle of traditional polymer products, the development of environmental-friendly, degradable polymeric materials has attracted extensive interest in the last decades. Among various materials, biorenewable polymers derived from different natural resources are considered as an almost inexhaustible source of sustainable materials in the world. Different kinds of sustainable polymers have been applied in many areas, such as the paper industry, water treatment, material reinforcement, and physiological hygiene. Biorenewable polymers can be easily obtained from abundant inexpensive renewable bioresources, and their properties of abundance, easy availability, and low cost have made them one of the most promising raw materials. Some of the most common sustainable polymers are cellulose, starch, chitin, chitosan, pectin, proteins, wheat gluten, gelatine, casein, polyhydroxyalkanoates, and polylactic acid. These biosourced and biodegradable materials are expected to gradually replace the currently existing family of oil-based polymers as they become competitive in terms of cost and performance.

Keeping in mind the immense advantages of sustainable polymers, this book primarily focuses on the processing and properties of different sustainable polymers procured from diverse biorenewable resources. Several critical issues and suggestions for future work are comprehensively discussed in this book with the hope that the book will provide a deep insight into the state of the art of sustainable polymers. We would like to thank the publisher for the invaluable help in the organization of the editing process.

Finally, we would like to thank our parents for their continuous encouragement and support.

Vijay Kumar Thakur
Manju Kumari Thakur

Chapter 1

Sustainable Polymers and Applications

Rita Mahanta[a] and Ranadeep Mahanta[b]

[a]*Department of Zoology, Cotton College, Guwahati, Assam, India*
[b]*Jorhat Medical College, Jorhat, Assam, India*

ritamahanta@yahoo.co.in

1.1 Introduction

Polymers are groups of macromolecules collectively recognized as polymer due to presence of repeating blocks of covalently linked atomic arrangement in formation of these molecules. The repetitive atomic arrangements forming the macromolecule by forming covalent links are the building block or constituent monomers. As the covalent bond formation between monomer units is the essence of polymer formation, polymers are organic or carbon compounds of either biological or synthetic origin.

The phenomenon or process of polymerization enables to create diverse forms of macromolecules with varied structural and functional properties and applications from few limited but selected sets of monomeric units covering multiples of structural

Handbook of Sustainable Polymers: Processing and Applications
Edited by Vijay Kumar Thakur and Manju Kumari Thakur

ISBN 978-981-4613-53-8 (Hardcover), 978-981-4613-54-5 (eBook)
www.panstanford.com

and functional utility and applications in both biotic and abiotic environment and state. Polymerization and polymers formed during the primordial world was the big bang for the organic evolution ultimately leading to the present form of the living world with concomitant development of non-biological polymerization augmenting the support system to the life forms in the universe. Basically or fundamentally, biological polymerization and polymers evolve the living system and non-biological polymers are utility products innovated to support living systems.

1.2 Basics of Polymers

Polymer is a high-molecular-weight molecule made up of a small repeat unit termed as monomer which is a low-molecular-weight compound that can be connected together to give a polymer. Short polymer chains usually containing less than or up to 10 monomer units is called or termed as an oligomer. When a polymer contains only one type of monomer it is a simple polymer or homopolymer. When it is formed by two or more types of monomers, it is termed as a copolymer or a heteropolymer. Heteropolymers may be of three types. They may be alternate, block, and graft variety.

When the two types of monomeric units are present in a heteropolymer in alternate repetitive sequence, it forms an **alternate co- or heteropolymer**. In **block copolymers**, the monomeric units are repeated in the sequence in the form of blocks of monomeric units. In some heteropolymer, blocks or groups of one or more types of monomeric units are grafted or linked to the primary and long polymer chain forming some short branches or side chains resulting in formation of **graft copolymers**. Graft copolymers can be of multiple combinations of oligomeric blocks forming very complex types of polymers and are mostly present in biological world as structural biopolymers with biophysical function and importance. However, the simple and other types of copolymers are common in both biological and synthetic polymer types. Another type of synthetic polymer is **random copolymer**, where the monomers are present randomly without any specific sequence.

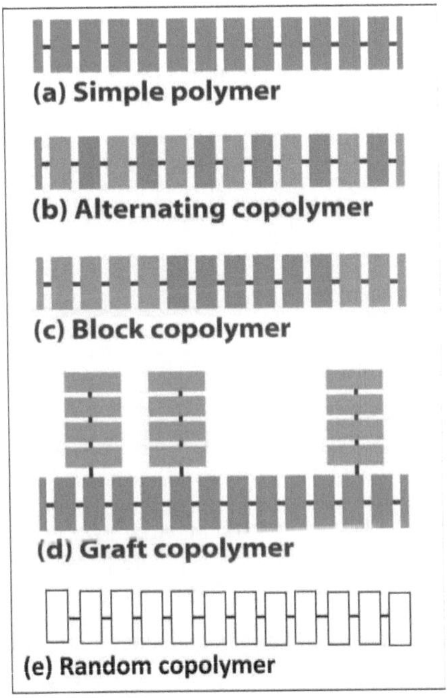

Types of polymer.

1.2.1 Polymer Families

Basically there are two families of polymers: synthetic and natural. Polymers formed by abiotic process by laboratory or industrial synthesis form the family of synthetic polymers and the polymers synthesized in biological systems by specific metabolic pathways form the family of natural polymers.

1.2.1.1 Synthetic polymers

Synthetic polymers are either polyolefins synthesized from olefin and alkene monomers by covalent bonding or polymers of esters, amides, urethanes, and some other functional group monomers. Polyethylene, polypropylene, polystyrene, polyvinyl chloride, and Teflon are some common examples of polyolefins and polyalkenes. PET, nylon, Kevlar, and Spandex are some examples of polyesters, amides, and urethanes.

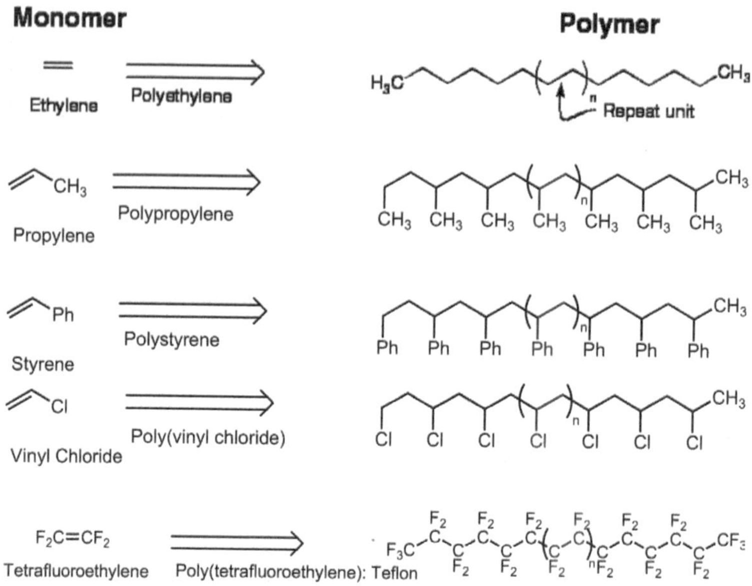

Some common polyolefin and polyalkenes synthetic polymers.

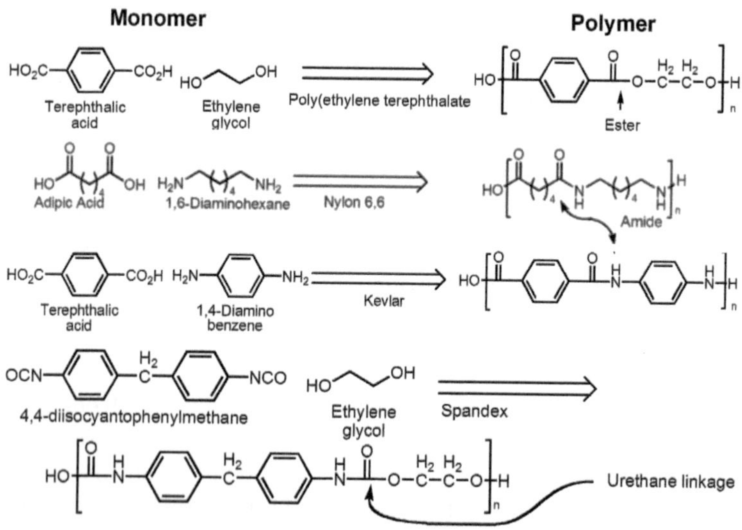

Examples of few common polyester, amide, and urethane synthetic polymers.

1.2.1.2 Natural polymers

Natural polymers are polymers synthesized by some biological system through specific metabolic pathways by polymerization of biological building block monomer groups by a variety of covalent links as ethylene, ester, amide, peptide, glycosidic, phosphosphodiester, etc., from monomeric units as isoprene, glycan, amino, purine, pyrimidine, etc., and their derivatives forming polymers as rubber, cellulose, proteins, nucleic acids, etc. These are only a very few examples of natural polymers from an almost infinite list considering the immense variety of living organism or living system synthesizing this group of polymer through individual and specific metabolic pathways with addition of the immense scope for post synthetic modification in vitro for target oriented applied utilization in diverse biomedical, industrial and ecological spheres.

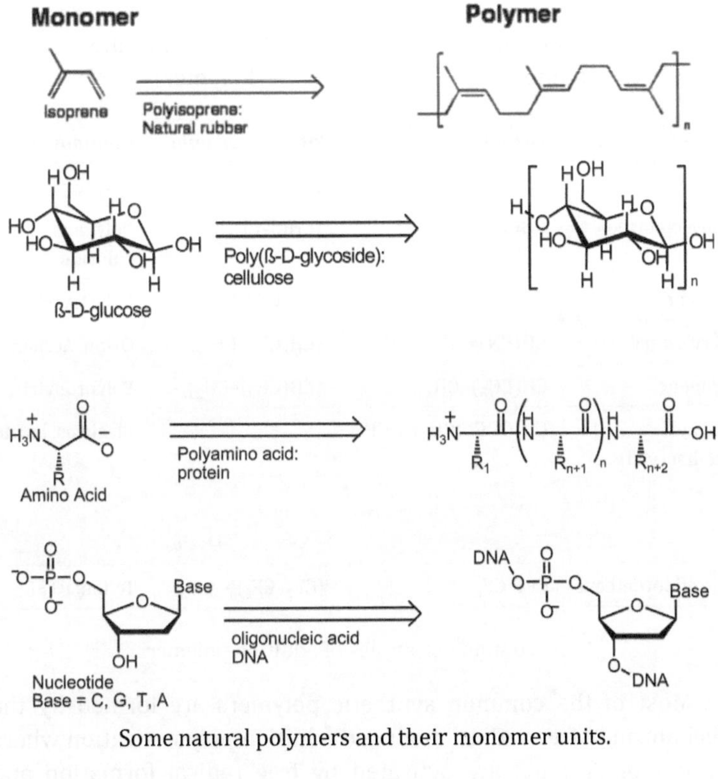

Some natural polymers and their monomer units.

1.2.2 Polymer Formation

Polymer synthesis or formation occurs by two major mechanisms. The mechanisms are known as **addition polymerization** and **condensation polymerization**.

1.2.2.1 Addition polymerization

In the mechanism for addition polymerization, the monomers react to form a polymer without loss of any atom. The total number of atoms in the formed polymer is simple multiples of the number of monomeric units used to form the polymer.

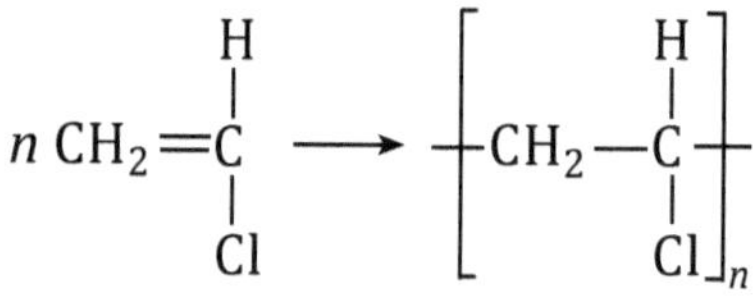

Ethylene chloride (Monomer) ⇨ Polyethylene chloride (Polymer)

Monomer name	Formula	Polymer formula	Common name
Ethene	$CH_2{=}CH_2$	$-(CH_2-CH_2)_n-$	Polyethylene
Vinyl chloride	$CHCl{=}CH_2$	$-(CHCl-CH_2)_n-$	Polyvinyl chloride
Styrene	$CH(C_6H_5){=}CH_2$	$-(CH(C_6H_5)-CH_2)_n-$	Polystryrene
Acrylonitrile	$CH(CN){=}CH_2$	$-(CH(CN)-CH_2)_n-$	Orlon, Acrilan
Propene	$CH(CH_3){=}CH_2$	$-(CH(CH_3)-CH_2)_n-$	Polypropylene
Methyl methacrylate	$CH_3OOCC(CH_3)C{=}CH_2$	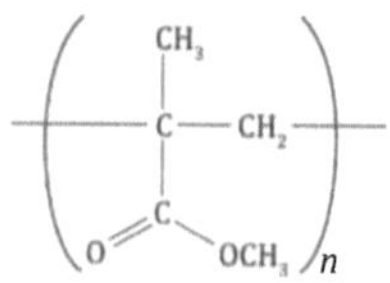	Plexiglas, Lucite
Tetrafluoroethene	$CF_2{=}CF_2$	$-(CF_2-CF_2)_n-$	Teflon, PTFE†

A few common examples of addition polymers.

Most of the common synthetic polymers are formed by the mechanism of free radical mediated addition polymerization where the monomer units are activated by free radical formation and subsequent prolongation of the polymer chain is maintained by free radical chain reaction at the activated end of the growing polymer chain.

1.2.2.2 Condensation polymerization

In the mechanism of condensation polymerization, the polymer is formed and grows with condensation reaction between monomer units by splitting off a small molecule such as water or carbon dioxide. The process of condensation is the mechanism for formation of only a few synthetic polymers as nylon but most of the natural polymers are formed by condensation polymerization.

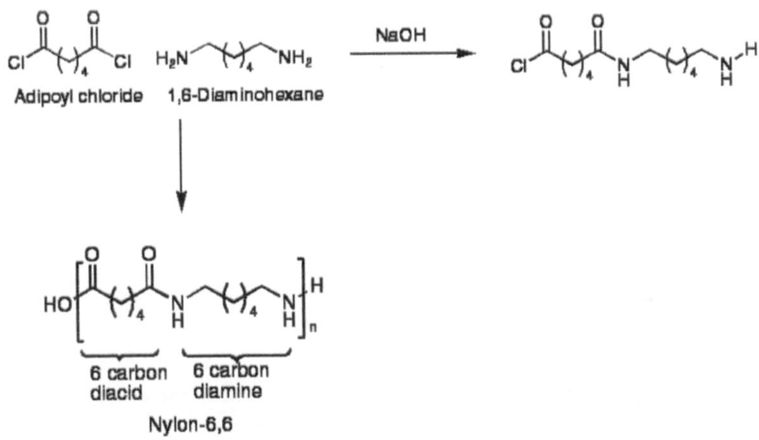

In condensation polymerization for biosynthesis of natural polymers, the building block monomer for polymer formation is initially converted to an activated form; then it is transferred to or condensed with the receiving unit of the developing polymer chain, releasing a low-molecular-weight compound such as water, carbon dioxide, or the activating fragment itself, as in case of synthesis of the natural polymer glycogen from the activated form of the monomer glucose such as UDP-glucose, forming the polymer glycogen and releasing UDP with condensation of each glucose units to the growing polymer chain. The processes of condensation polymerization for synthesis of natural or biopolymers are enzyme mediated and metabolically regulated. The fundamental role of enzymes in the process of biopolymerization by condensation, forming the natural polymers is to create a microenvironment for forming the condensing covalent bond between monomer units under physiological conditions compatible with the living process within moderate ranges of reaction conditions as temperature, pressure, hydrogen ion concentration, electrolyte and ionic environment, etc.

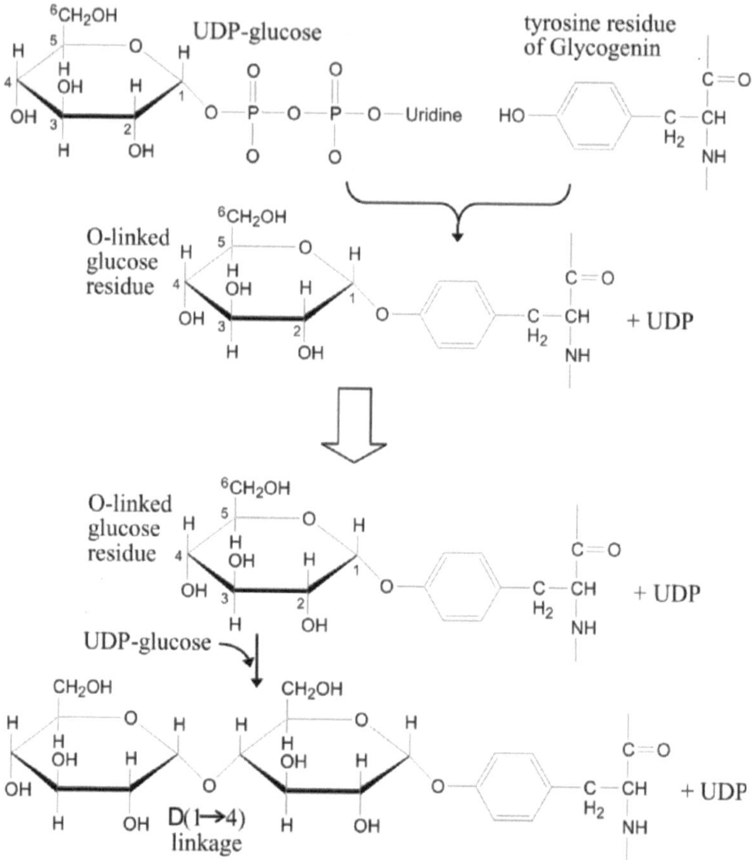

Step of initiation and prolongation during synthesis of the natural polymer glycogen by condensation polymerization.

It is to be remembered here that as in any chemical reaction involving either formation or breaking of a covalent bond, the energy of activation is to be achieved by any means, for formation of a free radical state in one of the reaction participant to initiate the reaction and once initiated, the reaction will continue until the free radical mediated electron imbalance is dissipated through formation of a new compound with requirement for a higher energy of activation and lower free energy. The same fundamentals of reaction conditions and mechanics are exploited or rather governing the reactions and mechanisms of polymerization equally well for both synthetic and natural polymers. In addition polymerization for synthetic polymer formation, the free radical generation and electrophilic attack is generated or initiated by drastic of reaction conditions as high and prolong temperature, pressure, extremes of pH, radiation, etc. During condensation polymerization of natural polymers, the free radical and electrophilic state is created and bond formation is mediated in the active sites of the enzymes in the form of a micro environment requiring reduced energy of activation for the polymerization reaction.

1.3 Classification

Depending on the thermal character, behavior under stress, formation, and origin, grossly the polymers may be classified as thermoplastics, thermosets, elastomers, and natural polymers. Thermoplastics such as polyethylene soften on heating. Thermosets or resins as epoxy harden when two components forming the polymer are heated together. Elastomers as rubbers change shape when stress is applied but regain original shape when the force is withdrawn. Natural polymer itself is a very diverse group of polymers that provide mechanical, structural, and functional basis of most plant and animal life.

Natural polymers are broadly classified as polysaccharides, proteins, polynucleotide, and natural rubber. This classification is based on the type of monomer building blocks for the polymer and type of covalent bonds for formation of the polymer. Polysaccharides that are carbohydrate polymers are formed by condensation of monomeric units called monosaccharides, proteins or peptide polymers are formed from amino acids and polynucleotide are synthesized from nucleotide condensation reaction forming glycosidic, peptide, and phosphodiester bonds, respectively. All natural polymers are condensation polymers.

Natural polymers with monomer units and type of linkage

Class of natural polymer	Monomer units	Type of linkage
Polyisoprene	Isoprene	Ethylene
Polysaccharide (carbohydrate)	Monosaccharide	Glycosidic
Protein and polypeptide	Amino acid	Peptide
Polynucleotide	Nucleotide	Phosphodiester

Polyisoprene (natural rubber)

Natural rubber is a hydrophobic polyisoprene formed by enzyme mediated condensation reaction between isopentenyl pyrophosphate and isoprene units. It is an elastomer and a thermoplastic polymer. During biosynthesis, rubber particles are formed in the cytoplasm of specialized latex-producing cells called laticifers within rubber producing plants. The particles of rubber are surrounded by a single phospholipid membrane with hydrophobic tails pointed inward toward the core of particles. The phospholipid monolayer membrane allows biosynthetic proteins involved with rubber synthesis to be sequestered at the surface of the growing rubber particle, which allows new monomeric units to be added from outside the biomembrane, but within the laticifer. The rubber particle is an enzymatically active entity that contains three layers of material, the rubber particle, a biomembrane, and free monomeric isoprene units. The biomembrane is held tightly to the rubber core by presence of high negative charge along the double bonds of the rubber polymer backbone. Free monomeric units and conjugated proteins make up the outer layer. The precursor of rubber is isopentenyl pyrophosphate, which elongates by enzyme mediated condensation reaction requiring presence of Mg^{2+} and the action of the enzyme rubber transferase. The monomer is added to the pyrophosphate end of the growing polymer. The process releases the terminal high-energy pyrophosphate. In this reaction a *cis* polymer is produced. The initiation step is catalyzed by prenyltransferase that converts three monomers of isopentenyl pyrophosphate into farnesyl pyrophosphate. The farnesyl pyrophosphate can bind to rubber transferase to elongate a new rubber polymer. The necessary isopentenyl pyrophosphate is derived from the mevalonate pathway, which originates from acetyl-CoA as cytosolic reactions. The relative ratio of the farnesyl pyrophosphate initiator unit and isoprenyl pyrophosphate elongation monomer is the primary factor for determination of the rate of new particle synthesis against elongation of existing rubber particles. The initial steps of rubber biosynthesis in rubber producing plants and cholesterol biosynthesis in animals share a common metabolic pathway.

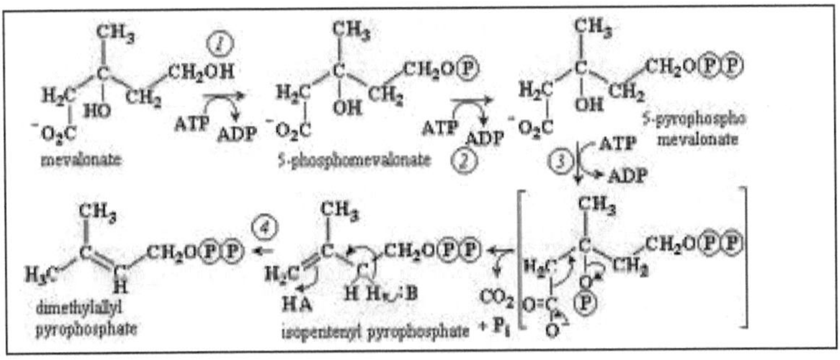

1.4 Introduction to Polysaccharides

Polysaccharide is a macromolecule consisting of a large number of monosaccharide residues joined to each other by glycosidic linkages. Polysaccharides that are composed of only one kind of monosaccharide are described as homopolysaccharides or homoglycans. Likewise, if two or more different kinds of monomeric unit or monosaccharides are present, the class of polysaccharide is known as heteropolysaccharides or heteroglycans. A general name for a homopolysaccharide is obtained by replacing the ending "-ose" of the monosaccharide name by "-an". As for example, it is xylan for polymers of xylose, mannan for polymers of mannose, and galactan for polymers of galactose. Cellulose and starch are both glucans, as they are composed of glucose residues. On this basis, the common name for polysaccharides is glycan, as they are formed by glycosyl units.

As the name of the polysaccharide merely indicates the monosaccharide unit in homopolysaccharides or homoglycans only and it does not carry any information about the type of glycosidic linkage forming it and as it is not informative for the names of heteropolysaccharides or heteroglycans, a system of "shorthand notation" may be or rather conveniently used for designating the glycans. In this system of shorthand notation for glycans, the monosaccharide glycosyl units are represented by the first three letters of their names, except for glucose, which is represented as **Glc**. When the monomer is a **D-sugar**, generally the D is omitted but for **L-sugars**, L is designated before the sugar monomer name. The ring form, that is, whether **pyran** or **furan** is indicated by ***p*** or *f* after the sugar names and **uronic acids,** is indicated by a suffix **A.** **α** and **β** are designated for the appropriate **anomeric** form of the glycosidic linkage.

1.4.1 Examples of Shorthand Notation for Polysaccharides

Cellobiose = β-Glc*p*(1→4)-Glc or βGlc*p*1,4Glc
Lactose = β-Gal*p*(1→4)-Glc or βGlc*p*1,4Glc
Maltose = α-Glc*p*(1→4)-Glc or αGlc*p*1,4Glc
Gellan = [3)β-D-Glc*p*(1→4)-β-D-Glc*p*A-(1→4)-β-D-Glc*p*(1→4)-α-L-Rha*p*(1→]$_n$ or [3βGlc*p*1,4βGlc*p*A1,4β Glc*p* 1,4αLRha**p**1,→]$_n$

1.4.2 Polysaccharide Classification

Polysaccharide carbohydrates are condensation polymers formed by a type of covalent linkage known as glycosidic bond between monomeric units of carbohydrates known as monosaccharides and so they are known as polysaccharides. The different types of polysaccharides are formed by the combination of variation in monosaccharide units forming them and the type of glycosidic linkage. The type of glycosidic bond during condensation reaction for formation or biosynthesis of polysaccharides can vary because the monosaccharides exhibit stereoisomerism due to presence of anomeric carbons in the monomer. This variability induces formation of two different types of polysaccharides with dramatic differences in character, property, and functional utilization from a single type of monomer monosaccharide. In addition to this aspect of variation in polysaccharides, another scope for introduction of variation in polysaccharide biosynthesis is branching during polymerization by condensation reactions between oligomeric blocks and the primary polymer chain producing block polymers of different character and function.

1.4.3 Scheme of Polysaccharide Classification

Due to diverse origin and structure of the polysaccharides, it is very difficult to classify the polysaccharides under a singles scheme. Depending on the purpose of utilization and information presentation, the polysaccharides are effectively classified on the basis of some fundamental commonness under different schemes.

1.4.3.1 General scheme for classification of polysaccharides

The following scheme is convenient for classification of polysaccharides where initially it is classified in four groups depending on source, structure, monomers, and charge followed by further sub groups as presented in the following scheme.

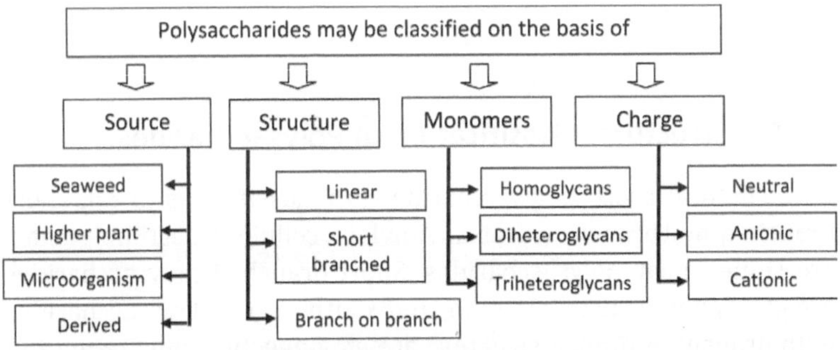

1.4.4 Sources of Polysaccharides and Related Classification

Polysaccharides are present in seaweeds, higher plants, and microorganisms, and derivatives are formed from polysaccharides present in these sources. Polysaccharides such as agars, alginates, and carrageenans are present and extracted from seaweeds. Higher plants contain soluble polysaccharide as pectin and insoluble polysaccharide as cellulose in their cell walls. Cereal starch, guar gum, and locust bean gum are polysaccharides present in seeds of higher plants. Tubers of potato and roots of tapioca contain starch. Gum arabic and gum tragacanth are polysaccharides of higher plant exudates. Microorganisms produce polysaccharides such as xanthan gum and gellan gum. Some derived polysaccharides are modified starch, carboxymethyl cellulose, propylene glycol, and alginate.

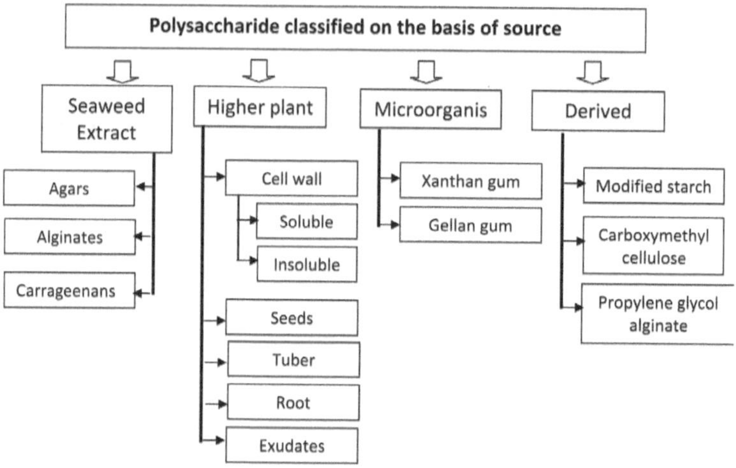

1.4.5 Structural Classification of Polysaccharides

On structural basis, polysaccharides are classified as linear, short branched, and branch on branch. Amylose, cellulose, pectin, alginates are some linear polysaccharides. Short-branched polysaccharides are guar gum, locust bean gum, and xanthan gum. Polysaccharides with branch-on-branch structure are amylopectin, gum arabic, and arabinoxylan.

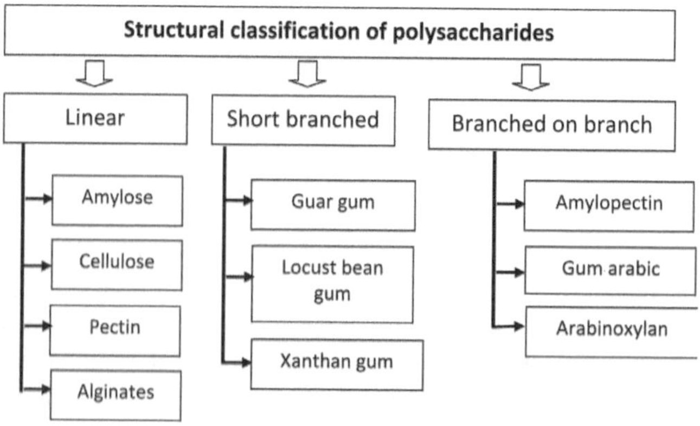

1.4.6 Classification of Polysaccharides on Monomer Composition

On the basis of monomer component composition, polysaccharides are classified as homoglycans, diheteroglycans, and triheteroglycans. Homoglycans are polysaccharides formed by polymerization of only a single type of monomer carbohydrate, i.e., monosaccharide. Diheteroglycans contain two different types of monosaccharides in a repetitive sequence. In triheteroglycans, the repetitive sequence is formed by three different monomer units of monosaccharides. Starch, cellulose, and inulin are homoglycans. Common diheteroglycans are agars, alginates, and carrageenans, whereas xanthan, gellan, and arabinoxylan are common forms of triheteroglycans.

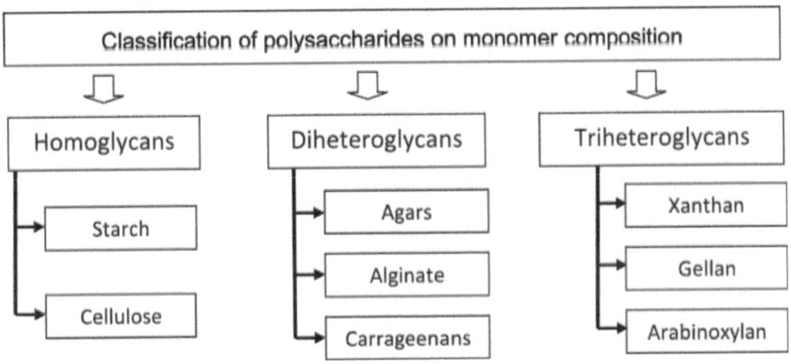

1.4.7 Polysaccharide Classification by Charge

There may be presence of some charged groups in the monomer units of polysaccharides producing charged polysaccharides. On the basis of surface charge, the polysaccharides are classified as neutral, anionic, and cationic. Amylose, amylopectin, cellulose, guar gum, etc., are neutral polysaccharides. Anionic polysaccharides are alginates, carrageenans, Gellan, gum arabic, xanthan. Chitosan is a cationic polysaccharide.

1.4.8 Characterization of Polysaccharides

Polysaccharides with diverse characters are obtained from a variety of plant, animal, and microbial sources. The factors determining the character of a polysaccharide are types of monosaccharides present, linkage, types of repetitive oligosaccharides, substituent groups present in the monomer units and their character, presence of branching, type, and frequency of branching, etc., and the combination of all these variable factors in the polysaccharide macro molecule. On the basis of these factors, the natural polysaccharides are characterized and the resultant character as a consequence of combination of these factors finally determines the biological as well as industrial and pharmaceutical applications of the polysaccharides.

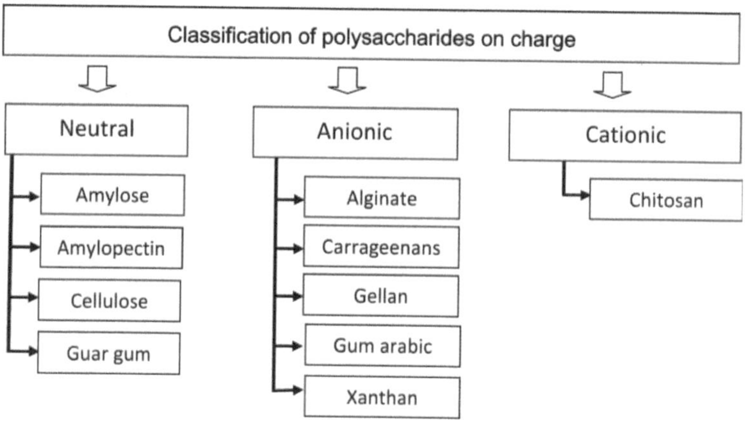

1.4.9 Characterization of Starch

Starch is a plant polysaccharide present in seeds, tubers, and roots. It is a major component of human diet and one of the most abundant carbohydrates. Starch from different sources has various levels of digestibility depending on the physical structure, which is genetically determined by the producing plant and tissue, and the digestibility can be modified by physical, chemical, and enzymatic treatment. It is a starting material for many foods, nonfood industrial material, and pharmaceutical products. Starch primarily comprises two types of molecules: amylose and amylopectin. Amylose is a linear molecule and amylopectin is branched. Starch can be subjected to acid or enzymatic hydrolysis to produce the oligomeric maltodextrins. Molecular formula of starch is $(C_{12}H_{22}011)_n$ with notation $[\alpha Glcp1,4Glc]_n$.

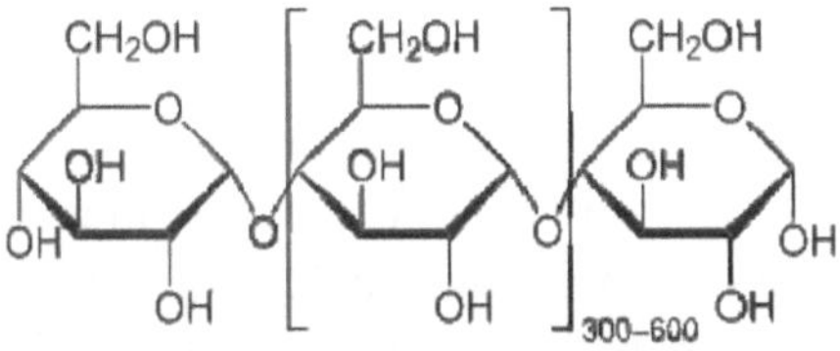

Structure of the amylose molecule.

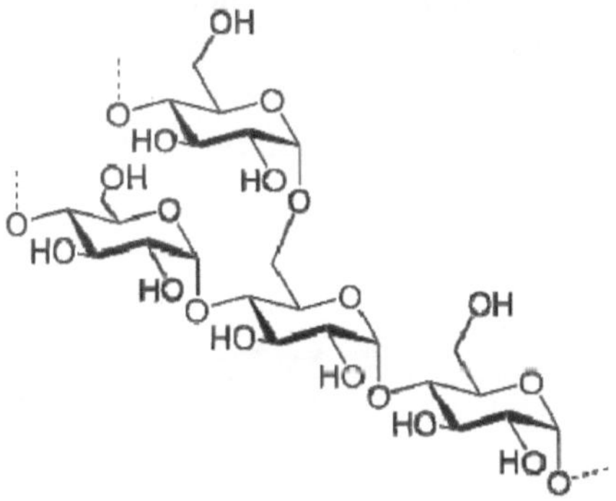

Structure of the amylopectin molecule.

Guar gum is a galactomannan obtained from the seed of guar plant, *Cyanaposis tetragonolobus,* mostly grown in India and Pakistan. Guar gum is formed by chains of (1→4)-linked β-D-mannopyranosyl units, with averagely 1.8 chain units attached with single α-D-galactopyranosyl units via (1→6)-linkages with relatively even distribution of side chains along the main chain. It is a neutral polysaccharide with molecular weight: 150,000–1,500,000. In solution, guar gum shows pseudoplastic behavior.

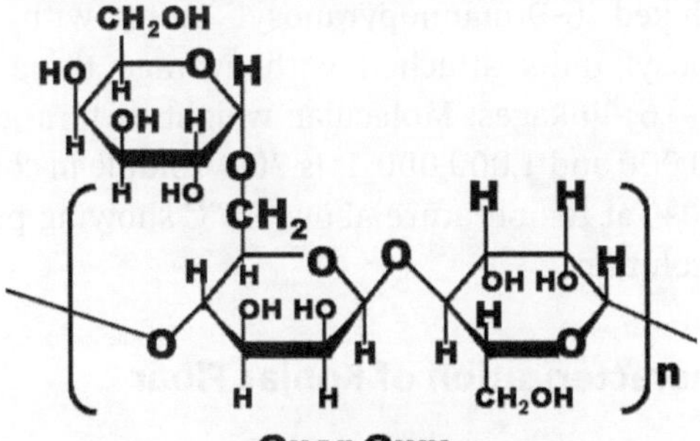

Guar Gum

Mannose backbone Galactose side chain

1.4.10 Characterization of Locust Bean Gum

It is a galactomannan obtained from the seed of carob tree, *Ceratonia siliqua*, grown around the Mediterranean Sea. The polysaccharide is formed by chains of (1→4)-linked β-D-mannopyranosyl units, with averagely 3.9 chain units attached with single α-D-galactopyronosyl units via (1→6)-linkages. Distribution of the side units is highly uneven along the main chain in this neutral polysaccharide with molecular weight: 400,000–1,000,000. Locust bean gum requires heating to hydrate completely and shows pseudoplastic behavior in solution.

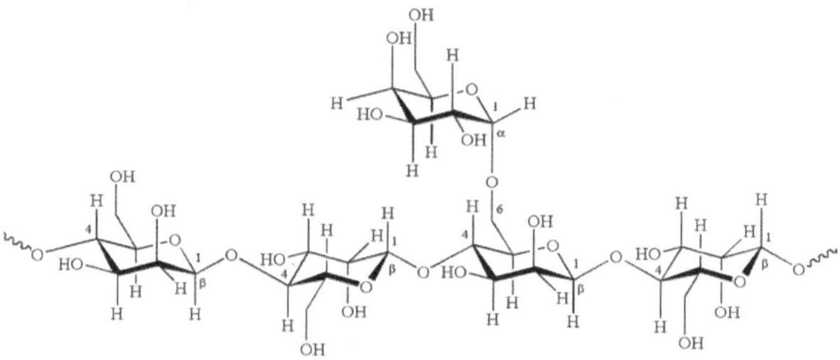

Locust bean gum

1.4.11 Characterization of Tara Gum

Tara gum is the galactomannan obtained from the seed of Tara shrub, *Caesalpinia spinosa*, mostly found in northern Africa and Peru of South America. The main polysaccharide chain is formed of (1→4)-linked β-D-mannopyranosyl units, with single α-D-galactopyranosyl units attached with average three main chain units via (1→6)-linkages. Molecular weight of tara gum may fall between 300,000 and 1,000,000. It is 70% soluble in cold water and hydrated 100% at temperature above 80°C showing pseudoplastic behavior in solution.

1.4.12 Characterization of Konjac Flour

Konjac flour is a glucomannan obtained from the tuber of *Amorphophallus konjac* grown in Asia. The polysaccharide chain is of mannose and glucose units in a molar ratio of 1.6:1 connected with β-(1→4)-linkages. In the main chain one acetyl group is present at the C-6 for about 6 to 20 sugar units. Konjac flour is a neutral polysaccharide with molecular weight 200,000–2,000,000 that swells in cold water and becomes fully hydrated after heating, showing pseudoplastic behavior in solution.

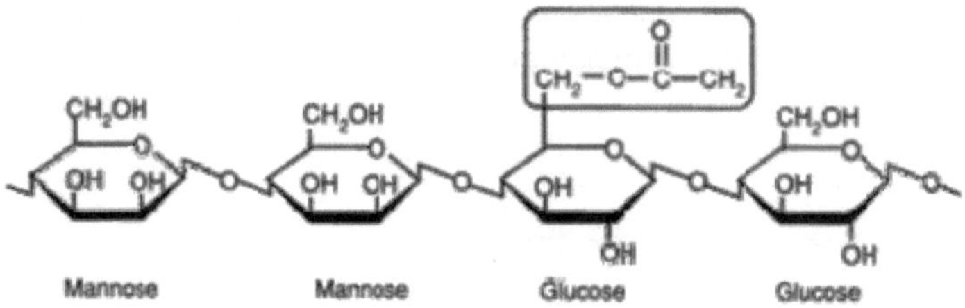

Konjac flour

1.4.13 Characterization of Gum Arabic

Gum arabic is the gummy exudate obtained from *Acacia* trees in Africa, primarily in Sudan. It is a highly branched polysaccharide containing both neutral monomers and uronic acids covalently attached with proteinaceous fractions, rendering its excellent function as an extensively used emulsifier with molecular weight about 250,000. It is highly soluble, producing low viscosity solution even at 40% concentration. The solution shows *Newtonian* behavior with constant viscosity at different shear rates.

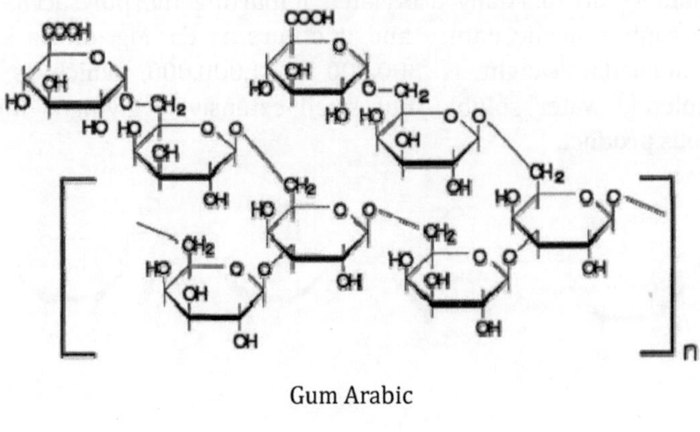

Gum Arabic

1.4.14 Characterization of Gum Tragacanth

Tragacanth gum is an exudate of *Astragalus*, a perennial short brush in Asia. It is slightly acidic and occurs as Ca, Mg, or Na salt. The polysaccharide chain is formed by neutral highly branched arabinogalactan and tragacanthic acid, which is a linear (1→4)-linked α-D-galacturonopyranosyl unit with some substitutions. The gum is highly viscous with some emulsification properties showing pseudoplastic behavior in solution.

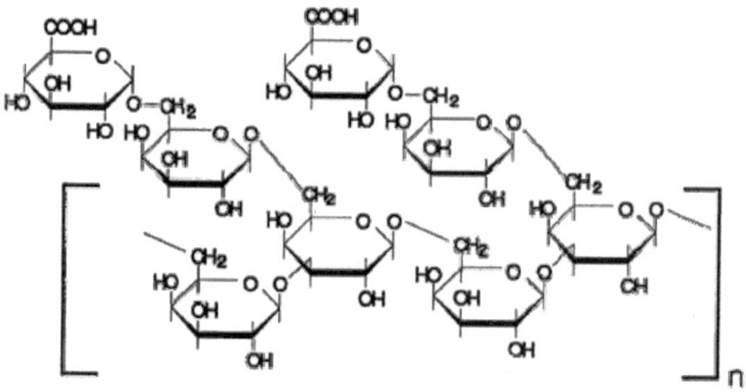

Gum Tragacanth

1.4.15 Characterization of Gum Karaya

This gummy polysaccharide is an exudate of *Sterculia urens*, a tree grown in India. The main polysaccharide chain composed of D-galactose, D-glucuronic acid, and L-rhamnose. The component monomers are partially acetylated imparting the polysaccharide with slightly acidic nature and it occurs as Ca, Mg, or Na salts. Its molecular weight is 300,000 to 1,000,000, which is not completely water soluble, but swell extensively giving a highly viscous product.

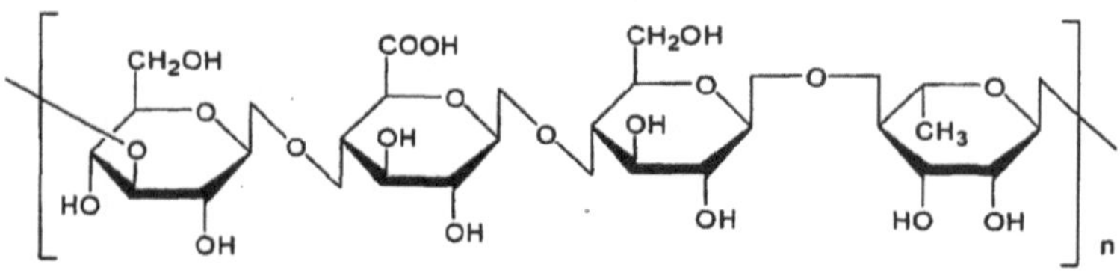

Gum Karaya

1.4.16 Characterization of Carrageenan

Carrageenan is obtained from the red seaweeds of the class Rhodophyceae, in which it is functions as a structural material. It is a group of linear galactan with 15–40% (w/w) ester sulfate content and is formed by alternating (1→3)-α-D and (1→4)-β-D-galactopyranosyl or 3,6-anhydro-α-D-galactopyranosyl linkages. There are three types of commercially available carrageenans: κ, ι, and λ. Carrageenans are anionic polysaccharides with molecular weight 100,000–1,000,000, and they form gels with potassium or calcium ions. The gel formation of carrageenans is more effective in milk than in water due to incorporation of casein micelles into the gel network.

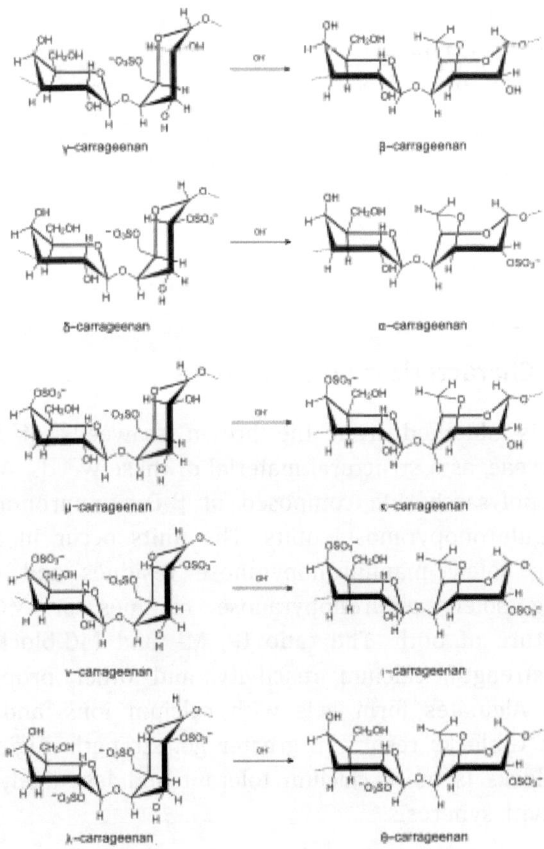

Different types of Carrageenans

1.4.17 Characterization of Agar

Agar is obtained from the red seaweeds of the class Rhodophyceae, as a structural material in the weeds. Agar has linear chain similar with carrageenans, consisting of repeating sections of (1→3)-linked β-D-galactopyranosyl units joined to 3,6-anhydro-α-L-galactopyranosyl units via (1→4)-linkages with very few sulfate groups. In agar, up to 21% of C-6 carbon on β-D-galactopyronosyl units may contain methoxyl groups, affecting the gelation properties of agar. The unique gelation property of agar is that, gelation occurs at temperature of 30–35°C, which is far below the gel melting temperature of 90–95°C.

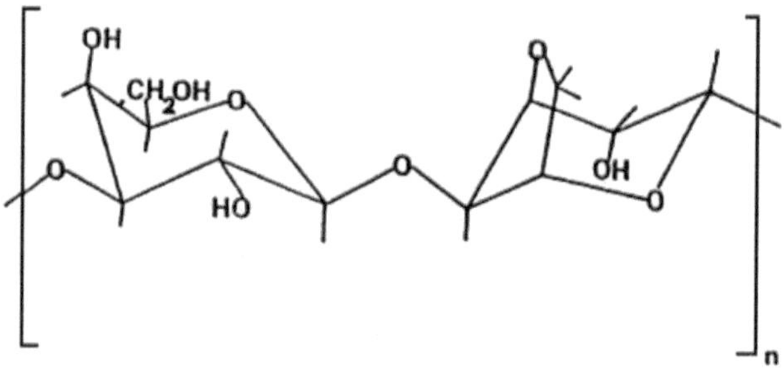

Agar

1.4.18 Characterization of Alginate

Alginate is obtained from the brown seaweeds of the class Phaeophyceae, as a structural material of these weeds. Alginate is a linear polysaccharide composed of β-D-mannuronopyranosyl and α-L-guluronopyranosyl units. The units occur in M blocks containing solely mannuronopyranose residues and G blocks containing solely guluronopyranose residues or MG blocks with mixture of both. The ratio G-, M-, and MG-blocks affects the gel strength, calcium reactivity, and other properties of alginates. Alginates form gels with calcium ions, and alginate with high G-blocks results in greater gel strength. Alginate with high M-blocks is more calcium tolerant and less likely to have problem with syneresis.

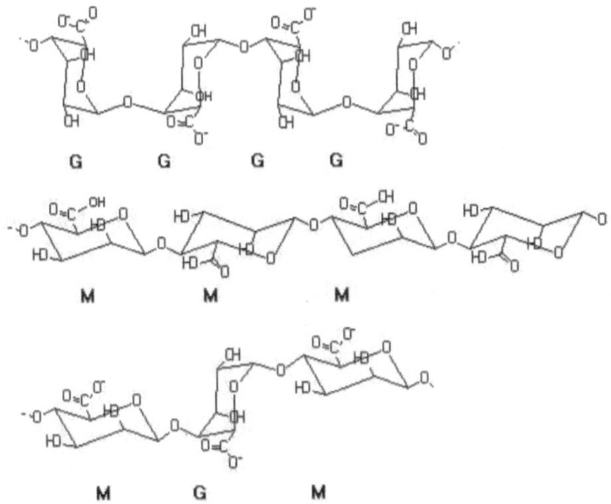

Alginates

1.4.19 Characterization of Pectin

Pectin is found in virtually all land-based plants as a structural material. Commercial pectin is extracted from citrus peel, apple pomace, sugar beet, or sunflower heads. It is a linear chain of galacturonic acid units with molecular weight about 110,000–150,000. In native pectin one free galacturonic acid unit is followed by five methyl esters of galacturonic acid, with degree of esterification (DE) of 83.3%. Esterification can be controlled during extraction processing, to obtain low methoxyl ester pectin with DE below 50% or high methoxyl ester pectin with DE above 50%. Low ester pectins gel with calcium, and high ester pectins gel with high sugar concentrations and solidify at low pH.

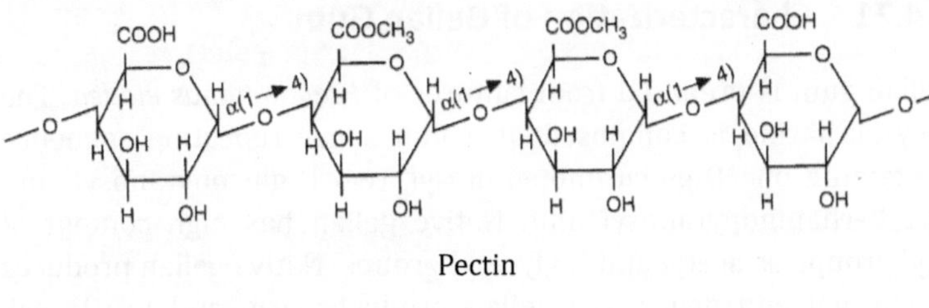

Pectin

1.4.20 Characterization of Xanthan Gum

Xanthan gum is prepared by culturing *Xanthomonas campestris,* which is a single-cell organism producing the gum as a protective coating. It has a backbone chain identical to cellulose. To the cellulose like backbone a trisaccharide side chain is attached to alternate D-glucosyl units at the C-3 position. The side chain consists of a D-glucuronosyl unit between two D-mannosyl units. Approximately 50% terminal mannosyl units contain a pyruvic acid moiety as a 4,6-cyclic acetal. The non-terminal mannosyl units are substituted at the C-6 position with an acetal group. Xanthan gum has molecular weight of about 2,000,000–3,000,000. It shows pseudoplastic shear thinning behavior in solution and is viscosity stable in a broad temperature and pH range.

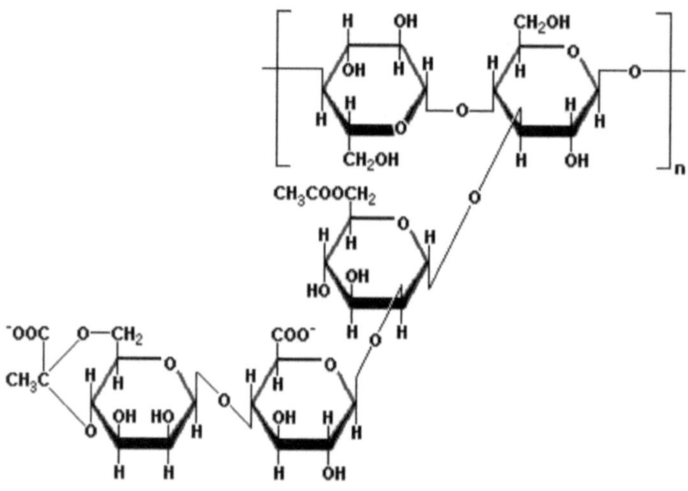

Xanthan gum

1.4.21 Characterization of Gellan Gum

Gellan gum is prepared from cultures of *Pseudomonas elodea.* The polysaccharide is composed of a four-sugar repeating sequence containing one D-glucuronopyranosyl, two D-glucopyranosyl, and one L–rhamnopyranosyl unit. Native gellan has high content of acyl groups as acetyl and L-glyceryl groups. Native gellan produces elastic gel and deacylated gellan produces firm and brittle gel.

Molecular weight of gellan gum is about 1,000,000–2,000,000. It requires either monovalent or divalent cations to form a gel. Gel strength is highly affected by calcium ions.

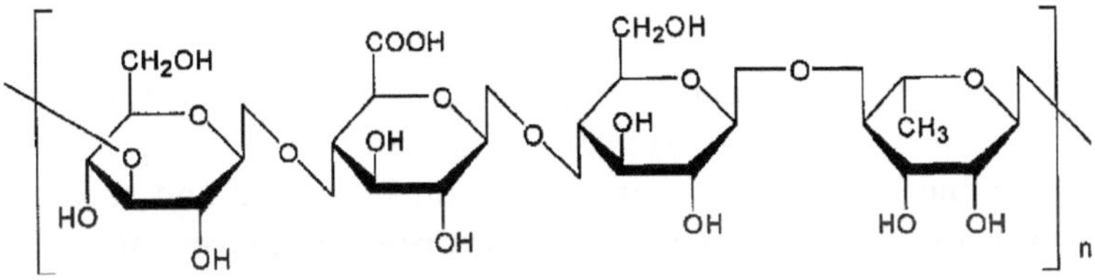

Gellan gum

1.4.22 Characterization of Cellulose

Cellulose is a linear homoglucan formed by β-1→4 linkage. It is the most abundant organic polymer. Due to the β-1→4 linkage in formation of the polysaccharide chain, the monomeric glucose units are such oriented with the adjacent monomer units in conformation that the linear polymer chain takes turn in the form of a helix. The helical conformation of the cellulose chain brings about the multiple -OH groups present in the glucose units in close association to form multiple hydrogen bonds. This phenomenon in the cellulose molecules imparts high tensile strength to the cellulose fibers. The hydroxyl groups (-OH) of cellulose can be partially or fully modified by reactions with various reagents to produce cellulose derivatives with useful properties. Many cellulose esters and cellulose ethers (-OR) are prepared by modification reactions.

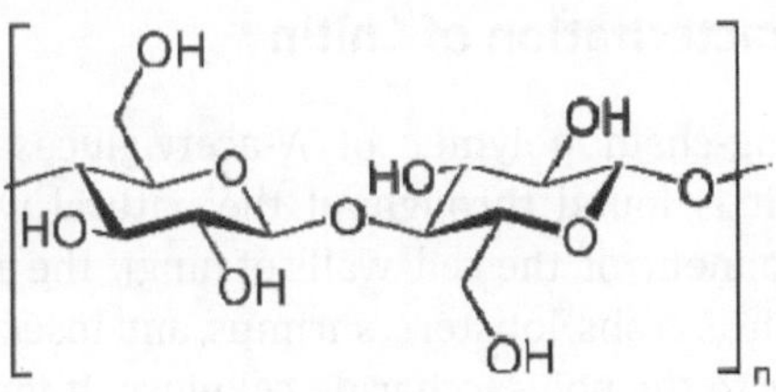

Cellulose

Microcrystalline cellulose is extracted from wood cellulose by removing amorphous regions under high shear and acid to release small crystalline bundles of cellulose fibers. Microcrystalline cellulose is stable in a relatively broad range of temperature and pH. **Carboxymethyl cellulose** (CMC) is prepared by soaking cellulose in aqueous sodium hydroxide and reacting with monochloroacetic acid. **Methylcellulose** (MC) is prepared by soaking cellulose in aqueous sodium hydroxide and reacting with methyl chloride. **Hydroxypropylcellulose** (HPC) is prepared by soaking cellulose in aqueous sodium hydroxide and reacting with propylene oxide. **Methylhydroxypropylcellulose** (MHPC) is prepared by soaking cellulose in aqueous sodium hydroxide and reacting with a mixture of methyl chloride and propylene oxide. CMC, MC, HPC, and MHPC solutions are optically transparent and pseudoplastic. MC and MHPC solutions form gel if heated to incipient gelation temperature at 50–90°C and revert to the liquid state upon cooling. HPC solution precipitates if heated to its cloud point and redissolves upon cooling.

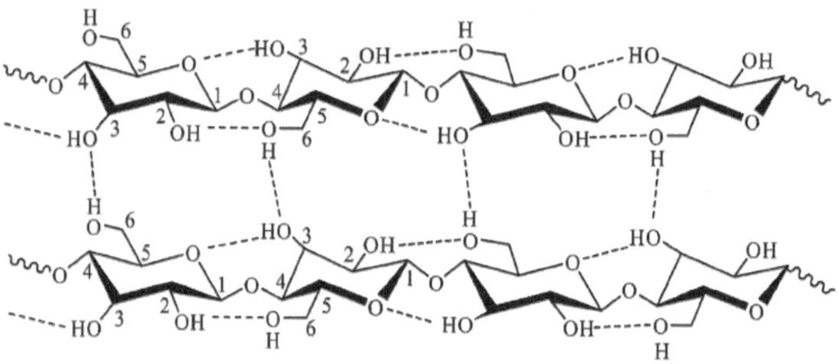

H-Bonds in cellulose

1.4.23 Characterization of Chitin

Chitin is a long-chain polymer of *N*-acetylglucosamine derived from glucose. It is found throughout the natural world. Chitin is the main component of the cell walls of fungi, the exoskeletons of arthropods such as crabs, lobsters, shrimps, and insects. Structurally chitin is similar to the polysaccharide cellulose. It forms crystalline nanofibrils or whiskers on deacetylization of chitin with alkali. The alkali treatment converts the *N*-acetyl groups in the *N*-acetylglucosamine monomers in chitin to glucosamine producing chitosan. Chitosan has multiple industrial and pharmaceutical uses.

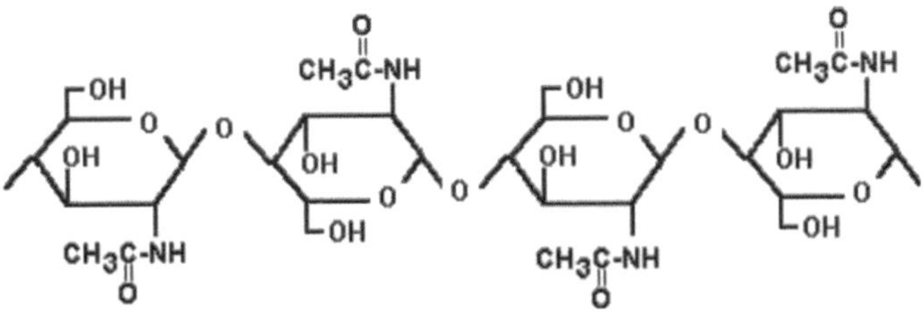

Chitin

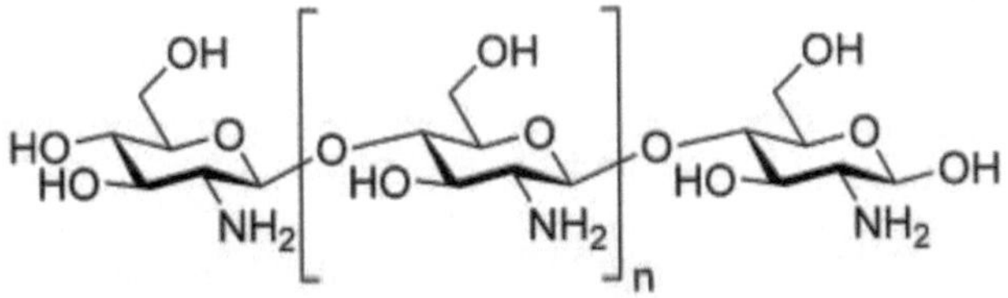

Chitosan

1.5 Proteins and Polypeptides

Proteins and polypeptides are a group of natural or biopolymer with diverse biological, structural, and functional characters. The uniqueness of this group of natural polymer is that all members of this polymer group are heteropolymer with monomeric units linked in specific sequences by a special type of amide bond called the peptide bond. The sequence of the monomeric units linked by the peptide bond is genetically determined and controlled during formation of the polymer. The genetically controlled and regulated event during polymerization of these groups of polymer is essential for their structural and functional integrity.

The monomeric units used for polymerization into proteins and polypeptides are **α-amino acids**. These amino acids are chemically carboxyl amine or amino carboxylate. They contain both amino and carboxyl groups covalently attached to a single carbon atom. One of the remaining two valances of that carbon is fulfilled by a hydrogen atom in all amino acids while the other valance is fulfilled by a functional group. The functional group differs in different amino acids and that makes amino acids different. There may be innumerable types of functional groups that can form as many amino acids; but through the long process of evolution, nature has selected only 20 types of functional groups for amino acids producing only 20 types of amino acids for making proteins and polypeptides.

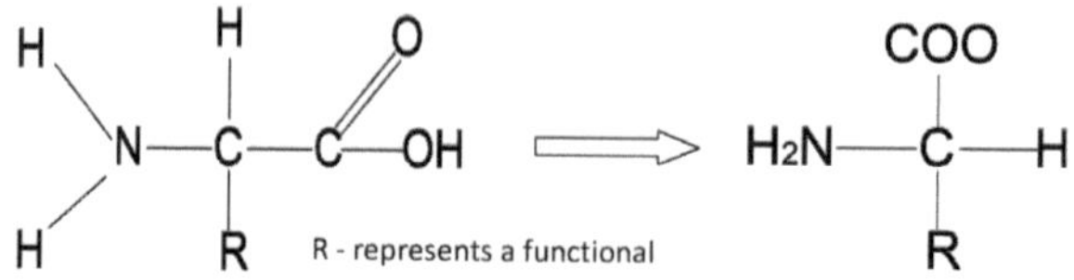

General structure of amino acid

During polymerization of amino acids for formation of proteins, the amino acids undergo a condensation reaction forming a peptide bond and the process is continued with subsequent addition of other amino acids and formation of peptide bongs producing elongation of the polypeptide chain until the desired polymer in the form of protein is formed. The entire process is highly regulated under biological system and the final character of the polymer in the form of a functional protein is primarily determined by the amino acid monomer sequence during polymerization and the process is precisely regulated and controlled by an interactive set of genetic and biomolecular system collectively forming a protein synthesizing machinery.

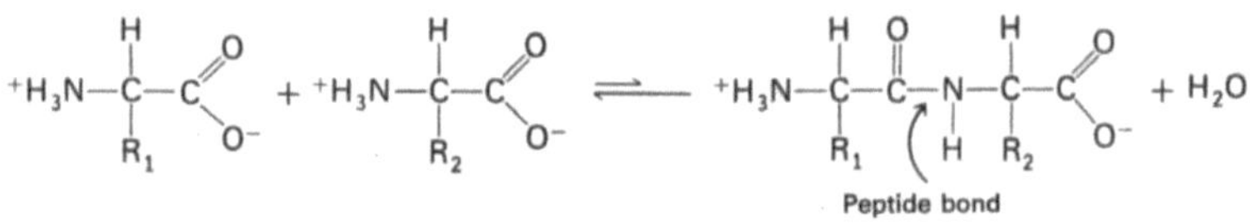

Formation of peptide bond

Proteins, the polymer of amino acids, are highly specific in structure and the overall structure of a protein is known as conformation. All proteins are conformation dependent for their biological function. Conformation of a protein is generated from the specific sequence of the amino acids during formation of the polymer in association with multiple intrinsic and extrinsic factors. The process of attaining the conformation after polymerization of amino acids in to polypeptide and acquiring the conformation is known as protein folding.

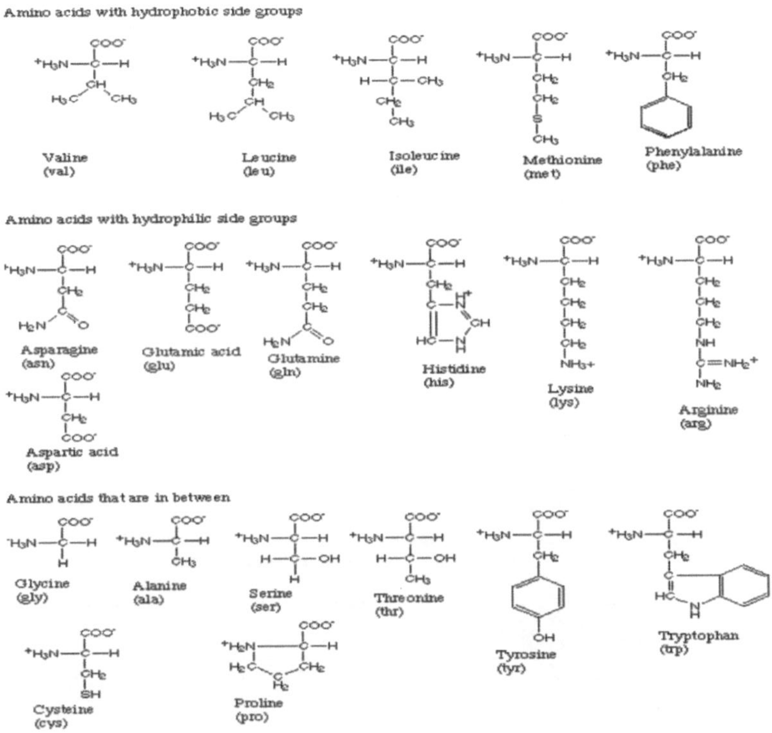

Amino acids for polypeptide and protein formation and their character

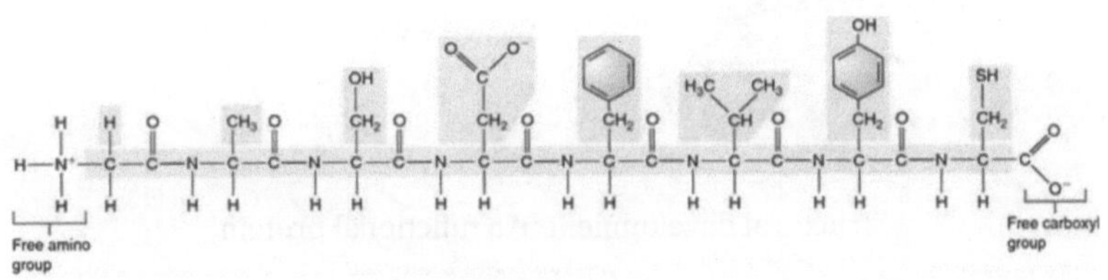

A polypeptide chain

Once the amino acids polymerize to form a polypeptide and ultimately folded in to a protein with specific conformation, the conformation of the entire polymer in the form of a functional protein is stabilized by combination of various internal and environmental forces and interactions. These interactive forces stabilize the polymer in the form of a functional protein.

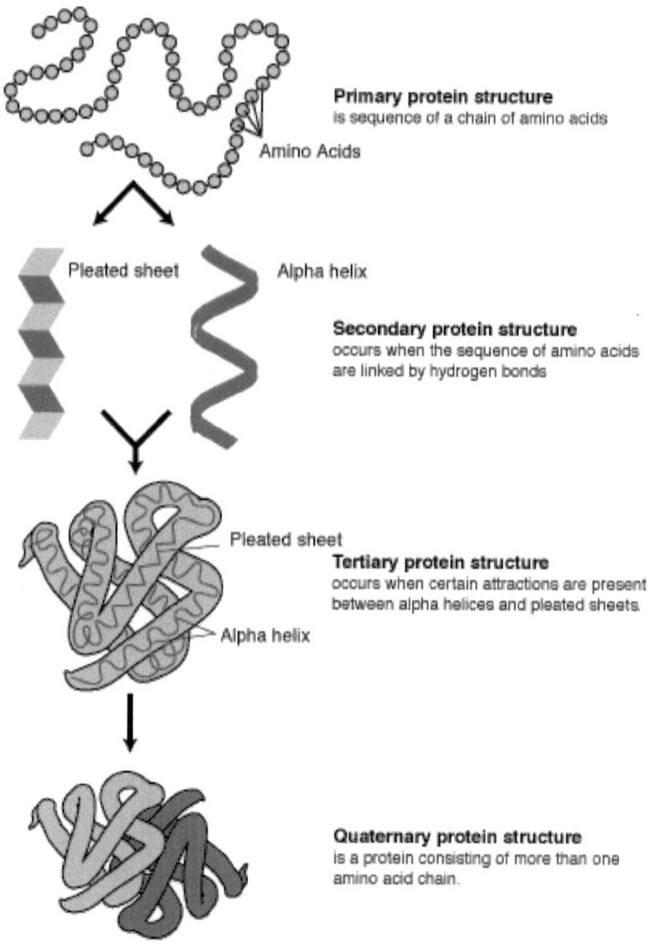

Structural development of a functional protein

1.6 Polynucleotides

Polynucleotides are the group of natural polymers in which the monomeric units are formed by a type of biomolecule known as nucleotides. Nucleotides are basically nucleoside phosphates. Nucleoside is the part of a nucleotide that is formed by glycosidic linkage between a monosaccharide and a type of nitrogenous compound called a nucleotide base. The nucleotide base may be of two types: a purine base or a pyrimidine base. Depending on the type of nitrogenous base, the nucleotides may be classed as purine nucleotides or pyrimidine nucleotides. The sugar or monosaccharide of the nucleoside component of nucleotides may be the pentoaldoses ribose or deoxyribose giving rise to the formation of ribonucleotides or deoxyribonucleotides. Nucleotides are synthesized by different biosynthetic pathways and polymerize to form the polynucleotides. In biological system, non-polymerized nucleotides are known as free nucleotides. Free nucleotides have extensive and different roles in metabolism and are the building block or raw material for formation of polynucleotides. Polynucleotides are the biological information molecules that store and transmit genetic information.

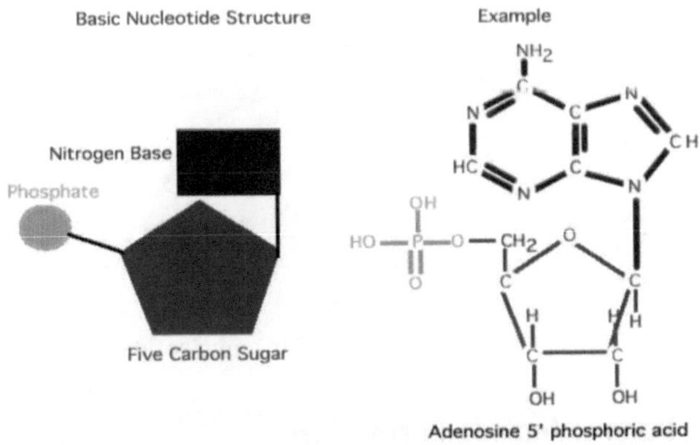

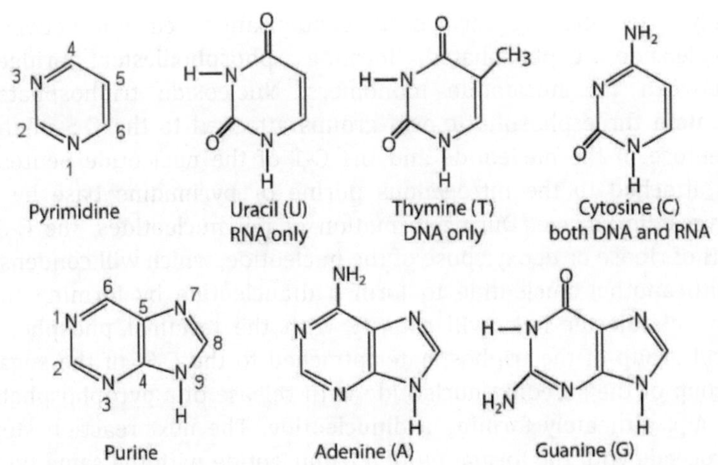

Purine and pyrimidine bases and their occurrence in polynucleotides

The covalent linkage between the base and sugar component of a nucleotide is a type of glycosidic linkage between the pentose sugar ribose or deoxyribose and the nitrogenous purine or pyrimidine base. The link between the nucleoside component and the phosphoric acid is an ester bond. However, it may be noted that, during biosynthesis the glycosidic linkage is initially formed between the phosphopentose and base initiation group and subsequently over that the base is developed to form the nucleotide.

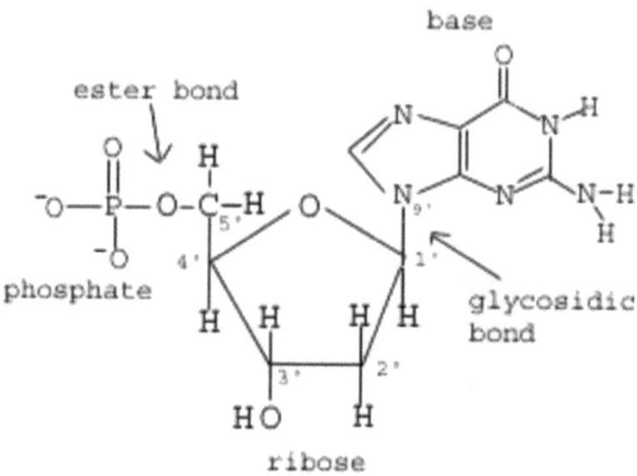

Ester and glycosidic bond in a nucleotide

1.6.1 Polynucleotide Formation

Polynucleotides are formed by condensation reaction between nucleoside triphosphates forming phosphodiester bridges between the nucleotide monomers. Nucleoside triphosphates contain three phosphoric acid groups attached to the C-5 of the pentose of the nucleotide and the C-1 of the nucleotide pentose is attached to the nitrogenous purine or pyrimidine base by a glycosidic linkage. During formation of polynucleotides, the C-3′ OH of ribose or deoxyribose of the nucleotide, which will condense with another nucleotide to form a dinucleotide by forming the phosphodiester link, will esterify with the terminal phosphoric acid group of the triphosphates attached to the C-5′ of the sugar group of the seceding nucleotide with release of a pyrophosphate group, ultimately forming a dinucleotide. The next reaction step proceeds with the formation of a trinucleotide with the same type of condensation reaction between the formed dinucleotide and a new nucleotide triphosphate. The polymerization of nucleotides proceeds with a similar mechanism and finally produces the polynucleotides with several thousand nucleotide monomer units.

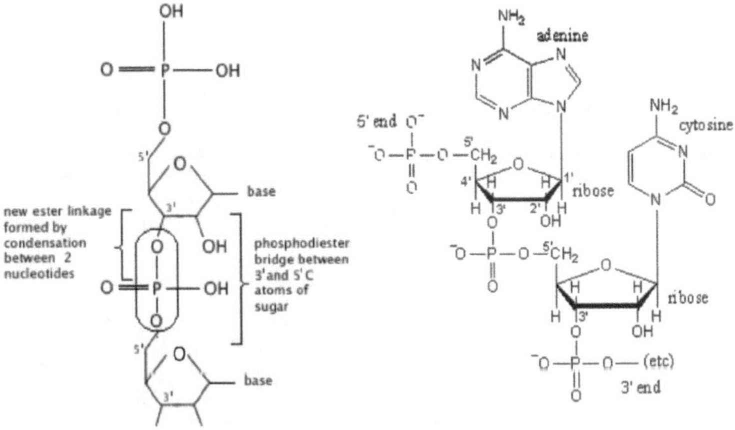

Phosphodiester link in a dinucleotide

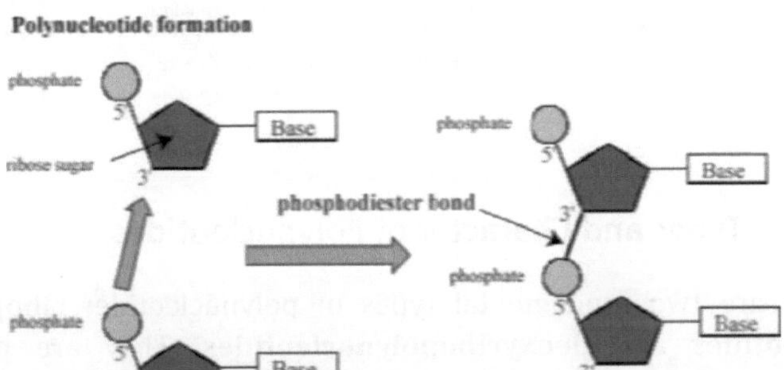

Formation of a dinucleotide

The natural process of polynucleotide synthesis is regulated with very high precision under systems controlled by enzymes and other biological factors and the sequences of nucleotides in the new polynucleotides are always guided by some available preformed sets of polynucleotides acting as guiding templates during the process. Basically the processes produce a polynucleotide by polymerizing nucleotide triphosphates, either by replicating a pre-existing polynucleotide or by taking pre-existing polynucleotides as guiding templates and synthesizing new polynucleotides by condensing the nucleotide triphosphates monomers.

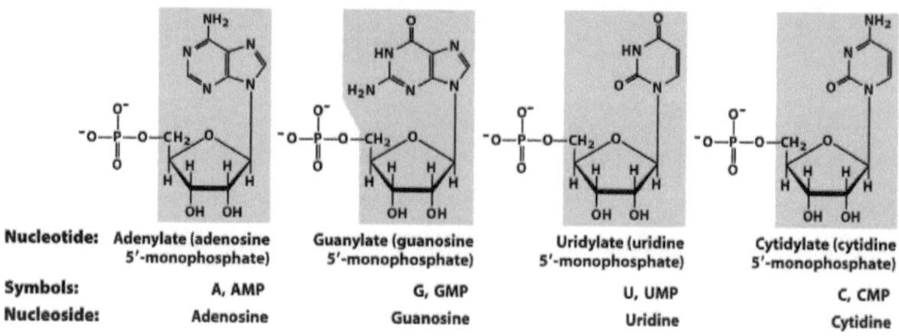

Ribonucleotides

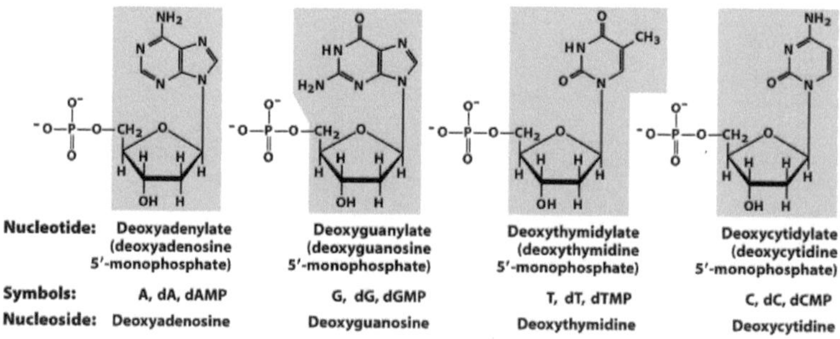

Deoxyribonucleotides

1.6.2 Types and Character of Polynucleotides

There are two fundamental types of polynucleotide: **ribopolynucleotides** and **deoxyribopolynucleotides**. They are popularly known as **ribonucleic acids** (RNA) and **deoxyribonucleic acids** (DNA) due to presence of the series of ionizable phosphate groups in the phosphodiester links producing the polynucleotide. Ribonucleic acids contain ribose as the pentose sugar in the constituent nucleotides and DNAs contain deoxyribose as the pentose sugar in the constituent monomer nucleotide. The difference between ribose and deoxyribose is in the atoms attached with the C-2 in these two pentoses. In ribose, the C-2 is attached with one H and one –OH apart from C-1 and C-3; but in deoxyribose, C-2 is attached with two hydrogen atoms and there is no oxygen containing –OH group in C-2 and so it is a deoxysugar producing a deoxynucleotide.

The minor difference in structure of ribose and deoxyribose produces some major and biologically very important differences in structure, character, and function of the two types of ribose and deoxyribose containing polynucleotides, in the form of RNAs and DNAs. The presence or absence of the oxygen atom at the C-2 of the pentoses in the formed polynucleotide determines how much near two polynucleotide chains can approach and orient with each other on polymerization and form hydrogen bonds between the nitrogenous purine and pyrimidine bases of adjacent polymer chain. Polynucleotide strands composed of deoxyribonucleotides approach each other easily and form hydrogen bonds between base pairs of each strands; forming double-stranded deoxypolynucleotides giving more structural stability to the double-stranded polymer in the form of DNA. Polyribonucleotides do not form hydrogen bonds easily due to steric hindrance and produce RNA, which is a single-stranded polynucleotide with relatively less structural stability.

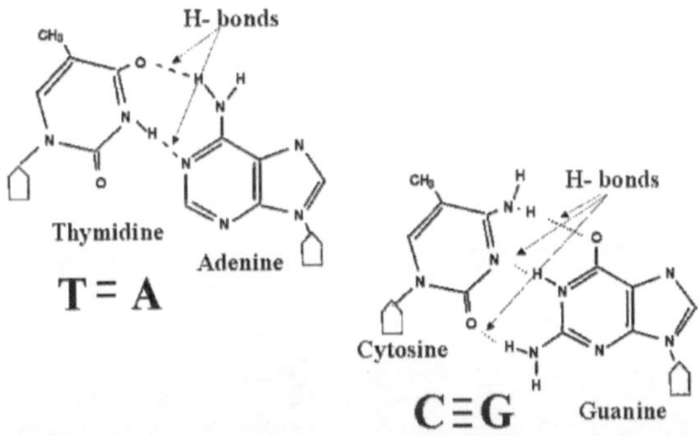

Hydrogen bonding and base pairing in deoxyribonucleotides

The capacity to form hydrogen bonds and base pairing by deoxyribo polynucleotides forming the DNAs imparts very high levels of stability to these polymers formed by nucleotide monomers. Due to its higher stability, DNA is naturally selected as the ideal polymer molecule for storage of information. The information in DNA is stored in the form of specific sequence of the four types of deoxyribonucleotides arranged in the linear chain of one strand of the polydeoxyribonucleotide for one strand of a DNA. There is another complimentary DNA forming base pairs with the other strand giving rise to the complete double-stranded DNA molecule. In RNAs formed by polymerization of ribonucleotides, the base pairing between strands of the polynucleotide is hindered and so RNAs usually remain single stranded. The single-stranded characters of RNAs make them relatively less stable than DNAs and they are more suitable for short-term functional existence. RNAs are commonly used as transporter of stored information from the DNA to locations of information dependent regulation and synthesis in metabolic processes of living systems. RNAs basically function as DNA transcripts.

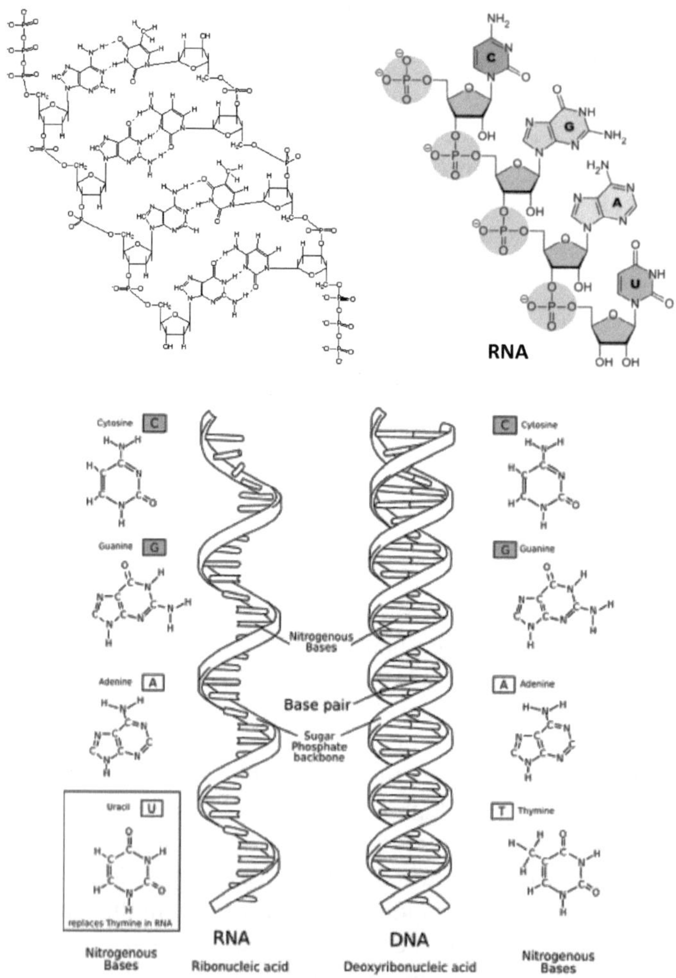

Structure and components of RNA and DNA

1.7 Synthesis of DNA

The basics of chemistry for the synthesis of DNA by the polymerization of deoxyribonucleoside triphosphates (dNTP) involve condensation of dNTPs or oligo-dNTP and dNTP monomer by phosphodiester linkages and maintenance of the chain elongation process to obtain the final DNA with specific numbers of nucleotides in specific sequences and appropriate base pairing through hydrogen bonds.

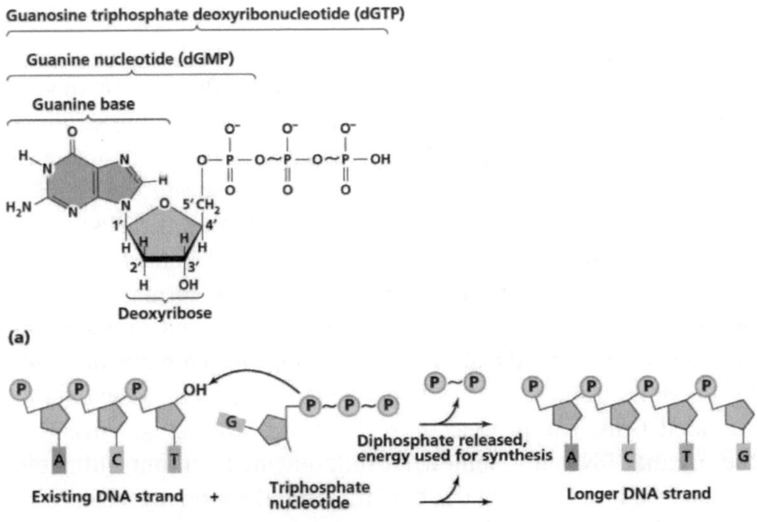

Basics of DNA synthesis

However, in biological systems, synthesis of these polymer macromolecules with maintenance of all its structural and functional integrity is a highly complicated and precisely regulated process with involvement of multiples of accessory molecules, molecular assemblies, and interrelated regulatory events. Structurally, DNA is a very large double-stranded polymer macromolecule with thousands of nucleotides arranged in a precise sequence, which is very neatly and compactly packed, and functionally, this packed molecule preserves the information necessary for the genesis of the entire proteome in the sequence of its nucleotides. This requires periodic unpacking of the compactly packed molecule either for synthesis of new DNA for transfer to a new living system or to transcribe a fragment of the total information content for some other biological work as synthesis of proteins. For both the processes, it requires systematic unpacking and repacking of the DNA molecules, which act as the primary information source for synthesis of a polynucleotide as either DNA or RNA. During the process of DNA synthesis, actually a pre-existing DNA is replicated and so it is called DNA replication.

1.8 Fundamentals of DNA Replication

For the replication of DNA, the DNA that is to be replicated is sequentially unfolded and uncoiled under influence of specific enzymes and factors. The strands of the uncoiled double-stranded DNA is than separated and kept separated by protein factors associated with the process. One of the single strand than acts as guide to select specific nucleoside triphosphates (dNTPs) to be condensed and polymerized by phosphodiester bond formations under catalytic activity of specific enzymes. On formation of the phosphodiester bonds, the base pairing is achieved between the original single strands of the DNA acting as template or guide for the new strand of DNA polymerized and forms the new double-stranded DNA. The process goes on in the other single strand of the original DNA in a somewhat different manner, but ultimately it also produces a new double-stranded DNA finally replicating a new double-stranded DNA from an original double-stranded

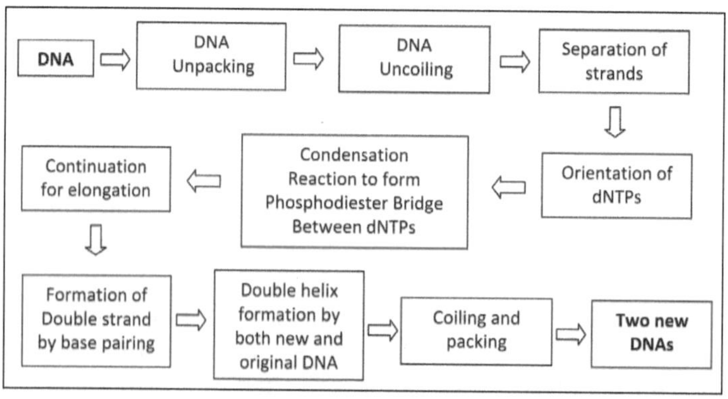

Fundamental steps of DNA replication

DNA where, in both the DNAs, one strand is new and one strand is the original.

In describing the process of replication of DNA, the original DNA that is replicated is called a mother or parent DNA and the new DNA formed is known as the daughter DNAs. The daughter DNAs has one strand from parent DNA and one strand newly formed during replication. Therefore, the process of DNA replication is semi-conservative where all replicated DNAs contain one original DNA strand. The stage of unwinding is initiated and maintained by **topoisomerase** and **helicase** enzymes. Topoisomerase regulate unwinding and over winding of DNA and helicase unwind the parental double helix and prepare the single strands for replication. The single DNA strands are maintained separated by a type of protein called single strand binding proteins (SSBP). The process of polymerization is continuous from the 3′ to 5′ direction of the parent template single strand DNA catalyzed by the enzyme **DNA polymerase**. This strand with continuous polymerization is called a **leading strand**. In the other strand, the polymerization is discontinuous and it is called a **lagging strand**. In the lagging strand, polymerization is mediated by the enzymes **primase** and a different type of polymerase (**pol-α**) producing discontinuous fragmented polymerization. The nascent DNA fragments are known as **Okazaki fragments**, which are subsequently linked or joined together to produce the continuous strand by the enzyme **DNA ligase**. Both the leading and the lagging strands finally form two double-stranded DNA.

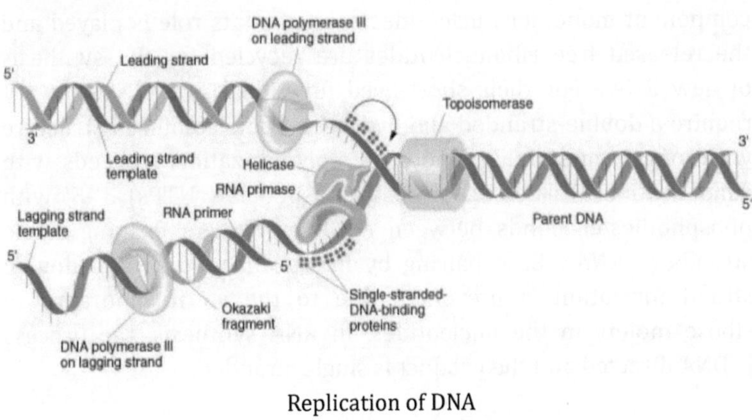

Replication of DNA

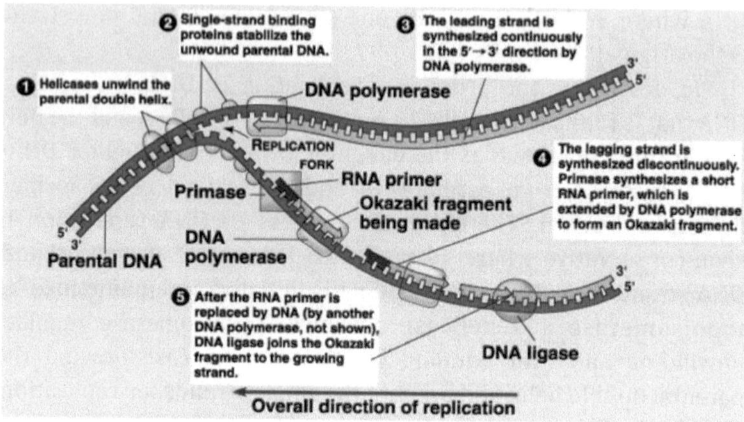

Summary of events in replication of DNA

1.9 RNA Synthesis

RNA is synthesized from ribonucleoside triphosphates as single-stranded polynucleotides using a strand of DNA as a template for maintaining the sequence of nucleotides in the specific RNA. The primary function of RNA is to carry a transcript of the information stored in DNA in the form of certain specific sequences of nucleotides in some segments of DNA to locations where these transcribed information is utilized in processes for expression of the genome. For serving its primary function RNAs are synthesized as and when necessary and degraded to its component monomer nucleotides soon after its role is played and the released free ribonucleotides are recycled for the synthesis of new RNAs. For their short-lived functional role, RNAs do not require a double-stranded and hydrogen bond–stabilized structure with long-term stability. The basic polymerization proceeds with condensation of ribonucleoside triphosphates (rNTPs) 3′-5′ with phosphodiester bonds between rNTP monomers forming single strands of RNAs. Base pairing by hydrogen bonding and double strand formation is prevented due to the steric hindrance of ribose moiety in the nucleotides. In RNA synthesis, the process is DNA directed and the product is single stranded.

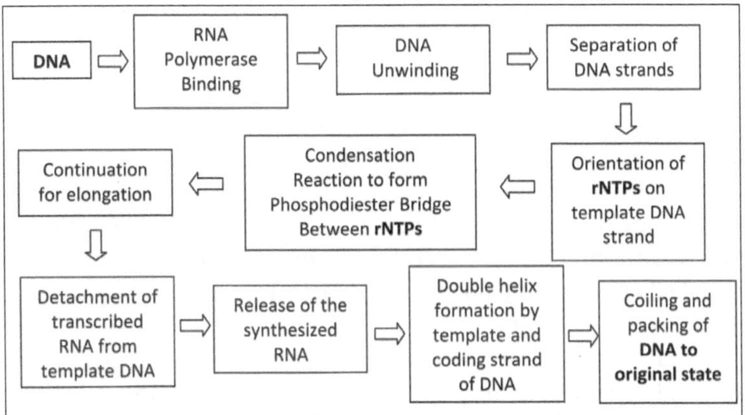

Fundamental steps of RNA synthesis by transcription

The process is initiated by interaction of RNA polymerase to a portion of DNA to be transcribed. The binding of RNA polymerase locally separates the two strands of DNA and one strand of the separated DNA act as a template for RNA synthesis and the other strand act as the coding strands for the nucleotide sequence in the transcribed RNA. When the required portion of the DNA is transcribed in to RNA, the synthesized RNA is detached as a single-stranded polymer and the separated DNA strands reunite by base pairing to form the original double-stranded DNA.

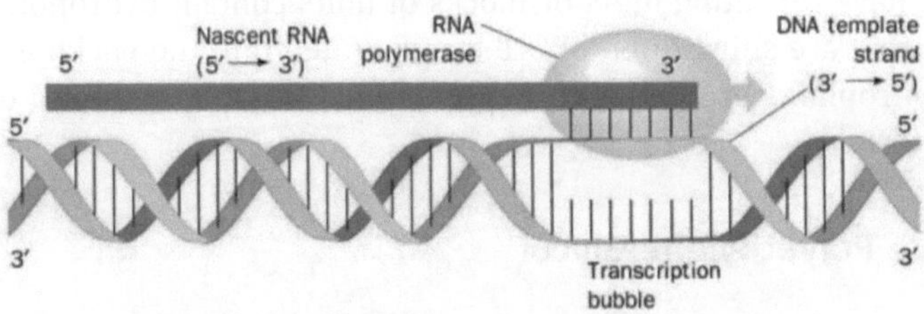

5′—GCGGCGACGCGCAGUUAAUCCCACAGCCGCCAGUUCCGCUGGCGGCAUUUU—3′ **mRNA**
3′—CGCCGCTGCGCGTCAATTAGGGTGTCGGCGGTCAAGGCGACCGCCGTAAAA—5′ **Template strand of DNA**
5′—GCGGCGACGCGCAGTTAATCCCACAGCCGCCAGTTCCGCTGGCGGCATTTT—3′ **Coding strand of DNA**

RNA synthesis and relationship among nucleotide sequences in transcribed RNA, template strand of DNA and coding strand of DNA through the process of transcription.

1.10 Applications of Polymers

With recent advancement in polymer science, there has been the innovation of some novel drug delivery systems using polymers. Either the polymers used for pharmaceutical application are obtained from natural resources, which may be chemically modified for various applications, or some are synthesized chemically for use. A large number of natural and synthetic polymers are water soluble. Water-soluble polymers for pharmaceutical application are synthetic and natural.

Initially polymers were used as solubilizers, stabilizers, and mechanical supports for sustained release of drugs. However, over a period of time, the functionalities of polymers have changed. The polymers have been synthesized and designed to suit specific needs or rather solve specific problems associated with development of drug delivery systems.

1.11 Synthetic Water-Soluble Polymers

Synthetic water-soluble polymers are substances that dissolve, disperse or swell in water and by these processes modify the physical properties of aqueous systems in the form of gelation, thickening or emulsification, and stabilization. These polymers usually have repeating units or blocks of units contain hydrophilic groups that are substituted or are incorporated into the backbone. The hydrophilic groups may be nonionic, anionic, cationic, or amphoteric.

1.11.1 Polyethylene Glycol

A low polydispersity index (PDI) is a criterion for the polymer to have pharmaceutical applications. Lower PDI value makes the polymer more homogenous to provide reliable residence time in the body system. This is fulfilled by polyethylene glycol (PEG) as it has PDI of 1.01. In addition, PEG shows a high solubility in organic solvents and so end-group modifications are relatively easy. PEG is suitable for biological applications because it is soluble in water and has low intrinsic toxicity. The high hydrophilic nature of PEG enhances the solubility of hydrophobic drugs or carriers when conjugated with them. It enhances the physical and chemical stability of drugs and prevents aggregation of the drugs in vivo and also during storage. PEG helps in reducing the aggregation of red blood cells and thereby improves the blood compatibility of PEG copolymers that are implanted as cardiovascular devices as stents. It is primarily used in storage of blood and organs. Temperature-responsive and chemically cross-linked hydrogels may be formed from PEG. Temperature-responsive systems have become increasingly affective as injectable drug delivery systems.

PEG–drug conjugates are investigated for a variety of molecules and drugs as insulin, daunorubicin camptothecin, peptides, and lipids. The main advantages of using PEG–drug conjugates are reduced protein immunogenicity, increased residence time in the body, and reduced enzymatic degradation. These ensure that the drug reaches the site of action and prevents clearance from the body because it is not recognized as a foreign body. Many of the conjugated drugs and liposomal and micellar formulations are PEG-containing products. Some drugs delivered as PEGylated products are adenosine deaminase, asparginase, doxorubicin, interferon, etc.

In antitumoral therapy, the conjugation of PEG with enzymes looks very promising as several enzymes have proven to be active against various types of cancer by acting through different mechanisms. Enzymes such as asparaginase, methioninase, and arginine deiminase are able to reduce plasma levels of corresponding tumor target amino acids in cancer therapy. The advantage of enzymes is their specificity.

Some PEGylated drugs and their use

PEGlylated drugs	Therapeutic use
Irinotecan	Brest, ovarian, and cervical cancer
Docetaxel	Hormone refractory prostate, lung, lung, gastric cancer
Camptothecin	Gastro-esophageal adenocarcinoma

1.11.2 Polyvinyl Pyrrolidone

Polyvinyl pyrrolidone is a water-soluble polymer having molecular weight ranging from 40,0000 to 360,000 synthesized by polymerization of vinnylpyrrolidone in water or isopropanol. Based on molecular weights **polyvinyl pyrrolidone** (PVP) is available in different grades and is used as a binder in tablet formulations. Compared to other binders, wet granulation with PVP having a molecular weight of 25,000–90,000 gives harder granulates with good flow ability. When the active substances have poor aqueous solubility, they have limited bioavailability. By adding PVP, their solubility is increased with resulting increase in bioavailability. Addition of PVP form water-soluble complexes with many active substances and increase the bioavailability. The soluble PVP grades are also useful for preparing solid solutions and dispersions because of their good hydrophilization properties. Soluble grades of PVP and polyvinyl pyrrolidone-vinyl acetate (PVP-VA) as copolymer can be used to improve the bioavailability of many water-soluble drugs with poor solubility such as indomethacin, tolbutamide, and nifedipine.

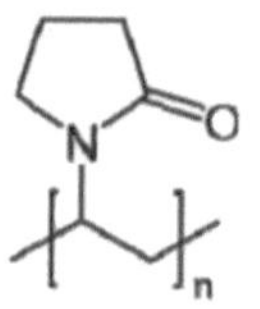

Polyvinyl pyrrolidone

Some common applications of PVP

Function of PVP as	Used in pharmaceutical form
Binder	Tablets, capsules, granules
Improved Bioavailability	Tablets, pellets, suppositories, transdermal systems
Film forming agent	Tablets, ophthalmic solutions
Solubilizing agent	Oral, parenteral and topical solutions
Taste masking	Oral solutions, chewing tablets
Lyophilizing agent	Injectables, oral lyophilisates
Stabilizer	Suspensions, dry syrups
Hydrophilizer	Sustained-release forms of suspensions
Adhesive	Transdermal systems, adhesive gels
Stabilizer	Enzymes in diagnostics, different forms
Toxicity reducer	Injectables, oral preparations, etc.

1.11.3 Polyvinyl Alcohol

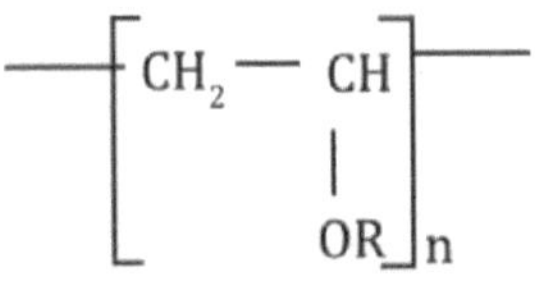

Where R = H or COCH3

Polyvinyl alcohol is synthesized by the polymerization of vinyl acetate to polyvinyl acetate (PVAc) which is hydrolyzed to get polyvinyl alcohol (PVA). PVA is soluble in highly polar and hydrophilic solvents as water, dimethyl sulfoxide, ethylene glycol, and N-methyl pyrrolidone. Water is the most common solvent for PVA. The solubility of PVA in water depends on the degree of polymerization, hydrolysis, and solution temperature. Any change in these three factors affects the degree solubility of PVA by changing the character of hydrogen bonding. PVA is mainly used in topical pharmaceutical and ophthalmic formulations. It is used as stabilizer in emulsions and viscosity increasing agent for viscous formulations as ophthalmic products.

1.11.4 Polyacrylic Acid

Polyacrylic acid (PAA) is a biodegradable water-soluble polymer with a wide range of medicinal applications and is considered pharmaceutically safe. They are used for oral and mucosal contact applications in the form of controlled release tablets, oral suspensions, and bioadhesives. The unique property of PAA is that it exists as a liquid at pH 5 and as a gel at pH 7. Permeation of cations into the gelled polymer converts the gel back to a liquid. It is an ideal polymer used for ocular delivery of ribozymes to the corneal epithelium as a drug delivery vehicle.

1.11.5 Polyacrylamide

Polyacrylamide is a synthetic polymer derived from acrylamide monomer introduced as a support matrix for electrophoresis. Polyacrylamide gels result from polymerization of acrylamide with a suitable bifunctional cross-linking agent, most commonly, *N,N'*-methylenebisacrylamide or bisacrylamide.

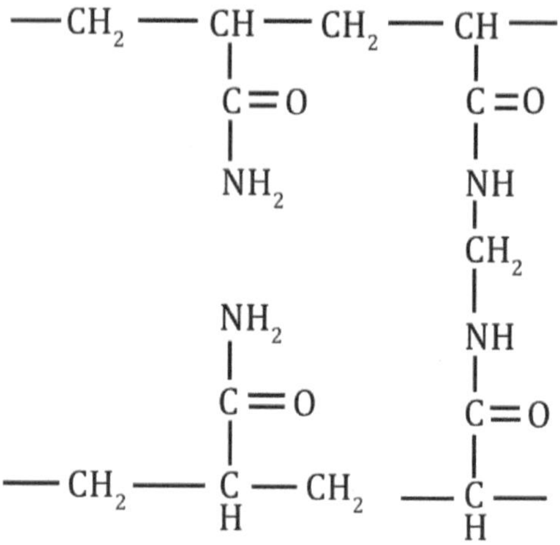

Cross link

Polyacrylamide is used in a wide range of cosmetic products such as moisturizers, lotions, creams, self-tanning products, etc. Polyacrylamides were first used as an implantable carrier for sustained delivery of insulin to lengthen the life of diabetic rats. Since then, various drug delivery systems based on polyacrylamide have been developed. It is also used as a carrier for other bioactive macromolecules and cells to produce the desired effects. Polyacrylamide–chitosan hydrogels are biocompatible and are used for sustained antibiotic release. Polyacrylamide is also used in extra corporeal toxin-removing devices, which remove unwanted toxic substances and subsequently returns the detoxified component to the circulation effectively. The function of polyacrylamide in an extracorporeal toxin removal modality is to provide a support matrix for immobilization of the functional parts or ligands.

1.11.6 Divinyl Ether-Maleic Anhydride

Divinyl ether-maleic anhydride (DIVEMA) is a water-soluble polymer that has antitumor activity against various types of tumors. The biological activities of DIVEMA are due to its ability to activate immunocompetent macrophages and natural killer cells. DIVEMA may be used as a drug carrier for superoxide dismutase and anticancer agents such as adriamycin and methotrexate.

1.11.7 Polyoxazoline

Polyoxazolines are structurally similar to peptides. Their physicochemical properties can be modulated by varying the alkyl substituent. Properties of polyoxazolins range from high hydrophilicity, which enables synthesis of hydrophilic water-soluble biocompatible polymers with good antibiofouling properties through thermal sensitivity of thermo responsive polymers, to hydrophobicity, typical for hydrophobic aromatic or aliphatic polymers. They have the ability to form nanostructures, and they act as versatile polymers. Polyoxazolines are used as adhesive and in coating formulations, and in various drug delivery applications. They may be used for the synthesis of lipo-polyoxazolines-poly (2methyl-2-oxazoline) and poly (2-ethyl-2-oxazoline)-based lipid conjugates as an alternative to PEG-based materials.

1.11.8 Polyphosphates

Biodegradable polyphosphoesters (PPE) such as polyphosphates and polyphosphonates are also used in drug delivery, gene delivery, and tissue engineering. These polymers have a backbone consisting of phosphorous atoms attached to either carbon or oxygen. The uniqueness of this class of polymer is in the chemical reactivity of phosphorous, which enables attachment of side chains to alter the biodegradation rates and molecular weight of the polymer.

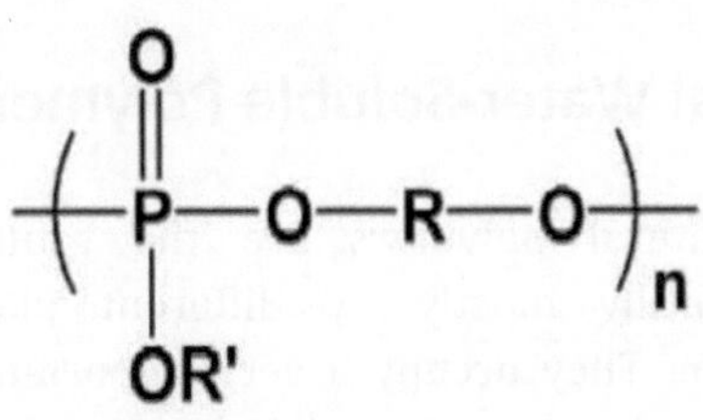

General structure of polyphosphoesters where R = divalent organic groups.

Water-soluble positively charged polymers are useful for gene delivery. Positively charged polymer interacts with negatively charged DNA by electrostatic interactions resulting in the formation of complexes and thereby they provide protection to DNA from enzymatic attack. This also enables greater cellular uptake of DNA. The commonly used cationic polymers for gene delivery are the ones having amide bond like poly(L-lysine) and vinyl bonds like in polyethyleneimine (PEI) as they show very good stability in aqueous solutions.

1.11.9 Polyphosphazenes

Polyphosphazene belongs to a class of polymers with inorganic moiety as the main chain and two active chloride groups on each repeat unit. Substitution of these chloride groups gives multifunctional polyphosphazenes with adjustable physicochemical and biological properties. These polymers are used to formulate nanofibers and hydrogels. They are used for delivery of water-soluble anticancer agent like Doxorubicin HCl. Some water-soluble polyphosphazenes containing ionic groups are used to formulate hydrogel microspheres or nanospheres for controlled release and drug delivery applications. These are ideal for protein encapsulation as they do not use organic solutions or heat. Polymers include polyphosphazene immune adjuvants, which are also formulated in microspheres and used for mucosal immunization study. Water-soluble polyphosphazene containing amino aryloxy and methyl amino side groups are synthesized and tried as an inert polymeric carrier for the covalent attachment of biologically active agents.

1.12 Natural Water-Soluble Polymers

Water-soluble natural polymers are the groups of polymer synthesized naturally mostly by different plants and lower forms of organism. They occupy a very important portion of the total pharmaceutical application of polymers. Some of the most frequently used polymers under this group are presented with a short description.

1.12.1 Xanthan Gum

Xanthan is a free-flowing powder soluble in both hot and cold water to give viscous solutions at low concentrations. Its industrial importance is based upon its ability to control the rheology of water-based systems. It is a very effective thickener and stabilizer as it gives highly viscous solutions even at low concentrations as compared to other polysaccharide solutions. Xanthan gum solutions exhibit pseudoplastic behavior as viscosity is regained immediately even at high shear rates. Its pseudoplastic nature enhances feeling effect of mouth and release of flavor sensation. Xanthan gum solutions offer very good stability. They are least affected by changes in pH and are stable in both alkaline and acidic conditions. The solution properties of xanthan are not affected in a pH range of 1–13. Xanthan is compatible with most commercially available thickeners such as sodium alginate, carboxymethyl cellulose, and starch.

Xanthan gum is widely used in cosmetics and in toothpastes. It can be easily extruded from the tube or dispenser because of the shear thinning flow behavior. It also ensures that the paste will keep a stable hold on the brush. The shear thinning characteristics also improve the dispersion on and the rinsing from the teeth. Toothpastes thickened with xanthan gum have a bright, shiny cord with short flow behavior. Xanthan gum is used as a thickener and stabilizer in personal care and cosmetic products like creams, eye gels, etc. Typical xanthan gels feel very gentle and soft due to their shear thinning low behavior. In emulsions or suspensions for pharmaceutical use xanthan gum prevents the separation of insoluble ingredients, e.g., barium sulfate in X-ray contrast media. Most of the ready to eat, semi-prepared foods and convenience foods would not be possible without stabilizers and thickeners. In order to adjust the desired flow behavior, xanthan gum is often used in combination with other hydrocolloids.

1.12.2 Pectins

Pectin is a made up of mixture of polysaccharides. Pectins are mainly obtained from citrus peel or apple pomades. Pectins are soluble in pure water. Monovalent cation alkali metal salts of pectinic and pectic acids are soluble in water; di- and tri-valent cations salts are weakly soluble or insoluble. Dry powdered pectin, when added to water, forms clumps. This clump formation can be prevented by dry mixing pectin powder with water-soluble carrier material or by the use of pectin having improved dispersibility. Other properties such as viscosity, solubility, and gelation are generally related. For instance, factors that increase gel strength will increase the tendency to gel, decrease solubility, and increase viscosity, and vice versa. These properties of pectins are a function of their structure, which is that of a linear polyanion polycarboxylate. Monovalent cation salts of pectins are highly ionized in solution, and the distribution of ionic charges along the molecule keeps it in an extended form by coulombic repulsion.

Pectin has been used in the pharmaceutical industry for a wide range of applications. Pure and standardized pectin has is used as a binding agent in tablets. High methoxy pectin is used as monolithic bioerodible system, preparation of directly compressible tablets. Low methoxy pectin is used to prepare beads by ionotropic gelation technique, sustained-release drug delivery using calcium pectinate gel beads. Pectin based microspheres are also prepared by emulsification technique. Film coated tablets may also be prepared using combination of HM-pectin and ethyl cellulose aqueous dispersion, HM or LM pectin with chitosan mixtures.

Pectin also has several unique properties that have enabled it to be used as a matrix for the entrapment and/or delivery of a variety of drugs, proteins, and cells. Pectin helps in reduction of blood cholesterol in a diverse group of subjects. At least 6 g/day of pectin is effective in reducing cholesterol levels but amounts less than 6 g/day of pectin are not effective. Pectin is also used as a thickening, stabilizing and gelling agent in the food and beverage industry. It effectively removes lead and mercury from the gastrointestinal tract and respiratory organs. Intravenous administration of pectin shortens the coagulation time of blood and help in controlling hemorrhage or local bleeding.

Pectin hydrogels are used as a binder in tablet formulations and have been used in controlled-release matrix tablet formulations.

Using an extruder, spherical pellets containing calcium pectate were prepared. These were then coated in pectin solution resulting in the formation of insoluble calcium pectinate gel around the pellets.

1.12.3 Chitosan Derivatives

Chitin and chitosan have been used extensively in many areas ranging from food processing to waste management, medicine, biotechnology and pharmaceutical industries. Chitosan in particular is used widely in pharmaceutical applications as a formulation excipient because of its biodegradable, biocompatible and less toxic nature. It has been used as a mucoadhesive, oral absorption enhancer, and in protein and gene delivery. The main drawback with chitin and chitosan is their difficulty in dissolving them in water and in neutral pH. Therefore, water-soluble derivatives of chitosan and chitosan have been synthesized by chemical modification. These chemical modifications result in the formation of hydrophilic chitin or chitosan which have more affinity to water or organic solvents. Carboxymethylation of chitosan results in formation of *N*-carboxymethylchitosan (N-CMC), which is soluble in a wide range of pH.

1.12.4 Dextran

Dextran is by fermentation of media containing sucrose by *Leuconostoc mesenteroides*. Dextran is an α-D-1,6-glucose-linked glucan with side-chains 1–3 linked to the backbone units of the dextran biopolymer. Fractions of dextran are readily soluble in water forming clear, stable solutions. The solubility of dextran is not affected by pH. They are also soluble in other solvents like methyl sulfide, formamide, ethylene glycol, and glycerol. Dextran fractions are insoluble in alcohols like methanol, ethanol and isopropanol, and also most ketones, as acetone and 2-propanone. Dextran is used as macromolecular carriers. Dextran hydrogels are obtained in various ways, based on either chemical or physical cross-linking. Dextrans cross-linked with methacrylate and hydroxyethylmethacrylate have been used as hydrogel implants and microspheres for scaffolds. Conjugation of thrombolytic enzymes with water-soluble dextran carrier produces a stabilized enzyme preparation with longer circulation time and reduced immunogenicity. These preparations are used for treatment of acute myocardial infarction, acute pulmonary artery thromboembolism, and peripheral arterial and deep vein thrombosis. In comparison with the native enzyme, streptodekaza, it has a prolonged life-time in the circulation.

1.12.5 Carrageenan

Three main types of carrageenan that are widely used in food industry are called iota, kappa, and lambda carrageenan. Carrageenan is considered to be a good substitute for gelatin in hard and soft gel capsules. The incorporation in glycerin–water mixture masks the chalkiness of antacid gels. It is used in both topical bases and suppository bases. The active ingredients are trapped inside the fibers by spinning the insoluble carrageenan fibers. This modification helps in wound healing by absorbing large quantities of water and thereby keeping the wound clean. The texture of any formulation or polyols can be controlled by utilizing the property of unique interactions between carrageenan and polyols. Carrageenan is used as a thickening agent in hand lotions and shampoos thus promoting healthy skin and hair. Many vaginal products have certain drawbacks such as leakage because of their inability to maintain gel-like structure when applied. Carrageenan gels can be modified suitably and can be used for quick rehealing and protection during intercourse.

Carrageenan has unique properties like viscosity, continuous phase gel formation and specific interactions with the abrasive. Combination of these properties helps in stabilizing the toothpaste preparations. The continuous phase gel matrix entraps the abrasive and flavor oil micelles within the gel matrix thereby enhance the emulsion stability. The gel structure also imparts short texture to the toothpaste providing a clean break on extrusion from the tube or pump. Carrageenan helps in dispersing and stabilizing the solids thus preventing hardening, caking and drying out. This is because of interaction between carrageenan and surface of the abrasives. These distinct properties of carrageenan make it unique as compared to other binders used in dentifrice industry. It can be safely used with CMC as it does not contain enzymes, while other binders such as xanthan gum contain enzymes that attack CMC making it unsuitable for use in combinations. Apart from this, carrageenan is widely used in food industry as a thickening agent, stabilizer, gelling agents for a wide range of products like juices, dressings and sauces, beer, and wine.

1.12.6 Guar Gum

Guar gum is used as a binder and disintegrant in tablet formulations. It also acts as a stabilizers, emulsifier, thickening, and suspending agent in liquid formulations. It is widely used for colonic drug delivery applications and for this purpose it may be used in nanoparticulate preparations to deliver the active ingredient to the clone where it slowly release the drug for a longer retention time. The swelling ability of guar gum is used in the retardation of drug release from the dosage forms. Its utility as a carrier for colon specific drug delivery is based on its degradation by colonic bacteria.

1.12.7 Cellulose Ethers

A very wide range of polymer products can be prepared using different cellulose ethers. They differ from each other with respect to type of substituent, substitution level, molecular weight, viscosity, and particle size. The common types of cellulose ethers are hydroxypropylmethyl cellulose (HPMC), hydroxypropyl cellulose (HPC), hydroxyethyl cellulose (HEC), Sodium carboxy methyl cellulose (Na-CMC). Pure cellulose is insoluble in hot or cold water due to strong intramolecular hydrogen bonding. When the cellulose is converted to cellulose esters or cellulose ethers, the derivatives become water soluble. These water-soluble cellulose derivatives are used in wide range of applications. Thus, modified cellulose derivatives enhance water retention capacity, pseudoplastic behavior, film-forming properties, and complexation. The advantages of cellulose ethers are that they are biocompatible and hence can be used for pharmaceutical purposes; cosmetics and food. They are mainly used as binders, coating agents, emulsifying, stabilizing, agents, and tablet disintegrants.

Sodium carboxymethyl cellulose (NaCMC) is used as an emulsifying agent in pharmaceuticals and in cosmetics. It is a polymer of choice because of its wide range of functional properties like binding, thickening, stabilizing agent. NaCMC can also be used in preparation of microspheres by using glutaraldehyde as a cross-linker. Ketorolac tromethamine, which is an anti-inflammatory and analgesic agent, is successfully encapsulated into these microspheres and drug encapsulation.

Hydroxypropyl cellulose (HPC) is a non-ionic water-soluble and pH insensitive cellulose ether. It can be used as thickening agent, tablet binder and modified release and film coating polymer. Buccal delivery formulations containing HPC and polyacrylic acid are used for many years and is used for mucoadhesive delivery systems for several drugs.

Hydroxypropylmethyl cellulose (HPMC) is a water-soluble cellulose ether that is mainly used in the preparation of controlled release tablets. Viscosity is the main variable responsible for controlling the release. The hydration and gel forming abilities of HPMC can be used to prolong the drug release of the active ingredient.

Hyaluronic acid (HA), the natural polyanionic polysaccharide distributed widely in the extracellular matrix and the joint liquid of mammalians is non-toxic, biocompatible mucoadhesive polysaccharide. It has negative charge and it is biodegradable. It is mainly distributed in the connective tissue, eyes, intestine and lungs. The expression of CD-44 receptor is an endogenous ligand for HA and it makes hyaluronic acid a good natural polymer for drug targeting. HA is composed of two sugar units—glucuronic acid and *N*-acetylglucosamine—and is polymerized into large macromolecules of over 30,000 repeating units. It is readily soluble in water and produces a gel. The high solubility of hyaluronic acid is problematic in the development of polymers for tissue engineering. This property of HA is more helpful in orthopedic surgery but it requires more chemical stabilization and structural stability. The length of the chain, degree of entanglement, cross-linking, pH, chemical variations affect the viscosity of the gel.

Hyaluronic acid polymers are used in the preparation of gels for delivery of drugs to eye and installation into other cavities. They are used along with other polymers like alginic acid, HPMC, poloxamers, etc., for achieving the desired property in drug delivery systems. Combination of these polymers influences the biophysical properties and alters the pharmacokinetics. HA-based corneal shields have more prolonged steroid delivery than by direct application. Insulin absorption from eye drops via the cornea is enhanced in the presence of HA. HA gel has been successfully used as a carrier mechanism for antibiotics to the eye as the gel prevents tears from washing away the drug and gives a more prolonged release. HA based nanosystems used for gene delivery, cancer, and asthma.

Other applications of HA are as follows wound healing by extracellular regeneration, epithelial regeneration, topical treatment of dry eye syndrome and Sjögren's syndrome. It is also used as a viscosity agent in pulmonary pathology for achieving alveolar patency and as a filler for wrinkles and cutaneous lines.

One interesting use of HA is in the field of viscosurgery and viscosupplementation. In reproductive medicine, HA enhances the retention of the mobility of cryo-preserved and thawed spermatozoa. This property is used to select viable spermatozoa and improve artificial insemination and other in vitro fertilization methods.

1.12.8 Starch or Starch-Based Derivatives

Starch is a natural polymer that is mainly used as a binder, diluent, and disintegrant and also as a sustained-release agent in matrix systems. It is also used as a thickening and gelling agent in food industry. Its low cost, biodegradability and renewability make it a suitable candidate for developing sustainable materials. Starch has many hydroxyl functional groups in its structure and so it is hydrophilic in nature. Native starch is not used because of its poor processability, and poor mechanical properties of the end products. Starch is either chemically or physically modified to improve the properties of starch. Such derivatives have physicochemical properties that are different from the parent while still maintaining the biodegradability. Starch is modified by hydroxypropylation to enhance starch clarity and cold storage stability because the presence of hydroxypropyl groups increases water holding and reduces reassociation of starch chains. This results in formation of a more stable gel. Starch-based biodegradeable polymers have good biocompatibility. Its degradation products are non-toxic and have good mechanical properties. These Starch-based biodegradeable polymers are widely used in bone tissue engineering scaffolds, in drug delivery as microspheres or hydrogels. Modified starches are used as functional ingredients in sustained-release applications because of their improved functionality over their native counterparts. Among them, cross-linked high amylose corn starch is the most extensively used one. The reasons for starch acting as a sustained-release agent is for its gel-forming ability, biodegradability and biocompatibility. The molecular structure of the gel layer and the mechanical and physicochemical characteristics of the matrix such as gel strength and porosity contribute to the sustained-release properties of the matrix.

1.12.9 Albumin

Albumin is a protein polypeptide. Albumin has a molecular weight of 66.5 kDa and is the most abundant plasma protein, which is present in the concentration of 35–50 g/L in human serum and is synthesized in the liver. Human serum albumin (HSA) has a half-life of 19 days. It acts as a solubilizing agent for long chain fatty acids and is therefore essential for the transport and metabolism of lipids. It binds very well to penicillins, sulfonamides, indole compounds, benzodiazepines, copper, and nickel in a specific and calcium and zinc in a relatively nonspecific manner. It is responsible for osmotic pressure of the blood.

Albumin is acidic and a very soluble protein. It is also soluble in 40% ethanol. It is stable in the pH range of 4–9, soluble in 40% ethanol, and highly thermostable even when heated at 60°C for up to 10 h. It is biodegradable in nature and lacks toxicity and immunogenicity. It is very well tolerated and taken up by the tumor tissues. All these properties make it an ideal candidate for drug delivery. It is a versatile protein carrier and is used in drug targeting for achieving better pharmacokinetic profile of peptide or protein based drugs. Purification of albumin is easy and it is soluble in water, which makes it convenient to delivery by injection, and hence it is considered an ideal candidate for nanoparticle preparation. Protein-based nanoparticles have the advantage of greater stability during storage and are easy to scale up as compared to other delivery systems. Covalent derivatization of albumin nanoparticles with drug targeting ligands is possible due to the presence of functional groups as amino and carboxylic groups on the nanoparticle surfaces. HSA based formulations such have good tolerability. So the efficacy of albumin formulations with minimum side effects is guaranteed. It is also suitable for gene delivery.

Chapter 2

Tara Gum, the New Biopolymer for Various Applications: Hydrodynamic Properties

Martin Alberto Masuelli

Laboratorio de Membranas, Instituto de Física Aplicada—CONICET, Área de Química Física, Departamento de Química, Facultad de Química, Bioquímica y Farmacia, Universidad Nacional de San Luis, Chacabuco 917, San Luis 5700, Argentina

masuelli@unsl.edu.ar

2.1 Introduction

Caesalpinia spinosa (Molina) Kuntze [1], commonly known as tara, is a leguminous shrub [2] native of Peru. It is cultivated as a source of tannins and is cultivated as an ornamental plant because of its colorful flowers and inflorescences. *C. spinosa* is in the family *Fabaceae*. It is distributed in the northern and central of South America. Some scientific synonyms are *Caesalpinia tinctoria*, *Poinciana spinosa*, *Caesalpinia pectinata*, *Coulteria tinctoria*, *Tara tinctoria*, and *Caesalpinia stipulata*. It is also known by various common names such as taya, tara, guarango, cuica, vinilo, vainillo, acacia amarilla, and divi divi. It contains polysaccharide derivatives of mannose (neutral mucilage). *C. spinosa* produces a galactomannan

Handbook of Sustainable Polymers: Processing and Applications
Edited by Vijay Kumar Thakur and Manju Kumari Thakur

ISBN 978-981-4613-53-8 (Hardcover), 978-981-4613-54-5 (eBook)
www.panstanford.com

(Tara gum) soluble with an intermediate galactose–mannose ratio of locust bean gum and guar gum. It is obtained by crushing the seed endosperm.

Among other uses, pulverized pods are used by its tanning properties. Certain gums extracted from legume seeds are very similar in terms of their rheological properties, such as gum Tara (*Caesalpinia spinosa*), guar gum (*Cyamopsis tetragonolobus*), and locust bean gum (*Ceratonia siliqua*), the latter two having been used since ancient times [3].

2.1.1 Botanical Aspect

Caesalpinia spinosa is in the family Fabaceae [4] and was discovered in northwestern and center of South America (Colombia, Ecuador, Peru, Bolivia, and north Argentina (Jujuy and Salta) [5–17].

Figure 2.1 *Caesalpinia spinosa* tree, fruit, and flowers [17].

C. spinosa reaches a height of 2–5 m; its bark is dark gray, with scattered thorns and hairy branches. The leaves are alternate, evergreen, lacking stipules, bipinnate, and lacking petiole and rachis glands (see Fig. 2.1). The sheets comprise 3 to 10 pairs of primary leaflets 8 cm long and 5 to 7 pairs of leaflets subsessile elliptic secondary, each approximately 1.5 to 4 cm long. The inflorescences are terminal 15–20 cm-long clusters with many flowers and covered with tiny hairs. The flowers are yellow to orange petals 6–7 mm, and the lowest sepal is shaped like a boat; many marginal teeth are colored stamens yellow, irregular in length and barely protruding. The fruit is an oblong flat, indehiscent, about 6–12 cm long and 2.5 cm wide, containing 4 to 7 round, black seeds, which turn red when mature [18–22].

The primary monosaccharide containing Tara gum is mannose (neutral mucilage), obtained by crushing the endosperm of the seeds. Soluble galactomannan of Tara gum has the galactose–mannose intermediate relation between LBG and guar. The Tara gum chiefly consists of polysaccharides of high molecular weight composed mainly of galactomannans [23]. The principal component consists of a linear chain of (1,4)-β-D-mannopyranose units with α-D-galacto-pyranose units attached by (1,6) linkages [24]. The ratio of mannose to galactose in Tara gum is 3:1; in carob bean gum, this ratio is 4:1 and in guar gum 2:1 [25].

2.1.2 Obtension Process

The gum is contained in a portion of the seed call endosperm. This is the food reserve for development of the embryo during germination. As the seed is dicotyledonous, you have two for each seed endosperm. The endosperm is surrounding the embryo and a shell of a blackish brown color surrounds these. In the severe case of impairment, the seed turns black and rubber and performance is very low when the gum contaminated with yellow spots to gray [26]. The Tara seeds measure 0.6 to 0.7 cm. and endosperm are 22–24% of the seed weight. The endosperm is crushed and sold as Tara gum.

In commercial processing of Tara gum, a variety of methods are used effectively to separate the endosperm from the husk and the germ or embryo. A thermal-mechanical differential grinding process separates the husk, germ, and endosperm, and there is a difference in the hardness of each component. Then white colour is classified by optical selector as a 98%; after finishing, the endosperm is cleaning 100% at the end of the process. The separated endosperm, which contains 80% galactomannan, is finally ground to a fine particle size and is sold as Tara gum. This grinding is done in several phases and sifting. Hammer mills, roller, or friction is used.

The characteristics of Tara gum produced depend on how it was processed and similar to other galactomannans [27–30] (see Fig. 2.2).

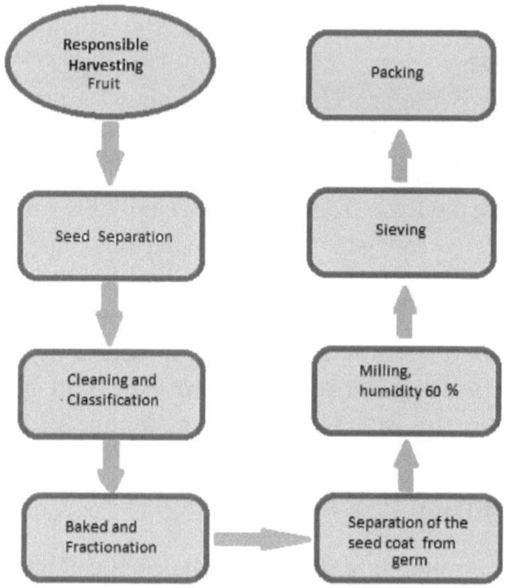

Figure 2.2 Schematic of Tara gum production [17].

2.1.2.1 Food grade

Food-grade Tara gum is obtained by the purification of the endosperm. The determination of moisture, protein, and acid-insoluble residue, generally regarded as the minimum monitoring requirement, tests the efficiency of the purification process: The presence of the germ is reflected in a high protein, shell residues show a high acid-insoluble residue. Moisture contents of above 15% or lower than 8%, excess moisture balance and indicate an abnormal treatment at some point or points from the harvest of the seed until the powder.

Determining insoluble crude fiber is performed in acid medium, resulting in a residue. Crude fiber, which is the residue insoluble in acid and alkaline hydrolysis, can generate impurities in natural food grade tara gum. The contents of the residue derived from mineral ash are extractable by ether fractions. It is common to determine the content of the low molecular weight sugars, fats and oil products, present in small amounts.

2.1.2.2 Industrial grade

Most of Tara gum sold for industrial applications is made from the endosperm and is as pure as food-grade Tara gum. Industrial-grade modifiers, however, particularly as applied chemical additives, such as moisturizing properties, viscosity and stability, can be added to control gelation and decreased viscosity [31, 32]. These methods and their combinations have resulted in increased use of Tara gum and its consumption.

The feature of Tara gum as water-binding agent makes it ideal for fast hydration-forming viscous colloidal solutions. It is versatile as a thickener or viscosity modifier. Tara gum is used in ice cream stabilizers, especially in high-temperature short-time processes where conditions require 80°C for 20 to 30 sec. Tara gum is also used in stabilizing sorbets [33]. It is used in a variety of products from soft cheese in cream cheese, pasteurized processed in production, to increase the yield of curd solids. It produces soft curds and compact, excellent texture. Cream cheeses are produced by mixing 1% to 2% of Tara gum with the other ingredients of cheese by melting and then cooling the homogeneous mixture. When applied to ice cream, it provides a good "overrun" (volume due to the incorporation of air during whipping).

2.1.2.3 Pharmaceutical biopolymer

In the past 40 years, polymers and biopolymers have been increasingly used as a support and as a tool for the controlled release of drugs or active substances that are in a given formulation. There are many applications of polymers for drug delivery due to the unique characteristics of these materials. Some of these are as follows: protection, support, and improved formulation stability; good processability; hydrophobic or hydrophilic character according to the requirement; rapid or controlled release of the active substance, improved bioavailability or acceptability of the medicament by the patient; and finally safe use.

Support applications in the pharmaceutical excipient comprise a polymer, which is inactive in the development and serves as a vehicle to enable the preparation of the drug and give consistency and stability, among other functions. A medicine may contain one or more than one active ingredient. The latter is called the active ingredient and its activity depends on its nature but is related to the amount ingested or absorbed. The pharmacologically active substance taken from a living organism, once purified and/or modified, is referred to as a drug or medicament. Morphologically, the polymer can be the matrix of the pharmaceutical composition or may be a shell or coating of the preparation.

A variety of polymers for pharmaceutical applications and the various embodiments of drugs or preparations are extruded or pelleted and some may be further coated or encapsulated.

Many drugs have low solubility, permeability, and bioavailability. The formulations can be improved by means of excipients that can be used as solubility promoters and bioavailability. Methods for improving the solubility of the drug include chemical modification using a pre-drug concept, physical modification through downsizing and surface modification, pH control system using a buffer, using co-solvents, and surfactant promoter absorption.

At present, trends worldwide are directed to the use of naturally occurring biopolymers for use in the pharmaceutical and food industry. These biopolymer-additives are substances intentionally incorporated in concentrations necessary in order to give the best product processing and preservation [34]. Within these biopolymers, additives are hydrocolloids that have a high water retention capacity and can form colloidal suspensions. The galactomannans are used in many formulations since they have great thickening viscosity and emulsion stability capacity. Tara gum obtained from the tree *Caesalpinia spinosa* is a great novel and promise in the pharmaceutical and food industry.

Polymers are becoming increasingly important in the field of drug delivery. The pharmaceutical applications of polymers range from their use as binders in tablets to viscosity and flow-controlling agents in liquids, suspensions, and emulsions. Polymers can be used as film coatings to disguise the unpleasant taste of a drug, to enhance drug stability, and to modify drug release characteristics. This review focuses on the use of pharmaceutical polymers for controlled drug delivery applications. Examples of pharmaceutical polymers and the principles of controlled drug delivery are outlined and applications of polymers for controlled drug delivery are described. The field of controlled drug delivery is vast; therefore, this review aims to provide an overview of the applications of pharmaceutical polymers. The reader will be directed, where necessary, to appropriate textbooks and specialized reviews [35]. Although polymers are used extensively as pharmaceutical packaging, this review is concerned with the use of polymers in the formulation of dosage forms. This review will be of interest to anyone who has an interest in the pharmaceutical use of polymers, whether as a researcher or as a manufacturer of medical devices [36].

Biopolymeric materials in contact with a drug product should fulfill a number of requirements besides the typical requirements such as adequate mechanical properties and suitability for mass production. The polymer should be chemically resistant to the excipients of the drug product. Moreover, it should be suitable for sterilization and have good barrier properties toward water, preservatives, and preferably gases. It should comply with the existing regulations regarding the amount and toxicity of leachables. The number of commercially available polymer materials that can be used is rather limited. Moreover, the biopolymeric materials are typically hydrophilic, which is known to be a disadvantage owing to water adsorption [37–44].

Tara gum is used as an appetite suppressant and as a binding agent and disintegrating agent in compressed tablets. It is also used to thicken various cosmetic products such as lotions and creams [45–49]. Some excipients that improve solubility are hydrophilic solvents [50, 51] such as propylene glycol and polyethylene glycol. Hydrophobic solvents are vegetable oils, triacetin, and diethyl phthalate. Tara gum is used as polymer emulsifer with polysorbates and vitamin E [52, 53].

There are four major sources of seed galactomannans for pharmaceuticals applications: locust bean (*Ceratonia siliqua*), guar (*Cyamopsis tetragonoloba*), Tara (*Caesalpinia spinosa, Kuntze*), and fenugreek (*Trigonella foenum-graecum L.*). Through keen references of reported literature on galactomannans, in this review, we have described the occurrence of various galactomannans, their physicochemical properties, characterization, and applications and an overview of some major galactomannans [54].

2.1.2.4 Food hydrocolloid

The commercial gum may be further specified in terms of viscosity and loss on drying. The color of Tara gum is white to white-yellow, it is nearly odorless powder, and its functional uses are as the thickening agent and stabilizer. Natural gums can be classified according to their origin and as uncharged or ionic polymers (polyelectrolytes).

Once the pod tree (*Caesalpinia spinosa*) is harvested, it is washed and dried, and then the seed is removed from the pod. Following the separation, the separately ground pit shell, which is an extraordinary product, is exported as a raw material for the production of tannic acid, which is widely used in the chemical, paint, and high-quality fur industries and pharmaceuticals. The seeds, or the pips, after undergoing a thermal-mechanical process, result in a gum from the endosperm, which is an alternative to the traditional rubber in the global food industry, pharmaceuticals, paints, and varnishes, among others. This gum has been approved by resolution of September 26, 1996 (ECC N°: E-417) by the European Community for use as a thickener and stabilizer in food for human consumption. Thus came into picture the global market for food hydrocolloids as an alternative product to the locust bean gum, produced in Spain and the Middle East.

Tara gum and locust bean gum are associated with metal ions such as Ca^{2+}, Na^{+}, K^{+}, and Mg^{2+}. These salt gums are classified as natural synthetic gums, modified or produced at low concentrations (less than 1% wt.); generate effects of gelling or viscous suspensions; and are used as adhesives, crystal inhibitors, and binding agents. A more frequent use is as an emulsion stabilizer in foods and ice, adjusting the viscosity of the aqueous phase. Hydrocolloids or gums have a wide application in the food industry as stabilizers, emulsifiers, or thickeners. Although they do not contribute to the aroma, flavor, or nutritional value of food, they may affect its acceptability by improving the texture or consistency. They are also used in the pharmaceutical, paper, and textile industries by improving the properties of different products. Other notable properties of hydrocolloids are clotting, lubricating, and film forming, still being in very low concentrations. Certain gums are extracted from legume seeds. For example, Guar gum (*Kyamopsio tetragonoloba*) and Locust bean gum (*Ceratonia siliqua*) have been used since the ancient times and still today are important as food additives because they give highly viscous solutions at low concentrations, even when the pH is low; are also compatible with other hydrocolloids, such as carrageenan, agar, and xanthan gum; and are able to reduce syneresis of some dairy products [55–79].

Creams

Functionality of Tara gum in creams:

- moisture retention, thus avoiding syneresis
- improvement in the appearance: sleek and smooth
- liberation clean taste, i.e., a definite flavor
- decreasing the rate of separation of fat
- providing body and creamy texture
- Critical adding cream in Tara gum.
  - Mix the stabilizer with xanthan gum (solids not fat) milk 1:4.
  - Add immediately after dispersing the xanthan gum it must allow proper hydration of the milk proteins to which an agitation of at least 15 min is recommended.
  - The stabilizer must also properly hydrated and dispersed, so it must be stirred for about 5 min before continuing the process.

Functional Tara gum mixtures with other hydrocolloids specific for different types of creams [80–84]:

- block syneresis
- creaminess, body and shine
- dosages of 0.4% to 0.8%
- prevent migration of fat
- viscosities according to the type of yogurt cream

Bakery Products

Among the additives, the most common are the gums or hydrocolloids. These products have a structure that makes them especially suitable to retain water and increase the viscosity of the masses, in fact are often referred to as thickeners. Their inclusion in the masses improves gas retention and makes less sticky mass. In some articles of bakery, products are called hydrocolloids gluten substitutes. Tara gum finds application in pastry and bakery, imparting smoothness and shape to the dough and makes cutting easier.

When added to different types of dough during kneading, Tara gum increases performance, gives more elasticity, and produces a longer and better handling properties softer texture, shelf life. In sponge cakes and pastries, Tara gum produces a smoother product that is easily removed from the molds and easily sliced without crumbling [17, 54].

The use level cake gum varies depending on the constituents of the mixture. Generally, the dose varies between 0.1% and 0.15% of the total weight of the dry ingredients.

Meat

Tara gum acts as a binder and lubricant in the manufacture of a variety of meat products such as sausages, meat products, and canned food. Tara Gum reduces weight loss during storage. Combined with xanthan gum, it makes the cutting of meat products easier.

The strong water retention properties of Tara gum, both cold and hot water, are very effective when used as a binder and lubricant in the manufacture of sausages and derivatives.

Tara gum process offers several advantages:

- rapid absorption of free and bound water for the preparation of products made from ground beef
- improves the speed of filling the bags
- linked free water, removing the separation and migration during operations of cooking and/or smoking
- allows viscosity when cooled, providing desirable firmness in the sausage; in the final product a uniform appearance is achieved, and when the product is sliced, it retains moisture longer, preserving a fresh and desirable appearance

The application tier is recommended 1 kg rubber per 1000 kg of free water. The total water added is a combination of water and the amount of water in wet-processed meat [17, 84].

Drinks

Tara gum is useful in thickening various fruit drinks and diet drinks without sugar. Tara gum and carrageenan are used to stabilize chocolate syrups and chocolate powder mixtures. Nectars consisting of fruit puree, fruit juice, sugar, citric acid, and ascorbic acid get a good texture and a stable viscosity [85, 86] by the addition of 0.2% to 0.8% of Tara gum.

Dressings and Sauces

Tara gum's property of thickening is used to maintain the stability [87, 88] and appearance of salad dressings, sauces, pickles, spicy seasonings, and barbecue sauces. Tara gum and guar gum are consistent with the very sour emulsions, with effective rates of 0.2% to 0.8% of the total weight.

2.2 Mineral-Petroleum Applications

2.2.1 Mineral Application

Polysaccharides are condensation polymers of high molecular weights based on simple monosaccharide sugar units. Various types of polysaccharides exist in nature, but only a small number of them have been used by the mineral industry, especially in flotation. These include starch, dextrin, guar gum, and carboxymethyl cellulose, and they are mainly used as flotation depressants [89]. Selective depressants are indispensable ingredients in any flotation reagent scheme targeting selective separation of different minerals. Various inorganic depressants have been routinely used. Many of these depressants, especially the ones used in differential sulfide flotation, are extremely toxic and environmentally unacceptable. Examples of such depressants are sodium cyanide, sodium dichromate, sulfur dioxide, arsenic trioxide, phosphorous pentasulfide, etc. Some of these inorganic depressants are reducing agents and thus are susceptible to oxidation in aerated flotation pulps [90], resulting in high reagent consumption.

On the other hand, polysaccharides are natural organic polymers that are non-toxic and biodegradable. They are also cheaper and more resistant to oxidation than the inorganic depressants. Not only do these properties make them ideal as flotation reagents, but they have also shown potential for use as selective depressants in many differential mineral flotation systems. Polysaccharides have been used in the mineral industry for almost 70 years, mainly as depressants for iron oxides, inherently hydrophobic minerals, and rock-forming gangue minerals [91]. They are also reported as being selective in the differential flotation of sulfide minerals. However, a general lack of understanding of the interaction mechanism between the polysaccharides and mineral surfaces has hindered the wider application of these polymers [92].

Guar gum is a branched polysaccharide belonging to the group of galactomannan. Structurally, it is made of a straight chain of β-D-mannose connected through C-1–C-4 linkage with attached α-D-galactose units through C-1–C-6 linkage every second β-D-mannose monomers. The α-D-galactose is similar to α-D-glucose except that the –OH on C-4 is in the axial position. The β-D-mannose is similar to β-D-glucose. However, the –OH group on C-2 is in axial position rather than equatorial. This makes the two –OH groups on C-2 and C-3 in A *cis* B conformation, i.e., they are on the same side of the monomer ring. It is known that the two –OH groups on C-2 and C-3, when in the *cis* axial-equatorial conformation or in the Trans diequatorial conformation, can participate in complexation reactions. Therefore, natural starch, dextrin, cellulose, and guar gum potentially have the ability to form complexes with foreign guests. [93–95].

The above descriptions show the monomeric structures of starch, dextrin, cellulose, and guar gum. In reality, these polysaccharides can be extracted from different sources and the chain length and configuration, molecular weights, and the contents of impurities may vary considerably. Generally, starches have been used mainly as flocculants or flotation depressants for iron oxide minerals and phosphate minerals while the associated silica is floated. Dextrin has been mainly tested as depressants for inherently hydrophobic minerals such as talc, molybdenite, and coal [96]. Applications of polysaccharides in other mineral systems, both in the laboratory and in commercial processes, have also been frequently reported. As can be seen, the polysaccharides have been used or tested as selective depressants in practically all types of mineral systems, ranging from oxides, sulfides, salt-type, and inherently hydrophobic minerals.

The predominant perception regarding the use of polysaccharides in mineral flotation is that their adsorption is non-selective primarily resulting from hydrogen bonding and that their functions are unpredictable. The applicability of the polysaccharides in such a wide range of mineral systems, however, seems to indicate that this general perception is probably oversimplified and incorrect. In fact, we will show that the adsorption mechanisms of the polysaccharides are more complicated than hydrogen bonding [97].

Polysaccharides contain a large number of hydroxyl groups, which is why these polymers can be used as flotation depressants. For pure polysaccharides, the hydroxyl groups are the only polar groups available in the polymer chain. However, naturally occurring polysaccharides always contain impurities, resulting in the presence of varying amount of carboxylate, phosphate, and amino acid groups. The concentrations of impurities depend on the type, source, and manufacturing process of the polysaccharides but are generally less than 1%. The impurities in the polysaccharides used in many of the mineral flotation studies were usually but not well characterized. The observed depressant functions were broadly attributed to the hydrogen bonding of the hydroxyl groups to mineral surfaces. Hydroxyl groups do not affect the validity of the conclusions since the impurity contents are usually very low and do not seem to contribute significantly to the adsorption [98]. Several hypotheses regarding polysaccharide adsorption mechanisms, principally hydrogen bonding and hydrophobic interactions, also proposed chemical complexation of starch with calcite and hematite surfaces. The current picture of the polysaccharide adsorption reflects the role of metal-hydroxylated species present on mineral surfaces. The direct consequence of the interaction of natural polysaccharides with metal-hydroxylated species is that the interaction, and thus the adsorption, is strongly dependent on pH [99]. The pH-dependence of polysaccharide adsorption was in fact first observed by measuring the adsorption of a corn starch on hematite and it was found that the adsorption was much higher at pH 6.8 than at pH 11.3; a higher starch and amylopectin adsorption on FeO at pH 6 was observed than at pH 8.2 studied the adsorption of a number of polysaccharides on pyrite and found that the maximum adsorption always occurred around pH 7. In fact, the isoelectric exchange point (IEP) values of various iron oxides are around 5.2 to 8.6. It follows that around the IEP, the iron oxide surfaces were fully hydroxylated. The observed adsorption maximum of natural polysaccharides around the IEP of iron-bearing minerals is in line with the postulation that the polysaccharides interact with surface metal-hydroxylated species [100].

2.2.2 Petroleum Applications

Fracking hydraulic stimulation is a technique to enable or enhance the extraction of gas and oil from underground. Hydraulic fracturing (fracking/HF) involves injecting wells at high pressure with water, proppants, radioactive tracers and chemical additives to fracture the formation and produce new cracks and pathways to help extract the gas. While chemical additives make up less than 2% of the fracking fluid, this still translates into large quantities. The procedure involves drilling a vertical well in which—once the desired depth is reached—some material is pressure injected in the soil with the aim of expanding existing fractures in the bedrock that contains the gas or oil and are typically less than 1 mm, thus favoring its exit. Usually the injected material is water with sand and chemicals, which seeks to promote cracking or even the dissolution of the rock [101–109].

It is estimated that in 2010 this technique was present in approximately 60% of the extraction wells in use. Due to the rising price of fossil fuels, which has made these methods economically profitable, their use has increased in recent years, especially in the United States [101, 105].

The injected component is based on a 99.51% water and 0.49% sand and supporting additives. The supporting additives are the ones that generate more controversy, as critics say that toxic substances, including carcinogens [110] and allergens, leave the subsoil conditions unrecoverable. However, the proponents of this extraction technique claim that the additives used can be found in household items. The purpose is to generate the necessary ways to extract shale gas, keep the channels open, and preserve the soil and prevent it from degrading during the operation. Hydraulic fracturing fluids usually include gelling agents to hold the proppant in suspension—e.g., mixtures of industrial guar gum or galactomannan (Tara gum and locust bean gum), diesel, and alkanes/alkenes) [111–116].

The decrease of gellant concentration would degrade the proppant carrying capacity of the fracturing fluid. Adding fiber in the base fluid could reduce the concentration of gellant under the premise of ensuring the proppant pack stability. Experiment results showed that the settling velocity of proppant decreased with increasing quantity and length of fiber. Rheological experiments showed that viscosity value increment reduced from 335.9 to 107.4 mPas when the dosage of fiber increased from 0 to 2.0%. Shear rate had smaller effect on the viscosity of fracturing fluid after adding fiber [117–120].

It is important to develop regenerable fracturing fluid with low cost for enhanced exploitation of oil-gas fields. The degraded hydroxypropyl Tara Gum (DHPTG) was synthesized by Tang et al. [121], and the influence of hydrogen peroxide dosage on its degradation degree was studied. The rheological properties results show that the gel systems with less than three regeneration times possess good viscoelasticity, thixotropy, and shear thinning properties, and the rheological properties of the gels before and after regeneration are not changed significantly [122–128].

Aimed at the problems of the low efficiency of fracturing fluid preparing in fracturing operation, the recent rising price of the raw materials of guar gum, many drainage liquids after construction, and the high cost of treatment, the instant guar gum fracturing fluid and the reutilization feasibility are very important. Based on a series of experiments, such as the rate of dissolution determination of the guar gum, gel breaking experiment and the performance assessment of repeated cross-linking, the performance of instant guar gum compared with that of the conventional guar gum and recycling feasibility of instant guar gum was researched. The experimental results show that the instant guar gum dissolves fast and has a good dispersion property and complete gel breaking performance, which can be as a thickening agent consisting of fracturing fluid system with shear resistant performance, certain reutilization performance, economy and environmental protection, and a broad application prospect [129, 130].

2.3 Experimental

2.3.1 Sample Preparation

Tara gum (TG) and Aglumix 01 were provided by INDUNOR from Argentina; 0.5 g of TG was dissolved in 200 ml of distilled water, and then the mixture was centrifuged to 7000 rpm for 5 min to separate supernatant from impurities; finally, purified TG was dried at 60°C for 48 h. Tara gum solutions were prepared by dissolving in 100 ml of distilled water 0.01, 0.02, 0.03, 0.04, 0.05, 0.07, and 0.1 g, with the addition of 1 ml HCl 1M (BIOPACK) at 60°C for 2 h.

2.3.2 Capillary Viscometry

Solutions and reference solvents were analyzed using an Ubbelohde 1 C viscometer (IVA, Argentina) at 25°C, under precise temperature control using thermostatic bath (Haake 1C, Germany).

2.3.3 Density Measurement

The densities of the solution and the solvent were measured with Anton Paar densimeter DMA5N, Austria.

2.4 Physics Chemistry Properties

2.4.1 Viscosity

If a solution tends to be independent of shear, then the measurement of viscosity (η in poise) is based on Poiseuille's law [131] and can be made easy by grouping all those terms related to a specific viscometer as a calibration constant A (cm^2/s^2).

$$\eta = A\rho t, \tag{2.1}$$

where ρ is the density of solution. If we divide both sides by ρ, we have

$$\upsilon = \frac{\eta}{\rho} = At, \tag{2.2}$$

where υ is the cinematic viscosity of solution in cm^2/s.

In macromolecular chemistry, the relative viscosity η_r is often measured. The relative viscosity is the ratio of the viscosity of the solution to that of the solvent:

$$\eta_r = \frac{\rho t}{\rho_0 t_0} \tag{2.3}$$

The specific viscosity η_{sp} is obtained from the relative viscosity by

$$\eta_{sp} = \eta_r - 1 \tag{2.4}$$

2.4.2 Density

In physicochemical terms, density (ρ) is a quantity relative to the amount of mass contained in a given volume of a substance. The measured density is the ratio of the mass of a body and the volume it occupies. Therefore, the density is equal to

$$\rho = \frac{m}{v}, \tag{2.5}$$

where m is mass in g, v is volume in cm^3, and ρ in g/cm^3.

There are different factors that can modify this magnitude, the overall density of a substance varies mainly when the pressure or temperature changes, and status changes:

- When the pressure increases, the density increases.
- As the temperature increases, the density decreases (if the pressure remains constant).
- By varying the mass, the density varies in the same direction to the change (if both the temperature and the pressure remain constant.)

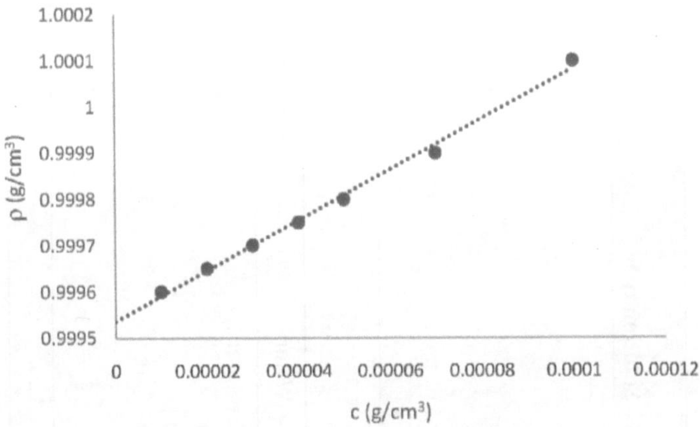

Figure 2.3 Density in function of concentration of Tara gum solutions.

The density values of Tara gum are shown in Fig. 2.3, which shows that with increasing gum concentration, the density increases linearly. From this plot, we can determine the value of the partial specific volume, data with which the hydration value is determined as long as the specific volume is known (see Table 2.1).

Table 2.1 Intrinsic viscosity and Mark-Houwink data from different authors

Galactomannan and botanical name	Solution type	$[\eta]$ (cm^3/g)	a	k (cm^3/g)	M_v (g/mol) × 10^{-5}	Reference
Carob	—	2710	0.75–0.80	0.104–0.67	3.19	[137]
Guar		3850			6.57	
Lotus		69–474			0.37–3.32	
Lucerna		1470			2.39	
Red clover		1920			4.16	
Sophora		1030			15.2	
Soybean		1760			1.66	
Guar gum	—	230–1170	1.0348	7.396 × 10^{-4}	3.8–20.2	[138, 139]
Locust bean gum		270–1120	0.9960	3.572 × 10^{-4}	11–13.8	
Guar gum	Water	450–1250	0.723	0.038	4.4–16.5	[140]
Mimosa scabrella	Water	—	—	—	9.72	[141]
Cyamopsis tetragonoloba (Guar gum)					13	
Schizolobium amazonicum					11	
Ceratonia Siliqua (Locust bean gum)					14	
Guar gum	—	—	0.72	0.0513	0.46–16	[142]
Mimosa scabrella	0.1 M $NaNO_3$, 30°C	—	—	—	20	[143]

Galactomannan and botanical name	Solution type	$[\eta]$ (cm^3/g)	a	k (cm^3/g)	M_v (g/mol) × 10^{-5}	Reference
Tara gum	—	—	—	—	9.4	[144]
Guar gum enzymatic modification	Water	—	1.05	0.00067	—	[145]
Guar gum	—	272–1600	0.747	0.0304	0.63–19.8	[146]
Locust bean gum Tara gum	Water	1103 1646	—	—	16.1 2.53	[147]
Guar gum	Water + alkali metal chloride, 24°C	1150	—	—	—	[131]
Guar gum Tara gum Locust bean gum	—	—	0.72–0.747	0.0304–0.06	—	[148]
Locust bean gum	Water	1330–1530	—	—	10–12	[149]
Gleditsia gum Sophora gum Adenanthera gum Pulcherrima gum	Water	1006 893–958 885 1091	0.01155	$a^* = 0.98$ $[(1-a)\, M_v]^{a^*}$ $a = 1/[(1$ $- M/G) + 1]$	16.2 12.5–13.3 18.1 17.5	[150]
Locust bean gum Tara gum	Water, 25°C	1260 1420	—	0.77 0.79	10.5 22.5	[151]

2.4.3 Intrinsic Viscosity

The intrinsic viscosity, denoted by $[\eta]$, is defined as

$$[\eta] = \lim_{c \to 0} \frac{\eta_{sp}}{c} = \lim_{c \to 0} \frac{1}{c} \ln \eta_r \quad (2.6)$$

or as

$$[\eta] = \lim_{\eta_{sp} \to 0} \frac{\eta_{sp}}{c}, \quad (2.7)$$

where c is the concentration of the polymer in grams per 100 cm^3 or grams per milliliter of the solution. The quantity η_{sp}/c is called the reduced viscosity. The unit of intrinsic viscosity is deciliters per gram (dl/g), (ml/g), and (cm^3/g) depending on the concentration unit of the solution. The intrinsic viscosity is also called the limiting viscosity number. The plot of η_{sp}/c versus c or $1/c \ln \eta_r$ versus c often gives a straight line, the intercept of which is $[\eta]$.

2.4.4 Huggins' and Kraemer's Methods

Huggins showed that the slope is

$$\frac{d\eta_{sp}/c}{dc} = k_H [\eta]^2. \quad (2.8)$$

Rearranging and integrating the resulting equation gives

$$\frac{\eta_{sp}}{c} = [\eta] + k_H [\eta]^2 c, \quad (2.9)$$

where k_H is a dimensionless constant, called the Huggins constant [131]. The value of k_H is related to the structures of polymers or biopolymers. Kraemer equations for the determination of $[\eta]$ is

$$\frac{1}{c} \ln \eta_r = [\eta] + k_K [\eta]^2 c. \quad (2.10)$$

For molecules of high intrinsic viscosity, a correction must be made for the effect of the rate of shear strain. For relatively low intrinsic viscosity, the rate of shear strain does not have any appreciable effect.

Figure 2.4 shows Huggins' and Kraemer's methods for Tara gum solutions, whose intrinsic viscosity data obtained by Huggins is 3595.9 cm^3/g, Kraemer is 3385.9 cm^3/g and Solomon-Ciuta single point is 3886.2 cm^3/g. Constants are k_H = 5.41 and k_K = 4.36, the sum of the $k_H + k_K$ = 9.77, far as value of 5 provides for a system of ideal solution.

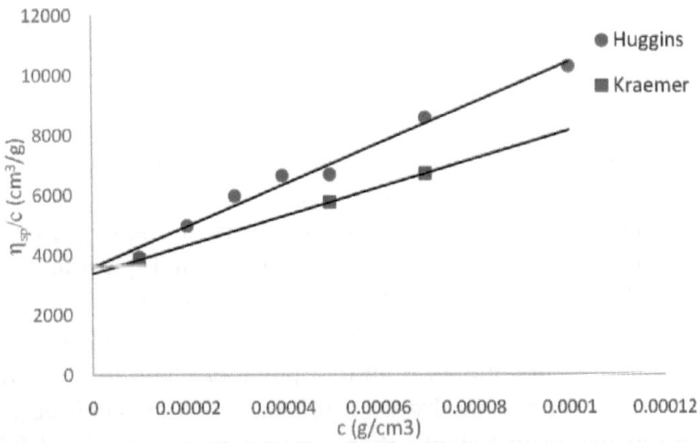

Figure 2.4 Huggins method for Tara gum solutions.

2.4.5 Hydrodynamic Properties: Molecular Weight and Mark-Houwink Parameters

Staudinger (1932) suggested that the viscometric molecular weight M_v of polymers is proportional to the reduced viscosity:

$$[\eta] = k_0 M_v, \quad (2.11)$$

where k_0 is proportionality constant.

Mark-Houwink correlated the intrinsic viscosity with molecular weight:

$$[\eta] = k M_v^a, \quad (2.12)$$

where both k and a are constants. The Mark-Houwink equation is applicable to many polymers and is extensively used to determine molecular weight. The constants k and a vary with polymers and solvents. Equation (2.12) describes the relationship between intrinsic viscosity and molecular weight. Molecular weight is related to the size of the polymer chain by modified Einstein´s equation.

The calculation of Mark-Houwink (M-H) parameters is carried out by the graphic representation of the following equation:

$$\ln[\eta] = \ln k + a \ln M_v, \quad (2.13)$$

where k and a are M-H constants; these constant depend on the type of polymer, solvent, and temperature of viscometric determinations. The exponent a is a function of polymer geometry, and varies from 0.5 to 2. These constants can be determined experimentally by measuring the intrinsic viscosity of several polymer samples for which the molecular weight has been determined by an independent method (i.e., osmotic pressure or light scattering). Using the polymer standards, a plot of the ln $[\eta]$ versus ln M_v usually gives a straight line. The slope is a value and the intercept is equal to ln k value. The M-H exponent bears the signature of a three-dimensional configuration of a polymer chain in the solvent environment. For a values from 0 to 0.5, the macromolecule acquires the conformation of a rigid sphere in an ideal solvent; from 0.5 to 0.8, the macromolecule acquires the conformation of a random coil (ellipsoid) in a good solvent; and from 0.8 to 2, the macromolecule acquires a rigid-rod or rod-like conformation (stiff chain). The fact that the intrinsic viscosity of a given polymer sample is different according to the solvent used gives an insight into the general shape of polymer molecules in solution. A long-chain polymer molecule in solution takes on a somewhat kinked or curled shape, intermediate between a tightly curled mass (coil) and a rigid linear configuration. All possible degrees of curling may be displayed by any molecule, but there will be an average configuration that will depend on the solvent. In a good solvent which shows a zero or negative heat of mixing with the polymer, the molecule is fairly loosely extended, and the intrinsic viscosity is high. The Mark-Houwink "*a*" constant is close to 0.75 or higher for these "good" solvents. In a "poor" solvent that shows a positive heat of mixing, segments of a polymer molecule attract each other in solution more strongly than attract the surrounding solvent molecules. The polymer molecule assumes a tighter configuration, and the solution has a lower intrinsic viscosity. The M-H "*a*" constant is close to 0.5 in "poor" solvents. For a rigid or rod-like polymer molecule that is greatly extended in solution, the M-H "*a*" constant approaches a value of 2.0 [132–136].

From the Huggins method, data are calculated for the Mark-Houwink parameters for the Tara gum solution used in this work, where $a = 0.8100$ and $k = 0.01646$ cm^3/g are obtained. Comparing these data with those of literature [137–144], the Mark-Houwink parameters are very similar. The calculated molecular weight is of 3770000 g/mol, greater than of other authors [137–144], where this increase may be due to the high viscosity of the solutions and protons presence from 0.1 M HCl. This is evidenced by the high value of intrinsic viscosity and its difficult Tara gum solution in water. As to the value of "*a*" this shows that Tara gum for this system behaves as rod-like, although others place it as an ellipsoid (random-coil) where "*a*" is 0.70 to 0.747 for galactomannans and 0.79 for Tara gum; the latter boundary between random-coil and rod-like is although closer to the rod-like conformation (see Table 2.1). This high molecular weight and intrinsic viscosity is special macromolecule for petroleum fracking (as galactomannan), for treatment by flocculation, and is excellent food thickener.

Morris et al. conducted a work studying hydrodynamics and chain flexibility of galactomannan (guar gum) [152], and *Mimosa scabrella* in 0.1 M $NaNO_3$ were realized by Mazeau and Rinaudo [153].

2.4.6 Hydration Values

It is well known that biopolymers adsorbed water during dry storage and its quality depends on water content. For example, the length of keratin depends on water content and therefore it is used as a hygrometer. The amount of adsorbed water depends on temperature and pressure of water vapor.

On the other hand, biopolymers in solution exhibit the phenomenon of hydration due to the polar properties of water molecules. The electronic formula of water shows that the center of charge of the negatively charged electrons is nearer to the oxygen atom to the positively charged hydrogen nuclei. The center of the positive charges is nearer to the two hydrogen atoms. Assuming a molecule in which the centers of the positive and negative charges do not coincide, a polar molecule or dipole, always, dipoles are attracted by ions. In the first phase, the ion attracts the opposite pole and repels the pole of same sign; in a second phase, attraction is stronger than repulsion because the attracted pole is nearer the ion than the repelled pole. For similar reasons, attraction takes place between two dipoles.

Hydration consists of the binding of water dipoles to ions or ionic groups, to dipoles, or polar groups. Hydration takes place in solid substances as well as in solution.

Since the components of a compound are linked to each other in such a way that they have lost some of their free translational mobility, the volume of the hydrated molecule is always smaller than the sum of the volumes of its components; the hydration is accompanied by a decrease of the total volume.

The amount of water bound to the proteins and polysaccharides depends primarily on the ratio of water to the biopolymer in the investigated system. The two extreme cases are the dry biopolymer (water content tend to zero) and highly diluted aqueous solutions of the biopolymers. The dry biopolymer undergoes hydration if is exposed to the water vapor of increased vapor pressure. The extent of hydration can be determined and measuring the increment in weight. It is much more difficult to determine the extent of hydration in aqueous solutions of biopolymers. Although hydration is accompanied by a volume contraction of the solute and the solvent, this change in volume is very small and difficult to measure directly. It is customary to measure the density of biopolymer solution.

The amount of hydrated biopolymer and of free water in the biopolymer–water system, the thermodynamic notion of partial specific volume has been introduced and is frequently determined.

$$[\eta]=[\eta]_0+\left\{\frac{1-\rho_0\bar{v}}{\rho_0}\right\} \tag{2.14}$$

$$\left(\frac{\eta_{sp}}{c}\right)=\left(\frac{\eta_{sp}}{c}\right)_0+\left\{\frac{1-\rho_0\bar{v}}{\rho_0}\right\} \tag{2.15}$$

Of course, if this latter is not known for the solvent conditions being used, or cannot be calculated from the chemical composition of the macromolecule, then solution density measurements are required:

$$\bar{v} = \left\{ \frac{1 - \partial\rho/\partial c}{\rho_0} \right\}, \tag{2.16}$$

where ρ_0 and ρ are density of solvent and solution, respectively, and can be measured using densimeter or picnometer.

The swollen specific volume v_{sp} (cm^3/g) is defined when anhydrous biomacromolecules essentially expand in suspended or dissolved in solution because of solvent association, and

$$v_{sp} = \frac{v_H M}{N_A}, \tag{2.17}$$

where v_H is swollen or hydrodynamic volume (cm^3), M the molecular weight (Da or g/mol), and N_A is Avogadro's number. This associated solvent, which we consider in more detail below, can be regarded as the one that is either chemically attached or physically entrained by the biopolymers; v_{sp} can be related to a popular term called the hydration value, δ, by the relation

$$v_{sp} = \bar{v} + \frac{\delta}{\rho_0}. \tag{2.18}$$

The corresponding value of the "hydration" δ of the molecule (see Table 2.2), defined by

$$\delta = (v_{sp} - \bar{v})\rho_0, \tag{2.19}$$

where v_{sp} is specific volume (cm^3/g). Although because of the approximations we have made, the actual numerical value must be treated with great caution, this treatment does suggest that polysaccharides are highly expanded, but perhaps not to the same extent as found for coil-like polysaccharide structures to be ~50g/g, and are important in food hydrocolloids [154, 155]. According to the above concentrations also there are many cases where this value can be much higher than previously described [156, 157].

2.4.6.1 Perrin number

Most biological polymers, such as proteins and nucleic acids and some synthetic polymers, have relatively inflexible chains. For rigid particles, the size is no longer of predominant importance, because the polymer chain is no longer in the form of a flexible random coil; instead, the shape becomes an important parameter. Following are some theoretical proposals for the estimation of the shape factor P from the viscosity measurement (Table 2.2). The term f/f_0 is sometimes denoted as P, the Perrin constant.

Combination of the Perrin function, P, often referred as the "frictional ratio due to shape" with the frictional ratio (f/f_0) enables the degree of expansion of the molecule $(v_H/\bar{v})$ to be estimated, where v_H, (cm^3/g) is the volume of the swollen molecule (polysaccharide or protein + associated solvent) per unit mass of polysaccharide and $\bar{v}$ is the partial specific volume (essentially the anhydrous molecule):

$$\frac{f}{f_0} = P\left(\frac{v_H}{\bar{v}}\right)^{1/3} \tag{2.20}$$

When the biopolymer is contracted, the term of expansion is negligible [158].

2.4.6.2 Einstein viscosity increment (Simha number)

There are two molecular contributions to the intrinsic viscosity: one from shape and the other from size or volume, as summarized by the following relation:

$$[\eta] = \nu_{a/b} v_{sp}, \tag{2.21}$$

where $\nu_{a/b}$ is a molecular shape parameter known as viscosity increment and v_{sp} is as defined in Eq. (2.19).

The viscosity increment $\nu_{a/b}$ is referred to as a universal shape function or Simha number (Table 2.2); it can be directly related to the shape of a particle independent of volume. For its experimental measurement, it does, however, require measurement of v_{sp}, $\bar{v}$, δ, ρ_0, and, of course, $[\eta]$.

Table 2.2 Hydrodynamic data of Tara gum solution

$\bar{v}$ (g/cm^3)	$\nu_{a/b}$	P
0.4541	11.12	15.69
R_H (nm)	δ(g/g)	v_{sp}(cm^3)
63.02	337.22	336.81

A study of the viscosity of a solution of suspension of spherical particles (colloids) suggested that the specific viscosity η_{sp} is related to a shape factor $\nu_{a/b}$ as follows:

$$\eta_{sp} = \nu_{a/b}\phi, \tag{2.22}$$

where ϕ is the volume fraction;

$$\phi = \frac{n\,\nu\,\nu_{a/b}}{V}, \tag{2.23}$$

where n is the number of identical particles that are not interacting, ν is the volume of each particle, and V is the volume of the solution or suspension. Assume that the molecules are of a spherical shape, rigid and large relative to the size of the solvent molecules, and that the particles are small enough to exhibit Brownian motion but large enough to obey the laws of macroscopic hydrodynamics [158–160]. Then $\nu_{a/b} = 2.5$ for spherical particle [158]. The Einstein equation is now used as a reference to estimate the shape of macromolecules. Any deviation can be interpreted as the fact that the molecules are not a sphere [160].

As seen in Table 2.2, the values of P and $\nu_{a/b}$ are far away from the spherical form and are accurate to form rod-like, as pectin [159], alginate [160]. The hydration value accounts for the high water adsorption capacity for this polysaccharide and its great industrial potential application in highly viscous and thick solutions. The R_H of Tara gum is between the values specified by Wu et al. [161].

Acknowledgments

The author thanks Universidad Nacional de San Luis (Project 2-81/11), FONCyT (PICT 2004-N°23-2548 and PICT 2008-N° 21-84), CONICET (PIP 6324 Res. 1905/05) and PROIPRO 2-2414 (Regional Polysaccharides: Purification and Physicochemical Characterization. Applications: Analytics, Separative Processes and Food Industry) for the financial support.

References

1. Kuntze Otto (Molina), 1898. Revisio Generum Plantarum, **3**(3): 54.
2. Burkart A. 1952. Las Leguminosas Argentinas, 569 pp.

3. Kapoor M., Khandal D., Gupta R., Arora P., Seshadri G., Aggarwal S., Khandal R. K. 2013. Certain rheological aspects of functionalized guar gum. *Int. J. Carbohydr. Chem.*, 2013, Article ID 463907, 15 pages.
4. Brack Egg A. 1999. Diccionario Enciclopédico de Plantas Útiles del Perú, Cusco, Perú: CBC.
5. McClintock E. (1996). Caesalpinia. In: Hickman J. C. (ed.), *The Jepson Manual: Higher Plants of California*. University of California Press.
6. De la Cruz Lapa P. 2004. An integral and rational utility of tara (Caesalpinia spinosa—Caesalpinia tinctoria). *Rev. Inst. Invest. Fac. Minas Metal Cienc. Geogr.*, **7**(14), 64–73.
7. Garro Galvez J. M., Riedl B., Conner A. H. 1997. Analytical studies on tara tannins. *Holzforschung*, **51**, 235–243.
8. Borzelleca J. F., Ladu B. N., Senti F. R., Egle, Jr. J. L. 1993. Evaluation of the safety of tara gum as a food ingredient: A review of the literature. *J. Am. Coll. Toxicol.*, **12**, 81–89.
9. Carrillo F. E. 1974. Las leguminosas del valle del Rimac (Sub-Familias: Mimosoideae y Caesalpinoideae). *Bol. Soc. Peruana Bot.*, **7**(1/2), 40–68.
10. De la Barra N. 1997. *Reconstr. Evol. Paisaje Veg. Cochabamba i–v*, 1–174. Thesis, Universidad Mayor San Simón, Cochabamba.
11. Hickman J. C. 1993. *The Jepson Manual: Higher Plants of California*, 1–1400. University of California Press, Berkeley.
12. Hokche O., Berry P. E., Huber O. 2008. Nuevo Cat. Fl. Vasc. Venezuela 1–860. Fundación Instituto Botánico de Venezuela, Caracas.
13. Idárraga-Piedrahita A., Ortiz R. D. C., Callejas Posada R., Merello M. 2011. Fl. Antioquia: Cat. 2, 9–939. Universidad de Antioquia, Medellín.
14. Jørgensen P. M., Ulloa Ulloa C. 1994. Seed plants of the high Andes of Ecuador: A checklist. *AAU Rep.*, **34**, 1–443.
15. Jørgensen P. M., León-Yánez S. (eds.) 1999. *Cat. Vasc. Pl. Ecuador, Monogr. Syst. Bot. Miss. Bot. Gard.*, **75**, i–viii, 1–1181. Missouri Botanical Garden, St. Louis.
16. Killeen T. J., García Estigarribia E., Beck. S. G. 1993. *Guía Árb. Bolivia*, 1–958. Herbario Nacional de Bolivia & Missouri Botanical Garden, La Paz.
17. Macbride, J. F. 1943. *Leguminosae*. **13**(3/1), 3–507. In: Macbride J. F. (ed.) *Fl. Peru. Publ. Field Mus. Nat. Hist., Bot. Ser. Field Museum*, Chicago.

18. Basurto Rodríguez L. 2003. Todo sobre la Tara, Caesalpinia Spinosa o Caesalpinia Tinctoria, ALNICOLSA del Perú S. A. C., Callao, Perú. http://taninos.tripod.com/#goma.
19. USDA, ARS, National Genetic Resources Program. Germplasm Resources Information Network (GRIN). National Germplasm Resources Laboratory, Beltsville, Maryland.
20. Coppen J. J. W. 1995. *Non-wood Forest Products. 6. Gums, Resins and Latexes of Plant Origin*. FAO, Rome.
21. Duke J. A. 1981. Caesalpinia spinosa. In: *Handbook of Legumes of World Economic Importance*. Plenum Press, New York. pp. 32–33.
22. Rogers J. S., Beebe C. W. 1941. Leaching and tanning experiments with Tara pods. *J. Am. Leather Chem. Ass.*, **36**, 525–539.
23. Wrann H. J., Arriagada B. M. 1988. Experimental plantations of tannin-producing species in the semi-arid zone of Chile. *Cien. Invest. Forestal*, **3**, 51–66.
24. Piet J. H., Daas Schols H. A., de Jongh H. H. J. 2000. On the galactosyl distribution of commercial galactomannans. *Carbohydr. Res.*, **329**, 609–619.
25. Tapie N., Malhiac C., Hucher N., Grisel M. 2008. Determination of galactose and mannose residues in natural galactomannans using a fast and efficient high-performance liquid chromatography/UV detection. *J. Chromatogr. A*, **1181**, 45–50.
26. Dakia P. A., Blecker C., Robert C., Wathelet B., Paquot M. 2008. Composition and physicochemical properties of locust bean gum extracted from whole seeds by acid or water dehulling pre-treatment. *Food Hydrocolloids*, **22**, 807–818.
27. Dakia P. A., Wathelet B., Paquot M. 2007. Isolation and chemical evaluation of carob (*Ceratonia siliqua L.)* seed germ. *Food Chem.*, **102**, 1368–1374.
28. Kök M. S., Hill S. E., Mitchell J. R. 1999. A comparison of the rheological behaviour of crude and refined locust bean gum preparations during thermal processing. *Carbohydr. Polymers*, **38**, 261–265.
29. Kök M. S., Hill S. E., Mitchell J. R. 1999. Viscosity of galactomannans during high temperature processing: Influence of degradation and solubilisation. *Food Hydrocolloids*, **13**, 535–542.
30. Kök M. S. 2007. A comparative study on the compositions of crude and refined locust bean gum: In relation to rheological properties. *Carbohydr. Polymers*, **70**, 68–76.

31. Kök M. S. 2010. Rheological study of galactomannan depolymerisation at elevated temperatures: Effect of varying pH and addition of antioxidants. *Carbohydr. Polymers,* **81**, 567–571.

32. Brummer Y., Cui W., Wang Q. 2003. Extraction, purification and physicochemical characterization of fenugreek gum. *Food Hydrocolloids,* **17**, 229–236.

33. Rincón F., Muñoz J., Ramírez P., Galán H., Alfaro M. C. 2014 Physicochemical and rheological characterization of *Prosopis juliflora* seed gum aqueous dispersions. *Food Hydrocolloids,* **35**, 348–357.

34. Cristina W., Vendruscolo C. W., Carmen Ferrero C., Edgardo Pineda E. A. G., Silveira J. L. M., Freitas R. A., Jiménez-Castellanos M. R., Bresolin T. M. B. 2009. Physicochemical and mechanical characterization of galactomannan from Mimosa scabrella: Effect of drying method. *Carbohydr. Polymers,* **76**, 86–93.

35. Zhang C., Chen J.-D., Yang F.-Q. 2014. Konjac glucomannan, a promising polysaccharide for OCDDS. *Carbohydr. Polymers,* **104**, 175–181.

36. Vipul D., Prajapati, Girish K., Jani, Naresh G., Moradiya, Narayan P., Randeria, Bhanu J., Nagar. 2013. Locust bean gum: A versatile biopolymer. *Carbohydr. Polymers,* **94**, 814– 821.

37. Kulkarni V., Butte K., Rathod S. 2012. Natural polymers–A comprehensive review. *Int. J. Res. Pharm. Biomed. Sci.,* **3**, 1597–1613.

38. Vinod K. R., Vasa S., Sandhya S. 2010. Emerging trends in pharmaceutical polymers. *Der Pharm. Lett.,* **2**(1) 172–180.

39. Ravi V., Mahendra C., Datta V. M., Gowda D. V., Shivakumar H. G., Bhargav E. 2013. Thiomers fresh drift of polymers and their prospective in pharmaceuticals: A review. *World J. Pharm. Pharm. Sci.,* **3**(1), 204–220.

40. Kadajji V. G., Betageri G. V. 2011. Water soluble polymers for pharmaceutical applications. *Polymers,* **3**, 1972–2009.

41. Jones D. 2004. Pharmaceutical applications of polymers for drug delivery. Volume 15 de RAPRA review reports: RAPRA Technology Limited, Rapra Review Reports. Rapra review reports; 15, 6, 174 (Rapra Technology Limited) Editor: iSmithers Rapra Publishing.

42. Oledzka E., Sobczak M. 2012. Polymers in the pharmaceutical applications: Natural and bioactive initiators and catalysts in the synthesis of biodegradable and bioresorbable polyesters and polycarbonates. In: *Innovations in Biotechnology*. Edited by Dr. Eddy C. Agbo. Publisher InTech.

43. Liu D.-Q., Chen L., Yan F.-Y., Wang B. 2012. Teaching exploration and study in polymers for pharmaceuticals. Education and educational technology, *Adv. Intel. Soft Comput.*, **108**, 95–98.

44. Park H., Park K. 1994. Polymers in pharmaceutical products. Chapter 1, 2–15. In: *Polymers of Biological and Biomedical Significance. ACS Symposium Series*, 540. American Chemical Society.

45. Labarre D., Ponchel G., Vauthier C. 2010. *Biomedical and Pharmaceutical Polymers*. Pharmaceutical Press, U.K.

46. Ogaji I. J., Nep E. I., Audu-Peter J. D. 2012. Advances in natural polymers as pharmaceutical excipients. *Pharm. Anal. Acta*, **3**, 146.

47. McGinity J. W., Felton L. 2008. *Aqueous Polymeric Coatings for Pharmaceutical Dosage Forms. Third Edition, Therapeutics and Novel Approaches*. 3rd ed. Informa Healthcare, USA.

48. Reddy D. M., Begum S. R., Jyothirmai K., Dileep Kumar G. V., Jayasree S., Jeevankumar J., Gita Samira A. V. S. 2006. A novel review on natural polymers used in formulation of pharmaceutical dosage forms. http://www. pharmaresearchlibrary.com.

49 Ostrovsky G. 2008. *Biodegradable Polymers Deliver Pharmaceuticals to Diseased Cells*. Intertek Pharmaceutical, Germany.

50. Gandhi K. G. J., Deshmane S. V., Biyani K. R. 2012. Polymers in pharmaceutical drug delivery system: A review. *Int. J. Pharm. Sci. Rev. Res.*, **14**(2), 57–66.

51. Chasin M. 1990. Biodegradable Polymers as Drug Delivery Systems. Volume 45 of *Drugs and the Pharmaceutical Sciences*. Editorial, Taylor & Francis.

52. Nep E. I., Odumosu P. O., Ngwuluka N. C., Olorunfemi P. O., Ochekpe N. A. 2013. Pharmaceutical properties and applications of a natural polymer from grewia mollis. *J. Polymers*, 2013, Article ID 938726, 8 pages.

53. Allender C. J., Richardson C., Woodhouse B., Heard C. M., Brain K. R. 2000. Pharmaceutical applications for molecularly imprinted polymers. *Int. J. Pharm.*, **195**(1–2), 39–43.

54. Chasin M., Langer R. 1990. *Biodegradable Polymers as Drug Delivery Systems*. Marcel Dekker.

55. Prajapati V. D., Jani G. K., Moradiya N. G., Randeria N. P., Nagar B. J., Naikwadi N. N., Variya B. C. 2013. Galactomannan: A versatile biodegradable seed polysaccharide. *Int. J. Biol. Macromol.*, **60**, 83–92.

56. Bresolin T. M. B., Milas M., Rinaudo M., Reicher F., Ganter J. L. M. S. 1999. Role of galactomannan composition on the binary gel formation with xanthan. *Int. J. Biol. Macromol.*, **26**, 225–231.

57. Bresolin T. M. B., Sander P. C., Reicher F., Sierakowski M. R., Rinaudo M., Ganter J. L. M. S. 1997. Viscometric studies on xanthan and galactomannan systems. *Carbohydr. Polymers,* **33**, 131–138.

58. Bresolin T. M. B., Milas M., Rinaudo M., Ganter J. L. M. S. 1998. Xanthan–galactomannan interactions as related to xanthan conformations. *Int. J. Biol. Macromol.,* **23**, 263–275.

59. Chandrasekaran R., Radha A. 1997. Molecular modeling of xanthan: Galactomannan interactions. *Carbohydr. Polymers,* **32**, 201–208.

60. Morris E. R., Foster T. J. 1994. Role of conformation in synergistic interactions of xanthan. *Carbohydr. Polymers,* **23**, 133–135.

61. Morris V. J., Brownsey G. J., Ridout M. J. 1994. Reply to Role of conformation in synergistic interactions of xanthan. *Carbohydr. Polymers,* **23**, 139–140.

62. Launay B., Cuvelier G., Martinez-Reyes S. 1997. Viscosity of locust bean, guar and xanthan gum solutions in the Newtonian domain: A critical examination of the master curves. *Carbohydr. Polymers,* **34**, 385–395.

63. Pinheiro A. C., Bourbon A. I., Rocha C., Ribeiro C., Maiad J. M., Goncalves M. P., Teixeira J. A., Vicente A. A. 2011. Rheological characterization of k-carrageenan/galactomannan and xanthan/galactomannan gels: Comparison of galactomannans from non-traditional sources with conventional galactomannans. *Carbohydr. Polymers,* **83**, 392–399.

64. Khouryieh H. A., Herald T. J., Aramouni F., Alavi S. 2006. Influence of mixing temperature on xanthan conformation and interaction of xanthan–guar gum in dilute aqueous solutions. *Food Res. Int.,* **39**, 964–973.

65. Khouryieh H. A., Herald T. J., Aramouni F., Alavi S. 2007. Intrinsic viscosity and viscoelastic properties of xanthan/guar mixtures in dilute solutions: Effect of salt concentration on the polymer interactions. *Food Res. Int.,* **40**, 883–893.

66. Kayacier A., Dogan M. 2006. Rheological properties of some gums-salep mixed solutions. *J. Food Eng.,* **8**(72), 261–265.

67. Higiro J., Herald T. J., Alavi S. 2006. Rheological study of xanthan and locust bean gum interaction in dilute solution. *Food Res. Int.,* **39**, 165–175.

68. Dolz M., Hernandez M. J., Delegido J., Alfaro M. C., Muñoz J. 2007. Influence of xanthan gum and locust bean gum upon flow and thixotropic behaviour of food emulsions containing modified starch. *J. Food Eng.,* **81**, 179–186.

69. Arocas A., Sanz T., Fiszman S. M. 2009. Improving effect of xanthan and locust bean gums on the freeze-thaw stability of white sauces made with different native starches. *Food Hydrocolloids,* **23**, 2478–2484.

70. Mandala I. G., Savvas T. P., Kostaropoulos A. E. 2004. Xanthan and locust bean gum influence on the rheology and structure of a white model-sauce. *J. Food Eng.*, **64**, 335–342.

71. Mannion R. O., Melia C. D., Launay B., Cuvelier G., Hill S. E., Harding S. E., Mitchel J. R. 1992. Xanthan/locust bean gum interactions at room temperature. *Carbohydr. Polymers,* 19, 91–97.

72. Zhan D. F., Ridout M. J., Brownsey G. J., Morris V. J. 1993. Xanthan-locust bean gum interactions and gelation. *Carbohydr. Polymers,* 21, 53–58.

73. Lundin L., Hermansson A.-M. 1995. Supermolecular aspects of xanthan-locust bean gum gels based on rheology and electron microscopy. *Carbohydr. Polymers,* 26, 129–140.

74. Gamier C., Schorsch C., Doublier J.-L. 1995. Phase separation in dextran/locust bean gum mixtures. *Carbohydr. Polymers,* 28, 313–317.

75. Schorsch C., Gamier C., Doublier J.-L. 1995. Microscopy of xanthan/galactomannan mixtures. *Carbohydr. Polymers,* **28**, 319–323.

76. Scborsch C., Gamier C., Doublier J.-L. 1997. Viscoelastic properties of xanthan/galactomannan mixtures: Comparison of guar gum with locust bean gum. *Carbohydr. Polymers,* **34**, 165–175.

77. Fernandes P. B. 1995. Influence of galactomannan on the structure and thermal behaviour of xanthan/galactomannan mixtures. *J. Food Eng.*, **24**, 269–283.

78. Tako M., Nakamura S. 1985. Synergistic interaction between xanthan and guar gum. *Carbohydr. Res.*, **138**, 207–213.

79. Tako M. 1991. Synergistic interaction between xanthan and Tara-bean gum. *Carbohydr. Polymers,* **16**, 239–252.

80. Cairns P., Morris V. J., Miles M. J., Brownsey G. J. 1986. Comparative studies of the mechanical properties of mixed gels formed by kappa carrageenan and tara gum or carob gum. *Food Hydrocolloids,* **1**(1), 89–93.

81. Cairns P., Miles M. J., Morris V. J. 1986. X-ray diffraction studies of kappa carrageenan-tara gum mixed gels. *Int. J. Biol. Macromol.*, **8**, 124–127.

82. Tabilo-Munizaga G., Barbosa-Canovas G. V. 2005. Rheology for the food industry. *J. Food Eng.*, **67**, 147–156.

83. Gomez-Diaz D., Navaza J. M. 2004. Rheology of food stabilizers blends. *J. Food Eng.*, **64**, 143–149.
84. Renaud M., Belgacem M. N., Rinaudo M. 2005. Rheological behaviour of polysaccharide aqueous solutions. *Polymer*, **46**, 12348–12358.
85. Gomez-Diaz D., Navaza J. M. 2003. Rheology of aqueous solutions of food additives effect of concentration, temperature and blending. *J. Food Eng.*, **56**, 387–392.
86. Rinaudo M. 2001. Relation between the molecular structure of some polysaccharides and original properties in sol and gel states. *Food Hydrocolloids*, **15**, 433–440.
87. Zhong W., Chang G.-S., Zhang H. 2008. Review on applications of the quick-freezing food stabilizers. *Grain Distribution Technol.*; 2008-01.
88. Hu G.-H., Zhai R.-W. 2002. The speciality of guar gum and its application in food industry. *Beverage Fast Frozen Food Ind.*; 2002-04.
89. Jianxiong X., Shan L., Zhendong G. 2004. The rheology influencing factors of guar gum. *Guangzhou Food Sci. Technol.*; 2004-04.
90. Zhu C., Xue H., Sun D., Zhang W., Shi J., Gu G. 2005. Progress in studies on amendatory guar gum; Chinese Wild Plant Resources; 2005-04.
91. Bicak O., Ekmekci Z., Bradshaw D. J., Harris P. J. 2007. Adsorption of guar gum and CMC on pyrite. *Minerals Eng.*, **20**, 996–1002.
92. Liu Q., Zhang Y., Laskowski J. S. 2000. The adsorption of polysaccharides onto mineral surfaces: An acid-base interaction. *Int. J. Miner. Process*, **60**, 229–245.
93. Wang W., Zhang J., Wang A. 2009. Preparation and swelling properties of superabsorbent nanocomposites based on natural guar gum and organo-vermiculite. *Appl. Clay Sci.*, **46**, 21–26.
94. Laskowski J. S., Liu Q., O'Connor C. T. 2007. Current understanding of the mechanism of polysaccharide adsorption at the mineral/aqueous solution interface. *Int. J. Miner. Process*, **84**, 59–68.
95. Liu Q., Zhang Y., Laskowski J. S. 2000. The adsorption of polysaccharides onto mineral surfaces: An acid-base interaction. *Int. J. Miner. Process*, **60**, 229–245.
96. Wang J., Somasundaran P., Nagaraj D. R. 2005. Adsorption mechanism of guar gum at solid–liquid interfaces. *Minerals Eng.*, **18**, 77–81.
97. Rath R. K., Subramanian S. 1997. Studies on adsorption of guar gum onto biotite mica. *Minerals Eng.*, **10**(12) 1405–1420.

98. Bicak O., Ekmekci Z., Bradshaw D. J., Harris P. J. 2007. Adsorption of guar gum and CMC on pyrite. *Minerals Eng.,* **20**, 996–1002.

99. Rath R. K., Subramanian S. 1999. Adsorption, electrokinetic and differential flotation studies on sphalerite and galena using dextrin. *Int. J. Miner. Process,* **57**, 265–283.

100. Rao H. K., Forssberg K. S. E. 1997. Mixed collector systems in flotation. *Int. J. Miner. Process,* **51**, 67–79.

101. Zhao M., Lu Y., Dun W. 2009. Research and application of guar gum mining technology. *Min. Metallurgy*; 2009-01.

102. Ma X., Pawlik M. 2005. Effect of alkali metal cations on adsorption of guar gum onto quartz. *J. Colloid Interface Sci.,* **289**, 48–55.

103. Peng H., Wang K., Wang L. 2013. Progress of research on guar gum derivatives for use in hydraulic fracturing. *Guangzhou Chem. Ind.*; 2013-18.

104. Tian N., Zheng R. Z. 2002. Study and performance appraisal of cationic guar gum. *J. Chengde Petroleum Coll.*; 2002-01.

105. Cheng W., Fang B., Lu Y., Yang Q., Cao D., Luo S. 2007. Rheological properties of the low molecular weight cationic guar gum cross-linked gel systems. *J. East China Univ. Sci. Tech.* (*Nat. Sci. Ed.*); 2007-05.

106. Zou S., Wang K., Yin Q., Jiang B. 2004. Preparation and characterization of hydroxypropyl guar gum. *Chem. Res. Appl.*; 2004-01.

107. Cui J., Wu W. 2010. Investigation of a water-soluble polymer fracturing fluid system. *J. Guangxi Univ.* (*Nat. Sci. Ed.*); 2010-03.

108. Lin S., Zhang J., Wang R., Xu L. 2013. Study on development and reutilization feasibility of instant guar gum fracturing fluid. *Fault-Block Oil Gas Field*; 2013-02.

109. Zhang H., Zhou J., Zhou Y., Zhang Y. 2013. Research and application of carboxymethyl hydroxypropyl guar (CMHPG) fracturing fluid. *Complex Hydrocarbon Reservoirs*; 2013-02.

110. Ma W., Luo Q., Shang Z., Li J. 1999. Study on polysaccharide in tara seed; *Chem. Ind. Forest Prod.*; 1999-04.

111. Lin S., Zhang J., Wang R., Xu L. 2013. Study on development and reutilization feasibility of instant guar gum fracturing fluid. *Fault-Block Oil Gas Field;* 2013-02.

112. Immig J. 2013. Toxic Chemicals in the exploration and production of gas from unconventional sources. National Toxics Network, Bangalow, Australia.

113. Cun Q., Chen L. 2003. Modification of vegetable gum and its application to drilling engineering. *J. Phenol. (Sci. Technol. Ed.)*; 2003-02.
114. Yi J., Li Z., Qiao W. 2005. Chemical modification of guar gum. *China Surf. Detergent Cosmetics*; 2005-02.
115. Hu Y. L. R. 2004. Progress in the modification of guar gum. *Guangdong Chem. Ind.*; 2004-06.
116. Zhao X., Feng H., Wang Y., Luo H. 2009. Synthesis of carboxyl methyl guar flocculant and its flocculating property. *Ind. Water Wastewater*; 2009-01.
117. Wang C., Huang G. 2008. Synthesis and characterization of hydroxy-propyl guar gum. *Speciality Petrochem.*; 2008-03.
118. Liu Y., Guan B., Liu P., Liang L., Chen X., Qiu X. 2012. Effects of fiber on the proppant carrying capacity of fracturing fluid. *Oilfield Chem.*; 2012-01.
119. Li W., Zhang J., Zhang X. 2000. The new advances in the study of the rheological property of fracturing fluid. *J. Xi' Petroleum Inst.*; 2000-02
120. Liu J., Zhou X., Lu H., Li Z., Guan B., Zhang X., Guo Z. 2008. Study on rheology of the new low molecular weight fracturing fluid; Well Testing; 2008-05
121. Tang M., Su C., Fang B. 2011. Rheological properties of the regenerable gel systems of low molecular weight carboxymethyl guar gum. *J. Chem. Eng. Chin. Univ.*; 2011-04.
122. Zhu G., Fang B. 2012. The preparation and rheological properties of regeneratable borax/degrated hydroxypropyl tara gum gel systems. *J. Chem. Eng. Chin. Univ.* 2012-05.
123. Tang M., Su C., Fang B. 2011. Rheological properties of the regenerable gel systems of low molecular weight carboxymethyl guar gum. *J. Chem. Eng. Chin. Univ.*; 2011-04.
124. Cheng W., Fang B., Lu Y., Yang Q., Cao D., Luo S. 2007. Rheological properties of the low molecular weight cationic guar gum cross-linked gel systems. *J. East China Univ. Sci. Technol. (Nat. Sci. Ed.)*; 2007-05.
125. Liu H., Xiao D. 2008. A novel low-damage vegetable gum-based fracturing fluid and its application in low-permeability reservoirs. *Acta Petrolei Sin.*; 2008-06.
126. Zhang J., Li Z., Xia D., Bao S., Luo X., Yang S. 2009. Investigating actuality and the vista of tara tannin and tara polysaccharide. *Nat. Prod. Res. Dev.*; 2009-01.
127. Fang B., Jiang Z., Lu Y., Cheng W., Guan C. 2008. Rheological property and constitutive equation of the oligomeric guar gum fracturing fluid system. *Nat. Gas Ind.*; 2008-02.

128. Guan B., Wang Y., He Z., Du B., Liu J. 2006. Type recoverable low molecular mass guar gum fracturing fluid. *Oilfield Chem.*; 2006-01.

129. Liu J., Zhou X., Lv H., Li Z., Guan B., Zhang X. 2008. Study on rheology of the new low molecular weight fracturing fluid. *Well Testing*; 2008-05.

130. Tang M., Su C., Fang B. 2011. Rheological properties of the regenerable gel systems of low molecular weight carboxymethyl guar gum. *J. Chem. Eng. Chin. Univ.*; 2011-04.

131. Zhang T., Liu P., Quan H., Dai S., Lu H. 2012. Modification of low molecular weight guar gum with etherifying agent. *Adv. Fine Petrochem.*; 2012-05.

132. Guo J., Zhu R., Yan F. 2007. Evaluation on the properties of a type of modified guar gum. *J. Southwest Petroleum Univ.*; 2007-S2.

133. Ma X., Pawlik M. 2007. Intrinsic viscosities and Huggins constants of guar gum in alkali metal chloride solutions. *Carbohydr. Polymers,* **70**, 15–24.

134. Masuelli M. A., Takara A., Acosta A. 2013. Hydrodynamic properties of tragacanthin. Study of temperature influence. *J. Argent. Chem. Soc.,* **100**, 25–34.

135. Martin A. Masuelli. 2013. Hydrodynamic properties of whole arabic gum. *Am. J. Food Sci. Technol.,* **1**(3), 60–66.

136. Masuelli M. A., Sansone M. G. 2012. Hydrodynamic properties of Gelatin. Studies from intrinsic viscosity measurements. In: *Products and Applications of Biopolymers.* Chapter 5, pp. 85–116. Editor Verbeek, C. J. R. INTECH.

137. Masuelli M. A. 2013. Study of bovine serum albumin solubility in aqueous solutions by intrinsic viscosity measurements. *Adv. Phys. Chem.,* 2013, Article ID 360239, 8 pages.

138. Curvale R., Masuelli M., Perez Padilla A. 2008. Intrinsic Viscosity of bovine serum albumin conformers. *Int. J. Biol. Macromol.,* **42**, 133–137.

139. Sharman W. R., Richards E. L., Malcom G. N. 1978. *Biopolymers,* **17**, 2817–2833.

140. Doublier J. L., Lounay B. 1976. *Proc. Inter. Congr. Rheology*, Gorhorbarg, 6th, pp. 532–533.

141. Doublier J. L., Launay B. 1981. Rheology of galactomannan solutions: Comparative study of guar gum and locust bean gum. *J. Text. Stud.,* **12**(2) 151–172.

142. Robinson G., Ross-Murphy S. B., Morris E. R. 1982. Viscosity molecular-weight relationships, intrinsic chain flexibility, and dynamic solution properties of guar galactomannan. *Carbohydr. Res.,* **107**, 17–32.

143. Petkowicz C. L. O., Reicher F., Mazeau K. 1998. Conformational analysis of galactomannans: From oligomeric segments to polymeric chains. *Carbohydr. Polymers*, **37**, 25–39.

144. Beer M. U., Wood P. J., Weisz J. 1999. A simple and rapid method for evaluation of Mark–Houwink–Sakurada constants of linear random coil polysaccharides using molecular weight and intrinsic viscosity determined by high performance size exclusion chromatography: Application to guar galactomannan. *Carbohydr. Polymers*, **39**, 377–380.

145. Petkowicz C. L. O., Milas M., Mazeau K., Bresolin T., Reicher F., Ganter J. L. M. S., Rinaudo M. 1999. Conformation of galactomannan: Experimental and modelling approaches. *Food Hydrocolloids*, **13**, 263–266.

146. Dolnik V., Gurske W., Padua A. 2001. Galactomannans as a sieving matrix in capillary electrophoresis. *Electrophoresis*, **22**, 707–719.

147. Wientjes R. H. W., Duits M. H. G., Bakker J. W. P., Jongschaap R. J. J., Mellema, J. 2001. Linear viscoelastic behavior of enzymatically modified guar gum solutions: Structure, relaxations, and gel formation. *Macromolecules*, **34**, 6014.

148. Cheng Y., Brown K. M., Prud'homme R. K. 2002. Preparation and characterization of molecular weight fractions of guar galactomannans using acid and enzymatic hydrolysis. *Int. J. Biol. Macromol.*, **31**, 29–35.

149. Sittikijyothin W., Torres D., Goncalves M. P. 2005. Modelling the rheological behaviour of galactomannan aqueous solutions. *Carbohydr. Polymers*, **59**, 339–350.

150. Picout D. R., Ross-Murphy S. B. 2007. On the Mark–Houwink parameters for galactomannans. *Carbohydr. Polymers*, **70**, 145–148.

151. Pollard M. A., Kelly R., Fischer P. A., Windha E. J., Eder B., Amado R. 2008. Investigation of molecular weight distribution of LBG galactomannan for flours prepared from individual seeds, mixtures, and commercial samples. *Food Hydrocolloids*, **22**, 1596–1606.

152. Cerqueira M. A., Pinheiro A. C., Souza B. W. S., Lima A. M. P., Ribeiro C., Miranda C., Teixeira J. A., Moreira R. A., Coimbra M. A., Gonçalves M. P., Vicente A. A. 2009. Extraction, purification and characterization of galactomannans from non-traditional sources. *Carbohydr. Polymers*, **75**, 408–414.

153. Picout D. R., Ross-Murphy S. B., Jumel K., Harding S. E. 2002. Pressure cell assisted solution characterization of polysaccharides. 2. Locust bean gum and tara gum. *Biomacromolecules*, **3**, 761–767.

154. Morris G. A., Patel T. R., Picout D. R., Ross-Murphy S. B., Ortega A., Garcia de la Torre J., Harding S. E. 2008. Global hydrodynamic analysis of

the molecular flexibility of galactomannans. *Carbohydr. Polymers,* **72**, 356–360.

155. Mazeau K., Rinaudo M. 2004. The prediction of the characteristics of some polysaccharides from molecular modeling. Comparison with effective behavior. *Food Hydrocolloids,* **18**, 885–898.

156. Sittikijyothin W., Sampaio P., Goncalves M. P. 2007. Heat-induced gelation of β-lactoglobulin at varying pH: Effect of tara gum on the rheological and structural properties of the gels. *Food Hydrocolloids,* **21**, 1046–1055.

157. Parvathy K. S., Susheelamma N. S., Tharanathan R. N. 2007. Hydration characteristics of guar gum samples and their fractions. *Food Hydrocolloids,* **21**, 630–637.

158. Masuelli, M. A. 2013. Dextrans in aqueous solution. Experimental review on intrinsic viscosity measurements and temperature effect. *J. Polymer Biopolymer Phys. Chem.,* **1**(1), 13–21.

159. Masuelli, M. A. 2011. Viscometric study of pectin. Effect of temperature on the hydrodynamic properties. *Int. J. Biol. Macromol.,* **48**, 286–291.

160. Masuelli, M. A., Illanes C. I. 2014. Review of the characterization of sodium alginate by intrinsic viscosity measurements. Comparative analysis between conventional and single point methods. *Int. J. BioMater. Sci. Eng.,* **1**(1), 1–11.

161. Wu Y., Li W., Cui W., Eskin N. A. M., Goff H. D. 2012. A molecular modeling approach to understand conformation functionality relationships of galactomannans with different mannose/galactose ratios. *Food Hydrocolloids,* **26**, 359–364.

Chapter 3

High-Value Utilization of Natural Cellulose: Cellulose-Based Biocomposite Materials

Yong Zhang and Juming Yao

The Key Laboratory of Advanced Textile Materials and Manufacturing Technology of the Ministry of Education,
College of Materials and Textiles, Zhejiang Sci-Tech University,
Hangzhou 310018, China

yaoj@zstu.edu.cn

3.1 Introduction

Cellulose, which is a naturally occurring polymer and considered as an almost inexhaustible source of organic raw material in the world, has been applied in many areas, such as the paper industry, water treatment, material reinforcement, physiological hygiene, and so on. Cellulose can be easily obtained from abundant inexpensive renewable biomass and its abundance, easy availability, and low cost have made it one of the most promising raw materials.

Cellulose is a linear polysaccharide with long chains, which consists of β-D-glucopyranose units joined by β-1,4-glycosidic linkage and contain a large number of hydroxyl groups (three –OH groups per AGU unit). In one repeating unit of cellulose molecule,

Handbook of Sustainable Polymers: Processing and Applications
Edited by Vijay Kumar Thakur and Manju Kumari Thakur

ISBN 978-981-4613-53-8 (Hardcover), 978-981-4613-54-5 (eBook)
www.panstanford.com

there are one methylol and two hydroxyl groups as functional groups. Although it is a linear polymer and contains two types of hydroxyl groups, primary hydroxyl in methylol group (–CH_2OH) at C-6 and secondary hydroxyl groups (–OH) at C-3 and C-4, both of which are hydrophilic, it does not dissolve in water and in common solvents due to strong hydrogen bonding between the cellulose chains. So it is important to modify the physical or chemical properties of surfaces in order to improve the flexibility and ensure the availability of cellulose. Now cellulose has been widely concerned and studied focusing on its excellent biological, chemical, and mechanical properties.

For a long time, cellulose together with hemi-cellulose, has played a key role in the paper industry as one of its traditional uses. Besides this, at present, cellulose has been formed into fibers, films, gels, micro- and nanoparticles for various applications, as pristine cellulose exhibits low toxicity and biodegradability and poses no serious environment concerns, providing impetus for its application in high value-added composites, such as food, fertilizers, biomaterials, and pharmaceuticals.

In this chapter, we review recent developments in the use of modified cellulose for emerging functional material applications, specifically superabsorbent, physiological, and medical materials, wastewater flocculating, material reinforcing and controlled drug delivery. We focus on the modification of cellulose with chemical and biological functionalities, reviewing practical considerations such as reusability, low toxicity, biodegradability, and scale-up capability.

3.2 Methods of Cellulose Graft Copolymerization

A graft copolymer generally consists of a long sequence of one monomer, referred to as the backbone polymer (main chain) with one or more branches (grafts) of long sequences of a different monomer [1]. Among the methods of modification of polymers, graft copolymerization offers an attractive and versatile means of imparting a variety of functional groups to a polymer [2]. The synthesis of cellulose graft copolymers is one of the key ways of modifying the physical properties and chemical properties of cellulose [3]. This is usually achieved by modifying the cellulose molecules through the creation of branches (grafts) of synthetic polymers that impart specific properties onto the cellulose substrate, without destroying its intrinsic properties [3]. Indeed, depending on the polymer that is grafted onto the cellulose, it is possible to attain properties such as dimensional-stability, resistance to abrasion and wear, wrinkle recovery, oil and water repellence, elasticity, sorbancy, ion exchange capacity [1] microbiological attack [4–10].

The graft copolymerization of many monomers onto cellulose and onto cellulose derivatives has been carried out by different methods that can be generally classified into three major groups: (i) chemical methods, (ii) radiation-induced grafting, and (iii) plasma-initiated grafting [11].

3.2.1 Chemical Methods

Chemical method of graft copolymerization involves the use of chemicals, which act as initiators to generate active sites on backbone. The use of various redox initiator systems like Lewis acids, strong bases, and metal carbonyls has been reported for chemical grafting. Grafting of vinyl monomers onto polymeric backbones has been accomplished by using a range of free radical initiators and redox systems like dibenzoyl peroxide, Azobis (isobutyronitrile) (AIBN), ceric ammonium nitrate, potassium persulfate, potassium permanganate, and Fenton's reagent [11–15]. Researchers are also developing new redox systems for the incorporation of desired properties into backbone. Zahran and Mahmoud [11, 16] investigated the use of potassium peroxydiphosphate–metal ion–cellulose thiocarbonate redox initiator system for the graft copolymerization of methyl methacrylate onto cellulosic fabrics. Atom transfer radical polymerization (ATRP) is a modern method to generate free radical sites on dormant chains. In this method, halogen atoms are capped onto inactive sites that are reversibly transferred to metal complexes in lower states. Vlcek [11, 17] reported the controlled grafting of cellulose diacetate with methyl methacrylate and butyl acrylate through ATRP. Tizzotti [11, 18] in a review article reported about ATRP graft copolymerization of polysaccharides along with other living polymerization techniques like RAFT and NMP.

3.2.2 Radiation-Induced Grafting

Graft copolymerization involving radiations as an initiating system allows a high degree of control over the number and length of grafted chains. This can be done by careful selection of dose and its rate. This is the most convenient method for the graft copolymerization. When electromagnetic radiations pass through backbone, it results in the formation of active sites for the reaction. Radiation technique is able to maintain the purity of product as it is free from contamination. It allows carrying out reaction at different depth of backbone, which depends upon the penetrating power of the radiation used. Moreover, the molecular weight of polymer formed can be regulated in case of radiation-induced grafting. The unique features of radiation method make it a preferred technique of graft copolymerization [11, 19, 20].

3.2.2.1 Low-energy radiation-induced grafting

Grafting can proceed by the use of either low-energy radiations or high-energy radiations. Irradiation of backbone with low-energy radiations like UV and visible light results in the homolytic fission to generate free radicals on the backbone. Khan [63–66] modified the jute yarn by graft copolymerization of vinyl monomers using UV radiations. In presence of low-energy radiations, grafting may proceed with or without photo initiator. The graft copolymerization of poly(ethylene glycol) dimethyl acrylate oligomers onto cotton was investigated under UV radiations using benzophenone as photo initiator. The photo grafting was found to increase the wrinkle resistance of cotton [11, 19].

3.2.2.2 High-energy radiation-induced grafting

High-energy radiation like X-rays, gamma rays or accelerated electrons having energy varying between 6.5 and 3.0 MeV has been used for graft copolymerization of various backbones. Gamma radiations are most commonly used radiations, and Co-60 is the most widely used source of gamma radiations due to its long half-life period of 5.3 years. Cellulose fabric was grafted with divinylbenzene, and the extent of grafting was found to increase with total gamma radiation dose [11, 20]. Chauhan [11, 21] carried out the graft copolymerization of styrene onto cellulose and studied the effect of some additives on grafting parameters in limited aqueous medium.

3.2.2.3 Microwave radiation-induced grafting

Microwave radiation is another important energy source and is emerging as an effective tool for graft copolymerization. Microwave radiations have an advantage of instantaneous in core processing of reactants in homogeneous manner. Microwave radiations can rapidly transfer the energy into bulk of reaction mixture, which results in the rapid interaction of the material in the reaction mixture [11, 22].

Microwave irradiation significantly reduces the use of toxic solvents, as well as the reaction time for almost all the grafting reactions of interest here, ensuring high yields, product selectivity and clean product formations. Microwave-synthesized polysaccharide copolymers exhibit better properties for commercial exploitation than their conventionally synthesized counterparts [11, 23].

3.2.3 Plasma-Initiated Grafting

Surface modification of polymers through plasma polymerization technique has received increasing interest. Plasma conditions attained through slow discharge offer about the same possibilities as with ionizing radiation [11, 24, 25]. The main processes in plasmas are electron-induced excitation, ionization and dissociation. Thus, the accelerated electrons from the plasma have sufficient energy to induce cleavage of the chemical bonds in the polymeric structure, which forms macromolecular radicals to initiate the graft copolymerization [26].

Plasma-initiated polymerization of grafting can be carried out by using polymerizing gases and precursors like fluorocarbons, hydrocarbons and silicone containing monomers (Fig. 3.1). Carrier gas plays important roles in these plasma–surface interactions, and usually inert gas like helium or argon is used as carrier gases [27].

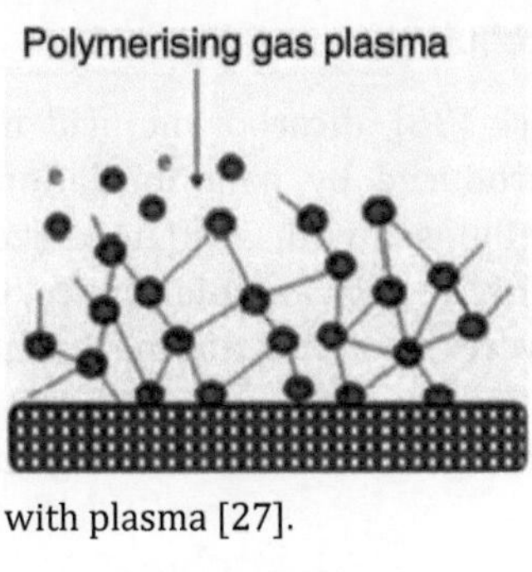

Figure 3.1 Grafting with plasma [27].

3.3 Water Treatment Application

With the rapid development of modern industry and living standard, the problem of environmental pollution is becoming more and more serious, as municipal wastewater, dying, printing, textile, domestic sewage, which has a serious influence on our live and been draw our attentions.

In recent years, a variety of methods, including flocculation, adsorption, filtration, oxidation, and electrolysis, have been carried out to remove contaminant from wastewater [28–32]. Therein, flocculation is one of the most widely applied methods in primary purification due to its low cost and easy operation [33–35].

Nowadays, great attention has been paid to natural polymer-based flocculants, since they are considered as wide-resourced, nontoxic, environment-friendly and high-efficient materials. They have been even acclaimed as "Green Flocculants of twenty-first Century" [34, 35]. Among them, cellulose is one of the most important natural polymers, and the first most abundant one in the world [34, 36]. In term of its novel characteristics, cellulose and its derivatives have a very wide application in many fields such as pharmaceutical, food, paper, and building material industries [34, 36, 37]. In addition, cellulose also has good water purification effects because it has abundant free -OH groups on the chain, enabling efficient removal of metal ions and organic matters from water for excellent chelating effect.

The utilization of cellulose as raw materials for flocculant is mainly divided into two major categories: (i) modified cellulose, (ii) cellulose-based flocculant.

3.3.1 Modified Cellulose

3.3.1.1 Anionic cellulose

In Suopajärvi's work [38], dicarboxylic acid nanocellulose (DCC) flocculants were produced by nanofibrillation of periodate and chlorite-oxidized celluloses with a homogenizer. The flocculation performance levels of five such anionic nanocelluloses with variable charge densities were examined in the coagulation-flocculation treatment of municipal wastewater and the results compared with the performance of a commercial coagulant and a synthetic polymeric flocculant. In addition, the aldehyde and carboxyl content, charge density (CD), size and stability of each DCC were determined in an aqueous solution. A schematic illustration of this cellulose oxidation and coagulation-flocculation process is shown in Fig. 3.2.

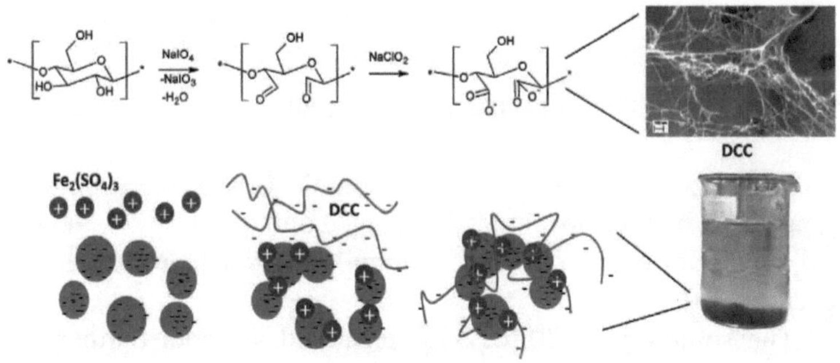

Figure 3.2 The synthesis of CDAC and coagulation-flocculation process [38].

The anionic nanofibrillated DCC bioflocculants synthesized here showed good performance in the coagulation-flocculation treatment of municipal wastewater. Their turbidity reduction efficiency fell short of that of the synthetic reference polymer, but their COD removal performance was as good as the latter. Combined treatment resulted in a lower residual turbidity and COD in settled suspension with highly reduced total chemical consumption relative to coagulation with ferric sulfite alone. DCC flocculants with a high charge density and high nanofibril content gave the best flocculation performance. The DCCs showed good stability over a long period in aqueous suspensions and good performance in the pH range chosen here [38].

3.3.1.2 Cationic cellulose

In Sirviö's work [39], a water-soluble cationic cellulose derivative was synthesized by cationizing dialdehyde cellulose (DAC) produced by periodate oxidation of birch cellulose pulp by means of a reaction between aldehyde groups and cationic Girard's reagent T ((2-hydrazinyl-2-oxoethyl)-trimethylazanium chloride, GT) to produce cationic dialdehyde cellulose (CDAC). A schematic illustration of this cellulose oxidation and DAC cationization processes is shown in Fig. 3.3.

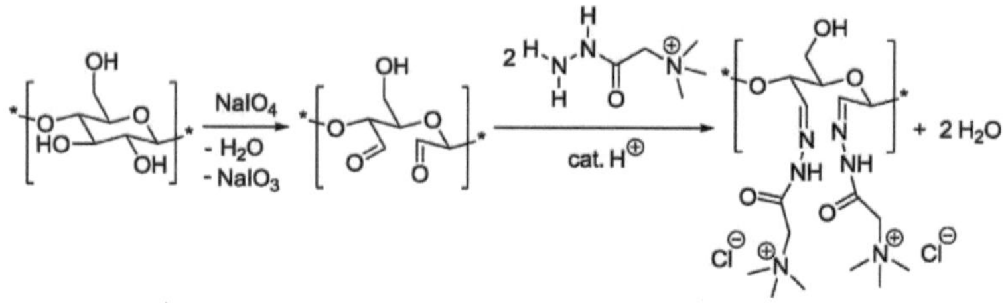

Figure 3.3 The synthesis of DAC and CDAC [39].

With promising results, Water-soluble CDACs were obtained from DACs having an aldehyde group content of 11.77 mmol/g and maximum cationicity of 4.27 mmol/g was achieved with 72 h reaction [39].

The solubility of CDACs as determined at room temperature was shown to be highly affected by the degree of oxidation of the DAC and the cationicity of the CDAC. CDACs produced from a DAC having an aldehyde content of 11.77 mmol/g or higher were soluble when the cationicity exceeded 3 mmol/g, whereas CDACs with the same cationic group content but a lower original aldehyde content in the native DAC only formed a gel-like material in water [39].

Preliminary studies using water-soluble CDACs as a biopolymeric flocculation agent indicate that they have a high potential for use as new, environment-friendly flocculation aids. Further studies on the use of CDAC as a flocculation agent are currently under way [39].

Likewise, in Zaman's work nanocrystalline cellulose (NCC) resulting from sulfuric acid hydrolysis of wood cellulose fiber [40], was rendered cationic by grafting with glycidyltrimethylammonium chloride (GTMAC). An optimization of the reaction parameters, such as water content, reactant mole ratio, and reaction media was performed. The presence of cationic GTMAC on the surface of NCC was confirmed by Fourier transform infrared spectroscopy (FTIR). The cationically modified NCC was characterized by surface charge density, degree of substitution, zeta potential, and particle size. It was found that the cationic surface charge density of NCC can be increased by controlling the water content of the reaction system. Surface cationization of NCC led to an increase in the surface charge density over the un-modified NCC. The cationically modified NCC was well dispersed and stable in aqueous media due to enhanced cationic surface charge density. Transmission electron microscopy (TEM) images showed the improvement in state of dispersion of cationically modified NCC over the un-modified NCC. The optimum water content was found to be 36 wt% for aqueous-based media and 0.5 of water to DMSO volume ratio for aqueous-organic solvent reaction media. The increased surface charge density of NCC also delayed the onset of gelation in aqueous system.

3.3.2 Cellulose-Based Flocculant

The flocculant based on cellulose can be generally classified into four major groups: (i) cationic cellulose-based-flocculant, (ii) anionic cellulose-based flocculant [41], (iii) amphoteric cellulose-based flocculant, (iv) nonionic cellulose-based flocculant [42] as follows:

3.3.2.1 Nonionic cellulose-based flocculant

Das [42] studied the development and application of a high-performance polymeric flocculant based on hydroxypropyl methyl cellulose grafted with polyacrylamide (HPMC-g-PAM). The flocculation characteristics were evaluated in two different synthetic effluents, namely kaolin and iron-ore suspensions using floc size measurement as well as traditional turbidity and settling velocity measurement. Based on the observed results, it can be concluded that polyacrylamide grafted hydroxypropyl methyl cellulose may be used as high-performance flocculant for treatment of synthetic effluents and mine wastewater. The flocculation kinetics demonstrates that it is in good agreement with both aggregation of particle and particle collision models. In addition, bridging flocculation is the major mechanism for this system. Finally, the rapid adsorption of polymers on suspended particles results in high settling rate. Being biodegradable and non-toxic, this novel graft copolymer-based flocculant has significance for the primary treatment of synthetic effluents and mine wastewaters.

3.3.2.2 Anionic cellulose-based flocculant

In Mishra's work [41], an environmentally benign and efficient route of synthesis of polyacrylic acid grafted carboxymethyl cellulose (CMC-g-PAA) is developed using microwave radiation alone to initiate the grafting reaction. The synthesis is optimized in terms of percentage grafting and intrinsic viscosity, by varying the microwave irradiation time and monomer (acrylic acid) concentration. The synthesized grades of the graft copolymer were characterized through various physicochemical techniques. FTIR spectroscopy provided first ever experimental proof of the proposed mechanism of the microwave effect involved in microwave initiated grafting. Flocculation efficacy of the graft copolymer was studied through standard "Jar test" procedure in river water and was compared to that of the starting material (CMC) and also with alum (coagulant). It was found that the CMC-g-PAA grade with highest hydrodynamic volume (i.e., intrinsic viscosity) showed the maximum flocculation efficacy, as predicted by "Singh's easy approachability model" and "Brostow, Pal and Singh model of flocculation." The high flocculation efficacy of polyacrylic acid grafted CMC (CMC-g-PAA) makes it a good candidate as flocculant for river water clarification, as confirmed by detailed study of water quality of the supernatants from jar test procedure.

3.4 Biomedical Material Application

In recent years, a rapid progress has been made in the field of biomedical materials, which combine natural and synthetic polymers and can be used in a variety of applications, including wound closure [43, 44], drug delivery systems [45, 46], novel vascular grafts [47–50] or scaffolds [51, 52] for in vitro or in vivo tissue engineering [50, 53, 54].

3.4.1 Wound Closure

Recent advances in the field of biomaterials and their medical applications indicate the significance and potential of various microbial cellulose in the development of novel classes of medical materials. This last, emerging area seems to be particularly important since many efforts have been devoted in recent years to exploring new skin substitutes and modern wound dressing materials using tissue engineering approaches. Various polymeric materials recently have been investigated for wound dressing application yielding many successful outcomes, but the search for an ideal skin graft substitute with properties and functionality similar to human skin is still continuing. Researchers believe that MC, while chemically the same as plant cellulose, displays novel physical properties determined by the particular genetics of the organism. In such a case, MC has a distinctive nanofibrillar structure that may become a perfect matrix as an optimal wound healing environment [43].

For example, a microbial cellulose membrane has been successfully used as a wound-healing device for severely damaged skin and as a small-diameter blood vessel replacement. The nonwoven ribbons of microbial cellulose microfibrils closely resemble the structure of native extracellullar matrices, suggesting that it could function as a scaffold for the production of many tissue-engineered constructs. In addition, microbial cellulose membranes, having a unique nanostructure, could have many other uses in wound healing and regenerative medicine, such as guided tissue regeneration (GTR), periodontal treatments, or as a replacement for dura mater (a membrane that surrounds brain tissue). In effect, microbial cellulose could function as a scaffold material for the regeneration of a wide variety of tissues, showing that it could eventually become an excellent platform technology for medicine. If microbial cellulose can be successfully mass produced, it will eventually become a vital biomaterial and will be used in the creation of a wide variety of medical devices and consumer products [44].

3.4.2 Drug Delivery System

Development of nanoparticle delivery systems for drugs, proteins, and enzymes remains an important challenge in the field of biomedical research on account of the effective delivery and release of the carrier molecules at the target site. Over the past few decades, extensive studies have been performed on both synthetic and natural polymer-based nano carriers and several of them are already in clinical use [46, 55–57]. Of the readily available natural polymers, polysaccharide nanoparticles, especially cellulose nanowhiskers (CNWs) have attracted increasing attention as a promising delivery system for drugs due to their outstanding properties such as nano dimension, high surface area, hydrophilicity, biocompatibility, and biodegradability [46, 58].

In Dash's work [46], they combined the aromatic linker and cellulose nanowhiskers to synthesize a novel delivery system for amine containing biologically active compounds and drug molecules. Based on their previous study focused on grafting of small chain primary amines to nanowhiskers employing a sequence of oxidation, condensation and *in situ* reduction reactions, c-aminobutyric acid, a spacer molecule, was grafted to the nanowhisker surface. C-aminobutyric acid was chosen as a spacer molecule because of its biocompatibility and ease of availability. In the next step, syringyl alcohol was used as an aromatic linker and attached to the carboxylic group of the amino acid. The effective grafting of spacer and linker molecules to the nanowhisker surfaces was confirmed by a series of characterization techniques including FT-IR, NMR, XPS and TEM. Further, their effective binding to the cellulose nanowhisker surface was evidenced by 2D-DOSY NMR spectroscopy. TEM observations showed no change in size and shape of the nanowhiskers after the series of reactions, which is substantial for a delivery system. Our future study involves exploring the conjugation of a small model amine drug (phenylpropanolamine) [59] to the nanowhisker delivery system. Finally, the concept of carrier molecule from a renewable, biocompatible and biodegradable resource provides a platform that could be widely adapted for the controlled delivery of enzymes, proteins and amine-containing drugs with the selection of desired linker molecules.

3.4.3 Vascular Grafts

Bacterial cellulose (BC) and fibrin have great potential uses for biomedical applications especially for artificial blood vessel (BV) applications [47, 48]. The emergence and increase of coronary heart diseases [60] brings about the need of replacement blood vessels. Synthetic BVs are available but have satisfactory success only with large-diameter vessels while they exhibit major failures as small-diameter vessels, those with diameter ~6 mm, e.g., coronary, carotid, femoral arteries [60, 61]. To date, the quest for a suitable artificial small-diameter blood vessels is still ongoing since none has yet immerged to give a satisfactory long-term success [62]. Both BC and fibrin are biocompatible and their surfaces are suitable for the growth of endothelial cells, cells that line the blood vessel lumen and prevent the formation of plaques that cause blood flow blockage [48, 63, 64]. BC has superior strength, is stable in the wet state [65, 66], and can be molded into tubular form with diameter of 6 mm [44, 54].

Fibrin on the other hand has remarkable elasticity and can extend up to 330% of its original length [67]. Also, cross-linked fibrin is part of the circulatory system that was tested to be relatively non-thrombogenic [68, 69]. Yet, BC falls short in elongation when compared to native BV [69] and fibrin reveals insufficient strength when subjected to physiological environments [47, 70].

In previous Brown's work [49], bacterial cellulose/fibrin composites were treated with glutaraldehyde in order to cross-link the polymers and allow better match of the mechanical properties with those of native small-diameter blood vessels. Glutaraldehyde-treatment of BC/fibrin composites produced materials that had ultimate tensile strength, modulus and time-dependent viscoelastic properties comparable to those of a native small-diameter blood vessel. The composite's strain at break, although considerably improved compared to that of untreated BC/fibrin blends, was significantly lower than that of the native BV. However, glutaraldehyde treatment changed the shape of the material's tensile stress strain curve by enhancing an initial low modulus plateau, resembling the strain hardening mechanism observed in native blood vessel. This strain hardening mechanism and the resulting improved properties were ascribed to cross-linking between the components, which allowed pronounced cellulose fiber alignment during loading as further observed by SEM of the fractured samples. Further investigations are needed to assess the viability, in particular cytotoxicity, and interest of glutaraldehyde-treated BC/fibrin composites for artificial blood vessel applications.

Similar to bacterial cellulose, NCC is biocompatible, stable, chemically inactive, and physiologically inert [58]. Moreover, NCC's nanoscale size allows for easy dispersion and its superior strength can provide effective reinforcement to a low strength matrix such as fibrin. These characteristics make NCC a promising nanobiomaterial for SDRVG application.

In Brown's subsequent work [50], the potential of synthesizing new nanocomposites from NCC and fibrin for small-diameter replacement vascular graft (SDRVG) application was demonstrated Fig. 3.4.

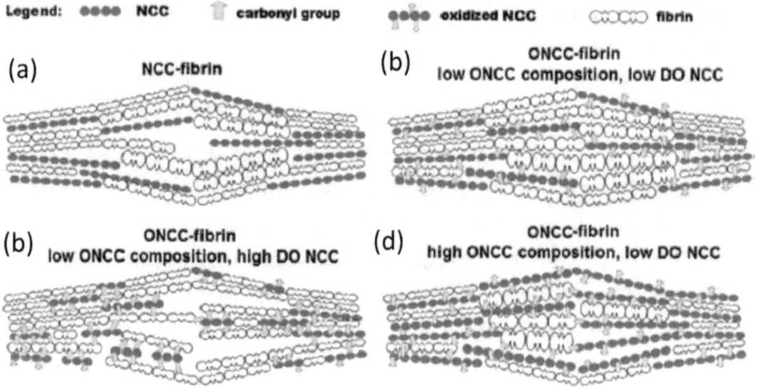

Figure 3.4 Postulated molecular interactions between ONCC and fibrin in nanocomposites [50].

Periodate oxidation of NCC can augment reactive carbonyl groups on NCC and facilitate its cross-linking with fibrin. Periodate oxidation of NCC from 4 to 12 h led to an augmentation of 2–7 carbonyl groups per 100 anhydroglucose units on NCC. NCC-fibrin nanocomposites were synthesized, composed of homogeneously dispersed oxidized NCC (ONCC) in a fibrin matrix, with fibrin providing elasticity and ONCC providing strength. The maximum strength and elongation of the nanocomposites were determined by atomic force microscopy (AFM) and compared with a native blood vessel. The manipulation of degree of oxidation of NCC and the NCC-to-fibrin ratio resulted in the variation of strength and elongation of the nanocomposites, indicating that the nanocomposites can be tailored to conform to the diverse mechanical properties of native blood vessels. A mechanistic understanding of the molecular interactions of ONCC and fibrin was illustrated. This study established fundamental information to utilizing NCC for SDRVG applications [50].

3.4.4 Scaffold

The biocompatibility of cellulose and its derivatives is well established [71]. The good match of their mechanical properties with those of hard and soft tissue has been demonstrated [72]. It has been employed in the form of membranes (e.g., dialyse, biosensors) [73] and bioadhesive cellulose gels as scaffold for growing functional cardiac cell constructs in vitro [51, 52].

In Müller's work [51], non-woven cellulose II fabrics were used as scaffolds for in vitro cartilage tissue engineering. The scaffolds were activated in a saturated $Ca(OH)_2$ solution and subsequently coated with a calcium phosphate layer precipitated from a supersaturated physiological solution. Chondrocyte cell response and cartilage development were investigated. The cell adherence was significantly improved compared to untreated cellulose fabrics, and the proliferation and vitality of the adhered chondrocytes were excellent, indicating the biocompatibility of these materials. A homogeneous distribution of the seeded cells was possible and the development of cartilageous tissue could be proved. In contact with a physiological chondrocyte solution, calcium is expected to be leached out from the precipitated layer, which might lead to a microenvironment that triggers the development of cartilage in a way similar to cartilage repair in the vicinity of subchondral bone.

3.5 Eco-Friendly Superabsorbent Polymer

Polymeric hydrogels have exhibited potential for application in many fields, such as hygiene [74], drug delivery formulations [75], wastewater treatment [76], agriculture [77] and tissue engineering scaffolds [78]. Those based on natural polymers with physical cross-links are especially interesting because of their unique advantages of abundance, nontoxicity, biocompatibility, and biodegradability [79, 80]. Cellulose hydrogels can be obtained by chemical cross-linking of water-soluble cellulose derivatives, such as cellulose acetate, hydroxyethylcellulose, and sodium carboxymethylcellulose [80–83]. Physically cross-linked cellulose hydrogels are recently prepared by developing intermolecular hydrogen bonds via hydroxyl groups on cellulose molecular chains [80, 84–86].

3.5.1 Superabsorbent Polymer in Agricultural Application

Superabsorbent polymer is a three-dimensionally cross-linked hydrophilic resin gel with excellent swelling and retaining capabilities. It could absorb liquids tens to thousands times its own mass and retain the liquids even under pressure [87, 88].

Two factors, water and fertilizer, play a crucial role in the growth of plants and their quality [88]. Therefore, it is very important to improve the utilization of water resources and fertilizer nutrients [88, 89]. However, about 40–70% of nitrogen, 80–90% of phosphorus, and 50–70% of potassium of the applied normal fertilizers is lost to the environment and cannot be absorbed by plants, which not only causes large economic and resource losses but also very serious environmental pollution [77, 90]. One method of reducing fertilizer nutrient losses involves the use of slow-release fertilizers. Slow-release fertilizers are made to release their nutrient contents gradually and to coincide with the nutrient requirement of a plant [77]. The release and dissolution rates of water-soluble fertilizers depend on the coating materials [77, 91–93]. Wu and Liu [77] have successfully prepared a novel cellulose acetate-coated compound fertilizer with controlled release and water retention, which possessed the three-layer structure. Its core was water-soluble compound fertilizer granular, the inner coating was cellulose acetate (CA), and the outer coating was poly(acrylic acid-co-acrylamide)/unexpanded vermiculite (P(AA-co-AM)/UVMT) superabsorbent composite.

Cellulose acetate is the classic membrane material used by the pioneers of modern membrane technology to create asymmetric membranes. It has been widely applied in the biomedical, pharmaceutical, and agricultural fields. In many of these applications, CA is extremely attractive due to its low price, good biodegradability, and nontoxicity. Therefore, the soluble compound fertilizer coated by CA would be an ideal slow-release formulation [77]. However, its application in this field has met some problems because most of these polymeric superabsorbents are based on pure poly(sodium acrylate), and then they are too expensive and not suitable for saline-containing water and soils [77, 94]. In the research of Wu [77], the water absorbency of CAFCW was 72 times its own weight if it was allowed to swell in tap water at room temperature for 90 min. The structure of CAFCW fertilizer granule was the three-layer, which the core is compound fertilizer (55.6 wt%), the inner coating is CA (6.7 wt%), and the outer coating is P(AA-co-AM)/UVMT (37.7 wt%), respectively, which was shown in Fig. 3.5 as follows [77]. This product with good controlled release and water retention capacity, being degradable in soil and environment friendly, could be especially useful in agricultural and horticultural applications.

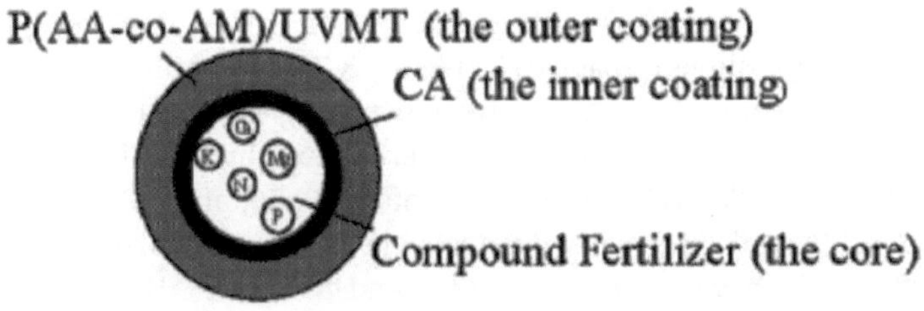

Figure 3.5 Cross-sectional schematic view of a CAFCW fertilizer granule [77].

From the research of Zhang [88], a novel flax yarn waste-g-poly(acrylic acid-co-acrylamide) (FYW/PAA) superabsorbent composite has been synthesized (Fig. 3.6). In order to develop an eco-friendly superabsorbent composite, flax yarn waste (FYW) was used as raw material. Acrylic acid (AA) and acrylamide (AM) were grafted onto the pretreated flax yarn waste (PFYW) by free-radical graft copolymerization in homogeneous aqueous solution. As a result, the prepared FYW/PAA attained the best water absorbency of 875 g/g in distilled water, 490 g/g in rainwater and 90 g/g in 0.9 wt% NaCl solution.

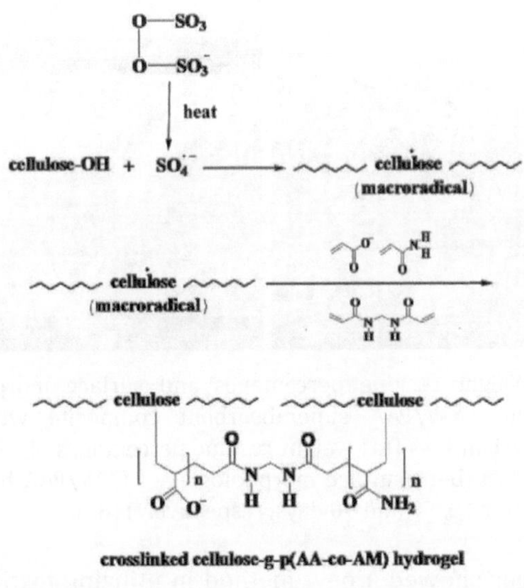

Figure 3.6 Synthetic principle of the FYW/PAA superabsorbent composite [88].

In this work, natural soil burial test was also completed [88]. Results were shown in Fig. 3.7. Figure 3.7a compares the weight residue percentages and surface morphologies of the FYW/PAA superabsorbent composite with different weight loss. The weight residues of FYW/PAA buried in soil after 10, 50, and 90 days were 94.3, 61.3, and 53.6 wt%, respectively. While the corresponding values of PAA were only 99.8, 93.1, and 87.0 wt%. Based on the flax cellulose skeleton material, the FYW/PAA superabsorbent composite presented an excellent biodegradability compared with PAA. Figures 3.7b–d shows the different surface morphologies of FYW/PAA buried in soil after 10, 50, and 90 days, respectively. The surfaces of the soil buried samples were not smooth and dense. Moreover, with the increase of burial time, more and more dots and cracks appeared on their surfaces.

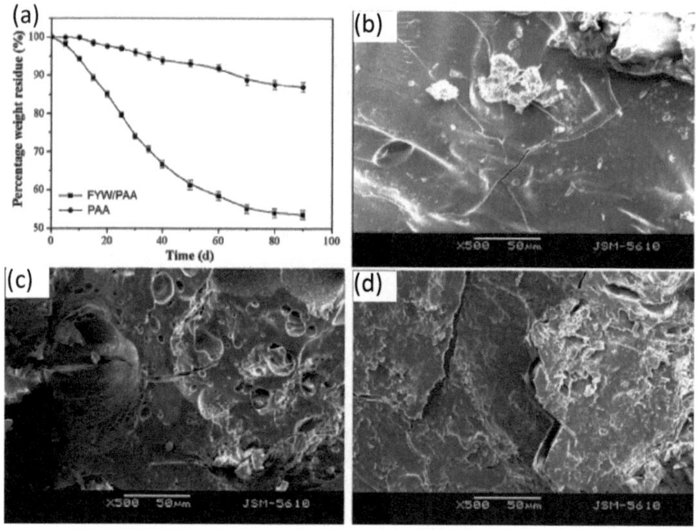

Figure 3.7 Weight residue percentages and surface morphologies of the FYW/PAA superabsorbent composite with different weight loss ((a) weight residue percentages of FYW/PAA and PAA; (b–d) surface morphologies of FYW/PAA buried in soil after 10, 50 and 90 days, respectively) [88].

This paper showed a new method in utilizing textile cellulose wastes such as cotton linter, cotton yarn waste and flax yarn waste, in the production of superabsorbent material. The introduction of them into superabsorbent could significantly reduce the production cost and make the technique quite environmental friendly, which is crucial to superabsorbent for agricultural and horticultural use [88].

In another work [95], differences in responses of corn subjected to SAP application under different irrigation levels were evident during our observations. Although the application of SAP had little effect on shoot dry mass (SDM) under adequate or moderate irrigation, the values increased remarkably by 133.5% under deficit irrigation. Correspondingly, water use efficiency under deficit irrigation was also increased by 97.1% with SAP. This study has shown that the application of superabsorbent polymer favored different physiological activities (intercellular carbon dioxide concentration, photosynthesis and leaf water potential) and increased the survivability of corn plant under drought stress. Thus, use of SAP could be an effective means for corn production under the arid conditions of northern China or the areas with similar ecologies.

It is mentioned that slow-release fertilizer provides an effective means of overcoming the high waste, low use efficiency, and environmental problems associated with the use of conventional fertilizers [96–98]. Coated fertilizer is the major category of slow-release fertilizer [96, 99], which is prepared by coating granules of conventional fertilizers with various materials that ensure a slow release of nutrients to soil by diffusion through the pores or by erosion and degradation of the coatings [96, 100–103].

In Xie's work [96], a new double-coated slow-release nitrogen and phosphorus fertilizer with water retention was prepared with the aim of improving fertilizer use efficiency and minimizing the negative impact on the environment. Wheat straw was introduced into the formulations as the basic coating material. Specifically, poly(acrylic acid-co-N-hydroxymethyl acrylamide)/wheat straw superabsorbent composite was used as the outer coating, and wheat straw/sodium alginate blends were used as the inner coating. The superabsorbent composite synthesized under the optimal conditions showed super water absorbency and excellent degradability. The results revealed that the coating material based on wheat straw imparted the product with high extent of degradation, low production cost, and broad application in general agricultures. The new approach shows promise in utilizing agro-waste and makes the technique quite environment friendly, simultaneously [96].

3.5.2 Superabsorbent Polymer in Hygiene Application

In the research results from Kamat and Malkani [74], it says that disposable diapers have almost become an indispensable item in the list of baby care products. Apart from providing convenience benefits, disposable diapers also offer health benefits. Furthermore, over the last couple of decades, the technology of diapers has also advanced with introduction of a super absorbent material (SAM) and better designs to ensure leakage protection. Compared to cloth nappy, disposable diapers offer benefits that help reduce the possibility of potential infections.

Disposable diapers have an absorbent core with SAM. SAM is a revolutionary material that can absorb water up to 100 times its weight i.e., 1 gm of SAM can absorb up to 100 gm of water. SAM is added to the inner core of the diaper and it absorbs urine and converts it into a gel. It wicks away the urine and prevents the mixing of urine and feces, further contributing to better containment of feces within the diaper. Disposable diapers have been significantly more effective in prevention of leakage of urine and feces as compared to cloth nappies (with or without overlying plastic/vinyl pants).

Another research from María [104] assesses the feasibility of degrading used disposable diapers, an important component (5–15% in weight) of urban solid waste in Mexico, by the activity of the fungus *Pleurotus ostreatus*, also known as oyster mushroom. Disposable diapers contain polyethylene, polypropylene, and a super absorbent polymer. Nevertheless, its main component is cellulose, which degrades slowly. *P. ostreatus* been utilized extensively to degrade cellulosic materials of agro-industrial sources, using *in situ* techniques. The practice has been extended to the commercial farming of the mushroom. This degradation capacity was assayed to reduce mass and volume of used disposable diapers. Pilot laboratory assays were performed to estimate the usefulness of the following variables on conditioning of used diapers before they act as substrate for *P. ostreatus*: (1) permanence versus removal of plastic cover, (2) shredding versus grinding, and (3) addition of grape wastes to improve structure, nitrogen, and trace elements content. Wheat straw was used as a positive control. This research indicates that growing. *P. ostreatus* disposable diapers could be a good alternative for two current problems: reduction of urban solid waste and availability of high-protein food sources.

In the work from Liu [105], an eco-friendly superabsorbent based on flax yarn waste for sanitary napkin applications has been successfully prepared. Till now, most of the waste cellulose materials from textile industry, such as cotton linter, cotton yarn waste and flax yarn waste, are still burned or land filled, causing both resource waste and environmental pollution [105, 106]. Reutilization of these waste materials could not only reduce the consumption of cotton or flax cellulose but also decrease the treatment of industrial waste. The major component (70%) of flax yarn waste is cellulose, which has been introduced as a basic skeleton of superabsorbent in previous studies [105, 106–109]. Fluff pulp, as a good absorbent material, is a kind of special pulp with villiform fibers. Nowadays, it is widely used as raw material for the production of sanitary napkin and paper diaper [105].

The objectives of this study were to investigate the preparation of an eco-friendly superabsorbent based on flax yarn waste under different synthesis conditions and to determine the properties of the resulting superabsorbent used for absorbent core layer of sanitary napkins. The results gained from this study strongly suggest that the prepared FYW/PAA superabsorbent has potential application in the field of eco-friendly sanitary napkins as a low-cost and eco-friendly synthetic material [105].

3.6 Nanocrystalline Cellulose

Polymer nanocomposites have attracted considerable attention in recent years as a result of their good performance, improved properties, design flexibility, lower life-cycle costs, and a unique large applicability range in various industrial fields [111]. Cellulose is an abundant and naturally occurring polymer that can be obtained from many sources. Nanometer-sized single-crystal cellulose, which is commonly referred to as nanocrystalline cellulose, nanowhiskers, or nanofibrils, can be obtained from various sources such as natural fibers and sea animals. The extraction of NCC from renewable sources has gained increasing attention in recent years, because of its exceptional mechanical properties (high specific strength and modulus), large specific surface area, high aspect ratio, environmental benefits, and low cost [112–114].

3.6.1 Introduction to Nanocrystalline Cellulose

Nanocrystalline cellulose, a rod-shaped nanoscale material with exceptional strength and physicochemical properties, can be prepared from inexpensive renewable biomass. Besides its potential use as a reinforcing agent for industrial biocomposites, pristine NCC exhibits low toxicity and poses no serious environmental concerns, providing impetus for its use in bioapplications [115].

Cellulose microfibrils contain crystalline and amorphous regions that are randomly distributed along their length. In the former, cellulose chains are packed closely, whereas the latter is more susceptible to chemical or enzymatic attack. NCC, pure cellulose in crystalline form with nanoscale dimensions, can be processed from different sources of biomass, under moderate conditions of acid or enzyme hydrolysis, to yield powder, liquid or gel forms by the removal of these amorphous regions [115]. The resulting NCC has a rigid rod-shape structure, 1–100 nm in diameter and tens to hundreds of nanometers in length [114–116]. As one of the strongest and stiffest natural materials available, it exhibits remarkable properties: high tensile strength (7500 MPa), high stiffness (Young's modulus of 100–140 GPa), high aspect ratio (70), large surface area (150–250 m^2/g) and other intriguing electrical and optical properties [117]. This green material has been advocated for diversified applications, such as a potential nanofiller for the preparation of industrial composites. The presence of surface hydroxyl groups on the NCC may permit further modification to alter its hydrophilicity, provide a stabilization matrix for anchoring metallic nanoparticles, or prepare the biomaterial for targeted applications [115]. There have been several reviews on NCC that deal with the synthesis, physical and optical properties, characterization and application of NCC as reinforcing agents for biocomposites [115, 118–122].

Depending on the desired application, chemical modification of NCC can be carried out at the hydroxyl groups of the glucose units on the crystalline backbone structure [115]. Some reactions to accomplish such modifications include sulfonation, oxidation [115, 123], cationization [115, 124], grafting [115, 125] (via acid chloride [115, 126], acid anhydride [115, 127], isocyanate [115, 128], and silylation [115, 129] (Fig. 3.8).

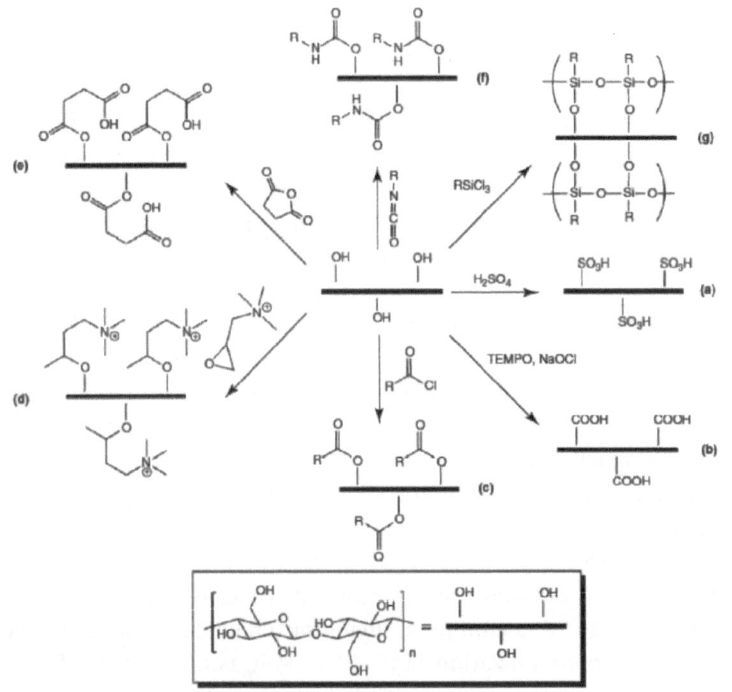

Figure 3.8 Some possible routes for chemical modification of NCC (clockwise from right): (a) sulfonation, (b) oxidation by TEMPO, (c) ester linkages via acid chlorides, (d) cationization via epoxides, (e) ester linkages via acid anhydrides, (f) urethane linkages via isocyanates, (g) silylation [115].

Recently, more homogeneous NCC has been prepared using ammonium persulfate (APS), a strong oxidant, instead of acid hydrolysis [115, 130, 131]. This versatile one-pot procedure can process a variety of cellulosic biomass without the need for pretreatments to remove non-cellulosic plant contents. The use of APS results in the formation of highly carboxylated NCC, as opposed to sulfonated NCC produced using H_2SO_4. As such, highly carboxylated NCC can be produced directly without the need to undergo an additional step of (2,2,6,6-tetramethylpiperidin-1-yl)oxyl (TEMPO) oxidation that is often required for NCC produced by acid hydrolysis [115].

Of interest is the synthesis of high-density metallic nanoparticles onto tunicate NCC surfaces using a cationic surfactant, cetyl trimethylammonium bromide (CTAB) [115, 138]. Metallic salts (Ag, Au, Cu and Pt) are reduced with $NaBH_4$ and the resulting nanoparticles are stabilized by CTAB. The hydrophilic ends of CTAB associate with the hydroxyl groups of NCC to produce metallic nanoparticle chains. However, the nanoparticle size distribution is large, particularly for Au and Cu nanoparticles, which would limit their effectiveness in size-specific applications such as catalysis [115]. A novel material has been prepared by the deposition of preformed carbonate-stabilized Au nanoparticles onto the surface of poly(diallyldimethyl ammonium chloride) (PDDA)-coated carboxylated NCC [115, 139]. A major advantage to this synthesis technique is the ability to deposit preformed metallic nanoparticles onto the surface of the NCC without agglomeration of the nanoparticles [115].

Based on their dimensions, functions, and preparation methods, nanocellulosic fibers can be classified into three types. The first type is referred to as microfibrillated cellulose (MFC) and is derived by treating microscale cellulose fibers using a variety of techniques, including chemical treatment and mechanical homogenization [140–142]. MFC is often also referred to as "microcrystalline cellulose," "microcrystallites," "cellulose nanofibrils," and "microfibrils" [140, 143]. They are characterized by being long and flexible and are composed of alternating crystalline and amorphous regions [140, 144]. The dimensions vary based on the type and amount of treatment; references in the literature cite diameters ranging from 10 to 100 nm and a wide range of lengths sometimes exceeding several thousand nanometers [140, 144]. The second type is referred to as NCC or nanowhiskers [140]. The name is derived from the long rod-like appearance of these entirely crystalline fibers [140, 143]. NCC is composed of the crystalline regions present in MFC, with these regions being liberated by mechanical, ultrasonic and chemical processes to cleave the amorphous regions [140, 145]. Their dimensions are 5–20 nm across and 100–500 nm in length [140, 143, 146]. The third type is bacterial nanocellulose (BNC), which also appears in the literature as "bacterial cellulose," "microbial cellulose," and "biocellulose" [140, 147]. These fibers contain about 99% water, which is added during biosynthesis and have diameters of 30 nm after swelling with water [140, 148].

Composites utilizing cellulose fibers have been prepared with many different materials, especially polymers. It has been well demonstrated that these fibers help to alter and in general enhance the physical properties of polymeric composites [140, 149–157]. Additionally, their bio-degradability and biocompatibility enables cellulose-reinforced materials to be suitable for bio-scaffolding in medical applications, if the polymeric component is also biocompatible [140, 158]. Some surface modifications have been performed on cellulose to add selected characteristics, such as antimicrobial properties to polymeric matrixes [140, 159].

3.6.2 Nanocrystalline Cellulose as Reinforce Phase

Cellulose nanowhiskers, rod-like nanocrystals with highly mechanical performance obtained from a range of renewable biosources, have been successfully used as reinforcing fillers in a series of synthetic and natural polymeric matrices [80, 160–163]. These nanowhiskers have a high tendency to form networks via surface hydroxyl group, which is advantageous for the formation of load-bearing percolating architectures within the host polymer matrix because the stress transfer can be facilitated by hydrogen bonding between the nanowhiskers [80, 164]. Since the cellulose nanowhisker fillers can also develop hydrogen bonds with the host polymer matrix, they may impact the gel formation and properties in addition to mechanical strength [80].

For example, in the Li's work [112], rod-shaped NCC was prepared from microcrystalline cellulose (MCC) using the purely physical method of high-intensity ultrasonication. The reinforcement capabilities of the obtained NCC were investigated by adding it to poly(vinyl alcohol) (PVA) via the solution casting method. The results revealed that the prepared NCC had a rod-shaped structure, with diameters between 10 and 20 nm and lengths between 50 and 250 nm. Because of the nanoscale size and large number of free-end chains, the NCC degraded at a slightly lower temperature, which resulted in increased char residue (9.6% to 16.1%), compared with that of the MCC (6.2%). The storage modulus of the nanocomposite films were significantly improved compared with that of pure PVA films. The modulus of PVA with 8 wt% NCC was 2.4 larger than that of pure PVA. SEM images of the MCC treated for different times are shown in Fig. 3.9. It can be seen that MCC particles had irregular shapes (Fig. 3.9a). After being treated by ultrasonication for 1 min, the MCC was damaged and fragmented into various shapes (Fig. 3.9b) and some erosion on the surface of the MCC was observed. After treatment for 2 min, some MCC particles began to show fluffiness and external fibrillation on their surfaces (Fig. 3.9c). After a 10 min treatment, the width of the NCC particles was in the range of 10–20 nm and a web-like network structure formed (Fig. 3.9d) [112]. Figure 3.10 shows the TEM images of the treated MCC. It can be seen that the obtained NCC samples had a regular rod shape [112].

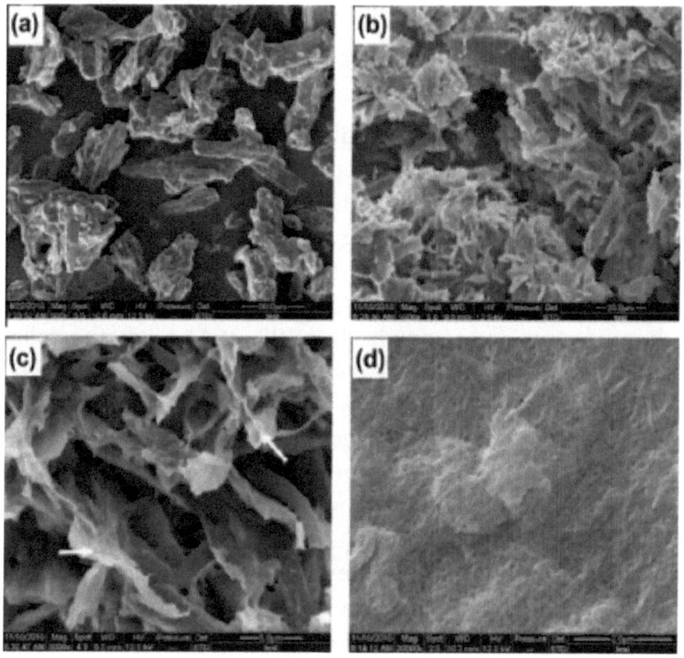

Figure 3.9 SEM images of MCC treated for different times: (a) 0 min, (b) 1 min, (c) 2 min, and (d) 10 min at an ultrasonication power of 1500 W [112].

The conclusions [112] are that the char residue amounts of all the NCC were significantly larger than that of the MCC. The DMA data showed a marked increase in storage modulus from 1.7 GPa for pure PVA to 4.2 GPa for PVA/NCC-8 at 30°C, because of the high modulus of NCC.

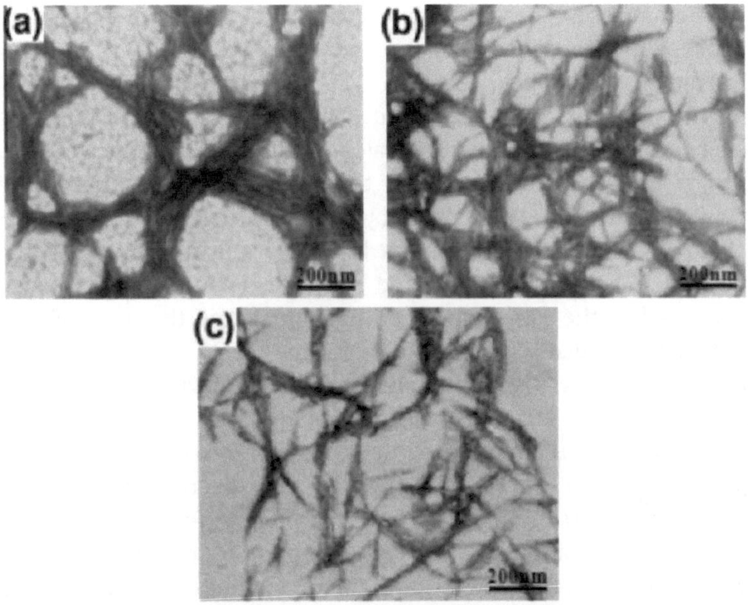

Figure 3.10 TEM images of NCC treated using different ultrasonication times: (a) NCC-5 min, (b) NCC-10 min and (c) NCC-15 min [112].

In another work of Ljungberg [165], nanocomposite films of isotactic polypropylene reinforced with cellulose whiskers highly dispersed with surfactant were prepared for the first time and compared with either bare or grafted aggregated whiskers. Evaluation of the crystallization behavior showed that the aggregated or surfactant-modified whiskers induced two crystalline forms (α and β) in the nanocomposites and also acted as nucleating agents for isotactic polypropylene. The linear mechanical properties above the glass-rubber transition were found to be drastically enhanced for all three of the nanocomposites as compared to the neat polypropylene matrix, and these effects were attributed to a mechanical coupling between the polypropylene crystallites and filler/filler interactions. For the mechanical experiments at large deformations, the quality of the whisker dispersion was found to play a major role. The nanocomposites obtained with the surfactant-modified whiskers exhibited enhanced ultimate properties when compared to the neat matrix or to the composites containing the other filler types [165].

Some researches on the use of cellulose nanowhiskers to enhance polylactic acid have been completed. In the work of Sanchez-Garcia and Lagaron [166], polylactic acid (PLA) nanocomposites were prepared using CNWs as a reinforcing element in order to assess the value of this filler to reduce the gas and vapor permeability of the biopolyester matrix. The nanocomposites were prepared by incorporating 1, 2, 3, and 5 wt% of the CNW into the PLA matrix by a chloroform solution casting method. The results indicated that the nanofiller was well dispersed in the PLA matrix, did not impair the thermal stability of this but induced the formation of some crystallinity, most likely transcrystallinity. CNW prepared by freeze drying exhibited in the nanocomposites better morphology and properties than their solvent exchanged counterparts. Interestingly, the water permeability of nanocomposites of PLA decreased with the addition of CNW prepared by freeze drying by up to 82% and the oxygen permeability by up to 90%. Optimum barrier enhancement was found for composites containing loadings of CNW below 3 wt%. Typical modeling of barrier and mechanical properties failed to describe the behavior of the composites and appropriate discussion regarding this aspect was also carried out. From the results, CNW exhibit novel significant potential in coatings, membranes, and food agro-based packaging applications [166].

3.7 Summary

Science and technology continue to pay considerable attention to renewable raw materials and more eco-friendly and sustainable resources and processes. In this context, it has been essential to enhance the existing knowledge of organic and polymer chemistries as well as in the chemistry of low-molecular-weight carbohydrates and other polysaccharides in the basic and application-oriented work in the field of cellulose. In addition, it has been significant to intensify the interdisciplinary interactions between pharmacy, medicine, biology, and chemical engineering.

In this chapter, the current state of development in the field of cellulose modification and application is demonstrated. It should be also pointed out that cellulose as a natural product belongs to the polymers, which hold an impressive future potential for fundamental knowledge as well as for large-scale production in a wide range of applications.

References

1. Odian, G. (2004). *Principles of Polymerization,* 4th ed. (John Wiley and Sons, Hoboken).
2. Bhattacharya, A., and Misra, B. N. (2004). Grafting: A versatile means to modify polymers: Techniques, factors and applications, *Prog. Polym. Sci.,* **29**, 767–814.
3. Hebeish, A., and Guthrie, J. T. (1981). *The Chemistry and Technology of Cellulosic Copolymers,* Springer-Verlag, Berlin, **12**, 432–1981.
4. Chauhan, G. S., Mahajan, S., and Guleria, L. K. (2000). Polymers from renewable resources: Sorption of Cu^{2+} ions by cellulose graft copolymers, *Desalination,* **130**, 85–88.
5. Gupta, K. C., and Khandekar, K. (2003). Temperature-responsive cellulose by ceric(IV) ion-initiated graft copolymerization of N-isopropylacrylamide, *Biomacromolecules,* **4**, 758–765.
6. McDowall, D. J., Gupta, B. S., and Stannett, V. T. (1984). Grafting of vinyl monomers to cellulose by ceric ion initiation, *Prog. Polym. Sci.,* **10**, 1–50.
7. Waly, A., Abdel-Mohdy, F. A., Aly, A. S., and Hebeish, A. (1998). Synthesis and characterization of cellulose ion exchanger. II. Pilot scale and utilization in dye–heavy metal removal, *J. Appl. Polym. Sci.,* **68**, 2151–2157.
8. El-Salmawi, K. M., Abu Zaid, M. M., Ibraheim, S. M., El-Naggar, A. M., and Zahran, A. H. (2001). Sorption of dye wastes by poly(vinyl alcohol)/ poly(carboxymethyl cellulose) blend grafted through a radiation method, *J. Appl. Polym. Sci.,* **82**, 136–142.
9. Hebeish, A., and El-Hilw, Z. H. (1998). Preparation of DEAE cotton-g-poly(methacrylic acid) for use as ion exchanger, *J. Appl. Polym. Sci.,* **67**, 739–745.
10. Roy, D., Semsarilar, M., Guthrie, J. T., and Perrier, S. (2009). Cellulose modification by polymer grafting: A review, *Chem. Soc. Rev.,* **38**, 2046–2964.
11. Kalia, S., and Sabaa, M. W. (2013). *Polysaccharide Based Graft Copolymers.* Kalia, S., Sabaa, M. W., and Kango, S., Chapter 1 "Polymer

grafting: A versatile means to modify the polysaccharides," (Springer-Verlag, Berlin) 1–14.

12. Hsu, S. T., and Pan, T. C. (2007). Adsorption of paraquat using methacrylic acid-modified rice husk, *Bioresour. Technol.*, **98**, 3617–3621.
13. Cho, C. G., and Lee, K. (2002). Preparation of starch graft copolymer by emulsion polymerization, *Carbohyd. Polym.*, **48**, 125–130.
14. Misra, B. N., and Dogra, R. (1980). Grafting onto starch. IV. Graft copolymerization of methylmethacrylate by use of AIBN as radical initiator, *J. Macromol. Sci. Chem. A.*, **14**, 763–770.
15. Chiang, W. Y., and Hu, C. H. (1996). The improvements in flame retardance and mechanical properties of polypropylene/Fr blends by acrylic acid graft copolymerization, *Eur. Polym. J.*, **32**, 385–390.
16. Zahran, M. K., and Mahmoud, R. I. (2003). Peroxydiphosphate–metal ion–cellulose thiocarbonate redox system-induced graft copolymerization of vinyl monomers onto cotton fabric, *J. Appl. Polym. Sci.*, **87**, 1879–1889.
17. Vlcek, P., Janata, M., Latalova, P., Kriz, J., Cadova, E., and Toman, L. (2006). Controlled grafting of cellulose diacetate, *Polymer*, **47**, 2587–2595.
18. Tizzotti, M., Charlot, A., Fleury, E., Stenzel, M., and Bernard, J. (2010). Modification of polysaccharides through controlled/living radical polymerization grafting—towards the generation of high performance hybrids, *Macromol. Rapid Commun.*, **31**, 1751–1772.
19. Kaur, I., Misra, B. N., Gupta, A., and Chauhan, G. S. (1998). Radiochemical grafting of methacrylonitrile and its binary mixture with methyl acrylate onto gelatin, *Polym. Int.*, **46**, 275–279.
20. Kaith, B. S., and Kalia, S. (2008). Preparation of microwave radiation induced graft copolymers and their applications as reinforcing material in phenolic composites, *Polym. Compos.*, **29**, 791–797.
21. Chauhan, G. S., Mishra, B. N., Dhiman, S. K., Guleria, L. K., and Kaur, I. (2000). Polymers from renewable resources: Kinetics of 4-vinyl pyridine radiochemical grafting onto cellulose extracted from pine needles, *Radiat. Phys. Chem.*, **58**, 181–190.
22. Deshayes, S., Liagre, M., Loupy, A., Luche, J. L., and Petit, A. (1999). Microwave activation in phase transfer catalysis, *Tetrahedron*, **55**, 10851–10870.
23. Singh, V., Kumar, P., and Sanghi, R. (2012). Use of microwave irradiation in the grafting modification of the polysaccharides—a review, *Prog. Polym. Sci.*, **37**, 340–364.

24 Wenzel, A., Yamgishita, H., Kitamoto, D., Endo, A., Haraya, K., Nakane, T., Hanai, N., Matsuda, H., Kamuswetz, H., and Paul, D. (2000). Effect of preparation condition of photoinduced graft filling polymerized membranes on pervaporation performance, *J. Membr. Sci.*, **179**, 69–77.

25. Yamaguchi, T., Yamahara, S., Nakao, S., and Kimura, S. (1994). Preparation of pervaporation membranes for removal of dissolved organics from water by plasma-graft filling polymerization, *J. Membr. Sci.*, **95**, 39–49.

26. Bhattacharyaa, A., and Misra, B. N. (2004). Grafting: A versatile means to modify polymers techniques, factors and applications, *Prog. Polym. Sci.*, **29**, 767–814.

27. Kale, K. H., and Desai, A. N. (2011). Atmospheric pressure plasma treatment of textiles using nonpolymerizing gases, *Indian. J. Fiber. Text.*, **36**, 289–299.

28. Beltran-Heredia, J., Sanchez-Martin, J., Davila-Acedo, M. A. (2011). Optimization of the synthesis of a new coagulant from a tannin extract, *J. Hazard. Mater.*, **186**, 1704–1712.

29. Dukkanci, M., Gunduz, G., Yilmaz, S., and Prihod'ko, R. V. (2010). Heterogeneous fenton-like degradation of Rhodamine 6G in water using CuFeZSM-5 zeolite catalyst prepared by hydrothermal synthesis, *J. Hazard. Mater.*, **181**, 343–350.

30. Fang, R., Cheng, X. S., and Xu, X. R. (2010). Synthesis of lignin-base cationic flocculant and its application in removing anionic azo-dyes from simulated wastewater, *Bioresource Technol.*, **101**, 7323–7329.

31. Rafatullah, M., Sulaiman, O., Hashim, R., and Ahmad, A. (2010). Adsorption of methylene blue on low-cost adsorbents: A review, *J. Hazard. Mater.*, **177**, 70–80.

32. Sabah, E., and Cengiz., I. (2004). An evaluation procedure for flocculation of coal preparation plant tailings, *Water Res.*, **38**, 1542–1549.

33. Bratby, J. (2006). *Coagulation and Flocculation in Water and Wastewater Treatment*, 2nd ed. (IWA Publishing, London).

34. Cai, T., Yang, Z., Li, H. J., Yang, H., Li, A. M., and Cheng, R. S. (2013). Effect of hydrolysis degree of hydrolyzed polyacrylamide grafted carboxymethyl cellulose on dye removal efficiency, *Cellulose*, **20**, 2605–2614.

35. Xiao, J., and Zhou, Q. (2005). *Natural Polymer Flocculants.* (Chemical Industry Press, Beijing).

36. Gao, J., and Tang, L. (1996) Cellulose science. Science Press, Beijing, Ghimici L., Nichifor M. (2010). Novel biodegradable flocculating agents based on cationic amphiphilic polysaccharides.

37. Klemm, D., Heublein, B., Fink, H. P., and Bohn, A. (2005). Cellulose: Fascinating biopolymer and sustainable raw material, *Angew. Chem. Int. Ed.*, **44**, 3358–3393.

38. Suopajärvi, T., Liimatainen, H., Hormi, O., and Niinimäki, J. (2013). Coagulation–flocculation treatment of municipal wastewater based on anionized nanocelluloses, *Chem. Eng. J.*, **231**, 59–67.

39. Sirviö, J., Honka, A., Liimatainen, H., Niinimäki, J., and Hormi, O. (2011). Synthesis of highly cationic water-soluble cellulose derivative and its potential as novel biopolymeric flocculation agent, *Carbohydrate Polymers*, **86,** 266–270.

40. Zaman, M., Xiao, H. N., Chibante, F., and Ni, Y. H. (2012). Synthesis and characterization of cationically modified nanocrystalline cellulose, *Carbohydrate Polymers*, **89,** 163–170.

41. Mishra, S., Rani, G. U., and Sen, G. (2012). Microwave initiated synthesis and application of polyacrylic acid grafted carboxymethyl cellulose, *Carbohydr. Polym.*, **87**, 2255–2262.

42. Das, R., Ghorai, S., and Pal, S. (2013). Flocculation characteristics of polyacrylamide grafted hydroxypropyl methyl cellulose: An efficient biodegradable flocculant, *Chem. Eng. J.*, **229**, 144–152.

43. Czaja, W., Krystynowicz, A., Bielecki, S., and Brown, R. J. (2006). Microbial cellulose—the natural power to heal wounds, *Biomaterials*, **27**, 145–151.

44. Czaja, W., Young, D. J., Kawecki, M., and Malcolm Brown, Jr., R. (2007). The future prospects of microbial cellulose in biomedical applications, *Biomaterials*, **8**, 1–12.

45. Talukdar, M. M., Michoel, A., Rombaut, P., and Kinget, R. (1996). Comparative study on xanthan gum and hydroxypropylmethyl cellulose as matrices for controlled-release drug delivery I. Compaction and in vitro drug release behaviour, *Int. J. Pharm.*, **129**, 223–241.

46. Dash, R., and Ragauskas, A. J. (2012). Synthesis of a novel cellulose nanowhisker-based drug delivery system, *RSC Adv.*, **2**, 3403–3409.

47. Jockenhoevel, S., Zund, G., Hoerstrup, S. P., Chalabi, K., Sachweh, J. S., Demircan, L., Messmer, B. J., and Turina, M. (2001). Fibrin gel—advantages of a new scaffold in cardiovascular tissue engineering, *Eur. J. Cardio. Thorac.*, **19**, 424–430.

48. Klemm, D., Schumann, D., Udhardt, U., and Marsch, S. (2001). Bacterial synthesized cellulose—artificial blood vessels for microsurgery, *Prog. Polym. Sci.*, **26**, 1561–1603.

49. Brown, E. E., Laborie, M. G., and Zhang, J. W. (2012). Glutaraldehyde treatment of bacterial cellulose/fibrin composites: Impact on morphology, tensile and viscoelastic properties, *Cellulose*, **19**, 127–137.

50. Brown, E. E., Hu, D. H., Lail, N. A., and Zhang, X. (2013). Potential of nanocrystalline cellulose–fibrin nanocomposites for artificial vascular graft applications, *Biomacromolecules,* **14**, 1063–1071.

51. Entcheva, E., Bien, H., Yin, L. H., Chung, C. Y., Farrella, M., and Kostov, Y. (2004). Functional cardiac cell constructs on cellulose-based scaffolding, *Biomaterials*, **25**, 5753–5762.

52. Müller, F. A., Müller, L., Hofmann, I., Greila, P., Wenzel, M. M., and Staudenmaier, R. (2006). Cellulose-based scaffold materials for cartilage tissue engineering, *Biomaterials*, **27**, 3955–3963.

53. Svensson, A., Niclasson, E., Harrah, T., Panilaitis, B., Kaplan, D., Brittber, M., and Gatenholm, P. (2005). Bacterial cellulose as a potential scaffold for tissue engineering of cartilage, *Biomaterials*, **26**, 419–431.

54. Bodin, A., Ba¨ckdahl, H., Gustafsson, L., Risberg, B., and Gatenholm, P. (2006). Manufacturing and characterisation bacterial cellulose tubes using two different fermentation techniques. In: Mendez-Vilas, A. (ed.) *Modern Multidisciplinary Applied Microbiology: Exploiting Microbes and Their Interactions*. Weinheim, Wiley VC Verlag GmbH & Co. KGaA, Weinheim, Germany, pp. 619–622.

55. Liu, Z. H., Jiao, Y. P., Wang, Y. F., Zhou, C. R., and Zhang, Z. Y. (2008). Polysaccharides-based nanoparticles as drug delivery systems, *Adv. Drug. Deliv. Rev.*, **60**, 1650–1662.

56. Zhang, L., Gu, F. X., Chan, J. M., Wang, A. Z., Langer, R. S., and Farokhzad, O. C. (2008). Nanoparticles in medicine: Therapeutic applications and developments, *Clin. Pharmacol. Ther.*, **83**, 761–769.

57. Ha, C. S., and Gardella, J. A. (2005). Surface chemistry of biodegradable polymers for drug delivery systems, *Chem. Rev.*, **105**, 4205–4232.

58. Habibi, Y., Lucia, L. A., and Rojas, O. J. (2010). Cellulose nanocrystals: Chemistry, self-assembly, and applications, *Chem. Rev.*, **110**, 3479–3500.

59. Zhu, L., Kumar, V., and Banker, G. S. (2001). Examination of oxidized cellulose as a macromolecular prodrug carrier: Preparation and characterization of an oxidized cellulose-phenylpropanolamine conjugate, *Int. J. Pharm.*, **223**, 35–47.

60. Rosamond, W., Flegal, K., Furie, K., Go, A., Greenlund, K., Haase, N., Halipern, S. M., Ho, M., Howard, V., Kissela, B., Kittner, S., Lloyd-Jones, D., McDermott, M., Meigs, J., Moy, C., Nichol, G., O'Donnell, C., Roger,

V., Sorlie, P., Steinberger, J., Thom, T., Wilson, M., and Hong, Y. (2008). Heart disease and stroke statistics–2008 update: A report from the American Heart Association Statistics Committee and Stroke Statistics Subcommittee, *Circulation*, **117**, 32–43.

61. Dardik, H., and Greisler, H. (1999). History of prosthetic grafts, *Semi. Vasc. Surg.*, **12,** 1–7.
62. Hamilton, D., and Vorp, D. (2004). Tissue engineering of blood vessel. In: *Encyclopedia of Bioengineering and Biomedical Engineering*, Marcel Dekker Inc., New York.
63. Laki, K. (1968). *Fibrinogen*. Dekker, M. New York.
64. Swartz, D. D., Russell, J. A., and Andreadis, S. T. (2005). Engineering of fibrin-based functional and implantable small-diameter blood vessels, *Am. J. Physiol.*, **288**, 1451–1460.
65. Astley, O. M., Chanliaud, E., Donald, A. M., and Gidley, M. J. (2003). Tensile deformation of bacterial cellulose composites, *Int. J. Biol. Macromol.*, **32**, 28–35.
66. Nakayama, A., Kakugo, A., Gong, J. P., Osada, Y., Takai, M., Erata, T., Kawano, S. (2004). High mechanical strength double-network hydrogel with bacterial cellulose, *Adv. Funct. Mater.*, **14**, 1124–1128.
67. Liu, W., Jawerth, L. M., Sparks, E. A., Falvo, M. R., Hantgan, R. R., Superfine, R., Lord, S. T., and Guthold, M. (2006). Fibrin fibers have extraordinary extensibility and elasticity, *Science*, **313**, 634.
68. Skarja, G. A., Brash, J. L., Bishop, P., and Woodhouse, K. A. (1998). Protein and platelet interactions with thermally denatured fibrinogen and crosslinked fibrin coated surfaces, *Biomaterials*, **19**, 2129–2138.
69. Backdahl, H., Helenius, G., Bodin, A., Nannmark, U., Johansson, B. R., Risberg, B., and Gatenholm, P. (2006). Mechanical properties of bacterial cellulose and interactions with smooth muscle cells, *Biomaterials*, **27**, 2141–2149.
70. Shaikh, F. M., Callanan, A., Kavanagh, E. G., Burke, P. E., Grace, P. A., and McGloughlin, T. M. (2008). Fibrin: A natural biodegradable scaffold in vascular tissue engineering, *Cells Tissues Organs*, **188**, 333–346.
71. Martson, M., Viljanto, J., Hurme, T., and Saukko, P. (1998). Biocompatibility of cellulose sponge with bone, *Eur. Surg. Res.*, **30**, 426–32.
72. Poustis, J., Baquey, C., and Chauveaux, D. (1994). Mechanical properties of cellulose in orthopaedic devices and related environments, *Clin. Mater.*, **16**, 119–124.

73. Laurence, S., Bareille, R., Baquey, C., and Fricain, J. C. (2005). Development of a resorbable macroporous cellulosic material used as hemostatic in an osseous environment, *J. Biomed. Mater. Res. A.*, **73**, 422–429.
74. Kamat, M., and Malkani, R. (2003). Disposable diapers: A hygienic alternative, *Indian. J. Pediatr.*, **70**, 879–881.
75. Sadeghi, M., and Hosseinzadeh, H. (2008). Synthesis of starch-poly (sodium acrylate-co-acrylamide) superabsorbent hydrogel with salt and pH-responsiveness properties as a drug delivery system, *J. Bioact. Compat. Pol.*, **23**, 381–404.
76. Yi, J. Z., and Zhang L. M. (2008). Removal of methylene blue dye from aqueous solution by adsorption onto sodium humate/polyacrylamide/ clay hybrid hydrogels, *Bioresource Technol.*, **99**, 2182–2186.
77. Lan, W., and Liu, M. Z. (2008). Preparation and characterization of cellulose acetate-coated compound fertilizer with controlled-release and water-retention, *Polym. Adv. Technol.*, **19**, 785–792.
78. Nguyen, K. T., and West, J. L. (2002). Photopolymerizable hydrogels for tissue engineering applications, *Biomaterials*, **23**, 4307–4314.
79. Liang, S., Wu, J., Tian, H., Zhang, L., and Xu, J. (2008). High-strength cellu-lose/poly(ethylene glycol) gels, *Chem. Sus. Chem.*, **1**, 558–563.
80. Wang, Y. X., and Chen, L. Y. (2011). Impacts of nanowhisker on formation kinetics and properties of all-cellulose composite gels, *Carbohyd. Polym.*, **83**, 1937–1946.
81. Andrews, G. P., Gorman, S. P., and Jones, D. S. (2005). Rheological characterization of primary and binary interactive bioadhesive gels composed of cellulose derivatives designed as ophthalmic viscosurgical devices, *Biomaterials*, **26**, 571–580.
82. Entcheva, E., Bien, H., Yin, L., Dhung, C. Y., Farell, M., and Kostov, Y. (2004). Functional cardiac cell constructs on cellulose-based scaffolding, *Biomaterials*, **25**, 5753–5762.
83. Fundueanu, G., Constantin, M., Esposito, E., Cortesi, R., Nastruzzi, C., and Menegatti, E. (2005). Cellulose acetate butyrate microcapsules containing dextran ion-exchange resins as self-propelled drug release system, *Biomaterials*, **26**, 4337–4347.
84. Bingöl, B., Strandberg, C., Szabo, A., and Wegner, G. (2008). Copolymers and hydrogels based on vinyl phosphonic acid, *Macromolecules*, **41**, 2785–2790.
85. Hoepfner, S., Ratke, L., and Milow, B. (2008). Synthesis and characterization of nanofibrillar cellulose aerogels, *Cellulose*, **15**, 121–129.

86. Klemm, D., Schumann, D., Udhardt, U., and Marsch, S. (2001). Bacterial synthesized cellulose-artificial blood vessels for microsurgery, *Prog. Polym. Sci.*, **26**, 1561–1603.
87. Ma, Z. H., Li, Q., Yue, Q. Y., Gao, B. Y., Xu, X., and Zhong, Q. Q. (2011). Synthesis and characterization of a novel super-absorbent based on wheat straw, *Bioresource Technol.*, **102**, 2853–2858.
88. Zhang, Y., Wu, F., Liu, L., and Yao, J. M. (2013). Synthesis and urea sustained-release behavior of an eco-friendly superabsorbent-based on flax yarn wastes, *Carbohyd. Polym.*, **91**, 277–283.
89. Wu, J. F., Hao, M. Y., and Guo, Z. G. (2008). Study on urea determining methods for feed, *Feed Industry*, **29**, 42–46.
90. Akelah, A. (1996). Novel utilization of conventional agrochemicals by controlled release formulations, *Mater. Sci. Eng.*, **C4**, 83–98.
91. Tomaszewska, M., and Jarosiewicz, A. (2002). Use of polysulfone in controlled-release NPK fertilizer formulations, *J. Agric. Food Chem.*, **50**, 4634–4639.
92. Al-Zahrani, S. M. (2000). Utilization of polyethylene and paraffin waxes as controlled delivery system for different fertilizers, *Ind. Eng. Chem. Res.*, **39**, 367–371.
93. Krysiak, M. D., Madigan, D. P. (2004). Granular fertilizer. US Patent 069032.
94. Lin, L., Wu, J. H., Yang, Z., and Pu, M. L. (2001). Synthesis and properties of poly(acrylic acid)/mica superabsorbent nanocomposite, *Macromol. Rapid Commun.*, **22**, 422–424.
95. Islam, M. R., Hu, Y. G., Mao, S. S., Mao, J. Z., Eneji, A. E., and Xue, X. Z. (2011). Effectiveness of a water-saving super-absorbent polymer in soil water conservation for corn (Zea mays L.) based on eco-physiological parameters, *J. Sci. Food Agric.*, **91**, 1998–2005.
96. Xie, L. H., Liu, M. Z., Ni, B. L., and Wang, Y. F. (2012). New environment-friendly use of wheat straw in slow-release fertilizer formulations with the function of superabsorbent, *Ind. Eng. Chem. Res.*, **51**, 3855–3862.
97. Trenkel, M. E. (1997). Controlled-release and stabilized fertilizers in agriculture, (International Fertilizer Industry Association, Paris).
98. Liu, C. H., Wu, J. Y., Chang, J. S. (2008). Diffusion characteristics and controlled release of bacterial fertilizers from modified calcium alginate capsules, *Bioresour. Technol.*, **99**, 1904–1910.
99. Liang, R., and Liu, M. Z. (2006). Preparation and properties of a double-coated slow-release and water-retention urea fertilizer, *J. Agric. Food Chem.*, **54**, 1392–1398.

100. García, M. C., Vallejo, A., García, L., and Cartagena, M. C. (1997). Manufacture and evaluation of coated triple superphosphate fertilizers, *Ind. Eng. Chem. Res.*, **36**, 869–873.

101. Ibrahim, A. A., and Jibril, B. Y. (2005). Controlled release of paraffin wax/rosin-coated fertilizers, *Ind. Eng. Chem. Res.*, **44**, 2288–2291.

102. Jarosiewicz, A., and Tomaszewska, M. (2003). Controlled-release NPK fertilizer encapsulated by polymeric membranes, *J. Agric. Food Chem.*, **51**, 413–417.

103. Shaviv, A., Raban, S., and Zaidel, E. (2003). Modeling controlled nutrient release from polymer coated fertilizers: Diffusion release from single granules, *Environ. Sci. Technol.*, **37**, 2251–2256.

104. Sylvie, T. M., Irma, D. A., and Alethia, V. M. (2011). Disposable diapers biodegradation by the fungus Pleurotus ostreatus Espinosa-Valdemar Rosa María, *Waste Manage.*, **31**, 1683–1688.

105. Liu, H. Y., Zhang, Y., and Yao, J. M. (2014). Preparation and properties of an eco-friendly superabsorbent based on flax yarn waste for sanitary napkin applications, *Fiber Polym.*, **15**, 145–152.

106. Aloulou, F., Boufi, S., and Labidi, J. (2006). Modified cellulose fibres for adsorption of organic compound in aqueous solution, *Sep. Purif. Technol.*, **52**, 332–342.

107. Poplin, J. H., Swatloski, R. P., Holbrey, J. D., Spear, S. K., Metlen, A., Gratzel, M., Nazeeruddin, M. K., and Rogers, R. D. (2007). Sensor technologies based on a cellulose supported platform, *Chem. Commun.*, **20**, 2025–2027.

108. Kelne, A., and Schacht, E. H. (2005). Tailor-made polymers for local drug delivery: Release of macromolecular model drugs from biodegradable hydrogels based on poly(ethylene oxide), *J. Control. Release*, **101**, 13–20.

109. Nogi, M., and Yano, H. (2008). Transparent nanocomposites based on cellulose produced by bacteria offer potential innovation in the electronics device industry, *Adv. Mater.*, **20**, 1849–1852.

110. Wu, F., Zhang, Y., Liu, L., and Yao, J. M. (2012). Synthesis and characterization of a novel cellulose-g-poly(acrylic acid-co-acrylamide) superabsorbent composite based on flax yarn waste, *Carbohyd. Polym.*, **87**, 2519–2525.

111. Fortunati, E., Rinaldi, S., Peltzer, M., Bloise, N., Armentano, L. V., Jiménez, A., Latterini, L., and Kenny, J. M. (2014). Nano-biocomposite films with modified cellulose nanocrystals and synthesized silver nanoparticles, *Carbohyd. Polym.*, **101**, 1122–1133.

112. Li, W., Yue, J. Q., and Liu, S. X. (2012). Preparation of nanocrystalline cellulose via ultrasound and its reinforcement capability for poly(vinyl alcohol) composites, *Ultrason. Sonochem.*, **19**, 479–485.

113. Habibi, Y., Lucia, L. A., and Rojas, O. J. (2010). Cellulose nanocrystals: Chemistry, self-assembly, and applications, *Chem. Rev.*, **110**, 3479–3500.

114. Lima, M. M. D. S., and Borsali, R. (2004). Rodlike cellulose microcrystals: Structure, properties and applications, *Macromol. Rapid Commun.*, **25**, 771–787.

115. Lam, E., Male, K. B., Chong, J. H., Leung, A. C. W., and Luong, J H. T. (2012). Applications of functionalized and nanoparticle-modified nanocrystalline cellulose, *Trends Biotechnol.*, **30**, 283–290.

116. Ruiz, M. M., Cavaillé, J. Y., Dufresne, A., Gérard, J. F., and Graillat, C. (2000). Processing and characterization of new thermoset nanocomposites based on cellulose whiskers, *Compos. Interface.*, **7**, 117–131.

117. Revol, J.-F. (1998). Solid films of cellulose with chiral nematic order and optically variable properties, *J. Pulp Pap. Sci.*, **24**, 146–149.

118. Habibi, Y., Lucia, L. A., and Rojas, O. J. (2010). Cellulose nano crystals: Chemistry, self-assembly, and applications, *Chem. Rev.*, **110**, 3479–3500.

119. Moon, R. J., Martini, A., Nairn, John., Simonsen, J., and Youngblood, J. (2011). Cellulose nanomaterials review: Structure, properties and nanocomposites, *Chem. Soc. Rev.*, **40**, 3941–3944.

120. Ramires, E. C., and Dufresne, A. (2011). A review of cellulose nanocrystals and nano composites, *Tappi J.*, **10**, 9–16.

121. Klemm, D., Kramer, F., Moritz, S., Lindström, T., Ankerfors, M., Gray, D., and Dorris, A. (2011). Nanocelluloses: A new family of nature-based materials, *Angew. Chem. Int. Ed.*, **50**, 5438–5466.

122. Holt, B. L. (2010). Novel anisotropic materials from functionalized colloidal cellulose and cellulose derivatives, *J. Mater. Chem.*, **20**, 10058–10070.

123. Habibi, Y., Chanzy, H., and Vignon, M. R. (2006). TEMPO-Mediated surface oxidation of cellulose whiskers, *Cellulose.*, **13**, 679–687.

124. Hasani, M., Cranston, E. D., Westman, G., and Gray, D. G. (2008). Cationic surface functionalisation of cellulose nanocrystals, *Soft Matter*, **4**, 2238–2244.

125. Habibi, Y. (2008). Bionanocomposites based on poly(ε-caprolactone)-grafted cellulose nanocrystals by ring-opening polymerization, *J. Mater. Chem.*, **18**, 5002–5010.

126. Junior de Menezes, A., Siqueiraa, G., Curvelob, A. A. S., and Dufresne, A. (2009). Extrusion and characterization of functionalized cellulose whiskers reinforced polyethylene nanocomposites, *Polymer*, **50**, 4552–4563.

127. Pandey, J. K., Chua, W. S., Kima, C. S., Leeb, C. S., and Ahn, S. H. (2009). Bio-nano reinforcement of environ mentally degradable polymer matrix by cellulose whiskers from grass, *Compos. Part B-Eng.*, **40**, 676–680.

128. Siqueira, G., Bras, J., and Dufresne, A. (2009). Cellulose whiskers versus microfibrils: Influence of the nature of the nanoparticle and its surface-functionalization on the thermal and mechanical properties of nano composites, *Biomacromolecules*, **10**, 425–432.

129. Goussé, C., Chanzy, H., Excoffier, G., Soubeyrand, L., and Fleury, E. (2002). Stable suspensions of partially silylated cellulose whiskers dispersed in organic solvents, *Polymer.*, **43**, 2645–2651.

130. Leung, C. W., John, H. T. L., Hrapovic, S., Lam, E., Liu, Y., Male, K. B., Mahmoud, K., and Rho, D. (2010). Cellulose nanocrystal from renewable biomass. National Research Council of Canada. CA 2010/0003 72.

131. Leung, A. C. W., Hrapovic, S., Lam, E., Liu, Y., Male, K. B., Mahmoud, K. A., and Luong, J. H. T. (2011). Characteristics and properties of carboxylated cellulose nanocrystals prepared from a novel one-step procedure, *Small*, **7**, 302–305.

132. Son, W. K., Youk, J. H., and Park, W. H. (2006). Antimicrobial cellulose acetate nanofibers containing silver nanoparticles, *Carbohyd. Polym.*, **65**, 430–434.

133. Shin, Y., Bae, I. T., Arey, B. W., and Exarhos, G. J. (2008). Facile stabilization of gold–silver alloy nano particles on cellulose nanocrystal, *J. Phys. Chem. C*, **112**, 4844–4848.

134. Shin, Y., Bae, I. T., Arey, B. W., and Exarhos, G. J. (2007). Simple preparation and stabilization of nickel nanocrystals on cellulose nanocrystal, *Mater. Lett.*, **61**, 3215–3217.

135. Cai, J., Kimura, S., Wada, M., and Kuga, S. (2009). Nanoporous cellulose as metal nanoparticle support, *Biomacromolecules*, **10**, 87–94.

136. He, J. H., Kunitake, T., and Nakao, A. (2003). Facile in situ synthesis of noble metal nanoparticles in porous cellulose fibers, *Chem. Mater.*, **15**, 4401–4406.

137. Maneerung, T., Tokura, S., and Rujiravanit, R. (2008). Impregnation of silver nanoparticles into bacterial cellulose for antimicrobial wound dressing, *Carbohyd. Polym.*, **72**, 43–51.

138. Padalkar, S., Capadona, J. R., Rowan, S. J., Weder, C., Won, Y. H., Stanciu, L. A., and Moon, R. J. (2010). Natural biopolymers: Novel templates for the synthesis of nanostructures, *Langmuir*, **26**, 8497–8502.

139. Lam, E., Hrapovic, S., Majid, E., Chonga, J. H., and Luong, J. H. T. (2012). Catalysis using gold nanoparticles decorated on nanocrystalline cellulose, *Nanoscale*, **4**, 997–1002.

140. Floros, M., Hojabri, L., Abrah, E., Josea, J., Thomasc, S., Pothanb, L., Leaod, A. L., and Narine, S. (2012). Enhancement of thermal stability, strength and extensibility of lipid-based polyurethanes with cellulose-based nanofibers, *Polym. Degarad. Stabil.*, **97**, 1970–1978.

141. Gardner. D. J., Oporto, G. S., Mills, R., and Samir, M. (2008). Adhesion and surface issues in cellulose and nanocellulose, *J. Adhes. Sci, Technol.*, **22**, 545–567.

142. Siró, I., and Plackett, D. (2010). Microfibrillated cellulose and new nanocomposite materials: A review, *Cellulose*, **17**, 459–494.

143. Siqueira, G., Bras, J., and Dufresne, A. (2010). Cellulosic bionanocomposites: A review of preparation, properties and applications, *Polymers*, **2**, 728–765.

144. Visakh, P., and Thomas, S. (2010). Preparation of bionanomaterials and their polymer nanocomposites from waste and biomass, *Waste Biomass Valorization*, **1**, 121–134.

145. Rosa, M., Medeiros, E., Malmonge, J., Gregorski, K., Wood, D., and Mattoso, L. (2010). Cellulose nanowhiskers from coconut husk fibers: Effect of preparation conditions on their thermal and morphological behavior, *Carbohyd. Polym.*, **81**, 83–92.

146. Elazzouzi-Hafraoui, S., Nishiyama, Y., Putaux, J. L., Heux, L., Dubreuil, F., and Rochas, C. (2007). The shape and size distribution of crystalline nanoparticles prepared by acid hydrolysis of native cellulose, *Biomacromolecules*, **9**, 57–65.

147. Klemm, D., Kramer, F., Moritz, S., Lindstrom, T., Ankerfors, M., and Gray, D. (2011). Nanocelluloses: A new family of nature-based materials, *Angew. Chem. Int. Edit.*, **50**, 5438–5466.

148. Paul, G., and Dieter, K. (2010). Bacterial nanocellulose as a renewable material for biomedical applications, *MRS Bull.*, 208–213.

149. Wik, V. M., Aranguren, M. I., and Mosiewicki, M. A. (2011). Castor oil-based polyurethanes containing cellulose nanocrystals, *Polym. Eng. Sci.*, **51**, 1389–1396.

150. Roohania, M., Habibia, Y., Belgacema, N. M., Ebrahimb, G., Karimib, A. N., and Dufresnea, A. (2008). Cellulose whiskers reinforced polyvinyl

alcohol copolymers nanocomposites, *Abstr. Papers Am. Chem. Soc.*, **44**, 2489–2498.

151. Favier, V., Canova, G. R., Cavaille, J. Y., Chanzy, H., Dufresne, A., and Gauthier, C. (1995). Nano-composite materials from latex and cellulose whiskers, *Polym. Adv. Technol.*, **6**, 351–355.

152. Auad, M. L., Mosiewicki, M. A., Richardson, T., Aranguren, M. I., and Marcovich, N. E. (2010). Nanocomposites made from cellulose nanocrystals and tailored segmented polyurethanes, *J. Appl. Polym. Sci.*, **115**, 1215–1225.

153. Ljungberg, N., Bonini, C., Bortolussi, F., Boisson, C., Heux, L., and Cavaille, J. Y. (2005). New nanocomposite materials reinforced with cellulose whiskers in atactic poly-propylene: Effect of surface and dispersion characteristics, *Biomacromolecules*, **6**, 2732–2739.

154. Pei, A. H., Malho, J. M., Ruokolainen, J., Zhou, Q., and Berglund, L. A. (2011). Strong nanocomposite reinforcement effects in polyurethane elastomer with low volume fraction of cellulose nanocrystals, *Macromolecules*, **44**, 4422–4427.

155. Hajji, P., Cavaille, J. Y., Favier, V., Gauthier, C., and Vigier, G. (1996). Tensile behavior of nano-composites from latex and cellulose whiskers, *Polym. Composite.*, **17**, 612–619.

156. Helbert, W., Cavaille, J. Y., and Dufresne, A., Thermoplastic nanocomposites filled with wheat straw cellulose whiskers. Part 1: Processing and mechanical behavior, *Polym. Composite*, **17**, 604–611.

157. Chazeau, L., Cavaille, J. Y, Canova, G., Dendievel, R., and Boutherin, B. (1999). Viscoelastic properties of plasticized PVC reinforced with cellulose whiskers, *J. Appl. Polym. Sci.*, **71**, 1797–1808.

158. Muller, F. A., Muller, L., Hofmann, I., Greil, P., Wenzel, M. M., and Staudenmaier R. (2006). Cellulose-based scaffold materials for cartilage tissue engineering, *Biomaterials*, **27**, 3955–3963.

159. Kenawy, E. R., Worley, S. D., and Broughton, R. (2007) The chemistry and applications of antimicrobial polymers: A state-of-the-art review, *Biomacromolecules*, **8**, 1359–1384.

160. Chen, Y., Liu, C., Chang, P. R., Cao, X., and Anderson, D. P. (2009). Bionanocomposites based on pea starch and cellulose nanowhiskers hydrolyzed from pea hull fibre: Effect of hydrolysis time, *Carbohyd. Polym.*, **76**, 607–615.

161. Lu, Y., Weng, L., and Cao, X. (2006). Morphological, thermal and mechanical properties of ramie crystallites-reinforced plasticized starch biocomposites, *Carbohyd. Polym.*, **63**, 198–204.

162. Paralikar, S. A., Simonsen, J., and Lombardi, J. (2008). Poly(vinyl alcohol)/cellulose nanocrystals barrier membranes, *J. Membrane Sci.*, **320**, 248–258.

163. Roohani, M., Habibi, Y., Belgacem, N. M., Ebrahim, G., Karimi, A. N., and Dufresne, A. (2008). Cellulose whiskers reinforced polyvinyl alcohol copolymers nanocomposites, *Eur. Polym. J.*, **44**, 2489–2498.

164. Capadona, J. R., Shanmuganathan, K., Trittschuh, S., Seidel, S., Rowan, S. J., and Weder, C. (2009). Polymer nanocomposites with nanowhiskers isolated from microcrystalline cellulose, *Biomacromolecules*, **10**, 712–716.

165. Ljungberg, N., Cavaille, J. Y., and Heux, L. (2006). Nanocomposites of isotactic polypropylene reinforced with rod-like cellulose whiskers, *Polymer*, **47**, 6285–6292.

166. Sanchez-Garcia, M. D., and Lagaron, J. M. (2010). On the use of plant cellulose nanowhiskers to enhance the barrier properties of polylactic acid, *Cellulose*, **17**, 987–1004.

Chapter 4

Bioprocessing of Natural Plant Cellulosic Polymer Materials

C. Vigneswaran,[a] P. Kandhavadivu,[a] and M. Ananthasubramanian[b]

[a]*Department of Fashion Technology and*
[b]*Department of Biotechnology, PSG College of Technology, Coimbatore 641004, Tamilnadu, India*

cvw@fas.psgtech.ac.in

4.1 Introduction

Biotechnology is the application of living organism and their components to industrial products and processes (Alat, 2001). It also offers the potential for new industrial processes that require less energy and are based on renewable raw materials (Anon, 2001). Advances in biotechnology and enzymology have brought new lines of research on organic cotton textiles and have accelerated the development of enzymatic applications in textile wet processing for sustainable process. Among the various stages of cotton preparation, textile wet processing is a highly energy-, water-, and chemical-consuming process (Naik and Paul, 1997; Warke and Chandratre, 2003). Enzymes are known for their specificity, high efficiency, and ability to work under mild conditions and provide a promising solution to eco-friendly processing

Handbook of Sustainable Polymers: Processing and Applications
Edited by Vijay Kumar Thakur and Manju Kumari Thakur

ISBN 978-981-4613-53-8 (Hardcover), 978-981-4613-54-5 (eBook)
www.panstanford.com

challenges (Carlier, 2001). The need of the organic cotton fabric for textile wet processing is required to process with minimum safe chemicals to health or alternative way to go into enzyme technology, because enzymes are substrate specific biocatalysts; they operate best at ambient pressures, mild temperatures and often at a neutral pH range. Enzymes are gaining an increasingly important role as a tool in various wet textile pre-treatment and finishing processes (Etters and Anis, 1998). Biocatalysts have proven to be a flexible and reliable tool in wet textile processing and a promising technology to fulfill the expected future requirements.

The sustainable processing is the need of the textile wet processing for organic cotton fabrics to obtain eco-friendly and environment-friendly agents and sustainable methods of processing in order to have environmentally safe processing with less chemicals inputs and problems in effluent disposals (Aiteromem, 2008). Presently, the pectinase enzymes are used in the bioscouring of cotton textiles, but its drawback is longer reaction time and it removes only pectin groups in the cellulosic structure of cotton fibers. Many research works carried out in the past have shown the removal of wax and oil substances by lipase and protease enzymes on the cotton fibers. To improve enzyme kinetic reaction on pectinolytic and proteinolytic activity of cotton materials, ultrasonic and air pressure system are identified techniques for boosting the enzyme activities on the bioscouring processes for meeting the industrial needs (Xia and Li, 2009). It is clear that advanced enzyme technology can be used to develop usable, more environment-friendly, economical competitive textile wet processing for organic cotton process to develop hygienic and value-added products in the coming days in the textile era.

4.2 Biotechnology: Scope and Importance

Biotechnology, in pure scientific terms, is defined as "application of biological organisms, systems, and processes to manufacturing and processing industries." A bioprocess is any process that uses complete living cells or their components (e.g., bacteria, enzymes, chloroplasts) to obtain desired products. It also offers the potential for new industrial processes that requires less energy and are based on renewable raw materials (Gubitz and Cavaco-Paulo, 2001; Opwis et al., 2006). It is important to note that biotechnology is not just concerned with biology, but it is a truly interdisciplinary subject involving the integration of natural and engineering sciences (Tyndall, 1996). Defining the scope of biotechnology is not easy because it overlaps with so many industries, such as the chemical industry and the food industry being major ones, but biotechnology has found many applications in the textile industry also, especially in genetic engineering, textile processing, and effluent management (Emilla Csiszar et al., 1998; Traore and Buschle-Diller, 2000).

4.3 Enzyme Structures and Mechanisms

Enzymes are generally globular proteins and range from 62 amino acid residues in size. Most enzymes are much larger than the substrates they act on, and only a small portion of the enzyme (around 3–4 amino acids) is directly involved in catalysis (Carlier, 2001). The region that contains these catalytic residues binds the substrate and then carries out the reaction is known as the "active site." Enzymes can also contain sites that bind cofactors, which are needed for catalysis. Some enzymes also have binding sites for small molecules, which are often direct or indirect products or substrates of the reaction catalyzed (Alat, 2001). This binding can serve to increase or decrease the enzyme's activity, providing a means for feedback regulation. The mechanism of enzyme catalysis is shown in Fig. 4.1.

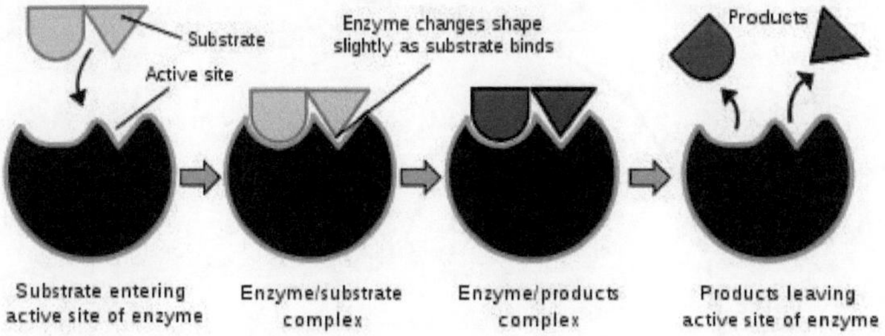

Figure 4.1 Mechanism of enzyme catalysis.

The active site continues to change until the substrate is completely bound, at which point the final shape and charge is determined (Etters and Anis, 1998).

4.3.1 Enzyme Kinetics and Their Reactions

The enzyme (E) binds a substrate (S) and produces a product (P). Enzyme kinetics is the investigation of how enzymes bind substrates and turn them into products. An enzymatic reaction occurs in two stages. In the first stage, the substrate binds reversibly to the enzyme, forming the enzyme–substrate complex. This is called the "Michaelis complex." The mechanism of single-substrate enzyme catalytic reaction is shown in Fig. 4.2.

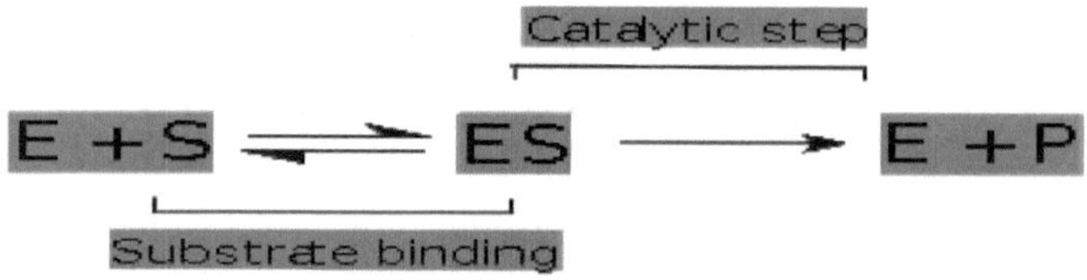

Figure 4.2 Mechanism of single-substrate enzyme catalytic reaction.

Enzymes can catalyze up to several million reactions per second. Enzyme rates depend on solution conditions and substrate concentration (Jayam et al., 2005). The maximum speed of an enzymatic reaction is based on the substrate concentration until a constant rate of product formation. This is shown in Fig. 4.3 for indicating the saturation curve. Michaelis-Menten constant (K_m) is the substrate concentration required for an enzyme to reach

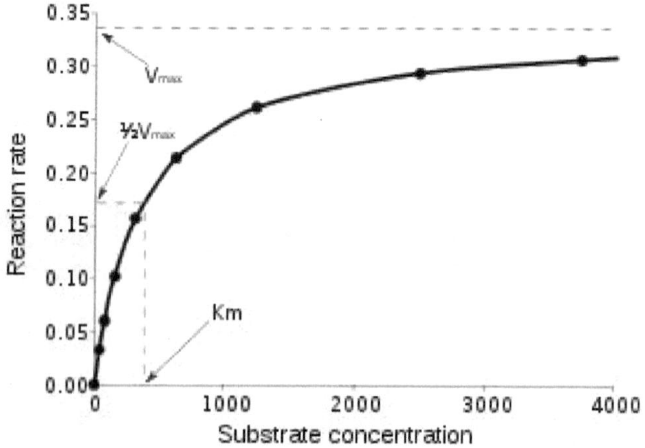

Figure 4.3 Saturation curve for an enzyme reaction showing the relation between the substrate concentration (S) and rate (v).

one-half its maximum velocity. Each enzyme has a characteristic K_m for a given substrate and this can show how tight the binding of the substrate is to the enzyme. Saturation happens because, as substrate concentration increases, more and more of the free enzyme is converted into the substrate-bound ES form. At the maximum velocity (V_{max}) of the enzyme, all the enzyme active sites are bound to substrate and the amount of ES complex is the same as the total amount of enzyme. However, V_{max} is only one kinetic constant of enzymes. The amount of substrate needed to achieve a given rate of reaction is also important.

4.3.2 Applications of Enzymes in Textile Wet Processing

Textile processing has benefited greatly on both environmental and product quality aspects through the use of enzymes. New enzymatic processes are being developed (cellulase, hemicellulase, pectinase, protease, xylanase, cutinase, and lipase), which offer the potential to replace the use of other chemicals in textile preparation processes for natural and synthetic materials (Warke and Chandratre, 2003). Environmental benefits are lower discharge of chemicals and wastewater and decreased handling of hazardous chemicals for textile workers. Many R&D works are being under progress in genetically modified micro-organisms and dyestuffs for the textile field. The use of enzymes in textile processing and after care is already the best established example of the application of biotechnology to textiles in the near- to medium-term future. The applications of hydrolase and oxidoreductase enzymes used in the textile wet processing are given in Tables 4.1 and 4.2, respectively.

Table 4.1 Application of hydrolase enzyme in fabric preparation

Enzyme name	Substrate	Textile application
Amylase	Starch	Starch desizing
Cellulase	Cellulose	Stone wash/Bio-polishing Biofinishing for hand modifications Carbonization of wood
Pectinase	Pectin	Bio scour replacing caustic soda
Catalase	Peroxides	In situ peroxide decomposition without any rinse in bleach bath

(*Continued*)

Table 4.1 (*Continued*)

Enzyme name	Substrate	Textile application
Protease	Protein molecules/ peptide bonds	Degumming of silk Bioantifelting of wool/cotton wax
Lipases	Fats and oils	Hydrophilicity of cotton and polyester

Table 4.2 Application of oxidoreductase in fabric preparation

Enzyme name	Substrate attacked	Textile application
Laccase	Color Chromophore and pigments	Discoloration of colored effluent Bio-bleaching of lignin containing fibers such as kenaf and jute. Bio-bleaching of indigo in denim for various effects
Peroxidases	Color Chromophore and pigments	Bio-bleaching of wood pulp
Glucose oxidases	Pigments	In situ generation of hydrogen peroxide for bleaching of cotton
AZO reductase	Color Chromophore and pigment	Discoloration of AZO dyes effluent
Peroxidase ostreatus	Color Chromophore and pigment	Discoloration of Remazol of basic dye effluent

4.4 Cotton Morphology and Chemistry

The cotton fiber is a single plant cell. Its cross section is oval, compared with the normal hexagonal plant cell. Cotton has a district cuticle, well-developed primary and secondary walls and a lumen. The cuticle is a waxy protective layer that provides water resistance to the fibers as they are growing. This layer is removed by scouring during processing before spinning. A mature cotton fiber has the following six parts: The "cuticle" is the outer waxy layer, which contains pectins and proteinaceous materials. It serves as a smooth, water-resistant coating, which protects the fiber. This layer is removed from the fiber by scouring. The "primary wall" is the original thin cell wall. Mainly cellulose, it is made up of a network of fine fibrils (small strands of cellulose). This makes for a well-organized system of continuous, very fine capillaries, which is shown in Fig. 4.4. It is well known that fine capillaries rob liquids from coarse capillaries. The fine surface capillaries of each cotton fiber contribute greatly to cotton's wipe-dry performance. The "winding layer" (also called the S1 layer) is the first layer of secondary thickening. It differs in structure from both the primary wall and the remainder of the secondary wall. It consists of fibrils aligned at 40° to 70° angles to the fiber axis in an open netting type of pattern. The "secondary wall" (also called the S2 layer) consists of concentric layers of cellulose, which constitute the main portion of the cotton fiber. After the fiber has attained its maximum diameter, new layers of cellulose are added to form the secondary wall. The fibrils are deposited at 70° to 80° angles to the fiber axis, reversing angle at points along the length of the fiber. The fibrils are packed close together, again, forming small capillaries. The "lumen wall" (also called the S3 layer) separates the secondary wall from the lumen and appears to be more resistant to certain reagents than the secondary wall layers. The "lumen" is the hollow canal that runs the length of the fiber. It is filled with living protoplast during the growth period. After the fiber matures and the boll opens, the protoplast dries up, and the lumen naturally collapses, leaving a central void, or pore space, in each fiber (Vigneswaran and Keerthivasan, 2008).

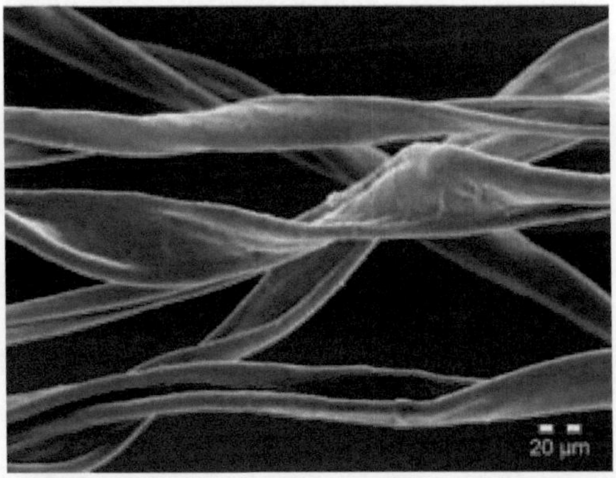

Figure 4.4 Morphological view of cotton fiber under microscope.

Under a microscope, a cotton fiber appears as a very fine, regular fiber, looking like a twisted ribbon or a collapsed and twisted tube. These twists are called convolutions; there are about 60 convolutions per centimeter. The convolutions give cotton an uneven fiber surface, which increases inter-fiber friction and enables fine cotton, yearns of squatted strength to be spun. The appearance of the cotton fiber's cross sections is referred to as being kidney-shaped. Cellulose is a macromolecule—a polymer made up of a long chain of glucose molecules linked by C-1 to C-4 oxygen bridges with elimination of water (glycoside bonds). The anhydroglucose units are linked together as beta-cellobiose; therefore, anhydro-beta-cellobiose is the repeating unit of the polymer chain (Fig. 4.5). The number of repeat units linked together to form the cellulose polymer is referred to as the "degree of polymerization." The cellulose chains within cotton fibers tend to be held in place by hydrogen bonding. These hydrogen bonds occur between the hydroxyl groups of adjacent molecules and are most prevalent between the parallel, closely packed molecules in the crystalline areas of the fiber.

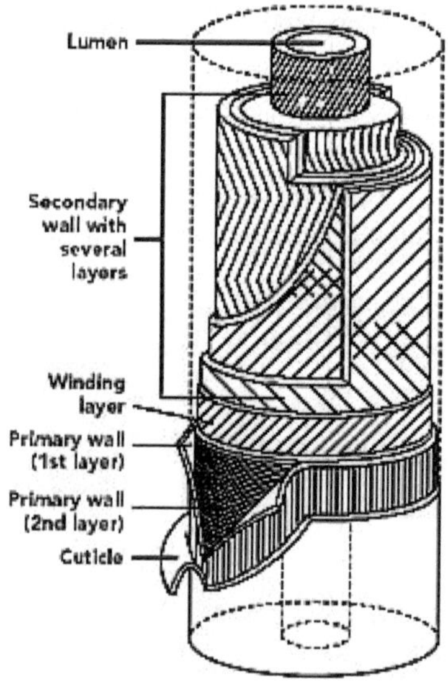

Figure 4.5 Microstructure of cotton fiber.

4.5 Textile Processing of Cotton Materials

Traditional method: In the cotton processing for improving the hydrophilic nature, cotton materials was treated with alkaline conditions for removal of wax, pectin, and fatty substances using sodium hydroxide and sodium carbonate under specific conditions. These processes are still followed in many textile industries, but the drawbacks are long incubation period, non-uniform treatments, and higher TDS, pH, and COD and BOD levels in textile preparatory processes.

Enzymatic methods: Many research organizations and scientists are working to improve the environment-friendly processing of cotton fabrics with enzymes. The many enzymes are today available for the removal of impurities in the cotton fibers. For example, alpha amylase is enzyme used for cotton desizing process, pectinase for pectin removal in cotton fiber and lipase, and protease for fatty substance removal in the cotton fiber treatments. Typical textile wet processing of cotton materials is shown in Fig. 4.6.

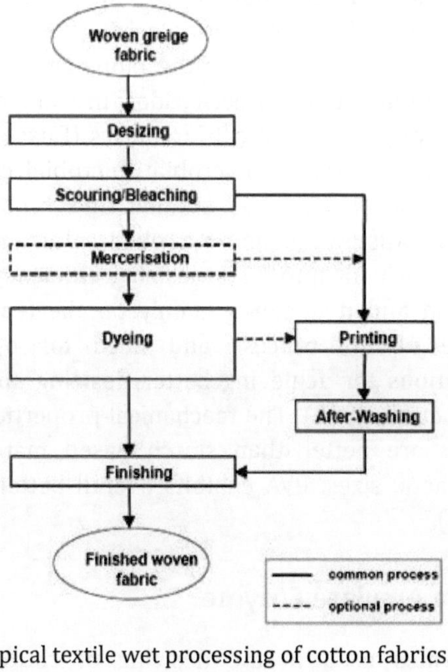

Figure 4.6 Typical textile wet processing of cotton fabrics.

4.6 Desizing of Cotton Fabric

Sizing: It is a process used for the application of a film forming polymer to provide temporary protection to the warp yarns from abrasive and other types of stresses generated on the weaving machines in order to reduce the warp breakages. Sizing helps in forming a coating that encapsulates the yarn, embeds the protruding fibers, and also causes some inter-fiber binding by penetration. The spun yarns being hairy usually require size add-on exceeding 8–10%; depending on the fabric to be woven. The small addition of synthetic binder to starch causes plasticization and increases the adhesion. A major drawback of the starch, brittleness of the film, can be reduced by providing internal plasticization (Moghe and Khera, 2005).

Traditional desizing: For cotton fabrics, traditional desizing is being carried out by high-temperature washing process and high concentrations of surfactants. As the process proceeds, the viscosity of the washing liquor rises rapidly because sizing agents dissolve. Therefore, large amounts of hot water are required. The use of industrial enzymes for desizing represents a major improvement because they cleave the biopolymers used a sizing agents into sub units as oligosaccharides. In some cases, this is a first step in biopreparation of cotton fabrics (Buschie-Diller et al., 1994; Tzanko et al., 2001). Anaerobial microbial cultures in the desizing of cotton fabrics was studied and reported that the bioreactor from wastewater and microbial culture for desizing of cotton was performed upto 73% desizing efficiency at 55°C and time of 60 min and it depends mainly on the temperature and reaction time of the process and need for optimizing the process conditions for achieving better desizing and low energy (Heikinheimo et al., 2003). The mechanical properties of synthetic size materials are better than starch based materials. Among different synthetic sizes, PVA exhibits overall better performance (Bayard, 1983).

4.6.1 Alpha Amylase Enzyme

Amylase enzymes used in the desizing of cotton fabrics are obtained from different sources with different activity levels. There are two kinds of amylases available in sources, namely endoamylase and exoamylases. Endoamylases cleave α-(1-4)-glycosidic bonds present in the inner parts of amylose, whereas exoamylases act on external glucose residues of starch molecules.

Exo enzymes include alpha amylases (1,4)-β-D-glucan, glucanohydrolase, EC 3.2.1.2), exomaltotriohydrolases (EC 3.2.1.95), exomaltohexohydrolases (EC 3.2.1.98), and glucoamylases (EC 3.2.1.3). Endo enzymes include alpha amylases (1,4)-β-D glucan, glucanohydrolase, EC 3.2.1.1), pullulanases (pullulan 6 glucanohydrolase EC 3.2.1.41), and isoamylases (glycogen 6 glucanohydrolase, EC 3.2.1.68).

4.6.2 Enzymatic Desizing of Cotton Fabric

Activity of amylases has been studied extensively using pure starch as a model compound (Opwis et al., 2000; Aranjo et al., 2004; Ibrahim et al., 2004; Lee et al., 2000; Liu et al., 2000; Reshmi et al., 2006; Shewale and Pandit, 2007), and attempts have been made to analyze the hydrolytic activities of pancreatic, malt, bacterial amylases, and cellulases in desizing (Hahn et al., 1998; Opwis et al., 2000; Feitkenhauer and Meyer, 2003). Amylase-assisted desizing of textile materials is carried out in machines such as jigger, jets, pad-batch, and pad-stream ranges, employing different levels of mechanical agitations. Among various steps involved in enzyme reaction, hydrolysis of starch needs longer time, depending upon the activity levels of enzyme and temperature conditions used in desizing. Composition, properties of starch (Moghe and Khera, 2005), ingredients added in size mix (Azevedo et al., 2003; Declerek et al., 2003), and process conditions employed in desizing (Ibrahim et al., 2004) have marked influence on the efficiency of desizing. Common waxes do not inactivate amylases but prevent quick wetting and penetration of enzymes, and other factors that affect the efficiency of size removal include viscosity of starch, amount of size applied, fabric construction, and method of washing off (Shamey and Hussein, 2005).

Enzymatic degradation of polyvinyl alcohol (PVA) was studied in the desizing of cotton fabrics. A mixture of two different PVA-degrading enzyme activities, including PVA oxidizing, was partially purified from the culture filtrate of a PVA-degrading mixed culture by ionic exchange chromatography. Optimum conditions for PVA-degradation by using this enzyme mixture were reported at pH 8.0 and temperatures ranging from 30°C to 55°C. Cotton fabrics sized with a PVA solution (25 $g\,dm^{-3}$) were desized with the enzyme mixture after a 1 h treatment at 30°C and pH 8.0. In this case, similar amounts of residual PVA in fabrics were detected in comparison with the conventional desizing process, which uses hot water at 80°C, 30 min (Mori et al., 1999; Vigneswaran, 2011). Several factors that affect the starch-size removal were studied and it was found that the effectiveness of enzymatic desizing can be enhanced by raising the desizing temperature up to 70°C; prolonging the desizing time up to 60 min; increasing the material-to-liquor ratio up to 20:1; increasing the Aquazym® 240-L dosage up to 6 g/L; and treating at pH 7 (Nabil et al., 2004; Csiszar et al., 2001). All enzymes work within a range of temperature specific to the organism. Increases in temperature generally lead to increase in reaction rates (Daniel et al., 2010). Most enzymes are sensitive to pH and have specific ranges of activity. The pH can stop enzyme activity by denaturating (altering) the three-dimensional shape of the enzyme. Most enzymes function between a pH of 6 and 10 (Li and Hardin, 1999).

4.6.3 Assessment of Enzyme Desizing

The assessment of alpha amylase enzyme based desized cotton fabric has been carried out by two methods, namely (a) Tegawa scale, a qualitative spotting test using iodine solution and (b) percentage weight loss (Shukla and Jaipura, 2004). Amylose binds ~20% of its weight of iodine at 20°C, shows deep blue color, violet and reddish brown, pale yellowish brown for undegraded starch, partially degraded dextrin, degraded dextrin, and completely hydrolyzed starch, while amylopectin binds 2% iodine only.

4.7 Scouring of Cotton Fabrics

The scouring process is to make the material hydrophilic, before it undergoes other processes like bleaching, dyeing, and printing in the textile wet processing (Holme, 2001). A desired hydrophilicity during the scouring can be achieved by removing non-cellulosic material from the cotton fabric, especially from the cuticle (waxes and fats) and the primary wall (e.g., pectin, protein, and organic acids). More precisely, scouring removes not only non-cellulosic material from cotton fibers but also substances that have adhered to the fibers during the production of the yarn or fabric. Substances such as dirt, lint, pesticides, oils, and any sizing agent applied to yarns to facilitate weaving (Eisisi et al., 1990; Ammayappan et al., 2003). Effective scouring is essential for subsequent processing of any cotton made substrate, regardless of its natural source. Even today, alkaline scouring of cotton is still the most widespread commercial technique for removing or rupturing the fiber cuticle to make the fiber absorbent for the cotton processing. Although sodium hydroxide is used generally for the scouring, sodium carbonate and calcium hydroxide are also mentioned as scouring agents (Hsieh and Cram, 1999). Scouring of cotton fabric is typically done with a hot solution (90–100°C) of sodium hydroxide (±1 mol/L) for up to 1 h (Emilla Csiszar et al., 1998). The concentration of alkali used and the time and temperature conditions needed depend on the desired quality of the scoured fabric. Other chemicals, for instance, wetting agents, emulsifying agents, and chelating agents (Nallankilli et al., 1992; Tyndall, 1996) are also included in typical preparation baths for scouring. Wetting agents act by reducing the surface tension of water enabling improved penetration of the chemicals into the cotton fabric. Emulsifying agents assist in removing waxy materials. Chelating agents remove polyvalent metal ions such as calcium, magnesium, iron, or other salts that can have a harmful effect on subsequent wet-processing operations (Vigneswaran et al., 2012a).

4.7.1 Drawbacks Associated with Alkaline Scouring

The scouring process requires large quantities of chemicals, energy, and water and is rather time consuming (Yonghua and Hardin, 1997). Owing to the high sodium hydroxide concentration and its corrosive nature, intensive rinsing is required, which leads to high water consumption. The use of high concentrations of sodium hydroxide also requires the neutralization of wastewater, which requires additional acid chemicals. Furthermore, the alkaline effluent requires special handling because of very high BOD and COD values. Apart from the above wet processing problems, the biggest drawback of alkaline scouring is a non-specific degradation of cellulose that produces fabrics of lower tensile strength and therefore of lower quality (Wang et al., 2007). Moreover, alkaline scouring is hazardous to workers and creates an unpleasant work atmosphere. Although, alkaline scouring is effective and the costs of NaOH are low, this process can be improved considerably to meet today's energy and environmental demands.

4.7.2 Overview of Enzymatic Scouring

Enzymes are substrate specific biocatalysts; they operate best at ambient pressures, mild temperatures, and often at a neutral pH range. Enzymes are gaining an increasingly important role as a tool in various wet textile pre-treatment and finishing processes (Alat, 2001; Anon, 2001; Carlier, 2001). Biocatalysts have proven to be a flexible and reliable tool in wet textile processing and a promising technology to fulfill the expected future requirements. Enzymatic scouring has been investigated extensively by various institutes and laboratories now for nearly one decade (Hartzell and Hsieh, 1998; Emilla Csiszar et al., 1998; Qiang Wang et al., 2006). Initial investigations explored the possibility of cotton scouring with enzymes, to see if cotton could be made hydrophilic in a reasonable time. Extracellular enzymes involved in the degradation of the plant cell wall's outer layer during the invasion of the plant, excreted by phyto-pathogenic fungi and by bacteria have been considered as candidates. Different enzymes like pectinases such as lyases (EC 4.2.2.2); polygalacturonase endo acting type (EC 3.2.1.15) and polygalacturonase exo acting type (EC 3.2.1.67), proteases (EC 3.4.21-25), cellulases such as endoglucanases (EC 3.3.1.4); cellobiohydrolases (EC 3.2.1.91), xylanases (EC 3.2.1.8), lipases (EC 3.1.1.3) and recently cutinases (EC 3.1.1.74) have been examined to degrade and subsequently remove the natural component present in the outer layer of cotton fibers (Buschie-Diller et al., 1998; Hartzell and Hsieh et al., 1998; Yonghua and Hardin, 1997). These studies incorporated staining tests, scanning electron microscopy (SEM), weight loss analysis, cotton wax residue, and nitrogen content analysis.

The scheme essentially contains the impregnation of cotton fabric with one or more enzymes in presence or absence of surfactants and chelators, followed by a high-temperature rinsing step. The enzyme incubation time used was up to 24 h depending on other process conditions and the density of the fabric. Lipases were found to be less effective in fulfilling this task (Dahod, 1987; El-Shafei and Rezkallah, 1997). Proteases were found to be efficient to improve whiteness rather than hydrophilicity (Ellaiah et al. 2002; Hsieh and Cram, 1999). Cellulases were the only enzymes reported to improve the wettability efficiently when applied without any other treatment or in combination with other enzymes. However, cellulase also causes a decrease in fiber strength and hence a decrease in fabric quality (Emilla Csiszar et al. 1998; Emre Karapinar and Merih, 2004). The best results have been obtained by alkaline pectinases or pectinases in combination with cellulase. Especially bacterial alkaline pectinase, a pectate lyase (EC 4.2.2.2) has been proven to be effective (Ghanem et al., 2000; Etters et al. 2003). Pectin acts as cement in the primary wall of cotton fibers. After enzymatic destabilization of a pectin structure, the different components present in the primary wall layer can be removed easily in subsequent rinsing steps (Hardin and Kim, 2000; Vigneswaran et al., 2012b). A proper interpretation of the enzymatic action on cotton fibers on a molecular basis was not possible because of the lack of structural knowledge of cotton fiber.

4.7.3 Pectinases in Bioscouring of Cotton

Pectinase is an enzyme that catalyzes the hydrolysis/depolymerization of the glycosidic bonds in the pectin polymers, classified according to their preferential substrates (high or low methylesterified pectin and polygalacturonic acid/pectate) and their reaction mechanism (Kristensen, 2001; Jayam et al., 2005). Pectinases are either endo-acting, cutting the polymer at random sites within the chain to give a mixture of oligomers, or exo-acting, attacking from one end of the polymer and producing monomers or dimers, identified by the rate of release of reducing sugars (Friend and Chang, 1982). Pectinases have been produced using solid-state and submerged fermentation (Pererra et al., 1993; Martin et al., 2004; Csiszar et al., 2007), with various bacterial (Bjo et al., 2002; Andersen et al. 2002) and fungal (Arguelles et al. 1995; Sharma and Gupta, 2001; Alves et al., 2002; Pyc et al., 2003; Solbak et al., 2005; Csiszar et al., 2007) sources, and different schools of thought exist to demonstrate their relative merits and demerits.

Alkaline pectinases are produced predominantly from *Bacillus* and *Pseudomonas* (Lawson and Hsieh, 2000), while *Aspergillus* appears to be the major acid pectinase producer (Csiszar et al., 2007). Comparisons have been made in the past, to assess the performance of acid pectinases (Pyc et al., 2003; Calafell and Garriga, 2004; Canal et al., 2004; Sahin and Gursoy, 2005) and alkaline pectinases (Bruhlmann, 1995; Almeida et al., 2003; Wan et al., 2007; Wang et al., 2007). Effects of various carbon sources, carbon supplements and other components in the culture medium have been well documented (Pererra et al., 1993; Kashyap et al., 2000; Maldonado and Saad, 2000; Miller et al., 2003; Presa, 2007).

4.7.4 Pectin

Pectin is a complex carbohydrate, which is found both in the cell walls of plants and between the cell walls; it helps regulate the flow of water in between cells and keeping them rigid. Pectins are considered intracellular adhesives, combining hydrophobic molecules like proteins, waxes, and hemicelluloses with various degrees of branching, and the three-dimensional structure of pectin breaks the cellulose array in cotton fibers (Potikha et al., 1999; Hardin and Kim, 2000; Brushwood, 2003; Gambler, 2003). Pectic acid, present in the primary wall of cotton fibers, contains negatively charged galacturonic acid residues and forms a bridge with calcium (pectates), thereby holding the inner primary wall of cotton that gives stability at high temperature. Pectolytic activities of pectin lyase, pectin methyl esterase, and polygalacturonase enzymes are shown in Fig. 4.7.

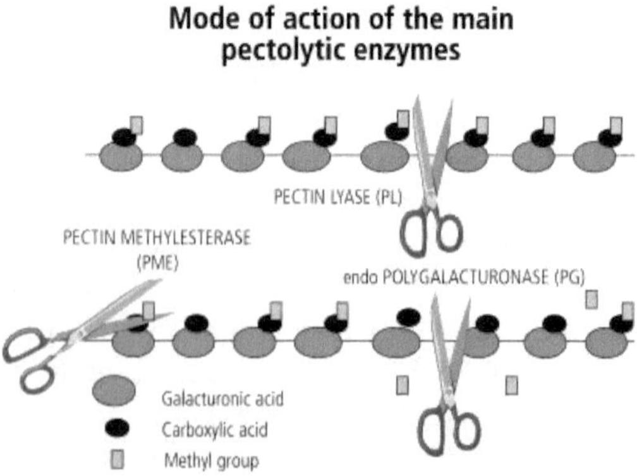

Figure 4.7 Pectolytic activity of pectin lyase, pectin methyl esterase and polygalacturonase enzymes.

4.7.5 Pectinase Treatments and Evaluation Methods

In enzymatic scouring of cotton, nonionic surfactants are used to overcome hydrophobicity of the substrate, which assist enzymes to penetrate through micro-pores or cracks and help them to orient themselves in favorable positions for catalytic actions (Sahin and Gursoy, 2005), while ionic surfactants complex with enzymes and disrupt their structure to different extents. Concentration of pectinase, pH, time, and temperature range used in scouring influence efficiency of the process (Etters et al., 1999; Canal et al., 2004). Microscopic observations (Sahin and Gursoy, 2005), absorbency (Li and Hardin, 1998; Lenting et al., 2002; Wang et al., 2007), selective staining methods (Calafell et al., 2005), weight loss (Li and Hardin, 1999; Lenting et al., 2002; Lenting and Warmoeskerken, 2004; Calafell et al., 2005; Schnitzhofer et al., 2006), residual cotton wax, and nitrogen content have been used for the assessment of bio-scoured fabrics (Lawson and Hsieh, 2000; Degari et al., 2002; Lenting et al., 2002; Brushwood, 2003; Chung et al., 2004).

4.7.6 Lipases in Processing of Cotton Fabrics

Lipases (triacylglycerol acylhydrolases, EC 3.1.1.3) are widely distributed among the animals, plants, and microorganisms, capable of hydrolyzing ester bonds of oils, fats and certain waxy substances. Lipases catalyze reactions, including synthesis and transesterification of glycerides and phosphoglycerides as well as a variety of non-glycerides and ester bonds (Sammour, 2005).

4.7.7 Applications of Lipases in Cotton Processing

The term cotton wax includes all lipids found in the cotton fiber surface, including waxes, fats, and oils (Lawson and Hsieh, 2000; Brushwood, 2003; Bargel et al., 2006). Cuticular lipids, of cotton, are complex mixtures of aliphatic and aromatic components, most of them resembling derivatives of nacyl alkanes. Benzene, chloroform, carbon tetrachloride, ethanol, isopropyl alcohol, alcohol-benzene mixture, and trichloroethylene have been recommended for extraction of the cotton wax. Lipase activity is often determined with reagents comprising a low concentration of buffer and a diglyceride dissolved therein (Obendorf et al. 2003; Varanasi et al., 2001; Pera et al. 2006). Lipases with activity of pH >6.5 at 30–60°C and ability to withstand many surfactants (linear alkyl benzene sulphonates) and proteases, meant for detergent formulations, can also withstand bleaching agents such as hydrogen peroxide and peracids. Lipases increase lipid removal from all morphological locations on the cotton fibers, including lumen and fiber surfaces.

4.8 Proteases in Cotton Fabric Processing

Proteins that occur (~1.3%) in the lumen and primary wall of the cotton fibers are the residual, dead protoplasm from biosynthesis, composed of several proteins and peptides, formed by various amino acids, rather than single protein. Proteases belong to the sub-class of peptide hydrolase or peptidase, and more conveniently, proteases are classified into serine, cysteine, aspartic, and metalloproteases (Gupta et al., 2002). Proteases are generally produced by submerged fermentation, though solid-state fermentation offers certain advantages in terms of reduced energy consumption during extraction of enzymes from the culture (Alves et al., 2002).

Most of the nitrogen-containing compounds of cotton can be removed by a mild alkali boil and a very low residue remains in scoured and bleached cotton (Najafi et al., 2005). Denaturation of protein substances by ageing, heating, and oxidation makes them less accessible to enzymatic degradation (Andrade et al., 2002). Pretreatments with boiling water enhance the accessibility of proteins in cotton beneath the waxes and improve scouring efficiency. Since protein contents are high among the non-cellulosics in cotton fiber, potential of proteolytic enzymes as scouring agents for raw cotton cannot be underestimated (Hartzell and Hsieh, 1998; Silva et al., 2006). Hydrolysates produced in the protease reactions depend on the source of proteases with 9–12 major amino acids. Nitrogen content of untreated cotton ~0.4% reduces to ~0.2% after treatment with proteases. Water absorption and retention capacity and *K*/*S* values have also been advocated for characterization of protease treated fabrics (Sae et al., 2007; Vigneswaran et al., 2012c). Though protease treatments in both scouring and detergency offer a range of advantages, there are certain limitations, also, in terms of proteolysis of enzymes in the crude cultures (Pererra et al. 1993).

4.8.1 Fabric Weight Loss and Strength Loss

Cotton fibers, free from pectic substances, show no damage in tensile strength and fluidity. Pectinase of *Aspergillus niger* at pH 4.2 and 40°C shows weight loss ~4–5.5%; however, lower weight losses in the range of 0.31% to 1.04% has also been reported. Strength loss of about 3.23% (warp) and 2.24% (weft) has been reported in fabrics, while the loss of single yarn strength has been reported at 2.6% (Calafell and Garriga, 2004; Calafell et al., 2005; Schnitzhofer et al., 2006).

4.8.2 Fabric Absorbency, Wettability, and Dyeability

Practically no difference in absorbency and wettability has been reported between alkali and bioscoured samples, alkaline pectinases and certain acidic pectinases (Etters, 1999; Lenting et al., 2002; Pyc et al., 2003; Canal et al., 2004; Calafell et al., 2005; Schnitzhofer et al., 2006). Drop absorbency similar to that of alkali scour (~1 sec) has been reported in most of the literature, except in certain cases where acidic pectinases have been used in the process (Calafell and Garriga, 2004). Degree of whiteness (CIE) of bioscoured fabrics shows lower value compared to alkali-scoured cotton fabrics (Canal et al., 2004; Calafell et al., 2005). When long-chain pectin (fully methylated pectic acid) is completely degraded to galacturonic acid, the iodine reducing value increases from zero to 8.95 (Karmakar, 1998). Dyeing of samples scoured with commercial pectinases shows no difference in absorbance values and color depth compared to the alkali-scoured samples; however, substantively of the dyes is lower on bioprepared fabric, due to "beneficial wax" that remains in the fibers (Etters, 1999; Calafell et al., 2005). Ruthenium red, a basic dye, selectively binds the carboxyl oxygen of galacturonide moiety and hydroxy oxygen of an adjacent galacturonide in the pectate chain, a characteristic reaction that helps to measure free carboxyl groups available in cotton fibers (Li and Hardin, 1998).

4.8.3 Effects of Process Parameters on Bioscouring

Besides the nature of substrates, efficiency of hydrolysis is also influenced by process conditions (Hemmpel, 1991; Tyndall, 1996; Paulo et al., 1996), co-reactants present in the process (Cavedon et al., 1990; Heikinheimo et al., 2003) and mechanical agitations employed in the reaction systems (Pendersen et al., 1992; Lee et al., 1996; Traore and Diller, 1999; Tzanko et al., 2001). The material-to-liquor ratio of process bath alters the efficiency of all the components exhibited by weight loss values, and little changes are observed in the range of 1: 10 to 1: 40 (Paulo et al., 1996). Mechanical actions, winch machines, and jet systems, between fabrics and equipment or surface-to-surface contact of fabrics, enhance reactivity of cellulases by improving two-way mass transfers and enhance weight loss and removal of weakened fibers from surfaces of yams and fabrics, thereby facilitating a clean surface to the fabrics (Traore and Diller, 1999; Ramkumar and Abdalah, 2001).

4.9 Cellulases on Cotton Fabrics

Cotton fabrics, using cellulases, is aimed to remove cellulosic impurities, individual and loose fiber ends that protrude from fabric surfaces, and to provide an enhanced appearance and handle, with or without the aid of mechanical agitations but without degrading the properties of the fabrics significantly. Cellulase enzymes are complex mixtures of three major constituents enzymes, namely endo-(1,4)-α-D-glucanases (ED) (EC 3.2.1.4), which randomly cleave internal glucosidic bonds, (1,4)-α-D-glucan cellobiohydrolases (CBH) (EC 3.2.1.91), which cleave them into cellobiases. Hydrolysis of cellobiose into the glucose end product is completed by β-glucosidases or cellobiases (EC 3.2.1.21), which split cellobiose units into soluble glucose monomers and complete hydrolysis of native celluloses, largely, depends on the synergistic actions of these three component enzymes. Figure 4.8 shows the structure of cellulose and catalysis of cellulase enzyme on cotton structure.

Cellulase-producing bacteria include *Cellulomonas, Clostridium, Pseudomonas, Streptomyces, Thermonospora,* and *Ruminococcus*, but bacterial cellulases digest cotton fibers to lower levels compared to amorphous celluloses (Ray et al., 2006). Cellulases have two functionally distinct domains in their structure, namely catalytic domain and cellulose substrate binding domain, linked by an interdomain, glycosylated linker peptide at either the N or the C terminal of the protein; cellulose binding domains exhibit different affinities, specificities, some binding to crystalline cellulose, while others restrict themselves to the disordered regions. Catalytic domain has an active site in the shape of a tunnel or cleft where hydrolytic reactions take place (Cavedon et al., 1990; Lee et al., 1996).

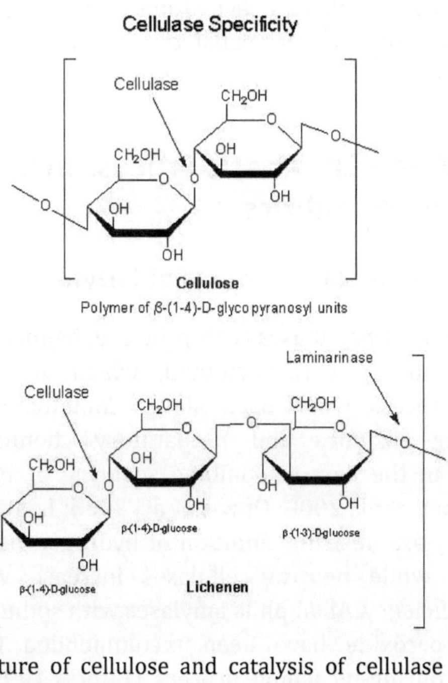

Figure 4.8 Structure of cellulose and catalysis of cellulase enzyme on cotton structure.

4.10 Bleaching of Cotton Fabrics

Conventional bleaching methods have been reviewed in many occasions, (Menezes and Chaudhari, 2005; Maekawa et al., 2007), oxidative chemical pretreatments are effective in degrading colorants and other impurities, though such methods often lead to oxidative degradation of substrates. However, peroxide bleaches are often referred to as the "color safe bleaches," due to minimal degradation of substrates.

4.10.1 Enzymatic Bleaching

Many attempts have been made to utilize various enzymes that belong to oxidoreductases, in bleaching of cotton fabrics (Nalankilli and Sundar, 2002; Tzanko et al., 2002; Opwis et al., 2006; Diller and Traore, 1998; Anis et al., 2008) and post-bleaching processes. Peroxidases are used to activate oxidizing agents like hydrogen peroxide; however, rapid deactivation of these agents in bleaching process does not guarantee satisfactory bleaching effects (Bernards et al., 2004).

4.11 Mixed Enzymatic Process in Scouring of Cotton Fabrics

4.11.1 Binary Combinations of Enzymes

Combinations of pectinases with protease, hemicellulase, cellulose, and lipase have been attempted, which are not efficient in scouring process when used alone. Combinations of amylases with other enzymes and preparatory chemicals have been attempted in the past to combine scouring or bleaching (Etters, 1999; Tzanov et al., 2000; Opwis et al., 2006; Lenting, 2008). In the case of amylase desizing, addition of hydrogen peroxide improves whiteness, while neutral cellulases increase weight loss and desizing efficiency. Alkaliphile amylases with sodium hydroxide and hydrogen peroxide have been recommended for a combined desizing-scouring-bleaching process (Etters, 1999; Csiszar et al., 2007). Integrated desizing and scouring using a-amylase and polygalacturonate lyase process involves two steps, in which fabrics are desized first, using a-amylases, then by a combination treatment of amylase and pectate lyase at 45–55°C at pH of 8.5 to 9.0, followed by washing in presence of chelating agent at 90–100°C (Lenting, 2008). Simultaneous desizing and scouring using amylase and pectinase obtained from single source, i.e.,

Bacillus and different sources have been attempted in the past (Lenting and Warmoeskerken, 2004; Dalvi et al., 2007). Higher weight loss values (7.0% to 15.3%), better drop absorbency (~1 sec), and dye absorption (KIS values of 7.45 against 6.9 of acid desized samples) have been reported with higher concentration of enzymes and longer incubation time. Whenever a very high whiteness in the fabrics is not required, desizing is combined with bleaching and scouring, incorporating protease, cellulose, and pectinase enzymes (Lange et al., 2001; Miller et al., 2003).

Combinations of pectinase or cellulase with hemicellulases like arabinases or pectinases with hemicellulase activities have been used for scouring. Crude mixtures of xylanase, cellulolytic, and pectinolytic enzymes destroy lignocellulosic structure of the seed coat fragments and produce soluble forms of lignin (Lange et al., 2001; Opwis et al., 2006). However, such combinations do not improve the drop absorbency (~290 sec), whereas no differences are observed in terms of strength loss and whiteness index compared to commercially scoured samples. Hemicellulases, in the pectinase scouring, are expected to give strong action against seed coat fragments, but their combinations with cellulase prove to show detrimental effects in terms of strength loss (Csiszar et al., 2001). Pectinase and cellulase act synergistically to improve the wettability within shorter process time, in which addition of non-ionic surfactants and mechanical agitations enhance their reactions. The presence of cellulase does not increase the pectin removal but results in higher weight losses (Schnitzhofer et al., 2006). Cellulase-containing pectinases result in flat ridges and concave grooves with polished surface in presence of surfactants, whereas alkali scoured fabrics often show fuzzy and blurred surface. In cellulase- assisted protease scouring, hydrolysis of cellulosic chains in primary wall is expected to enhance scouring action of protease synergistically (Diller et al., 1999; Guha and Shah, 2001).

Scouring using acid and alkaline pectinases and combined bleaching together with peracetic acid have been attempted in one bath, as one-step or two-step processes to yield commercially acceptable whiteness (Liu et al., 2000; Patra et al., 2004; Wan, 2007). Combined scouring and bleaching process using peroxide in a single step or two steps has also been advocated, by adding hydrogen peroxide at the end of pectinase scouring at higher pH (Tavcer et al., 2005) and such process is capable of retaining fabric strength up to ~90% with Hunter whiteness index and yellowness index of 84 and 16, respectively. An attempt has been made to utilize desize bath for bleaching using immobilized glucose oxidase in an aerated system at pH 10–11 and temperature of 90°C (Diller et al., 1999; Tzanko et al., 2000). However, such combined desizing and bleaching using glucose oxidase, often, results in non-uniform wetting properties, though wicking (~2.0 cm) and average drop absorbency (~1 sec) show similar values as that of commercial processes. Single-bath dyeing and biopreparation either simultaneously or sequentially using pectate lyase, at alkaline pH (>8), suitable for reactive dyes which result in wash fastness, rubbing fastness grades at least 3.5 to 4.0 (Liu et al., 2000).

4.11.2 Ternary and Quadruplet Combinations of Enzymes

In many situations, combinations of three or four different enzymes, including amylase, pectinase, protease, cellulase, glucose oxidase, and various hemicellulases, have been attempted since pectinases alone is often ineffective in the removal of impurities from cotton fibers and to improve absorbency of the scoured samples (Diller et al., 1999; Traore and Diller, 2000; Degari et al., 2002). Presence of xylanase and pectinase in commercial cellulase preparations also facilitates the removal of seed coats up to 70–85% (Cziszar et al., 1998). Protease and lipase are used together with pectate lyase to remove spinning, coning, and slashing lubricants from the fabrics (Miller et al., 2003; Wang et al., 2007). Lipase or protease treatments, alone, do not improve wetting or water retention values (Buchert et al., 2000). Research has been carried out (Traore and Diller, 2000) to study the effects of combined enzymes on the efficiency of scouring, using lipase, pectinase, xylanase, and cellulase, with strong agitation levels that result in weight loss up to 13.9%. Higher wickability is observed in the case of pectinase and xylanase combinations than pectinase and cellulase, pectinase and lipase combinations and the highest wicking observed in all-enzyme combination (Kim et al., 2005; Vigneswaran et al., 2012d).

Two-step scouring of cotton has also been suggested, with lipase and protease in the first step and cellulase in the second step.

Pectinase scouring produces 18-fold higher amounts of reducing sugars and galacturonic acid than any of the two-step processes, whereas lipase/proteases/cellulase scouring produces fivefold higher amounts of amino acids than the pectinase scouring (Sae et al., 2007; Karapinar and Sariisik, 2004). Pectinase–cellulase or, pectinase–cellulase with protease, or pectinase–cellulase–xylanase produces better scouring results than those individual enzymes, in terms of dyeability, *K/S*, and fastness properties. However, after hydrogen peroxide bleach, all the combinations (Wang et al., 2007) show similar CIE whiteness index (68–70).

4.12 Ultrasonic Treatment

The ultrasonic technique holds a promise in applications in the field of textiles. Ultrasonic represents a special branch of general acoustics, the science of mechanical oscillations of solids, liquids and gaseous media. Ultrasound can enhance a wide variety of chemical and physical processes, mainly by generating cavitation in liquid medium. The sonicator used is of 20 kHz frequency, which is found to be suitable for inducing cavitation (Abramov, 1998; Kamel et al., 2005). It is well known that cavitation, which causes the formation and collapse of microbubbles, is most effective for better fiber opening, which enhances water absorbency and dye uptake. This activated state causes chemical reaction between the fabric and the enzyme by forming waves and severe shear force capable of breaking chemical bonds. Ultrasound energy has great potential in industrial processes as it offers reduction in cost, time, energy, and effluents. Ultrasound reduces processing time and energy consumption, maintains or improves product quality, and reduces the use of auxiliary chemicals (Yachmenev, 2005).

4.12.1 Sonication: Basic Principle

In a solid, both longitudinal and transverse waves can be transmitted, whereas in gas and liquids, only longitudinal waves can be transmitted. In liquids, longitudinal vibrations of molecules generate compression and refractions, i.e., areas of high pressure and low local pressure. The latter gives rise to cavities or bubbles, which expand and finally, during the compression phase, collapse violently generating shock waves. The phenomena of bubble formation and collapse (known as cavitations) are generally responsible for most of ultrasonic effects observed in solid/liquid or liquid/liquid systems.

4.12.2 Ultrasonic Application: Textile Wet Processing

New biopreparation processes that utilize highly specific enzymes instead of conventional organic/inorganic chemicals are becoming increasingly popular in the textile industry (Yachmenev et al., 2005). The major shortcoming of this new technology is that the processing time is much longer than the conventional method. This limitation was overcome by use of ultrasound energy in combination with enzyme processing (Yachmenev et al., 2005). The combined enzyme/ultrasound biopreparation of greige cotton offers significant advantages such as less consumption of expensive enzymes, shorter processing time, better uniformity of treatment, and a notable decrease in the amount and toxicity of the resulting textile wastewater effluents. Sonolysis, enzyme treatment, and a combination of the two processes were tested for the degradation of phenol in aqueous medium. Degradation of starch followed by ultrasonic desizing could lead to considerable energy saving as compared to conventional starch sizing and desizing. Desizing of the pure cotton fabric with alpha amylase was conducted by the ultrasonic wave method and the traditional one respectively (Wan Qingyu, 2004). Research on the effect of ultrasonic wave on the enzyme desizing of the fabric indicated that it can improve desizing percentage, wettability, and whiteness of the fabric, reduce the strength loss, treatment time, and temperature, and save energy. Therefore, it has bright prospect of application. Introduction of ultrasonic energy during enzymatic biopreparation/bio-finishing of cotton textiles significantly improved enzyme performance but did not contribute to a decrease in fabric strength. Experimental data indicate that the maximum benefit provided by sonication of enzyme-processing solution occurred at lower enzyme concentrations.

The hydrolysis of maltoheptaose by α-amylase, and the resulting reaction was followed by the continuous monitoring of changes in ultrasonic velocity. As the reaction proceeds, ultrasonic velocity increases because the hydration level of the product is higher than that of the starting substrate, as shown in Fig. 4.9. It is simple to recalculate the ultrasonic curve to give the time dependence of the amount of substrate that has been hydrolyzed, providing the kinetic profile of the reaction, and allowing the enzyme's activity faster, which is shown in Fig. 4.10.

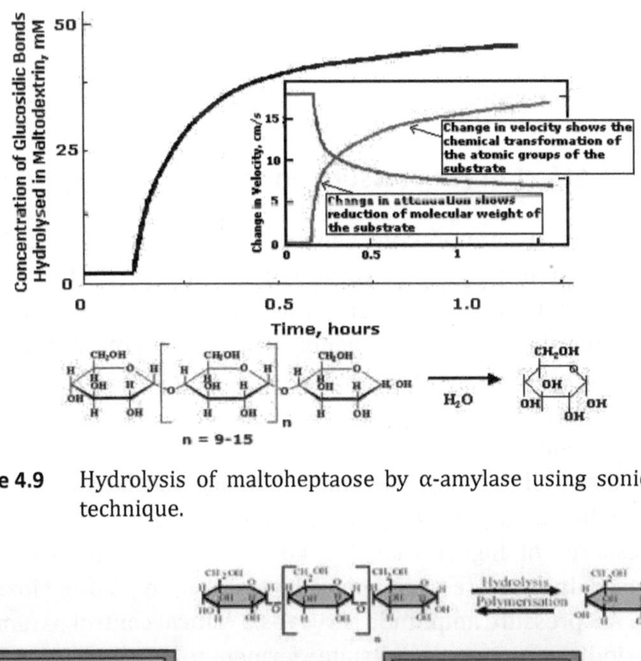

Figure 4.9 Hydrolysis of maltoheptaose by α-amylase using sonication technique.

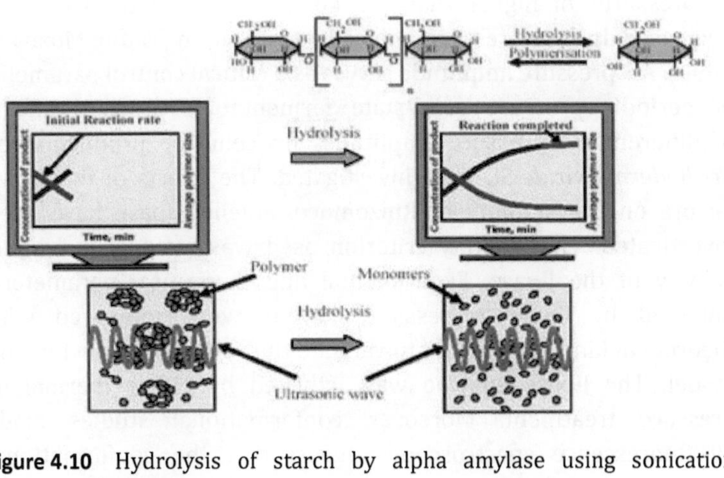

Figure 4.10 Hydrolysis of starch by alpha amylase using sonication technique.

4.13 Aerodynamic System

Aerodynamic technique has been studied and used for a variety of applications in liquids, dispersions, and polymers (Pinheiro, 2000). Limited research works have been reported to acceleration of enzyme kinetics through aerodynamic system (air pressure) to improve the reaction of substrate and enzyme binding to high quality and standardization of process parameters (Xia and Li, 2009). The aerodynamic system of enzyme acceleration has great potential in industrial processes as it offers reduction in cost, time, energy, and effluents.

4.13.1 Effect of Air Pressure on Enzyme Activity

Elevated hydrostatic pressure has been used to increase catalytic activity and thermal stability of enzymes. An increase in pressure at 20°C results in an exponential acceleration of the hydrolysis rate catalyzed by cellulase reaching a 6.5-fold increase in activity at 4700 atm (4.7 kbar). Due to a strong temperature dependence of the enzyme, the acceleration effect of high pressure becomes more pronounced at high temperatures. At 50°C, under a pressure of 3.6 kbar, cellulase enzyme shows activity that is more than 30 times higher than the activity at normal conditions (20°C, 1 atm). At pressures of higher than 3.6 kbar, the enzymatic activity is decreased due to a pressure-induced denaturation (Vadim Mozhaev, 1996). Air pressure amplitude serves as a critical control parameter of periodic pressure solid-state fermentation process. Effects of different air pressure amplitudes on cellulase production by *Trichoderma viride*-SL were investigated. The effects of these two factors on the stability of Rhizomucor miehei lipase have been investigated. The stability criterion used was residual hydrolytic activity of the lipase. Experimental and theoretical parameters, obtained by linear regression analysis were compared with theoretical kinetics in order to validate the series-type inactivation model. The lipase enzyme was activated by either thermal or pressure treatment. Moreover, conformational studies made by fluorescence spectroscopy suggest that the conformational changes induced by pressure were different from those induced by temperature. In addition, they show that after thermal deactivation there were less intermolecular hydrogen bonded structures formed than was the case for deactivation by high pressure.

4.14 Summary

A comprehensive review of biotechnology in the pretreatment of textiles reveals that enzymatic processing is environmental safe and a replacement of conventional harsh chemicals. The textile industry has plenty of opportunities for adapting biotechnology, but current awareness of biotechnology in this field is low. Enzymes are catalyzed and the processes operate under mild conditions of pH, temperature, and pressure than their chemical counterparts. Utilization of highly specific enzymes for various textile-processing applications is becoming increasingly popular because of their ability to replace harsh organic/inorganic chemicals currently used by the textile industry. It was established that ultrasound does activate the complex structures of enzyme molecules, and as a consequence, there was significant improvement in the performance of both cellulase and pectinase enzymes in the previous research work. The combined enzyme/sonication treatment of cellulosic textiles offers significant advantages such as less consumption of expensive enzymes, shorter processing time, less fiber damage and better uniformity of treatment. Researchers and scientists recommend and identified that the enzymes can be activated fast in reaction by sonication and aerodynamic techniques, which will result in shorter processing time and better quality of fabrics in textiles.

References

Abramov, O. V. (1998). *High Intensity Ultrasound: Theory and Industrial Applications*, Gordon and Breach Publications, London.

Aiteromem, A. L. (2008). Sustainable textile manufacturing with new enzyme based processes, *Int. Dyer*, **7**, 15–19.

Alat, D. V. (2001). Recent developments in the processing of textiles using enzymes, *Colourage*, **48**(2), 33–36.

Almeida, C., Branyik, T., Ferreira, P. M., and Teixeira, J. (2003). Continuous production of pectinase by immbolized yeast cells on spent grains, *J. Biosci. Bioeng.*, **96**(6), 513–518.

Alves, M. H., Takaki, G. M. C., Porto, A. L. F., and Milanez, A. I. (2002). Screening of mucor spp for the production of amylase, lipase, polygalacturonase and protease, *Braz. J. Microbiol.*, **33**(4), 325–330.

Ammayappan, L., Muthukrishnan, G., and Saravana P. C. (2003). A single stage preparatory process for woven cotton fabric and its optimization, *Man Made Text. India*, **47**(1), 29–35.

Andersen, L. N., Schulein, M., Lange, N. E. K., Ranvan, M. E. B., Fler, S. M., Glad, S. O., Kauppinen, M. S., Schnorr, K., and Kongsbak, L. (2002). Pectate lyases. US Patent No. 6 368843.

Andrade, V. S., Sarubho, L. A., Fukushima, K., Myaji, M., Nishimura, K., and Takaki, G. M. C. (2002). Production of extracellular proteases by mucor circinelloides using D-glucose as carbon source substrate, *Braz. J. Microbiol.*, **33**(2), 106–110.

Anis, P., Davulcu, A., and Erien, H. A. (2009). Enzymatic pretreatment of cotton. Part 2—Peroxide generation in desizing liquor and bleaching, *Fibres Text. East. Eur.*, **17**(2), 87–90.

Anon. (2001). Cotton preparation—A new enzymatic concept, *Novozyme Catalogue*, 23105–01.

Aranjo, M. A., Cunha, A. M., and Mota, M. (2004). Enzymatic degradation of starch based thermoplastic compounds used in protheses: Identification of the degradation products in solution, *Biomaterials*, **25**(13), 2687–2693.

Arguelles, M. E. A., Rojas, M. G., Gonzauz, G. V., and Torres, E. F. (1995). Production and properties of three pectinolytic activities produced by *Aspergillus niger* in submerged and solid state fermentation, *Appl. Microbiol. Biotechnol.*, **43**(5), 808–814.

Azevedo, H. S., Gama, F. M., and Reis, R. L. (2003). In vitro assessment of the enzymatic degradation of several starch based biomaterials, *Biomacromolecules*, **4**(6), 1703–1712.

Bargel, H., Koch, K., Cerman, Z., and Neinhuis, C. (2006). Structure function relationship of the plant cuticle and cuticular waxes—A smart materials, *Funct. Plant Biol.*, **33**(10), 893–910.

Bayard, J. (1983). Desizing with enzymes, *Canadian Text J.*, **100**(6), 168–169.

Bernards, M. A., Summerhurst, D. K., and Razen, F. A. (2004). Oxidase, peroxidase and hydrogen peroxide: The suberin connection, *Phytochem. Rev.*, **3**(1), 113–126.

Bjo, M. E., Anderssen, J. T., Schnorr, K., Schulein, M., and Kongsback, L. (2002). Pectate lyases. US Patent No. 6 399 351.

Bruhlmann, F. (1995). Purification and characterization of an extracellular pectate lyase from an *Amycolata* Sp, *Appl. Environ. Microbiol.*, **61**(10), 3580–3585.

Brushwood, D. E. (2003). Non-cellulosic constituents on raw cotton and their relationship to fiber physical properties, *Text. Res. J.*, **73**(10), 912–916.

Buchert, J., Peter, J., Puolakka, A., and Nousiainen, P. (2000). Scouring of cotton with pectinases, proteases and lipases, *Text. Chem. Colorists Am. Dyestuff Reporter*, **32**(5), 48–52.

Buschie-Diller, G., Zeronian, S. H., Pan, N., and Yoon, M. Y. (1994). Enzymatic hydrolysis of cotton, linen, ramie and viscose fabric, *Text. Res. J.*, **64**, 240–279.

Buschie-Diller, G., Mogahzy, E. Y., Inglesby, M. K., and Zeronian, S. H. (1998). Effect of scouring with enzymes, organic solvents and caustic soda on the properties of hydrogen peroxide bleached cotton yarn, *Text. Res. J.*, **68**(12), 920–929.

Calafell, M., and Garriga, P. (2004). Effect of some process parameters in the enzymatic scouring of cotton using acid pectinases, *Enzyme Microb. Tech.*, **34**(3), 326–331.

Calafell, M., Santner, B. K., Guebitz, G., and Garriga, P. (2005). Dyeing behaviour of cotton fabric bioscoured with pectate lyase and polygalacturonase, *Coloration Tech.*, **121**(6), 291–297.

Canal, J. M., Navarro, A., Calafell, M., Rodriguez, C., Caballero, G., Vega, B., Canal, C., and Paul, R. (2004). Effect of various Bioscouring systems on the accessibility of dyes into cotton, *Coloration Tech.*, **120**(6), 311–315.

Carlier, F. (2001). Enzymes, *Industrila Text.*, **54**(3), 121–123.

Cavedon, K., Leschine, S. B., and Parola, E. C. (1990). Characterization of the extracellular cellulases from a mesophilic clostridium (strain C7), *J. Bacteriol.*, **172**(8), 4231–4238.

Chung, C., Lee, M., and Choe, E. K. (2004). Characterization of cotton scouring using FT-IR ATR spectroscopy, *Carbohydr. Polymers*, **58**(4), 417–420.

Csiszar, E., Losonsczi, A., Szakacs, G., Rusznak, I., Bezur, L., and Reicher, J. (2001). Enzymes and chelating agents in cotton pretreatment, *J. Biotech.*, **89**(3), 271–279.

Csiszar, E., Szakacs, G., and Koczka, B. (2007). Biopreparation of cotton fabric with enzymes produced by solid state fermentation, *Enzyme Microb. Tech.*, **40**(7), 1765–1771.

Csiszar, E., Szakacs., G., and Rusznak, I. (1998). Combining traditional cotton scouring with cellulase enzymatic treatment, *Text. Res. J.*, **68**(3), 163–167.

Csiszar, E., Urbanszki, K., and Szakacs, G. (2001). Biotreatment of desized cotton fabric by commercial cellulase and xylanase enzymes, *J. Mol. Catal. B Enzymes*, **11**(4), 1065–1072.

Dahod, S. K. (1987). Improved process for conducting lipase enzyme hydrolysis. European Patent No. 257 716.

Dalvi, P., Anthappan, P., Darade, N., Kanoonge, N., and Adivarekar, R. (2007). Amylase and pectinase from single source for simultaneous desizing and scouring, *Indian J. Fibre Text. Res.*, **32**(4), 459–465.

Declerek, N., Machius, M., Joyet, P., Wiegand, G., Huber, R., and Gaillardin, C. (2003). Hyper thermostabilisation of *Bacillus licheniformis* a amylase and modulation of its stability over a 50°C temperature range, *Protein Eng.*, **16**(4), 287–293.

Degari, O., Gepstein, S., and Dosoretz, G. (2002). Potential use of cutinase in enzymatic scouring of cotton fiber cuticle, *Appl. Biochem. Biotech.*, **102**(1), 277–289.

Diller, G. B., Walsh, W. K., and Radhakrishnaiah, P. (1999). Effect of enzymatic treatment on dyeing and finishing of cellulose fibres: A study of the basic mechanisms and optimization of the process. National Textile Center, Annual Report No. **C96**, 1–9.

Eisisi, F. F., Hafiz, S. A. A., Rafie, M. H., and Hebeish, A. (1990). Development of a one-step process for desizing, scouring, bleaching cotton based textiles, *Am. Dyestuff Reporter.*, **79**(10), 39–43.

Ellaiah, P., Adinarayana, K., Pardhasaradhi, S. V., Srinivasulu, B. (2002). Isolation of alkaline protease producing bacteria from Visakhapatnam soil, *Indust, J. Microbiol.*, **42**(3), 173–175.

El-Shafei, H. A., and Rezkallah, L. A. (1997). Production, purification and characterization of Bacillus lipase, *Microbiol. Resource*, **152**(2), 199–208.

Etters, J. N. (1999). Cotton preparation with alkaline pectinase: An environmental advance, *Text. Chem. Colorists Am. Dyestuff Reporter*, **1**(3), 33–36.

Etters, J. N., and Anis, P. A. (1998). Textile enzyme use: A developing technology, *Am. Dyestuff Reporter*, **77**(5), 18–23.

Feitkenhauer, H., and Meyer, U. (2003). Anaerobic microbial cultures in cotton desizing—Efficient combination of fabric and waste water, *Text. Res. J.*, **73**(2), 93–97.

Friend, D. R., and Chang, G. W. (1982). Simple dye release for determining endopectinase activity, *J. Agric. Food Chem.*, **30**(5), 982–985.

Gambler, G. R. (2003). Effects of elevated temperatures on chemical properties on cotton fibre pectin, *Text. Res. J.*, **73**(2), 157–160.

Ghanem, E. H., Al Sayed, H. A., and Saleh, K. M. (2000). An alkalophilic thermostable lipase produced by a new isolate of Bacillus alkalophilus, *World J. Microbiol. Biotech.*, **16**(4), 459–464.

Gubitz, G. M., and Cavaco-Paulo, A. (2001). Biotechnology in the textile industry: Perspectives for the new millennium. *J. Biotech.*, **89**(2), 91–94.

Guha, S. B., and Shah, S. R. (2001). Enzymatic scouring of cotton fabrics, *J. Text. Assoc.*, **61**(2), 215–218.

Gupta, R., Beg, Q. K., and Lorenz, P. (2002). Bacterial alkaline proteases: Molecular approaches and industrial applications, *Appl. Microbiol. Biotech.*, **59**(1), 15–32.

Hahn, W., Axel, S., Riegels, M., Koch, R., and Pirkotsch, M. (1998). Enzyme mixtures and processes for desizing textiles sized with starch. US Patent No. 5 769 900.

Hardin, I. R., and Kim, J. (2000). The cuticular structure of cotton: What does enzymatic scouring do?, *Colourage Ann.*, **47**(12), 43–48.

Hartzell, M., and Hsieh, Y. L. (1998). Enzymatic scouring to improve cotton fabric wettability, *Text. Res. J.*, **68**(4), 233–241.

Heikinheimo, L., Oinonen, A. M., Paulo, A. C., and Buchert J. (2003). Effect of purified *Trichdenna reesei* cellulases on formation of cotton powder from cotton fabric, *J. Appl. Poly. Sci.*, **90**(7), 1917–1922.

Hemmpel, W. H. (1991). The surface modification of woven and knitted cellulosic fibre fabrics by enzymatic degradation, *Int. Text. Bull.—Dye./Printing/Finishing.*, **3**(1), 5–14.

Holme, L. (2001). Biopreparation: The eco-friendly alternative, *Int. Dyer*, **2**, 8–9.

Hsieh, Y. L., and Cram, L. (1999). Protease as scouring agents for cotton, *Text. Res. J.*, **69**(8), 590–597.

Ibrahim, N. A., El-Hossamy, M., Morsy, M. S., and Basma, M. E. (2004). Optimization and modification of enzymatic desizing of starch-size, *Polymer Plast. Tech. Eng.*, **43**(2), 519–538.

Jayam, R. S., Saxena, S., and Gupta, R. (2005). Microbial pectinolytic enzymes: A review, *Process Biochem.*, **40**(9), 2931–2944.

Kamel, M. M., El-Shishtawy, R. M., Yussef, B. M., and Mashaly, H. (2005). Ultrasound in textile dyeing and the decolouration/mineralization of textile dyes, *Dyes Pigments*, **65**(2), 103–110.

Karapinar, E., and Sariisik, M. O. (2004). Scouring of cotton with cellulase, Pectinases and proteases, *Fibres Text. East. Eur.*, **12**, 79–84.

Karapinar, E., and Sariisik, M. O. (2004). Scouring of cotton with cellulases, pectinases and proteases, *Fibres Text. East. Eur.*, **12**(3), 79–82.

Karmakar, S. R. (1998). Application of biotechnology in the pretreatment process of textiles, *Colourage Ann.*, **45**(12), 75–86.

Kashyap, D. R., Chandra, S., Kaul, A., and Tewari, R. (2000). Production, purification and characterisation of pectinase from a *Bacillus sp DTT, World J. Microbiol. Biotechnol.*, **16**(3), 277–282.

Kim, J., Choe, E. K., Kim, S. Y., and Nam, S. W. (2005). Optimization of enzymatic scouring, *J. Nat. Fibres*, **3**(2), 155–168.

Kristensen, H. (2001). Enzymatic treatment method. US Patent No. 6296671.

Lange, N. K. (1997). Lipase assisted desizing of woven cotton fabrics, *AATCC Rev.*, **29**(6), 23–26.

Lawson, M. M. H., and Hsieh, Y. L. (2000). Characterizing the non cellulosics in developing cotton fibers, *Text. Res. J.*, **70**(9), 810–819.

Lee, L., Evans, B. R., and Woodward, J. (2000). The mechanism of cellulase action on cotton fibres: Evidence from atomic force microscopy, *Ultra microsc.* **82**(1), 213–221.

Lee, L., Evans, B. R., Lane, L. M., and Woodward, J. (1996). Substrate enzyme interactions in cellulase systems, *Bioresource Tech.*, **58**(22), 163–169.

Lenting, H. B. M., and Warmoeskerken, M. M. C. G. (2004). A fast, continuous enzyme based pretreatment process concept for cotton containing textiles, *Biotransformation*, **22**(5), 361–368.

Lenting, H. B. M., Zwier, E., and Nierstrasz, V. A. (2002). Identifying important parameters for a continuous bioscouring process, *Text. Res. J.*, **72**(9), 825–831.

Li, Y., and Hardin, I. R. (1998). Enzymatic scouring of cotton surfactants, agitation, and selection of enzymes, *Text. Chem. Colorists,* **30**(9), 23–29.

Li, Y., and Hardin, I. R. (1999). Enzymatic scouring of cotton: Effects on structure and properties, *Text. Chem. Colorists,* **29**(8), 71–76.

Liu, J., Condon, B., and Showmaker, H. L. (2000). Single bath biopreparation and dyeing of textiles. US Patent No. 6 162260.

Maekawa, M., Hashimoto, A., and Tahara, M. (2007). Effects of pH in hydrogen peroxide bleaching of cotton fabrics pretreated with ferrous sulphate., *Text. Res. J.*, **77**(4), 222–226.

Maldonado, M. C., and Saad, A. M. S. (2000). Production of pectin esterase and polygalacturonase by *Aspergillus niger* submerged and solid state systems, *J. Indust. Microbiol. Biotech.*, **20**(1), 34–38.

Martin, N., De Souza, S. R., Silva, R., and Gomes, E. (2004). Pectinase production by fungal strains in solid state fermentation using agro industrial byproduct, *Braz. Arch. Biol. Tech.*, **47**(5), 813–819.

Menezes, E., and Chaudhari, M. (2005). New approach to textile bleaching, *Colourage*, **52**(1), 39–42.

Miller, C. A., Jorgensen, S. S., Otto, E. W., Lange, N. E. K., Condon, B., and Liu, J. (2003). Alkaline enzyme scouring of cotton text. US Patent No. 6 551 358.

Moghe, V. V., and Khera, J. (2005). Desizing parameters, *Colourage*, **52**(7), 85–87.

Mori, T., Sakimoto, M., Kagi., T., and Saki, T. (1999). Enzymatic desizing of polyvinyl alcohol from cotton fabrics, *J. Chem. Tech. Biotech.*, **68**(2), 151–156.

Mozhaev, V. V., Lange., R., Kudryashova., E. V., and Balny, C. (1996). Application of high hydrostatic pressure for increasing activity and stability of enzymes, *Biotechnol. Bioeng.*, **52**(2), 320–331.

Naik, S. R., and Paul, R. (1997). Application of enzymes in textile processing, *Asian Text. J.*, **6**(2), 48–55.

Najafi, M. F., Deabagkar, D., and Deepti, D. (2005). Potential application of protease isolated from *Psedomonas aeruginosa* PD 100, *Electron. J. Biotech.*, **8**(2), 197–203.

Nalankilli, G., and Sundar, A. E. (2002). Biobleaching, *Asian Text. J.*, **2**(7), 33–34.

Nalankilli, G., Saravanan, D., Govindaraj, N., Harish, P., and Ramachandran, T. (2008). Efficacy of solvent, alkali and pectinase on removal of non-cellulosics from cotton fibres, *Indian J. Fibre Text. Res.*, **33**(4), 438–442.

Nallankilli, G. (1992). Enzymes in Textile wet processing, *Text Indust. Trade J.*, **30**, 51–55.

Obendorf, S. K., Varanashi, A., Mejldal, R., and Nielsen, V. S. (2003). Lipid distribution of cotton textiles in relation to washing with cellulase, *J. Surfactants Detergents*, **6**(1), 1–5.

Opwis, K., Knittel, D., Doffler, C., and Koppe, A. (2006). Combined use of enzymes in the pretreatment of cotton, *Melliand Int.*, **12**(2), 130–136.

Opwis, K., Knittel, D., Kele, A., and Schollmeyer, E. (2000). Enzymatic recycling of starch containing desizing liquor, *Starch*, **51**(10), 348–353.

Patra, A. K., Agrawal, G., and Ray, S. (2004). Enzymatic scouring and bleaching compatibility of cotton, *Indian Text. J.*, **114**(4), 25–27.

Paulo, A. C. (1998). Mechanism of cellulase action in textile processes, *Carbohydr. Polymers*, **37**(3), 273–277.

Paulo, A. C., and Almeida, L. (1994). Cellulase hydrolysis of cotton cellulose: The effects of mechanical action, enzyme concentration and dyed substrates, *Biocatalysis*, **10**, 353–360.

Paulo, A. C., and Almeida, L. (1996). Kinetic parameters measured during cellulase processing of cotton, *J. Text. Inst.*, **1**(4), 227–233.

Pendersen, G. L., Screws, G. A., and Cedroni, D. M. (1992). Biopolishing of cellulosic fabrics, *Canadian Text. J.*, **109**(12), 31–35.

Pera, L. M., Romero, C. M., Baigori, M. D., and Castro, G. R. (2006). Catalytic properties of lipase extracts from *Aspergillus niger, Food Technol. Biotech.*, **44**(2), 247–252.

Pererra, S. S., Torres, E. F., Gonzalez, G. V., and Rojas, M. G. (1993). Effects of different carbon sources on the synthesis of pectinase by *Aspergillus niger* in submerged and solid stage fermentation, *Appl. Microbiol. Biotech.*, **39**(1), 36–41.

Pinheiro, R., Belo, I., and Mota, M. (2000). Air pressure effects on biomass yield of two different Kluyveromyces strains, *Enzyme Microb. Tech.*, **26**(9), 756–762.

Potikha, T. S., Collins, C. C., Johnson, D. L., and Delmer, D. P. (1999). The Involvement of hydrogen peroxide in the differentiation of secondary walls in cotton fibres, *Plant Physiol.*, **119**(3), 849–858.

Presa, P., and Forte Tavcer, P. (2007). Pectinases as agents for bioscouring, *Tekstilec*, **50**, 16–34.

Pyc, R., Ledakowics, J. S., and Bratkowska, H. (2003). Biosynthesis of enzymes by *Aspergillus niger* IBT-90 and an evaluation of their application in textile technologies, *Fibres Text. East. Eur.*, **11**(4), 71–77.

Ramkumar, S. S., and Abdalah, G. (2001). Surface characterization of enzyme treated fabrics, *Colourage*, **48**(4), 15–16.

Ray, S. (2006). A statistical tool for process optimization, *Indian Text. J.*, **116**(12), 24–30.

Reshmi, R., Sanjay, G., and Sugunan, S. (2006). Enhanced activity and stability of a amylase immobilized on alumina, *Catal. Commun.*, **7**(7), 460–465.

Sae, B. P., Sangwatanaroj, U., and Punnapayak, H. (2007). Analysis of the products from enzymatic scouring of cotton, *Biotechnol. J.*, **2**(3), 316–325.

Sahin, U. K., and Gursoy, N. C. (2005). Low temperature acidic pectinase scouring for enhancing textile quality, *AATCC Rev.*, **5**(1), 27–30.

Sammour, R. H. (2005). Purification and Pal1ial characterisation of an acid lipase in germinating lipidobody linseedlings, *Turkish J. Botany*, **29**(3), 177–184.

Schnitzhofer, W., Kandelbauer, A., Klug-Santner, B., Onos, M., Calafell, M., and Glibitz, G. M. (2006). Production of cellulase-free polygalacturonase preparations by sclerotium rolfsii for bioscouring of cotton, *J. Nat. Fibres*, **3**(2), 19–38.

Shamey, R., and Hussein, T. (2005). Critical solutions in the dyeing of cotton textile materials, *Text. Prog.*, **37**(1), 15–17.

Sharma, A., and Gupta, M. N. (2001). Purification of pectinase by three phase partitioning, *Biotechnol. Lett.*, **23**(19), 1625–1627.

Shewale, S. D., and Pandit, A. B. (2007). Hydrolysis of soluble starch using bacillus lichenifonnis a amylases immobilized on super porous celbeads, *Carbohydr. Res.*, **342**(8), 997–1008.

Shukla, S. R., and Jaipura, L. (2004). Estimating amylase activity for desizing by DNSA, *Text Asia*, **35**(11), 15–20.

Silva, C. J. S. M., Gubitz, G., and Paulo, A. C. (2006). Optimization of a serine protease coupling to *Eudrajit* S 100 by experimental design techniques, *J. Chem. Tech. Biotech.*, **81**(1), 8–16.

Solbak, A. I., Richardson, T. H., Mc Cann, R. T., Kline, K. A., Bartnek, F., Tomlinson, G., Tan, X., Gessert, T. P., Frey, G. J., Podar, M., Luginbuhl, P., Gray, K. A., Mathur, E. J., Robertson, D. E., Burk, M. J., Hazlewood, G. P., Short, J. M., and Kerovuo, J. (2005). Discovery of pectin degrading enzymes and directed evolution of a novel pectate lyase for processing cotton fabrics, *J. Biol. Chem.*, **280**(10), 9431–9438.

Tavcer, P. F., Krizman, P., and Presa, P. (2005). Combined bioscouring and bleaching of cotton fibres, *J. Nat. Fibres,* **3**(2), 83–97.

Traore, M. K., and Buschle-Diller, G. (2000). Environmentally friendly scouring process, *Text. Chem. Coloration,* **32**, 40–43.

Traore, M. K., and Diller, G. B. (1999). Influence of wetting agent and agitation on enzymatic hydrolysis of cotton, *Text. Chem. Colorists Am. Dyestuff Reporter,* **1**(4), 21–56.

Tyndall, R. M. (1996). Improving the softness and surface appearance of cotton fabrics and garments by treatment with cellulase enzymes, *Text. Chem. Colorists,* **24**(6), 23–26.

Tzanko, T., Calafell, T. M., Guebitz, G. M., and Cavaco-Paulo, A. (2001). Bio-preparation of cotton fabrics, *Enzyme Microb. Technol.,* **29**(6), 357–362.

Tzanko, T., Costa, S., Calafell, M., Guebitz, G. M., and Paulo, A. C. (2000). Enzymes for cotton fabrics preparation and recycling of waste water for dyeing, *Colourage Ann.,* **47**(12), 65–72.

Varanasi, A., Obendorf, S. K., Pedersen, L. S., and Mejldal, R. (2001). Lipid distribution on textiles in relation to washing with lipases, *J. Surfactants Detergents,* **4**(2): 135–146.

Vigneswaran, C., Keerthivasan, D. (2008). Bio-processing of cotton fabrics with commercial enzymes, *Melliand Int.,* **14**(5), 308–311.

Vigneswaran, C., Ananthasubramanian, M., and Anbumani, N. (2011). Biovision in textile wet processing industry: Technological challenges, *J. Text. Apparel Tech. Manag.,* **7**(1), 1–12.

Vigneswaran, C., Ananthasubramanian, M., Anbumani, N., and Rajendran, R. (2012a). Prediction of optimum process parameter to achieve eco-friendly desizing of organic cotton fabrics with indigenously produced alpha-amylase and their enzyme kinetics, *J. Text. Inst.,* **103**(4), 422–433.

Vigneswaran, C., Ananthasubramanian, M., and Anbumani, N. (2012b). Prediction and process optimization of bioscouring of organic cotton fabrics through specific mixed enzymatic system using artificial neural network (ANN), *J. Nat. Fibres,* **9**(1), 1–22.

Vigneswaran, C., Anbumani, N., Ananthasubramanian, M., and Rajendran, R. (2012c). Prediction and process optimization of pectinolytic reaction on organic cotton fabrics for bioscouring with alkaline pectinase, *Indian J. Fibre Text. Res.,* **37**(2), 183–190.

Vigneswaran, C., Ananthasubramanian, M., and Anbumani, N. (2012d). Neural network approach for optimizing the bioscouring performance of organic cotton fabric through aerodynamic system, *J. Text. Apparel Tech. Manag.*, **7**(3), 1–14.

Wan, Q., Fan, X., Hua, Z., Gao, W., and Chen, J. (2007). Degradation kinetics of pectins by an alkaline pectinase in bioscouring of cotton fabrics, *Carbohydr. Polymers*, **67**(4), 572–575.

Wang, Q., Fan., X., Gao., W., and Chen, J. (2006). Characterization of bioscoured cotton fabrics using FT-IR ATR spectroscopy and microscopy techniques, *Carbohydr. Res.*, **341**(12), 2170–2175.

Warke, V. V., and Chandratre, P. R. (2003). Application of biotechnology in textiles, *Man-Made Text. India*, **26**(4), 142–146.

Xia, Y.-J., and Li, Z.-Y. (2009). Effects of high pressure treatment on peroxidase (POD) activity in orange juice, *J. Nanjing Univ. Tech.* (*Natural Science Edition*), **5**, 124–130.

Yachmenev, V. G. (2005). The effects of ultrasound on the performance of industrial enzymes used in cotton biopreparation/biofinishing applications, *J. Nat. Fibres*, **3**(2), 99–112.

Yonghua, L., and Hardin, I. R. (1997). Enzymatic scouring of cotton: Effects on structure and properties, *Text. Chem. Colorist.*, **29**(8), 71–76.

Chapter 5

Biodegradable Polymers: Definition, Classification and Application as an Alternative to Plastic

Neha Patni and Shibu G. Pillai

Chemical Engineering Department,
Institute of Technology, Nirma University, India

neha.patni@nirmauni.ac.in

5.1 Introduction

The term "plastic" is derived from the Greek word "plastikos" meaning fit for molding, and "plastos" meaning molded (Kent and Riegel, 2007). It refers to the material's malleability, or plasticity, during manufacture that allows it to be cast, pressed, or extruded into a variety of shapes such as films, fibers, plates, tubes, bottles, boxes, and much more. Plastics have versatile qualities of strength, lightness, durability, and resistance to degradation. They have become an important commodity to enhance the comfort and quality of life. They are an essential part of almost all industries and have replaced glass and paper in packaging, but these very desirable properties have now become their greatest problem. Solutions to plastic waste management include source reduction,

Handbook of Sustainable Polymers: Processing and Applications
Edited by Vijay Kumar Thakur and Manju Kumari Thakur

ISBN 978-981-4613-53-8 (Hardcover), 978-981-4613-54-5 (eBook)
www.panstanford.com

incineration, recycling, and bio- or photo-degradation. However, most of these have problems associated with them. Incineration of plastics is potentially dangerous (it yields toxic emissions e.g., dioxin). and can be expensive. Also, during the combustion of plastic waste, hydrogen cyanide can be formed from acrylonitrile-based plastics and may cause potential health hazards. Recycling can be done but is very tedious. The sorting of the wide variety of discarded plastic material is also a very time-consuming process. Moreover, the presence of a wide variety of additives such as pigments, coatings, and fillers limits the use of the recycled material. Plastic recycling, though a necessity, yet shows a negative eco-balance, in nearly all cases, to wash the plastic waste as well as the energy consumption during the recycling process phases (waste grinding and plastic processing).

Nowadays a great amount of awareness is created in the society encouraging the green consumerism as an initial approach to curb the increasing amount of pollution. Rapid progress in materials, science, and technology has created new plastic products with favorable mechanical integrity and excellent durability. The total global capacity of commodity plastic production dramatically increased from 1.5 million tons in 1950 to 245 million tons in 2008, an annual growth rate of 9%. The biggest potential growth area is in the rapidly developing Asian countries. It is very difficult to reduce the consumption of plastic products due to their versatile properties and it is required to replace petroleum-based plastics with alternative materials that have polymer-like properties and that degrade after being discarded. In such a scenario, biodegradable plastics offer the best solution to the environmental hazard posed by conventional plastics. Biodegradable plastics of renewable resources origin also help to preserve the non-renewable resources and contribute to sustainable development. Biopolymers are renewable, largely biodegradable and can have very similar properties to conventional petroleum-based polymers.

Plastic materials are currently considered very important owing to their exceptional properties and performance over other materials such as metal and wood. It is projected that the demand for plastics will continue to rise following a trend that has increased since the 1950s (Azapagic et al., 2003; Rosato et al., 2003), as almost every product we purchase come encased in plastic.

Plastic packaging provides excellent protection for the product, it is cheap to manufacture and seems to last forever. Non-biodegradability is a major environmental problem. Plastics have very high volume-to-weight ratio and since those are non-biodegradable, their frequent use and consumption causes "White Pollution." Accumulated plastic film residues in soil have caused significant decrease in fertility of the soil and hence affect the overall quality of soil.

More than 99% of plastics are of fossil fuel origin, and their excess use will put pressure on the already limited natural oil reserves (Ren, 2003). Bio-based plastics are "plastics in which 100% of the carbon is derived from renewable agricultural and forestry resources such as corn starch, soybean protein and cellulose." In reality, most bio-based plastics on the market today are blends of bio-based and petroleum-based materials. The best examples of biopolymers based on renewable resources are cellulosic plastics, polylactides (PLA), starch plastics, and soybased plastics (Mohanty et al., 2000). Originally, biopolymers were intended to be used in packaging industries, farming, and other applications with minor strength requirements. Performance limitations and high cost are the major barriers to widespread acceptance of biopolymers as substitutes for traditional non-biodegradable polymers (Wang et al., 2008).

5.2 Plastic Spectrum

Over the past decade, researchers and citizen advocates have developed several tools to assist in decision making about plastics selection. The plastics pyramid (Fig. 5.1) developed by Thorpe and Van der Naalde in 1998 was an early attempt to visually display the life cycle hazards of different plastics to assist in materials selection. This ranking focused on the toxicity of the material, considering production hazards, use of harmful additives, hazards in use, and disposal hazards. In this pyramid, bio-based polymers form the bottom of the pyramid, indicating they are most preferable, as they are made from renewable resources, and theoretically are biodegradable and compostable (Rosalía et al., 2012).

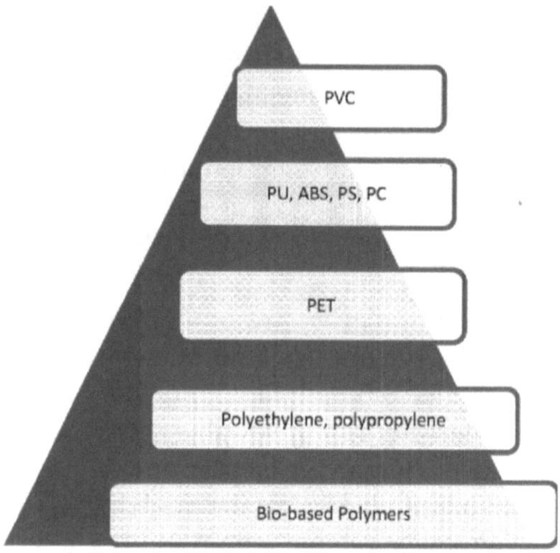

Figure 5.1 The plastics pyramid. Reproduced with permission from Rosalía et al. (2012).

Since the Plastics Pyramid (Fig. 5.1) was developed, bio-based plastics are much further along in their commercial development. These materials need to be evaluated carefully for sustainability. Later on the Plastic Spectrum (Table 5.1) and Plastics Scorecard were also made in order to rate different types of plastics based on their life cycle impacts and hazards to human health and environment.

Table 5.1 Preference spectrum of plastics

Avoid	→				Prefer
Polyvinyl chloride	Plastics with highly hazardous additives	Acrylonitrile butadiene Styrene, ethylene vinyl acetate, polycarbonate, polystyrene	Polyethylene cross-linked; polyethylene terepthalate	Polyethylene polypropylene thermoplastic polyolefin	Bio-based plastics sustainably grown

Development and commercialization of bio-based plastics for a variety of uses in products and packaging is also of great interest as manufacturers are looking for safer and healthier materials as substitutes for chemicals and materials of concern in consumer products and their production process may be more energy efficient than petroleum-based plastics processing (Rosalía et al., 2012). Sustainable materials are those that reduce the impact on occupational and public health as well as to the environment throughout their life cycles (Geiser, 2001).

5.3 Biodegradable Polymer

Definition: Degradation of polymer is the process of polymer chain scission by the cleavage of bonds between the monomers in the polymer backbone. Accordingly, degradation leads to a size reduction of the polymer chain. Degradation of the degradable polymers is measured in terms of duration of the application or in relation to our human lifetime rather than on absolute time. If a polymer degrades not within the human lifetime it is usually not considered degradable. A distinction between degradable and nondegradable polymers could be made by defining a Deborah number. Debohrah numbers are dimensionless. A Deborah number to distinguish degradable from non-degradable polymers could be defined as shown in Eq. 5.1.

$$D = \frac{\text{Time of degradation}}{\text{Human lifetime}} \tag{5.1}$$

Degradable polymers would then have small values for D ($D \rightarrow 0$) and non-degradable polymers have large ones ($D \rightarrow \infty$).

Biodegradation is the degradation of an organic material caused by biological activity (biotic degradation), mainly microorganisms' enzymatic action. Bacteria and fungi are the main participants in the process of biodegradation in the natural world. The end-products are new biomass, carbon dioxide, and water (aerobic conditions) or methane (anaerobic conditions), as defined in the European Standard EN 13432-2000. Depending on the type of standard to follow (ASTM or EN), different composting conditions (humidity and temperature cycle) must be realized to determine the compostability level (Steinbuchel, 2003).

$$\text{Polymer} + O_2 \rightarrow CO_2 + H_2O \tag{5.2}$$

According to ASTM standard D-5488-94d and European norm EN 13432, "biodegradable" means "capable of undergoing decomposition into carbon dioxide, methane, water, inorganic compounds, and biomass." The predominant mechanism is the enzymatic action of microorganisms, which can be measured by standard tests over a specific period of time, reflecting available disposal conditions. There are different media (liquid, inert, or compost medium) to analyze biodegradability. Material biodegradability using compost medium is known as compostability.

Biodegradation is thus nothing more than catabolism. With the possible exception the biological world readily degrades what it builds. Some synthetic polymers can be microbially degraded (Trozzolo, 1972; Roylance and Roylance, 2013), but the process is normally slow. Most have chemical compositions resisting enzymatic attack. This is not the case for polymers of biological origin, many of which have thermoplastic properties. But it is not only biodegradability that makes these polymers so fascinating; it is also their synthesis from renewable carbon sources, based on agriculture or even on industrial wastes, allowing a sustainable closed cycle process for the production and use of such polymers instead of the end-of-the-pipe technologies connected to production and use of conventional plastics.

Plastics are resistant against microbial attack, since during their short time of presence in nature, could not design new enzyme structures capable of degrading synthetic polymers. Many polymers that are claimed to be "biodegradable" are in fact "bio-erodable," "hydro-biodegradable," or "photo-biodegradable." These different polymer classes all come under the broader category of "environmentally degradable polymers." The "biodegradability" of plastics is dependent on the chemical structure of the material and on the constitution of the final product, not just on the raw materials used for its production. Hence, the polymer that undergoes biodegradation, as per the standards given, is known as biodegradable polymer.

Biodegradable plastics can be based on natural or synthetic resins. Natural biodegradable plastics are based primarily on renewable resources (such as starch) and can be either naturally produced or synthesized from renewable resources. They are coming under polysaccharides (starch, cellulose, lignin, etc.), proteins (gelatine, wool, silk, etc.), lipid (fats and oil), polyesters produced by plant or microorganisms (PHA), polyesters derived from bio-derived monomers (polylactic acid) and miscellaneous polymers like natural rubbers, composites, etc. Non-renewable synthetic biodegradable plastics are petroleum-based. Many natural biodegradable plastics are blended with synthetic polymers to produce plastics that meet the functional requirements.

5.4 Classification of Plastics

Similarly, there are three primary classes of polymer materials usually referred to in the general class of plastics by consumers and industry. Their design is often that of a composite, where a polymer matrix (plastic material) forms a dominant phase around a filler material. The filler is present in order to increase mechanical properties, and decrease material costs.

(1) Conventional plastics: They are resistant to biodegradation, as the surfaces in contact with the soil in which they are disposed are characteristically smooth. Microorganisms within the soil are unable to consume a portion of the plastic, which would, in turn, cause a more rapid breakdown of the supporting matrix. This group of materials usually has an impenetrable petroleum-based matrix, which is reinforced with carbon or glass fibers (Katarzyna et al., 2010).

(2) Partially degradable: They are designed with the goal of more rapid degradation than that of conventional synthetic plastics. Production of this class of materials typically includes surrounding naturally produced fibers with a conventional (petroleum based) matrix. When disposed of, microorganisms are able to consume the natural macromolecules within the plastic matrix. This leaves a weakened material, with rough, open edges. Further degradation may then occur (Katarzyna et al., 2010).

(3) Completely biodegradable: Here the polymer matrix is derived from natural sources (such as starch or microbially grown polymers), and the fiber reinforcements are produced from common crops such as flax or hemp. Microorganisms are able to consume these materials in their entirety, eventually leaving carbon dioxide and water as by-products (Katarzyna et al., 2010).

5.5 Classification of Polymers

5.5.1 On the Basis of Degradability

Biodegradable materials are generally of three types (Fig. 5.2):

(1) mixtures of synthetic polymers and substances that are easy digestible by microorganisms (inter alia, modified natural polymers, natural polymers such as starch and cellulose),

(2) synthetic materials with groups that are susceptible to hydrolytic microbial attack (for example, polycaprolactone), and

(3) the biopolyesters from bacterial sources. The most popular and important biodegradable polymers are aliphatic polyesters (e.g., polylactide, poly(ε-caprolactone) polyethylene oxide, poly(3-hydroxybutyrate), polyglycolic acid) and thermoplastic proteins (Liao, 2009; Katarzyna et al., 2010).

Agro-polymers		Bio-polyesters		
From Agro-resources (biomass)		From microorganisms (obtained by extraction)	From biotechnology (conventional synthesis from bio-monomers)	From petroleum-based Products
Polysaccharides	Proteins	PHA (polyhydroxy-alkanoates)	PLA (polylactic acid)	PCL (polycaprolactone)
Starches: wheat, potato, corn, etc.	Animal proteins: casein, gelatin, collagen, etc.			Polyesters
Cellulose: fibers	Plant proteins: gluten, corn, soya, etc.			
Others: chitosan/chitin/pectin, etc.				

Figure 5.2 Classification of biodegradable polymers, also on the basis of their origin (Avérous and Pollet, 2012).

5.5.2 From BioMass

5.5.2.1 From polysaccharides

Polysaccharides are the most abundant macromolecules in the biosphere. These complex carbohydrates constituted of glycosidic bonds are often one of the main structural elements of plants and animals exoskeleton (cellulose, carrageenan chitin). The polysaccharides covered in this chapter are cellulose, starch, chitin, chitosan, and pectins.

5.5.2.2 Cellulose

Cellulose is the most abundant of naturally occurring polymers comprising at least one-third of the vegetable matter in the world, which is the highly polar, main structural component of flax and hemp fibers. Natural cellulose fibers are low cost and biodegradable and have strong mechanical properties. Cellulosic materials are commonly used in textiles, fibers and packaging and are subdivided into two groups: regenerated and modified cellulose. Cellulose is readily biodegradable and is mineralized by variety of microorganisms (Chauhan, 2012).

These characteristics make cellulose fibers the most common choice for natural fillers in plastic materials. It was concluded that the presence of 25% w/w of cellulose fibers in a polypropylene matrix causes a significant increase in tensile modulus. Cellulose has a very long molecular chain, which is infusible and insoluble in all but the most aggressive solvents. Therefore, it is most often converted into derivatives to increase solubility, which further increases adhesion within the matrix (Kolybaba et al., 2001).

5.5.2.3 Starch

Starch is an agricultural feedstock hydrocolloid biopolymer found in a variety of plants and is mainly extracted from cereals (wheat, corn, rice, etc.) and from tubers (potatoes, manioc, etc.). It is usually utilized in the form of granules, which are composed of two glucopyranose homopolymers, the amylose and the amylopectin. Amylose, the linear polymer, comprises approximately 20% w/w of starch, whereas amylopectin, the branched polymer, constitutes the remainder. Depending on the botanical origin of the plant, starch granules can have different shapes (sphere, platelet, and polygon) and size (from 0.5 to 175 lm). Their proportions in the granules depend directly on the botanical source (Guilbot and Mercier, 1985). Often, granular starch is added to polyethylenes in order to increase the degradation rate of the plastic material (Kolybaba et al., 2001).

Natural filler materials may be incorporated into synthetic plastic matrices as a rapidly biodegradable component. Heating the starch in the presence of water during extrusion or injection molding causes the formation of a thermoplastic material that may be deformed during blending. This starch-based product is then blended with either natural or synthetic materials. Glycerol is often used as a plasticizer in starch blends, to increase softness and pliability. Starch granules that have been plasticized with water and glycerol are referred to as plasticized starches. These materials are widely used in polymer blends and are available in commercial grades suitable for extrusion, injection/blow molding and thermoforming processes (Chauhan, 2012).

5.5.2.4 Chitin and chitosan

Chitin is the second most abundant agro-polymer produced in nature after cellulose. It appears in nature as ordered crystalline microfibrils forming structural components in the exoskeleton of arthropods or in the cell walls of fungi and yeasts (Rinaudo, 2006). It is an acetylated polysaccharide composed of N-acetyl-D-glucosamine.

Chitosan is obtained from the deacetylation of chitin, which is found in marine environments. Because it is insoluble in water, chitosan is dissolved in acidic solutions before being incorporated into biodegradable polymer films. It can also be plasticized with glycerol to obtain a kind of thermoplastic material like, for instance, plasticized starch (Epure, 2011).

5.5.2.5 Pectin

Pectin is a linear macromolecule composed of D-galacturonic acid. This monomer unit could be partially replaced by L-rhamnose leading to a new structure named rhamnogalacturonan I. A third pectin structural type is rhamnogalacturonan II, which is a less frequent, but complex and highly branched polysaccharide (Thakur, 1997; Avérous and Pollet, 2012).

5.5.2.6 Proteins

Proteins are agro-polymers. They are an important renewable resources produced by animals, plants, and bacteria. The term "protein" comes from the Greek, proteios, for "primary, first and foremost." A certain number of proteins have received much attention as biodegradable polymers but few have led to actual industrial scale-up due to the high production cost and the low product performance. In terms of potential sources, soy protein, corn protein (zein) and wheat proteins (gluten) are among the main plant proteins. Casein, collagen protein or gelatin, and keratin are important animal proteins. Proteins can be processed with the same kinds of approaches developed for starch-based materials, e.g., formulation with plasticizers, cross-linkers.

5.5.2.7 Soybean proteins

Typically soybean proteins contain 38–42% crude protein, 16–20% triglycerides, and around 33% carbohydrates, on dry basis. Soy proteins are the isolated proteins from soybean. It is made of dehulled, defatted soybean meal. The structural characteristics of soy proteins give them potential for industrial applications in plastics and reinforced composite materials.

5.5.2.8 Corn proteins

Proteins form about 9% of the dry weight of corns. They are mainly composed of zein (a highly hydrophobic protein, soluble in alcohols), glutelin (soluble in aqueous alkaline solutions), albumins, and globulins. Zein comes from the alcohol soluble protein of corn, classified as a prolamin. It is the principle storage protein of corn and constituting 45–80% of endosperm protein. Zein is a mixture of proteins varying in molecular weight and solubility, which can be separated by differential solubility to give four related zeins named a, b, c and d (Zhang and Zeng, 2008; Avérous and Pollet, 2012).

5.5.2.9 Wheat gluten and casein

Wheat gluten and casein are covered in detail later in the chapter.

5.5.2.10 Gelatin

Gelatin is a natural material based on animal proteins. It is derived from collagen, which is elongated fibrils and mostly found in fibrous tissues such as tendon, ligament, and skin. It is commonly used for biomedical applications due to its biodegradability and biocompatibility in physiological environments, in contact with living tissues (Ofokansi et al., 2010).

5.5.3 Polymers Obtained by Microbial Production

5.5.3.1 Polyhydroxyalkanoates (bacterial polyesters)

Polyhydroxyalkanoates (PHAs) are a family of intracellular biopolymers, a class of bacterial polyesters, and are synthesized by a variety of naturally occurring bacteria by the process of fermentation of sugars and lipids. They are products of carbon assimilation employed by microorganisms as a form of energy storage molecule to be metabolized when other common energy sources are not available. A wide variety of prokaryotic organisms accumulate PHA from 30 to 80% of their cellular dry weight.

Polyhydroxybutyrate (PHB) is the most common type of polymer that falls in the category of polyhydroxyalkanoates. PHAs are generally classified into short-chain-length PHA (sCL-PHA) and medium-chain-length PHA (mCL-PHA) by the different number of carbons in their repeating units.

Several types of bacterial polyesters that are produced by biosynthesis are poly-3-hydroxybutyrate, poly-4-hydroxybutyrate, poly-3-hydroxyvalerate, poly-3-hydroxyhexanoate, poly-3-hydroxyheptanoate, etc., and their respective copolymer combinations. Due to their ability to degrade naturally in variety of environments, they will find a lot of applications in disposal items, short-term packaging, and also considered biocompatible in contact with living tissues and can be used for biomedical applications (e.g., drug encapsulation, tissue engineering) (Chauhan, 2012).

5.5.4 Conventionally and Chemically Synthesized Biodegradable Polymers with Monomers Obtained from Agro-Resource

5.5.4.1 Polylactic acid

Polylactic acid (PLA) is the second common biopolymer that is produced by microbial fermentation. It is thermoplastic aliphatic polyester that can be synthesized from biologically produced lactic acid polymerized by ring opening polymerization. Lactic acid is a chiral molecule existing as two stereoisomers, L- and D-lactic acid, which can be produced by different ways, i.e., biologically or chemically synthesized (Averous, 2008).

Bacterial fermentation is used to produce lactic acid from corn starch or cane sugar which is further processed to produce lactide monomer. Because lactic acid is difficult to polymerize directly to high polymers in a single step on a commercial scale, most companies used a two-step process. Lactic acid is first oligomerized to a linear chain with a MW of less than 3,000 by removing water. The oligomer is then depolymerized to lactide, a cyclic dimer. This six-membered ring is purified and subjected to ring-opening polymerization to produce a PLA with a MW of greater than 50,000–110,000. The enantiomeric monomers (D and L) are polycondensed via its cyclic dimer (lactide) by ring-opening polymerization (ROP) to a high molecular weight polymer. Thus, this material can be made stereo specific or racemic to yield different properties. Compared to the other biodegradable polyesters, PLA is the product that at the present time has one of the highest potentials due to its availability in the market and its low price. PLA and its copolymers have been used widely in medicine and surgery for the controlled release of drugs, biodegradable surgical sutures and implants for fixation of fractures, primarily because of their high biocompatibility. It is considered to be fully biodegradable and generally undergoes hydrolytic degradation (Chauhan, 2012).

PHA and PLA are both considered synthetic polymers, as they are not found in nature. However, they are wholly biodegradable (Kolybaba et al., 2001).

5.5.4.2 Petroleum-based polyesters

A large number of biodegradable polyesters are based on petroleum resources, obtained chemically from synthetic monomers (Okada, 2002; Albertsson and Varma, 2002; Vert et al., 1995; Sinclair, 1996; Lunt, 1998; Steinbuchel, 2003; Bigg, 1996). According to the chemical structures, we can distinguish polycaprolactone, aliphatic copolyesters, and aromatic copolyesters. All these polyesters are soft at room temperature (Avérous and Pollet, 2012).

5.5.4.3 Polycaprolactone

It is usually obtained by ROP of e-caprolactone in the presence of metal alkoxides (aluminium isopropoxide, tin octoate) (Okada, 2002; Albertsson and Varma, 2002; Chiellini and Solaro, 1996). It is a semi-crystalline water-soluble synthetic polymer made from polyvinyl acetate. PCL is widely used as a PVC solid plasticizer or for polyurethane applications, as polyols. However, it finds also some applications based on its biodegradable character in domains such as biomedicine (e.g., drugs controlled release) and environment (e.g., soft compostable packaging).

5.5.4.4 Aliphatic copolyesters

A large number of aliphatic copolyesters based on petroleum resources are biodegradable copolymers. They are obtained by the combination of diols such as 1,2-ethanediol, 1,3-propanediol, and 1,4-butadenediol and of dicarboxylic acids such as adipic, sebacic, and succinic acid.

5.5.4.5 Aromatic copolyesters

Compared with totally aliphatic copolyesters, aromatic copolyesters are often based on terephthalic diacid. BASF markets a product, Ecoflex, which is a copolyester of butanediol, adipic acid, and dimethyl terephthalate. Eastman's Eastar Bio Copolyester is similar aliphatic/aromatic copolyester. DuPont markets a modified PET known as Biomax (Chauhan, 2012).

5.5.4.6 Polyvinyl alcohol

Polyvinyl alcohol (PVA) is a semi-crystalline water-soluble synthetic polymer made from hydrolysis of polyvinyl acetate. This material is also found to be biodegradable and mineralized in various environments. The degree of solubility and biodegradability as well as other physical attributes can be controlled by varying the MW and the degree of hydrolysis of the polymer (Chauhan, 2012).

5.6 Mechanism of Biodegradability of Polymers

The biodegradation of polymers is a very complex process and can occur in a number of ways. According, to the American Society for Testing of Materials (ASTM) and the International Standards Organization (ISO), the degradable plastics undergo a significant change in chemical structure under specific environmental conditions (Chauhan, 2012). Polymers do not rust, but they can exhibit environmental degradation through a variety of mechanisms (Roylance and Roylance, 2013). Both the chemical and physical properties of plastics play an important role in the mechanism of biodegradation. The surface conditions such as surface area, hydrophobic properties and hydrophilic, and the first-order structures such as chemical structure, molecular weight distribution and molecular weight, also the high-order structures (glass transition temperature, melting temperature, crystallinity, modulus of elasticity, and crystal structure) of polymers affect the biodegradation processes. A general statement regarding the breakdown of polymer materials is that it may occur by microbial action, photodegradation, or chemical degradation. All three methods are classified under biodegradation, as the end products are stable and found in nature (Ashwin, 2011).

5.6.1 Thermal Depolymerization

Although long-chain single-bonded polymer molecules are chemically stable, they can be depolymerized back to the low-molecular weight state if sufficient thermo-chemical driving energy is available. This degradation is usually negligible at ambient temperatures, but in some polymers, it may be a problem during processing. One important example is a condensation polymer such as nylon 66. During polymerization, nylon 66 converts two monomers—adipic acid and hexamethylene diamine—to three products: nylon polymer, water, and heat. At sufficiently high extents of reaction, as the concentration of product rises, the probability of back reaction to monomer increases, lowering the yield of polymer. This is handled industrially by removing water and keeping the temperature low during polymerization. However, later, when the polymer is melted during extrusion or injection molding, especially if water is present, conditions will be right for depolymerization. This degradation mechanism—a consequence of Le Chatilier's principle—is avoided by making sure the polymer resin is dried before melt processing (Roylance and Roylance, 2013).

5.6.2 Photolytic Oxidation

The photon energy in solar radiation is the most damaging component of the outdoor environment, serving to initiate a wide variety of chemical changes in polymeric materials (Trozzolo, 1976).

Photodegradable polymers undergo degradation from the action of sunlight. In many cases, polymers are attacked photochemically and broken down to small pieces. Further microbial degradation must then occur for true biodegradation to be achieved (Ashwin, 2011).

In order to induce chemical change, the photon must first be absorbed by the material. UV-absorbing groups may be present naturally in the polymer; or they may be introduced adventitiously by any of several means, such as oxidation during fabrication, polymerization anomalies, introduction of various additives, etc. The energy contained in photon-excited high-energy orbitals may be dissipated harmlessly in the form of heat, but a certain fraction of these excited states may relax by initiating chemical change. It is this latter process that produces deterioration. The deterioration process is frequently some form of photo-initiated oxidation.

Typically, the process is initiated when a sufficiently energetic photon takes away a labile proton from the polymer, leaving behind a free radical:

$$RH \xrightarrow{h\nu} R\bullet + H\bullet \tag{5.3}$$

The radical is highly unstable, and reacts easily with atmospheric oxygen to form a peroxy radical, which further reacts to form hydroperoxide:

$$R\bullet + O_2 \longrightarrow ROOH \tag{5.4}$$

The oxidation process may now be repeated, so that the initial abstraction by the photon serves to set up a chain reaction, which may include thousands of steps. The hydroperoxides formed are generally felt to be the direct cause of degradation in polymer properties, as they decompose via mechanisms which cause polymer chain scission and/or cross-linking.

Scission and cross-linking may be regarded as competitive mechanisms, with one or the other dominating in a particular situation. Cross-linking may also be dangerous, as it reduces molecular mobility, leads to the sort of brittleness often seen in elastomers that have been exposed to sunlight. Also, it may be the production of chromophoric chemical species that may impart an unacceptable discoloration to the polymer if they absorb in the visible range of light, and an autocatalytic UV degradation may be established if UV-absorbing chromophores are produced, which in turn serve to capture more UV photons (Roylance and Roylance, 2013).

Polyolefins (a type of petroleum-based conventional plastic) are the polymers found to be most susceptible to photodegradation. Further approaches includes incorporating additives that accelerate photochemical reactions (e.g., benzophenone), modifying the composition of the polymers to include more UV absorbing groups (e.g., carbonyl), and synthesizing new polymers with light sensitive groups (Ashwin, 2011).

5.6.3 Moisture Absorption

Although polymers do not rust in the presence of environmental moisture, most polymers absorb some moisture through diffusion in high humidity environments. Absorbed water is generally less damaging to polymeric materials than is sunlight, but its effects may be important in certain cases. Cases in which water causes permanent degradation by hydrolysis of chemical bonds are not common, and the plasticizing effects can generally be reversed upon drying the material (Roylance and Roylance, 2013).

Water is of special concern in the case of glass-fiber reinforced polymer matrix composites, since glass is known to be subject to permanent hydrolytic damage, especially when simultaneously exposed to stress. Water causes significant reductions in the strengths of fiberglass composites, and it has been common practice in the marine industry to design to "wet" rather than dry strength limits. The nature of the water-induced damage is not well understood, since its effect is often but not always reversible on drying (Hertz, 1973; Roylance and Roylance, 2013).

5.6.4 Microbial Action

Many biopolymers are designed to be discarded in landfills, composts, or soil. The materials will be broken down, provided that the required microorganisms are present. Normal soil bacteria and water are generally all that is required, adding to the appeal of microbially reduced plastics. Polymers that are based on naturally grown materials (such as starch or flax fiber) are susceptible to degradation by microorganisms (Fig. 5.3). The material may or may not decompose more rapidly under aerobic conditions, depending on the formulation used, and the microorganisms required (Ashwin, 2011). In the case of materials where starch is used as an additive to a conventional plastic matrix, the polymer in contact with the soil and/or water is attacked by the microbes. The microbes digest the starch, leaving behind a porous, and sponge like structure. When the starch component has been depleted, the polymer matrix begins to be degraded by an enzymatic attack. Each reaction results in the scission of a molecule, slowly reducing the weight of the matrix until the entire material has been digested (Ashwin, 2011).

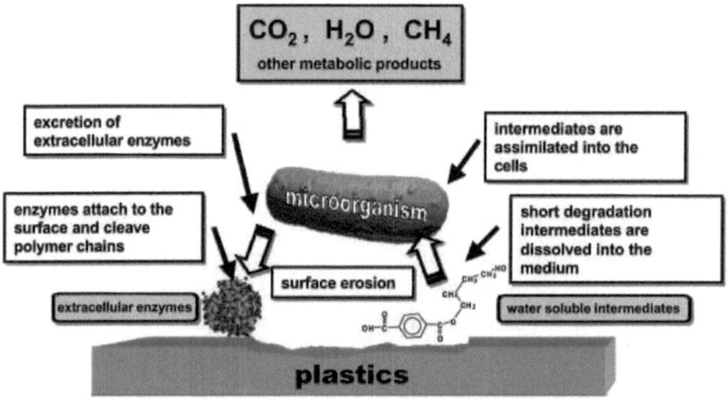

Figure 5.3 General mechanism of biodegradation by microbial action.

Another approach to microbial degradation of biopolymers involves growing microorganisms for the specific purpose of digesting polymer materials. This is a more intensive process that ultimately costs more. Although this method reduces the volume of waste, it does not aid in the preservation of non-renewable resources (Ashwin, 2011).

Some biodegradable polymer materials experience a rapid dissolution when exposed to particular (chemically based) aqueous solutions. Biopolymer materials that disintegrate upon exposure to aqueous solutions are desirable for the disposal and transport of biohazards and medical wastes. Industrial "washing machines" are designed to dissolve and wash away the aqueous solutions for further microbial digestion (Ashwin, 2011). Two types of microorganisms are of particular interest in the biodegradation of natural and synthetic polymers: bacteria and fungi.

5.6.4.1 Fungi

Eumycetes, or true fungi, are microorganisms of particular importance in causing the degradation of materials. More than 80,000 species are known. True fungi are present everywhere. Their importance as deteriorative agents is a result of the production of enzymes that break down nonliving substrates in order to supply nutrient materials present in polymer compositions. Certain environmental conditions are essential for optimum growth and degradative activity. These include an optimal ambient temperature, the presence of nutrient materials, and high humidity.

5.6.4.2 Bacteria

Schizomycetes, a bacteria, have played an undetermined role in relation to fungi in polymer deterioration. Bacteria can be single-cell rods, cocci, or spirilla; others are chainlike or filamentous. It can either be aerobic or anaerobic; in contrast, fungi are necessarily aerobic. Some bacteria are motile; bacteria are predominantly nonchlorophyllous. Their degradative action is also chiefly a result of enzyme production and resultant breakdown of the nonliving substrate in order to obtain nutrient materials. Bacteria present in soil are important agents for material degradation. Particularly affected are cellulosic plant life, wood products, and textiles subject to cellulytic degradation.

5.6.4.3 Enzymes

Enzymes are essentially biological catalysts, with the same action as chemical catalysts.

The vast majority of enzymes are proteins having a polypeptide chain with a complex three-dimensional structure. Enzyme activity is closely related to conformational structure with characteristic primary structures (i.e., specific amino acid sequences) which form an active site. At the active site, the interaction between the enzyme and substrate takes place leading to a chemical reaction, giving a particular product.

For optimal activity, certain enzymes must associate with cofactors, which can be metal ions, e.g., sodium, potassium, magnesium, calcium, or zinc. Organic cofactors are also called coenzymes and they can vary in structure. Some are derived from different B-vitamins (thiamine, biotin, etc.), while others are important compounds in metabolic cycles such as nicotinamide adenine dinucleotide (NAD^+), nicotinamide adenine dinucleotide phosphate ($NADP^+$), flavin adenine dinucleotide (FAD^+), and adenosine triphosphate (ATP). An enzyme plus a cofactor is called a holoenzyme while an enzyme lacking a cofactor is called an apoenzyme. For enzymes with absolute specificities, the "key-and-lock" theory, which implies an unchangeable rigid conformation, is a plausible model. The initial contact between an enzyme and substrate forms an optimal orientation at the active site giving good possibilities for maximum bonding (enzyme–substrate), often the cofactor induces these changes when binding to the enzyme.

5.7 Tests for Biodegradable Polymers

Biodegradable plastics, as novel materials are environment friendly. A first generation of biodegradable plastics consisted simply of polyethylenes blended with starch. Initially, these were sold as biodegradable plastics, but in practice they did not fulfill the expectations of the users. Arguments for claiming these blends as biodegradable included the growth of microorganisms on the material's surface, or a certain loss in mechanical properties (e.g., tensile strength) when they were exposed to the environment. However, the evaluation methods used were unsuitable to characterize biodegradable materials. At the time, the failure of these polyethylenes led to a generally negative image of biodegradable plastics; however, the subsequent development of suitable testing methods and evaluation criteria for biodegradable plastics has resulted in the definition of standards by various national and international standardization bodies during the past 10 years. Indeed, this process is ongoing, as the number of different environments in which plastics may be degraded has made necessary the establishment of a complex and extensive battery of test methods and evaluation criteria for these materials (Dutta et al., 2004).

5.7.1 Testing Methods

Nowadays, the evaluation of the degradability of chemicals in the environment as one important aspect of the ecological impact of a compound has become very important when attempting to bring a new chemical product to the marketplace. For this reason, a large number of standardized tests have been developed for different environments, and with the use of different analytical methods, as shown in Fig. 5.4.

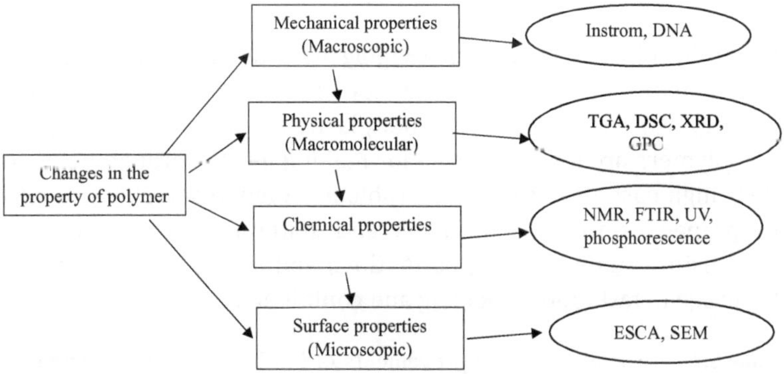

Figure 5.4 Different levels of investigation on polymer biodegradation.

These characterization techniques are meant to study the physical, chemical, mechanical properties of the polymer before and after degradation. The study of mechanical properties comprises tensile strength, elongation at fail and modulus of polymer by using instron. The physical properties of the polymer monitored are (micro cracks, embrittlement using SEM, transmission optical microscopy), density, contact angle, viscosity, molecular weight distribution (using GPC) melting temperature(T_m), glass transition temperature (T_g) (doing TGA and DSC) and changes in crystalline and amorphous region (X-ray diffraction, SAXS, and WAXS). The changes in chemical properties that could be measured are including formation and disappearance of functional groups as determined by FTIR. The molecular weight and molecular weight distribution of the degraded products or intermediates are characterized by techniques such as TLC, GC, GCMS, CL, MALDITOF, and NMR. The level of information as derived from each technique, as shown in Fig. 5.4, increases as one moves downward, thus understanding the mechanism of biodegradation. CO_2 level is measured by using GC titrating with barium hydroxide. The metabolic activity of the cells in the cultures can be seen by ATP assays, protein analysis and FDP analysis. Thermally stimulated current spectra obtained from electro-thermal analysis reveals the electric polarization properties of polymers which is used for investigating biodegradability (Lammers and Kromer, 2002).

5.8 Current Trend of Polymer Usage

Generally plastics used in packaging field are not biodegradable and their discards occupy tremendous space, there is a growing urgency to develop novel bio-plastic. However, the typical biopolymers are less employed for making plastic, probably due to their high cost and processing problems. A few researchers started to prepare starch-based foams and natural fiber/starch foam composites, (Ganjyal et al., 2004; Guan and Hanna, 2004) targeted for usage in fast food packaging and containers.

Table 5.2 Sustainability improvements of bio-based plastics relative to petroleum-based plastics (PBP) (Rosalía et al., 2012)

Bioplastic	Sustainability improvement
Polyhydroxyalkanoates	Highly biodegradable
Polylactic acid	Production uses 30–50% less fossil energy and generates 50–70% less CO_2 emissions than PBP. Competitive use of water with the best performing PBP, recyclable, compostable at temperatures above 60°C
Cellulose and Lignin	The biological degradation of lignin is lower than cellulose, compostable
Corn zein and soy protein	Biodegradable and compostable

Source: Reproduced with permission from Rosalía et al. (2012).

There are various substitutes available for the conventional petroleum-based plastics, which are manufactured using renewable resources and are capable of biodegrading in some fixed duration. Some of the most used substitutes can be listed as follows (Patni et al., 2012):

(1) wheat gluten
(2) PHA: Polyhydroxyalkanoates
(3) wood powder material
(4) liquid wood
(5) cellulosic plastics
(6) PLA: polylactides
(7) soy-based plastics
(8) plastic from casein

All of the above substitutes can be compared based on the biodegradability as shown in Table 5.2.

5.9 Biodegradable Polymers as an Alternative to Plastic

5.9.1 Biodegradable Polymer from Casein

Casein is a naturally occurring macromolecule that accounts for approximately 80% of the protein content of cow's milk; it is a phosphoprotein that can be separated into various electrophoretic fractions, such as α_s-casein, κ-casein, β-casein, and γ-casein in which each constituent differs in primary, secondary, and tertiary structure, amino acid composition, and molecular weight (Ghosh et al., 2009; Audic et al., 2003; Barreto et al., 2003). It finds use in making adhesives and paper coatings.

5.9.1.1 Why casein aerogel composites?

Considering that most foam plastics used in packaging field are not biodegradable and their discards occupy tremendous space, there is a growing urgency to develop novel biofoams. However, the typical biopolymers mentioned above are less employed for making foam plastic, probably due to their high cost and processing problems. A few researchers started to prepare starch-based foams and natural fiber/starch foam composites, targeted for usage in fast food packaging and containers.

Starch-based biodegradable polymers are thermally unstable and difficult to process into packaging products. When heated, they degrade before reaching the melting point. The casein aerogel composites can be used at high temperatures. High production costs remain a major barrier to wide use of microbially produced biodegradable polymers such as poly-3-hydroxybutyrate (PHB). There is no involvement of any kind of microorganisms in the production of this polymer product, and so it is relatively less costly.

5.9.1.2 Presence of casein in milk

Milk contains 3.3% total protein. There are two major categories of milk protein that are broadly defined by their chemical composition and physical properties. The casein family contains phosphorus and will coagulate or precipitate at pH 4.6. The serum (whey) proteins do not contain phosphorus, and these proteins remain in solution in milk at pH 4.6. The principle of coagulation, or curd formation, at reduced pH is the basis for cheese curd formation. In cow's milk, approximately 82% of milk protein is casein and the remaining 18% is serum, or whey protein.

The casein family of protein consists of several types of caseins (α-s1, α-s2, and β-6). Each of them has its own amino acid composition, genetic variations, and functional properties. The caseins are suspended in milk in a complex called a micelle.

5.9.1.3 Casein structure

Casein is a protein that is found in milk and is used independently in many foods as a binding agent. Its structure comprises the presence of amino acids. Amino acids have a variety of chemically reactive groups like phenolic hydroxy groups, presence of peptide bonds. Casein also includes amino groups, ketones and hydrazine groups. Its structure is shown in Fig. 5.5.

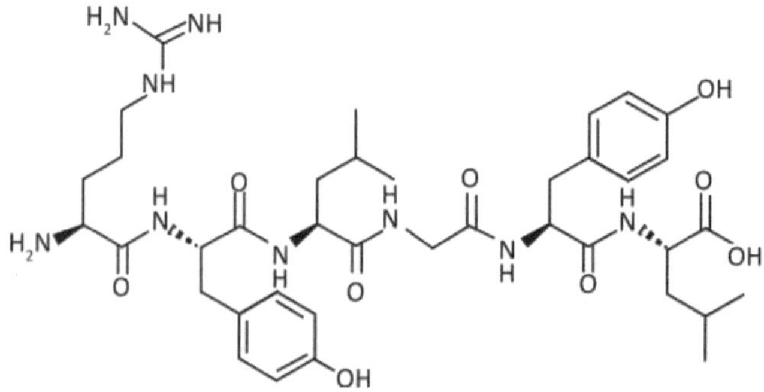

Figure 5.5 Structure of casein.

5.9.1.4 Extraction of casein from milk

When a sample of non-fat milk is precipitated using acetic acid after heating milk, the casein separates from the sample leaving a clear solution. By stirring the solution while adding acid, the casein separates from the milky solution as an amorphous mass which could be removed using suction filtration. The separated casein is then mixed with the ether solution for a few minutes, the ether is allowed to settle, and the process is repeated with a second portion of ether. After the second washing with ether, suction filters the product. The ether washings remove any small quantities of fat that may have precipitated with the casein (Agarwal et al., 2009).

5.9.1.5 Biodegradability of casein polymer composite

Casein is not very strong, and water can wash it away. To beef it up and boost its resistance to water, scientists blended in a small amount of clay and a reactive molecule called glyceraldehyde, which links casein's protein molecules together.

The polymer product formed satisfies almost all the properties that a conventional petroleum-based polymer can as far as strength and elasticity is compared. Biodegradability studies in compost media revealed distinct differences in biodegradation rates between the wheat starch and aerogels. The aerogels did in fact exhibit useful rates of biodegradation, rates that were enhanced by chemical cross-linking of the casein polymer, with the attendant changes in its protein structure (Pojanavaraphan et al., 2010). They also concluded that it is strong enough for commercial uses, and biodegradable, with almost a third of the material breaking down within 30 days (Pojanavaraphan et al., 2010).

5.9.1.6 Applications of biodegradable polymer

This foam-like material holds promise for a wide range of applications where the low density and environmental friendliness are of great importance; the ultra-low-density layered architectures result in favorable mechanical and thermal insulation properties (Pojanavaraphan et al., 2010).

In addition to cost savings and efficiency, the polymer adds excellent strength, toughness, and flexibility to help improve products. It also finds following applications:

- hose and tubing
- ice and frozen food bags
- food packaging
- stretch wrap blended with other polymers, etc.

5.9.2 Polymer from Wheat Gluten

Plant proteins from wheat show the advantage for use as films and plastics because of their abundant resources, low cost, good biodegradability, and suitable properties such as rheological, water sensitivity, sound absorption, and thermal behavior, etc. Plant proteins from soy (Lodha and Netravali, 2005), corn (Gioia and Guilbert, 1999; Romero et al., 2004), cottonseed (Grevellec et al., 2001), and wheat (Cherian et al., 1995; Irissin et al., 2001; Mangavel et al., 2002; Sanchez et al., 1998) show the advantage for use as films and plastics because of their abundant resources, low cost, good biodegradability, and suitable properties.

Plant protein from wheat, i.e., wheat glutenins are inexpensive and abundant co-product derived from renewable resources and is biodegradable. Industrial wheat gluten (WG) containing more than 75% protein with 40–50% gliadins and 35–45% glutenins has been studied as film former due to its cohesive and elastic properties (Payne and Corfield, 1979). Gliadins are single chain polypeptides, whilst glutenins are multiple chain polymeric proteins interlinked by intermolecular disulfide and hydrogen (Dong, 2010; Zárate et al., 2011; Kunanopparat et al., 2008; Yuan et al., 2010). Gliadins are low-molecular-weight, single-chain, compact and globular-shape proteins with high molecular weight of 20–50 kDa. Glutenins are linear and have high molecular weight up to 10^6. Thus, a part of glutenins belongs to the largest proteins in nature. Disulfide bonds play an important role in determining the structure and properties of gluten proteins.

Wheat gluten is a large extended polypeptide polymer without globular structure (Shewry et al., 2002). In practice, the term "gluten" refers to the proteins and contains hundreds of protein components that are present either as monomers or, linked by interchain disulfide bonds, as oligo- and polymers. They are unique in terms of their amino acid compositions, which are characterized by high contents of glutamine and proline and by low contents of amino acids with charged side groups.

WG is an ideal candidate for development of biodegradable materials because WG plastics can fully biodegrade without releasing toxic products (Domenek et al., 2004).

5.9.2.1 Conversion of WG into Composites

Gluten can be readily prepared by gently washing dough under a stream of running water. The common way to prepare gluten from wheat flour dough is either by hand washing or by automatic gluten washing devices. Gluten comprises some 75% protein on a dry weight basis, with most of the remainder being starch and lipids. When water-soluble starch has been washed away from wheat flour, the water insoluble protein remains and this removes the bulk of the soluble and particulate matter to leave a protenaceous mass that retains its cohesiveness on stretching. The protein fraction varies from different wheat quality and growth environment. WG can be readily processed into films and plastics through convenient method of mechanical mixing and thermal molding (Mangavel et al., 2004; Pommet et al., 2005).

WG is able to form network upon thermosetting so that plasticized WG can be processed into films and plastics conveniently through thermal molding. Several factors, including plasticizer content, molding temperature, and cross-linking type, might influence moisture absorption, mechanical and dynamic mechanical properties of the final products (Song et al., 2011).

In general, plasticizer agents are required to prepare protein-based plastics through thermal processing such as kneading, molding, and extrusion. Without it, WG materials are brittle and are difficult to handle (Gontard et al., 1993). Hydrophilic compounds such as water, polyols, oligosaccharides and lactic acid can be used as plasticizer to proteins.

Silica is widely used as a reinforcing material in polymer composites for improving mechanical properties and reducing cost. Additive like Xylan (Kayserilioðlu et al., 2003), filler like hydro ethyl cellulose (Reddy and Yang, 2011), lubricator like sulfonamide, salicyclic acid (Angellier et al., 2011), binder like urea and sodium hydroxide (Reddy and Yang, 2011) and reinforcing material like jute fibers (Reddy and Yang, 2011) can also be incorporated into gluten to form biodegradable composite films. L-cysteine, glutaraldehyde or formaldehyde (Sun et al., 2007),

1-ethyl-3-(3-dimethylaminopropyl) carbodiimide (EDC), and *N*-hydroxysuccinimide (NHS) are also used to promote the formation of a cross-linked protein network.

5.9.2.2 Biodegradability of WG

WG, which constitutes the protein by-product of the starch fabrication, is an interesting raw material for the development of biopolymers. A large variety of WG-based bioplastics, plasticized with glycerol, were subjected to biodegradation performed in the liquid medium (modified Sturm test) and in farmland soil (Reddy and Yang, 2011). All gluten materials were fully degraded after 36 days in aerobic fermentation and within 50 days in farmland soil. Domenek et al. (2004) carried out the Sturm test on five different gluten samples, for that cellulose was chosen to be the reference material. The results obtained showed that the degradation of all materials was faster than the degradation of the reference material cellulose. Gluten materials degraded considerably faster than well-known materials like paper (t_1 = 10 days) or uncoated cellophane (t_1 = 30 days). New materials based on starch/ polycaprolactone mixtures, e.g., degrade in times more comparable to cellophane (t_1 = 36 days). Also the protein bulk of gluten materials even strongly modified technologically shows no toxic effect on the microorganisms in the test vessel during the degradation process (Domenek et al., 2004).

5.10 Summary and Future Perspectives

Nowadays a great amount of awareness is created in the society encouraging the green consumerism as an initial approach to curb the increasing amount of pollution. Developments in science and technology, especially over the last two decades, have increased the amount of synthetic polymers viz. plastic, produced worldwide each year and it is very difficult to reduce their consumption. If plastic, which has numerous properties and widespread utilization in everyday life, is produced from some renewable resources and is biodegradable, it can immensely help in achieving the aforementioned goal without creating any harm to the existing ecology of Earth. The term biodegradable means that a substance is able to be broken down into simpler substances by the activities of living organisms and therefore is unlikely to persist in the environment. Degradation of the degradable polymers is measured in terms of duration of the application or in relation to our human lifetime rather than on absolute time. There are many different standards used to measure biodegradability, with each country having its own. The reason traditional plastics are not biodegradable is because their long polymer molecules are too large and too tightly bonded together to be broken apart and assimilated by decomposer organisms. With the observation that natural cellulose- and protein-based materials are biodegradable, the importance of natural products for replacing petroleum-derived materials becomes very clear with increasing emphasis on the environmental issues.

Originally, biopolymers were intended to be used in packaging industries, farming and other applications with minor strength requirements. Performance limitations and high cost are the major barriers to widespread acceptance of biopolymers as substitutes for traditional non-biodegradable polymers. Development and commercialization of bio-based plastics for a variety of uses in products and packaging is also of great interest as manufacturers are looking for safer and healthier materials as substitutes for chemicals and materials of concern in consumer products. Bio-based materials are promising as their feedstocks are renewable, theoretically they can be composted or recycled, and their production process may be more energy efficient than petroleum-based plastics processing.

Some of the various biopolymers and biodegradable substitutes of plastics are cellulosic plastics, polylactides (PLA), starch plastics, soy plastics, liquid wood (bioplastics), etc., which are nontoxic, biodegradable, environment friendly. Starch-based biodegradable polymers are thermally unstable and difficult to process into packaging products. When heated, they degrade before reaching the melting point. The casein aerogel composites can be used at high temperatures. This foam like material holds promise for a wide range of applications where the low density and environmental friendliness are of great importance; the ultra-low-density layered architectures result in favorable mechanical and thermal insulation properties flexibility to help improve products. It also finds numerous applications.

Extensive efforts are being made to develop such polymers from renewable resources as the matrix and/or reinforcing materials. Both partially green composites where either the reinforcing or the matrix materials are biodegradable and completely green composites where both the reinforcing and matrix materials are biodegradable and derived from renewable resources have been developed. Plant proteins from wheat show the advantage for usage as films and plastics because of their abundant resources, low cost, good biodegradability and suitable properties. Gluten-based materials can now be converted into a usable plastic with properties approaching those of commercial polymeric materials such as polypropylene and epoxies. WG can easily be biodegraded compared with plastic in aerobic conditions within 40 days. It is water insoluble and elastic in nature and can be used to make environment friendly thermosetting composites by blending it in different weight ratios with plasticizer followed by compression molding of the mixture in order to cross-link the proteins. Cross-linking is carried out to improve film properties such as water sensitivity and tensile strength.

Hence, the authors conclude the chapter by stating that making our planet Earth green has nowadays become the ultimate goal of all revolutionary ideas around, today, in this era of technological revolution developing a new technology has become less valuable than bringing out innovations in an existing technology, approaching towards a greener earth.

References

Agarwal O. P. edition 26 (2009). Isolation of casein and lactose from milk. *Adv. Pract. Org. Chem.*, 371–372.

Albertsson, A. C., and Varma, I. K. (2002). Aliphatic polyesters: Synthesis, properties and applications. *Adv. Polym. Sci.*, **157**, 1–40.

Kumar, A., Karthick, K., and Arumugam, V. (2011). Biodegradable Polymers and Its Applications, *Int. J. Biosci. Biochem. Bioinformatics*, **1**, 173–176.

Averous, L. (2008). Polylactic acid: Synthesis, properties and applications. In: Belgacem, M. N., Gandini, A., eds., *Monomers, Oligomers, Polymers and Composites from Renewable Resources*, Elsevier, Amsterdam, pp. 433–450.

Audic, J. L., Chaufer, B., and Daufin, G. (2003). *Lait*, **83**, 417–438.

Avérous, L., and Pollet, E., eds. (2012). *Environmental Silicate Nano-Biocomposites, Green Energy and Technology*, DOI: 10.1007/978-1-4471-4108-2_2, Springer-Verlag London, pp. 1–14.

Azapagic, A., Emsley, A., and Hamerton, I. (2003). *Polymers: The Environmental and Sustainable Development*. John Wiley and Sons, England, p. 219.

Barreto, P. L. M., Pires, A. T. N., and Soldi, V. (2003). Thermal Degradation of Edible Films Based on Milk Proteins and Gelatin in Inert Atmosphere. *Polym. Degrad. Stab.*, **79**, 147–152.

Bigg, D. M. (1996). Effect of copolymer ratio on the crystallinity and properties of polylactic acid copolymers, *J. Eng. Appl. Sci.*, **2**, 2028–2039.

Chauhan, A. (2012). Environment-friendly biodegradable polymers and their applications, *Malays. Polymer J.*, **7**, 62–67.

Chiellini, E., and Solaro, R. (1996). Biodegradable polymeric materials, *Adv. Mater.*, **8**, 305–313.

Dutta, P. K., Tripathi, S., Mehrotra, G. K., and Dutta, J. (2004). Perspectives for chitosan based antimicrobial films in food applications. *Food Chem.*, **114**, 1173–1182.

Epure, V., Griffon, M., Pollet, E., and Averous, L. (2011). Structure and properties of glycerolplasticized chitosan obtained by mechanical kneading, *Carbohyd. Polym.*, **83**, 947–952.

Ganjyal, G. M., Reddy, N., Yang, Y. Q., and Hanna, M. A. J. (2004). Biodegradable packaging foams of starch acetate blended with corn stalk fibers. *Appl. Polym. Sci.*, **93**, 2627–2633.

Geiser, K. (2001). *Materials Matter: Toward a Sustainable Materials Policy First*. MIT Press, p. 479.

Ghosh, A., Ali, M. A., and Dias, G. J. (2009). Effect of cross-linking on microstructure and physical performance of casein protein. *Biomacromolecules*, **10**, 1681–1688.

Guan, J., and s Hanna, M. A.(2004). Functional properties of extruded foam composites of starch acetate and corn cob fiber. *Ind. Crops Prod.*, **19**, 255–269.

Guilbot, A., and Mercier, C. (1985). The polysaccharides. In: Aspinall, G. O., ed., *Molecular Biology*, Academic Press Incorporation, New York, **3**, pp. 209–282.

Leja, K., and Lewandowicz, G., (2010). Polymer biodegradation and biodegradable polymers—A review, *Polish J. Environ. Stud.*, **19**, 255–266.

Kent and Riegel's, (2007). *Handbook of Industrial Chemistry and Biotechnology*, 623–688.

Kolybaba, M., Tabil, L. G., Panigrahi, S., Crerar, W. J., Powell, T., and Wang, B. (2003). Biodegradable polymers: Past, present, and future, paper number: RRV03-0007, St. Joseph, Mich.: ASAE, pp. 1–15.

Lammers, P., and Kromer, K. (2002). Competitive natural fiber used in composite materials for automotive parts. In Proceeding of 2002 Annual International Meeting, ASAE Paper, Chicago, USA.

Lawton, J. W., Shogren, R. L., Tiefenbacher, K. F., IndVert, M., Schwarch, G., and Coudane, J., (1995). Present and future of PLA polymers, *J. Macromol. Sci. Pure Appl. Chem. A*, **32**, 787–796.

Liao, H. T., and Wu, C. S. (2009). Preparation and characterization of ternary blends composed of polylactide, poly(ε-caprolac tone) and starch. *Mater. Sci. Eng. A,* **515**, 207.

Lunt, J. (1998). Large-scale production, properties and commercial applications of polylactic acid polymers, *Polym. Degrad. Stabil.*, **59**, 145–152.

Madison, L. L., and Huisman, G. W. (1999). Metabolic engineering of poly(3-hydroxyalkanoates): From DNA to plastic, *Microbiol. Mol. Biol. Rev.*, **63**, 21–53.

Mohanty, A. K., Misra, M., and Hinrichsen, G. (2000). Biofibers, biodegradable polymers and biocomposites: an overview. *Macromol. Mater. Eng.*, **276**, 1–24.

Ofokansi, K., Winter, G., Fricker, G., and Coester, C. (2010). Matrix-loaded biodegradable gelatin nanoparticles as new approach to improve drug loading and delivery, *Eur. J. Pharm. Biopharm.*, **76**, 1–9.

Okada, M. (2002). Chemical syntheses of biodegradable polymers. *Prog. Polym. Sci.* (Oxford), **27**(1): 87–133 90. Albertsson, A.-C., Varma, I.K. (2002) Aliphatic polyesters: Synthesis, properties and applications, *Adv. Polym. Sci.*, **157**, 1–40.

Patni, N., and Tripathi, N. (2012). Bio based plastics as viable and biodegradable substitutes for petroleum based counterparts, *GIT-J. Eng. Technol.*, **1**, 1–5.

Pojanavaraphan, T., Magaraphan, R., Chiou, B. S., and Schiraldi, D. A. (2010). Development of Biodegradable Foamlike Materials Based on Casein and Sodium Montmorillonite Clay, *Biomacromolecules*, **11**, 2640–2646.

Ren, X. (2003). Biodegradable plastics: A solution or a challenge?, *J. Cleaner Prod.*, **11**, 27–40.

Rinaudo. M. (2006). Chitin and chitosan: Properties and applications, *Prog. Polym. Sci.*, **31**, 603–632.

Rosalía, C., Edwards, S., Rafael, M. E., and Geiser, K. (2012). Sustainability of bio-based plastics: General comparative analysis and recommendations for improvement, *J. Cleaner Prod.*, **23**, 47–56.

Rosato, D. V. (2003). *Plastics Engineered Product Design*, Elsevier, USA, pp. 1, 2, 20–22, 588.

Roylance, M., and Roylance, D. (2013). *Environmental Degradation of advanced and Traditional Enginering Materials. Forms of Polymer Degradation: Overview*, CRC press, pp. 216–280.

Sinclair, R. G. (1996). The case for polylactic acid as a commodity packaging plastic, *J. Macromol. Sci. Pure Appl. Chem.*, **33**, 585–597.

Thakur, B. R., Singh, R. K., and Handa, A. K. (1997). Chemistry and uses of pectin—a review, *Crit. Rev. Food Sci. Nutr.*, **37**, 47–73.

Trozzolo, A. M. (1972). In: Hawkins, W. L., ed. *Photooxidation of polyolefins, Polymer Stabilization*, Wiley-Interscience, pp. 153.

Wang, H. J., Rong, M. Z., Zhang, M. Q., Jing, Hu, Chen, H. W., and Czigany, T. (2008). Biodegradable foam plastics based on castor oil, *Biomacromolecules*, **9**, 615–623.

Zhang, L., and Zeng, M. (2008). Proteins as sources of materials. In: Belgacem, M., Gandini, A., eds., *Monomers Polymers and Composites from Renewable Resources*, Elsevier, Amsterdam, pp. 479–493.

Chapter 6

Potential Applications of Sustainable Polymers in Functionalization of Cellulosic Textile Materials

Nabil A. Ibrahim and Basma M. Eid

Textile Research Division, National Research Center, El-Behouth St., Dokki, Giza 12622, Egypt

nabibrahim49@yahoo.co.uk

6.1 Introduction

In the recent years, the increasing environmental concerns and demands for eco-friendly textile processes and highly functional cellulosic textile products with higher added value and better durability have increased significantly. New co-applications of certain sustainable polymers with the advancement of emerging technologies such as nano and plasma technologies for the functionalization of cellulose-based textiles impart novel or significantly enhance functional properties. Such properties include antibacterial, UV protection, self-cleaning, flame retardancy, and easy care.

The first section of this chapter discusses the classification of cellulosic textile materials, pretreatment, and coloration processes required to achieve successful functional finishes as

Handbook of Sustainable Polymers: Processing and Applications
Edited by Vijay Kumar Thakur and Manju Kumari Thakur

ISBN 978-981-4613-53-8 (Hardcover), 978-981-4613-54-5 (eBook)
www.panstanford.com

well as environmental concerns. The most widely used sustainable polymers, their chemical compositions, as well as the most promising approaches and proper emerging technologies for enhancing the extent of loading the nominated polymers onto the selected substrates are highlighted in Section 6.2. Some key evaluation methods of the imparted functional properties and forecasting future developments are demonstrated in Section 6.3.

6.2 Cellulose

Cellulose is the most abundant renewable biopolymer. It is a natural linear polysaccharide 1,4-glucan polymer. Sources of cellulose include seed fibers (cotton), wood fibers, bast fibers (flax, jute), grasses (bamboo), bacteria and algae. Cellulose component represents about 90–99% in purest cotton form, 70–75% in flax, 60–65% in jute, 40–55% in bamboo, and 40–50% in wood. Its degree of polymerization (DP) is dependent on its source. On the other hand, its degree of crystallinity and the availability and accessibility of the CH_2OH and -OH groups in the amorphous areas have an important role in its chemical functionalizations, i.e., etherification, esterification, cross-linking, and graft–copolymerization. Reactivity of its one primary (at C_6) and the two secondary (at C_2 and C_3) hydroxyl groups is governed by the nature of chemical functionalization. In case of etherification in an alkaline medium, it follows the descending order -OH at C-2> $-CH_2OH$ at C-6> -OH at C-3, while the primary-CH_2OH group is the most active in esterification. Further, its degree of substitution (DS) is determined by the average number of hydroxyl groups replaced by substituent [48, 86, 122, 145].

6.2.1 Cellulose Fibers

Cellulosic fibers are characterized by favorable properties such as renewability, biodegradability, environment friendly, excellent affinity for chemical functionalization as well as potential applicability [149]. Cellulosic fibers may be natural, such as cotton, flax, and jute, or regenerated fibers such as lyocell, using NMMO (N-methyl, morpholine-N-oxide) as a solvent for cellulose pulp, viscose, via, more environment-friendly viscose process, as well as bamboo viscose fibers [24, 142, 149].

6.2.1.1 Pretreatment processes

Unlike regenerated cellulosic fibers, natural cellulosic fibers contain different non-cellulosic inherent impurities, e.g., wax, fats, pectin, lignin, and seed husks, as well as added ones, e.g., spinning oils, sizes, and lubricants, during spinning, weaving, or knitting processes. Pretreatment processes include the following (Fig. 6.1):

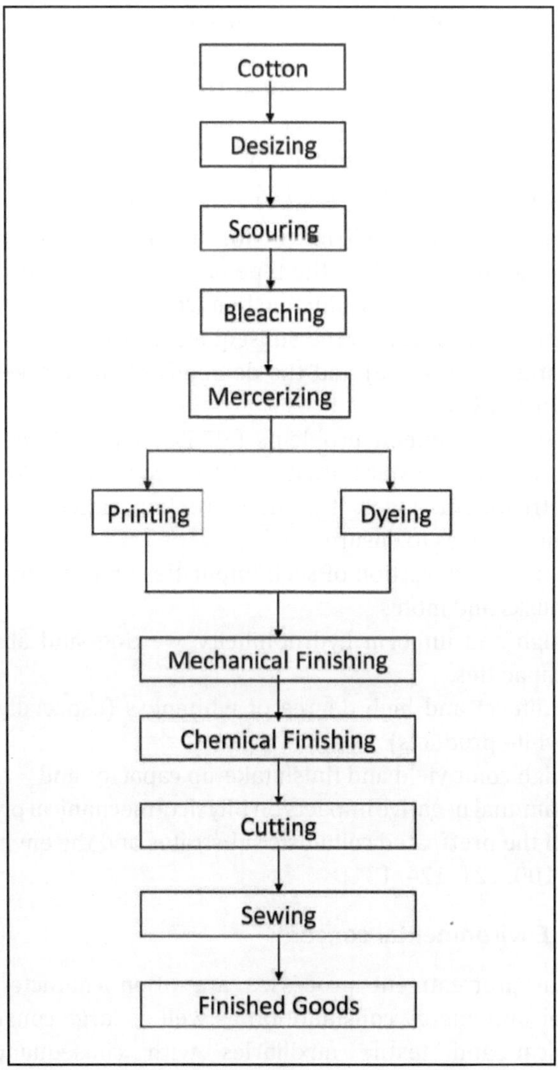

Figure 6.1 Flow chart of woven cotton fabric wet processing.

- singeing to remove projecting fiber ends and get smooth fabric surface
- desizing to get rid of sizing materials, e.g., starch, PV, OH and CMC
- scouring to swell the cellulosic fibers and seed husks, as well as extract natural and other applied substances
- bleaching to destruct and remove the pre-swell seed husks, oxidize natural coloring matters and get rid of any remnant size, e.g., starch and polyvinyl alcohol
- mercerizing to improve dyeability, tensile strength, and dimensional stability as well as to increase luster, especially in case of using chain mercerization
- optical brightening especially in case of seeking of high degree of whiteness [123, 124, 131, 146].

Number, sequence, chemical formulations, and treatment conditions are dependent on the type of cellulosic substrate, fabric form (woven or knitted), production line (continues, semi-continues or discontinues), as well as the subsequent wet processes (dyeing, printing and/or finishing) and the demands of the market for the final product [121].

Most of the common problems (~75%) encountered in subsequent wet processes are attributed to incorrect and/or in adequate pretreatment stages. It is therefore, imperative to select the pretreatment stages to ensure

- uniform extraction of such impurities and removal of seed husks and motes,
- high and uniform hydrophilicity, wetting, and absorption capacities,
- uniform and high degree of whiteness (especially for full white products)
- high color yield and finish take-up capacity, and
- minimal negative impacts on physico-mechanical properties of the pretreated cellulosic substrates and the environment [100, 121, 124, 131].

6.2.1.2 Environmental concerns

Traditional pretreatment processes are often characterized by high water and energy consumption as well as large consumption of chemical and textile auxiliaries with consequence high environmental negative impacts. In recent years, there have been increasing demands and concerns to develop the conventional pretreatment processes to

- comply with the more stringent environmental legislation/regulations/norms and reduce environmental impacts,
- reduce heat energy, water and chemicals consumptions,
- upgrade both the product and environment quality, and
- fulfill the ever-growing consumer needs and cope with his awareness.

The use of waste minimization approach to pollution control in pretreatments offers environmental and economical advantages over effluent treatment namely [88]

- reduction of environmental impacts,
- minimization of production costs,
- financial payback,
- positive impacts on the subsequent wet processes, and
- upgrade of cellulosic products and environment quality.

The key opportunities for pollution prevention in pretreatment of cellulosic substrates are as follows [100, 88, 162]:

- rationalization and optimization of pretreatment stages,
- continuous pretreatment processes instead of batchwise one,
- use of combined process, e.g., desizing/scouring, scouring/bleaching or desizing/scouring/bleaching in one step,
- replacement of hazardous chemicals by eco-friendly ones such as NaOCl by H_2O_2, acetic acid by formic acid, non-biodegradable auxiliaries by biodegradable ones,
- reduction of water and energy consumption by using ultra liquor ratio equipment (liquor to material ratio 5:1), as well as automation of machinery,
- use of proper enzymes in pretreatment processes like α-amylase in desizing, alkaline pectinase in bio-scouring, lipase in removal of vegetable fats and oils, catalase for bleach clean-up, as alternative to the conventional harsh chemicals, led to considerable environmental improvements as well as energy and water saving [45, 52, 53, 57, 63, 107, 130],
- combined enzymatic treatment in one bath and in one step will help to save more water, energy and processing time and ultimately costs [53], and

- potential applications of plasma technologies, an environment-friendly way for surface modification, enhancing fiber hydrophilicity and wettability as well as dyeability, as well as in desizing, led to energy water and time savings [40, 58, 74, 135, 138].

6.2.2 Coloration Process

Highly efficient pretreatment regime is essential for optimum coloration of cellulosic substrates. Coloration of cellulosic substrates is carried out by dyeing and/or printing processes. Dyeing of cellulosic substrates with reactive, direct, vat and sulfur classes, in aqueous medium, are widely used to get uniform coloration. The dyeing process comprises four steps namely: adsorption from the dye bath to the cellulosic substrate surface, penetration into the interior fabric structure, followed by subsequent interaction and fixation via chemical linkages, e.g., reactive, hydrogen and van der Waals forces, e.g., direct, or via reduction followed by in-situ oxidation, e.g., vat and sulfur [88, 124, 131]. Conventional dyeing process demands the use of a large number of chemicals and textiles auxiliaries along with an intensive water and energy consumption for a prolonged process time. Improved dyeing quality along with reduced environmental impacts have promoted the wide spread use of bifunctional reactive and vat dyes in cellulosic substrates dyeing.

On the other hand, printing of cellulosic fabrics is a localized form of dyeing, using the same dyestuff for their dyeing, to get clear/well-defined design on the fabric surface with the help of using proper thickeners, binders especially for pigment printing. In recent years, eco-friendly pigment printing using green thickening and binding agents is one of the most important and commercially used methods of printing cellulosic substrates [54, 67–69].

6.2.2.1 Environmental Concerns

Features of modern and innovative coloration processes of cellulosic substrates trends, taking in consideration pollution prevention/waste minimization/cost saving/as well as product and environment quality, include the following:

- processes optimization and recipes rationalization
- polyfunctional reactive dyes for promoting high extent of exhaustion and fixation [87, 88]
- ultra low liquor or waterless dyeing equipments
- dyeing in super critical carbon dioxide for dyeing instead of water [11, 35]
- right first time dyeing [112]
- chemical modification, e.g., cationization, of cellulosic substrates for zero-salt dyeing and higher extent of dye exhaustion and fixation [43, 101]
- innovative biodegradable textile auxiliaries for coloration [87, 147]
- automation in textile coloration, e.g., central coloration kitchen, computerized color measurements and matching, computerized management system [87]
- dyeing in ionic liquids to reduce or completely replace the aqueous medium [7]
- utilization of plasma and/or biotechnologies for surface modifications and efficient coloration [107, 130]
- scaling up digital ink-jet printing processes [39]
- application of environmental management system [EMS], REACH system, Eco-labels certifying [103]
- use of natural dyes in textile coloration for their better biodegradability and high compatibility with the environment [141].

6.2.3 Finishing Process

It is well known that efficient and proper pretreatment stages are essential for uniform and successful coloration processes as well as for optimum finishing results. A number of mechanical and/or chemical finishes were developed to impart different desirable properties to the post-finished substrates.

Mechanical finishing processes such as compacting, calendaring, raising, polishing, etc., are performed to alter fabrics appearance or function like fabric luster, smoothness, softness, residual shrinkage, etc., by using mechanical means alone or accompanied by application of heat, steam pressure, etc [96].

Chemical finishes mean the use of proper chemical finishing formulations to impart a wide variety of desired properties. Application of a chemical finish is governed by type of the cellulosic fabric, the demanded fabric properties, and the active ingredients of the finishing formulations, available machinery and application method, i.e., in batch by exhaustion or by continuous process such as padding technique [127].

The main requirements of applying functional finishes of cellulosic substrates to impart specific functional properties are to [12, 47]

- develop innovative textile products with high value-added to maintain a competitive edge;
- meet increasing awareness of environmental concerns;
- minimize the total production cost to remain competitive;
- avoid any negative impacts on the inherent physico-mechanical and comfort properties of cellulosic substrates;
- cope with the market demands and the growing needs of individual consumer.

Consequently, the need for innovative, cleaner, durable, comfortable, cost-efficient, and value-added functionalized cellulosic textile products, e.g., antimicrobial, UV protection, self-cleaning, easy care, soft-handle, insect repellent, water/oil repellent, antiodor and fragrance finished, etc., are still a challenge [12, 47, 91, 111, 127].

6.3 Biopolymers Classifications and Applications for Functional Cellulose Textiles

6.3.1 Sustainable Biopolymers

In recent years biodegradable polymers have widely attracted considerable attention as an eco-friendly alternative for the currently used hazardous textile chemicals and auxiliaries for textile modification and functionalization [10, 18, 37, 72, 132]. As a consequence of their unique properties such as abundance, renewability, biocompatibility, biodegradability, and environment friendliness (thereby leading to economic, environmental, and social positive impacts), a vast number of biodegradable polymers have been used to add or enhance desired functionalities within textiles as well as to find a sustainable competitive edge.

Biodegradable polymers can be classified according to their chemical composition, resource, synthesis and application methods, etc. Figure 6.2 demonstrates classification of the main biodegradable polymers according to their resource. On the other hand, Fig. 6.3 shows the chemical structure of some of the most important biodegradable polymers.

Biodegradable polymers

A. Natural

1. By direct extraction from bio-mass

1.1. Polysaccharides, e.g. starch, cellulose, chitin, chitosan, alginate, lignin, etc., and their derivatives

1.2. Proteins, e.g., collagen, gelatin, corn zein, sericin, casein, soy protein

2. By microbial production, e.g.,

- polyhydroxyalkanoates (PHAs)
- Poly-3-hydroxbutyrate (PHB)
- Bacterial Cellulose (BC)
- Polylactic acid (PLA)
- Polyglyconic acid (PGA)

B. Synthetic

1. Aliphatic Polyester, e.g.,

- PLA& PGA and their copolymers
- Polyvinyl acohol PVOH

2. Aromatic Polyester

- Often based on terephthalic acid, e.g., poly (butylenes-adipate-co-terephthalate) (PBAT)
- An increase in terephthalic acid content results a decrease in the biodegradation rate.

Figure 6.2 Classification of biodegradable polymers based on their origin. Adapted from [10, 18, 37, 132].

6.3.2 Applications of Sustainable Biopolymers in Functionalization of Cellulosic Textiles

Recently, the field of functional finishes of cellulose based textiles has been revolutionized by developing and applying new, advanced, innovative, and emerging green-process technologies, e.g., nano-, bio-, plasma technologies, etc., and bio-nanocomposite materials for creating new, safe, and innovative socially acceptable textile products with extra functional properties, highly value-added, and more appealing taking in consideration the possible negative impacts on the environment (E), human health (H) and safety (S) and sustainability (S), i.e., (EHS/S), during product life cycle [1, 37, 41, 45, 59, 104, 117, 119, 139].

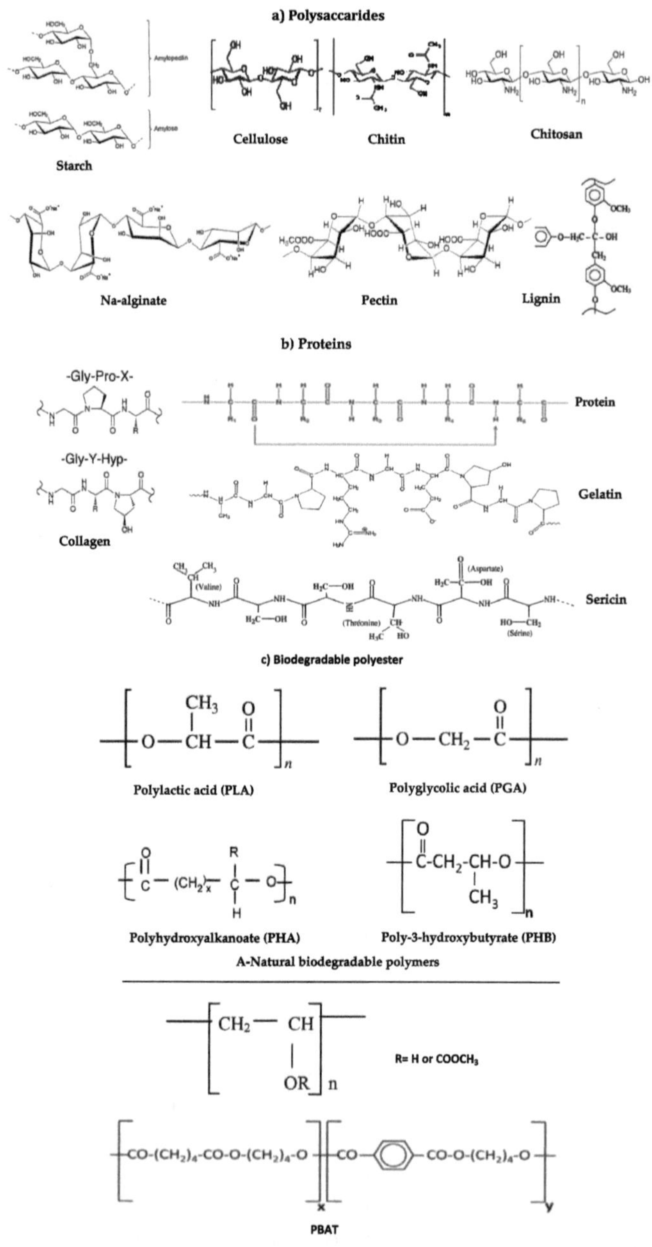

Figure 6.3 Chemical structure of some of the most important biodegradable polymers.

6.3.2.1 Options for enhancing functionalization of textiles

In order to cope with the future demands for high quality and functionality textile products taking in consideration the economic, environmental, and ecological concerns without adversely affecting the inherent physico-mechanical and comfort properties of the textile material, many emerging and innovative approaches have focused on surface modification of fabric surface, before subsequent functional finishing, for increasing and improving the availability and accessibility of its active sites and functional groups. This surface modification may results in enhancing the extent of functionalization and creation of new and effective functional properties such as antimicrobial, anti-UV, self-cleaning, water repellent, insect repellent, and flame retardant. The promising results have been obtained by applying the following emerging techniques:

- Plasma treatment for development of durable functional textiles via surface activation, deposition of active materials, and/or polymerization of coating onto the textile surface for various functionalities [1, 38, 41, 59, 72, 117, 152].
- Enzyme treatment as a cost effective and environmentally benign tool for surface modification [1, 9, 40, 108, 109]. On the other hand, the applicability and efficiency of immobilized enzymes to increase the activity and enhance the functionality of textiles are governed by the nature of enzyme, its activity and stability, immobilization method and accessibility [41, 56, 72].
- Nanocoatings on textiles to create textile surface with innovative functional properties and without adversely affecting the fabric feel via using various techniques such as nanosol, polymer dispersions, and chemical vapor deposition [12, 41, 60, 65, 69–71].
- Layer-by-layer assembly method of deposition to impart durable multifunctional properties to textiles without negative impact on strength properties [41, 76, 144, 148].

6.3.3 Antimicrobial Functional Finishes

Cellulosic fibers are prone to host and support the growth of pathogenic or potential pathogenic microorganisms such as bacteria and fungi. Consequently, there is increasing demand for cellulosic textile equipped with highly efficient antimicrobial agents to protect the textile user from pathogenic or odor generating microorganisms as well as to protect textile itself from undesirable aesthetic change or deterioration, i.e., off-quality and strength loss. Moreover, antibacterial agent can be classified into biocidal, which kills, or biostatic, which inhibits the growth of microorganisms, as well as can be divided into controlled-release, i.e., leaching-type, or bound-type agent [36, 47, 66, 133, 150].

An ideal antimicrobial functional finish should fulfill the following requirements [36, 47, 66, 128, 133, 150]:

- should be effective against a broad range of harmful microorganisms without adversely affecting resident flora of non-pathogenic bacteria on the user skin,
- should be safe to both the manufacturer, in plant handling and the consumer,
- should be durable to repeated laundering,
- should be eco-friendly and cope with standards in compatibility tests,
- should be compatible with other desired functions, and
- should be cost-effective and have no negative impacts on the inherent properties of the treated fabric.

Additionally, antimicrobial finish kill microorganism or inhibit their growth by [12, 36, 47]

- cell wall damage,
- inhibition of cell wall synthesis,
- alteration of cell wall permeability,
- inhibition of proteins and nucleic acids synthesis, and/or
- inhibition of enzyme action thereby killing or inhibiting the microorganism.

To cope with textile-user demands for cellulosic textile products that offer comfort along with other desired functional properties taking in consideration both the ecological and environmental concerns, there is an urgent need for and a great motivation in developing a new generation of antimicrobial agents based on sustainable biopolymers, alone or in combination with other active ingredients, as alternatives for the harmful ones, in modification and functionalization of textile substrates. This part highlights the application of most important of biodegradable polymers such as chitosan, β-cyclodextrin, etc., in antimicrobial finishing of cellulosic textiles.

6.3.3.1 Chitosan and chitosan derivatives

Main characteristics of chitosan, deacylated derivatives of chitin, include polysaccharide, biodegradable, biocompatible, non-toxic, polycationic nature, $-NH_2$ and $-OH$ active sites, soluble in weak acid solutions, versatile chemical and physical properties, antimicrobial/wound healing/antistatic activities, deodorizing and film forming abilities [47, 72, 73, 133]. Most of chemical reactions with chitosan occur primarily on amino groups on the atom C_2. Reactivity of its hydroxyl groups follows the decreasing order: $-CH_2OH$ on the C-6 atom > $-CHOH$ of atom C-3 [136].

Chitosan can be chemically modified by etherification and esterification of its hydroxyl groups, by the N-substitution of its nucleophilic amino group (C-2), or by graft copolymerization to get a broad range of chitosan derivatives that can easily dissolve in the entire range of pH [93].

The brief examples of chitosan derivatives, their chemical structure, characteristics features, and modes of action are highlighted in Table 6.1.

6.3.3.2 Application methods

The most currently used approaches for efficient and permanent fixation of chitosan and chitosan derivatives onto/within the cellulosic fabrics to impart antimicrobial functional properties are summarized in Table 6.2.

6.3.3.3 Sericin protein

Sericin biopolymer, an amorphous glue-like substance, is made up of 18 amino acids most of which ~70% have a polar terminal functional group, such as $-OH$, $-COOH$, and $-NH_2$ groups, which are responsible for its solubility as well as water absorption and desorption [72, 143]. Biocompatibility, biodegradability, excellent antimicrobial activity, resistance to UV radiation, and moisture absorption ability inherent properties along with its availability have facilitated its potential applications in different fields, including textiles [72, 77, 143].

Rajendran et al. [118] have studied antimicrobial properties of cotton fabric coated with extracted sericin, using ice cold ethanol, by pad-dry-cure method. They found that the sericin-coated fabric showed an efficient bactericidal activity against G+ve (*S. aureus*) and G-ve (*E. coli*) bacteria.

Table 6.1 Some of the most important chitosan ($HO\text{-}CH\text{-}NH_2$) derivatives, their chemical structure, and characteristic features

Derivatives	Chemical structure	Characteristic feature	Ref.
Chitosan salts e.g., chitosan acetate	$\overset{+}{HO\cdot CH\cdot NH_3}\overset{-}{O}OC\text{-}CH_3$	Strong antimicrobial activity, precipitate in an alkaline medium, antimicrobial activity is pH dependent	93, 99
Quaternized chitosan derivatives	$[HO\cdot CH\cdot \overset{+}{N}(CH_3)_3]^-$ Cl (TMCC) N,N,N-triemethyl chitosan chloride $[HO\cdot CH\cdot NH\cdot CH_2\text{-}CH(OH)\text{-}CH_2\text{-}N(CH_3)\text{-}CH_3]^{+-}Cl$ (with CH_3 above N, OH below CH, CH_3 below N) (HTCC): N[(2-hydroxy-3-triemethyl ammonium) propyl] chitosan chloride $[HO\cdot CH\cdot \overset{+}{N}(CH_2)_2]^-I$ with CH_2R on N N-alkyl chitosan derivatives	–High water solubility, improved antimicrobial activity –Number and accessibility of the cationic actives sites as well as alkyl chain length strongly affect the antimicrobial efficacy of the quaternized chitosan derivatives	25, 79, 93, 94, 102, 143
Carboxy-alkylated	$HO\cdot CH\cdot NH\ CH\cdot COOH$ (NCMC): N-carboxmethyl chitosan	–Improved solubility and antimicrobial activity compared to chitosan	29, 42, 80, 93, 106

Derivatives	Chemical structure	Characteristic feature	Ref.
chitosan derivatives	HO·CH·NH $(CH_2)_3$-COOH (NCPC): N-carboxypropyl chitosan HO·CH·NH $(CH_2)_4$-COOH (NCBC): N-carboxybutyl chitosan	–Antimicrobial activity is governed by the number of available –NH_2 groups, degree of carboxyalkylation and the amphoteric properties of carboxyalkyl chitosan –Improved solubility is attributed to the contribution of high degree of carboxyalkylation and decreased crystallinity of chitosan	
N-sulfonated and N-sulfobenzyol chitosans	HO·CH·NH-SO_3H (SC): N-sulfonated chitosan 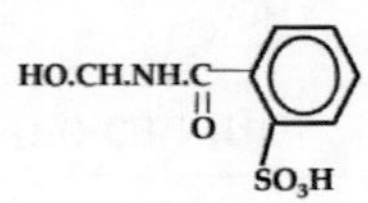 (SBS): N-Sulfobenzyol Chitosan	–Superior antimicrobial activity and higher water solubility than unmodified chitosan –The excellent antibacterial activity, at certain sulfur content, was attributed to the better change interaction between the negatively charged cell walls and the chitosan molecules	21, 22
Carbohydrate branched chitosan derivative	D-Glucosamine-branched chitosan 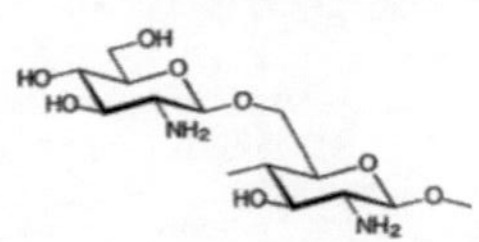	–Higher growth suppression against numerous types of micro-organisms compared to unmodified one –Water soluble (DS = 0.45), biodegradable chitosan derivative	89, 93

(*Continued*)

Table 6.1 *(Continued)*

Derivatives	Chemical structure	Characteristic feature	Ref.
Chitosan nanoparticles		–Antimicrobial activity of chitosan depends on type of chitosan, degree of polymerization, molecular weight, pH and solvent –Chitosan nanoparticles could exhibit superior antimicrobial effect against various micro-organism than chitosan itself –The unique character of chitosan nanoparticles for their positive charge and small particle size is responsible for their potential antibacterial activity and acceptable biocompatibility	113
Unmodified chitosan	[HO-CH-NH_2] 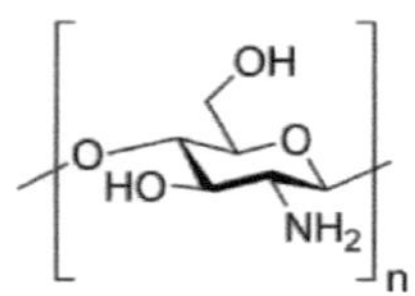	–N-deacetylated derivative of chitin –Both degree of deacetylation (DD) and pH determine its charge density which in turn affects the level of antimicrobial activity –The increase in DD and decrease in pH (<6.5) enhance the antimicrobial activity of chitosan –Its antimicrobial activity arises from its polycationic nature –Its mode of action comprises the following step: interaction with negatively charged bacterial surface → disruption of the cell membrane → increase in its permeability → leakage of intracellular substances → interruption of all essential functions and finally death of these cells –Can be used as multifunctional textile finishing agent.	72, 73, 93, 133, 136

Table 6.2 Antimicrobial finishing of cellulosic substrates using chitosan and its derivatives

Bioactive agent	Substrate	Finishing formulation and finishing technique	Comments	Ref.
Chitosan	Cotton fabric	–Using BTCA or Arkofix NEC as a cross-linker –Chitosan. HCl as a finishing agent and as a catalyst for Arkofix NEC-pad-dry-cure	–Broad spectrum antimicrobial against G+ve and G-ve bacteria and fungi-tested –The extent of improvement in the imparted properties (antimicrobial/durable press) was determined by type and concentration of cross-linker, chitosan molecular weight and conc. As well as finishing condition.	31
Chitosan	Cotton fabric	–Chitosan along with epichlorohydrin as a finishing agent –Padding in finishing formulation → pre-drying at 160°C → followed by mercerization.	–Good dyeability along with high antimicrobial properties. –The retained antimicrobial properties over 91% even after 25 washing cycles	92
	Cotton and cotton/polyester (50/50)	–Chitosan or chitosan/DMDHEU –Surface treatment –Pretreatment with chitosan followed by subsequent treatment with malonic acid or DMDHEU/$MgCl_2.6H_2O$	–A sufficiently high level of antimicrobial activity along with a sufficient durability to wash, especially in case of after treatment with DMDHEU followed by drying and steaming at 150°C/4 min.	154

(*Continued*)

Table 6.2 (*Continued*)

Bioactive agent	Substrate	Finishing formulation and finishing technique	Comments	Ref.
	Cotton and cotton/polyester (50/50)	–Chitosan or chitosan/DMDHEU –Surface treatment –Pretreatment with chitosan followed by subsequent treatment with malonic acid or DMDHEU/$MgCl_2 \cdot 6H_2O$	–A sufficiently high level of antimicrobial activity along with a sufficient durability to wash, especially in case of after treatment with DMDHEU followed by drying and steaming at 150°C/4 min.	154
	Cotton, viscose, linen and cotton/polyester blended fabrics	–Incorporation of chitosan into pigment printing formulation as a bioactive ingredients –Simultaneous drying and fixation of pigment printed substrates was carried out using microwave oven at output power of 386W/5 min.	–High-quality pigment prints with remarkable antimicrobial properties, durability to wash as well as better coloration properties	67, 68
	Cotton fabric	–Using carboxymethyl chitosan alone by pad-dry-cure method	–Good antibacterial activity, improved wrinkle recovery, dyeability with cationic dyes –Durable for 5 laundering cycles	42

Bioactive agent	Substrate	Finishing formulation and finishing technique	Comments	Ref.
Carboxymethyl chitosan (CMCTS)	Cotton fabric	–Post treatment of pre-cationized cotton fabric with CMCTS by padding technique followed by batching at 60°C/over night	–Ionic cross-linking of cotton fabric results in enhancement in easy care properties along with a moderate improvement in antibacterial activity	29
HTCC	Cotton fabric	–HTCC along with different cross-linking agents including citric acid (CA), BTCA and glutaraldehyde (GA) in the presence of a proper catalysts (NaH_2PO_2 or $MgCl_2$) –Pad-dry-cure technique	–Treated fabric samples with CA and BTCA show good antimicrobial properties, improved wrinkle recovery after repeated laundering –Using BTCA as a cross-linker resulted in functionalized cotton fabrics without yellowing and unpleasant odor	102
	Mercerized cotton fabric	–HTCC along with DMDHEU and BTCA or CA as a cross-linker and in the presence of Na-acetate, NaH_2PO_2 or $MgCl_2$ as a catalyst using pad-dry-cure technique	–BTCA proved to be the proper one for imparting prolonged antimicrobial activity along with durable press properties	84
(NMA-HTCC) O-acryl-amidomethyl HTCC	Cotton fabric	–Aqueous solution of NMA-HTCC along with the alkaline catalyst –By pad-dry-cure or by cold pad-batch method	–Remarkable and durable antibacterial activity against *S. aureus*	95

(*Continued*)

Table 6.2 (*Continued*)

Bioactive agent	Substrate	Finishing formulation and finishing technique	Comments	Ref.
Cationic hyperbranched dendritic poly amidoamine (PAMaM) chitosan	Cotton fabric	–By pad-dry-cure method	–Enhanced antimicrobial activity against *S. aureus* (G+ve) even at neutral pH	85
Chitosan metal complex	Jute fabric	–By padding jute fabric samples with chitosan metal complex solutions containing Ag^+, Zn^{2+} or Zr^{2+} ions followed by drying and curing	–The imparted antibacterial properties are governed by the type of Chitosan complex as follow: chitosan-Zn > chitosan-Zr > chitosan-Ag.	46
Chitosan cyanuric chloride (CHI-CNC)	Cotton fabric	–CHI-CNC solution along with Na_2CO_3 as an alkaline catalyst –By exhaustion technique at 75°C for 9 min. followed by neutralization with acetic acid	–80% reduction of bacteria after 20 launderings due to its durable fixation	82
Chitosan/Ag oxide nanocomposite	Cotton fabric	–The inclusion of nanoparticles was used to treat cotton fabric by pad-dry-cure method	–The imparted remarkable wash fast antimicrobial activity was attributed to the synergy of chitosan and Ag-oxide, as well as to the large surface of the nanocomposite onto the cotton surface	51

Bioactive agent	Substrate	Finishing formulation and finishing technique	Comments	Ref.
ZnO/CMCTS bio-nanocomposite	Cotton fabric	–By padding cotton fabric in aqueous solution of the prepared composite followed by drying and curing at 160°C/3 min	–An improvement in both antimicrobial activity and UV protection functionality of finished cotton	30
ZnO nano-particles chitosan composite	Cotton fabric	By coating technique using a lab padder (immersing → padding → drying at 130°C/30 min.)	Decreasing the molecular weight of chitosan component resulted in an improvement in antibacterial activity	33
TiO_2–chitosan composite	Cotton fabric	Immobilization of TiO_2–CS composite on cotton fabrics using epichlorohydrin as a cross-linking agent and dodecyl sodium sulfate and applying a 30 min cross-linking reaction time	–Immobilization of TiO_2 chitosan composite onto cotton fabric resulted in a significant improvement in its bacterial and mildew-proof (BMF) properties –The imparted properties attributed to the TiO_2-photcatalysis and the bactericidal activity of chitosan.	115

Doakhan et al. [28] have studied the feasibility of using sericin/TiO_2 nanocomposite as antimicrobial finishing agent to impart antimicrobial functionality to cotton fabric, using the pad-dry-cure technique in the presence and absence of polycarboxylic acid as a cross-linker, e.g., BTCA, CA. The obtained results showed that

- the imparted antibacterial activity followed the decreasing order: *S. aureus* > *E. coli*;
- the extent of bacterial reduction is directly related to type and concentration of cross-linker, pH of the finishing formulation as well as concentration of the composite components;
- finishing with nanocomposite did not adversely affect the strength properties of treated cotton;
- the proper processing conditions for attaining the optimal antibacterial activity were 50 g/L TiO_2-NPs, 25 g/L sericin, and 50 g/L cross-linker 30 g/L NaH_2PO_2;
- the remarkable antibacterial activity of the functionalized cotton was attributed to both photocataytic activities of TiO_2-NPs together with polycationic nature of sericin.

6.3.3.4 Cyclodextrins

Cyclodextrins (CDs) are polysaccharides produced during biodegradation of starch using cyclodextrin glucanotransferase (CG Tase) enzyme [14]. CDs are of three types, namely α-CD, β-CD, and υ-CD, which are composed of six, seven, and eight D-glucose units, respectively (Fig. 6.4). Among them, β-CD is the most important and widely used [16, 27]. The unique feature of CDs is their ability to host guest molecules in their internal hydrophobic cavities, thereby forming non-covalent host-guest inclusion complexes. Inclusion in CDs has positive impacts on the physico-chemical properties of guest molecules such as solubility enhancement, stability improvement against the degradation effects, volatility and sublimation, controlled release of active ingredients, masking mal-order, etc. [72, 98, 114]. Chemical modifications of β-CD through its –OH groups on the exterior rims have been carried out to alter their undesirable inherent properties, e.g., poor water solubility, relatively small cavity size, etc., as well as to enable permanent fixation onto the cellulosic matrix. These unique properties of fiber-reactive β-CD derivatives make them appropriate candidates and greener textile auxiliaries for potential applications in functional finishing of cellulosic materials [26, 114].

Table 6.3 Antimicrobial finishing of cellulosic substrates using fiber reactive β-CD derivatives

β-CD or its derivatives	Cellulosic substrate	Finishing formulation and finishing technique	Comments	Ref.
Antibiotic/CPβ-CDs inclusion complexes: —Triclosan/ CPβCDs –Butylparaben /CPβCDs –CPβCD: cationic β-CD polymer	Cellulose fiber	–By exhaustion technique	–Cationization of β-CD increased its water solubility –Inclusion of antibiotic with CPβCDs significantly increased their water solubility –Antibiotic/CPβCDs inclusion complexes imparted excellent antimicrobial activity to the cellulose fibers. –Antimicrobial effect was attributed to their negative effects on the metabolism of the bacteria instead of damaging the cell membrane	30
β-CD	Cotton fabric	Grafting of β-CD onto cotton by ester-cross-linking using citric acid as a cross-linker and NaH_2PO_4 as a catalyst along with β-CD at 180°C/10 min. followed by subsequent treatment with $AgNO_3$ solution at 27°C for 24 h to entrap Ag^+ ions into cavities of loaded β-CD	–The slow release of Ag^+ ions present within the internal cavities was attributed to the hydro-phobic nature of these cavities along with binding of Ag^+ ions with oxygen atoms of glucopyranose rings within the cavity –According to their results, the loaded Ag^+ ions into the grafted β-CD cavities exhibit excellent antibacterial functionality with slow release mechanism	13

(*Continued*)

Table 6.3 (*Continued*)

β-CD or its derivatives	Cellulosic substrate	Finishing formulation and finishing technique	Comments	Ref.
MCT-βCD (monochloro-triazine-βCD)	Cotton and viscose fabric	Reactive printing and β-CD loading in one step followed by post treatment with neem oil as an eco-friendly antimicrobial agent	–Inclusion of neem oil into βCD cavities resulted in a significant improvement in the antibacterial activity of the reactive prints. –The imparted antibacterial activity was attributed to neem oil-functional ingredients especially azadirachtin	64
β-CD	Knitted organic cotton	–Grafting or β-CD onto organic cotton using CA/NaH_2PO_2 cross-linking system by padding → drying and curing, followed by loading of thymol into the grafted β-CD cavities by immersing method	–Inclusion of thymol into grafted fabric enhanced its antibacterial functionality as well as durability to wash –The imparted antibacterial efficacy against G+ve and G-ve bacteria was attributed to capacity of thymol to destroy bacterial cells	125
MCT-βCD	Cotton fabric	Grafting of MCT-βCD onto cotton in alkaline medium using the pad-dry-cure fixation method followed by subsequent treatment with triclosan in the ethanol	–The functionalized cotton fabric samples had excellent antibacterial properties	17
MCT-βCD	Knitted organic cotton	–Biopolishing of organic cotton, to enhance the grafted yield, followed by grafting of MCT-βCD, using Na_2CO_3 as a catalyst and pad-dry-cure method, and finally post treated with thymol solution by soaking	–Eco-friendly durable antibacterial functionalized cotton fabric was obtained. –Biopolishing pre-treatment enriches the extent of grafting of MCT-βCD onto cotton fabric and subsequent accommodation of the antimicrobial agent into the βCD cavities	140

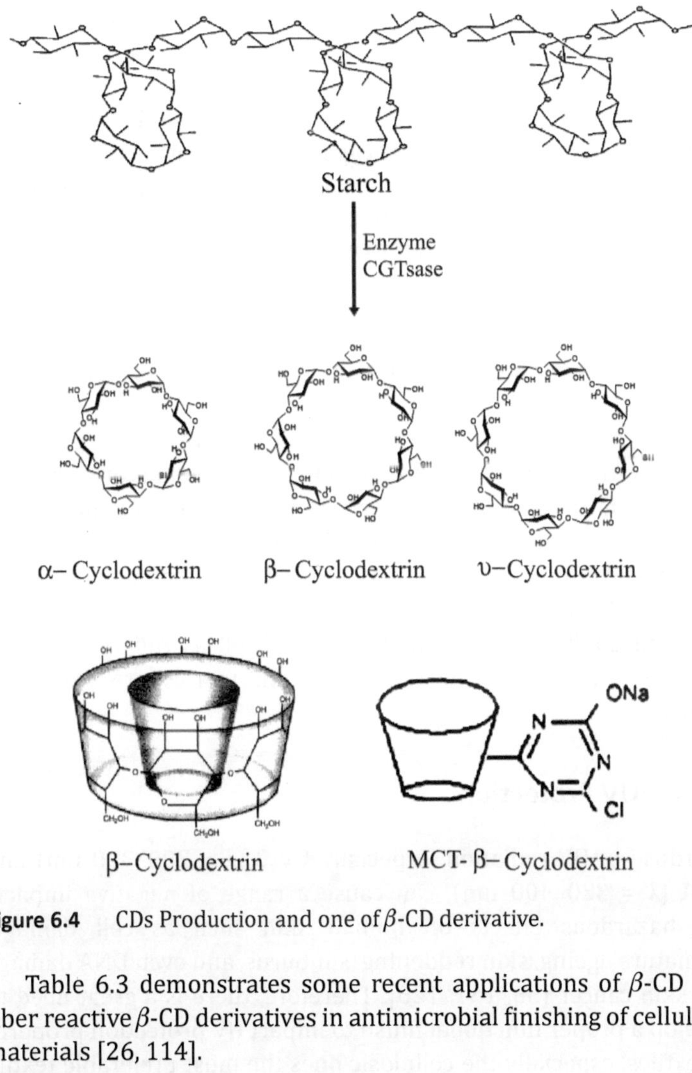

Figure 6.4 CDs Production and one of β-CD derivative.

Table 6.3 demonstrates some recent applications of β-CD and fiber reactive β-CD derivatives in antimicrobial finishing of cellulosic materials [26, 114].

6.3.3.5 Na-Alginate

Na-alginate is natural polysaccharide extracted from seaweeds, with a large availability in nature. Chemically, alginate composed of block polymers namely polyguluronate (poly-G), poly-mannuromate (poly-M) and their copolymer in random sequence [8, 72].

Alginate is an excellent gel forming materials capable of holding a huge amount of water, in presence of polyvalent cations, e.g., Ca^{2+},

Zn^{2+}, Al^{3+} etc., alginic acid becomes water insoluble. Its industrial utilization depends on its uronic acid composition. Being nontoxic, biocompatible, biodegradable and bioabsorber has given it a wide range of potential applications such as in the wound management, in controlled release and in encapsulation of various biocides, in biosorption of antimicrobial acting ions Ag^{+}, Zn^{2+}, Cu^{2+}, etc., in antimicrobial finishing as well as a stabilizer for the radiolytic synthesis of Au-NPs [8, 81, 97, 133, 137].

6.3.3.6 Aloe vera gel

Aloe vera is a natural plant contains wide range of physiologically active constituents such as anthraquinones, saccharides, enzymes, vitamins, and low-molecular-weight components that have antimicrobial, antioxidant, UV protection and anti-inflammatory activities in addition to wound and burn healing promoting properties [23, 73]. These active ingredients and multifunctional properties give special attention to Aloe vera gel and its potential applications in the textile finishing.

Some applications of Na-alginate as well as Aloe vera gel in antimicrobial functionalization of cotton fabrics are shown in Table 6.4.

6.3.4 UV Protection

Overdose of UV radiation, especially UV-B (λ = 280–320 nm) and UV-A (λ = 320–400 nm), can cause a range of negative impacts and hazardous effects on human skin such as cell damage, premature ageing, skin reddening, sunburns, and even DNA damage and skin cancer [55, 151, 126]. Therefore, there is a great need to develop a proper functional finish to impart UV protection property to textiles, especially the cellulosic ones the most preferable textile type during the summer time. UV protection factor (UPF) indicates the extent of protection against the harmful UV radiation, and the minimum UPF for protective clothing should be $40–50^{+}$. On the other hand, the enhancement in the UPF value is governed by the fiber type, chemical composition, fabric construction, moisture content, color and depth of shade, presence of UV protection additives such as optical brighteners, UV absorbers, etc., as well as finishing treatments [36, 90].

Table 6.4 Antimicrobial functionalization of cotton fabrics using some bio-active materials based on Na-alginate and Aloe vera

Biopolymer or its derivatives	Substrates	Finishing formulation and finishing technique	Comments	Ref.
Alginate micro capsules containing herbal extract	Cotton denim	–(herbal extract + Na-alginate)→spraying into $CaCl_2$ solution→microcapsules –Application of microcapsules on the fabric by exhaustion method using citric acid as a cross-linker, followed by drying and curing of the treated fabric at 120°C/2 min	–Using microcapsules technique, by ionic gelatin, enhanced the durability of the imparted antimicrobial functionality even after 15 washes –The antimicrobial activity against both *E. coli*, and *S. aureus* bacteria followed the decreasing order *S. aureus* > *E. coli*	137
Alginate-quaternary ammonium complex nanoparticles [SA-TSA, with an average size of 99 nm]	Cotton fabrics	Sodium alginate (SA)/3-(trimethoxysilyl) propyl-octadecyl-dimethylammonium chloride (TSA) complex nanoparticles were used for antimicrobial finishing of cotton fabrics by pad-dry-cure technique	–The functionalized cotton fabrics showed a remarkable durable antimicrobial activity against G+ve and G-ve bacteria –The enhancement in durability was attributed to the chemical binding of silanol groups, due to the partial hydrolysis of trimethoxysilyl groups in SA-TSA nanoparticles, and the cotton hydroxyl groups during thermo-fixation	81
Aloe vera extract	Cotton fabric	Finishing of cotton fabric was carried out using the exhaust	–Aloe vera-treated fabric samples showed a reduction in the growth of G+ve and G-ve bacteria	78

(*Continued*)

Table 6.3 (*Continued*)

Biopolymer or its derivatives	Substrates	Finishing formulation and finishing technique	Comments	Ref.
		method according to the following recipe: Aloe vera extract (6% owf), nonionic wetting agent (1.5% owf) at pH 5.5, LR 15:1 at 80°C/1 h	up to 90% before wash and 58% after four washes –Addition of Na-benzoate enhanced the stability of Aloe vera for about 6 months	
Aloe vera extract	Semi—bleached cotton fabric	Aloe vera, chitosan, and curcumin were applied in alone and in combination on cotton by exhaustion method	–Peroxide pretreatment of cotton fabrics showed better antimicrobial activity that their corresponding intact ones –The imparted antimicrobial activity was governed by the type of bio-active material and followed the decreasing order: Aloe vera > chitosan > curcumin –Combination of the above-mentioned bio-active agent resulted in an enhancement in both the activity and durability of the imparted functional properties to cotton fabrics.	6
Aloe vera extract	Cotton/wool (50/50) Viscose/wool (50/50) blends	–Inclusion of Aloe vera in the pigment printing formulation along with other constituents. –Printed fabric samples were dried/fixed in a one step process using microwave oven	–Inclusion of Aloe vera in the printing paste resulted in significant improvement in the depth of the obtained pigment prints along with a remarkable improvement in their antibacterial functionality even after 15 consecutive washing cycles	70

Some new applications of biopolymers or their derivatives in the enhancement in UV-protective functionality of cellulosic-containing fabrics are given in Table 6.5.

Table 6.5 The positive role of biopolymers or their derivatives in enhancing the UV-protective functional property of cellulose-containing fabrics

Biopolymer or its derivatives	Substrate	Finishing formulation and finishing technique	Comments	Ref.
β-CD	Cotton containing fabrics	–Cross-linking of cotton–containing fabrics using citric acid along with βCD as a reactive additive –Post-treated of cross-linked-grafted fabrics with Cu-acetate	–Easy care finished cotton containing fabrics with higher UPF level	55
Chitosan (as mordant)	Cotton	–Using chitosan as a mordant for natural dyeing with green tea extract. –Premordanting with chitosan followed by subsequently natural dyeing.	–Using chitosan as a mordant resulted in an improvement in dyeing and UV-protecting properties. –This was attributed to the positive effect of chitosan on enhancing the extent of uptake of phenoloic moiety, i.e., catechin	83
MCT-β-CD	Cotton containing fabrics	–Alkaline disperse dyeing of cotton/poly ester in the absence and presence of MCT-β-CD as a reactive additive	–Improving disperse dyeing properties along with a significant enhancement in the UPF level of the obtained dyeing	61

(Continued)

Table 6.5 *(Continued)*

Biopolymer or its derivatives	Substrate	Finishing formulation and finishing technique	Comments	Ref.
Dialdehyde chitosan	Cotton fabrics	–Improving –Pre-modification of cotton fabric with dialdehyde chitosan solution containing 3, aminopropyltriethoxy-silane (APETS) and 1,2 ethandiamine (EDA) using pad-dry-cure technique followed by after treatment with Na-borohydride to reduce the forming Schiff base	–Modification of cotton fabrics resulted in a significant improvement in its UV protection, wrinkle recovery as well as a marked increase in the color depth shade	44

6.3.5 Flame Retardancy

Flame-retardant textiles are textiles or textile-based materials that inhibit or resist the spread of fire. Factors affecting flammability and thermal behavior of textile include fiber type, fabric construction, thermal behavior of textile polymer and its composition as well as the presence or absence of flame additives. On the other hand, flame-retardant additives can be classified by their chemical composition or by mode of action, i.e., gas phase action or by the formation of protective barrier [49, 50]. Moreover, flame-retardant functional finishes of cellulose-based textiles can be accomplished by: (i) using inorganic phosphates, (ii) with organophosphorous compounds, (iii) with sulfur-derivatives or (iv) by grafting flame retardants monomers [49, 50].

In the last few decades, with the growing public concerns about fire hazardous, environmental risks, cleaner surrounding and safe/healthy life style, considerable efforts have been made to develop an eco-friendly/green new generation of flame retardants to replace the restricted hazardous or banned ones [3, 4, 49, 50].

Some of the new generations of flame-retardant additives used to confer retardant features to the cellulose containing fabrics are given in Table 6.6.

Table 6.6 Potential applications of some biopolymers to cotton textiles for flame retardancy

Additive	Substrate	Finishing formulation and finishing technique	Comments	Ref.
Chitosan	Cotton fabric	–Pretreatment with Na-poly-phosphate and subsequently with chitosan solution –Post-treatment of poly-phosphate treated samples with chitosan solution along with ethylenediamine tetramethacrylate as a cross-linker followed by drying and curing at 60°C for 5 min.	–Co-application of chitosan with polyphosphate resulted in enhancement in durability and flame retardancy compared with the untreated one, formation of a carbonaceous char and suppression of after glow	20
Chitosan	Cotton fabric	–The cotton fabric samples were padded with finishing bath containing diammonium hydrogen phosphate (DAHP), chitosan along with BTCA as a cross-linker and $NaHPO_4$ as a catalyst using the pad-dry-cur technique	–Incorporation of chitosan as well as increasing its concentration in the finishing formulation improved the durability and flame retardancy property. –The enhancement in the aforementioned properties was attributed to its synergetic effects as well as to the enhancement in the extent of phosphorylation	32

Table 6.6 (*Continued*)

Additive	Substrate	Finishing formulation and finishing technique	Comments	Ref.
Whey protein isolate (WPI)	Cotton fabric	–Folded and unfolded WPI products were deposited on cotton fabric by padding technique followed by drying	–The presence of protein coating on cotton fabric surface increased total burning time along with a reduced burning rate most probably due to its good oxygen barrier properties and great water vapor thereby affecting the thermal degradation of cotton and modify its combustion	13
Casein and hydrophobins (phosphorous and sulfur rich proteins)	Cotton fabric	–Casein suspension and hydrophobin solution were used without any additives –Treatment was carried out in a climatic chamber (30°C and 30% R.H.) followed by gently pressing and drying	–Presence of protein coatings changed the flammability feature of cotton, enhanced its flame resistance as well as modified the resistance to a heat flow of the treated fabrics –This was attributed to the ability of this surface treatment to favor to dehydration of cellulose instead of its depolymerization	5

6.3.6 Aroma Finishing

Inclusion of fragrance oils into the inner cavity of the grafted CDs onto the cellulosic substrates, via formation of inclusion complexes, and the gradual/control release of aroma during use, has been reported to impart new functional property, i.e., perfumed textiles [75, 118]. Using some biopolymers in aroma finishing of cellulosic fabrics, as binding or encapsulating agents, are given in Table 6.7.

Table 6.7 Aroma finishing of cotton cellulose using textiles auxiliaries based on biopolymers

Additives	Substrate	Finishing formulation and finishing technique	Comments	Ref.
Chitosan	Cotton fabric	-Chitosan solution along with an essential oil (rosemary)/Tween® 80 emulsion. -By padding technique, to wet pick up of 110%, followed by drying at 60°C/30 min.	-Treatment of the cotton fabric with rosemary essential oil and chitosan imparted aroma therapeutic properties to the finished substrate	105
β-CD	Cotton fabric	-Combined reactive dyeing and β-CD loading in one step. -Post-treatment for loading of essential oil (in dilute form as fragrance) by immersion or by spraying	-Incorporation of β-CD into dyeing bath enhanced the extent of fixation and life time of the fragrance on dyed cotton, and has no noticeably effect on fastness properties of the obtained perfumed dyeing.	19
MCT β- CD	Cotton fabric	-Pre-modification of cotton fabric using MCT β-CD along with Na_2CO_3 as a catalyst using pad-dry-cure technique -Post-treatment with alcoholic solution of sandal wood oil by spraying	-Fixation of MCT β-CD onto the cellulose cotton enhanced the fragrance retention for 21 days without negative impact on tensile strength of the treated fabric	134

6.3.7 Other Functional Properties

Table 6.8 shows the potential applications of biopolymers or their derivatives in imparting other functional properties such as insect-repellent, waterproof/breathable coating, hand building, easy care finishing/hand building to the treated cotton-containing fabrics.

Table 6.8 The positive role of certain biopolymer or their derivatives in imparting other functional properties to cotton-containing fabrics

Functional finish	Additives	Sub-strate	Finishing formulation and finishing technique	Comments	Ref.
Insect repellent	MCT β-CD	Cotton fabric	–Grafting of MCT β-CD onto cotton fabric by pad-dry-cure technique –Subsequent immersing in emulsion of pyethroid insecticide permethrin (PERM) for 12 h to include it into internal cavities of β-CD	Inclusion of PERM insect repellent into the free β-CD cavities resulted in enhancing the durability of the imparted insecticide/irritate functionality for a longer time compared with the current impregnation method	120
Waterproof breathable finish	PVOH (synthetic bio-degradable polymer)	Cotton fabric	–Coating cotton fabric with PVOH and partially neutralized poly (acrylic acid) as a cross-linker solution followed by drying at 90°C/10 min and curing 160°C/25 min	–Optimal conditions for attaining optimum breathable properties were: PVOH/poly (acrylic acid) molar ratio of 25 with coating add-on 18% –The coating was continuous without cracks and durable even after a harsh washing conditions	110

Functional finish	Additives	Sub-strate	Finishing formulation and finishing technique	Comments	Ref.
Hand-building	PVOHs	Cotton and C/PET blend (50/50)	Treatment of cotton containing solution using pad-dry-cure technique	–Fixation of PVOH onto/within cotton containing fabrics, via formation of semi inter penetrated network, resulted in remarkable increase in fabric stiffness as well as pilling resistance	153
Easy care and hand-building	PVOH CMS CMC	Cotton fabric	Treatment of cotton fabric with DMDHEU alone and in the presence of PVOH, CMS, CMC, PVOH/CMC or PVOH/CMS and ammonium salts as a catalysts using pad-dry-cure technique	–Incorporation of these additives in finishing formulation resulted in enhancing fabric stiffness, oil stain release, and affinity of basic dyeing along with a reasonable fabric resiliency. –The extent of functionalization was governed by type and concentrate-ion of additives as well as type of catalyst	2

6.4 Evaluation of Functional Finishes

Evaluation tests of functionalized cellulosic textiles are summarized in Table 6.9.

Table 6.9 List of standard test method for functionalized cellulosic textiles

Functional finishes	Test method	Comments
Antimicrobial	AATCC 147-1993 (parallel streak method)/ antibacterial assessment of textile method	–Agar diffusion tests semi-quantitative –Demonstrated zone of inhibition and around the treated fabric samples
	AATCC 90-1982 (agar plate method)/ antibacterial activity of fabrics	
	JISL 1902-1998 antibacterial of textiles	
	AATCC 100-1993 assessments of antibacterial finishes on textiles	
Insect repellent	AATCC 24-1999	
UV protection	Australia/New Zealand AS/NES 4399-1996	UPF Classification Grade >40 Excell Protection III 30–40 V. good Protection II 20–90 good Protection I
	AATCC183-2004	
Flame retardancy [129]	16CFR 1610 sponsor: CPSC (consumer product safety commission	–Fabric at 45°C angle to flame for 1 s –For general apparel
	NFPA 1971 sponsor: NFPA (national firefighters protection association)	–Fabric held vertical to flame for 12 s –For protection clothing
	ASTMD-2863 sponsor: ASTM (LOI: limiting oxygen index)	–Fabric held vertical –Oxygen/nitrogen atmosphere –Ignition from the top –Determines minimum oxygen level to support combustion

Functional finishes	Test method	Comments
	ISO 6940/6941 sponsor: ISO	–Vertically held specimens –Ease of ignition/flame spread properties
Fragrance retention and release	Assessment of the smell intensity of the samples of by: (i) sensory evaluation by a panel of experimenters at regular intervals [110], or (ii) by the GC-MS measurements (with solid-phase micro-extraction technique) [153]	–A blind method
Hand building	Fabric stiffness and drapability using Cantilever method: BS 3356-1961 and ASTM 1388-55T	
Waterproof breathable coating	AATCC 127-1977 on a Shirley hydrostatic-head tester	Resistance to water penetration
	BS 7209-1990	Water vapor permeability
	ASTM-D 737-96	Resistance to air penetration
Easy care	AATCC test method 66 → wrinkle recovery (woven fabric)	Dry and wet
	AATCC test method 124 → DP rating	–Finished fabrics are washed and dried in a standard manner –Fabric appearance is compared to standard scale (1–5) –Aim is a DP rating of > 3.5

6.5 Future Scope

To ensure further development and application of sustainable biopolymers in multi-functionalization of cellulose textile products, the following challenges should be taken into consideration.

- The future potential for application of sustainable biopolymers and derivatives to add a wide range of functionalities to cellulosic textiles, requires an economical and uniform production as well as evaluation of cytotoxicity of these products for full-scale implementation to assure the reproducibility and durability of the imparted functions together with a consumer safety
- Use of emerging technologies such as plasma, nano- and biotechnology along with sustainable biopolymers in functional finishing taking in consideration human, production, and disposal ecology concerns to develop more efficient and highly value-added cellulosic textile products are likely to continue in the future.
- Biopolymeric nanoparticles still need more R&D efforts for their development and large-scale production and implementation.
- Consumer demands for functional, smart, intelligent, interactive, and eco-friendly textile products using new generation and innovative textile auxiliaries based on sustainable bionanopolymers and other nanomaterials are undoubtedly growing.

References

1. Abdel-Aziz, M. S., Eid, B. M., and Ibrahim, N. A. (2014). Biosynthesized silver nanoparticles for antibacterial treatment of cellulosic fabrics using O_2-plasma, *AATCC J. Res.*, **1**, 6–12.
2. Abo-Shosha, M. H., Ibrahim, N. A., Fahmy, H. M., and Hebeish, A. (1995). Utilizing water soluble size additives in easy-care finishing, *Am. Dyestuff Reptr.*, **84**(7), 44–47.
3. Alongi, J., Carletto, R. A., Di Blasio, A., Corosio, F., Bosco, F., and Malucelli, G. (2013). DNA: A novel green, natural flame retardant and suppressant for textiles, *J. Mater. Chem. A*, **1**, 4779–4785.

4. Alongi, J., Corosio, F., and Malucelli, G. (2013). Current emerging technique to impart flame retardancy to fabrics: An overview, polym. *Degrad. Stabil.*, http://dx.doi.org/10.1016/j.polymdegradstab.2013.07.012.
5. Alongi, J., Carletto, R. A., Bosco, R., Carosio, F., Di Blasio, A., Cuttica, F., Antonucci, V., Giordano, M., and Malucelli, G. (2014). Caseins and hydrophobins as novel green flame retardants for cotton fabrics, *Polym. Degrad. Stabil.*, **99**, 111–117.
6. Ammayappan, L., and Moses, J. J. (2009). Study of antimicrobial activity of aloe vera, chitosan, and curcmin on cotton, wool, and rabbit hair, *Fiber Polym.*, **10**, 161–166.
7. Anderson, K. (2010). Ionic liquid: An environmentally friendly alternative. Available at: http://www.techexchange.com/library/Ionic%20Liquids%20%20An%20Environmentally%20Friendly%20Alternativc.pdf.
8. Anh, N. T., Phu, D. V., Dury, N. N., Du, B. D., and Hien, N. Q. (2010). Synthesis of alginate stabilized gold nanoparticles by γ- and seed particles enlargement, *Radiat. Phys. Chem.*, **79**, 405–408.
9. Araujo, R., Casal, M., and Cavaco-Paulo, A. (2008). Application of enzymes for textiles fibres processing, *Biocatal. Biotransform.*, **26**, 332–349.
10. Averous, L., and Pollet, E. (2012). Biodegradable polymers, in *Environmental Silicate Nano-Bioçomposite* (Averous, L., and Pollet, E., eds.), chapter 2, Springer-Verlarg London, pp. 13–39.
11. Bach, E., Cleve, E., and Schollmeyer (2002). Past, present and future of supercritical fluid dyeing technology—An overview, *Rev. Prog. Color.*, **32**, 88–102.
12. Bajaj, P. (2002). Finishing of textile materials, *J. Appl. Polym. Sci.*, **83**, 631–659.
13. Bajpai, M., Gupta, P., and Bajpai, S. K. (2010). Silver (I) ions loaded cyclodextrin-grafted-cotton fabric with excellent antimicrobial property, *Fiber Polym.*, **11**, 8–13.
14. Biwer, A., Antranikian, G., and Heinzle, E. (2002). Enzymatic production of cyclodextrins, *Appl. Microbiol. Biotechnol.*, **59**, 609–617.
15. Bosco, F., Carletto, R. A., Alongi, J., Marmo, L., Di Blasio, A., and Malucelli, G. (2013). Thermal stability and flame resistance of cotton fabrics treated with whey proteins, *Carbohyd. Polym.*, **94**, 373–377.
16. Buschmann, H.-J., Knittle, D., and Schollmeyer, E. (2001). New textile applications of cyclodextrins, *J. Incl. Phenom. Macrocycl. Chem.*, **40**, 169–172.

17. Cabrales, L., Abidi, N., Hammond, A., and Hamood, A. (2012). Cotton fabric functionalization with cyclodextrins, *J. Mater. Environ. Sci.*, **3**, 561–574.

18. Chandra, R., and Rustgi, R. (1998). Biodegradable polymers, *Prog. Polym. Sci.*, **23**, 1273–1334

19. Chao-Xia, W., and Shui-Lin, C. (2004). Anchoring β-cyclodextrin to retain fragrances on cotton by means of heterobifunctional reactive dyes, *Color. Technol.*, **120**, 14–18.

20. Charuchinda, S., Srikulkit, K., and Mowattana, T. (2005). Co-application of sodium polyphosphate and chitosan to improve flame retadancy to cotton fabric, *J. Sci. Res. Chula. Univ.*, **30**, 97–107.

21. Chen, C. S., Su, J. C., and Tsai, G. J. (1998). Antimicrobial effect and physical properties of sulfonated chitosan, in *Advances in Chitin Science* (Chen, R. H., and Chen, H. C., eds.), Vol. 3, Rita Advertising Co. Ltd, Taiwan, pp. 278–282.

22. Chen, C.-S., Liau, W.-Y., and Tsai, G.-T. (1998). Antibacterial effects of N-sulfonated and N-sulphbenzoyl chitosan and applications to oyster preservation, *J. Food Protect.*, **61**, 1124–1128.

23. Choi, S., and Chung, M.-H. (2003). A review on the relationship between aloe vera component and their biological effects, *Semin. Integration Med.*, **1**, 53–62.

24. Cook, J. G. (2001). *Handbook of Textile Fibres, Vol. 1: Natural Fibres*, 5th Ed. (Woodhead, Cambridge).

25. Daly, W. H., and Guerrini, M. M. (1998). Antimicrobial properties of quaternary ammonium cellulose and chitosan derivatives, *Polym. Mater. Sci. Eng.*, **79**, 220–221.

26. Dawson, T. L. (2008). Nanomaterials for textile processing and photonic applications, *Color. Technol.*, **124**, 261–272.

27. Diaz, D., Vurgas-Baca, I., and Gracia-Mora, J. (1994). β-cyclodextrin inclusion complexes with iodine, *J. Chem. Educ.*, **71**(8), 709–714.

28. Doakhan, S., Montazer, M., Rashidi, A., Moinri, R., and Moghadam, M. B. (2013). Influence of sericin/TiO_2 nanocomposite on cotton fbaric: Part 1. enhanced antibacterial effect, *Carbohyd. Polym.*, **94**, 737–648.

29. El-Shafei, A. M., Fouda, M. G., Knittle, D., Schollmyers, E. (2008). Antibacterial activity of cationically modified cotton fabric with carboxymethyl chitosan, *J. Appl. Polym. Sci.*, **110**, 1289–1296.

30. El-Sahfei, A., and Abou Okeil, A. (2011). ZnO/caroxymethyl chitosan bio nano-composite to impart antibacterial and UV-protection for cotton fabric, *Carbohyd. Polym.*, **83**, 920–925.

31. El-Tahlawy, K. F., El-Bendary, M. A., El-hendawy, A. G., and Hudson, S. M. (2005). The antimicrobial activity of cotton fabrics treated with different crosslinking agents and chitosan, *Carbohyd. Polym.*, **60**, 421–430.

32. El-Tahlawy, K. (2008). Chitosan phosphate: A new way for production of eco-friendly flame-retardant cotton textiles, *TJTI*, **99**, 185–191.

33. Farouk, A., Moussa, S., Ulbricht, M., and Textor, T. (2012). ZnO-nanoparticles-chitosan composite as antibacterial finish of textiles, *Int. J. Carbohyd. Chem.*, **2012**, article ID 693629, 8 pp. doi:10.1155/2012/693629, available at: http://dx.doi.org/10.1155/2012/693629.

34. Farouk, A., Textor, T., Schollmeyer, E., Tarbuk, A., and Grancacic, A. M. (2009). Sol-gel derived inorganic-organic hybrid polymers filled with ZnO nanoparticles as ultraviolet protection finish for textiles, *Autex Res. J.*, **9**, 114–120.

35. Fernandez Cid, M. V., van Sprosen, J., van der Kraan, M., Veuglers, W. J. T., Woerlee, G. F., and Witkamp, G. J. (2005). Excellent dye fixation on cotton dyed in supercritical carbon dioxide using fluorotriazine reactive dyes, *Green Chem.*, **7**, 669–616.

36. Gao, Y., and Cransion, R. (2008). Recent advance in antimicrobial treatments of textiles, *Text. Res. J.*, **78**, 60–72.

37. Ghanbarzadeh, B., and Almasi, H. (2013). Biodegradable polymers, in *Biodegradation: Life of Science* (Chamy, R., and Rosenkranz, F., eds.), Chapter 6, InTech, Croatia, pp. 142–185. ISBN 978-953-51-1154-2.

38. Goddard, J. M., and Hotchjkiss, J. H. (2007). Polymer surface modification for the attachment of bioactive compound, *Prog. Polym. Sci.*, **32**, 698–725.

39. Gorgani, A. S., and Shakib, N. (2013). Single-phase ink-jet printing onto cotton fabric, *Color. Technol.*, **129**, 109–113.

40. Gorjanc, M., Gorensek, M., Jovancic, P., and Mozetic, M. (2013). Multifunctional textiles: Modification by plasma, dyeing and nanoparticles, in *Eco-Friendly Textile Dyeing and Finishing*, (Gunay, M., ed.), Chapter 1, InTech, Croatia, pp. 1–31.

41. Gulrajani, M. L., and Gupta, D. (2011). Emerging technique for functional finishing of textiles, *IJFTR*, **36**, 388–397.

42. Gupta, D., and Haile, A. (2007). Multifunctional properties of cotton fabric treated with chitosan and carboxymethyl chitosan, *Carbohyd. Polym.*, **69**, 164–171.

43. Hashem, M., Hauser, P., and Smith, P. (2003). Wrinkle recovery for cellulosic fabric by means of ionic crosslinking, *Text. Res. J.*, **73**, 762–766.

44. He, X., Tao, R., Zhou, T., Wang, C., and Xie, K. (2014). Structure and properties of cotton fabrics treated with functionalized dialdehyde chitosan, *Carbohyd. Polym.*, **103**, 558–565.

45. Hebeish, A., and Ibrahim, N. A. (2007). The impact of frontier sciences on textile industry: Review article, *Colourage Ann.*, **54**, 41–55.

46. Higazy, A., Hashem, M., El-Shafie, A., Shaker, N., and Abdel-Hady, M. (2010). Development of antimicrobial jute packaging using chitosan and chitosan-metal complex, *Carbohyd. Polym.*, **79**, 867–874.

47. Holme, I. (2007). Innovative technologies for high performance textiles, *Color. Technol.*, **123**, 59–73.

48. Hon, D. N. S. (1996). *Chemical Modification of lignocellulosic Materials.* (Marcel Dekker, New York).

49. Horrocks, A. R., Kandola, B. K., Davies, P. J., Zhang, S., and Padbury. S. A. (2005). Developments in flame retardant textiles: A review, *Polym. Degrad. Stabil.*, **88**, 3–12.

50. Horrocks, A. R. (2011). Flame retardant challenges for textile and fibres: New chemistry versus innovatory solutions, *Polym. Degrad. Stabil.*, **96**, 377–392.

51. Hu, Z., Chan, W. L., and Szeto, Y. S. (2008). Nanocomposite of chitosan and silver oxide and its antibacterial property, *J. Appl. Polym. Sci.*, **108**, 52–56.

52. Ibrahim, N. A., El-Hossamy, M., Morsy, M. S., and Eid. B. M. (2004). Development of new eco-friendly options for cotton wet processing, *J. Appl. Polym. Sci.*, **93**, 1825–1836.

53. Ibrahim, N. A., Abd Allah, S. Z., Hassan, T. M., and Borham, H. A. T. (2005). Economical and ecological bio-treatment/half bleaching of cotton-containing knit fabrics on industrial scale, *Polym. Plastic Tech. Eng.*, **44**, 881–899.

54. Ibrahim, N. A., El-Zairy, M. R., Zaky, S., and Borham. H. A. (2005). Environmentally sound pigment printing using synthetic thickening agents, *Polym. Plastic Technol. Eng.*, **44**, 111–132.

55. Ibrahim, N. A. Refaie, R., Yossef, M. A., and Ahmed, A. F. (2005). Proper finishing treatment for sun protective cotton containing fabrics, *J. Appl. Polym. Sci.*, **97**, 1024–1032.

56. Ibrahim, N. A, Gouda, M., El-shafei, A. M., and Abdel-Fatah O. M. (2007). Antimicrobial activity of cotton fabrics containing immobilized enzymes, *J. Appl. Polym. Sci.*, **104**, 1754–1761.

57. Ibrahim, N. A., El-Hossamy, M., Hashem, M. M., Refai, R., and Eid, B. M. (2008). Novel pretreatment processes to promote linen-containing fabrics properties, *Carbohyd. Polym.*, **74**, 880–891.

58. Ibrahim, N. A., Hashem, M. M., Eid, M. A., Refai, R., El-Hossamy, M., and Eid, B. M. (2010). Eco-friendly plasma treatment of linen-containing fabrics. *J. Text. Inst.*, **101**, 1035–1049.

59. Ibrahim, N. A., Eid, B. M., Hashem, M. M., Refaie, R., and El-Hossamy, M. (2010). Smart options for functional finishing of linen-containing fabrics, *J. Ind. Text.*, **39**, 233–265.

60. Ibrahim, N. A., Refaie, R., and Ahmed, A. F. (2010). Novel approach for attaining cotton fabric with multi functional properties, *J. Ind. Text.*, **40**, 65–83.

61. Ibrahim, N. A., Zairy, W. R., and Eid, B. M. (2010). Novel approach for improving disperse dyeing and UV-protective function of cotton-containing fabrics using MCT-βCD, *Carbohyd. Polym.*, **79**, 839–846.

62. Ibrahim, N. A. (2011). Dyeing of textile fibre blends, in handbook of textile and industrial dyeing, in *Application of Dyes. vol.* 2 (Clark, M., ed.), Chapter 4, Woodhead, Cambridge, pp. 147–172.

63. Ibrahim, N. A., EL-Badry, K., Eid, B. M., and Hassan, T. M. (2011). A new approach for biofinishing of cellulose–containing fabrics using acid cellulases, *Carbohyd. Polym.*, **83**, 116–121.

64. Ibrahim, N. A., Eid, B. M., and El-Zairy, E. R. (2011). Antibacterial functionalization of reactive-cellulosic prints via inclusion of bioactive neem oil/*β*CD complex, *Carbohyd. Polym.*, **86**, 1313–1319.

65. Ibrahim, N. A., Amr, A., Eid, B. M., Mohammed, Z. E., and Fahmy H. M. (2012). Poly (acrylic acid)/poly (ethylene glycol) adduct for attaining multifunctional cellulosic fabrics, *Carbohyd. Polym.*, **89**, 684–660.

66. Ibrahim, N. A., Eid, B. M., and El-Batal, H. (2012). A novel approach for adding smart functions to cellulosic fabrics, *Carbohyd. Polym.*, **87**, 744–751.

67. Ibrahim, N. A., Eid, B. M., Abou Elmaaty, T. M., and Abd El-Aziz, E. (2013). A smart approach to add antibacterial functionality to cellulosic pigment prints, *Carbohyd. Polym.*, **94**, 612–618.

68. Ibrahim, N. A., Abou Elmaaty, T. M., Eid, B. M., and Abd El-Aziz, E. (2013). Combined antimicrobial finishing and pigment printing of cotton/polyester blends, *Carbohyd. Polym.*, **95**, 379–388.

69. Ibrahim, N. A., Eid, Abdel Aziz, E., and Abou Elmaaty, T. M. (2013). Functionalization of linen/cotton pigment prints using inorganic nano structure, *Carbohyd. Polym.*, **97**, 537–545.

70. Ibrahim, N. A., Khalil, H. M., El-Zairy, E. M. R., and Abdalla, W. A. (2013). Smart options for simultaneous functionalization and pigment coloration of cellulosic/wool blends, *Carbohyd. Polym.*, **96**, 200–210.

71. Ibrahim, N. A., Amr, A., Eid, B. M., Almetwally, A. A., and Mourad, M. M. (2013). Functional finishes of stretch cotton fabrics, *Carbohyd. Polym.*, **98**, 1603– 1609.

72. Islam S. U., Shahid, M., and Mohammed, F. (2013). Green chemistry approaches to develop antimicrobial textiles based on sustainable biopolymers: A review, *Ind. Eng. Chem. Res.*, **52**, 5245–5260.

73. Joshi, M., Ali, S. W., and Purwar, R. (2009). Eco-friendly antimicrobial finishing of textiles using bioactive agents based on natural products, *IJFTR*, **34**, 295–304.

74. Kale, K. H., and Desai, A. N. (2011). Atmospheric pressure treatment of textiles using non-polymerizing gases, *IJFTR*, **36**, 289–299.

75. Kant, A., Linforth, R. S. T., Hort, J., and Tylor, A. J. (2004). Effect of β-cyclodextrin on aroma release and flavor perception, *J. Agr. Food Chem.*, **52**, 2028–2035.

76. Kathirvelu, S., D'Souza, L., and Dhurai, B. (2009). A study on functional finishing of the cotton fabrics using nano-particles of zinc oxide, *Mater. Sci. (Medziagotyra)*, **15**(1), 75–79.

77. Khalifa, I. B., Ladhari, N., and Touay, M. (2012). Application of sericin to modify textile supports. *J. Text. Inst.*, **103**, 370–377.

78. Khan, A. F. (2012). Extraction, stabilization and application of antimicrobial agents from aloe vera, PTJ, April, pp. 42–44. Available at: http://www.ptj.com.pk/Web-2012/04-2012/April-2012-PDF/Finishing-AVM.pdf.

79. Kim, C. H., Choi, J. W., Chun, H. J., and Choi, K. S. (1997). Synthesis of chitosan derivatives with quaternary ammonium salt and their antimicrobial activity, *Polym. Bull.*, **38**, 387–393.

80. Kim, C. H., and Choi, K. S. (1998). Synthesis and properties of carboxyalkyl chitosan dervatives, *J. Ind. Eng. Chem.*, **4**, 19–25.

81. Kim, H. W., Kim, B. R., and Rhee, Y. H. (2010). Improving durable antimicrobial properties to cotton fabrics using alginate-quaternary ammonium complex nanoparticles, *Carbohyd. Polym.*, **79**, 1057–1062.

82. Kim, J. Y., Chung, Y. S., and Kim, S.-W. (1990). Preparation of chitosan derivative with cyanuric chloride and antimicrobial finish of cotton fabric, *J. Korean Fibre Soc.*, **36**, 618–623.

83. Kim, S.-H. (2006). Dyeing characteristics and UV-protection property of green tea dyed cotton fabrics-focusing on the effect of chitosan mordanting condition, *Fiber Polym.*, **7**, 255–261.

84. Kim, Y. H., Nam, C. W., Choi, J. W., and Jang, J. (2003). Durable antimicrobial treatment of cotton fabrics using N-(2-hydroxy) propyl-3-trimethylammonium chitosan chloride and polycarboxylic acids, *J. Appl. Polym. Sci.*, **88**, 1567–1572.

85. Klaykruayat, B., Siralertmukful, K., and Sirkulkit, K. (2010). Chemical modification of chitosan with cationic hyperbranched denderitic polyanidoamine and its antimicrobial activity on cotton fabric, *Carbohyd. Polym.*, **80**, 197–207.

86. Klemm, D., Heublein, B., Fink, H. P., and Bohn, A. (2005). Cellulose: Fascinating biopolymers and sustainable raw material, *Angew. Chem. Int. Ed.*, **44**, 2–37.

87. Koh, J. (2011). Dyeing of cellulosic fibers, in *Handbook of Textile and Industrial Dyeing* (Clark, M., ed.), Chapter 3, Woodhead, Cambridge, pp. 129–148.

88. Kumar, A., and Roy Choudhury, A. K. (2013). Green chemistry and the textile industry, *Text. Prog.*, **45**, 66–68, 91–92.

89. Kurita, K., Kojima, T., Nishiyama, Y., and Shimojoh, M. (2000). Synthesis and some properties of non-natural amino polysaccahrides: Branched chitin and chitosan, *Macromolecules*, **33**, 4711–4716.

90. Kursun, S., and Ozcan, G. (2010). An investigation of UV-protection of swimwear Fabrics, *Text. Res. J.*, **80**, 1811–1818.

91. Lam, Y. L., Kan, C. W., and Yuen, C. W. M. (2012). Development in functional finishing of cotton fibers-wrinkle resistance, flame-retardant and antimicrobial treatments, *Text. Prog.*, **44**, 175–249.

92. Lee, S. H., Kim, M. J., and Park, H. (2010). Characteristics of cotton fabrics treated with epichlorohydrin and chitosan, *J. Appl. Polym. Sci.*, **117**, 623–628.

93. Lim, S. H., and Hudson, S. M. (2003). Reviews of chitosan and its derivatives as antimicrobial agents and their uses as textile chemicals, *Polym. Rev.*, **43**, 223–269.

94. Lim, S. H., and Hudson, S. M. (2004). Application of a fibre-reactive chitosan derivative to cotton fabric as an antimicrobial textile finish, *Carbohyd. Polym.*, **56**, 227–234.

95. Lim, S. H., and Hudson, S. M. (2004). Synthesis and antimicrobial activity of a water-soluble chitosan derivative with a fibre-reactive group. *Carbohyd. Res.*, **339**, 313–319.

96. Lockett, A. P. (2003). Mechanical finishing: Traditional and modern, in *Textile Finishing* (Heywood, D., ed.), Chapter 4, *Society of Dyers and Colourists,* Bradford, pp. 114–134.

97. Mahltig, B., Soltmann, U., and Haase, H. (2013). Modification of algae with zinc, copper and silver ions for usage as natural composite for antibacterial applications, *Mater. Sci. Eng.*, **C33**, 979–983.

98. Martin Del Valle, E. M. (2004). Cyclodextrins and their uses: A review, *Process Biochem.*, **39**, 1033–1046.

99. Mellegard, H., Strand, S. P., Christensen, B. E., Granum, P. E., and Hardy, S. P. (2011). Antibacterial activity of chemically defined chitosans: Influence of molecular weight, degree of acetylation and test organism, *Int. J. Food Microbiol.*, **148**, 48–54.

100. Menezes, E., and Choudhari, M. (2011). Pretreatment of textile prior to dyeing, in *Textile Dyeing* (Hauser P., ed.), Chapter 11, Intech, Croatia, pp. 221–240.

101. Montazer, M., Malek, R. M. A., and Rahimi, A. (2007). Salt free reactive dyeing of cationized cotton, *Fibers Polym.*, **8**, 762–766.

102. Monatzer, M., and Afjeh, M. G. (2007). Simultaneous X-linking and antimicrobial finishing of cotton fabric, *J. App. Polym. Sci.*, **103**, 178–185.

103. Moore, S. E., and Wentz, M. (2009). Eco-labelling for textiles and apparel, in *Sustainable Textile-Life Cycle and Environmental Impact* (Blackburn R. S., ed.), Chapter 10, Woodhead, Cambridge, pp. 214–230.

104. Morose, G. (2010). The 5 principles of "design" for safer nanotechnology, *J. Clean. Prod.*, **18**, 285–289.

105. Muresan, A., Cereemplel, A. Duncu, S., Muresan, R., and Butnaru, R. (2009). Aromatherpeutic charcteristics of cotton fabrics treated with rosmary essential oil, *Cell. Chem. Technol.*, **43**, 435–442.

106. Muzzarelli, R. A. A., Muzzarelli, C., Tarsi, R., Miliani, M., Gabbanelli, F., and Cartolari, M. (2001). Fungistic activity of modified chitosan against saprolegnia parasitica, *Biomacromolecules*, **2**, 165–169.

107. Nielsen, P. H., Kuilderd, H., Zhau, W., and Lu., X. (2009). Enzyme biotechnology for sustainable textiles, in *Sustainable Textiles: Life Cycle and Environmental Impact*, Chapter 5, Woodhead, Cambridge, pp. 113–138.

108. Nithya, E., Radhai, R., Rajendran, R., Shalini, S., Rajendran, V., and Jayakumar, S. (2011). Synergetic effect of DC air plasma and cellulase enzyme treatment on the hydrophilicity of cotton fabrics, *Carbohyd. Polym.*, **83**, 1652–1658.

109. Nithya, E., Radhai, R., Rajendran R., Jayakumar, S., and Vaideki, K. (2012). Enhancement of antimicrobial propertiy of cotton fabric using plasma and enzyme pretreatments, *Carbohyd. Polym.*, **88**, 986–991.

110. Palankkumaran, M., Agrawal, A. K., and Jassal, M. (2008). Waterproof breathable coating based on poly (vinyl alcohol) for cellulosic fabric, *J. Ind. Text.*, **38**, 151–166.

111. Pan, N., and Sun, G. (2011). Fibre protection (Chapter 3); Improved water, oil and soil protection (Chapter 4), in *Functional Textiles for Improved Performance, Protection and Health*, The Textile Institute, Woodhead, Cambridge, pp. 54–84.

112. Park, J., and Shore, J. (2009). Evolution of right-first-time dyeing production, *Color. Technol.*, **125**, 133–140.

113. Qi, L., Xu, Z., Jiang, X., Hu, C., and Zou, X. (2004). Preparation and antibacterial activity of chitosan nanoparticles, *Carbohyd. Res.*, **339**, 2693–2700.

114. Qian, L., Guan, Y., Ziaee, Z., He. B., Zheng, A., and Xiao, H. (2009). Rending cellulose fibres antimicrobial using cationic *β*-cyclodextrin-based polymers include with antibiotic, *Cellulose*, **16**, 309–317.

115. Qian, T., Su, H., and Tan, T. (2011). The bacterial and mildew-proof activity of a TiO_2-chitosan compsoite, *J. Photochem. Photobiol. A: Chem.*, **218**, 130–136.

116. Qin, Y. (2008). Review alginate fibres: An overview of the production processes and applications in wound management, *Polym. Int.*, **57**, 171–180.

117. Radetic, M., Jovancic, P., Puac, N., and Petrovic, Z. L. (2007). Environmental impact to plasma application of textiles, *J. Phys. Conf. Ser.*, **71**, 1–11, doi:10.1088/1742-6596/71/1/012017.

118. Rajendran, R., Balakumar, C., Sivakumar, R., Amruta, T., and Devaki, N. (2012). Extraction and application of natural silk sericin from bombyx mori as antimicrobial finish for cotton fabrics, *J. Text. Inst.*, **103**, 458–462.

119. Rhim, J. W., Park, H. M., and Ha, C. S. (2013). Bio-nanocomposites for food packaging applications, *Prog. Polym. Sci.*, **38**, 1629–1652.

120. Romi, R., Nistro, P. L., Bocci, E., Ridi, F., and Baglioni, P. (2005). Bioengineering of a cellulosic fabric for insecticide delivery via grafted cyclodextrin, *Biotechnol. Prog.*, **21**, 1724–1730.

121. Roulette, H.-K. (2000). *Encyclopedia of Textile Finishing* (Springer, Berlin), pp. 184–192.

122. Rowell, R. M., Young, R. A., and Rowell, J. (1996). Chemical composition of agro-based fibres, in *Paper and Composites from Agro-based Resources* (Hans, J. S., and Rowell, J. S., eds.), Chapter 5, CRC-Lewis, Florid, pp. 83–134.

123. Roy Choudhury, A. K. (2006). Chemistry of textile materials, in *Textile Preparation and Dyeing*, Chapter 1, Science Publishers, USA, pp.1–16.

124. Roy Choudhury, A. K. (2006). Prepartory processes I: cleaning and preparatory (Chapters 4); Prepartory processes II: washing (Chapters 5), in *Textile Preapration and Dyeing*, Science Publishers, USA, pp. 131–207, 255–302.

125. Rukmani, A., and Sundrarajan, M. (2011). Inclusion of antibacterial agent thymol on β-cyclodextrin-grafted organic cotton, *J. Ind. Text.*, **42**, 132–144.

126. Saravanan, D. (2007). UV-protection textile materials, *AUTEX Res. J.*, **7**, 53–62.

127. Schindler, W. D., and Hauser, P. (2004). Introduction to chemical finishing, in *Chemical Finishing of Textiles*, Chapter 1, Woodhead, Cambridge, pp. 1–6.

128. Schindler, W. D. and Hauser, P. J. (2004). *Chemical Finishing of Textiles*, Chapter 15 "Antimicrobial Finishes" (Woodhead, Cambridge), pp. 165–174 .

129. Schindler, W. D., and Hauser, P. J. (2004). Flame retardant finishes, in *Chemical Finishing of Textiles*, Chapter 8, Woodhead, Cambridge, p. 113.

130. Shaikh, M. A. (2010). Enzymes: A revolution textile processing, *Pakistan Text. J.*, April, pp. 48–51.

131. Shamy, R., and Hussein, T. (2005). Critical solutions in the dyeing of cotton textile materials, *Text. Prog.*, **37**, 1–24.

132. Simkovic, I. (2013). Unexplored possibilities of all-polysaccharide composites, *Carbohyd. Polym.*, **95**, 697–715.

133. Simoncic, B., and Tomsic, B. (2010). Structure of novel antimicrobial agents for textiles: A review, *Text. Res. J.* **80**, 1721–1737.

134. Sricharussin, W., Spajaree, C., Manee, Rung, T., and Sangsuriya, N. (2009). Modification of cotton fabrics with β-cyclodextrin derivative for aroma finishing, *J. Text. Inst.*, **100**, 682–687.

135. Stegmaier, T., Linke, M., Dinkelmann, A., Von Arnim, V., and Plank, H. (2009). Environmentally friendly plasma technologies for textiles, in *Sustainable Textiles: Life Cycle and Environmental Impact*, Chapter 7, Woodhead, Cambridge, pp. 155–178.

136. Strand, S., Sauperl, O., and Fras-Zemljic, L. (2010). Cellulose fibres functionalizated by chitosan, in *Characterization and Application in Biopolymers,* (Elnashar, M., ed.), Chapter 9, InTech, Croatia, pp. 181–200. ISBN 978-953-307-109-1.

137. Sumithra, M., and Raaja, N. V. (2012). Micro-encapsulation and nano-encapsulation of denim fabric with herbal extract, *IJFTR*, **37**, 321–325.

138. Sun, D., and Stylios, G. K. (2005). Investigating the plasma modification of natural fibre fabrics-the effect on fabric surface and mechanical properties, *Text. Res. J.*, **75**, 639–644.

139. Sundar, S., Kundu, J., and Kundu, S. C. (2010). Biopolymeric nanotechnology, *Sci. Technol. Adv. Mater.*, **11**(1), 1–13, doi:10.1088/1468-6996/11/1/014104.

140. Sundrarajan, M., and Rukmani, A. (2013). Biopolishing and cyclodextrin derivative grafting on cellulosic fabric for incorporation of antibacterial agent thymol, *J. Text. Inst.*, **104**, 188–196.

141. Talreja, D., Talerja, P., and Mathur, M. (2003). Eco-friendliness of natural dyes, *Colourage*, **50**, 35–44.

142. Taylor, J. M., and Mears, P. (1991). Synthetic fibres in the dyehouse: The manufactures role, *J. Soc. Dyers Colour.*, **107**, 64–69.

143. Tsurugui, K., and Hiradie, T. (1994). Preparation and antibacterial activities of N-trimethyl-ammonium salt of chitosan, *Sen-I Gakkaishi*, **50**, 215–220.

144. Ugur, S. S., Sariisik, M., Aktas, A. H., Ucar, M. C., and Erden, E. (2010). modifying of cotton fabric surface with nano-zno multilayer films by layer-by-layer deposition method, *Nanoscale Res. Lett.*, **5**, 1204–1210.

145. Varshney V. K., and Naithani, S. (2011). Chemical functionalization of cellulose derived from nonconventional sources, in *Cellulose Fibers: Bio- and Nano-Polymeric Composites, Green Chemistry and Technology* (Kaila, S., Kaith, B. S., and Kaur, I., eds.), Chapter 2, Springer, Berlin, pp. 43–60.

146. Vigo, T. L. (1994). Preparatory processes, in *Textile Processing and Properties: Preparation, Dyeing, Finishing and Performance*, Chapter 1, Elsevier, Netherlands, pp. 1–43.

147. Wang, H.-J., and Chen, K.-M. (2005). Preparation and surface activity of biodegradable polymeric surfactants. I. preparation and surface activity of dextrin derivatives, *J. App. Polym. Sci.*, **98**, 711–717.

148. Wang, Q., and Hauser, P. T. (2010). developing a novel UV protection process for cotton based on layer-by-layer self-assembly, *Carbohyd. Polym.*, **81**, 491–496.

149. Wesolowska, E., and Wawro, D. (2009). An introduction to cellulosic fibres, in *Handbook of Textile Fibre Structure, Vol.2: Natural, Regenerated* (Eichhorn, S. J., Hearle, J. W. S., Jaffe, M., and Kikutani, T., eds.), Chapter 1, Woodhead, Cambridge, pp. 3–61.

150. Windler, L., Height, M., and Nowack, B. (2013). Comparative evaluation of antimicrobials for textile applications—a review, *Environ. Int.*, **58**, 62–73.

151. Xin, J. H., Daoud, W. A., and Kong, Y. Y. (2004). A new approach to UV-blocking treatment of cotton fabrics, *Text. Res. J.*, **74**, 97–100.

152. Zemljic, L. F., Persin, Z., and Stenius, P. (2009). Improvement of chitosan adsorption onto cellulosic fabric by plasma treatment, *Biomacromolecules*, **10**, 1181–1187.

153. Zia, K. M., Zuber, M., Rizwan, A., Jamil, T., Tubasum, S., and Shahid, M. (2012). Modification of cellulosic fabric using polyvinyl alcohol-part 1: physicochemical properties, *Carbohyd. Polym.*, **87**, 2063–2067.

154. ktem, T. (2003). Surface treatment of cotton fabrics with chitosan, *Color. Technol.*, **119**, 241–246.

Chapter 7

Research Progress in Polymer Complexes and Their Applications

Altaf H. Basta and Houssni El Saied

Cellulose and Paper Department, National Research Centre, Dokki-12622, Cairo, Egypt

Altaf_Basta2004@yahoo.com

7.1 Introduction

Chelating polymers have been employed successfully in some areas, such as removal of harmful trace metal ions, because of the highly selective adsorptive for heavy metal ions. They have also been given much attention with respect to the recovery of the adsorbed metal ions [1, 2]. In other words, the utilization of cellulose materials in their as-received state is facing some problems, for application as valuable new materials, such as dissolution of cellulose and coloring, and lacks its microorganisms and fire-resistant behaviours. Cellulose–metal complexes have been used to initiate graft copolymerization onto cellulose [3, 4] and dissolution of cellulose [5]. Functionalization of cellulose by either esterification or etherification reactions requires one approach for enhancing its solubilization in organic or water and, moreover, motivates its applications [5]. Cellulose derivatives

Handbook of Sustainable Polymers: Processing and Applications
Edited by Vijay Kumar Thakur and Manju Kumari Thakur

ISBN 978-981-4613-53-8 (Hardcover), 978-981-4613-54-5 (eBook)
www.panstanford.com

can react with metal ions, such as alkaline earth elements and heavy metal ions.

Sodium carboxymethyl cellulose (CMC) is considered to be one of the most important water-soluble cellulose derivatives. It is widely used in paper, textiles, food, pharmaceutical, petroleum, and other important industries [6]. CMC can react with metal ions to form precipitates; in this case, the polyvalent ions functions acts as a cross-linking agent [7]. Lepri and Desideri [8] investigated the chromatographic behavior of some metal ions on thin layers of CMC and Dowex. They estimated the stability constant of the resulting complexes as well as the selectivity coefficients of the ions for these two exchangers. CMC also has positive effect for flotation of nickel sulfide transition ore [9]. Despite highly promising applications of cellulose derivative–metal complexes, the knowledge about their compositions and structures had not been clearly established before year 1993. Therefore, since 1993 the elucidations the structures of polymers complexes have been a subject of interest to the authors of this chapter and others [10–19]. The structure, and consequently the characteristics of polymer complexes, helped us to specify their field of applications, in preparation of valuable products, e.g., functional paper and agro-composite [20–23]. In this chapter, we present our previous publications in the subject of structure elucidation and field of application in production functional of paper sheets [20–22].

7.2 Structures of Cellulose Derivative–Metal Complexes

Our published articles focused on using carboxymethyl cellulose and hydroxyethyl cellulose, and cellulose acetate, as ligands for cellulose derivative–metal complexes. These derivatives are examples of water-soluble cellulose ethers and solvent-soluble cellulose esters, respectively. The micro-analyses, electronic and IR-spectra, and magnetic susceptibility measurements were used as tools for studying the nature of the chemical structures of cellulose ethers complexes with some transition metals [10–16], while electrical and thermal analyses were carried out to identify the functional properties of cellulose ether provided from chelation with metal ions [17–20].

For cellulose ether–metal complexes [10–13, 17]

This study was carried out using CMC (Hercules Inc. product), with different degrees of substitution (DS 0.46, 0.67 and 1.3) and hydroxyethyl cellulose (HEC) of Natrosol 250 type (D.S. 1.5, MS. 2.5), as ligands and Cr(III), Cu(II), Ni(II), and Fe(III) chloride salts, Fe(III) sulfate, nitrate salts, as metal ions. These complexes were prepared by dissolving equimolar amounts of CMC (with respect to glucopyranose unit) and metal salt in the smallest possible volume of deionized water. The unreacted metals were separated by filtration or washing the casted films, several times by deionized water, followed by drying in vacuum oven. The final complexes were subjected to elemental analyses, magnetic moment measurements (at r.T., using Gouy's method), electronic absorption and IR-spectra measurements [24–26]. The data obtained are listed in Tables 7.1–7.5.

Table 7.1 Microanalytical data, magnetic moment and spectrophotometric determination of Cr(III) content of Cr(III)–CMC complexes

		Calculated (found)			
Sample	DS	C (%)	H (%)	μ_{eff} (BM)	Cr^{3+}
L_1	0.46	41.770 (43.4)	5.268 (4.6)	—	—
L_2	0.67	40.853 (41.4)	4.949 (4.7)	—	—
L_3	1.31	38.771 (40.3)	4.239 (4.9)	—	—
Cr–L_1	—	36.049 (35.7)	4.543 (4.9)	3.72	16.3
Cr–L_2	—	36.368 (36.6)	4.759 (4.5)	3.56	44.8
Cr–L_3	—	37.115 (38.1)	4.058 (3.8)	3.23	46.3

Table 7.2 Energy transitions and ligand field parameters of Cr(III)-CMC complexes

Sample	$^4T_{2g}\leftarrow{}^4A_{2g}$ ν_1 (cm^{-1})	$^4T_{2g}\leftarrow{}^4A_{2g}$ ν_2 (cm^{-1})	$^4T_{2g}\leftarrow{}^4A_{2g}$ ν_3 (cm^{-1})	D_q (cm^{-1})	B (cm^{-1})	β	C (cm^{-1})
Cr-L_1	16633	22512	37235	1663	616	0.67	2772
Cr-L_2	15285	21115	37230	1528	566	0.62	2547
Cr-L_3	15249	21017	37147	1525	554	0.60	2493

Table 7.3 Main IR frequencies of the CMCs and their Cr(III) complexes

L_1 (DS = 0.46)		L_2 (DS = 0.67)		L_3 (DS = 1.31)		Cr(III)–L_1		Cr(III)–L_2		Cr(III)–L_3	
ν (cm^{-1})	$E°$	ν (cm^{-1})	$E°$	ν (cm^{-1})	$E°$	ν (cm^{-1})	$E°$	ν (cm^{-1})	$E°$	ν (cm^{-1})	$E°$
—	—	—	—	3828	Very weak	—	—	—	—	3951	Very weak
—	—	—	—	3811	Very weak	—	—	—	—	3856	Very weak
3472	2.7002	3428	2.1478	3446	4.2243	3431	3.5250	3425	3.1429	3441	4.8004
—	—	—	—	—	—	1747	1.4318	1745	1.4272	1746	1.5876
1608	2.4141	1619	2.8706	1615	3.0587	1620	2.6835	1620	2.5099	1622	2.0603
1425	1.4340	1426	1.7295	1423	1.9520	1425	1.6878	1424	1.3431	1429	1.9808
1271	0.2275	—	—	1270	0.0449	—	—	—	—	—	—
—	—	1114	2.2051	1118	1.6694	—	—	—	—	1114	2.2812
1061	1.9469	1065	2.3459	1059	1.8144	1066	3.1179	1063	2.7157	1066	2.7213
906	0.8452	902	1.2384	902	0.0892	901	0.2648	902	0.2072	900	Very weak
715	1.2494	717	1.2913	710	0.6862	726	0.9383	721	0.3802	727	0.9225

Table 7.4 Nephlauxetic parameter of metal complexes (β) and electronegativity of metal cations

Complex sample	β	Electronegativity of metal cation
$[2(CMC)Cr \cdot 2H_2O]Cl$	0.62	1.6
$[(CMC)FeCl \cdot H_2O]Cl \cdot 2H_2O$	0.43	1.6
$[2(CMC)Cu]$	—	1.8
$[(CMC)Co \cdot 2H_2O]Cl$	0.78	1.7
$[(CMC)Ni \cdot 4H_2O]Cl$	0.80	1.8
$[(HEC)Cr \cdot 2Cl \cdot 2H_2O]$	0.72	1.6
$[(HEC)FeCl]Cl$	0.46	1.6
$[(HEC)Cu \cdot 2Cl] \cdot 2H_2O$	—	1.8
$[(HEC)Co \cdot 2Cl \cdot 2H_2O]$	0.73	1.7
$[(HEC)Ni \cdot Cl \cdot 3H_2O]$	0.69	1.8

Table 7.5a IR spectral measurements of cellulose ethers (CE) and their metal complexes

	CMC			HEC		
Complex	Δv (cm^{-1})	MHBS	C_rD	Δv (cm^{-1})	MHBS	C_rD
—	—	1.272	2.092		1.173	1.607
CE-Cr(III)	+3 (red)	1.786	2.901	+53 (red)	1.248	1.048
CE-Fe(III)	−76 (blue)	1.160	1.221	+42 (red)	1.451	1.628
CE-Cu(II)	−20 (blue)	1135	1.192	−10 (blue)	1.385	1.145
CE-Co(II)	−11 (blue)	1.380	1.789	−20 (blue)	1.278	1.058
CE-Ni(II)	−20 (blue)	1.517	2.697	46 (red)	2.863	2.258

Note: Red shift: the band maximum was shifted to a lower wave number; blue shift: the band maximum was shifted to a higher wave number.

Table 7.5b Relative absorbancies of hydroxyl and carbonyl bands of CE–Fe(III) complexes

Sample	v_{OH} (cm^{-1})	E_{OH}	$v_{C=O}$(cm^{-1})	$E_{C=O}$
CMC	3361	1.271	1595	0.785
CMC-Fe(III)	3437	0.502	1601	1.589
HEC	3416	1.576	1653	0.459
HEC-Fe(III)	3374	0.635	1636	0.617

Note: $E = Av/A_{1160\,cm^{-1}}$ where $A_{1160\,cm^{-1}}$ is the absorbance of the band assigned C=O of the glucopyranose ring, which is constant in all samples.

The studies showed that the degree of substitution and the type of ligand and metal ions played a significant role on the mode of chelation. For example, for CMC with Cr(III) ions, it was observed that changing the degree of substitution of CMC (L_1 = 0.46, L_2 = 0.67, and L_3 = 1.3) was accompanied by changing the elemental analyses, ligand field parameters, magnetic moment values, and IR-spectra (Tables 7.1–7.3). The Racah (B) was reduced compared with that of the free ion value (β_o = 918 cm^{-1}), which indicated the covalent character of CMC. The (β) value decreased with increasing DS of CMC. According to Jørgensen [24] a decreased nephlauxetic parameter (β) is associated with a reduction in nuclear charge on the cation. The reduction is more marked with the increase in carboxymethyl groups, leading to an increase in the degree of covalence in the metal–ligand–σ bond. Therefore, the nephlauxetic parameter values decreased as a function of the degree of substitution.

Based on the values of transition ratios (2.44, 1.76, 1.38), as well as the magnetic moments values 3.56 and 3.25 BM, the structures are nearly in agreement with the octahedral geometry around the Cr(III) ions. Moreover, the IR-spectra indicated that for Cr(III)-L_1 and Cr(III)-L_3 CMC **l** complexes, the chelating sites are not only the carboxymethyl group but also the hydroxyl group (Str. I). However, in the case of the Cr(III)-L_3 complex, the chelation provided through two carboxymethyl groups (Str. II) [10]. Also, the changes in the metal ions associated the change in structure geometry of CMC and HEC complexes. e.g., the side structures of HEC–Cr(III) and HEC–Ni(II) complexes [17]. The spectral data reported in Tables 7.4 and 7.5 [10–13, 17] reveal that anions of the CMC Ligands were chelated with metal ions corresponding to the formula $[(CMC)_2\cdot Cr\cdot 2H_2O]Cl$, $[(CMC)\cdot Co\cdot 2H_2O]Cl$, $[(CMC)Ni\cdot 4H_2O]Cl$, $(CMC)_2\cdot Cu$, $[(CMC)Fe\cdot H_2O]Cl.2H_2O$, $[(CMC)Fe\cdot H_2O]H_2O$, and $[(CMC)Fe, 2NO_3]\cdot 3H_2O$. While, HEC –metal complexes had formula $(HEC)\ Cr\cdot 2Cl\cdot 2H_2O$, $(HEC)Co\cdot 2Cl\cdot 2H_2O$, $(HEC)Ni\cdot Cl\cdot 3H_2O$, $[(HEC)Cu\cdot 2Cl]\cdot 2H_2O$ $[(HEC)FeCl$, $[(HEC)Fe\cdot H_2O]H_2O$ and $[(HEC)Fe]NO_3$.

For cellulose ester metal complexes [14–16]

Abdel-Hadi and Helmy [15] prepared Cr(III) complexes using acetylated bagasse dissolving pulp (I) and acetylated bagasse kraft paper pulp (II) as ligands. These complexes were also characterized by means of electronic and vibrational data. The electronic absorption spectra of Cr III with I and II exhibit absorption bands at 16,666 and

23,324 cm^{-1}, which are assigned in octahedral symmetry to the transitions $^4T_{2g} \leftarrow {}^4A_{2g}$ and $^4T_{1g} \leftarrow {}^4T_{2g}$, respectively. The third highest band, associated with the $^4T_{1g} \leftarrow {}^4T_{2g}$ transition, is not observed, being obscured by charge transfer absorption. The ligand field parameters D_q and B have been calculated. The nephelauxetic parameter, β, has the value 0.67 suggesting strong covalence in the metal–ligand–σ bond. Cr III–II complexes have higher optical absorbance than Cr III–I complexes. Since I has a greater acetyl content and lower degree of polymerization (D.P.) than II, II has more unacetylated hydroxyl groups, then the hydroxyl groups must be more involved in chelation. The secondary cellulose acetate (SCA) acts as uninegatively charged bidentate ligand and coordinates via the O atom of the hydroxyl group and the O atom of the ester group to form stable five-membered rings.

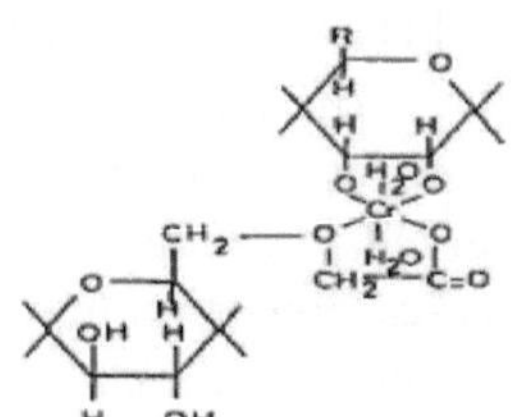

I

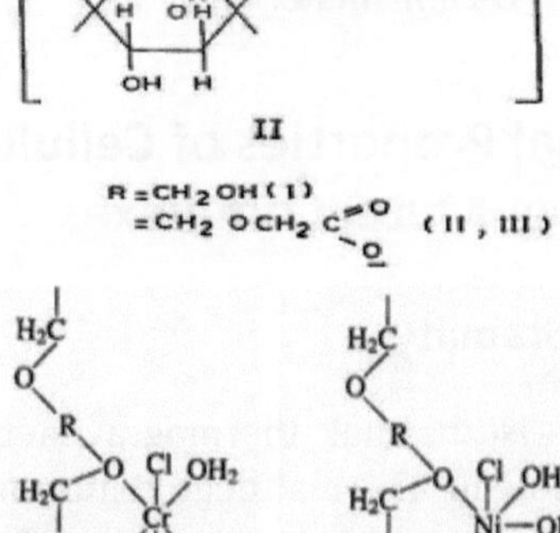

II

R = CH_2OH (I)
= $CH_2OCH_2C(=O)O$ (II, III)

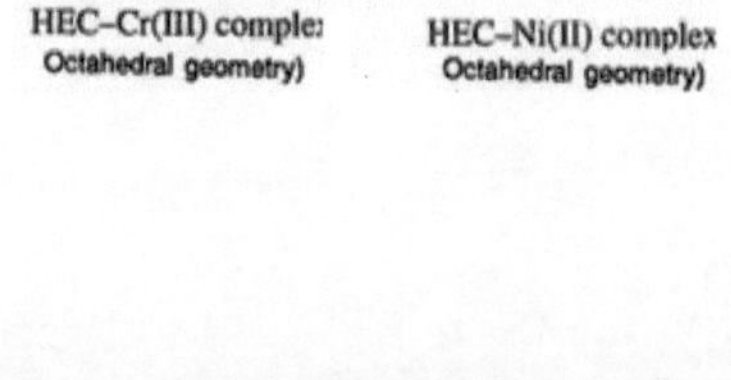

R = CH_2–CH_2

HEC-Cr(III) comple: Octahedral geometry) HEC-Ni(II) complex Octahedral geometry)

For our publication in this subject [14], we focused on elucidating the structure of cellulose acetate (CA)-UO_2(VI), because despite the importance of using cellulose derivatives as chelating resins for removing rare metal ions in waste- and sea-water, before year of publishing of this study, there was a lack of information regarding the structure of cellulose acetate-UO_2 (VI) complexes. We had prepared cellulose acetate-UO_2(VI) complex samples from three degrees of substitution of viscose pulp with uranyl acetate. The role of acetyl content of CA on the amount of chelated uranium was also evaluated (Table 7.6).

Probable structures of the cellulose acetate complexes were inferred from the elemental analyses data, conductance measurements, IR, electronic, and H NMR spectra. The results obtained show that the formula of UO_2(V1) complex with cellulose acetate of DS = 2.2 and 2.45 [(CA)·UO_2] is more probable than [(CA)·UO_2]. 2(CH_3COO), while the reverse is true for the case of a UO_2 complex with CA of DS = 2.86. For the former formula, cellulose acetate acts as a uninegatively charged bidentate ligand and reacts with UO^{+2} through the ether–carbon–oxygen of the secondary acetylated hydroxyl group of the anhydroglucose unit and the oxygen atom of the residual secondary unacetylated hydroxyl group, forming a five-membered chelate ring. For the later formula, cellulose acetate of DS = 2.86 acts as a neutral bidentate chelating agent through the two ether oxygen atoms of the vicinal ester groups of secondary acetylated hydroxyl groups in anhydroglucose units also forming a five-membered chelate ring. The uranium atom in these complexes is 8-coordinate.

7.3 Functional Properties of Cellulose Derivative–Metal Complexes

7.3.1 Thermal Stability

In this study, non-isothermal thermogravimetric analysis was quite helpful in following thermal degradation of cellulose ethers and esters and also for understanding how the metal chelation affected the thermal stability of cellulose ethers and cellulose ester [14, 17].

Table 7.6a Elemental analysis, uranium content and molar conductivity of cellulose acetate (CA) samples and their complexes with uranyl acetate

Sample	DS	Micro-analytical Values calculated (found) C (%)	H (%)	Ash (%)	Uranium (%)	λ_M (Ω^{-1} cm^2 mol^{-1})
CA(I)	2.20	44.06 (48.5)	5.66 (5.9)	0.113	—	—
CA(2)	2.45	49.38 (49.1)	5.63 (5.8)	0.106	—	—
CA(3)	2.86	49.89 (49.29)	5.57 (5.55)	0.113	—	—
CA(1)–UO_2 complex	—	37.32 (37.5)	4.31 (5.05)	22.243	19.277	19.40
CA(2)–UO_2 complex		37.92 (38.35)	4.32 (4.65)	25.525	22.131	21.10
CA(3)–UO_2 complex	—	35.12 (34.70)	3.99 (4.3)	34.467	29.922	46.80

Table 7.6b Electronic spectral data (nm) of cellulose acetate–dioxouranium (VI) complexes

Complex sample	^1E_g $^2\Pi_a$ transition	Π–Π* transition
CA(1)–UO_2 complex	405 Sh., 415 Sh., 440 Sh. and 450 Sh.	305
CA(2)–UO_2 complex	420 Sh., 425 Sh., 440 Sh. and 450 Sh.	303
CA(3)–UO_2 complex	405 Sh., 410 Sh., 420 Sh. and 450 Sh.	325

The CMC, HEC, and their complexes with the forgoing metal ions samples were heated in pure nitrogen (flow rate 50 mL/min) at 10°C/min, and within the typical temperature range 35–600°C, i.e., until no additional weight loss was observed. Measurements were made using calcined alumina as reference material. Differential thermogravimetric (DTG) peaks were examined for evidencing different behaviors between the samples, and also the kinetic parameters based on the weight loss data of TG curve analysis were determined according to the method described elsewhere [27, 28]. The TGA and DTGA of CMC and its complex with Cr(III) are illustrated in Fig. 7.1. The TGA measurements of all investigated cellulose derivative–metal complexes are illustrated in Tables 7.7–7.9.

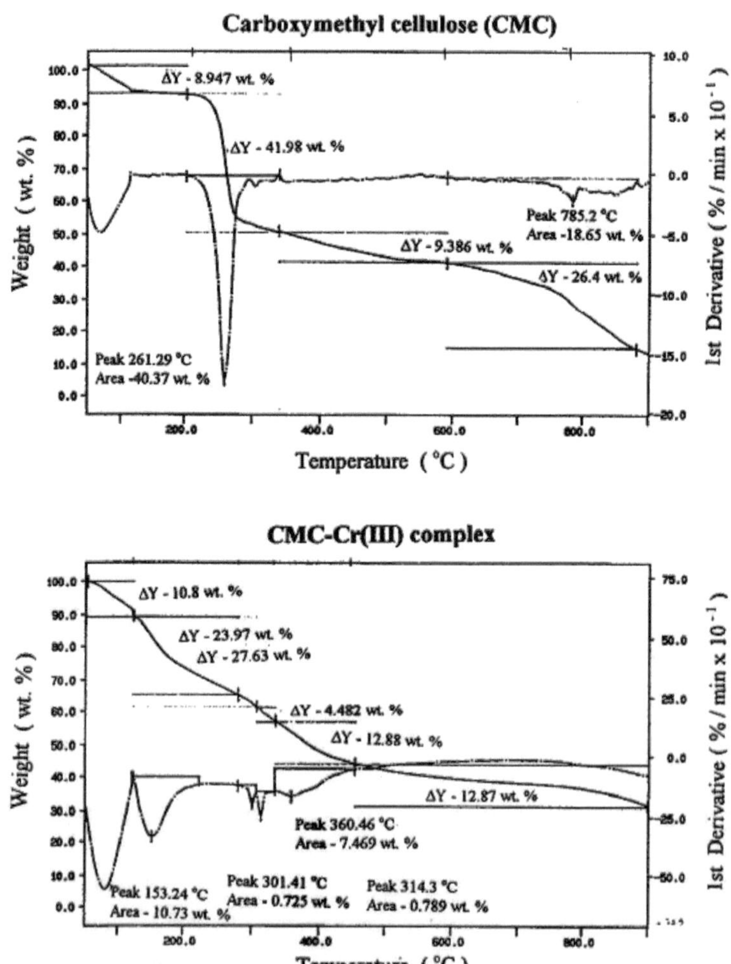

Figure 7.1 TGA and DTGA curves of CMC and its complexes with Cr(III).

Table 7.7 Temperature range, DTGA peak temperature, statistical parameters, and activation energies of the degradation stages of CMC and its complexes with some transition metals

Sample	Stage	Temp. range (°C)	DTGA peak (°C)	Order n	$-r$	SE	E_a (kJ/mol)
CMC	1st	204–340	261	1.0	0.9994	0.018	289.65
	2nd	592–882	785	2–0	0.9973	0.057	482.17
							ΣE_a = 771.8
CMC-Cr(II) complex	Ist	125–224	153	2.0	0.9957	0.068	100.72
	2nd	250–307	301	0.5	0.9915	0.061	347.73
	3rd	307–335	314	1.5	0.9935	0.074	550.06
	4th	385–445	361	2.0	0.9849	0.129	169.59
	5th[a]	789–>1000	—	—	—	—	—
							ΣE_a >1168.1
CMC-Fe(III) complex	1st	131–230	168	2.0	0.9892	0.110	135.5
	2nd	230–315	239	2.0	0.9911	0.101	177.72
	3rd	315–442	322	1.5	0.0986	0.103	113.63
	4th	442–679	617	1.5	0.9963	0.052	128.0
	5th	679–881	837	1.0	0.9987	0.028	333.5
							ΣE_a = 888.4
CMC-Cu(II) complex	1st	148–289	254	0	0.9889	0.052	36.304
	2nd	323–422	348	2.0	0.9923	0.095	282.94
	3rd*	726–>1000	—	—	—	—	—
							ΣE_a = >541.2
CMC-Co(II) complex	1st	175–227	199	1.5	0.9981	0.039	208.881
	2nd	227–37	269	2.0	0.9942	0.080	136.04
	3rd	398–454	416	1.5	0.9936	0.073	341.8
	4th	765–870	834	1.5	0.9939	0.070	451.44
							ΣE_a = 1138.2
CMC-Ni(II) complex	1st	137–202	186	1.5	0.9978	0.043	179.5
	2nd	202–351	258	1.5	0.9952	0.060	100.53
	3rd	351–389	370	1.5	0.9903	0.091	501.89
	4th	722–866	795	1.0	0.9974	0.038	385.67
							ΣE_a = 1167.59

[a]This stage not completed; it needs temperature >1000°C.

Table 7.8 Temperature range, DTGA peak temperature, statistical parameters, and activation energies of the degradation stages of HEC and its complexes with some transition metals

Sample	Stage	Temp. range (°C)	DTGA peak (°C)	Order n	$-r$	SE	E_a (kJ/mol)
HEC	1st	151–16	328	1.0	0.9988	0.0252	113.29
	2nd	416–695	542	0.5	0.9845	0.0765	117.93
	3rd	695–886	821	1.5	0.9968	0.0516	520.67
							$\sum E_a = 751.9$
HEC-Cr(II) complex	1st	114–221	144	2.0	0.9946	0.0756	206.232
	2nd	221–350	270	1.5	0.9854	0.1030	88.997
	3rd	350–613	393	2.0	0.9806	0.1379	91.992
	4th	613–893	808	2.0	0.9806	0.1465	288.30
							$\sum E_a = 676.0$
HEC-Fe(III) complex	1st	104–119	112	1.0	0.9972	0.0418	416.02
	2nd	119–304	170	2.0	0 9980	0.0447	65.606
	3rd	304–409	358	1.0	0.9955	0.0502	172.290
	4th	409–480	498	1.5	0.9945	0.0633	138.91
	5th	606–800	784	2.0	0.9924	0.0904	215.634
							$\sum E_a = 1008.5$
HEC-Cu(II) complex	1st	113–161	123	2.5	0.9787	0.1873	93.19
	2nd	161–388	228	2.0	0.9823	0.1297	62.54
	3rd	430–693	576	1.5	0.9952	0.0578	118.42
	4th	700–833	790	1.5	0.9817	0.1001	257.16
							$\sum E_a = 53.13$
HEC-Co(II) complex	1st	129–307	189	2.0	0.993 8	0.0807	87.51
	2nd	307–413	380	1.0	0.9913	0.0684	120.50
	3rd	413–466	457	1.0	0.9921	0.0679	332.12
	4th	466–649	502	2.0	0.9932	0.0864	191.36
	5th	715–879	796	2.0	0.9888	0.935	176.889
							$\sum E_a = 908.4$
HEC-Ni(II) complex	1st	150–346	230	2.0	0.9975	0.0513	86.448
	2nd	346–406	361	2.5	0.9859	0.1538	443.30
	3rd	406–597	530	1.5	0.9969	0.0477	138.91
	4th	597–726	669	1.5	0.9967	0.0518	320.22
	5th	726–851	796	1.5	0.9862	0.1183	325.58
							$\sum E_a = 1314.5$

Table 7.9 The average activation energy and order of degradation calculated from the DTA of the samples

Sample	Degradation stage	E_a (kJ/mol)	n
CDA	1st Stage	51.7278	1.0
	2nd Stage	39.9387	0.5
CTA	1st Stage	81.4183	1.5
	2nd Stage	60.9334	0.5
CDA-UO_2	1st Stage	33.1757	0.0
complex	2nd Stage	71.0970	1.0
	3rd Stage	84.2932	0.5
CTA-UO_2	1st Stage	51.9928	0.0
complex	2nd Stage	142.1134	1.0
	3rd Stage	53.5447	0.5

The promising finding was summarized in the following points:

- Complexation of cellulose ether with some transition metals does not behave in the same manner as that observed for parent cellulose ether (i.e., the number of degradation stages was increased and order of degradation were varied). This is due to the elimination of coordinated water and decomposition of cellulose ether–metal coordination bonds.
- Chelation of cellulose ethers with metal ions increased their thermal stability, except for the case of CMC–Cu(II), HEC–Cr(III), and HEC–Cu(II) complexes. The extent of increasing the thermal stability depends on the type of cellulose ether (anionic or neutral), electronegativity of metal ion, and the nephlauxetic parameter (degree of covalence in the metal-ligand-σ bond).
- Good relation existed between the thermal stability of cellulose ether complexes (Tables 7.7 and 7.8) and IR measurements data (mean hydrogen bond strength [MHBS] and degree of crystallinity [CrD]; Table 7.5).
- From the values of total activation energies (ΣE_a) in Tables 7.7 and 7.8, it was evident that the thermal stabilities of CMC–metal

complexes follow the sequence CMC–Cr (III) > CMC–Ni(II) > CMC–Co(II) > CMC–Fe(III) > CMC (Table 7.7; [17]).

However, for the case of HEC–metal complexes, the thermal stabilities follow the sequence HEC–Fe(III) > HEC–Co(II) > HEC > HEC–Cr(III) > HEC–Cu(II) (Table 7.8; [17]).

- Procedure for DTA peak analysis of dioxouranium [UO, (VI)] complexes with cellulose acetate (CA), using the equation of Prout and Tompkins (Table 7.9) showed that the increase of DS of cellulose acetate (from 2.2 to 2.86) and its chelation with uranium (VI) ions increased the activation energies for degradation. The increase in the thermal stabilities (activation energy) of cellulose acetate complexes can be attributed to the coordination bonds between dioxouranium and acetyl of cellulose acetate, i.e., to the formation of five-membered rings [14, 29].

7.3.2 Electrical Properties

Dielectric properties are related to the capability of a material to be polarized under the influence of an externally applied electrical field. The polarizability of a material depends on its structure and molecular properties and therefore dielectric measurements can provide information in this respect. The study of dielectric properties of various substances is an important tool for investigating their molecular structure and a rapid method to follow up the compatibility of the blends before and after the addition of different types of compatibilizers. Moreover, the materials with a high dielectric constants are good insulators [30, 31].

In continuation to our previous reported studies, the dielectric properties of the cellulose ether-transition metal complexes were examined [18]. The aim of this investigation was to find the relation between the dielectric measurements (permittivity and relaxation time) with the previous proposed structures, and ligand field parameters. The polymer complexes chosen for this study were prepared from cellulose ethers (CMC and HEC) with transition metals $CuCl_2$, $NiCl_2$, $CoCl_2$ and $FeCl_3$. The dielectric properties were studied over a frequency range 0.1–80 kHz, at ~25°C. Example of the relation between the permittivity (ε') and dielectric loss (ε″), we illustrated here the case of CMC and its metal complexes (Fig. 7.2). Figure 7.3 represents the analyses for CMC and CMC–Co(II) complexes. The dielectric properties in relation of metal contents and nephlauxetic parameter are given in Tables 7.10 and 7.11.

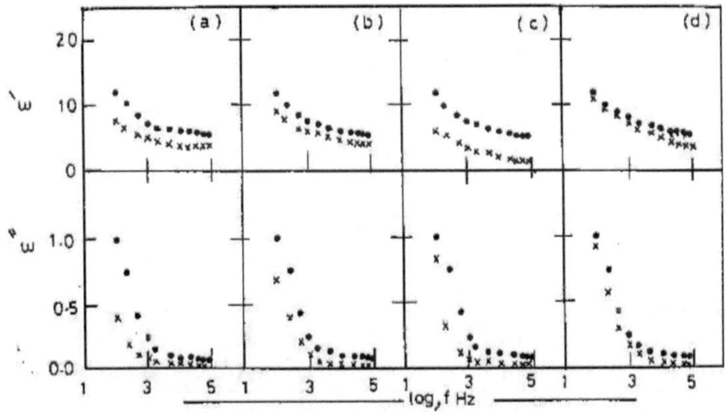

Figure 7.2 Relation between the permittivity (ε′) and dielectric loss (ε″) for carboxymethyl cellulose and its metal complexes: (a) copper chloride, (b) nickel chloride, (c) cobalt chloride, and (d) ferric chloride. (•) CMC × CMC–Metal complex.

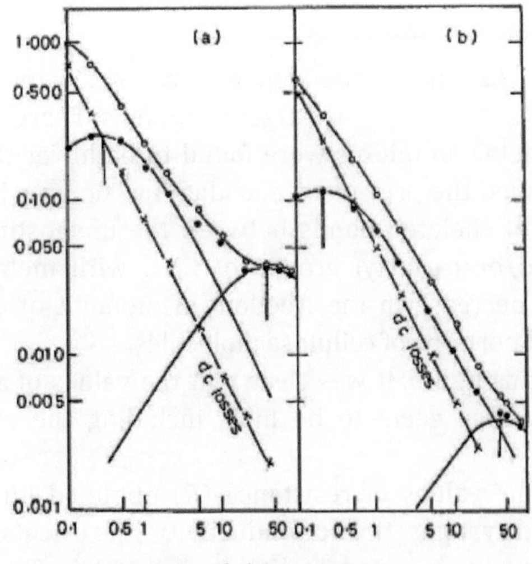

Figure 7.3 Absorption curves of (a) CMC, (b) CMC–Co(II) complex and (ε″ exp. × ε″ dc) • (ε″ exp. – ε″ dc).

Table 7.10 Dielectric and spectral data of carboxymethyl cellulose (CMC) and their complexes

Material	$\sigma_{dc} \times 10^{11}$ Ω^{-1} cm^{-1}	$\tau_1 \times 10^4$ s	$\tau_2 \times 10^6$ s	Metal, %	β
CMC	4.3	6.4	5.3	—	—
CMC–Cu(II)	2.0	8.4	5.0	6.9	—
CMC–Ni(II)	3.4	8.6	5.3	9.1	0.80
CMC–Co(II)	2.8	8.4	5.0	11.5	0.78
CMC–Fe(III)	4.2	9.9	5.0	15.8	0.43

Table 7.11 Dielectric and spectral data of hydroxyethyl cellulose (HEC) and their complexes

Material	$\sigma_{dc} \times 10^{11}$ Ω^{-1} cm^{-1}	$\tau_1 \times 10^4$ s	$\tau_2 \times 10^6$ s	Metal, %	β
HEC	5.2	5.3	4.0	—	—
HEC-Cu(II)	3.0	—	4.0	88	—
HEC-Ni(II)	4.0	6.2	4.0	7.1	0.73
HEC-Co(II)	3.3	6.3	4.0	8.3	0.69
HEC-Fe(III)	5.0	7.2	4.0	15.1	0.46

Figure 7.2 shows the dependence of ε'' upon the applied frequency for the above investigated systems. These measurements for CMC–metal complexes were found to be lower than CMC. This was confirmed the previously elucidated structures [11–13], to the formation of chelated bonds between the unsubstituted hydroxyl groups and/or carboxyl groups of CMC with metal ions, which leads to a decrease in the freedom of moment of carboxymethyl groups and portions of cellulose molecules.

From this figure, it was clear that the values of ε' in the lower frequency range seem to be high, including the presence of dc conductivity.

From the values of resistance (R) obtained for the different investigated systems, the dc conductivity was calculated, according to the equation reported in [32] and listed in Table 7.10. After subtracting the dc losses from the measured ε'' values, the results showed a well-defined absorption region (Fig. 7.3) according to the Frohlich equation [33]. Table 7.11 shows these values in the case of HEC and its complexes. Table 7.10 shows that the structure of CMC–metal complexes, either neutral or electrolyte, as well as the initial value of the dc conductivity of metal chloride salts, both play an important effect on the value of σ_{dc} of the complexes. However, the neutral structure CMC–Cu(II) complex prepared from relatively low conducting salt possesses a low σ_{dc} value, compared with other complexes [CMC–Ni(II) and CMC–Co(II)], which have electrolyte structures [11, 12]. On the other hand, CMC–Fe(III) complexes had relatively high σ_{dc} value and near to the of CMC and relaxation time (τ).

For the case of HEC–metal complexes (Table 7.11), the relaxation times (τ) were found to be less than that detected for different CMC–metal complexes. This view was realized to the presence of polyhydroxyethyl groups on the cellulose backbone, which result in inter-chain spaces and consequently reduce the effect of volume expansion due to the chelation of HEC with metal ions [18].

On comparison of the dielectric measurements with the previously calculated nephlauxetic parameter β [11–13], Tables 7.10 and 7.11 show a good relation between the increase in the metal content, which is considered to be a measure for complex formation, and degree of covalence of metal–ligand–σ bonds (β-value) of cellulose ether complexes with Ni(II), Co(II) and Fe(III) ions. We excluded the complexes with Cu(II), because the geometry structure proposed for these complexes square planner Cu(II) complexes could not be calculated the β-value. The decrease in the value of β indicates considerable overlap with a strong covalent metal.

Based on the definition of dielectric relaxation time, it could be recommended that CMC provided higher semiconductor and insulators than HEC. Therefore, our further studies were concerned with CMC–metal complexes as semiconductor materials and their application in paper production. Depending on distinguishing characteristics of the previous investigated cellulose derivative-metal complexes the field of their applications can be specified.

7.3.3 Electrical Conductivity

In [19], we studied the electrical conductivity (AC) of CMC and its CMC–Cu(II) complex. Moreover, the role of carboxymethyl content (DS) of CMC and origin of Cu(II) ion on semiconductor behavior of resulted complex materials were studied. Three degree of substitution of CMC (0.43, 0.67 and 1.3), and Cu(II) ion originated from $CuCl_2 \cdot 2H_2O$ and $CuSO_4 \cdot 5H_2O$ salts were used in preparing these complexes. The AC electrical conductivity measurements were carried out on the complex samples in the form of pellets (~1.25 cm diameter and 0.2 cm thick.). The pellet and copper electrode were connected to high-frequency oscillator, through a suitable standard resistance, across which voltage values were measured by means of an oscilloscope type Trio-10 MHz/Cs 1562 A (Japan). The temperature dependence of the conductivity was obtained in the temperature range 25–400°C. The variation of the electrical conductivity, σ, of three grades of CMCs and their complexes with Cu(II) in temperature range 25–100°C was determined and illustrated in Figs. 7.4–7.6. Also, the calculated activation energies [33, 34] of CMCs and their complexes were registered in Table 7.12.

Table 7.12 Values of logσ, activation energy (ΔE), and temperature for the First and second peaks of CMCs and CMC–Cu(II) complexes

	First Peak				Second Peak			
Sample	-log σ (Ω^{-1} cm^{-1})	t (°C)	ΔE_1 (eV)	$-r''$	-log σ (Ω^{-1} cm^{-1})	t (°C)	ΔE_2 (eV)	$-r''$
				CMC of DS ~0.46 (L_1)				
L_1	4.31	110	0.37	0.98	2.20	352	2.71	0.97
L_1–Cu(II)[b]	5.40	112	0.03	0.91	4.95	311	0.53	0.97
L_1–Cu(II)[c]	4.53	218	035	0.94	4.80	290	0.33	1.00
				CMC of DS ~0.67 (L_2)				
L_2	4.50	118	0.33	0.99	2.10	352	3.14	0.97
L_2–Cu(II)[b]	5.43	121	0.04	0.93	4.45	330	1.35	0.99
L_2–Cu(II)[c]	4.79	158	0.25	0.97	4.55	285	0.76	0.95
				CMC of DS ~1.31 (L_3)				
L_3	4.45	120	0.30	0.99	2.08	342	4.10	0.973
L_3–Cu(II)[b]	5.46	120	0.03	0.93	4. 83	315	0.76	0.955
L_3–Cu(II)[c]	4.45	153	0.12	0.96	5.10	271	0.13	0.986

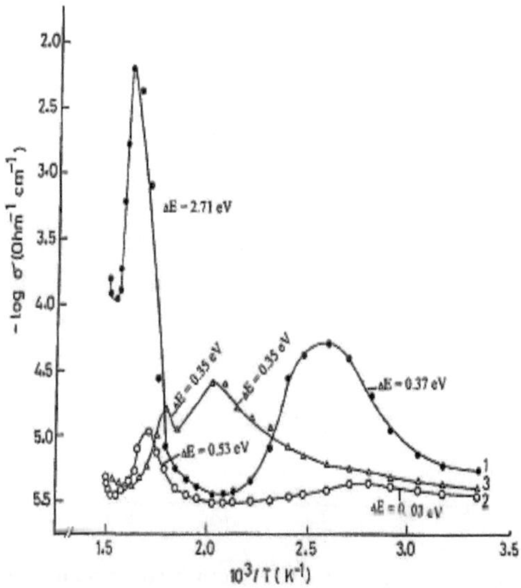

Figure 7.4 Variation of electrical conductivity with reciprocal temperature at 100 kHz for (1) CMC (L_1) and its complexes with (2) $CuCl_2$ and $CuSO_4$.

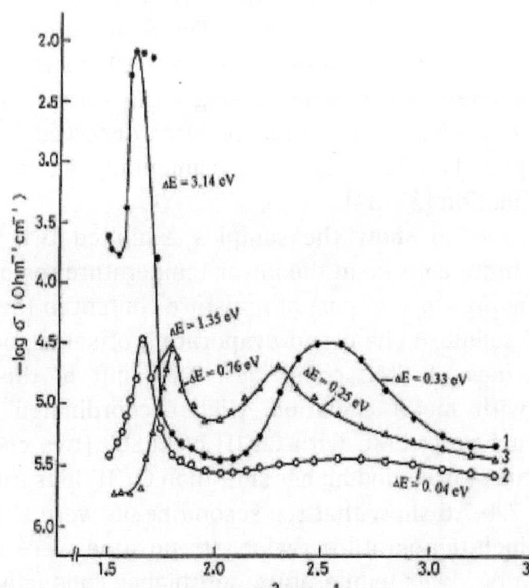

Figure 7.5 Variation of electrical conductivity with reciprocal temperature at 100 kHz for (1) CMC (L_2) and its complexes with (2) $CuCl_2$ and $CuSO_4$.

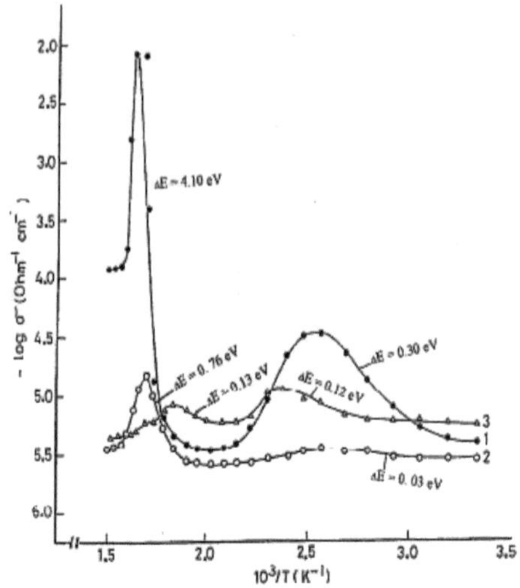

Figure 7.6 Variation of electrical conductivity with reciprocal temperature at 100 kHz for (1) CMC(L_3) and its complexes with (2) $CuCl_2$ and $CuSO_4$.

From Figs. 7.4–7.6, it is clear that the shape of the plots have more or less similar trends in behavior. For all investigated samples, the values of resistivity lie within the range 10^5 to 10^2 Ω cm, and these values decreased with increasing temperature. This meant that we were dealing with a behavior often observed in semiconductor compounds, which varies experimentally according to the Arrhenius equation [33, 34].

Figures 7.4–7.6 show the samples exhibited two peaks. In general, the first peaks lie in the lower temperature region and are related to the presence of part of moisture content in the capillary structure of cellulose chain and evaporating of some coordinated water molecules of CMC complexes. The shift of this peak is associated with metal chelation. Higher coordinated water in complex structure of CMC with Cu(II) of $CuSO_4$ (two coordinated water molecules) provided higher shift than Cu(II) ions from $CuCl_2$.

Figures 7.4–7.6 show that the second peaks were sharper and lied in the high-temperature region. Its position were shifted in the direction of lower temperature and higher conductivity when the DS increased from 0.46 to 1.31. Metal chelation of CMC also led to a shift of the peak maxima to a lower temperature and conductivity according to the sequence CMC–Cu(II) of $CuSO_4$ < CMC

–Cu(II) of $CuCl_2$ < CMC (Table 7.12). It was also observed that the trend of variation of activation energy values (ΔE_1 and ΔE_2) of different examined samples was similar to the corresponding trend for the –log σ peak values (Table 7.12). The increase of the percentage of the Cu(II) content in the complex as a result of the increase of the DS value of the ligand caused a decrease in activation energy.

The observed increase in the values of the activation energies, in the high-temperature region (second peak) means that the samples were converted from low to high semiconductor materials. In other words, it was motivated by the semiconductor behavior of the materials. However, due to the combined effect of the conduction mechanism and the number of metal and chelating groups, the decreased in activation energy was observed.

7.4 Applications of Cellulose Derivative–Metal Complexes for Paper Production (Functional Paper Sheets)

In this section, we summarize our published articles on how we can utilize the previously investigated cellulose-ether-metal complexes to overcome some problems faced with respect to the quality of paper, as functional paper sheets.

Trials have been made by numerous investigators to improve the quality of paper sheets, to meet the different applications, e.g., coloration, as insulating materials for electrical equipment, which need higher voltages, as well as durable and fire-, magnetic-, and biological-resistant paper sheets.

For providing electrical paper sheets (reduce the dielectric loss and increase the dielectric properties), literature attempts include the modification of paper substrate by oxidation [35], impregnation of cellulose paper with insulating oils [36], and replacing the cellulose fibers by some synthetic materials that exhibit better dielectric properties than cellulose, e.g., polypropylene fibers and other synthetic fibers [37, 38], or by treating with ferric chloride [39].

For flame-retardant paper, the flame prevented from the presence of some elements such as phosphorous, halogen, nitrogen, and boron in the retardant [40, 41]. Phosphorous compounds strongly affect pyrolysis and char formation, whereas the halogens considerably influence polymer breakdown and combustion. Nitrogen in combination with phosphorus has a synergistic effect. Phosphoric acid and aluminum chloride probably increased the rate of dehydration, while boric acid may have inhibited levoglucosan formation. Dehydration reactions of polysaccharides, catalyzed by many inorganic salts, lead to a higher portion of charring with a reduced amount of evolved levoglucosan. The drawback of lignocellulosic products treated by these salts resulted from the unfixation and leaching of these salts when exposed to humidity and consequently leads to the loss of their fire retardancy property together with creating environmental impact.

For magnetic paper, maghemite (γ-Fe_2O_3) and magnetite Fe_3O_4 are the widely used magnetic pigments in the production of magnetic recording and information storage media. Magnetic fibers are also obtained by synthesizing ferrites in situ using suitable natural cellulose fibers possessing appropriate functional groups or chemically modified cellulose fibers to introduce these functional groups, e.g., carboxyl, sulfonyl, and sulfate groups [42, 43].

For coloring the paper sheets, there are a number of requirements for colored paper considered in certain applications, especially as cable and laminate base paper for the furniture industry and fine paper. Strength, quality, longevity, and heat resistance are the major requirements for such types of paper. There are two parts to a coloring system, (1) The selection of the actual coloring agents, and (2) the method by which the colorants are attached to the substrate [44].

Currently, the manufacture of colored paper is achieved by dyeing the paper stock, surface coloring in the size press, or surface coloring by coating, using dyes or pigments as coloring agents. Dyes are more affected by IR, UV, and visible light due to photochemical reactions, causing fading of the color. The coloring agents that provide the best longevity are all pigments (not dyes). Unfortunately, pigments are insoluble and have no affinity to the substrate and require the use of binders. A common and economic way of retaining additives in paper pulp is by means of alum and rosin. This often results in an acidic paper and causes its deterioration and aging [44–46]. When flame retardancy and biological resistance are required, it is necessary to provide such properties together with longevity to the colored paper.

Our trials through published articles [20–22] to provide functional paper sheets with serving the previous drawbacks, and based on the characteristics of cellulose derivative–metal complexes, the CMC- and HEC–metal complexes were used as beater additives during the formation of paper sheets from wood pulp. The thermal degradation and magnetic and electrical conductivity properties were tested to evaluate the performance of paper product besides the strength properties.

7.4.1 Application of Cellulose Ether–Fe(III) Complexes

For applying CMC–Fe(III) and HEC–Fe(III) complexes, their ligands complexes with Zn(II), and metal oxides of Fe_2O_3 and ZnO were added for comparison. In this study, we examined the effect of ligand type, cations and anions of metal salts, and pH during sheet formation. The electrical and strength measurements are illustrated in Figs. 7.7 and 7.8 and summarized in Table 7.13.

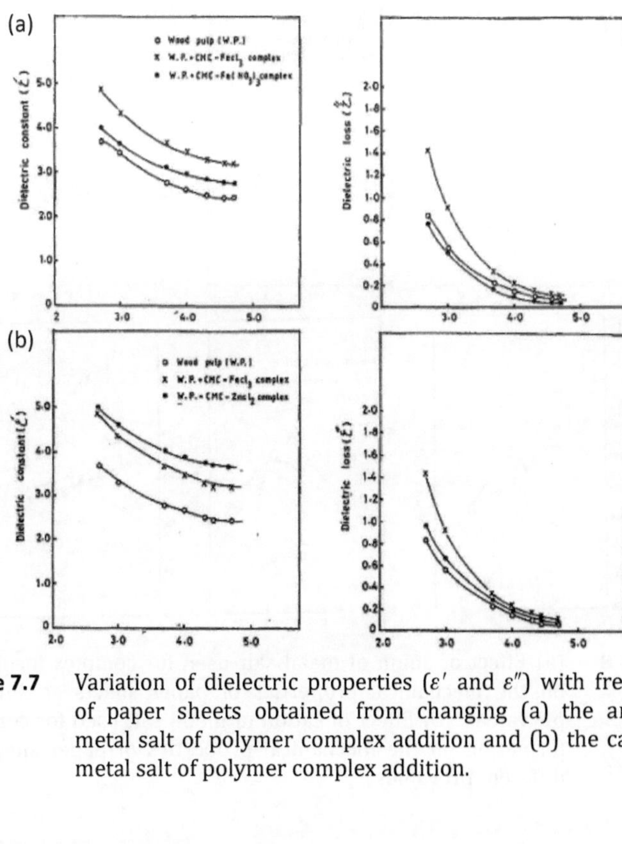

Figure 7.7 Variation of dielectric properties (ε′ and ε″) with frequency of paper sheets obtained from changing (a) the anion of metal salt of polymer complex addition and (b) the cation of metal salt of polymer complex addition.

Table 7.13 Effect of $FeCl_3$ and polymer complexes addition on mechanical and dielectric properties of produced paper sheets

			Mechanical properties			Dielectric properties*	
Additive	**pH**	**Ash%**	**Breaking length, *m***	**Burst factor**	**Tear factor**	ε'	ε''
—	6.53	0.21	4539.35	33.07	108.07	3.68	0.84
$FeCl_3$	6.65	7.42	5044.49	45.55	148.55	4.25	1.72
HEC-$FeCl_3$ complex	7.08	8.92	4646.17	35.17	129.84	3.89	0.69
CMC-$FeCl_3$ complex	6.36	10.40	6122.38	54.42	153.48	4.90	1.44
CMC-$Fe(NO_3)_3$ complex	7.32	7.69	6657.32	57.98	159.90	4.00	0.78
CMC-$ZnCl_2$ complex	6.62	5.27	7049.34	60.42	136.12	4.99	0.97

*Dielectric properties at frequency 0.5 kHz.

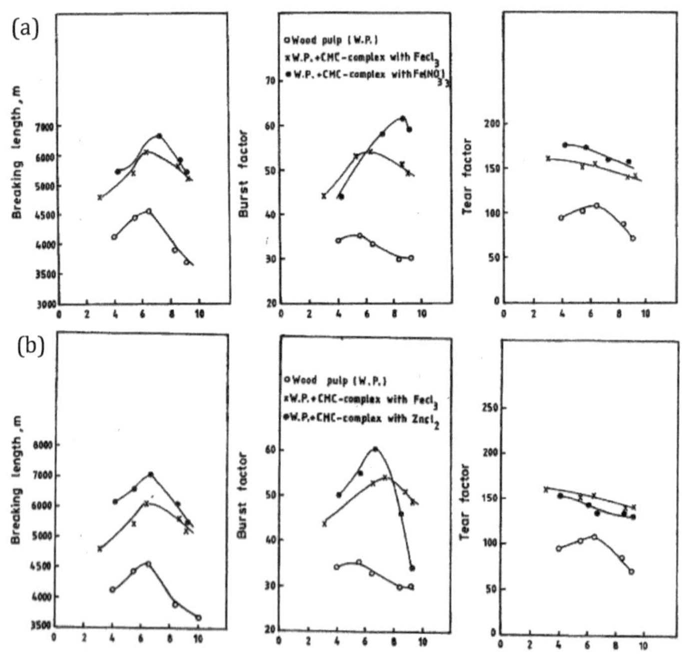

Figure 7.8 (a) Effect of anion of metal salt used for complex formation on the mechanical properties of paper sheets at different pH values. (b) Effect of cation of metal salt used for complex formation on the mechanical properties of paper sheets at different pH values.

From Fig. 7.7, it can be noticed that the variation of electrical properties of paper sheets with changing the type of ligand polymer was greatly noticed at low frequency (0.5 kHz). Therefore, the values of dielectric measurements at 0.5 kHz were recorded in Table 7.13, together with their mechanical properties.

From Table 7.13 it can be seen that the dielectric constant values obtained with $FeCl_3$, CMC–Fe(III) complexes at a given frequency were greater than that obtained from untreated wood pulp. The improvement in dielectric properties (ε' and ε'') in case of polymer complex addition was higher than $FeCl_3$ addition paper sheets.

The increase of (ε') in the case of $FeCl_3$ addition paper sheets can be explained, if we take into consideration the polarization of the adsorbed ferric ions, due to their displacement under the effect of the electric field [39]. However, the increase in such property in case of polymer complex addition is probably related to molecular polarization of side group of unchelated polymer (COO^- and OH), in addition to the polarization of adsorbed (chelated) ferric ions.

Figure 7.7a shows that the electrical properties of polymer complex paper sheets were greatly influenced by the type of anion of ferric salt used for complex formation, whereas the Fe(III) cation of $FeCl_3$ greatly improves the ε'' of produced sheet, compared with that resulted from Fe(III) of $Fe(NO_3)_3$. These results reverse to some extent the trend of mechanical properties (Fig. 7.8a, Table 7.13). The latter complex provided greater strength properties of produced paper sheet. This observation refers to higher bonded OH groups of cellulose pulps, and consequently, less oscillation and polarizability of OH group occur.

For the effect of cation of metal salt of polymer complex, it was observed that (Figs. 7.7b and 7.8b), CMC–Zn(II) complex provided paper sheet with greater dielectric properties than that produced by CMC–Fe(III) complex. Both type of complexes produced relatively high insulating paper than paper made from untreated wood pulp. The results obtained are probably due to the change of the type of combination between metal and cellulose. However, in case of Zn(II), the ionical bond is formed, while chelated covalent coordinated bond is formed in case of Fe(III) [47].

From the above results, it could be concluded that the addition of polymer complexes to wood pulp during beating process was successful for producing new insulating paper sheet with high strength properties. This surprising data persuaded us to conduct further studies, e.g., on thermal, magnetism, and biological activities.

For thermal stability study, Fig. 7.9 shows that the addition of oxides, especially Fe_2O_3 with CMC during beating stage improves the resistivity of wood pulp paper sheet to thermal decomposition, compared with beating addition of cellulose ether–Fe(III) complexes in wood pulp fibers. The addition of cellulose ether-Fe(III) complexes led to shift the decomposition temperature, especially at 10%–40% weight loss, to a lower value, compared with untreated sheets. The extent of shift behaves the following order; paper sheet included HEC–$FeCl_3$ complex > CMC–$FeCl_3$> CMC–$Fe(NO_3)_3$ > CMC–$ZnCl_2$.

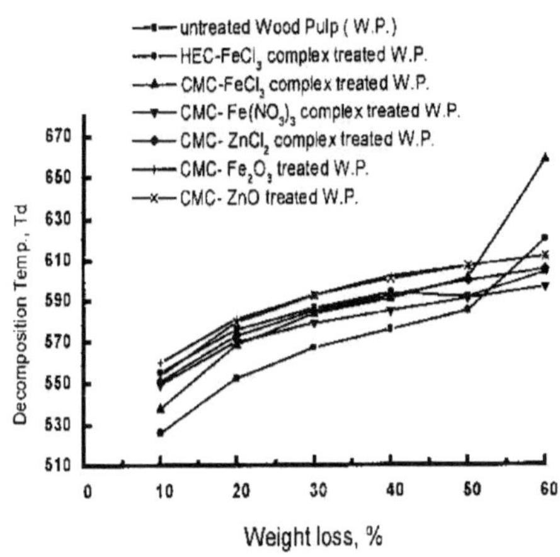

Figure 7.9 Temperature of decomposition, T_d, of untreated and wood pulp–treated paper sheets at different percentages of weight losses.

From the values of DTG peaks and specific rate constant (Table 7.14), it was observed that beater adding CE–metal complexes increased the rate of depolymerization and dehydration of the prepared paper samples, in addition to increasing the percentage of residue, compared with addition of CMC–metal oxides. This means they decreased the formation of levoglucosan regards the catalyze for formation of fuel for the flame [48]. In other words, these adding complexes gain the produced paper sheets a relatively high flame-retardancy behavior compared also with adding CMC–$ZnCl_2$ complex. The improving in fire retardance of resulted paper sheets follows the following sequence: HEC–$FeCl_3$ > CMC–$FeCl_3$ > CMC–$Fe(NO_3)_3$ > CMC–$ZnCl_2$ > CMC–ZnO > CMC–Fe_2O_3.

Table 7.14 Activation energies, frequency factors, specific rate residues and order of degradation of the main degradation stages of untreated and treated paper sheets

Type of additive	Temp. range (°C)	DTG peak maxima (°C)	E_a (kJ/mol)	A (sec^{-1})	$K \times 10^{-6}$ (min^{-1})	Residue, %	"n"
—	231–377	332.6	98.552	2.66×10^5	22.53	19.229	0.5
HEC-$FeCl_3$, complex	193–358	305.0	60.922	4.81×10^2	69.40	43.138	0.0
CMC-$FeCl_3$, complex	203–354	320.6	75.860	7.92×10^3	36.26	42.820	0.0
CMC-Ft $(NO_3)_3$ complex	206–331	319.7	81.761	2.68×10^4	31.82	28.047	0.0
CMC-$LnCl_2$ complex	221–380	327.2	87.806	5.52×10^4	27.11	27.800	0.0
CMC-Fe_2O_3	245–354	336.4	120.133	1.51×10^7	11.20	24.900	0.5
CMC–ZnO	213–350	334.6	90.983	1.07×10^5	16.30	27.600	0.0

For the magnetic and electrical properties of paper sheets, the mass susceptibility (X_g, c.g.s.) and AC conductivity (σ ohm^{-1}cm^{-1}) of paper samples were measured and recorded in Table 7.15. It was observed that this investigated CMC–Fe(III)-based beater additive improved to less extent the semiconductor property of produced paper sheets; however, the trend was reversed for the case of HEC–Fe(III) and CMC–Zn(II) complexes. These observations are probably related to the magnetic moment, molar conductivity and electrical conductivity of metal and metal complexes [19, 49].

Table 7.15 Magnetic susceptibility and electrical conductivity of paper samples

Type of additive	Mass susceptibility (c.g.s.) $\times 10^{-6}$	Electrical conductivity (Ohm^1 am^1) $\times 10^{-10}$
—	—	2.85
HEC-$FeCl_3$ complex	0.483	7.03
CMC-$FeCl_3$ complex	9.386	2.31
CMC-$Fe(NO_3)_3$ complex	7.939	1.38
CMC–$ZnCl_3$ complex	0.269	4.01
CMC-Fe_2O_3	5.578	4.166
CMC-ZnO	2.871	5.61

It was surprising to notice that (Tables 7.16 and 7.17). Paper sheets containing metal complexes had relatively high air permeability values (porous structure) and water resistance property together with high strength properties. These promising data, in comparison with those obtained from previously investigated paper treatments, e.g., acetylation or adding hydrophobic PAN [50, 61], confirm that this new type of paper sheets can be used for the production of nursery-plant boats.

Table 7.16 Change percentages of strength properties, water absorption and air permeability of paper samples

Type of additive	Change percentage Breaking, length, *m*	Burst factor	Tear factor	Water absorption, %	Air permeability ml/min
—	0.000	0.000	0.000	0.000	180
HEC-FeO_3 complex	2.354	6.350	20.144	-10.856	230
CMC-$FeCl_3$ complex	34.874	64.500	42.019	-28.013	450
CMC-$Fe(NO_3)_3$ complex	46.68	75.325	47.960	-28.703	800
CMC-$ZnCl_2$ complex	55.294	82.703	25.955	-34.561	85
CMC-Fe_2O_3	28.187	83.950	47.358	—	320
CMC-ZnO	-9.9435	0.148	5.506	—	0.00

Table 7.17 Change percentages of strength properties and water absorption of previously treated paper samples [24, 25]

Paper samples	Change percentage Breaking length, *m*	Burst factor	Tear factor	Water* absorption, %
Untreated wood pulp (WP)	0.000	0.000	0.000	0.000
Carboxymethylated W.P (D.S. 0.046)	-13.13	-74.07	55.54	-3.047

Paper samples	Change percentage			
	Breaking length, *m*	Burst factor	Tear factor	Water* absorption, %
Wood pulp + CMC (D.S. 0.046)	14.77	-7.711	12.70	10.33
Acetylated wood pulp (Acetyl content 18.38%)	-55.31	-76.119	-75.06	-8.05
Wood pulp + 9% homopolymer (PAN)	-23.30	-6.99	—	4.33
PAN-grafted pulp (25 g AN/100WP)	-84.68	-85.63	-97.22	8.391
Wood pulp + 9% homopolymer (PMA)	-13.93	-38.47	—	9.964
PMA-grafted pulp (25 g MA/100WP)	-46.30	26.16	-20.65	4.251

*Immersing time for determination of water absorption % is 2 hrs., at 23°C.
PAN = polyacrylonitrile,
PMA = polymethyl acrylate.

7.4.2 Application of Carboxymethyl Cellulose–Cu(II) Complexes [22]

This work focused on studying the used of CMC–Cu(II) complexes as a new approach for the preparation of high-performance colored paper. In this respect, the paper strength, thermal stability, biological resistance, magnetic properties, as well as the durability of aged paper were evaluated.

For strength properties, it was found that (Table 7.19) using carboxymethyl cellulose–copper complexes [CMC–Cu(II)] as paper additive enhanced the strength properties of wood pulp-paper sheet, and depended on the anion of the copper salt used and the pH value during the preparation process. The best polymer complex was that produced from using copper sulfate as the origin of copper ion at pH 5.4.

The observed improvement may be ascribed to the attraction of neutral or positively charged polymer complexes sphere to the negative charges known to exist on the fibers, i.e., cellulose fibers bridged through CMC–Cu(II) complex. Also, the presence of

Table 7.18 Anti-microbial action of paper samples on some microorganisms*

Paper sample with the used additives	*Salmonella typhi*	*Penicillum chrysogen*	*Candida albicous*	*Staphylococcus aureus*	*Escherichia coil*	*Pseudomoas aeruginosa*	*Aspergillus niger*	*Bacillus stabtilis*	*Saccoro-myces cervisiae*
Wood pulp	—	—	—	—	—	—	—	—	—
HEC-$FeCl_3$ complex	—	—	—	0.45	—	—	—	0.45	—
CMC-$FeCl_3$ complex	—	0.6	—	0.55	—	—	—	0.45	—
CMC-$Fe(NO_3)_3$ complex	—	—	—	0.45	—	—	—	—	0.55
CMC-$2nCl_2$ complex	—	—	—	1.05	—	—	0.55	—	0.45

*Diameter of unloaded diskette = 0.4 cm.

unchelated carboxymethyl cellulose in the produced sheets causes an increase in van der Waals bonding, which supports the strength properties. However, the change in the strength properties as a function of pH value and the origin of Cu(II) ion is probably the result of the pH dependence in the adsorption or desorption of copper (II) ions, in $Cu(Ac)_2$ via the carboxymethyl groups and exhibits the formula $[Cu(CMC)_2]$ and [Cu(CMC)AcO], respectively; it acts as a binegatively charged bidentade ligand when it is chelated with $CuSO_4$ via the carboxymethyl and secondary hydroxyl groups and exhibit the formula $[Cu(CMC)\cdot 2H_2O]$ [19]. These formed bonds are low stable in acidic medium [52].

Table 7.19 Change percentages of strength properties of paper sheets from beater adding CMC–Cu(II) complexes

	Breaking length, *m*			Burst factor			Tear factor		
pH-value	CMC-$CuCl_2$	CMC-$CuSO_4$	CMC-Cu$(AcO)_2$	CMC-$CuCl_2$	CMC-$CuSO_4$	CMC-Cu$(AcO)_2$	CMC-$CuCl_2$	CMC-$CuSO_4$	CMC-$(AcO)_2$
Blank	0	0	0	0	0	0	0	0	0
3.65	23.8	16.41	12.91	6.78	30.34	14.91	-24.74	-15	-14.92
4.7	24.6	24.92	23.19	24.73	35.14	18.23	-1.34	-7.37	3.43
5.45	28.63	30.46	15.82	11.37	38.43	11.98	9.13	9.10	4.84
6.5	15.48	15.05	12.11	-4.99	-12.52	-11.46	-10.54	-11.56	7.65
8.5	-12.99	4.43	10.54	-15.02	-18.89	-15.28	-19.87	-12.80	-8.79
9.2	-18.30	-5.30	2.86	-36.17	-31.32	-19.10	-27.75	-30.20	-26.08

As can be seen that elevating the pH value during paper sheet preparation by dilute NaOH to alkaline range, pH 8–10, was accompanied by the improvement or deterioration in strength properties of polymer complex-wood pulp sheets. The explanation of these data is probably ascribed to the competition of Na ions with Cu(II) ions or complex sphere to attract on the negative charge of cellulose fibers surface. The univalent sodium ions that are present have no power to form a bridge between the cellulose fibers, as the case of CMC–Cu(II) complexes, and in the same time it diminishes the formation of hydrogen bond between the hydroxyl and carbonyl groups. In other words, the presence of sodium ions reduces the probability of bridging of the cellulose fibers through complex sphere. To confirm this view, further measurements were carried out, such as roughness, permeability, oil, penetration and absorption as well as wax pick.

These measurements are recorded in Table 7.20.

Table 7.20 Physical measurements of $Cu(CMC) \cdot 2H_2O$ complex-containing paper sheets

	Permeability	Roughness		Oil penetration sec.		Wax
pH-value	ml/min.	Side a	Side b*	Side a	Side b	no.
3.65	300	590	680	9	18	14
~6.0	190	560	695	22	28	14
9.2	260	615	760	4	5	10

It is obvious that elevating the pH values by sodium hydroxide during sheet formation leads to weakness of the bond possibly formed (intra and inter) between CMC–Cu(II) complexes and hydroxyl group on wood pulp constituents. This was accompanied by increasing the voids between fiber chains, roughness, and fasting migration of the oil inside paper, compared with that made in neutral pH value.

For thermal study, the behavior of copper complexes containing paper sheets on the thermogram shape showed difference compared with wood pulp-sheet alone. Incorporating [CMC–$CuCl_2$] and [CMC–$CuSO_4$] complexes (Fig. 7.10) showed additional peaks at relatively low DTGA peak temperatures for the stages responsible for the formation of levoglucosan (main stage). In the same time, these stages were started at a relatively high temperature and the weight loss of volatilization stage near completion at relatively low temperature resulted in relatively high percentage of the weight remaining, compared with the wood pulp sheet, i.e., reduction of formation of levoglucosan occurred (Table 7.21).

It was interesting to note that not only the three investigated copper complexes lead to increasing the thermal stability (increased activation energy and onset temperature of degradation) of the paper, but these complexes, especially based on $CuCl_2$ and $CuSO_4$, also retard the rate of mass loss with the remaining values of about 29–31.9%. In other words, these complexes provided additional fire resistance property through minimizing the formation of levoglucosan, and consequently reduce the formation of toxic gases.

These data demonstrated that the presence of copper complexes enhanced the dehydration stage than depolymerization and formation of levoglucosan, and thus result in high thermal stability.

Table 7.21 Thermal degradation measurements of copper complex-containing paper sheets in comparison to wood pulp sheet (main degradation stage)

Complex	Temperature range (°C)	DTG peak (°C)	Order "*n*"	E_a (kJ/mol)	Weight remaining (%)
—	234.7–349.3	318.5	0.0	89	24.52
CMC–$CuCl_2$	241–276	277.7	1.0	273.34	29.17
	277.7–341.8	317.2	1.0	203.15	
				Σ 476.49	
CMC–$CuSO_4$	241.9–293.1	292.1	0.5	161.85	31.904
	293.1–341.9	320.7	1.5	352.39	
				Σ 514.24	
CMC–Cu(AcO)$_2$	244.2–356.7	329.1	0.5	102.66	23.46

Note: E_a is the activation energy of the main degradation stages.

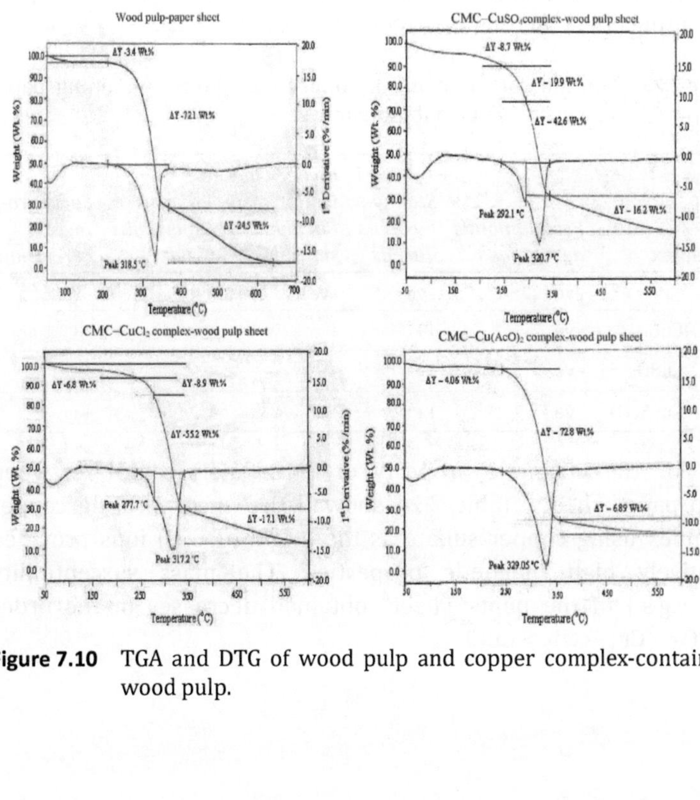

Figure 7.10 TGA and DTG of wood pulp and copper complex-containing wood pulp.

The thermal measurements were in agreement with the data of FTIR spectra [crystallinity index (C.I.) and (MHBS)] (Table 7.22).

Table 7.22 IR measurements of copper complexes–containing paper sheets

Complex	E_{OH}	$E_{C=O}$	Asymmetric index	MHBS	Cr.I.
—	1.48	0.81	1.7	1.5	1.2
CMC–$CuCl_2$	1.37	0.49	2.0	2.3	2.5
CMC–$CuSO_4$	1.24	0.44	2.1	2.0	2.6
CMC–u$(AcO)_2$	1.43	0.35	2.0	1.9	2.2

Note: Asymmetric index of the bands correspond to the stretching vibration of OH groups, MHBS is the mean hydrogen bond strength, and Cr.I. is the crystallinity index (A_{1430}/A_{900}).

For biological resistance of CMC–Cu(II) complex wood pulp paper sheets, the data obtained in Table 7.23 show that incorporating CMC–Cu(II) complexes to wood pulp has an inhibiting effect to all tested microorganisms except bacteria of the type *pseudomonas aeruginosa*. Also, Cu(II)-complex originated from using $CuCl_2$ no inhibiting effect on fungus (*Aspergillus niger*).

Table 7.23 Antimicrobial action of copper complexes containing paper sheets on some microorganisns

Complex	*Pseudomonas aeruginosa*	*Staphy-lococcus aureus*	*Escherichia Coli Bacillus subtilis*	*Aspergillus niger*	*Sacchoro-myces cerevisiae*
—	−ve	−ve	−ve	−ve	−ve
CMC–$CuCl_2$	−ve	++	+	−ve	++
CMC–$CuSO_4$	−ve	+++	+	+	++
CMC–$Cu(AcO)_2$	−ve	++	++	+	++

For the magnetic property of CMC–Cu(II) complexes-wood pulp paper sheets, Table 7.24 showed that incorporating copper complex using copper sulfate as the origin of C(II) ions provided relatively high magnetic properties. The mass susceptibility (X_g; c.g.s.) of the paper sheets obtained decreases in the order $CuSO_4 > Cu(AcO)_2 > CuCl_2$.

Table 7.24 Magnetic behavior of copper complexes and magnetic filler-containing paper sheets

Complex	Mass susceptibility (X_g; c.g.s.)
—	3.99×10^{-7}
CMC–$CuCl_2$	4.65×10^{-5}
CMC–$CuSO_4$	1.27×10^{-6}
CMC–$Cu(AcO)_2$	5.59×10^{-6}

The relatively low mass susceptibility in case of $CuCl_2$ was ascribed to the formation of CMC–Cu(II) complex [$Cu(CMC)_2$], where the lone pair of electrons of aliphatic ether and the carboxyl groups of carboxymethyl cellulose were coordinated together with Cu(II) ions of $CuCl_2$ salt. However, the liberation of hydrogen atoms from the second hydroxyl group of CMC to coordinate with Cu(II) was more possible than the coordination of the lone pair electron on oxygen of ether linkage in complexes from copper salts of sulfate and acetate anions. It was interesting to note that the mass susceptibility of the investigated paper when using $CuSO_4$ is higher than that produced from adding known magnetic pigment, such as Fe_2O_3.

7.4.3 Evaluation of CMC–Cu(II) Complexes in the Production of Durable Paper Sheets

The foregoing encouraging data led us to study the longevity of these high-quality colored papers. Previous studies revealed that the main parameters of the deterioration of paper documents are the chemical composition of the paper and ink [53]. Other contributing factors are the storage conditions, such as heat and humidity. In this work, we applied the dry heat treatment at 100°C in the investigation of accelerated ageing for wood pulp- and copper complex wood pulp paper sheets (at optimum pHs). It is considered that three days under these conditions correspond to 25 years of natural ageing [54]. The changing of strength and color strength of all investigated paper sheet against the time of accelerated ageing are illustrated in Fig. 7.11. Durability was estimated according to equation mentioned elsewhere [55, 56].

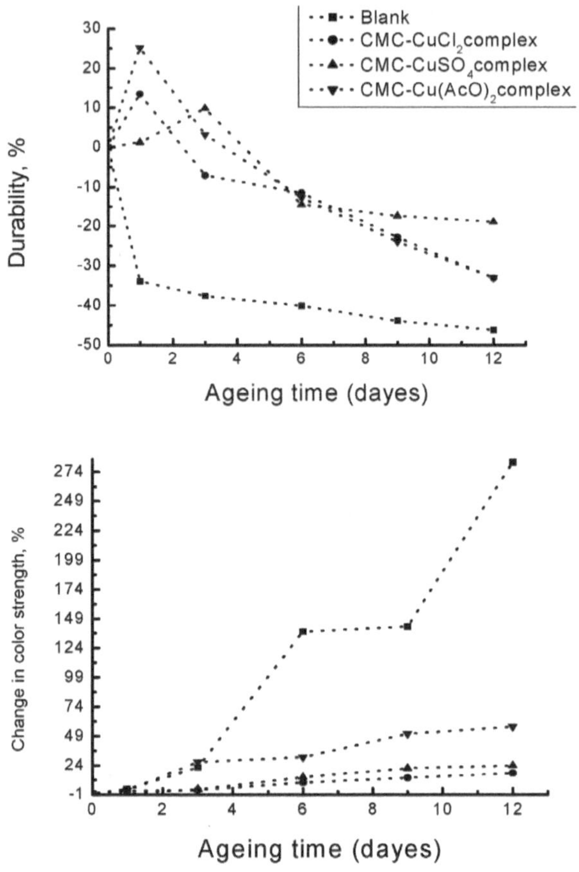

Figure 7.11 Durability and change in color strength of CMC–Cu(II) wood pulp sheets toward accelerated ageing.

Generally, a significant decrease in paper strength and change in color strength were noticed upon accelerated ageing. The relatively high change was observed in the case of free complex paper sheet compared to copper complexes containing paper sheets. The deterioration rate in the former paper is caused by one day of ageing, while incorporating the complexes allows the deterioration rate to start after 3–6 days. The loss of the strength of wood pulp-paper upon accelerated ageing is related to a decrease in the fiber strength as a result of cellulose depolymerization and oxidation of lignin and hemicelluloses, which leads to the formation of additional carboxylic groups.

There is an exponential relation between the reduction of carbonyl groups (chromophores formed) and discoloration [53, 55, 56]. In our investigated paper samples, the polymer complexes scavenged the hydroxyl groups of wood pulp constituents through bond formation and consequently protected it to some extent from oxidation.

As can be seen that at ageing time from 1 to 6 days, the [CMC–Cu(AcO_2)] complex achieved high durability (Fig. 7.11) compared with other complexes, and the (CMC–$CuCl_2$) complex had the lowest durability. However, the reverse trend was observed at higher ageing time (9–12 days), where paper prepared from the acetate complex had the same durability as that produced from chloride complex, whereas the copper complex-based on copper sulfate achieved high durability. This may be ascribed to cleavage of the acetate group contained in the polymer complex upon accelerated ageing, which consequently increases the acidity of the paper from its reaction with absorbed water in the sheet.

7.5 Conclusions

- All the investigated CMC–Fe(III) and CMC–Cu(II) complexes, as new beater paper additives, had positive effects for improving the quality of obtained paper sheets.
- Improvement in strength properties is dependent on the anion of metal salts and pH value during the preparation process. Incorporating the CMC–Fe(III) complex, using $Fe(NO_3)_3$ as the origin of Fe(III) ions, achieved higher magnetic property besides improvement in the strength, dielectric insulation and biological and fire-resistant properties. However, CMC–Cu(II) complex, using $CuSO_4$ as the origin of Cu(II) ions, at pH 5.4 achieved the best improvement in strength quality, thermal stability, and biological resistance. Moreover, it provided less change in strength color strength upon accelerated ageing, compared with other Cu(II) salts (chloride and acetate).
- Beater adding CMC–$Fe(NO_3)_3$ provides high-permeability and water-resistant paper sheets, together with high strength properties, making it convenient for the production of nursery-plant boats.

References

1. Sugasaka, K., Katoh, S., Takai, N., Takahashi, H., and Umezawa, Y. (1981). *Sep. Sci. Technol.*, **16**, 971.
2. Sakuragi, M., Ichimura, K., Fugishige, S., and Katoh, M. (1981). *Bul. Res. Inst. Polym. Text*, **130**, 43.
3. Samal, R. K., Satrusallya, S. C., and Sahoo, P. K. (1984). *Appl. Polym. Sci.*, **29**, 318.
4. Muller, L. L., Hebert, J. J., and Rollins, M. L. (1971). *Appl. Polym. Sci.*, **15**, 1425–1435.
5. Whistler, R. L. (1963). *Methods in Carbohydrate Chemistry*, Vol. III., Academic Press, London, pp. 193–198.
6. Casey, J. P. (1960). *Pulp and Paper*, 2nd ed., Interscience Publishers, Inc., New York, Vols. I–II.
7. Klug, E. D. (1965) in *Encyclopedia of Polymer Science and Technology*, Vol. 1, Interscience Publishers, Inc. New York, p. 520.
8. **Lepri**, L., and **Desideri**, P. G. J. (1973). *Chromatogr. A.*, **84**(1) 155.
9. Wellham, E. J., Elber, L., and Yan, D. S. (1992). *Miner. Eng.*, **5**(3–5), 381.
10. El-Saied, H., Basta, A. H., Abdel-Hadi, A. K., and Hosny, W. M. (1994). *J. Polym. Int.*, **35**, 27–33.
11. Abdel-Hadi, A. K., Hosny, W. M., Basta, A. H., and El-Saied, H. (1994). *Polym. Plast. Technol. Eng.*, **33**, 781–791.
12. Hosny, W. M., Abdel-Hadi, A. K., El-Saied, H., and Basta, A. H. (1995). *J. Polym. Int.*, **37**, 93–96.
13. Hosny, W. M., Basta, A. H., and El-Saied, H. (1997). *Polym. Int.*, **42**, 157.
14. Basta, A. H., and Hosny, W. M. (1998). *Polymer Deg. Stab.*, **60**, 239.
15. Abdel-Hadi, A. K., and Helmy, S. A. (1993). *Polym. Degrad. Stab.*, **40**, 121.
16. Hanna, A. A., Basta, A. H., El-Saied, H., and Abadir, I. F. (1998). *Angew. Makromol. Chem.*, **260**, 1.
17. Basta, A. H., and El-Saied, H. (2000). *Polym. Plast. Technol. Eng.*, **39**, 887.
18. Abd-El-Messieh, S. L., Basta, A. H., and El-Saied, H. (2001). *J. Polym. Mater.*, **18**, 279.
19. El-Saied, H., Basta, A. H., Hanna, A. A., and El-Sayed, A. M. (1999). *Polym. Plast. Technol. & Eng.*, **38**, 1095.
20. El-Saied, H., Basta, A. H., and Abde-El-Nowr, K. N. (1997). *Int. J. Polym. Mater.*, **36**, 131.
21. Basta, A. H., and El-Saied, H. (2001). *Int. J. Polym. Mat.*, **50**, 185.

22. Basta, A. H., and El-Saied, H. (2008). *Carbohydr. Polym.*, **74**, 301.
23. Basta, A. H., El-Saied, H., Gobran, R. H. (2004). *Polym.-Plastic Technol. & Eng.*, **43**(3), 743.
24. Jørgensen, C. K. (1962). *Prog. Inorg. Chem.*, **4**, 73.
25. Risen, W. M., Jr, et al. *Coulombic Interactions in Macromolecular Systems*, ACS Symposium Series, 1988, pp. 302, 368.
26. Haggin, J. (1989). *Chem. Eng. News*, **40**, 175.
27. Coat, A. W., and Redfern, J. P. (1964). *Nature*, **201**(4914), 68–69.
28. Basta, A. H. (1999). *Int. J. Polym. Mater.*, **44**, 1.
29. Tanaka, H., and Tokumitsu, M. (1984). *J. Thermal Anal.*, **29**, 87.
30. Selim, I. Z., Basta, A. H., Mansour, O. Y., and Atwa, A. I. (1994). *Polym. Plast. Technol. Eng.*, **33**, 173.
31. Younan, A. F., Abd-El-Messieh, S. L., and Gasser, A. A. (1998). *J. Appl. Polym. Sci.*, **70**, 2068.
32. Abd El-Messieh, S. L. M. (1989). Science. thesis, Cairo Univ.
33. Frohlich, H. (1949). *Theory of Dielectric*, Oxford University Press, London.
34. Khilla, M. A., and Hanna, A. A. (1981). *Thermochim. Acta*, **51**, 335.
35. Clark, F. M., and Montsinger, V. M. (1925). *Gen. Elec. Rev.*, **28**, 286.
36. Taiel, F. M. (1984). *Egypt J. Phys.*, **15**, 279.
37. Poppel, E. M., Boutros, D., and Ciobanu, R. (1989). *Cell. Chem. Technol.*, **23**, 761.
38. **Jawaid**, M., and **Abdul Khalil**, H. P. S. (2011). *Carbohydr. Polymers*, **86**, 1–18.
39. Abdel-Motelb, M. M., and El-Shafee, F. (1994). *Cellulose*, **1**, 197.
40. Gordon, L. N. (1990). *Fire and Polymers, Hazards Identification and Prevention*, ACS, Washington, DC, p. 335.
41. Singh, R., Arora, S., and Lal, K. (1996). *Thermochim. Acta*, **289**, 9.
42. Ziolo, R. F. (1985), US Patent 4474866 (1984). *Chem. Abstr.*, **2**, 15114.
43. Raymond, **L.**, Revol, **J.-F.**, Ryan, **D. H.**, and **Marchessault**, R. H. (1994). **Chem. Mater.**, **6**(2), 249.
44. *Ullmann's Encyclopedia of Industrial Chemistry*, 5th ed., VCH veriagsgesellschaft mbH, D-6940 Weinheim, 1991.
45. El-Saied, H., Basta, A. H., and Abdou, M. M. (1998). *Restaurator*, **19**, 155.
46. Koretsky, K. (1983). *Color for the Hand Papermaker*. Carriage House Press, Brookline, MA.

47. Hojo, H. S., and Arakawa, N. (1996). In: Paper presented at 15th High Polymer Conference (Japan), Nagoya, Abstracts, p. 130.

48. Shatizadeh, F., Lai, Y. Z., and Melntyre, C. R. (1978). *J. Appl. Polym. Sci.*, **22**, 1183.

49. Lee, J. D. (1991). *A Text Book of "Concise Inorganic Chemistry,"* 5th edition, London.

50. Fahmy, Y., and El-Saied, H. (1974). *Holzforschung*, **28**, 29.

51. Fahmy, Y., and El-Saied, H. (1974). *Holzforschung*, **28**, 61.

52. Pasteka, M. (1992). *Acta Polymer*, **43**, 227.

53. Havlimova, B., Babiakova, D., Brezova, V., Durovic, M., Novotna, M., and Belanyi, F. (2002). *Dyes Pigment*, **54**, 173.

54. Kirova, Kn. L., Stepanova, L. A., Flyate, D.M., and Shul'man, I. S.: Problemy Sokhraunosti Documental-nykh Materialov. Leningrad (Sankt Peterburg) 1977: 13.

55. Basta, A. H. (2003). *Resturator*, **24**, 106.

56. Basta, A. H. (2004). *Resturator*, **25**(2), 129.

Chapter 8

Hemicellulose: Isolation and Its Application in Pharmacy

Muchlisyam, Jansen Silalahi, and Urip Harahap

Faculty of Pharmacy, University of North Sumatra, Medan, Sumatera Utara, Indonesia

muchlisyam@gmail.com

8.1 Introduction of Hemicellulose

Polysaccharides are biopolymers commonly found in living organisms and are known to reveal the physiological functions by forming a specific conformation. However, polysaccharide chains are still in their premature state with respect to their structure in solid and in solution. Structural analysis may offer the most fundamental knowledge to understand the functions of polysaccharides, but the diversity and irregularity of polysaccharide chains make it a formidable task. Polysaccharide chains are partly organized but are considered to be mostly amorphous [9].

Hemicelluloses are heterogeneous polysaccharides, which are located between the lignin and cellulose fibers. Depending on wood species, hemicelluloses constitute about 20–30% of the naturally occurring lignocellulosic plant biomass [19]. Lignocellulosic biomass represents a renewable, widespread, and

Handbook of Sustainable Polymers: Processing and Applications
Edited by Vijay Kumar Thakur and Manju Kumari Thakur

ISBN 978-981-4613-53-8 (Hardcover), 978-981-4613-54-5 (eBook)
www.panstanford.com

low-cost source of sugars, which can be used as substrate for the production of special chemicals. Various agricultural residues contain about 20–30% hemicellulose, the second most abundant biopolymer found in nature. They are heterogeneous polymer of pentoses (xylose and arabinose), hexoses (mannose, glucose and galactose) and sugar acids. Xylan is the major component of hemicellulose and is heteropolysaccharide with homopolymeric backbone chains of 1,4 linked β-D-xylopyranose units [7, 9].

Non-starch polysaccharides (NSP) is a lignocellulosic biomass that located as byproduct of the agricultural material processing which is amounted to around 30–50% of the total number of processed agricultural products. The largest component of the NSP is the fiber, consisting of non-polysaccharide and heteropolysaccharides such as cellulose (20–35%), hemicellulose (20–35%), pectin, alginate, gum, mucilage, and lignin (10–25%). By conducting the process of extraction and careful fractionation of biomass byproduct of this material, it will result in a various compounds, which can be converted into highly valuable products [9, 26, 39, 57].

For thousands of years, nature has provided humankind with a large variety of materials for the most diversified applications for its survival, such as food, energy, medicinal products, protection and defense tools, and others. The pharmaceutical industry has benefitted from such diversity of biomaterials and has exploited the use of natural products as sources of both drugs and excipients [27, 28]. One example of a promising biomaterial for pharmaceutical use is xylan, a hemicellulose largely found in nature, being considered the second most abundant polysaccharide after cellulose [57].

These polysaccharides, as natural polymers, are by far the most abundant renewable resource on the earth with an annual formation rate that sustainable materials synthesized by the sun's energy and fully biodegradable in the original state. The most important biological role of hemicelluloses is their contribution to strengthening the cell wall by interaction with cellulose and, in some walls, with lignin. Hemicelluloses include xyloglucans, xylans, mannans and glucomannans, and β-(1$\rightarrow$3, 1$\rightarrow$4)-glucans are generally classified according to the main sugar residue in the backbone that has β-(1$\rightarrow$4)-linked backbones with an equatorial configuration. Hemicellulose can be any of several heteropolymers (matrix polysaccharides), present in almost all plant cell walls along with cellulose. They differ from cellulose by the composition of their sugar units, chain length, and branching of the chain molecules [9].

The different subtypes of hemicellulose can be grouped into two general categories based on the hydration of the fibers [57]:

(1) Low-hydration polysaccharides include the arabinoxylans, glucuronoxylans, xyloglucans, and linear mannans. With the exception of the linear mannans, which serve mainly as a seed storage compound, this class of hemicelluloses functions primarily to stabilize the cell wall through hydrogen-bonding interactions with cellulose and covalent interaction with lignin. They are water-soluble in their native state generally due to their branched construction.

(2) Other hemicelluloses, composed mainly of hydrocolloids, are used primarily as an extracellular energy, raw materials storage system, and as a water retention mechanism in seeds. Galactoglucomannans, glucomannans, galactomannans, and h-glucans tend to be heavily hydrated and have fewer, if any, ester-linked side chains [57].

Potential of hemicellulose in the pharmaceutical field could potentially be developed as drug formulation excipients such as binders, disintegrator, thickeners, and stabilizers. In addition, applications of hemicellulose have been done by making the derivatives that have pharmacological effects such as lowering cholesterol and inhibitors of HIV [26–28, 41, 57].

Several researchers have investigated the functional group in organic compounds that are present in nature, such as amine and hydroxyl groups in chitosan, hydroxyl, carboxylate groups of sodium alginate, and pectin-containing carboxylic and hydroxyl groups found to adsorb various metals such as aluminum, iron, lead and copper [39, 53, 56].

Hemicellulose is one of the most abundant polymers after cellulose. However, its potential has yet to be completely recognized. This polymer comprises a variety of monomers, including xylose, mannose, arabinose, glucose, and glucoronic acid having hydroxyl (OH) and carboxylate (COOH) group on each monomer [9]. In the qualitative chemical analysis of organic compounds, the OH and carboxylate group in both cyclical aliphatic, for example, citric acid will be able to bind to the metal to form complex salts.

Several studies have been conducted for the separation of hemicellulose from different plant byproduct with different methods of isolation. Lignification process conducted among others by NaOH in 70% ethanol, chlorine, sodium hypochlorite and 30% H_2O_2. Isolation of hemicellulose used alkaline compounds such as KOH and NaOH with various concentrations, whereas HCl and 90% ethanol is used for the purification of hemicellulose [39, 40, 57].

The aim of this chapter is to summarize some of the research findings on hemicellulose, a natural polymer extracted from corn cobs, which presents a promising application in the development of binding ability for cadmium and plumbum in rabbit's blood [33, 34].

8.2 Sources and Structure of Hemicellulose

Hemicellulose is a group of polysaccharides that can be found in the plant's cell wall that surrounds the cellulose, protein, lignin, and other materials that covalently bonded, hydrogen bonds, ionic bonds and interactions hydrophoby [35]. In plants, hemicellulose is a second polymer after cellulose that abundance in wood and cereals. It consists of approximately one-quarter to one-third of the plant material. This amount varies according to the type of plants, such as corn steins (28.0%), wheat straw (38.8%), rice straw (35.8%), and rye straw (36.9%). Hemicellulose serves as a supporter to cell wall and as an adhesive between single cells in plants [10].

These polysaccharides are one of the renewable resources that are most widely found in nature. These polymers present in plants together with cellulose and lignin. Hemicellulose is different from other celluloses in terms of the monomer composition, and has a long chain and molecular chain branching. The content of these compounds differ from species to species [9, 10].

Chemical and thermal stability of hemicellulose is generally lower than other celluloses, due to the lack of crystallinity and the lower degree of polymerization. This compound is very easily hydrolyzed, especially in the presence of hydroxyl groups on each unit of the chain. The results of hydrolysis include glucose, xylose, mannose, galactose, arabinose, glucuronic acid, and mannuronic acid in different amounts [9].

Hemicelluloses are heteropolysaccharides. Their structure and polymeric properties vary depending on the origin, i.e., plant species, industrial processing, isolation and purification procedure. They are a heterogeneous group of polysaccharides. The term was coined at a time when the structures were not well understood and biosynthesis was completely unknown [6].

These polysaccharides are shorter, or branch five-carbon sugars such as xylose or six-carbon sugar such as galactose and glucose. They are readily hydrolysable predominantly to xylose in the case of hardwood, and to mannose in the case of softwood. Hemicelluloses differ from cellulose in that they consist, for the most part, of pentose and hexose sugars others than glucose. They are usually branched and have much lower degrees of polymerization. They are not crystalline so that they do not present the same barriers to accessibility as cellulose does. Again, the hydroxyl groups are the most abundant [9, 11].

Hemicelluloses have relatively low molar mass compared with other wood polymers. The degree of polymerization of hemicelluloses is around 80–200. The general units of hemicelluloses are pentosans and hexosans. Their general formulas are $(C_5H_8O_4)_n$ and $(C_6H_{10}O_5)_n$, respectively. Hemicellulose does not have a repeated structure. There are many variations in the structure that are not well known. The detailed structure of the hemicelluloses and their abundance vary.

The chemical structure of hemicelluloses is the most important factor that influences their properties [1, 4, 7].

Polysaccharides present large differences solubility in solution and gel properties. Their chemical structure determines the shapes of the molecules adopted both in aqueous system and in the solid state. Depending on the plant species, developmental stage, and tissue type, various subclasses of hemicellulose may be found, including glucuronoxylans, arabinoxylans, linear mannans, glucomannans, galactomannans, galactoglucomannans, h-glucans, and xyloglucans [9, 42].

Hemicellulose can be found in hardwood and softwood. Softwood hemicelluloses consist of glucomannan, galactoglucomann, arabinan, and a small amount of arabino-(4-*O*-methylglucurono)-xylan. Softwood xylans do not contain acetyl groups. They are separated as a gel from acidic solutions, one of them, in methylated form being presented as a strong film, indicating xylans' linear structure. Galactoglucomannans are divided into two main fractions: totally water soluble, with a galactose:glucose: mannose ratio of 1:1:3, and partially water soluble, with a galactose:glucose: mannose ratio of 3:1:0.1. Both are constituted by D-glucose and D-mannose units linked by (1,4) glycosidic bonds. The ratio of mannose, glucose, and galactose depends on the mode of separation [9].

Hardwood hemicelluloses are represented by xylans and a small proportion of glucomannans. Hardwood xylans are linear polymers, constituted of (1,4)-linked β xylanopyranosyl units that constitute the main skeleton. Every tenth D-xylanopyranosyl unit is substituted by a 4-*O*-methyl-D-glucuronic acid residue, linked to the birch xylan chain by (1,2) linkages, that has been found to retard the alkaline peeling reaction. Analysis of partially hydrolyzed xylan indicates that 4-*O*-methyl-D-glucuronic acid units linked to the C_2 position are randomly distributed along the birch xylan backbone [9].

Based on the current stage of knowledge, hemicelluloses can be divided into four general classes of structurally different cell-wall polysaccharide types:

(a) xylans
(b) mannans
(c) β-glucans with mixed linkages
(d) xyloglucans

Hemicelluloses occur in structural variations differing in side-chain types, distribution, localization, and/or types and distribution of glycoside linkages in the macromolecular backbone. Polysaccharides might be differentiated from various points of view, such as function in plant tissues (reserve, supporting), primary structure, occurrence in plant tissues, etc. [15].

Hemicelluloses are synthesized by glycosyltransferases located in the Golgi membranes. Many glycosyltransferases needed for the biosynthesis of xyloglucans and mannans are known. In contrast, the biosynthesis of xylans and β-(1$\rightarrow$3, 1$\rightarrow$4)-glucans remains very elusive [15].

This compound has a non-crystalline nature, is not fiber, easy to inflate, slightly soluble in water, very hydrophilic, and easily soluble in alkali. High content of hemicellulose can contribute to the bonding between the fibers, as hemicellulose acts as an adhesive in every single fiber. At the time of the ripening process takes place, hemicellulose will be softened and at that time, most of the fibers will be easily separated [29].

8.2.1 Structure and Chemical Composition

Hemicellulose is a carbohydrate polymer that is colorless and relatively stable. Polymers branches with a number of units of sugar. This material is found in three groups, which are xylan, mannan, and galactans. Xylan presents in the form of arabinoxylan or arabino glucurunoxylan. Mannan is contained in the form of glucomannan and galactomannan. Galactans while relatively rare are in the form of arabinogalactans [39].

These polysaccharides are a polymer of pentose or xylose with ties of β-1,4 and the number of monomer ranging from 150 to 200 units. Hemicellulose itself is a polymer of sugar monomers that can be grouped according to its constituent, namely hexose (glucose, mannose, and galactose), pentoses (xylose, arabinopyranose, and arabinofuranose), hexuronide acid (glucuronic acid, methylglucoronate, and galacturonate), and deoxyhexose (rhamnose and fructose). The main chain of hemicellulose usually only consists of only one kind of monomer (homopolymer), for example, xylose, or may consist of two or more monomers (heteropolymer), for example, glucomannan [9].

Hemicellulose has substituents that are around the ring of the core structure. In general, the substituents that are found on the rear ring are an acetyl ring, arabinoxyl, and glucoronoxyl. Based on the structure, the hemicellulose contained in *angiospermae* or *gymnospermae* plants. Hemicellulose of seed plants is *O*-acetyl-4-*O*-methyl glucoronoxylan. Every 10 xylose carry a 4-*O*-methylglucoronide acid located in two positions of xylose [39].

The composition of hemicellulose from *gymnospermae* plants has more 4-*O*-methylglucoronoxylan and 4-*O*-metylglucoronide acid contained in the two-carbon chain. An example of *gymnospermae* plants that contains hemicellulose is open pine, while the *angiospermae* are seed rice, corn, and other cereals [39]. These heteroglycan compounds contain different types of sugar units, arranged in different proportions and with different structures. Hemicellulose derived from monomers consists of five and six carbon, i.e., pentose and hexose. Hemicellulose content of renewable resources in 20–35% of dry weight. The type and amount of monomer of the hemicellulose depends on the wood tissues [9, 40]. Sugar units that are part of the hemicelluloses' structure are D-xylose, L-arabinose, D-glucose, D-galactose, D-mannose, D-glucuronic acid, 4-*O*-methyl-D-glucuronide acid, D-galacturonic acid, and in small amounts of L-rhamnosa, L-fructose, and various neutral sugars. Hemicellulose is different from other celluloses, as its composition comprises various monomer units, the molecular chain is short and there is branching of the molecular chain. The monomer units that form hemicellulose can be divided into complexes such as pentose, hexose, hexuronic acid, and deoxy-hexoses [9, 40].

Hemicellulose can be used as a source of raw material for several industrial processes. This polymer is applied in the biotechnology and pharmaceutical uses by formulation of various derivatives through a variety of chemical reactions [33, 34, 40].

8.2.2 Isolation of Hemicellulose

Wood is in general composed of the linear polymer cellulose, different mainly linear hemicelluloses, and the network polymer lignin. The isolation of hemicelluloses from various lignocellulosic materials especially woods has been already studied a long time ago. Because the hemicelluloses are embedded in the cell wall, their isolation is difficult. Methods for the isolation of hemicelluloses from wood are too numerous to be discussed completely, but the most important methods will be described. Many researchers have studied different methods, solvents, operating conditions, and processes for the extraction of hemicelluloses. Different properties and solubility of the wood hemicelluloses require different separation methods. They can be roughly divided into methods using a high-energy input, which involves the formation of organic acids from the wood itself to dissolve the hemicelluloses, and methods where chemicals are added for the dissolution and the temperature is then usually only modestly raised [9, 26, 57].

Proper extraction process needs special requirements to fractionate a certain hemicellulose or mixture of hemicelluloses. However, the presence of lignin network as well as ester and ether lignin–carbohydrate linkages limits the hemicellulose extraction from the cell wall matrix of wood. The extensive hydrogen bonds between the individual polysaccharides components also impede their separation [1, 9, 46, 53, 56, 57].

The isolation of hemicellulose from various materials that have lygnocellulose has been done by several researchers. Numerous researchers have studied the different procedures, solvents, working conditions, and hemicellulose extraction processes. Various procedures have been developed for the isolation of hemicellulose, using sodium and potassium hydroxide, peroxide hydrogen and a base, or with a solution of barium or calcium hydroxide with a wide range of temperatures [7, 9, 14, 22, 25, 39, 40].

Different separation and purification procedures have been proposed to isolate hemicelluloses from different raw materials. Extraction of hemicelluloses from different resources has been studied since a long time ago. Isolation of hemicelluloses using cost efficient extraction methods would be beneficial to increase the utilization of hemicelluloses. Alkaline aqueous solutions and organic solvents have been used to extract hemicelluloses from the original or delignified wood (holocellulose) [1,9,21,25,31,33,34,57].

Different materials used in the isolation procedure both the plant and the chemicals and the soluble will be different, therefore hemicellulose requires a different design of separation procedures. However, the presence of lignin and lignin–carbs relationship will affect the extraction of hemicellulose from the lygnocellulose cell wall matrix. The presence of hydrogen bonds between the individual components of the polysaccharides also inhibits the process of isolation. Therefore, the proper insulation process is required in order to separate it from the other components of polymers such as cellulose and lignin [9, 16, 22, 26, 49, 50].

Isolation process of hemicellulose has been in conformity with the principles of green chemistry. The latter sets guidelines for the chemical industry in order to secure sustainable development, while increasing process economy. Briefly, green chemistry and the related green engineering call for an increase in and/or upgrading of

- process economy, by preventing waste generation; this represents a much superior approach to waste treatment;
- atom economy, by incorporating all reagents employed in the final product; this also contributes to reduction and/or elimination of waste;

- process safety, e.g., by using non-toxic, non-inflammable solvents and reagents;
- process efficiency, e.g., by material recycling into the process; where possible, use of catalytic pathways; use of catalysts that can be regenerated/recycled; rational use of energy, and reduction of the number of intermediate steps [17].

Green chemistry also calls for design for biodegradable end products, principally, by employing chemicals from renewable sources, and dictates the use of real-time, on-line analysis for better process control [17].

Muchlisyam et al. [33, 34] report green method has been carried out to isolation method of hemicellulose from corn cobs. The method used produce environment-friendly waste. That research was done to corn cobs from local corn in Medan, Indonesia.

The procedure is carried out with 50 g of corn cobs powder, which was added to 500 mL of 0.03 M NaOH in 70% ethanol and heated at 60°C and stirred for 2 h to dissolve the lignin. The suspension was allowed to cool to room temperature and filtered through Whatman filter paper. The precipitate was added to 500 mL of 0.2 M NaOH and stirred for 8 h at room temperature to dissolve hemicellulose, and then filtered. The filtrate was heated at a temperature of 65°C, and 137 mL of 3% H_20_2 was added in stages. Each addition of 1 mL 3% H_20_2 to the filtrate was made and stirred constantly. Stirring was continued until a clear solution formed.

The solution of 10% acetic acid in 95% ethanol with a ratio of 1:4 (v/v) was added to the sample solution and left at room temperature for 6 h until the precipitate formed. The suspension was centrifuged at 10,000 rpm for 15 min, the filtrate was discarded, and the precipitate was washed with 96% ethanol and dried in vacuum dryer. The washed precipitate was hemicellulose [33, 34].

The isolation of corn cobs hemicellulose used 500 cm^3 of 0.03 M NaOH in 70% ethanol and precipitation used the solution of 10% acetic acid in 95% ethanol with a ratio of 1:4 (v/v) and corn cobs hemicelluloses reached the levels of 12.04% [33, 34]. This method performs removal of lignin by using 3% H_20_2, while hemicellulose powder is used to obtain 10% acetic acid in 95% ethanol.

The deposition process of alkaline solution conforms to some isolation methods that have been conducted by various other researchers who use hydrochloride acid and chlorine. Replacement is intended to prevent the occurrence of waste in the form of dioxin. This compound is an organic material that is toxic to human health. Due to the use of chlorine, it will occur between lignin reaction with chlorine to form dioxins that are harmful to health and the environment. In addition, the delignification step was shown to be more acceptable in practice than the hazardous delignification with chlorine. The delignification step produce hemicelluloses that free of toxic waste such as dioxin [33, 34].

8.3 Characterization of Corn Cobs Hemicelluloses

The purpose of characterization of corn cobs hemicellulose using infrared spectrophotometry (FTIR) is to determine the functional groups of the hemicellulose and to observe a vibration of specific fingerprint. However, the purpose of characterization of corn cobs hemicellulose using high performance liquid chromatography is to determine the retention time, the peak height, and the symmetrical area of the chromatograph [3, 5].

8.3.1 Characterization of Hemicellulose by High-Performance Liquid Chromatography

The method of characterization by high-performance liquid chromatography (HPLC) is as follows: 25 mg of hemicellulose was put into 50 mL volumetric flask and pure water was added to the mark. The mixture was shaken and then filtered (first few milliliters of filtrate was discarded). The solution was then filtered through a membrane filter of 0.2 μm cellulose nitrate. Then about 100 mL solution was injected into the HPLC system via a loop injector with a 20 mL, using an isocratic elution system with distilled water with a mobile phase, flow rate 0.8 mL/min. The detection was made using a UV detector at a wavelength of 280 nm [34].

The chromatogram recorded and performed a qualitative analysis of the area, peak height, and retention time corn cobs hemicellulose.

From Fig. 8.1 it can be seen that the yield of hemicellulose is a hydrophilic compound as it is soluble in the distilled water. The HPLC testing using aquabidest as the mobile phase, using a C18 column, with a flow rate of 0.8 cm^3/min and ultraviolet light detector at a wavelength of 280 nm produces retention time of 1.8, peak height of 5.11558 and symmetrical area of 0.82 [34].

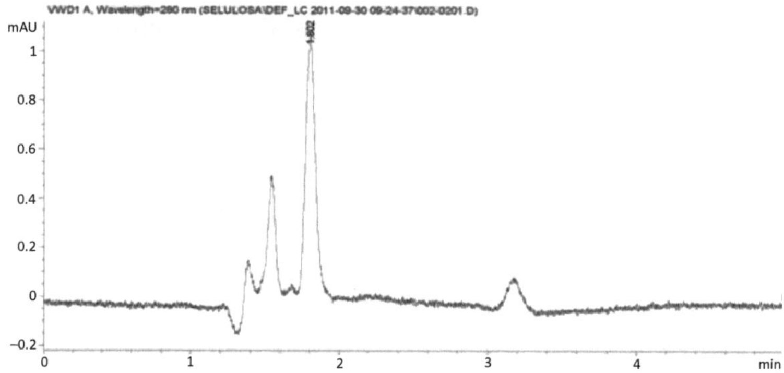

Figure 8.1 Characterization corn cobs hemicellulose by high-performance liquid chromatography.

8.3.2 Characterization of Corn Cobs Hemicellulose Using Infrared Spectrophotometry

To a weighed amount of 1 mg corn cobs hemicellulose, 200 mg of potassium bromide was added. The mixture was ground to a homogeneous form and then analyzed the vibration in the range of wave number 500–4000 cm^{-1}. Infrared spectra recorded the fingerprint spectrum and the spectrum of functional groups resulting from the hemicelluloses [5]. The results can be seen in Table 8.1.

FTIR characterization of the hemicellulose in Table 8.1 shows that the sample of the insulation has a C=O group and OH group. The functional group is present in the monomer on the hemicellulose and the right catchment area of 500–1500 cm^{-1} is an absorption area of the fingerprint region [5]. The sample gives the vibrations in the fingerprint region, which is shown in Fig. 8.2.

The FT-IR spectrum of corn cob hemicellulose is shown in Fig. 8.2. Typical signals of hemicellulose at 3282, 3076, 2041, and 895 cm^{-1} were clearly observed in the sample [34]. Hemicellulose exhibited a broad stretching intense characteristic peak at around 3282 cm^{-1} for the hydroxyl group and a weak band at 2929 cm^{-1} due to CH_2 and CH_3 groups [5].

Table 8.1 Characterization of corn cob hemicelluloses with solubility, FTIR and HPLC

Solubility	Aquadest	Poorly soluble
	Hot water	Dissolve
	1% HCl	Poorly soluble
	1% NaOH	Dissolve
FTIR		Hydroxyl
	Functional groups	Carbonyl
		Carboxylate
HPLC	Retention time	1.802
	Peak height	1.06418
	Area	5.11558
	Symmetric	0.82

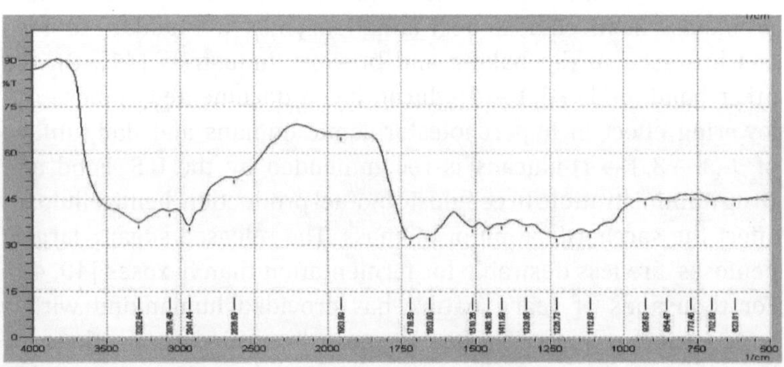

Figure 8.2 Characterization infrared vibration of corn cobs hemicelluloses.

A large peak around 1653 cm^{-1} was assigned to the carboxyl group of glucuronic acid. Hemicellulose has a band in the 750–1250 cm^{-1} region, which is dominated by ring vibrations overlapped with stretching vibrations of C–OH side groups and C–O–C glycosidic band vibration [5].

8.4 Chemical Structure and Properties of Hemicelluloses

The chemical structure of hemicelluloses is the most important factor that influences their properties. That is why polysaccharides present large differences in solubility in solution and gel properties. Their chemical structure determines the shapes of the molecules adopted both in aqueous system and in the solid state [9]. In industrial applications, hemicelluloses are used to control water and the rheology of aqueous phases. Thus, they may be used as food additives, thickeners, emulsifiers, gelling agents, adhesives, and adsorbents [9, 12]. Hemicelluloses have commercial significance. Seed storage hemicelluloses are used directly as products in the food industry, e.g., guar and locust bean gums (galactomannans), gum (glucomannan), and tamarind gum (xyloglucan) [41].

In addition, hemicelluloses give important properties to many food and feed products. In the baking industry, the insoluble arabinoxylans affect the baking quality. β-(1→3,1→4)-glucans and arabinoxylans are well-known anti-nutritional compounds in animal feed. They can cause filtering and haze problems in the brewery industry due to their viscosity. To alleviate these problems, hemicellulose-degrading enzymes are added to feed and are used in the baking and brewery industries [41]. On the other hand, β-(1→3,1→4)-glucan has a documented cholesterol-lowering effect in hypercholesterolemic humans and daily intake of β-(1→3,1→4)-glucans is recommended by the U.S. Food and Drug Administration. In cellulosic biofuel production, hemicelluloses affect the saccharification of biomass. The released sugars, largely pentoses, are less desirable for fermentation than hexoses [40, 43]. For thousands of years, nature has provided humankind with a large variety of materials for the most diversified applications for its survival, such as food, energy, medicinal products, protection and defense tools, and others [2, 9, 21]. In the pharmaceutical field, the hemicelluloses has a lot of potential and used for many purposes such as the producer of xylitol from xylan; as a binder; as a disintegrator in drug formulation; and as a thickener and stabilizer in syrup preparation. Moreover, several important applications of the xylan have been made by formulating its derivatives that have pharmacological effects such as cholesterol lowering and HIV inhibition in addition to being the dietary fiber [2, 25, 40].

Specifically, some hemicelluloses from plants and higher plants are a potential source of industrial polysaccharides. The pharmaceutical industry has benefitted from such diversity of biomaterials and has exploited the use of natural products as sources of both drugs and excipients. One example of a promising biomaterial for pharmaceutical use is xylan, a hemicellulose largely found in nature and considered the second most abundant polysaccharide after cellulose [41, 43]. Melo et al. [31] have reported research on the activity of antioxidant, antimicroba, and anticoagulant from corn cobs xylan.

Xylan also has drawn considerable interest due to its potential for packaging films and coating food, as well as for its use in biomedical products [28]. Because it is referred to as a corn fiber gum with a sticky behavior, xylan has been used as an antithrombotic activity, adhesive, drug delivery system, anti-cholesterol, thickener, and additive to plastics [2, 7, 9, 13, 36, 44, 45].

It increases their stretchability and breaking resistance as well as their susceptibility to biodegradation. Xylan has also been studied because of its significant mitogenic and comitogenic properties, which enable it to be compared to the commercial immunomodulator Zymosan. Another interesting application of xylan may be found in the food industry as an emulsifier and protein foam stabilizer during heating [11, 47, 49, 54].

Previous papers have investigated the suitable use of xylan in papermaking [10] and textile printing [21]. In the drug delivery field, xylan extracted from birch wood has been used for the production of nanoparticles after structural modification by the addition of different ester moieties, namely those with furoate and pyroglutamate functions [21]. On the other hand, the esterification of xylan from beech wood via the activation of the carboxylic acid with *N,N'*-carbonyldiimidazole has been carried out in order to produce prodrugs for ibuprofen release [10, 21, 22, 37].

Egito and colleagues have been working for over a decade on the extraction of xylan from corn cobs and its use for the development of microparticles as drug carriers for colon-specific delivery of anti-inflammatory and toxic drugs, such as sodium diclofenac (SD), 5-aminosalycilic acid (5-ASA), and usnic acid (UA). Xylan-coated microparticles have also been developed by Egito and co-workers in order to deliver magnetite particles [44]. Different microencapsulation techniques have been used for the production of xylan-based microparticles. Coacervation, interfacial cross-linking polymerization, and spray drying have been shown to be the most successful methodologies for that purpose [21, 26, 29].

Xylan degradation occurs by the action of hydrolytic enzymes named xylanases and *O*-xylosidases. Those enzymes are produced by a number of organisms, such as bacteria, algae, fungi, protozoa, gastropods, and arthropods. The degradation of xylan in ruminants has been well reported, while some human intestinal bacteria have been investigated for their ability to produce xylan-polymer–degrading enzymes. Among the intestinal species that are able to degrade complex carbohydrates, lactobacilli, bacteroides, and nonpathogenic clostridia have demonstrated that ability. Because of the presence of those bacteria in the human colon whether by induction of prebiotics or not, it is believed that xylan is a promising polymer for the composition of biodegradable drug carriers for colonic delivery. Hemicellulose is not decomposed in the upper gastrointestinal and is degraded by xylanases enzyme when reaching the colon [41].

Muchlisyam et al. [33, 34] worked for the isolation of hemicellulose from corn cobs to made eco-friendly waste and its use to reduce cadmium and plumbum from rabbit's blood [33, 34].

8.5 Application of Corn Cobs Hemicellulose as Chelating Agent for Toxic Metal

Chelating agents can be defined as organic compounds that complex or sequester metal ions. The chelating agent removes a metallic ion from a solid salt and holds it in solution. By forming a soluble complex from an insoluble compound it is possible to remove unwanted material, washing it away with water [8,30]. Common functional groups that are electron donors according to Ref. [30] among others are carboxylate, *N*-heterocyclic, aliphatic amines, aromatic amines, and phosphates.

Lewis Law developed by Pearson in 1963 [38] states that the hydroxyl group is monodentate ligands that are hard base. According to it, the hydroxyl groups in inorganic and organic compounds are strong base groups. They can also act as a ligand to the metal compounds that are weak acids and that are borderline acids [37]. This means that Lewis Law developed by Pearson in [20] not only does apply to the interaction of the same class, such as a strong acid with a strong base, but also can interact with a different class acid or base such as a borderline acid or weak base. [20]. Based on the above, this suggests that the corn cob hemicellulose can be used as ligand for toxic metals.

8.5.1 Test to Corn Cobs Hemicellulose Power Strap Respect for Cadmium (Cd) Ion and Plumbum (Pb) Ions with Titrimetric in the pH Bases

The corn cobs hemicelluloses weighed 100 mg. Then, the hemicelluloses were inserted into 12 erlenmeyers, dissolved in 1 cm^3 of 0.05 N NaOH, and diluted with distilled water to 10 cm^3. Six erlenmeyers of corncobs hemicelluloses solution titrated with a solution of 3 mg/cm^3 cadmium ion. However, the other six erlenmeyers of hemicelluloses solution titrated with a solution 10 mg/cm^3 plumbum ions. The end point of the titration was indicated by the formation of a white precipitate [3, 33].

The test of corn cobs hemicellulose's ability as a chelating agent metal titrimetric with titration was done indirectly, and as a solvent, 0.05 N NaOH as much as 1 cm^3 was used. This is based on the hemicellulose of hydroxyl groups that react with the sodium salt into alkaline salts that are soluble in water. This is in accordance with the legal principles that reactions between Lewis acids and bases will form water-soluble salts [20, 38].

The titration results of corn cobs hemicellulose bond between the metal ions cadmium and plumbum ions are shown in Tables 8.2 and 8.3. Based on Table 8.2, it appears that 100 mg of corn cobs hemicellulose was bond cadmium at (46.88 ± 0.5094) mg. According to Table 8.3, it appears that 100 mg of corn cobs hemicellulose was bond plumbum at (67.82 ± 0.4280) mg. This implies that the cadmium ions and plumbum ions can release sodium from hemicelluloses [33, 34]. Because based on their group in the periodic system, sodium was contained in class 1A and had a low electronegativity compared to cadmium ions in group II or plumbum ions in group IVA.

Table 8.2 Titration results of corn cobs hemicelluloses with cadmium 3 mg/cm^3 as a titrant

No	Corn cobs hemicellulose (mg)	Volume 3 mg/cm^3 Cd(cm^3)	Percentage (%)
1	100.4	15.65	46.51
2	100.7	16.05	47.46
3	101.4	16.20	47.23
4	101.7	15.90	46.11
5	101.9	16.35	47.20
6	101.7	16.15	46.80
		Average	**46.88**

Table 8.3 Titration results of corn cobs hemicellulose with plumbum 10 mg/cm^3 as a titrant

No	Corn cobs hemicellulose (mg)	Volume 3 mg/cm^3 Pb (cm^3)	Percentage (%)
1	100.2	67.63	67.51
2	100.5	68.34	68.00
3	100.4	68.07	67.80
4	101.2	69.32	68.50
5	100.3	67.70	67.50
6	101.5	68.61	67.60
		Average	**67.82**

This is reinforced by laws developed by [38], the law of the hard soft acid base, which states that the bonding that occurs in hemicellulose also occurs at hydroxyl groups contained in the carboxylate groups and hydroxyl corn cobs hemicellulose. Although the reaction between sodium—a strong acid—and a hydroxyl compound of hemicellulose—a strong base—can be replaced by other cations, there are soft acids such as cadmium ions and borderline acids such as plumbum ions [20]. This indicates that the titration performed in the alkaline pH of the hemicellulose can bind cadmium with endpoint white precipitation of cadmium hydroxide or endpoint white precipitation plumbum hydroxide from plumbum [3].

8.5.2 The Identification of Cadmium Corn Cobs Hemicellulose Bond and Plumbum Corn Cobs Hemicellulose Bond with Color Reaction and FTIR Vibration

The results of titration with cadmium corn cobs hemicelluloses and plumbum corn cobs hemicelluloses were centrifuged at 10,000 rpm for 15 min. Then, the filtrate was separated from sediment and the sediment was washed with aquadest. After that, it was suspended in 1 cm^3 of concentrated nitric acid, diluted with 10 cm^3 of distilled water, stirred for 2 min, then centrifuged and separated [24, 33].

8.5.2.1 Identification with the color reaction

The results of hemicellulose filtrate titration appear and the first reaction solution of potassium sulfide is added to cadmium forming yellow color and plumbum forming black brown color [33].

The final titration is showed by the existence of perfect reaction between hemicelluloses with cadmium and plumbum. Based on the result of the reaction, it appears that there has been a reaction between cadmium and plumbum with hemicellulose as indicated by the specific reaction of cadmium and plumbum in the filtrate titration results. Filtrate titration results indicate sediment that has specific color for cadmium and lead [3].

This confirms there has been an ionic bond between the cadmium or plumbum with corn cobs hemicellulose and it forms water-soluble complexes. This is evidenced also by the sediment test titration results showing the results of the same reaction. This means that the deposits were formed in the titration of a precipitate of cadmium hydroxide or plumbum hydroxide.

8.5.2.2 Identification by FTIR

Titration results with fried dryer and dry powder weighed as much as 1 mg and were identified by FTIR. The results were compared to the yield of vibrational corn cobs hemicellulose with infrared [24, 34].

The comparison of infrared vibrational forms corn cobs hemicellulose, the reaction corn cobs hemicellulose vibrations can be seen in Figs. 8.2 and 8.3a for cadmium corn cobs hemicellulose vibrational forms and Fig. 8.3b for plumbum vibrational forms. Figures 8.3a and 8.3b indicate a change in the form of vibrations of the corn cobs hemicellulose before and after the reaction with cadmium and plumbum.

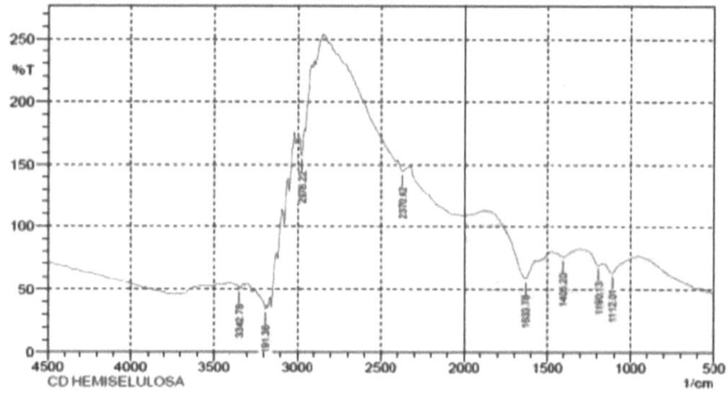

Figure 8.3a Characterization of the infrared vibration of corn cobs hemicellulose and cadmium complex.

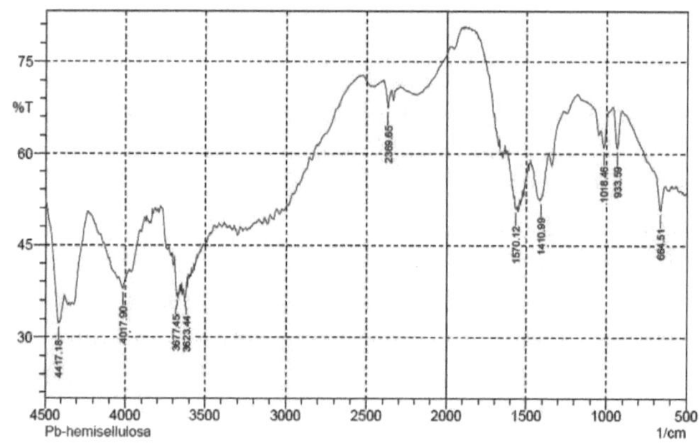

Figure 8.3b Characterization of the infrared vibration of corn cobs hemicellulose and plumbum complex.

From Figs. 8.3a and 8.3b, it can be seen that there are a shifting in the location of the vibration of the corn cob hemicellulose prior to reaction with cadmium and plumbum. This shows that the inclusion of cadmium and plumbum ions on corncob hemicellulose, will change the location of groups vibrations contained in corn cob hemicellulose. In the FTIR vibration, hydroxyl vibration of corn cobs hemicellulose in the region 3282 cm^{-1} changes to 3382 cm^{-1} with an addition of cadmium in the form complexes of corn cobs hemicellulose cadmium; or to 3623 and 3677.45 cm^{-1} with an addition of plumbum in the form complexes of corn cobs hemicellulose plumbum. This means there is a reaction between corn cob hemicellulose with cadmium or with plumbum then formed complexes that resulted in a change of vibration of the hydroxyl group into vibration corn cob hemicellulose complex shapes with cadmium or plumbum [34].

8.6 Cadmium and Plumbum Absorption

The hemicellulose and pectin, weighed at 100, 200, and 300 mg. Then each was dissolved into erlenmeyer containing 25 cm^3 solution 30 mg/cm^3 of cadmium in 0.1 N nitric acid and the pH adjusted to acid pH was performed at pH 2. The solution was stirred with a magnetic stirrer at room temperature for 2 h. Then, each was centrifuged at 10,000 rpm for 30 min. The supernatant was taken and diluted with 0.5 to 50 cm^3 aquabidest. Then 0.25 cm^3 of dilution was taken and diluted to 25 cm^3 with aquabidest. The cadmium levels were measured [34]. The cadmium in the supernatant was estimated using flame atomic absorption spectroscopy (Hitachi Analyst 100) at 228.3 nm [23, 32, 34, 51].

The results are presented as mean values ± standard deviation. The same procedure was applied to the 25 mL solution 40 mg/ml of plumbum in 0.1 N nitric acid. The plumbum in the supernatant was estimated using graphite furnace atomic absorption spectroscopy (Hitachi Analyst 100) at 283.8 nm [23, 32]. The results are presented as mean values ± standard deviation [18, 48]. The chelating ability of corn cobs hemicellulose and pectin toward cadmium and plumbum ion is shown in Tables 8.4 and 8.5. The administration of the chelating agent to bind cadmium and plumbum at pH 2, the higher the amount of chelating agent showed an increased chelating ability, but in hemicellulose from corn cobs, the increase was not proportional to the amount of hemicellulose. This is in accordance with the principle of adsorption that the adsorption process will end if there is equilibrium between the adsorbent, so the weight does not affect a significant absorbance with the increasing amount of hemicellulose for each treatment [34].

Tables 8.4 and 8.5 show that there is the influence of the type of chelating agent and the weight of the chelating agent applied. Compared to hemicellulose, 100 mg pectin has the lowest binding capacity of 21.10 ± 1.08 mg of plumbum ion or 52.75%. However, 200 and 300 mg pectin will give a greater ability in binding capacity of 32.48 ± 0.71 mg or 81.20% and 37.39 ± 0.22 mg or 93.47% of plumbum ion, respectively, which shows that the corn cob hemicellulose has a greater chelating effect than pectin [34].

Table 8.4 Chelating effect of corn cobs hemicellulose and pectin for 30 mg/25 mL cadmium ions

Chelating agent	Weight	Cadmium content in chelating agent (mg)	Percentage cadmium ion in chelating agent (%)
Hemicellulose	100 mg	24.41 ± 0.6185	81.36
	200 mg	25.65 ± 0.4279	85.50
	300 mg	26.68 ± 0.1490	88.93
Pectin	100 mg	20.86 ± 0.3924	69.53
	200 mg	25.30 ± 0.6599	84.33
	300 mg	26.89 ± 0.5817	89.63

Table 8.5 Chelating effect of corn cobs hemicellulose and pectin for 40 mg/25 mL plumbum ions

Chelating agent	Weight	Plumbum content in chelating agent (mg)	Percentage plumbum ion in chelating agent (%)
Hemicellulose	100 mg	35.97 ± 0.25	89.92
	200 mg	37.33 ± 1.90	93.32
	300 mg	39.53 ± 0.13	98.82
Pectin	100 mg	21.10 ± 1.08	52.75
	200 mg	32.48 ± 0.71	81.20
	300 mg	37.39 ± 0.22	93.47

Absorption characteristics between hemicelluloses and pectin are different where hemicelluloses react with plumbum ions. This suggests that the hydroxyl group of the hemicellulose causes attraction of plumbum ions present in solution. Thus, hemicellulose is more polar and has a better absorption capacity than the less polar substances. The uptake mechanism that occurs between the-hydroxyl function from hemicellulose monomer, for example, pentosan group attached to the surface with the positively charged metal ion (cation), is an ion exchange mechanism as follows [33]:

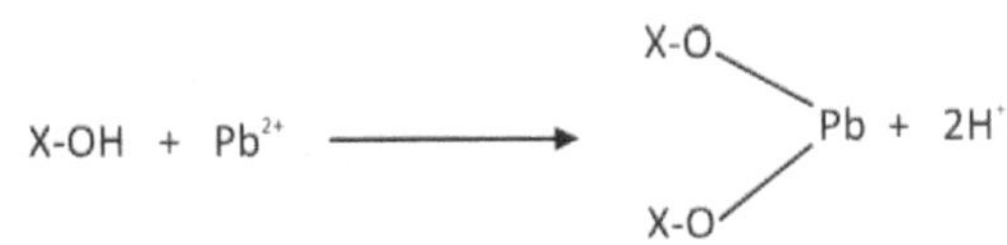

Metal ions in the form of plumbum ions, –OH groups are hydroxyl groups present on the C atom of the monomer compound in the hemicellulose and X is a matrix of hemicellulose place the –OH group attached. Interactions of the –OH group with the plumbum ion are also possible through the mechanism of formation of coordination complexes, as atomic oxygen (O) with the –OH group has a lone pair, while the plumbum ion has an empty orbital. The free electron pair will occupy an empty orbital that is owned by a metal ion, thus forming a compound or complex ion [8, 20, 33]. Chemical bonding that occurs between the active groups in the molecules of organic substances can be explained as the behavior of Lewis acid–base interactions, which produce a complex on the surface of solids [38]. In the system of binding of metal ion solution, for example, the interaction is written in general form [33]:

$$2[FH] + M^{z} \longrightarrow [F2M^{(Z-2)}] + 2H^{+}$$

where FH is a functional group found in organic matter, and M is the ion of *z* valency. This means that the in vitro hemicellulose, which is an organic compound having a hydroxyl group, can bind to metal of two valency in the dissolved state and acidic conditions [30, 33, 34].

The chelating ability of corn cobs hemicellulose, and pectin toward cadmium ion and plumbum is shown in Tables 8.4 and 8.5. The administration of chelating agent to bind distinct cadmium and plumbum in the pH 2 gives different results. From Tables 8.4 and 8.5, it can be seen that there is an influence of the type and weight of chelating agent applied. However, in contrast to pectin, with the addition of pectin, the concentration will also increase properly the strap to cadmium ions [34].

On the 100 mg pectin, there is a binding ability of (20.86 ± 0.3924) mg cadmium ion or 69.53% and (21.10 ± 1.08) mg plumbum ions or 52.74%, but the 200 mg pectin will give a greater ability around (25.30 ± 0.6599) mg cadmium ion or 84.33% and (32.48 ± 0.71) mg plumbum ions or 81.20%. Three hundred milligram of pectin bonds by an increase of (26.89 ± 0.5817) mg of cadmium ions, or 89.63%, and (37.39 ± 0.22) mg plumbum ions or 93.47% [34].

Compared with corn cobs hemicellulose, the average increase of the binding ability is around 4.33–6.61%. The overall number of increased binding ability of pectin is still below the binding ability. This means that a chelating effect of the corn cobs hemicellulose is greater than pectin. The higher amount of the chelating agent showed the higher increase of chelating agent ability. However, in corn cobs hemicellulose, the increase was not proportional to its amount [34]. It is in accordance with the principle of absorption that the absorption process will end if there is equilibrium between the absorbents combination [30]. Therefore, the weight does not affect a significant absorbance with the hemicellulose for each treatment. Statistical calculations [18, 48] show that the addition of corn cobs hemicellulose will give an increase to the binding effect for cadmium ion. However, it can be proved here that the corn cobs hemicellulose given as binding ability to a solution of 30 mg in 25 cm^3 of cadmium causes a decline in cadmium content in the solution, meaning that there is a chemical function of the corn cobs hemicellulose that can bind cadmium ions. This suggests that the hydroxyl and carboxylic groups of the corn cobs hemicellulose cause attraction for cadmium ions to be present in solution [34].

Based on Tables 8.4 and 8.5, it can be stated that the levels of cadmium and plumbum are tied with the chelating agent in the treatment of the type of interaction and the variation of weight. The chelating agent type shows the same thing with the sole influence on the differences that in chelating agent type. There is also difference in the binding of cadmium levels and plumbum levels can be maximized by a given weight of 300 mg [34].

The profile of chelating effect of corn cobs hemicelluloses and pectin in Figs. 8.4 and 8.5 shows that the weight will increase the ability of the chelating agent to corn cobs hemicellulose and pectin as a whole. Overall, the corn cobs hemicelluloses have greater ability than pectin except at the weight of 300 mg, pectin has the ability to absorb cadmium ions greater than the corn cobs hemicellulose. However, in the terms of pharmacology, pectin is used as anti-diarrhea only in doses below 165 mg, and if given in doses of 300 mg weight, constipation most likely occurred. Besides as a fiber, corn cobs hemicellulose has pharmacological effects to improve the digestive system and is not toxic [34].

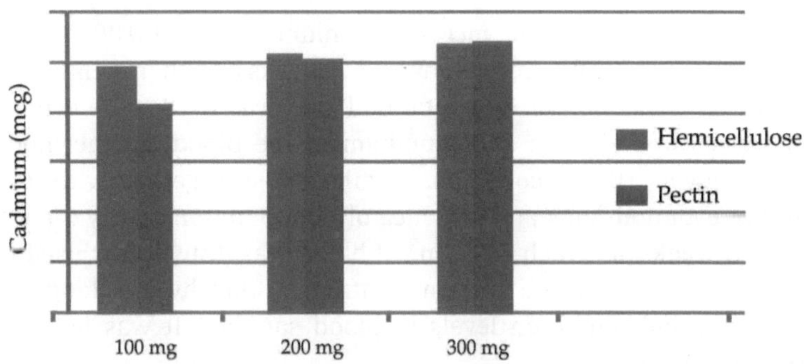

Figure 8.4 Charts comparison of hemicellulose and pectin bound to cadmium ions.

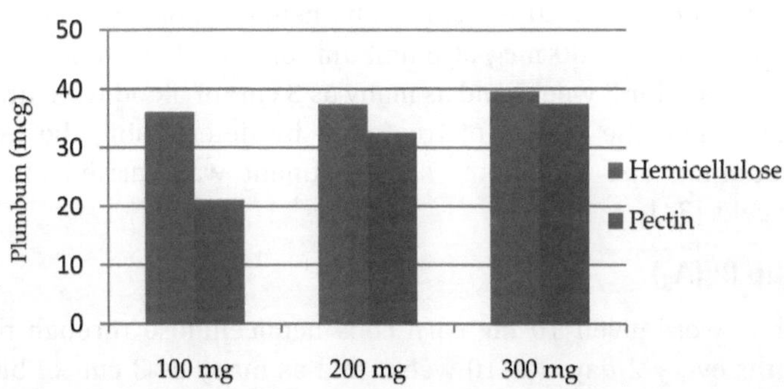

Figure 8.5 Charts comparison of hemicellulose and pectin bound to plumbum ions.

8.7 The in vivo Corn Cobs Hemicellulose Testing in Rabbits

This study was done on rabbits for 10 weeks (W). Measurements were made on cadmium and plumbum levels in rabbit's blood in every 2 weeks [34]. The test is performed with the following steps:

(1) As many as 15 rabbits were weighed and their blood was drawn to determine the initial level of cadmium and plumbum in blood.
(2) Rabbits were divided into three groups and each group consisted of 4 rabbits. They were given 3 cm^3 aquabides to drink and then treated as follows:

Group I (A_1)

Rabbits were given 100 mcg of cadmium ions and 100 mcg of plumbum ions orally every 2 days for 2 weeks and then as much as 3 cm^3 of it was drawn. The total of 1 cm^3 was used to determine the levels of cadmium and plumbum in the blood. Furthermore, it was given 10 mg corn cobs hemicellulose together with 100 mcg of cadmium ions and 100 mcg of plumbum ions every 2 days. After 2 weeks, as much as 3 cm^3 of blood was drawn. It is used to test the results of the implementation of this by checking the cadmium and plumbum levels in blood samples. It was held for 10 weeks [34].

Group II (A_2)

Rabbits were given 10 mg corn cobs hemicellulose and 100 mcg of cadmium and 100 mcg of plumbum ions through their mouths every 2 days for 2 weeks and as many as 3 cm^3 of blood were drawn to determine the results of treatment by determining the levels of cadmium and plumbum. The treatment was carried out for 10 weeks [34].

Group III (A_3)

Rabbits were given 10 mg corn cobs hemicellulose through their mouths every 2 days for 10 weeks and as many as 3 cm^3 of blood were taken every 2 weeks to determine the results of treatment by determining the levels of cadmium and plumbum. The treatment was carried out for 10 weeks [34].

8.7.1 Determination of Cadmium and Plumbum Levels in Blood of Rabbits

One cm^3 of blood was added to 5 cm^3 of concentrated nitric acid and heated on a hot plate at a temperature of 150°C for 30 min. Then, it was added 0.2 cm^3 of 50% perchloric acid solution and

0.4 cm^3 of concentrated sulfuric acid and heated in a row at a temperature of 150°C, 200°C and 250°C, respectively within 15 min. Once it was heated at a temperature of 320°C for 20 min. The results were obtained by the destruction of the white and absorbed in 6 N HNO_3 (v/v) at a temperature of 90°C for 30 min until dissolved, cooled and diluted with mineral-free water [32]. The determination of cadmium levels in a sample by flame AAS at a wavelength of 228.3 nm and plumbum level in a sample by graphite furnace AAS at 283.8 nm [32, 52, 55].

8.7.2 Binding Ability Corn Cobs Hemicellulose toward Cadmium in vivo

The corn cobs hemicellulose test with three treatments was carried out and the results of the treatments are shown in Tables 8.6, 8.7.

Tabel 8.6 Effect of treatment for 10 weeks on the average blood cadmium levels

Week	Treatment	N	Weight (ppm)	Standard of deviation
0	A_1	4	2.42	0.2902
	A_2	4	2.37	0.1127
	A_3	4	3.09	0.1676
2	A_1	4	4.32	0.5048
	A_2	4	3.96	0.3347
	A_3	4	4.46	0.2936
4	A_1	4	3.08	0.2927
	A_2	4	3.34	0.7862
	A_3	4	2.69	0.1706
6	A_1	4	0.39	0.1578
	A_2	4	0.79	0.3126
	A_3	4	0.59	0.3624
8	A_1	4	0.21	0.0150
	A_2	4	0.50	0.2435
	A_3	4	0.42	0.3738
10	A_1	4	None	None
	A_2	4	0.35	0.3070
	A_3	4	0.04	0.0526

Tabel 8.7 Effect of treatment for 10 weeks on the average blood plumbum levels

Week	Treatment	N	Weight	Standard of deviation
0	A_1	4	2.16	0.3867
	A_2	4	0.67	0.4478
	A_3	4	1.23	0.6678
2	A_1	4	1.30	0.3030
	A_2	4	0.89	0.8020
	A_3	4	1.15	0.7242
4	A_1	4	1.04	0.9785
	A_2	4	0.84	0.8218
	A_3	4	0.98	0.0709
6	A_1	4	0.23	0.2956
	A_2	4	0.39	0.3293
	A_3	4	0.04	0.0750
8	A_1	4	0.08	0.0980
	A_2	4	0.13	0.1618
	A_3	4	None	None
10	A_1	4	0.12	0.1698
	A_2	4	0.10	0.1611
	A_3	4	None	None

Note: A_1: 100 mcg of lead every 2 days for 2 weeks and then continued 100 mcg of plumbum and 10 mg hemicellulose A_2: 100 mcg of plumbum and 10 mg every 2 days hemicellulose A_3: hemicellulose 10 mg every 2 days.

Based on Tables 8.6 and 8.7, it appears from the blood samples taken prior to treatment (W_0) that the average blood cadmium level at (2.63 ± 0.1902) ppm/cm^3 and plumbum level at (1.35 ± 0.5008) ppm/cm^3 and after the third treatment every other day for 2 weeks there was a decrease in blood cadmium levels as proved in treatment A_1, A_2, and A_3. This is caused by the influence of carrots—eaten an average of 200 g per day—which contains cadmium and plumbum at 0.02 ppm and 0.05 ppm. Thus, it affects the content of cadmium and plumbum in the blood of rabbits. Under the same conditions, the examination of the fourth week of a decline in cadmium and plumbum levels in all treatments showed a decrease of blood cadmium and plumbum levels in rabbits. In the same conditions, results showed a decline in cadmium and plumbum levels and continued until measurements in 10 weeks.

The treatment was then stopped because the results of the analysis of blood cadmium levels in some rabbits were not detected and the mean cadmium levels were (0.13 ± 0.2299) ppm/cm^3 and plumbum was not detected [34].

This means that the administration of 10 mg corn cobs hemicellulose for 10 weeks can reduce cadmium levels of 2.50 ppm or 95.05% and plumbum levels of 1.24 ppm or 88.49%. On the other hand, F test [17, 47] results show that the statistical analysis before treatment (W_0) to check on W_{10} shows no significant difference. This means that the process of the treatment of three types of treatment for 10 weeks can reduce blood cadmium and plumbum levels in rabbits. In other words, corn cobs hemicelluloses can be used as an ingredient to reduce cadmium and plumbum levels in the blood as shown graphically in Figs. 8.6 and 8.7.

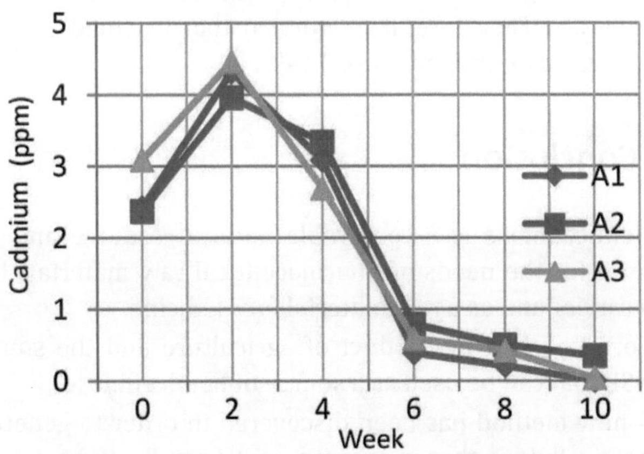

Figure 8.6 Graphics the relationship between time of administration and reduction of cadmium content in the blood.

Based on Figs. 8.6 and 8.7, it appears that the duration of treatment affects the levels of cadmium and plumbum in the blood. The results of blood tests at the end of W_2 show an increase in levels of cadmium and plumbum on A_1, A_2, and A_3 treatment. However, in the week W_4, there had been a decrease in the blood and the decline continued until W_{10} [34].

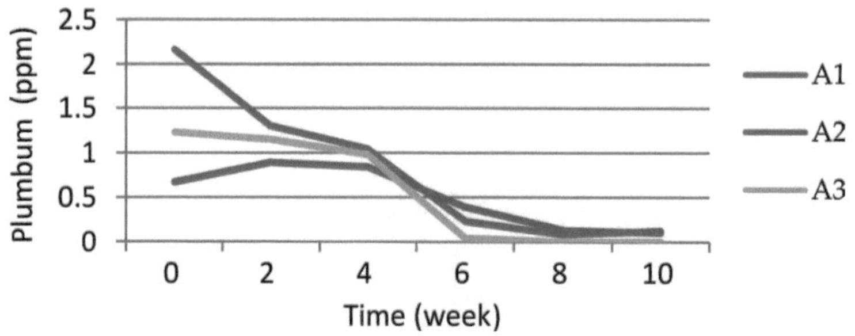

Figure 8.7 Graphics the relationship between time of administration and reduction of plumbum content in the blood.

Treatment was stopped due to blood cadmium and plumbum levels in several rabbits are not detected meaning that the administration of 10 mg corn cobs hemicellulose with a period of 10 weeks can reduce exposure to cadmium and plumbum levels in the blood. This also means when corn cobs hemicellulose is given in greater weight, it will affect the essential minerals needed by the body in large quantities such as calcium, iron, zinc, sodium, potassium, etc. Therefore, it is worried that it would also reduce levels of these essential minerals [34].

8.8 Conclusion

(1) Hemicellulose is a renewable natural resource and can be used for the needs of pharmaceutical raw materials both as a carrier and as a raw material for medicine.

(2) Corn cob is a by-product of agriculture and the sources of NSP that can be used as a source of hemicellulose.

(3) A new method has been discovered in order to generate the hemicellulose that is environment friendly and free of toxic compounds.

(4) Hemicellulose can be used to reduce the contamination of cadmium and plumbum in the body of rabbits.

References

1. Aspinall, G. (1959). Structural chemistry of hemicelluloses. *Adv. Carbohydr. Chem.*, **14**, 429–468.

2. Barroso, E. M. A., Costa, L. S., Medeiros, V. P., Cordeiro, L. S., Costa, M. S. S. P., Franco, C. R. C., Nader, H. B., Leite, E. L., Rocha, H. A. O. (2008). A non-anticoagulant heterofucan has antithrombotic activity in vivo. *Planta Med.*, **74**, 712–718.

3. Basset, J., Denny, R. C., Jeffrey, B. H., Mendham, J. (1994). *Vogel's Textbook of Quantitative Inorganic Analysis Including Elementary Instrumental Analysis*, Translation, EGC Book Medical Publishers, Jakarta.

4. Bikova, T., Treimanis, A. (2002). Solubility and molecular weight of hemicelluloses from *Alnus incana* and *Alnus glutinosa*. Effect of tree age. *Plant Physiol. Biochem.*, **40**, 347–353.

5. Britthin, H. G. (2006). *Spectroscopy of Pharmaceutical Solids*, Center for Pharmaceutical Physics Taylor and Francis Group, New York, doi:10.1201/9780849361333.

6. Caprita, R., Caprita, A., Julean, D. C. (2010). Biochemical aspect of non-polysaccharides. *Anim. Sci. Biotechnol.*, **43**(1), 1–4.

7. Carvalheiro, F., Duarte, L.C., Girio, F. M. (2008). Hemicellulose biorefineries: A review on biomass pretreatments. *J. Sci. Indus. Res.*, **67**, 849–864.

8. Das, N., Vimala, R., Karthika, P. (2008). Biosorption of heavy metals: An overview. *Indian J. Biotechnol.*, **7**, 159–169.

9. Dumitriu, S. (2005). Polysaccharides: Structural diversity and functional versatility. In: Hemicelluloses: Structure and Properties (ed. Spiridon, I., Popa, V. I.,), Marcel Dekker, New York. Chapter 18, pp. 493–507.

10. Ebringerova, A., Hromadkova, Z., Kacurakova, M., Antal, M. (1994). Quaternized xylans: Synthesis and structural characterization. *Carbohydr. Polym.*, **24**(4), 301–308.

11. Ebringerova, A., Hromadkova, Z., Alfodi, J., Hribalova, V., (1995). Structure and mitogenic activities of corn cob heteroxylans. *Int. J. Biol. Macromol.*, **17**(6), 327–331.

12. Ebringerova, A., Hromadkova, Z., Alfodi, J., Hribalova, V. (1998). The immunologicallyactive xylan from ultrasound-treated corn cobs: Extractability, structure and properties. *Carbohydr. Polym.*, **37**(3), 231–239.

13. Ebringerova, A., Hromadkova, Z. (1999). Xylans of industrial and biomedical importance, In: *Biotechnology and Genetic Engineering Reviews*, Intercept Ltd. Scientific, Technical & Medical Publishers, ISBN 0264-8725, pp. 325–346.

14. Ebringerova, A., Heinze, T, (2000). Xylan and xylan derivatives: Biopolymers with valuable properties, 1. Naturally occurring xylans structures, procedures and properties. *Macromol. Rapid Commun.*, **21**(9), 542–556.

15. Ebringerova, A., Hromadkova, Z., Heinze, T. (2005). Hemicellulose. *Adv. Polym. Sci.*, **186**, 1–67.

16. Ebringerová, A. (2005). Structural diversity and application potential of hemicelluloses. *Macromol. Symp.*, **232**(1), 1–12.

17. El Seoud, O. A., Heinze, T. (2005). Organic esters of cellulose: New perspectives for old polymers. *Adv. Polym. Sci.*, **186**, 103–149.

18. Ermer, J., Miller, J. H. (2005). *Method Validation in Pharmaceutical Analysis*, Wiley-VCH Verlag GmbH & Co. KgaA, Weinheim, pp. 3, 5, 16.

19. Gilbert, H. J., Hazlewood, G. P. (1993). Bacterial cellulases and xylanases. *J. Gen. Microbiol.*, **139**, 187–194.

20. Haryadi, W. (1990). *Analytical Chemistry*, PT Gramedia, Jakarta.

21. Heinze, T., Petzold, K., Hornig, S. (2007). Novel nanoparticles based on xylan. *Cellulose Chem. Technol.*, **41**(1), 13–18.

22. Hromadkova, Z., Kacurakova, M., Ebringerova, A. (1999). Study of the classical and ultrasound-assisted extraction of the corn cob xylan. *Ind. Crops Prod.*, **9**(2), 101–109.

23. Jickel, S., Negrusz, A. (2008). *Clarke's Analytical Forensic Toxicology*, 3th ed., Pharmaceutical Press, Chicago.

24. Kacuráková, M., Wellner, N., Ebringerová, A., Hromádková, Z., Wilson, R. H., Belton, P. S. (1999). Characterisation of xylan-type polysaccharides and associated cell wall components by FT-IR and FT-Raman spectroscopies. *Food Hydrocoll.*, **13**, 35–41.

25. Karaaslan, A. M., Tshabalala, M. A., Buschle, D. G. (2010). Wood hemicellulose/chitosan-based semi-interpenetrating networks hydrogels: Mechanical swelling and control drug release properties. *J. Bioresources*, **5**(2), 1036–1054.

26. Kumar, M. (2000). Nano and microparticles as controlled drug delivery devices. *J. Pharm. Pharm. Sci.*, **3**(2), 234–258.

27. Kumar, V., de la Luz Reus-Medina, M., Yang, D. (2002). Preparation, characterization, and tabletting properties of a new cellulose-based pharmaceutical aid. *Int. J. Pharm.*, **235**(1–2) 129–140.

28. Kumar, P., Barrett, D. M., Delwiche, M. J., Stroeve, P. (2009). Methods for pretreatment of lignocellulosic biomass for efficient hydrolysis and biofuel production. *Ind. Eng. Chem. Res.*, **48**(8), 3713–3729.

29. Li, X., Shi, X., Wang, M., Du, Y. (2011). Xylan chitosan conjugate: A potential food preservative. *Food Chem.*, **126**(2), 520–525.

30. Manahan, S. E. (2003). *Toxicological Chemistry and Biochemistry,* 3rd ed., Lewis Publishers, A CRC Press Company, New York, pp. 233–239.

31. Melo-Silveira, R. F., Fidelis, G. P., Costa, M. S. S. P., Telles, C. B. S., Elias, S. D. O., Ribeiro, V. B., Barth, A. L., Macedo, A. J., Leite, E. L., Rocha, H. A. O. (2012). In vitro antioxidant, anticoagulant and antimicrobial activity and in inhibition of cancer cell proliferation by xylan extracted from corn cobs. *Int. J. Mol. Sci.,* **13**, 409–426.

32. Moffat, C. A., Osellton, D. M., Widdop, B. (2005). Clarke's *Analysis of Drugs and Poisons*, Pharmaceutical Press, London.

33. Muchlisyam, Harahap, U., Silalahi, J., Alfian, Z. A. (2013). Isolation and utilization of corn cobs hemicellulose as chelating agent for lead ions. *Malays. J. Anal. Sci.,* **17**(1), 71 - 74.

34. Muchlisyam, Harahap, U., Silalahi, J., Alfian, Z. B. (2013). Binding ability of corn cobs hemicellulose toward cadmium. *Am. J. Anal. Chem.*, **4**, 86–93.

35. Muralikrishna, G., Rao, S. M. V. S. S. T. (2007). Cereal non-cellulosic polysaccharides: Structure and function relationship: An overview. *Crit. Rev. Food Sci. Nutr.,* **47**(6), 599–608.

36. Nagashima, T., Oliveira, E. E., Silva, A. E., Marcelino, H. R., Gomes, M. C. S., Aguiar L. M., Araujo, I. B., Soares, L. A. L., Oliveira, A. G., Egito, E. S. T. (2008). Influence of the lipophilic external phase composition on the preparation and characterization of xylan microcapsules: A technical note. *AAPS Pharm. Sci. Tech.*, **9**(3), 814–817.

37. Oliveira, E. E., Silva, A. E., Nagashima Jr., T., Gomes, M. C. S., Aguiar, L. M., Marcelino H. R., Araujo, I. B., Bayer, M. P., Ricardo, N. M. P. S., Oliveira, A. G., Egito, E. S. T. (2010). Xylan from corn cobs, a promising polymer for drug delivery: Production and characterization. *Bioresource Technol.*, **101**(14), 5402–5406.

38. Pearson, R. G. (2005). Chemical hardness and density functional theory. *J. Chem. Sci.*, **117**(5), 369–377.

39. Purwoningsih, E. (2008). Effect of molecular weight chitosan against levels plumbum (Pb) blood and enzyme activity-Alad δ (delta aminolevulinic acid dehydratase) Albino mice (Mus Musculus L) healthcare, Master Thesis, University of North Sumatra, Medan.

40. Richana, N., Irawadi, T. T., Nur, A. M. (2007). The extraction of hemicellulose from corn cobs. *Postharvest J.*, **4**(1), 38–43.

41. Saha, C. B. (2003). Hemicellulose bioconversion. *J. Microbiol. Biotechnol.*, **30**(16), 2003, pp. 279–291.

42. Scheller, H. V., Ulvskov, P. (2010). *Annual Review of Plant Biology*, vol. 61, pp. 263–289, volume publication date June 2010, DOI: 10.1146/annurev-arplant-042809-112315.

43. Shallom, D., Shoham, Y. (2003). Microbial hemicellulases. *Curr. Opin. Microbiol.*, **6**(3), 219–228.

44. Silva, A. E., Marcelino, H. R., Gomes, M. C. S., Oliveira, E. E., Nagashima, Jr, T., Egito, E. S. T. (2008). Xylan, a promising hemicellulose for pharmaceutical use. *Prod. Appl. Biopolym. J.*, www.intechopen.com./download/PDF/31734.

45. Silva, A. K. A., Silva, E. L., Oliveira, E. E., Nagashima, J. T., Soares, L. A. L., Medeiros, A. C., Araujo, J. H., Araujo, I. B., Carriço, A. S., Egito, E. S. T. (2007). Synthesis and characterization of xylan-coated magnetite microparticles. *Int. J. Pharm.*, **334**(1–2), 42–47.

46. Simó, C., Cifuentes. A., Gallardo, A. (2003). Drug delivery systems: Polymers and drugs monitored by capillary electromigration methods. *J. Chromatogr. B*, **797**(1–2), 37–49.

47. Sjostrom, E. (1993). *Wood Chemistry Fundamentals and Applications*, 2nd ed., Academic Press, San Diego.

48. Steel, R. G. D., Torrie, J. H. (1991). *Principles and Procedures Statistics: A Biometric Approach*, PT Gramedia Pustaka Husada, Jakarta.

49. Stuart, M. A. C. (2008). Supramolecular perspectives in colloid science. *Colloid Polym. Sci.*, **286**(8–9), 855–864.

50. Sun, R. M., Fang, J., Goodwin, A. M., Lawther, J. J., Bolton, A. (1998). Fractionation and characterization of polysaccharides from abaca fibre. *Carbohydr. Polym.*, **37**(4), 351–359.

51. Szkoda, J., Zmudzki, J. (2005). Determination of lead and cadmium in biological material by graphite furnace atomic absorption spectrometry method. *Bull. Vet. Inst.*, **49**, 89–92.

52. Tarigan, I. M. (2008). Analysis of the applicability of chitosan and chitosan beads as adsorbents standard solution to lower levels of iron (Fe) and aluminum (Al) by atomic absorption spectrophotometry, Master Thesis, University of North Sumatra, Medan, 2008.

53. Timell, T. (1965). Wood hemicelluloses II. *Adv. Carbohydr. Chem.*, **20**, 409–483.

54. Ünlu, C. H., Günister, E., Atici, O. (2009). Synthesis and characterization of NaMt biocomposites with corn cob xylan in aqueous media. *Carbohydr. Polym.*, **76**(4), 585–592.

55. Wong, W. W., Abbas, F. M. A., Liong, M. T., Azhar, M. (2008). Modification of Durian ring for improved biosorbent ability. *Int. Food Res. J.*, **15**(3), 363–365.

56. Wyman, C. E., Decker, S. R., Himmel, M. E., Brady, J. W., Skopec, C. E. (2005). *Hydrolysis of Cellulose and Hemicellulose*, Copyright by Marcel Dekker.

57. Yadav, M. P., Johnston, D. B., Hicks, K. B. (2007). Structural characterization of corn fiber gums from coarse and fine fiber and a study of their emulsifying properties. *J. Agric. Food Chem.*, **55**(15), 6366–6371.

Chapter 9

Optical Anisotropy of Cellulose Esters and Its Application to Optical Functional Films

Kultida Songsurang, Hikaru Shimada, Shogo Nobukawa, and Masayuki Yamaguchi

School of Materials Science, Japan Advanced Institute of Science and Technology, 1-1 Asahidai, Nomi, Ishikawa 923-1292 Japan

songsurang.k@gmail.com, m_yama@jaist.ac.jp

Cellulose is the most abundant natural biopolymer and is readily available from renewable resources. Esterified cellulose is a highly flexible material as its properties can be varied by controlling the type and amount of the ester substituents during the chemical manufacturing process. Some cellulose esters have been applied as optical films for decades by virtue of their excellent properties such as high transparency and heat resistance. The cellulose ester used is mainly cellulose acetate, while the applications are rather limited to photographic films and protective films.

In this study, in-plane and out-of-plane birefringences in various types of cellulose esters, including those prepared by the solution-cast method was examined. Birefringence is an important property that defines the optical behavior of optical films. The information on the molecular orientation and the birefringence

Handbook of Sustainable Polymers: Processing and Applications
Edited by Vijay Kumar Thakur and Manju Kumari Thakur

ISBN 978-981-4613-53-8 (Hardcover), 978-981-4613-54-5 (eBook)
www.panstanford.com

of a solution-cast film for cellulose triacetate was investigated. Incorporation of low-mass compound was also proposed as the means to modify the orientation birefringence in cellulose esters. Moreover, the effects of the type and amount of ester groups were investigated. It is expected that this chapter will contribute to a profound understanding regarding the mechanism for in-plane and out-of-plane birefringences in cellulose esters. This would open up possibilities of developing biopolymer-based optical films with birefringence suited for various functions in optical devices.

9.1 Optical Properties of Polymeric Materials

In recent years, conventional plastics have been widely used, which cause severe environmental problems with the increase in their waste. With mounting the environmental and legislative pressure to reduce plastic and packaging wastes, there has been an increased demand for biodegradable or biomass-based plastics that are compatible with the environment. Therefore, alternative materials that can be biodegraded have emerged [1, 2].

Cellulose derivative, as one of the biomass-based plastics, is also studied to substitute glassy petroleum polymers. Further, the excellent transparency and heat resistant property of cellulose esters, such as cellulose triacetate (CTA), cellulose acetate propionate (CAP), and cellulose acetate butyrate (CAB), encourage the study on the optical anisotropy especially for the application of optical display such as liquid crystal display. In this application, various types of polymer films are required to construct a display, such as anti-glare film, retardation film, protective film, brightness enhancement film, and diffusion film. Particularly, polarization is a significantly important property for protective and retardation films [3–8]. For the optical application, control of retardation, i.e., product of birefringence and film thickness ($R = d\Delta n$; d is the film thickness and Δn is birefringence), is the most important aspect in material design, and thus various methods have been proposed.

High-performance displays such as large-size TVs, tablet PCs, and smart phones usually have more demanding requirements for viewing angle and contrast and, as a result, require special in-plane (R_{in}) and out-of-plane (R_{th}) retardations for optical films, as shown in Table 9.1 [9]:

Table 9.1 Requirements for LCD retardation films [9]

Applications	LCD modes	R_{in}	R_{th}
Large size TV	VA mode	50 nm	50 nm
	IPS mode	0 nm (<10 nm)	0 nm (<10 nm)
PC, laptops	TN mode	125–570 nm	125–570 nm
Tablet PC, Smart phones	IPS mode	0 nm (<10 nm)	0 nm (<10 nm)

In the application as polarizer protective films, the films cover a polarizing film to protect the polarizer from mechanical damage, moisture, and oxidation. In order to keep the polarizing condition of the light after passing through the polarizer, the films have to be free from birefringence so that the polarized light will not be disturbed.

Another important optical film used in displays is the retardation films (compensation films), in which not only retardation at a specific wavelength but also those in broad range of visible light have to be controlled. For example, the absolute value of birefringence must show an increase with increasing the wavelength for wave plates such as quarter-wave plate and half-wave plate. This property is known as "extraordinary wavelength dispersion" of birefringence. Recently, a multi-band quarter-wave plate, one of the retardation films that give a retardation of a quarter of wavelength, has been required intensively and is often employed with a polarizing film to generate a circular-polarizing light for electro-luminescent panels and touch panels to prevent the reflection of ambient light, because it is used for advanced display devices, i.e., glasses for a 3D display and anti-reflection films for an electro-luminescent display.

Conventionally, optical films in LCD application are produced using the solution-cast method, in which solvent is evaporated to obtain a film, to avoid the occurrence of irregular birefringence associated with extrusion process [5, 10]. Therefore, the information on the molecular orientation and the birefringence of solution-cast films is significantly important.

For optical anisotropic films, three refractive indices, n_x, n_y, and n_z, along three principal axes have to be taken into consideration. The *x*-axis is the direction showing the maximum refractive index within the film plane in general, the *y*-axis is the direction perpendicular to the *x*-axis within the film plane, and the *z*-axis is the thickness direction that is normal to the *x*–*y* plane.

It is well known that the solution-cast method provides films without molecular orientation in the film plane, i.e., $n_x = n_y$ [11]. Therefore, a solution-cast film is preferably employed for a protective film rather than a melt-extruded one. However, the other component of birefringence, namely out-of-plane birefringence, is generally not zero. Therefore, it has to be precisely controlled to provide a high quality display. In this chapter, the in-plane birefringence or orientation birefringence (Δn_{in}) and out-of-plane birefringence (Δn_{th}) are defined by the following equations:

$$\Delta n_{\text{in}} = n_x - n_y \tag{9.1}$$

$$\Delta n_{\text{th}} = \frac{n_x + n_y}{2} - n_z \tag{9.2}$$

In the case of solution-cast films, the molecular orientation is caused by the stress induced by solvent removal. The alignment of polymer molecules during a solution-cast process has been investigated by several researchers [12–19]. According to these studies, polymer molecules generally tend to align in a film plane. Consequently, the in-plane refractive indices (n_x and n_y) are higher than the out-of-plane refractive index (n_z) when a material shows positive intrinsic birefringence which will be explained later. However, it is known to be difficult to control the out-of-plane birefringence, which can affect the performance of the displays.

Sosnowski and Weber [12] reported that the optical anisotropy in a solution-cast film of polystyrene is given by a result of the stress developed in a coated film during drying process, which makes polymer chains align in the film plane. Prest and Luca [14] also demonstrated the same results using polystyrene, polycarbonate and poly(2,6-dimethylphenyleneoxide). They showed that the sign of birefringence depends upon the orientation of the dominating polarizable group relative to the backbone chain. Furthermore, the out-of-plane birefringence was found to be pronounced for a thin film. Prest and Luca [14, 15], Cohen and Reich [16], and Machell et al. [17] studied the effect of polymer formulations with various casting conditions such as coating thickness, drying temperature and polymer concentration on birefringence values. They found that the out-of-plane birefringence is determined by the competition between the normal stress induced by the drying process and the Brownian motion leading to random conformation by entropic force. Therefore, it is necessary to predict the stress applied by evaporation accurately to evaluate the out-of-plane birefringence. The growth of stress in solution-cast films was studied in detail by several investigators. Croll [13] developed a simple elastic model to predict the stress, which was modified by Lei et al. [18]. Greener and Chen [19] calculated the out-of-plane birefringence by using the model. They found that the out-of-plane birefringence occurs when the solvent concentration is beyond a critical value. At this stage, the compression stress applied in the normal direction to the film plane starts to build up and induces molecular orientation in the film by overcoming the entropic force.

Conventional polymers, in general, show ordinary dispersion, in which the orientation birefringence increases with the wavelength. This behavior can be expressed by the following equation, which was proposed based on the Sellmeier relation [20–22], as follows:

$$n(\lambda)^2 = 1 + \sum_i \frac{A_i}{\lambda^2 - B_i}, \tag{9.3}$$

where λ and $n(\lambda)$ are the wavelength of light and the average refractive index at λ, respectively. A and B are the Sellmeier coefficients.

The wavelength dispersion of orientation birefringence must be controlled precisely for color display. Based on the Kuhn and Grün model proposed for the stress-optical behavior of cross-linked rubbers, the orientation birefringence $\Delta n_{\text{in}}(\lambda)$ of an oriented polymer is expressed as follows [20, 23–27]:

$$\Delta n_{\text{in}}(\lambda) = \frac{2\pi}{9} \frac{(n(\lambda)^2 + 2)^2}{n(\lambda)} N\, \Delta\alpha(\lambda) \left(\frac{3\langle \cos^2\theta \rangle - 1}{2} \right), \tag{9.4}$$

where N, $\Delta\alpha(\lambda)$, and θ are the number of chains in a unit volume, the polarizability anisotropy, and the angle that a segment makes with the stretch axis, respectively. The last bracketed term is identically equal to the Hermans orientation function [25, 28]. The other part in the right-hand-side term is called intrinsic birefringence, which is independent of the molecular weight and determined by the chemical structure of monomer unit. Consequently, Eq. 9.4 can be written as follows:

$$\Delta n_{\text{in}}(\lambda) = \Delta n_{\text{in}}^{\circ}(\lambda) F, \tag{9.5}$$

where $\Delta n_{\text{in}}^{\circ}(\lambda)$ is the intrinsic birefringence and F is the Hermans orientation function. Regarding the wavelength dispersion, the following simple relation is derived:

$$\frac{\Delta n_{\text{in}}(\lambda)}{\Delta n_{\text{in}}(\lambda_0)} = \frac{\Delta n_{\text{in}}^{\circ}(\lambda)}{\Delta n_{\text{in}}^{\circ}(\lambda_0)} = \text{const.} \tag{9.6}$$

where λ_0 is the arbitrary standard wavelength.

According to Eq. 9.6, the wavelength dispersion of birefringence, i.e., normalized birefringence $\Delta n_{\text{in}}(\lambda)/\Delta n_{\text{in}}(\lambda_0)$, is a constant and determined by the chemical structure.

Both Eqs. 9.3 and 9.6 demonstrate that it is significantly difficult to obtain a polymeric film having extraordinary wavelength dispersion of the orientation birefringence. For example, the orientation birefringence of stretched films of poly(methyl methacrylate) (PMMA) is shown in Fig. 9.1. As seen in the figure, the absolute values of the birefringence decrease with the wavelength. The wavelength dispersion of the birefringence is, therefore, contrary to that of an ideal wave plates. Furthermore, the wavelength dispersion of birefringence is also independent of the draw ratio as demonstrated in Fig. 9.1, in which both curves of the normalized orientation birefringence fall on one line.

However, some stretched films of cellulose esters are known to show extraordinary wavelength dispersion. In this chapter, wavelength dispersion of in-plane and out-of-plane birefringences for cellulose esters is mentioned in detail after a brief introduction on the concept of controlling the optical anisotropy. The molecular orientation and the birefringence of a solution-cast film for CTA are explained. Finally, an advanced method to obtain the extraordinary dispersion of the in-plane and out-of-plane birefringences is proposed.

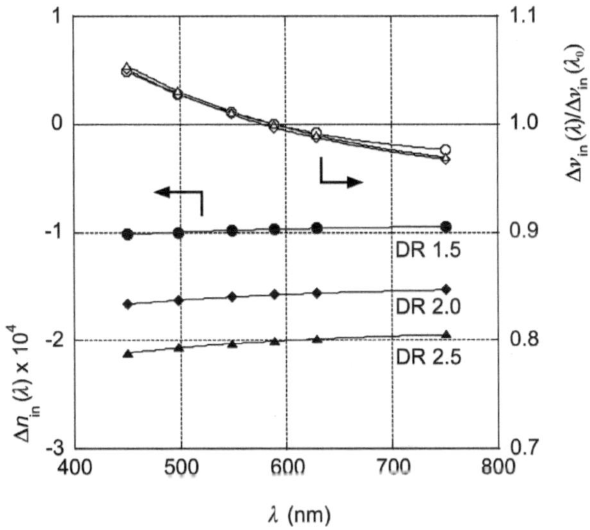

Figure 9.1 Wavelength dependence of orientation birefringence (closed symbols) and normalized orientation birefringence (open symbols) for stretched PMMA films at draw ratios of 1.5 (circles), 2.0 (diamonds), and 2.5 (triangles).

9.2 Wavelength Dispersion of Orientation Birefringence

An ideal wavelength phase difference plate shows a consistent retardation value, e.g., quarter wavelength ($\lambda/4$), over a wide range of visible light. As retardation is given by the product of birefringence and thickness, a retardation film must show an increase in birefringence with increasing wavelength. Industrially, the so-called extraordinary dispersion of the orientation birefringence is achieved by piling together two or more polymer films having different wavelength dispersion, with their fast axes set to be perpendicular to each other, as illustrated in Fig. 9.2. By rotating one of the films at 90°, its fast direction is converted into the slow one ($B_{\parallel} \rightarrow B_{\perp}$), and the combination of the birefringence from these films yields extraordinary wavelength dispersion of orientation birefringence. However, as two or more films are stuck together to produce the required retardation, it leads to poor cost-performance due to the complicated assembling process and results in a thick display. Thus, having a single material that shows extraordinary dispersion of orientation birefringence is desired as the cost production. Further, thickness can be reduced.

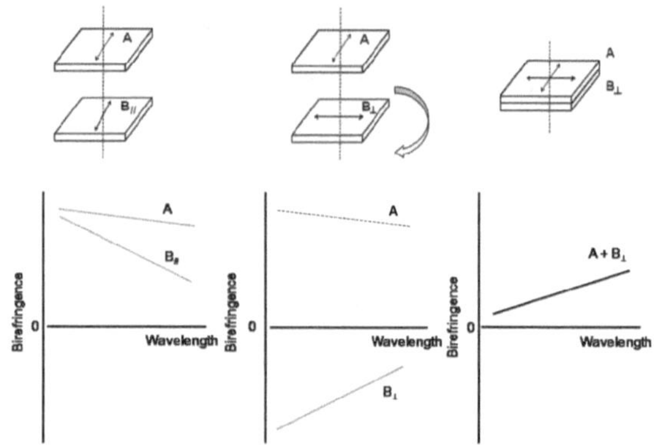

Figure 9.2 Combination of two films having different wavelength dispersion to produce extraordinary wavelength dispersion of orientation birefringence.

In order to produce the optical films with high performance, various techniques such as blending with other polymers [29–32] or small molecules [33–36] and copolymerization [37–40] have been studied.

In case of the polymer blend method, Uchiyama and Yatabe [30, 31] have successfully produced polymer films with extraordinary wavelength dispersion of orientation birefringence by using a miscible blend of poly(2,6-dimethyl-1,4-phenylene oxide) (PPO) and atactic polystyrene (PS). In the blend system, PS shows negative birefringence with strong wavelength dependence, whereas PPO exhibits a positive one with weak wavelength dependence. Although both polymers show ordinary wavelength dispersion, blends at certain compositions exhibit extraordinary wavelength dispersion as the summation of the contributions from both polymers as follows:

$$\Delta n_{\text{in}} = \Delta n_{\text{F}} + \sum_i \varphi_i \Delta n_{\text{in-}i}, \tag{9.7}$$

where i refers to the i-th component and φ_i is the volume fraction and Δn_{F} is the birefringence arising form or deformation effects.

Another study on the polymer blend technique was performed by Kuboyama et al. [32]. They managed to control the wavelength dispersion of a miscible blend comprising polynorbornene (NB) and poly(styrene-co-maleic anhydride) (SMA) with appropriate blend ratios. As predicted from the chemical structure, NB has positive birefringence with weak wavelength dispersion, while SMA has negative birefringence with strong wavelength dispersion [32]. A similar method was proposed by employing a copolymer of two monomers such as 2,2-bis(4-hydroxyphenyl)propane (BPA) and 9,9-bis(4-hydroxy-3-methylphenyl) fluorine (BMPF) [38]. Similar to the blend method, one monomer (BPA) has positive birefringence with weak wavelength dispersion, while the other (BMPF) has negative with strong dispersion.

In principle, the extraordinary wavelength dispersion of birefringence produced by the blend and copolymerization techniques is achieved by using a similar concept to that of the piling method (Fig. 9.2) as illustrated in Fig. 9.3.

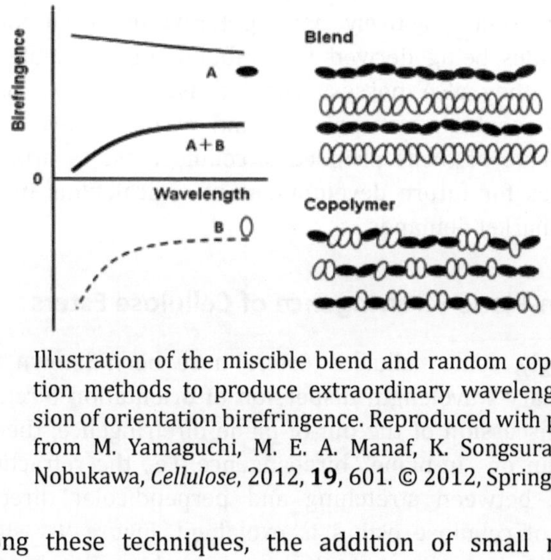

Figure 9.3 Illustration of the miscible blend and random copolymerization methods to produce extraordinary wavelength dispersion of orientation birefringence. Reproduced with permission from M. Yamaguchi, M. E. A. Manaf, K. Songsurang, and S. Nobukawa, *Cellulose*, 2012, **19**, 601. © 2012, Springer [41].

Among these techniques, the addition of small molecules is relatively easier than the others due to good miscibility with polymers. In contrast, the combination of species is limited for the polymer blend and copolymerization technique. Therefore, addition of small molecules is becoming a good candidate these days. Addition of needle-shaped solid particles having strong optical anisotropy also modifies the orientation birefringence and its wavelength dispersion as reported by Koike and co-workers [35]. Tagaya and co-workers reported that a small amount of rod-like molecules having large polarizability anisotropy can reduce the negative birefringence of PMMA. According to them, birefringence is controlled by only the concentration of the anisotropic molecules in the PMMA matrix [33]. Another study on the addition of small molecules was performed by Manaf and co-workers [36]. They also found that tricresyl phosphate (TCP) enhances birefringence of cellulose ester films. These results suggest that the rod-like molecule and TCP orient parallel to the chain segments of matrix polymers due to the intermolecular orientation coupling which is called the nematic interaction.

9.3 Optical Anisotropy of Solution-Cast Cellulose Ester Films

In optical device application, biomass-derived cellulose esters are being studied actively as a potential material for optical films. Besides being derived from abundantly available biomass resources, they also possess characteristics suitable as optical films such as high transparency and excellent heat resistance [8]. By virtue of these properties, cellulose esters provide huge possibilities for future development of optical films in order to meet the market demands.

9.3.1 In-Plane Birefringence of Cellulose Esters

Interestingly, some cellulose esters have been known to show extraordinary wavelength dispersion of orientation birefringence. Prior to discussion of the out-of-plane birefringence, the effect of ester group on "in-plane" birefringence, i.e., the refractive index difference between stretching and perpendicular directions in the film, of cellulose esters is explained. Figure 9.4 shows the wavelength dispersion of the in-plane birefringence for the stretched films at a draw ratio of 2.0. The numerals in the CAB samples represent the weight fraction (wt%) of the butyryl group. All samples were uniaxially stretched at a temperature where the storage modulus at 10 Hz is 10 MPa in order to provide the same stress level for all samples having different glass transition temperature. As seen in the figure, the in-plane birefringence seems to be dependent on the species and the contents of the ester groups. Except for CTA, all CAB, CAP and cellulose diacetate (CDA) samples employed in this experiment exhibit positive birefringence with extraordinary wavelength dispersion. In the case of CAB, the in-plane birefringence increases with the butyryl content. The trend of the birefringence indicates that the butyryl group has positive polarizability anisotropy to the stretching direction. Furthermore, CAP exhibits positive birefringence, as similar to CAB. These experimental results indicate that the acetyl group contributes negative birefringence, whereas the propionyl and butyryl groups show positive birefringence. The flexible and bulky methylene part in ester groups would lead to the parallel orientation of the polarizability anisotropy along the flow direction. Furthermore, the interaction between the ester groups and the main chains may be important because the higher-order structure

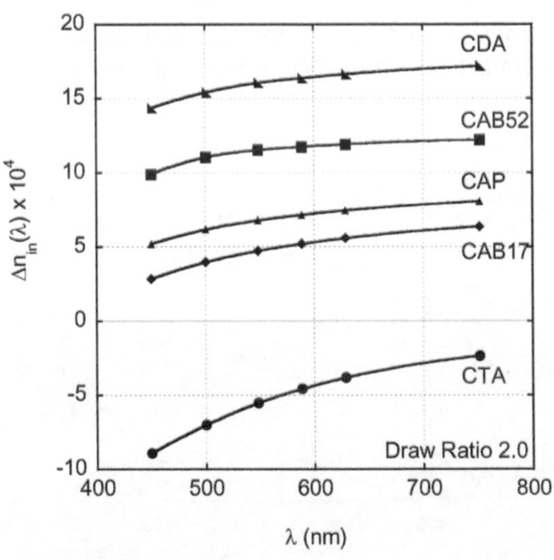

Figure 9.4 Wavelength dependence of orientation birefringence for stretched films of various cellulose esters. The numerals in the CAB represent the weight concentration of the butyryl group. The degree of substitution of propionyl group in CAP is 2.58, and those of butyryl group are 0.73, 1.74, and 2.64 for CAB17, and CAB52, respectively. Reproduced with permission from M. Yamaguchi, K. Okada, M. E. A. Manaf, Y. Shiroyama, T. Iwasaki, and K. Okamoto, *Macromolecules*, 2009, **42**, 9034. © 2009, ACS Publications [44].

of cellulose esters [3, 5, 42] is strongly dependent on the species of the ester groups. Furthermore, the data for CAB52 and CAP represent that the positive contribution of the butyryl group is stronger than that of the propionyl group. Apart from the other samples, CTA shows negative birefringence with ordinary dispersion, in agreement with the result by El-Diasty and co-workers [43]. The result demonstrates that the acetyl group provides negative birefringence. Finally, CDA shows the largest value of orientation birefringence among them, indicating the significant contribution of the hydroxyl group to positive birefringence.

The contributions of hydroxyl and acetyl groups in CDA were calculated from the experimental results of CTA and CDA using the same relation of Eq. 9.6, assuming the followings; (i) orientation relaxation is ignored; (ii) contribution of the main chains is ignored; (iii) contribution of the crystalline part of CTA is ignored; and (iv) both hydroxyl and acetyl groups have the same orientation function in CTA and CDA at a constant draw ratio. As seen in Fig. 9.5, great contribution of the hydroxyl group is confirmed, although the molar content in CDA is lower than that of the acetyl group.

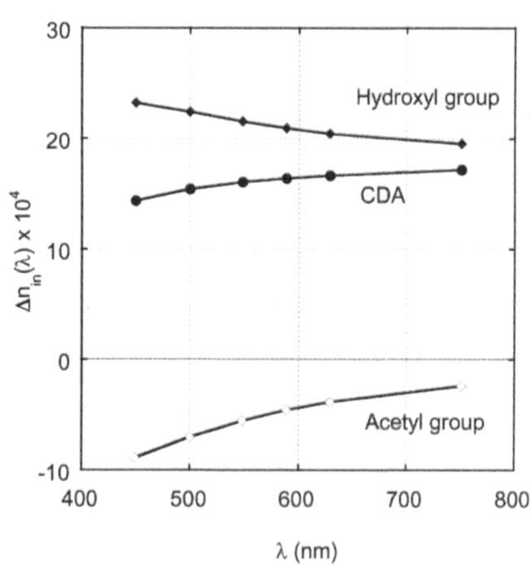

Figure 9.5 Contribution of (closed diamonds) hydroxyl and (open diamonds) acetyl groups to the orientation birefringence of (closed circles) CDA at a draw ratio of 2.0. The birefringence values of both groups are derived from the experimental results of CTA and CDA. Reproduced with permission from M. Yamaguchi, K. Okada, M. E. A. Manaf, Y. Shiroyama, T. Iwasaki, and K. Okamoto, *Macromolecules*, 2009, **42**, 9034. © 2009, ACS Publications [44].

As well known, CAP has at least two ester groups having different polarizability anisotropies, that is, acetyl and propionyl groups. Since the orientation birefringence of CAP is determined not by the orientation of main chains, but rather by the species and the amount of the ester groups, it is expressed by the summation of the birefringence arising from the acetyl group and that from the propionyl group as expressed in the following relation:

$$\Delta n(\lambda) = \Delta n_{\mathrm{A}}^{0}(\lambda)F_{\mathrm{A}} + \Delta n_{\mathrm{P}}^{0}(\lambda)F_{\mathrm{P}}, \tag{9.8}$$

where $\Delta n_{\mathrm{A}}^{0}(\lambda)$ and $\Delta n_{\mathrm{P}}^{0}(\lambda)$ are the intrinsic birefringences, and F_{A} and F_{P} are the orientation functions of the acetyl and propionyl groups, respectively. The extraordinary wavelength dispersion of the orientation birefringence with positive sign for CAP is explained by the illustration in Fig. 9.6. In the figure, the polarizability anisotropy of ester groups is represented by ellipsoids. As depicted in the figure, the addition of both components, denoted by a bold line, gives the extraordinary dispersion with positive orientation birefringence, when the wavelength dependence of the propionyl group is weaker than that of the acetyl group.

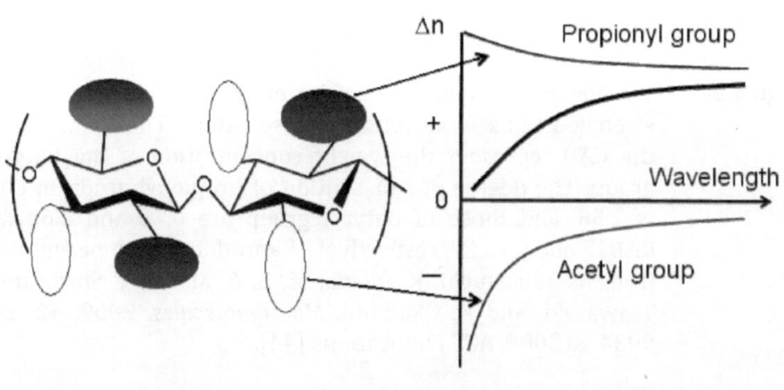

Figure 9.6 Illustration of CAP molecules and the contribution of each ester group to the orientation birefringence. Reproduced with permission from M. Yamaguchi, K. Okada, M. E. A. Manaf, Y. Shiroyama, T. Iwasaki, and K. Okamoto, *Macromolecules*, 2009, **42**, 9034. © 2009, ACS Publications [44].

The normalized orientation birefringence is shown in Fig. 9.7. It is demonstrated that the wavelength dependence of the normalized orientation birefringence becomes weak as the butyryl content is increased. This is reasonable because the butyryl group has weak wavelength dependence.

Figure 9.7 also indicates that the slope of the orientation birefringence for CAP is larger than that for CAB52, although both contain a similar level of the degree of substitution of ester groups such as propionyl and butyryl ones. It suggests that the birefringence of the propionyl group shows strong dependence on the wavelength as compared with that of the butyryl group.

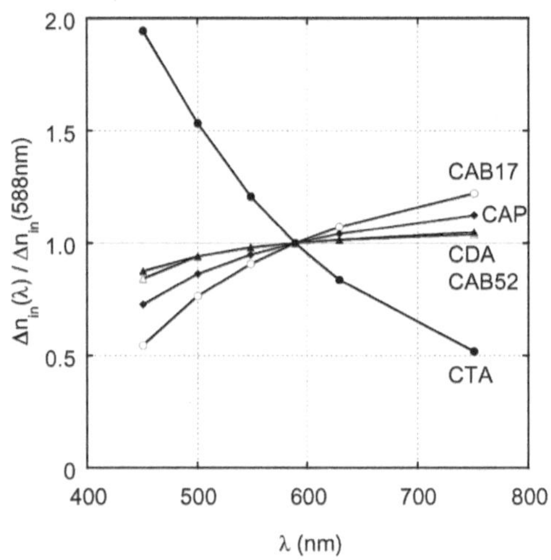

Figure 9.7 Wavelength dependence of orientation birefringence for stretched films of various cellulose esters. The numerals in the CAB represent the weight concentration of the butyryl group. The degree of substitution of propionyl group in CAP is 2.58, and those of butyryl group are 0.73 and 2.64 for CAB17 and CAB 52, respectively. Reproduced with permission from M. Yamaguchi, K. Okada, M. E. A. Manaf, Y. Shiroyama, T. Iwasaki, and K. Okamoto, *Macromolecules*, 2009, **42**, 22, 9034. © 2009, ACS Publications [44].

As a result, the following relation is expected:

$$-\left.\frac{d(\Delta n_{\text{in}}(\lambda))}{d\lambda}\right|_{\text{Acetyl}} > \left.\frac{d(\Delta n_{\text{in}}(\lambda))}{d\lambda}\right|_{\text{Propionyl}} > \left.\frac{d(\Delta n_{\text{in}}(\lambda))}{d\lambda}\right|_{\text{Butyryl}} \quad (9.9)$$

In the case of cellulose esters, Eq. 9.9 is predictable because λ_{ab}, which is the coefficient having the relation with the wavelength of a strong vibrational absorption peak in ultraviolet region of the ester groups, is in the following order: acetyl > propionyl > butyryl.

The wavelength dispersion of the orientation birefringence in cellulose esters is also dependent upon the draw ratio as shown in Fig. 9.8 [44]. It is obvious that the extraordinary dispersion becomes pronounced with increasing the draw ratio, which is not observed in a conventional polymer as discussed earlier (Fig. 9.1). Considering that the acetyl group shows the strong wavelength dispersion, the result indicates that the contribution of acetyl group, i.e., $F_A/(F_A + F_P)$, increases with the draw ratio. From the industrial point of view, the phenomenon indicates that the wavelength dispersion can be controlled by stretching conditions such as degree of stretching (draw ratio), stretching temperature, and stretching speed, which is considerably important information for the actual application.

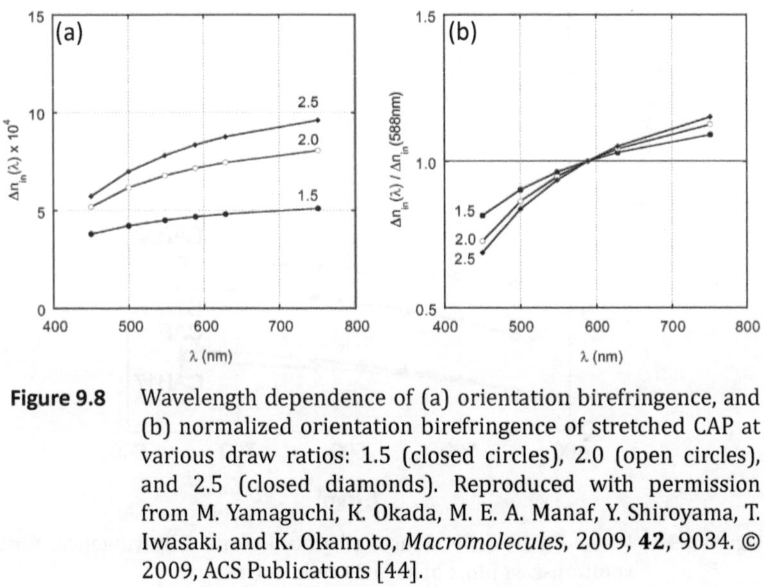

Figure 9.8 Wavelength dependence of (a) orientation birefringence, and (b) normalized orientation birefringence of stretched CAP at various draw ratios: 1.5 (closed circles), 2.0 (open circles), and 2.5 (closed diamonds). Reproduced with permission from M. Yamaguchi, K. Okada, M. E. A. Manaf, Y. Shiroyama, T. Iwasaki, and K. Okamoto, *Macromolecules*, 2009, **42**, 9034. © 2009, ACS Publications [44].

9.3.2 Out-of-Plane Birefringence of Cellulose Esters

The wavelength dispersion of the out-of-plane birefringence of the samples is shown in Fig. 9.9. The solution-cast films, including CTA (butyryl content is zero), show positive birefringence although CTA shows negative in-plane birefringence. The out-of-plane birefringence increases with the butyryl content. Moreover, the out-of-plane birefringence of CAP is significantly smaller than that of CAB52, implying that the propionyl group exhibits lower birefringence. Moreover, CDA shows the largest value of orientation birefringence indicating the significant contribution of the hydroxyl group to positive birefringence. The results obtained for CAP, CAB and CDA correspond to the trend of in-plane birefringence in Fig. 9.4, suggesting that the out-of-plane birefringence is simply originated from the chain orientation in the film plane. In contrast, the birefringence of CTA in Fig. 9.9 is not consistent with that in Fig. 9.4. In order to investigate the origin of the out-of-plane birefringence of CTA, the acetyl orientation is evaluated by Fourier-transform infrared spectroscopy (FT-IR).

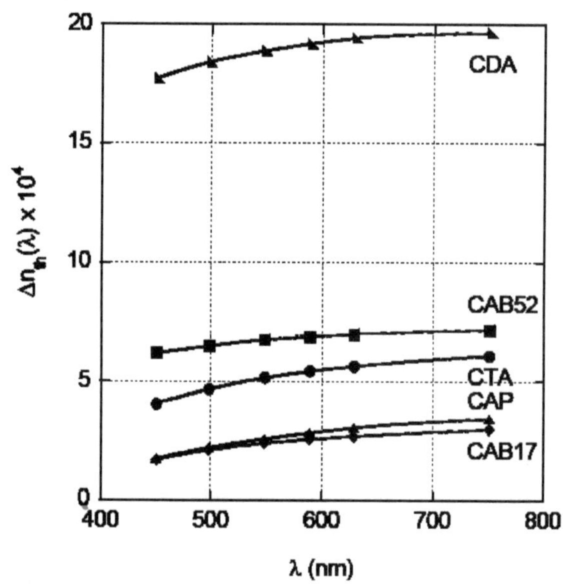

Figure 9.9 Wavelength dependence of out-of-plane birefringence for solution-cast films of various cellulose esters.

Figure 9.10 compares the FT-IR spectra for carbonyl and ether groups in CTA films by using the attenuated total reflection (ATR) method. CTA films were prepared at two different evaporation rates (fast and slow), leading high and low chain orientations, respectively. The chain orientation that affects the intensity of ether band in the pyranose ring is enhanced by the prompt evaporation. Additionally, the carbonyl band becomes strong with evaporation rate, indicating that acetyl orientation in the plane also increases. Since the birefringence of CTA is mostly determined by the acetyl orientation, the positive birefringence for the out-of-plane is attributed to the in-plane alignment of acetyl group generated by the solution-cast method.

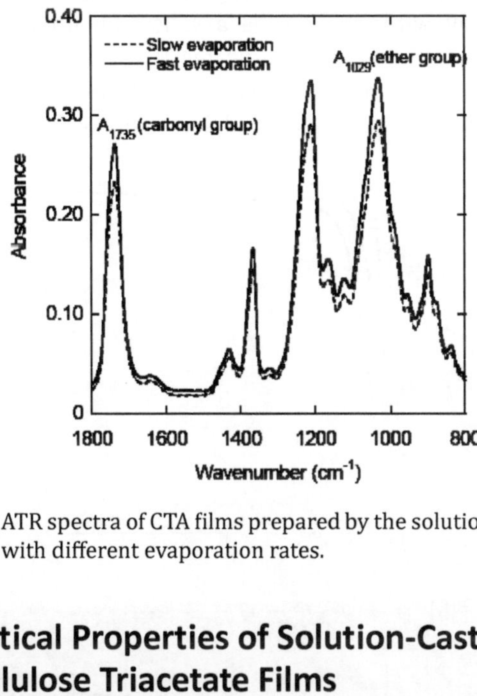

Figure 9.10 ATR spectra of CTA films prepared by the solution-cast method with different evaporation rates.

9.4 Optical Properties of Solution-Cast Cellulose Triacetate Films

The molecular orientation in the film occurs when the relaxation time of the solution becomes longer than the characteristic time for the biaxial deformation applied by the compressional stress due to the solvent evaporation, which was quantitatively calculated by Croll [13]. Therefore, the evaporation rate has to be considered to discuss the birefringence in a solution-cast film.

9.4.1 Effect of Film Thickness

Figure 9.11 shows the growth curves of the weight loss in percent for the CTA solutions to obtain films with various thicknesses. Since all films are prepared using the same Petri dish, the thickness is adjusted by the initial volume of the solution. The weight loss is proportional to the exposure time at first, and the slope in the figure of weight loss versus exposure time is a constant for all samples. Then, the slope becomes low in the final stage, as CTA retards the solvent evaporation. Considering that the initial slope is the same for all samples, the exposing area of the solution determines the evaporation rate. The result suggests that evaporation occurs homogeneously without creating a solid film on the top of the solution. Moreover, Fig. 9.11 indicates that the stress applied by the reduction of the solution increases with decreasing thickness of the final film because of the rapid drying process.

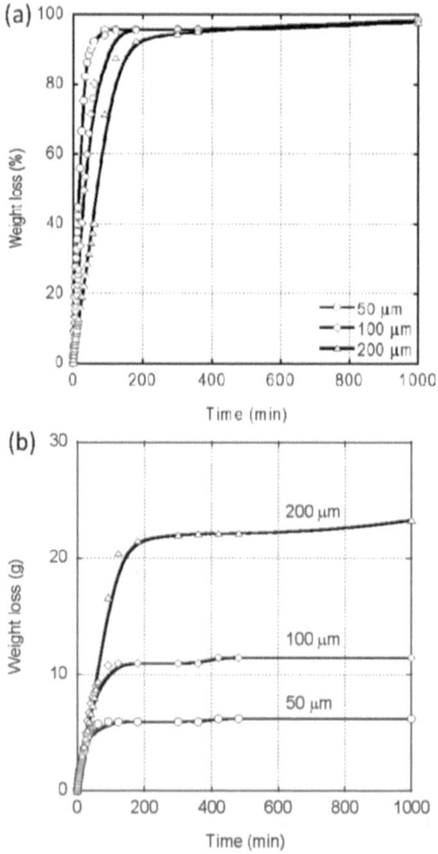

Figure 9.11 (a) Growth curves of the weight loss (%). (b) Growth curves of the weight loss (g) for CTA solutions to obtain films with various thicknesses.

The wavelength dispersion of the out-of-plane birefringence is shown in Fig. 9.12. It is found that the solution-cast films show positive birefringence ($n_z < n_x, n_y$) that increases with increasing wavelength, i.e., extraordinary wavelength dispersion. This is an anomalous phenomenon for CTA. The in-plane birefringence is, on the other hand, negligible for all films. Generally, the orientation birefringence of CTA is determined by the contribution of acetyl and hydroxyl groups considering the previous researches at the best of our knowledge [43–45]. Since the direction of polarizability anisotropy associated with the acetyl group is perpendicular to the main chain, the refractive index in the oriented direction is the lowest, i.e., negative orientation birefringence. Moreover, it is known that the absolute value of birefringence decreases with increasing the wavelength, i.e., ordinary wavelength dispersion, for CTA, as similar to most conventional polymers [8, 38, 41, 45]. However, the contribution of the hydroxyl group cannot be ignored, which provides positive and ordinary wavelength dispersion [44].

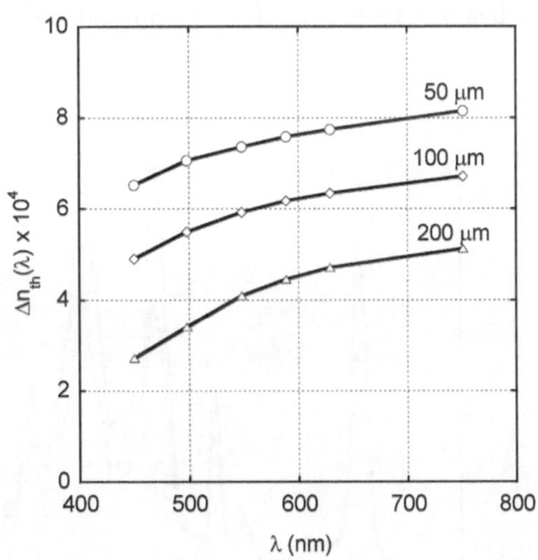

Figure 9.12 Wavelength dispersion of out-of-plane birefringence for cast films with various thicknesses.

Figure 9.12 also demonstrates that the birefringence decreases with increasing film thickness, indicating that the refractive index in the in-plane direction decreases with increasing the film thickness. The decrease in the molecular orientation for a thick film is reasonable, because the rate of solvent removal is slow. Since the solvent can be entrapped in a thick film for a long time as shown in Fig. 9.10, the molecules are less oriented in the film plane. Similar results have been reported by another research group, Greener et al. [46]. According to them, a thin film shows a high value of birefringence because the stress builds up faster in the drying process than the stress relaxation [46].

In order to clarify the effect of film thickness on the birefringence, ATR measurements are performed focusing on the C–O–C stretching vibration in the pyranose ring (A_{1029}) and C=O stretching vibration in the carbonyl group (A_{1735}). It should be noted that the same spectra were obtained for both surfaces (air and glass sides), indicating that the skin layer is not well-developed on the free surface (or the contribution of the skin layer on the birefringence can be ignored). As seen in Fig. 9.13, the absorbances of the pyranose ring and the carbonyl group decrease with increasing the film thickness. Considering that the penetration depth of IR beam into the sample is approximately 2.2 μm at

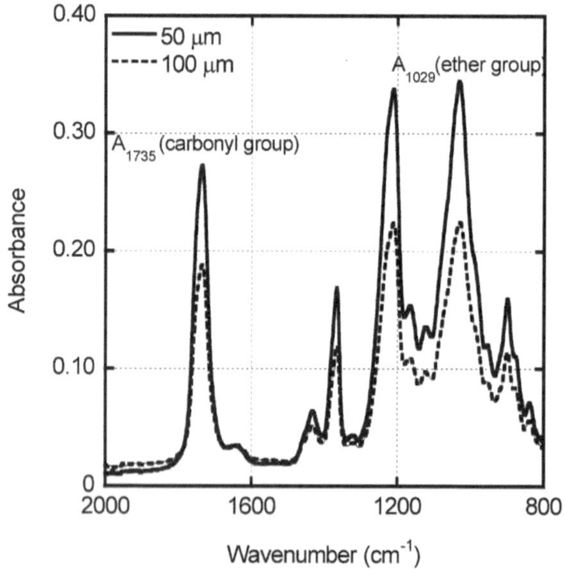

Figure 9.13 ATR spectra of CTA films with different film thicknesses: 50 μm (solid line) and 100 μm (dashed line).

1029 cm^{-1} and 1.2 μm at 1735 cm^{-1}, the film thickness itself does not affect the absorbance. The results indicate that the pyranose ring and the carbonyl group are aligned in the film plane, which is pronounced in a thin, i.e., rapid evaporation film. The in-plane orientation of the carbonyl group will be responsible for the positive out-of-plane birefringence. In the case of a hot-stretched film of CTA, the carbonyl group orients perpendicular to the stretching direction. Consequently, the refractive index in the perpendicular direction is always higher than that in the stretching direction, leading to negative orientation birefringence. On the contrary, the carbonyl group preferably exists in a film plane for a solution-cast film, which is attributed to the in-plane orientation of the pyranose ring. As a result, the refractive index in the film plane is larger than that in the thickness direction, although the backbone chains of CTA are also in the film plane.

According to the previous paper [44], the extraordinary wavelength dispersion for CDA is attributed to the contributions of the polarizability anisotropy of both hydroxyl and acetyl groups, in which the hydroxyl group provides positive birefringence with weak wavelength dispersion and the acetyl group gives negative one with strong wavelength dispersion. In this study, CTA also shows extraordinary wavelength dispersion suggesting that the contribution of the hydroxyl group cannot be ignored even if the amount is small.

In the case of CTA, the crystallization state has to be considered, because it is well known that CTA is a crystalline polymer [5, 47]. As shown in Table 9.2, it is found that the degree of crystallization is almost constant irrespective of the film thickness. Furthermore, the degree of the crystallization is calculated to be 32–40 wt% based on the literature data [47], which seems to be extremely high for pure CTA. Therefore, WAXD measurements were also carried out as shown in Fig. 9.14. Basically, CTA shows three sharp crystal peaks at 2θ of 9°, 13°, and 17°. As seen in the figure, it is found that any distinct peak is not detected in the diffraction pattern except for a broad peak ascribed to amorphous region for a solution-cast film, although a hot-stretched film shows high crystallinity. The result supports that the crystallites have a negligible effect on the out-of-plane birefringence of a cast film.

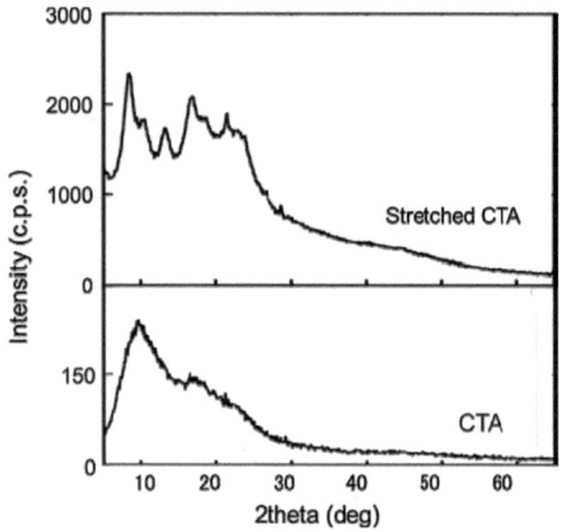

Figure 9.14 WAXD patterns for CTA films: (Bottom) A cast film with a thickness of 100 μm obtained from CH_2Cl_2/CH_3OH at a standard condition, and (top) a film stretched at a draw ratio of 1.5. Reproduced with permission from K. Songsurang, A. Miyagawa, M. E. A. Manaf, P. Phulkerd, S. Nobukawa, and M. Yamaguchi, *Cellulose*, 2013, **20**, 83. © 2013, Springer Link [11].

Table 9.2 Thermal properties of solution-cast films obtained at various conditions

Conditions				Heat of Fusion (J/g)	T_m (°C)
Thickness (μm)	**Evaporation Rate**	**Solvent**	**TCP**		
50	Standard	CH_2Cl_2/CH_3OH	Not included	13.8	302
100	Standard	CH_2Cl_2/CH_3OH	Not included	12.7	303
200	Standard	CH_2Cl_2/CH_3OH	Not included	11.5	304
100	Slow	CH_2Cl_2/CH_3OH	Not included	12.1	302
100	Very slow	CH_2Cl_2/CH_3OH	Not included	11.3	303
100	Standard	$CHCl_3/CH_3OH$	Not included	12.1	302
100	Standard	CH_2Cl_2/CH_3OH	Included	10.2	294

It can be concluded from Figs. 9.11–9.13 that the achievable anisotropy is found to be a function of the evaporation rate. Therefore, further study is performed focusing the effect of evaporation rate.

9.4.2 Effect of Evaporation Rate

The evaporation rate is controlled by covering the Petri dishes with an aluminum foil. The growth curves of the weight loss are shown in Fig. 9.15. In the figure, the "standard" represents the cast film obtained from uncovered Petri dishes as the same with the samples in Fig. 9.13. The "slow" and "very slow" denote the cast films obtained with an aluminum foil having large and small holes, respectively. The thickness of all films is approximately 100 μm. As seen in the figure, the evaporation rate can be controlled by this technique. The initial slope of "standard" is three times larger than that of "very slow."

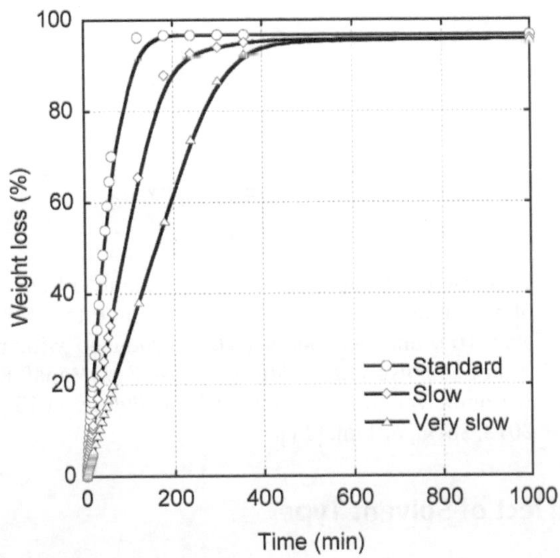

Figure 9.15 Growth curves of the weight loss (%) for CTA solutions at various evaporation rates; standard (○), slow (◇) and very slow (Δ). Reproduced with permission from K. Songsurang, A. Miyagawa, M. E. A. Manaf, P. Phulkerd, S. Nobukawa, and M. Yamaguchi, *Cellulose*, 2013, **20**, 83. © 2013, Springer Link [11].

The wavelength dispersion of out-of-plane birefringence of the samples is shown in Fig. 9.16. All films show positive birefringence with extraordinary wavelength dispersion irrespective of the evaporation rate. However, the magnitude of the birefringence increases with increasing the evaporation rate. Therefore, a similar situation with a thick film occurs for the film evaporated slowly.

The decrease in the birefringence for a film obtained by the prolonged evaporation process suggests that stress, i.e., molecular orientation, is relaxed. It is reasonable because the polymer chains are able to move randomly during the cast process.

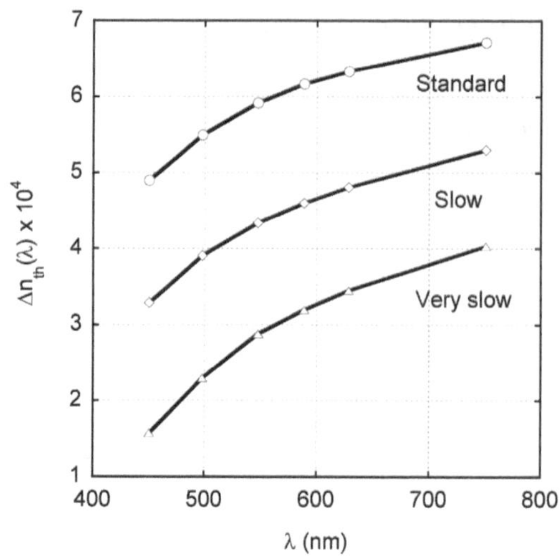

Figure 9.16 Wavelength dispersion of out-of-plane birefringence for films obtained at various evaporation rates; standard (○), slow (◇), and very slow (Δ). Reproduced with permission from K. Songsurang, A. Miyagawa, M. E. A. Manaf, P. Phulkerd, S. Nobukawa, and M. Yamaguchi, *Cellulose*, 2013, **20**, 83. © 2013, Springer Link [11].

9.4.3 Effect of Solvent Type

The species of solvents affects the evaporation rate and thus the out-of-plane birefringence as shown in Figs. 9.17, 9.18, respectively. As seen in Fig. 9.17, the evaporation rate of the mixed solvent of CH_2Cl_2 and CH_3OH is faster than that of $CHCl_3$ and CH_3OH, because the vapor pressure of CH_2Cl_2 (47 kPa at 20°C) is higher than that of $CHCl_3$ (21 kPa at 20°C) at room temperature. Moreover, it is identified that the whole curve of the $CHCl_3/CH_3OH$ solution is almost the same as that of the CH_2Cl_2/CH_3OH solution with an aluminum foil having small holes ("slow" in Fig. 9.15). The result demonstrates that the retardant effect of the evaporation by CTA is almost the same for both solvents. This would be attributed to similar molecular interaction with CTA for both solvents.

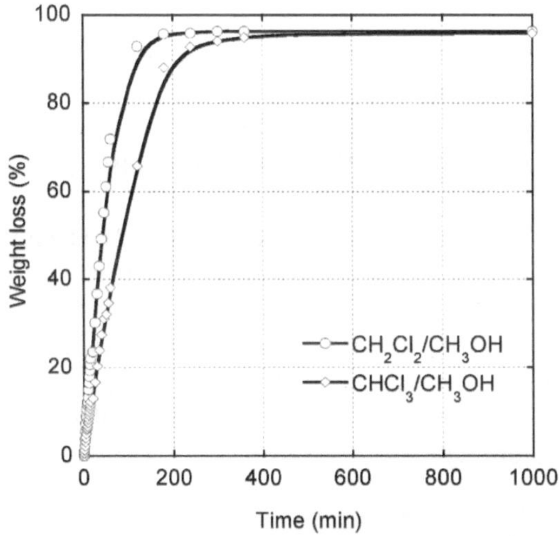

Figure 9.17 Growth curves of the weight loss (%) for CTA solutions using CH_2Cl_2/CH_3OH (○) and $CHCl_3/CH_3OH$ (◊) as solvents. Reproduced with permission from K. Songsurang, A. Miyagawa, M. E. A. Manaf, P. Phulkerd, S. Nobukawa, and M. Yamaguchi, *Cellulose*, 2013, **20**, 83. © 2013, Springer Link [11].

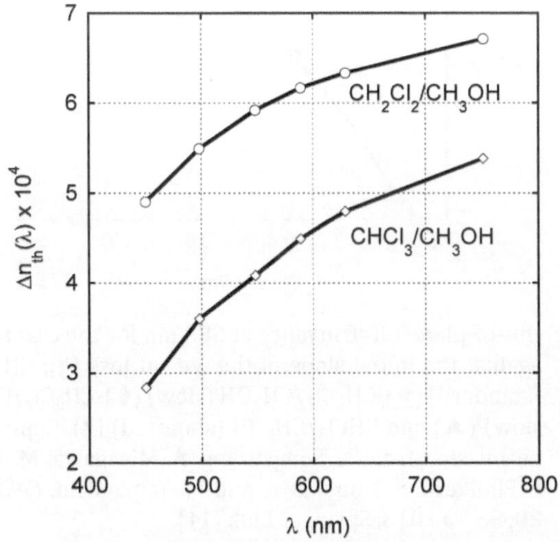

Figure 9.18 Wavelength dispersion of out-of-plane birefringence for the cast films obtained from CH_2Cl_2/CH_3OH (○) and $CHCl_3/CH_3OH$ (◊). Reproduced with permission from K. Songsurang, A. Miyagawa, M. E. A. Manaf, P. Phulkerd, S. Nobukawa, and M. Yamaguchi, *Cellulose*, 2013, **20**, 83. © 2013, Springer Link [11].

Because of the difference in the evaporation rate, the relaxation of molecular orientation of CTA is pronounced for the $CHCl_3$/CH_3OH solution as long as the evaporation condition is the same. Therefore, the CTA film prepared by CH_2Cl_2/CH_3OH shows higher out-of-plane birefringence than the film by $CHCl_3$/CH_3OH, as seen in Fig. 9.18.

The out-of-plane birefringence at 588 nm is plotted against the initial slope of the weight loss in Fig. 9.19. As seen in the figure, the birefringence of the film obtained from the $CHCl_3$/CH_3OH solution is located on the line of the data from the CH_2Cl_2/CH_3OH solution. The result demonstrates that the out-of-plane birefringence is independent of the species of solvents but depends on the evaporation rate.

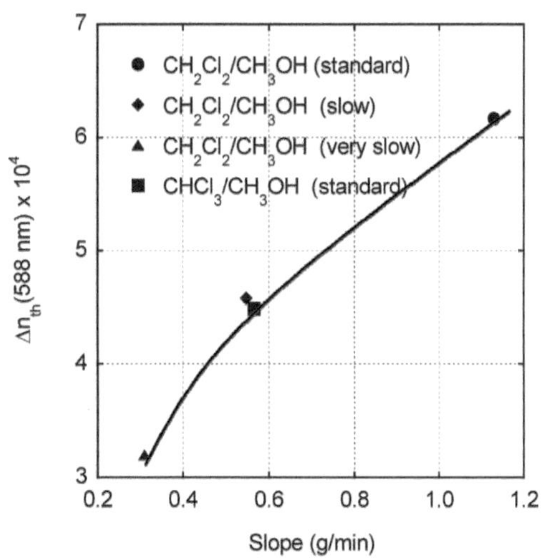

Figure 9.19 Out-of-plane birefringence at 588 nm for the cast films plotted against the initial slope of the weight loss (g); CH_2Cl_2/CH_3OH (standard) (●), CH_2Cl_2/CH_3OH (slow) (♦), CH_2Cl_2/CH_3OH (very slow) (▲) and $CHCl_3$/CH_3OH (standard) (▪). Reproduced with permission from K. Songsurang, A. Miyagawa, M. E. A. Manaf, P. Phulkerd, S. Nobukawa, and M. Yamaguchi, *Cellulose*, 2013, **20**, 83. © 2013, Springer Link [11].

9.4.4 Effect of Plasticizer

Recently, it has been clarified that addition of specific plasticizers can control the orientation birefringence of a stretched film. In particular, it was found that orientation birefringence is enhanced by TCP for cellulose esters [36]. The phenomenon is explained by nematic interaction, i.e., intermolecular orientation correlation [48, 49]. Although TCP is a low mass compound in a liquid state, the molecules are forced to orient in the same direction with the polymer chains. Considering that the nematic interaction occurs when the size of a low mass compound is comparable with the segmental size of a matrix polymer [50–52], TCP has an appropriate size for CTA. To the best of our knowledge, however, the nematic interaction in solution-cast films has not been studied yet.

Figure 9.20 shows the out-of-plane birefringence of a CTA film containing 5 wt% of TCP with the data of pure CTA before and after methanol immersion. As seen in the figure, addition of TCP greatly enhances the out-of-plane birefringence of CTA. Since the evaporation rate is not affected by the addition of TCP, the enhancement is attained by the orientation of TCP molecules in the film plane accompanied with CTA chains.

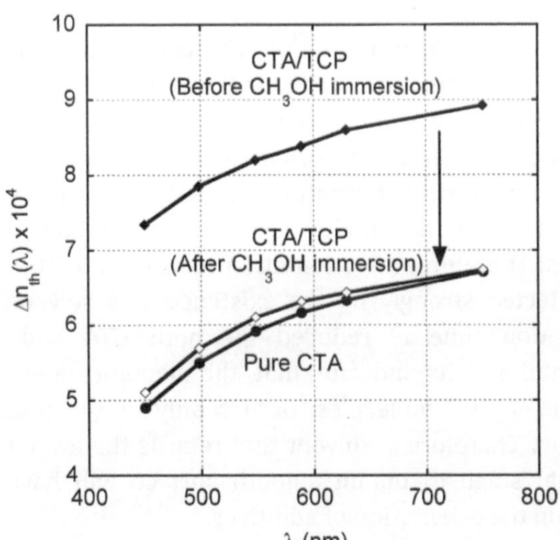

Figure 9.20 Wavelength dispersion of out-of-plane birefringence for CTA (●) and CTA/TCP before (♦) and after (◊) methanol immersion. The cast films with a thickness of 100 μm were prepared by CH_2Cl_2/CH_3OH at a standard condition.

In order to confirm the contribution of TCP to the birefringence, the CTA/TCP film is immersed in methanol for 24 h to remove

TCP. Then, the out-of-plane birefringence is measured again after drying at room temperature under a vacuum condition. Prior to the measurement, it is confirmed that the methanol immersion does not affect the birefringence of the pure CTA. The removal of TCP is checked by FT-IR spectra, and the following experiment is performed after keeping the sample in the temperature-and-humidity controller for 24 h. The FT-IR spectrum for CTA/TCP shows a strong absorption peak around 780 cm^{-1}, which is not detected in CTA but appears in TCP spectrum. This peak is attributed to the vibration of C–H bond in meta-disubstituted benzene of TCP [36]. Therefore, the lack of this peak suggests that TCP has been completely removed out. After methanol immersion, the birefringence of CTA/TCP decreases approaching to that of pure CTA, as illustrated in Fig. 9.20. The result demonstrates that TCP molecules oriented by nematic interaction enhance the out-of-plane birefringence. In other words, the out-of-plane birefringence can be controlled by not only evaporation rate but also the species and amount of additives.

As mentioned in introduction, the competition of molecular motion and deformation applied by the solvent evaporation determines the molecular orientation in a solution-cast film. In other words, CTA chains orient in the film plane by applied uniaxial compression deformation due to the solvent evaporation. At the same time, TCP molecules orient accompanied with the CTA chains. Therefore, the orientation relaxation of TCP molecules will be affected strongly by the existence of a solvent, although the relaxation time is reduced for both CTA and TCP. The experimental results indicate that the nematic interaction, i.e., orientation of TCP molecules, occurs only at the final stage of evaporation. Therefore, a solvent that retards the evaporation rate at the final stage to obtain smooth surface will have a strong influence on the orientation of additives.

9.4.5 Effect of Hot-Stretching

The hot-stretching was performed at the temperature where the tensile storage modulus at 10 Hz was 10 MPa (212°C). The stress–strain curve at a strain rate of 0.05 s^{-1} is shown in Fig. 9.21. The curve is a typical one for a viscoelastic body in a rubbery region.

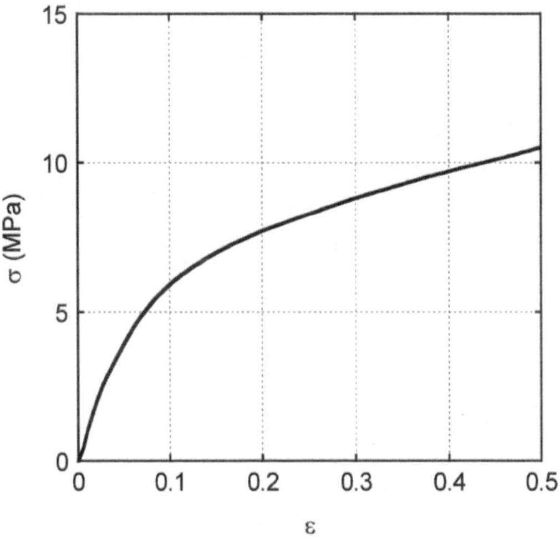

Figure 9.21 Stress–strain (α–ε) curve at 212°C for CTA. Reproduced with permission from K. Songsurang, A. Miyagawa, M. E. A. Manaf, P. Phulkerd, S. Nobukawa, and M. Yamaguchi, *Cellulose*, 2013, 20, 83. ©2013, Springer Link [11].

The refractive indices at 588 nm along three principle axes for films with various draw ratios are illustrated in Fig. 9.22. In this experiment, the hot-stretching was performed in the x direction.

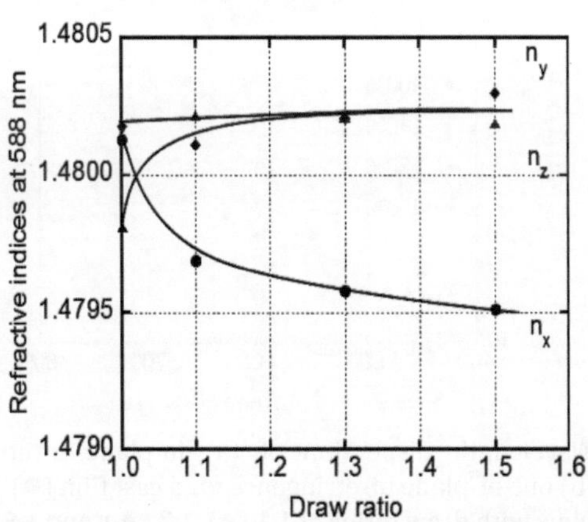

Figure 9.22 Refractive indices along three principle axes; n_x (●), n_y (♦) and n_z (▲) for CTA films stretched at various draw ratios. Reproduced with permission from K. Songsurang, A. Miyagawa, M. E. A. Manaf, P. Phulkerd, S. Nobukawa, and M. Yamaguchi, *Cellulose*, 2013, **20**, 83. © 2013, Springer Link [11].

As mentioned, the solution-cast film is randomly oriented in the film plane, i.e., $n_x = n_y$, but the principle refractive index in the thickness direction n_z is lower than n_x and n_y, which results in the positive out-of-plane birefringence. On the contrary, a hot drawn film shows negative in-plane birefringence.

The wavelength dispersion of in-plane and out-of-plane birefringences for the stretched films is shown in Fig. 9.23. As seen

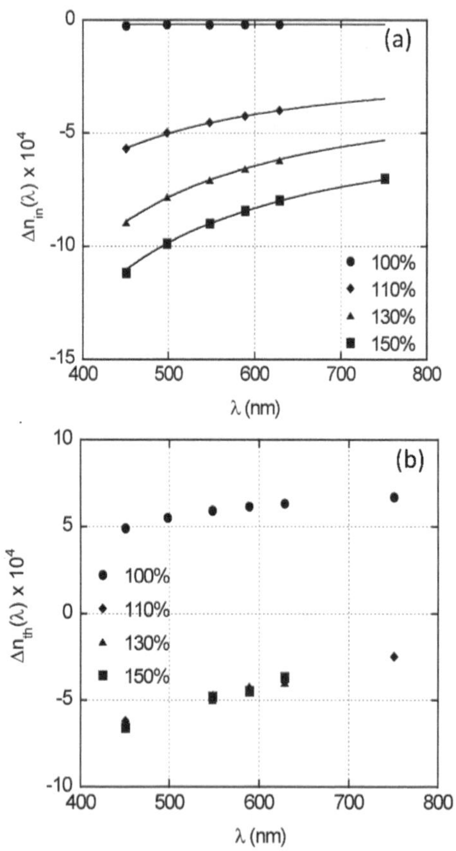

Figure 9.23 Wavelength dispersion of (a) in-plane birefringence and (b) out-of-plane birefringence for a cast film (●) and stretched films with draw ratios of 1.1 (♦), 1.3 (▲), and 1.5 (■). The cast films with a thickness of 100 μm were prepared by CH_2Cl_2/CH_3OH at the standard condition. Reproduced with permission from K. Songsurang, A. Miyagawa, M. E. A. Manaf, P. Phulkerd, S. Nobukawa, and M. Yamaguchi, *Cellulose*, 2013, **20**, 83. © 2013, Springer Link [11].

in the figure, the order in the refractive indices changes with the draw ratio. When the draw ratio is around 1.035, n_z is almost the same as $(n_x + n_y)/2$ at 588 nm. Beyond this draw ratio, the out-of-birefringence of the solution-cast films shows extraordinary wavelength dispersion, whereas in-plane and out-of-plane birefringences for the stretched films exhibit ordinary wavelength dispersion. Moreover, the out-of-birefringence of the stretched films is almost independent of the applied strain level, when the draw ratio is larger than 1.1.

The negative in-plane birefringence in CTA suggests that the direction of the polarizability anisotropy associated with the acetyl group is perpendicular to the main chain. This result corresponds to the previous reports [36, 43, 44, 53].

9.4.6 Effect of Crystallization

Because the hot-stretching is carried out beyond the glass transition temperature, the degree of crystallinity has to be considered more seriously. The top pattern in Fig. 9.14 shows the WAXD profile of a stretched film at a draw ratio of 1.5. As seen in the figure, several distinct peaks are clearly detected, which are attributed to the thermal history beyond the glass transition temperature with the flow induced crystallization during stretching. Therefore, the contribution of crystalline phase cannot be ignored especially after hot-stretching. It was found that the extraordinary wavelength dispersion of CAP becomes pronounced with increasing the draw ratio [44]. The result indicates that the acetyl group plays a more important role in the total birefringence with increasing the draw ratio. A similar situation is expected also for CTA, because the acetyl group is in the crystalline structure with the strong polarizability anisotropy in the direction perpendicular to the chain axis. In other words, the wavelength dispersion and the magnitude of birefringence for CTA, including solution-cast films, can be modified by controlling the crystalline state.

9.5 Advanced Method to Control Orientation Anisotropy

It has been revealed that orientation (in-plane) birefringences of a polymer film is modified by addition of a low-mass compound having strong optical anisotropy. Low-mass compounds that are dissolved into a polymer matrix are forced to orient in the same direction to the polymer chains by intermolecular orientation correlation called "nematic interaction" [48, 51, 54]. This intermolecular interaction was originally used to explain the orientation relaxation of a short chain in a polymer melt with broad molecular weight distribution. Then it has been clarified that the nematic interaction is expected to occur in a polymer blend system as long as the system is miscible. Quantitative studies on the nematic interaction in miscible polymer blends have been carried out by Urakawa and co-workers [50–52, 55] based on the orientation coupling theory [54]. This technique using miscible blends can be employed to modify the wavelength dispersion of orientation birefringence.

Yamaguchi and co-workers studied the effect of the small addition of poly(lactic acid) (PLA) on the orientation birefringence of CAP. They found that PLA is miscible with CAP on a molecular scale when the amount of PLA is less than 3 wt% [8]. Moreover, the orientation birefringence of CAP is greatly enhanced even with a small addition of PLA as shown in Fig. 9.24. Considering that

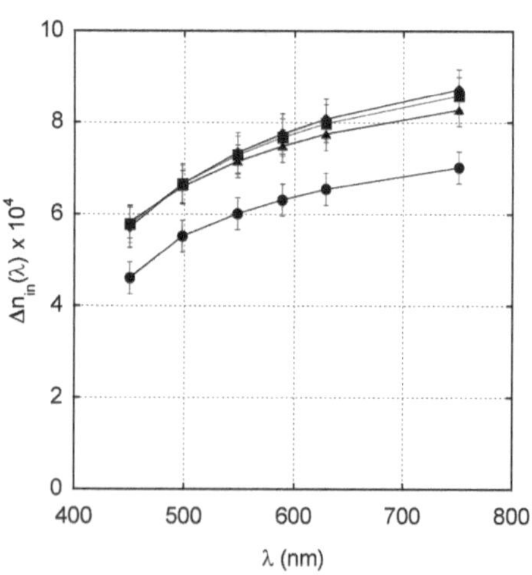

Figure 9.24 Wavelength dependence of orientation birefringence for CAP and CAP/PLA blends stretched at a draw ratio of 2.0: PLA (circles), 1 wt% of PLA (diamonds), 3 wt% of PLA (triangles), and 5 wt% of PLA (squares). Reproduced with permission from M. Yamaguchi, S. Lee, M. E. A. Manaf, M. Tsuji, and T. Yokohara, *Eur. Polym. J.*, 2010, **46**, 12, 2269. © 2010, Elsevier [8].

the relaxation times are 0.045 s for PLA and approximately 1000 s for CAP at the stretching temperature, the high level of orientation of PLA chains without relaxation is attained by the cooperative alignment (nematic interaction) with CAP chains. The study indicates that a retardation film having high level of birefringence, leading to a thin film, can be designed by polymer blends composed of only biomass-based materials.

Furthermore, the blends with more than 3 wt% of PLA exhibits large haze values, indicating the existence of phase-separated morphology as previously reported by Tatsushima and co-workers [56]. As a result, the blends lose transparency and show similar level of orientation birefringence to the blend with 1 wt% of PLA. This is reasonable because PLA chains in the dispersed phase have no nematic interaction with CAP chains, thus the orientation will relax immediately as compared to the CAP. As demonstrated, much attention has to be focused on the miscibility for this method.

Yamaguchi and Masuzawa found that CAP is miscible with poly(vinyl acetate) (PVAc), which has negative orientation birefringence. Although they only showed that the birefringence at a specific wavelength decreases with the PVAc content, the wavelength dispersion must be modified. They also revealed that poly(epichlorohydrin), having negative birefringence, is miscible with CAP [57].

Besides polymeric materials, plasticizers such as diethyl phthalate (DEP), bisphenoxyethanolfluorene (BPEF), and tricresyl phosphate (TCP) are effective in modifying the orientation birefringence of cellulose esters [11, 53, 57]. This phenomenon is also attributed to the nematic interaction. The degree of modification depends on the intrinsic birefringence of the plasticizer as well as the nematic interaction coefficient between the plasticizer and the host polymer. Such phenomenon is also observed in a solution-cast film [11]. Moreover, some low-mass compounds contribute negative orientation birefringence, when the optical slow axis is perpendicular to the long axis of the low-mass compound [58]. Using a low-mass compound like a plasticizer has a great benefit, because there are numerous miscible systems due to large mixing entropy. Because both TCP and DEP contribute to positive orientation birefringence, their incorporation into CTA, which has negative birefringence with ordinary dispersion, results in the blends having extraordinary wavelength dispersion. Figure 9.25

shows the orientation birefringence of pure CTA and its blends with 5 and 10 wt% of TCP. Apparently, CTA with 10 wt% of TCP shows extraordinary wavelength dispersion, of which the orientation birefringence shows almost an ideal dispersion (indicated by the straight dotted line) as required for a multi-band quarter-wave plate [36]. Since the molecular shape of TCP is rather disk, more effective low-mass compounds having rod-shape with strong optical anisotropy will be found in near future. Preliminary experimental results have been reported by our research group [59].

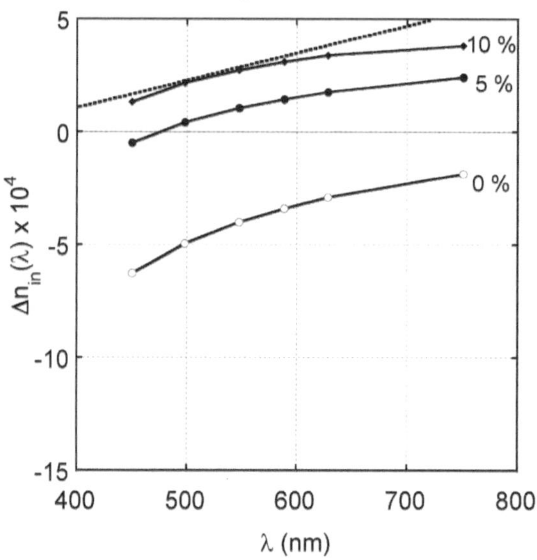

Figure 9.25 Wavelength dependence of orientation birefringence for (open circles) CTA, (closed circles) CTA/TCP (95/5) and (closed diamonds) CTA/TCP (90/10), stretched at a draw ratio of 1.5. In the figure, the straight dotted line represents the ideal wavelength dispersion for a multi-band quarter-wave plate with a thickness of 400 μm. Reproduced with permission from M. E. A. Manaf, M. Tsuji, Y. Shiroyama, and M. Yamaguchi, *Macromolecules*, 2011, **44**, 10, 3942. © 2011, ACS Publications [36].

Although the orientation of the main chain does not directly affect the orientation birefringence for cellulose esters, it has an impact on the orientation of the ester groups as well as the hydroxyl group. Furthermore, the orientation of the main chain strongly affects the nematic interaction with low-mass compounds.

Therefore, the main chain orientation, which is proportional to the stress, should be considered even for cellulose esters. For the better understanding, the effect of stress relaxation on the orientation birefringence was studied by holding the stretched sample in a hot chamber at various residence times [36]. As seen in Fig. 9.26, the orientation birefringence decreases monotonically with increasing holding time after stretching. Furthermore, there is no further decrease in the orientation birefringence above 30 min of the holding time. The results suggest that the TCP molecules relax the orientation when left at a high temperature without being quenched.

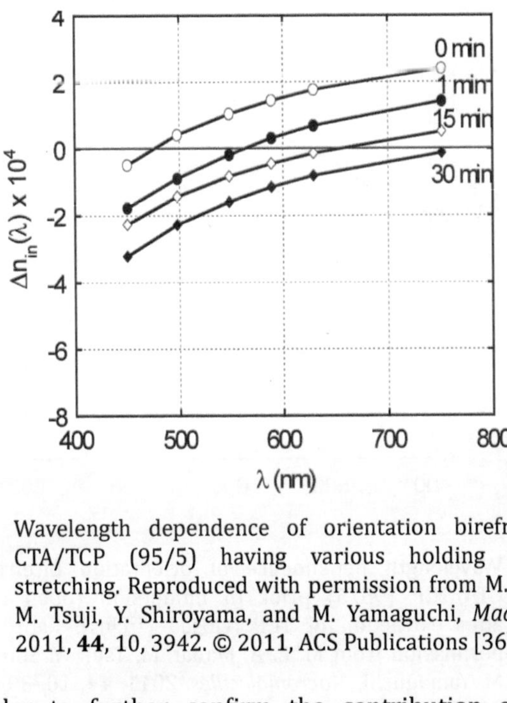

Figure 9.26 Wavelength dependence of orientation birefringence for CTA/TCP (95/5) having various holding times after stretching. Reproduced with permission from M. E. A. Manaf, M. Tsuji, Y. Shiroyama, and M. Yamaguchi, *Macromolecules*, 2011, **44**, 10, 3942. © 2011, ACS Publications [36].

In order to further confirm the contribution of TCP, the stretched sample of CTA/TCP (95/5) blend was immersed in methanol for 24 hours to remove TCP, and the orientation birefringence was measured. No considerable change in the dimension of the sample after 24 hours of immersion is observed, suggesting there is almost no alteration in the degree of stretching upon the methanol immersion. Furthermore, it is affirmed by IR spectra analysis that TCP has been completely removed.

Orientation birefringence of CTA/TCP (95/5) before and after the methanol immersion is shown in Fig. 9.27 [36]. After the immersion, the orientation birefringence of the CTA/TCP blend decreases and approaches to that of pure CTA. Therefore, the removal of TCP is reflected in the change of the orientation birefringence from positive to negative values. The result suggests that the positive birefringence in the CTA/TCP blend is strongly attributed to the orientation of TCP.

Nobukawa and co-workers studied the effect of the aromatic compounds having various alkyl chains on orientation birefringence of CAP. The aromatic additives (C_mPh) are synthesized from alkyl diol (*m*-carbons) and terephthalic acid [60, 61].

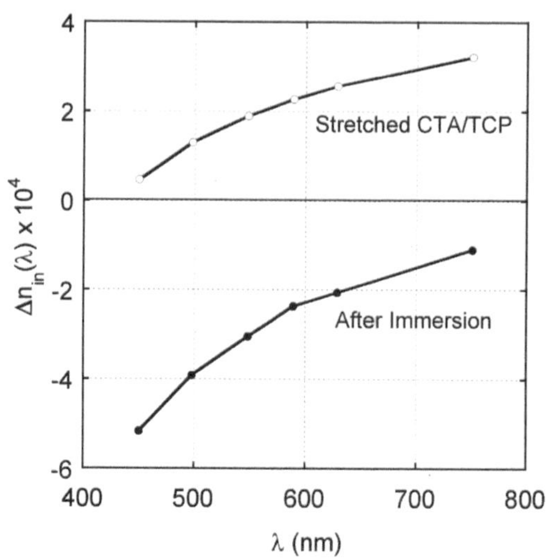

Figure 9.27 Wavelength dependence of orientation birefringence for CTA/TCP (95/5) prior to immersion (open circles) and after extraction of TCP (closed circles). Reproduced with permission from M. E. A. Manaf, M. Tsuji, Y. Shiroyama, and M. Yamaguchi, *Macromolecules*, 2011, **44**, 10, 3942. © 2011, ACS Publications [36].

As shown in Fig. 9.28, the additive improved the orientation birefringence of CAP, suggesting that the additive orientation is parallel to CAP chains via the "nematic interaction." The strength of nematic interaction was evaluated by using the theory concerning the intermolecular orientation correlation. The result shows the increase of orientation birefringence is dependent on the alkyl chain length, meaning that the alkyl chain length in the additive molecule strongly affects the orientation correlation.

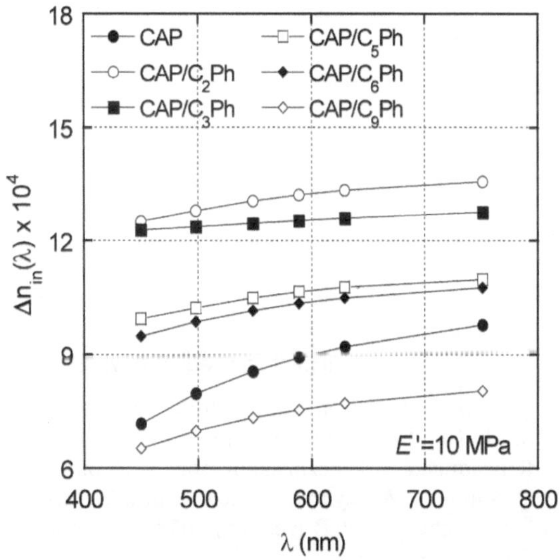

Figure 9.28 Wavelength dependence of orientation birefringence for CAP and CAP/additive blends (100/10 wt/wt) drawn at temperature where E' is 10 MPa. Reproduced with permission from S. Nobukawa, Y. Aoki, H. Yoshimura, Y. Tachikawa, and M. Yamaguchi, *J. Appl. Polym. Sci.*, 2013, **130**, 5, 3645. © 2013, Wiley Online Library [60].

Although the intrinsic birefringence monotonically decreases with the alkyl chain length, the nematic interaction shows the highest value for the additive with propionyl group (m = 3). Nobukawa et al. evaluated the nematic interaction parameters between CAP and aromatic additives, C_2Ph and C_2Np. Here, Ph and Np represents phenyl and naphthalate groups, respectively in the molecule. As shown in Fig. 9.29, naphthalate group generates the nematic interaction more strongly than phenyl group. Especially, C_2Np shows almost perfect orientation correlation with CAP chain. The results demonstrate that the contribution of birefringence from the aromatic molecule, which is determined by the intrinsic birefringence and orientation function, needs to be considered to design the optical films with high performances.

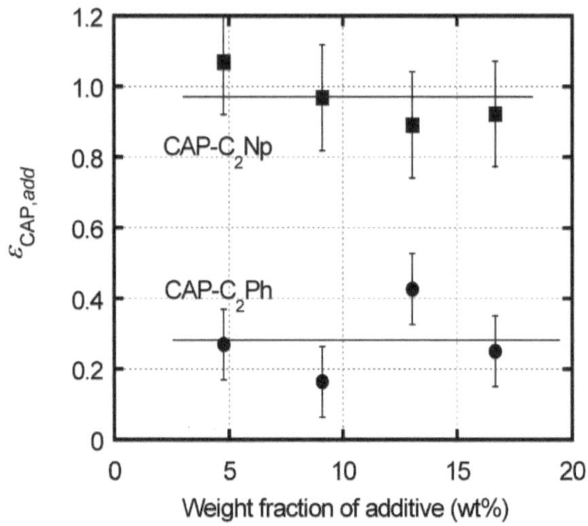

Figure 9.29 Nematic interaction parameters between CAP and PET or PEN, εCAP, add in CAP/PET and CAP/PEN blends. Reproduced with permission from S. Nobukawa, H. Hayashi, H. Shimada, A. Kiyama, H. Yoshimura, Y. Tachikawa, and M. Yamaguchi, *J. Appl. Polym. Sci.*, 2014, in press. © 2014, Wiley Online Library [61].

9.6 Conclusions

Cellulose esters are biomass-derived materials with great potential as optical films. Besides their relatively good cost-performance, cellulose esters also possess excellent physical properties suitable for optical film applications, such as excellent transparency and heat resistance. Some cellulose esters such as CAP and CAB show extraordinary dispersion of in-plane and out-of-plane birefringences, i.e., the magnitude of birefringence increases with the wavelength, a property essential for specific retardation films such as quarter-wave plate. Although this property can be provided by various techniques—through piling of two films, blending of two polymers having opposite birefringence or by random copolymerization, these techniques involve an adhesion or combination of two different types of polymers. In the case of cellulose esters, the orientation birefringence is mainly determined by the type and amount of the substitution groups. Therefore, it is possible to tailor the retardation/birefringence by manipulating the degree of substitution of the ester groups with intense intention to the hydroxyl group. We found that certain ester groups such as propionyl and butyryl as well as hydroxyl group contribute to positive birefringence, while acetyl group contributes to negative birefringence. Furthermore, addition of a suitable plasticizer is also effective in modifying the in-plane and out-of-plane birefringences of cellulose esters, provided the blend is miscible. This will be a key technology in the field of high-performance optical films to improve the contrast of LCD, because there are numerous low-mass compounds having strong polarizability anisotropy.

References

1. Shen, L., Worrell, E., and Patel, M. (2010). Present and future development in plastics from biomass, *Biofuel Bioprod. Bior.*, **4**, 25–40.
2. Pan, H. (2011). Synthesis of polymers from organic solvent liquefied biomass: A review, *Renewable Sustainable Energy Rev.*, **15**, 3454–3463.
3. Edgar, K. J., Buchanan, C. M., Debenham, J. S., Rundquist, P. A., Seiler, B. D., Shelton, M. C., and Tindall, D. (2001). Advances in cellulose ester performance and application, *Prog. Polym. Sci.*, **26**, 1605–1688.
4. Glasser, W. G. (2004). Prospects for future applications of cellulose acetate, *Macromol. Symp.*, **208**, 371–394.
5. Sata, H., Maruyama, M., and Shimamoto, S. (2004). Properties and applications of cellulose triacetate film, *Macromol. Sym.*, **208**, 323–333.
6. Kamide, K. (2005). *Cellulose and Cellulose Derivatives: Molecular Characterization and Its Application* (Elsevier Science, Amsterdam).
7. Yamaguchi, M. (2010). Optical properties of cellulose esters and their blends, in *Cellulose: Structure and Properties, Derivatives and Industrial Uses* (Lejeune, A., Deprez, T., eds.), Nova Science Publishers, Inc.: New York, pp. 325–340.
8. Yamaguchi, M., Lee, S., Manaf, M. E. A., Tsuji, M., and Yokohara, T. (2010). Modification of orientation birefringence of cellulose ester by addition of poly (lactic acid), *Eur. Polym. J.*, **46**, 2269–2274.
9. Wang, Z., Folkenroth, J., Zhou, W., Zhang, Y., and Shi, X. (2012). Cellulose ether polymers as optical compensation films for LCDs-high birefringence and tunable optics, *IDW/AD'12*, Kyoto, Japan, pp. 499–502.
10. Kimura, T., Yamato, M., Endo, S., Kimura, F., Sata, H., Kawasaki, H., and Shinagawa, Y. (2001). Orientation of cellulose triacetate films cast

from solution in high magnetic field, *J. Polym. Sci., Polym. Phys.*, **39**, 1942–1947.

11. Songsurang, K., Miyagawa, A., Manaf, M. E. A., Phulkerd, P., Nobukawa, S., and Yamaguchi, M. (2013). Optical anisotropy in solution-cast film of cellulose triacetate, *Cellulose*, **20**, 83–96.
12. Sosnowski, T. P., and Weber, H. P. (1972). Thin birefringent polymer films for integrated optics, *Appl. Phys. Lett.*, **21**, 310–312.
13. Croll, S. G. (1979). The origin of residual internal stress in solvent-cast thermoplastic coatings, *J. Appl. Polym. Sci.*, **23**, 847–858.
14. Prest, W. M., and Luca, D. J. (1979). The origin of the optical anisotropy of solvent cast polymeric films, *J. Appl. Phys.*, **50**, 6067–6072.
15. Prest, W. M., and Luca, D. J. (1981). The alignment of polymers during the solvent-coating process, *J. Appl. Phys.*, **51**, 5170–5175.
16. Cohen, Y., and Reich, S. (1981). Ordering phenomena in thin polystyrene films, *J. Polym. Sci. Polym. Phys. Ed.*, **19**, 599–608.
17. Machell, J. S., Greener, J., and Contestable, B. A. (1990). Optical properties of solvent cast polymer films, *Macromolecules*, **23**, 186–194.
18. Lei, H., Payne, A., McCormick, A. V., Francis, L. F., Gerberich, W. W., and Scriven, L. E. (2001). Stress development in drying coatings, *J. Appl. Polym. Sci.*, **81**, 1000–1013.
19. Greener, J., Chen, J. (2005). Optical properties of solvent-cast polarizer films for liquid crystal displays, *IDMC Taipei, Taiwan*.
20. Harding, G. F. (1986). In *Optical Properties of Polymers* (Meeten, G. H. ed.), Elsevier Applied Science, London, Chapter 2, p. 63.
21. Ghosh, G. (1999). Dispersion-equation coefficients for the refractive index and birefringence of calcite and quartz crystals, *Opt. Commun.*, **163**, 95–102.
22. Scharf, T. (2006). *Polarized Light in Liquid Crystals and Polymers*, Wiley: New York.
23. Kuhn, W., and Grün, F. (1942). Beziehungen zwishen elastischen konstanten und dehnungsdoppelbrechung hochelastischer stoffe, *Kolloid-Z*, **101**, 248–271.
24. Treloar, L. R. G. (1958). *The Physics of Rubber Elasticity* (Clarendon Press: Oxford).
25. Read, B. E. (1975). In *Structure and Properties of Oriented Polymers* (Ward, I. M. ed.), Applied Science Publishers: London, Chapter 4.
26. Jenkins, F. A., and White, H. E. (1981). In *Fundamentals of optics*, 4th edn., McGraw-Hill, New York.

27. Marks, J. E., and Erman, B. (1988). In *Rubberlike Elasticity: A Molecular Primer*. Wiley, New York.

28. Hermans, P. H., Platzek, P. (1939). Beitrage zür kenntnis des deformationsmechanimus und der feinstruktur der hydratzellulose X. die Kratky'sche kette als rechenmodell für deformationsmechanismus der hydratzellulosegele, *Kolloid-Z*, **88**, 68–72.

29. Saito, H., Inoue, T. (1987). Chain orientation and intrinsic anisotropy in birefringence-free polymer blends, *J. Polym. Sci. Polym. Phys. Ed.*, **25**, 1629–1636.

30. Uchiyama, A., and Yatabe, T. (2003). Analysis of extraordinary birefringence dispersion of uniaxially oriented poly(2,6-dimethyl 1,4-phenylene oxide)/atactic polystyrene blend films, *Jpn. J. Appl. Phys.*, **42**, 3503–3507.

31. Uchiyama, A., and Yatabe, T. (2003). Control of birefringence dispersion of uniaxially oriented poly(2,6-dimethyl 1,4-phenylene oxide)/atactic polystyrene blend films by changing the stretching parameters, *Jpn. J. Appl. Phys.*, **42**, 5665–5669.

32. Kuboyama, K., Kuroda, T., and Ougizawa, T. (2007). Control of wavelength dispersion of birefringence by miscible polymer blends, *Macromol. Symp.*, **249–250**, 641–646.

33. Tagaya, A., Iwata, S., Kawanami, E., Tsukahara, H., and Koike, Y. (2001). Anisotropic molecule dopant method for synthesizing a zero-birefringence polymer, *Jpn. J. Appl. Phys.*, **40**, 6117–6123.

34. Tagaya, A., Ohkita, H., Harada, T., Ishibashi, K., and Koike, Y. (2006). Zero-birefringence optical polymers, *Macromolecules*, **39**, 3019–3023.

35. Koike, Y., Yamazaki, K., Ohkita, H., and Tagaya, A. (2006). Zero-birefringence optical polymer by birefringent crystal and analysis of the compensation mechanism, *Macomol. Symp.*, **235**, 64–70.

36. Manaf, M. E. A., Tsuji, M., Shiroyama, Y., and Yamaguchi, M. (2011). Wavelength dispersion of orientation birefringence for cellulose esters containing tricresyl phosphate, *Macromolecules*, **44**, 3942–3949.

37. Cimrová, V., Neher, D., Kostromine, S., and Bieringer, T. (1999). Optical anisotropy in films of photoaddressable polymers, *Macromolecules*, **32**, 8496–8503.

38. Uchiyama, A., and Yatabe, T. (2003). Control of wavelength dispersion of birefringence for oriented copolycarbonate films containing positive and negative birefringent units, *Jpn. J. Appl. Phys.*, **42**, 6941–6945.

39. Iwasaki, S., Satoh, Z., Shafiee, H., Tagaya, A., and Koike, Y. (2012). Design and synthesis of zero–zero-birefringence polymers in a quaternary copolymerization system, *Polymer*, **53**, 3287–3296.

40. Uchiyama, A., Ono, Y., Ikeda, Y., Shuto, H., and Yahata, K. (2012). Copolycarbonate optical films developed using birefringence dispersion control, *Polym. J.*, **44**, 995–1008.

41. Yamaguchi, M., Manaf, M. E. A., Songsurang, K., and Nobukawa, S. (2012). Material design of retardation films with extraordinary wavelength dispersion of orientation birefringence: A review, *Cellulose*, **19**, 601–613.

42. Zugenmaier, P. (2004). Characteristics of cellulose acetates-Characterization and physical properties of cellulose acetates, *Macromol. Symp.*, **208**, 81–166.

43. El-Diasty, F., Soliman, M. A., Elgendy, A. F. T., and Ashour, A. (2007). Birefringence dispersion in uniaxial material irradiated by gamma rays: Cellulose triacetate films, *J. Opt. A Pure Appl. Opt.*, **9**, 247–252.

44. Yamaguchi, M., Okada, K., Manaf, M. E. A., Shiroyama, Y., Iwasaki, T., and Okamoto, K. (2009). Extraordinary wavelength dispersion of orientation birefringence for cellulose esters, *Macromolecules*, **42**, 9034–9040.

45. Yamaguchi, M., Iwasaki, T., Okada, K., and Okamoto, K. (2009). Control of optical anisotropy of cellulose esters and their blends with plasticizer, *Acta Mater.*, **57**, 823–829.

46. Greener, J., Lei. H., Elman, J., and Chen, J. (2005). Optical properties of solvent-cast polarizer films for liquid-crystal displays: A viscoelastic modeling framework, *J. SID.*, **13**, 835–839.

47. Takahashi, A., Kawaharada, T., and Kato, T. (1979). Melting temperature of thermally reversible gel. V. heat of fusion of cellulose triacetate and the melting of cellulose diacetate-benzyl alcohol gel, *Polym. J.*, **11**, 671–675.

48. Doi, M., and Watanabe, H. (1991). Effect of nematic interaction on the Rouse dynamics. *Macromolecules*, **24**, 740–744.

49. Watanabe, H., Kotaka, T., and Tirrell, M. (1991). Effect of orientation coupling due to nematic interaction on relaxation of Rouse chains, *Macromolecules*, **24**, 201–208.

50. Urakawa, O., Ohta, E., Hori, H., and Adachi, K. (2006). Effect of molecular size on cooperative dynamics of low mass compounds in polystyrene, *J. Polym. Sci. Polym. Phys. Ed.*, **44**, 967–974.

51. Nobukawa, S., Urakawa, O., Shikata, T., and Inoue, T. (2010). Evaluation of nematic interaction parameter between polymer segments and low-mass molecules in the mixture, *Macromolecules*, **43**, 6099–6105.

52. Nobukawa, S., Urakawa, O., Shikata, T., and Inoue, T. (2011). Cooperative dynamics in polystyrene and low-mass molecule mixtures, *Macromolecules*, **44**, 8324–8332.

53. Manaf, M. E. A., Tsuji, M., Nobukawa, S., and Yamaguchi, M. (2011). Effect of moisture on the orientation birefringence of cellulose esters, *Polymers*, **3**, 955–966.

54. Zawada, A. F., Fuller, G. G., Colby, R. H., Fetters, L. J., and Roovers, J. (1994). Measuring component contributions to the dynamic modulus in miscible polymer blends, *Macromolecules*, **27**, 6851–6860.

55. Urakawa, O., Nobukawa, S., Shikata, T., and Inoue, T. (2010). Dynamics of low mass molecules dissolved in polymers, *Nihon Reoroji Gakkaishi*, **38**, 41–46.

56. Tatsushima, T., Ogata, N., Nakane, K., and Ogihara, T. (2005). Structure and physical properties of cellulose acetate butyrate/poly(butylene succinate) blend, *J. Appl. Polym. Sci.*, **96**, 400–406.

57. Yamaguchi, M., and Masuzawa, K. (2007). Birefringence control for binary blends of cellulose acetate propionate and poly(vinyl acetate), *Eur. Polym. J.*, **43**, 3277–3282.

58. Manaf, M. E. A., Miyagawa, A., Nobukawa, S., Aoki, Y., and Yamaguchi, M. (2013). Incorporation of low-mass compound to alter the orientation birefringence in cellulose acetate propionate, *Opt. Mater.*, **35**, 1443–1448.

59. Songsurang, K., Shimada, H., Nobukawa, S. Y., Yamaguchi, M. (2014). Control of three-dimensional refractive indices of uniaxially-stretched cellulose triacetate with low-molecular-weight compound, *Eur. Polym. J.*, **89**, 105–112.

60. Nobukawa, S., Aoki, Y., Yoshimura, H., Tachikawa, Y., and Yamaguchi, M. (2013). Effect of aromatic additives with various alkyl groups on orientation birefringence of cellulose acetate propionate, *J. Appl. Polym. Sci.*, **130**, 3465–3472.

61. Nobukawa, S., Hayashi, H., Shimada, H., Kiyama, A., Yoshimura, H., Tachikawa, Y., and Yamaguchi, M. (2014). Strong orientation correlation and optical anisotropy in blend of cellulose ester and poly(ethylene 2,6-naphthalate) oligomer, *J. Appl. Polym. Sci.*, **131**, 40570.

Chapter 10

Mechanical Performance of *Eulaliopsis binata* Biofiber–Based Green Composites

Manju Kumari Thakur[a] and Vijay Kumar Thakur[b]

[a]*Division of Chemistry, Government Degree College Sarkaghat, Himachal Pradesh University, Shimla 171005, India*
[b]*School of Mechanical and Materials Engineering, Washington State University, USA*

shandilyamn@gmail.com

10.1 Introduction

Various kinds of materials derived from biorenewable resources have been in use for the past two decades because of their eco-friendly advantages [1–4]. Among these materials, natural fibers such as hibiscus sabdariffa, cotton, pine needles, jute, flax, kenaf, banana, grewia optiva, sisal, oil palm, and pineapple leafs have attracted a considerable interest for multifunctional applications during the last few decades [5–8]. The advantageous properties of natural fibers over traditional synthetic materials include their low density, high toughness, biodegradability, low cost, environment friendliness, acceptable specific properties, easy availability, and better thermal and insulating properties [5, 9–13]. One of the biggest assets of natural fibers is their neutrality toward the

Handbook of Sustainable Polymers: Processing and Applications
Edited by Vijay Kumar Thakur and Manju Kumari Thakur

ISBN 978-981-4613-53-8 (Hardcover), 978-981-4613-54-5 (eBook)
www.panstanford.com

emission of CO_2; thus, their impropriation into the polymers may lead to the reduction of greenhouse emission [14, 15]. *Eulaliopsis binata* fibers are one such type of natural fibrous materials that have remained unexplored so far. These fibers are obtained from the *Eulaliopsis binata* perennial plant [16]. This plant, abundantly grown in India, China, and Nepal, belongs to the family Poaceae [16]. The rapid expansion of this plant has resulted in serious concerns about its waste management as it is considered the waste biomass and most frequently burned to get rid of it. Limited information is available in the existing literature about the effective utilization of this material as reinforcement in the polymer composites for structural applications [17, 18].

Considerable efforts have been devoted during the past few years to prepare composites using different types of natural fibers as the reinforcement such as oil palm, flax, jute, henequen, grewia optiva, hibiscus sabdariffa, and pine needles in various polymer matrices procured from renewable/petroleum resources [2, 19–21]. Significant amount of research work has been published on the exploration of the mechanical properties of natural fiber–based polymer composites [22]. These materials have been reported to be competitive with the synthetic fiber–reinforced composites for some selected applications [5, 23–25]. Recently we have reported some of our preliminary results on the use of *Eulaliopsis binata* fibers as particle reinforcement in green composites [16]. Inspired by these results, in the present work, polymer composites were prepared using short *Eulaliopsis binata* as the *reinforcing* material. The green composites were prepared employing the compression molding method and were aimed at determining the tensile, compression, flexural and wear resistance behavior of these composites.

10.2 Materials and Methods

10.2.1 Materials

Eulaliopsis binata fibers were obtained from the local resources of the Himalayan region. Phenol, formaldehyde solution, and sodium hydroxide of Qualigens make were used as received from their vendors [26–29]. Phenol-formaldehyde (PF) resin to be used as the polymer matrix for the preparation of the composites was prepared using the standard method reported in the literature

[30, 31]. Prior to their use as the reinforcement, the *Eulaliopsis binata* fibers were thoroughly washed and purified as per the standard method and dried in a hot air oven [26–29]. These *Eulaliopsis binata* fibers were then converted into short fibers form and were used as reinforcement in the preparation of the polymer composites [26–29].

10.2.2 Preparation of Polymer Composites

The short *Eulaliopsis binata* fibers were used in the preparation of polymer composites. The phenol-formaldehyde resin used in the current research is a liquid resin self-prepared in the laboratory [27–29]. During the preparation of the polymer composite samples, the short fibers were impregnated with phenolic resin matrix using the mechanical stirrer. Polymer composites with different weight percentage loadings starting from 10–40% were prepared. A neat phenolic resin sample was also prepared to compare the mechanical strength results. For testing the mechanical properties, the rectangular-shaped sheets of size 150 mm × 150 mm × 5.0 mm were prepared [26–29].

10.2.3 Characterization of Green Polymer Composites

The analysis of mechanical properties such as tensile, compressive, flexural, and wear resistance of the polymer composites was done as per ASTM D 3039, ASTM D 3410, ASTM D 790, and ASTM D 3702 methods, respectively [26–29]. Thermal stability of the fiber-reinforced polymer composites was determined using thermogravimetric analysis [TGA] in nitrogen atmosphere on a thermal analyzer. The heating rate during the experiment was maintained at 10°C/min. The morphological images of the polymer composites were analyzed using scanning electron microscopy (SEM). Prior to the surface morphological evaluation, the samples to be analyzed were mounted on an aluminum stub employing a conductive silver paint followed by sputtering a gold platinum mixture.

10.3 Results and Discussion

Mechanical properties of polymer composite systems determine their actual potential in numerous applications, especially in structural and automotive [27–29, 32]. Both the reinforcement and the polymer matrix used to encapsulate the reinforcement determine the overall properties of the composites [27–29, 32]. Compared to the synthetic fibers that are frequently used as the reinforcement in the polymer matrices, natural cellulosic fibers are much cheaper and offers a number of advantages as discussed in Section 10.1 [27–29, 32]. These natural fibers were initially used in the polymer composite systems to replace the synthetic fibers such as of glass fiber in polymer composites. The cellulosic fiber-reinforced polymer composites are also advantageous from the health point of view as they do not cause health problems such as skin irritation [33]. In addition to their eco-friendly advantages and health benefits, natural fibers have been reported to exhibit significantly better elongation at break [27–29, 32]. This superior elongation at break results in the better composite damage tolerance [33]. In the present work the polymer resin matrix was of thermosetting origin and was synthesized using the condensation reaction of phenol with formaldehyde [27–29, 32]. The polymer resin contains lots of hydroxyl groups that may interact with the hydroxyl groups present in the *Eulaliopsis binata* fibers reinforcement.

The mechanical properties of the synthesized *Eulaliopsis binata* fibers/thermosetting polymer composites have been found to depend on the extent of fiber–polymer matrix bonding and the transfer of load from thermosetting polymer matrix to reinforcement similar to the earlier reported literature [16, 27–29, 32]. It has been reported in the existing literature that the higher magnitude of bonding between the matrix and fibers used as the reinforcing materials facilitates the load transfer between the matrix and reinforcement [16, 27–29, 32]. The transfer of load contributes to higher mechanical properties in the resulting *Eulaliopsis binata* fibers polymer composites. It has been observed in all the composite systems studied in the present work that the polymer composites with 30 wt% fiber loading exhibit the maximum mechanical properties and this behavior was attributed to the maximum fiber/matrix bonding as well as load transfer at this *Eulaliopsis binata* fibers loading [16, 27–29, 32]. Figure 10.1 shows the tensile strength results of short *Eulaliopsis binata* fiber-reinforced polymer composites. Figure 10.1 clearly demonstrates that the tensile strength of the *Eulaliopsis binata* fiber-reinforced composites is higher than that of the pure polymer matrix [16, 27–29, 32].

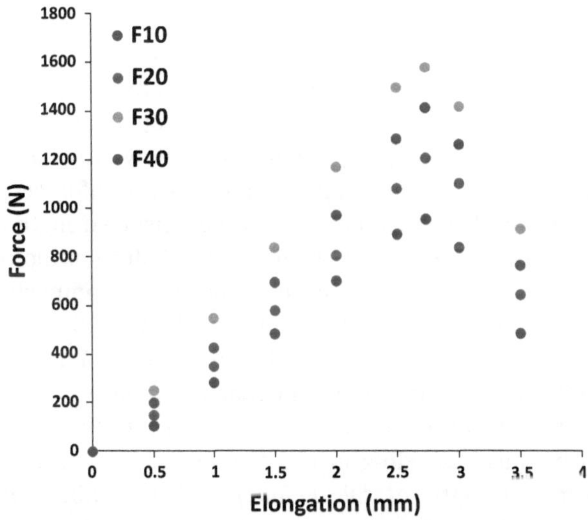

Figure 10.1 Load elongation curve of *Eulaliopsis binata* fiber-reinforced green composites.

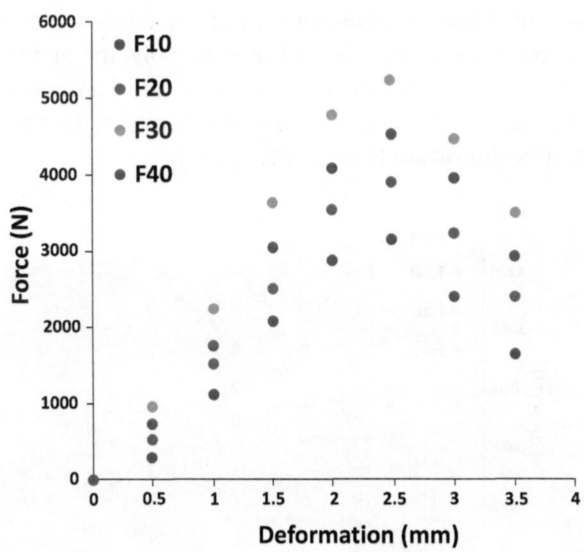

Figure 10.2 Load deformation curve of *Eulaliopsis binata* fiber-reinforced green composites.

This increment in tensile strength of the *Eulaliopsis binata* fiber-reinforced composites can be attributed the strong reinforcing effect of the fibers used as the reinforcement in the polymer matrix.

The decrease in the tensile strength properties of the *Eulaliopsis binata* fiber–reinforced composites beyond 30% can be attributed to the poor fiber–matrix adhesion as well as breaking of fibers at the weaker point followed by further propagation under the applied load [16, 27–29, 32]. Figure 10.2 shows the compressive strength results of the *Eulaliopsis binata* fiber–reinforced polymer composites and these results followed the tensile strength [16, 27–29, 32]. In case of the compression test, the failure of the resulting *Eulaliopsis binata* fiber–reinforced polymer composites was attributed to the sudden and dramatic buckling of the fibers used as the reinforcement [16, 27–29, 32]. From the different polymer composites samples, it was found that the failure of composites under compressive load can be attributed to the buckling of columns or micro buckling in the composites. Figure 10.3 shows the flexural strength of the *Eulaliopsis binata* fiber–reinforced green composites and these results also followed the trend of the tensile and compressive strength results [16, 27–29, 32]. The wear resistance results as demonstrated in Fig. 10.4 shows that the reinforcement of the *Eulaliopsis binata* fibers in the polymer matrix improves the wear resistance of the pure polymer matrix. These results of the wear resistance behavior of the *Eulaliopsis binata* fiber–reinforced composites are in conformity with the results reported in the literature [16, 27–29, 32, 34].

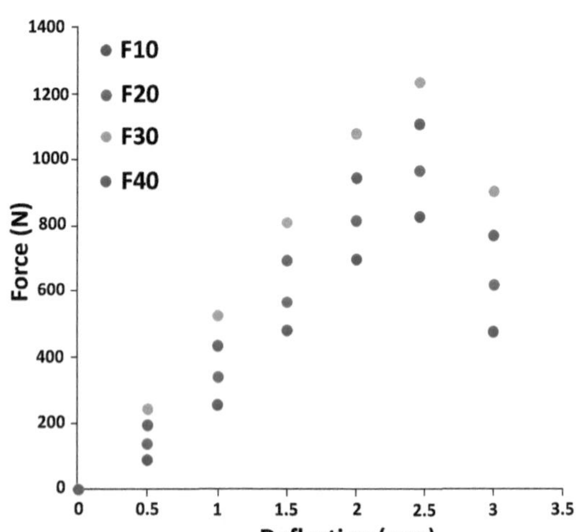

Figure 10.3 Load deflection curve of *Eulaliopsis binata* fiber–reinforced green composites.

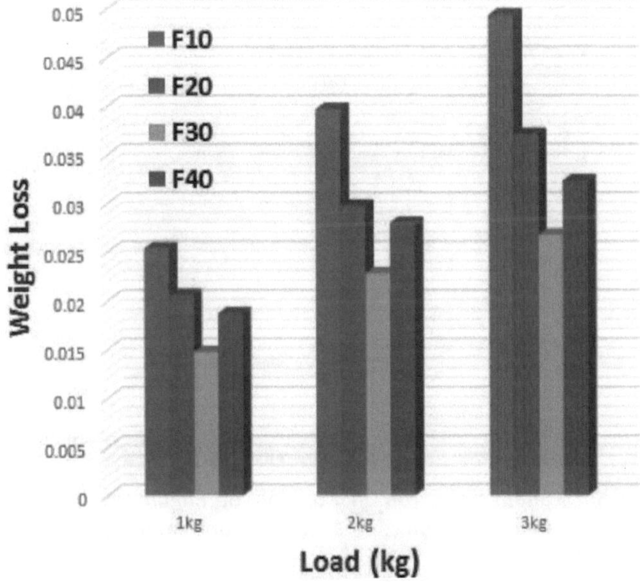

Figure 10.4 Wear resistance curve of *Eulaliopsis binata* fiber-reinforced green composites.

10.3.1 Morphological and Thermal Study of the Green Composites

The morphological images of the *Eulaliopsis binata* fiber-reinforced composites were analyzed by a scanning electron microscope. It was revealed from the micrographic images that the different *Eulaliopsis binata* fibers reinforcement in terms of weight have different impacts on the morphology of resulting composites. Figure 10.5 shows the morphological image of the polymer composite with optimized (30 wt% loading). From the figure, it is clear that the *Eulaliopsis binata* fibers were intimately mixed at this loading [34]. The thermal stability of the *Eulaliopsis binata* fiber-reinforced composite was studied using TGA analysis. It has been reported in the literature that natural fibers are thermally less stable than the synthetic fibers [34, 35]. In case of *Eulaliopsis binata* fiber-reinforced composites, the polymer composite with 30 wt% loading was found to exhibit intermediate behavior between the pure matrix and the fibers. This behavior was similar to the particle *Eulaliopsis binata* fiber-reinforced composite [16].

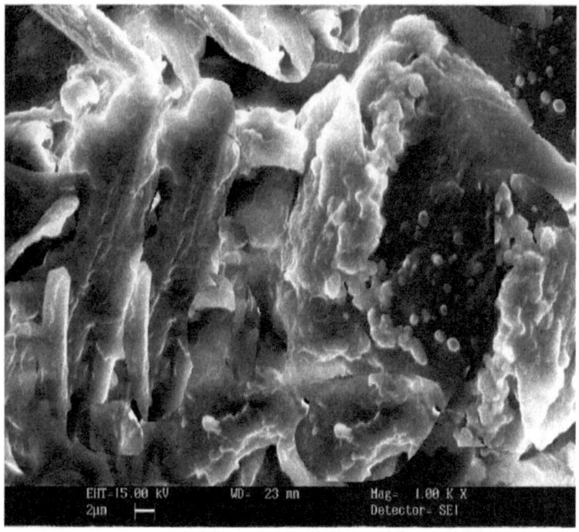

Figure 10.5 Scanning electron micrograph of *Eulaliopsis binata* fiber-reinforced green composites with optimum (30 wt%) loading.

10.4 Conclusions

Eulaliopsis binata fiber–reinforced polymer composites are more eco-friendly and cost effective compared to the traditional synthetic fiber–reinforced composites. The aim of the present work was to study the reinforcing potential of the *Eulaliopsis binata* fibers in the short fiber form. The mechanical performance of these *Eulaliopsis binata* fiber polymer composites was found to be higher than that of the pure polymer. However challenges still exist in further improving the mechanical properties of these composites to make them competitive to their synthetic counterparts.

Acknowledgments

The authors wish to thank their parental institutes for providing the necessary facilities to accomplish the present research work.

References

1. Ramanaiah, K., Prasad, A. V. R., Reddy, K. H. C. (2013). Mechanical and thermo-physical properties of fish tail palm tree natural fiber-reinforced polyester composites, *Int. J. Polym. Anal. Charact.*, **18**(2), 126–136.

2. Singha, A. S., Thakur, V. K., Mehta, I. K., Shama, A., Khanna, A. J., et al. (2009). Surface-modified *hibiscus sabdariffa* fibers: Physicochemical, thermal, and morphological properties evaluation, *Int. J. Polym. Anal. Charact.,* **14**(8), 695–711.
3. Bodirlau, R., Teaca, C.-A., Spiridon, I. (2010). Influence of ionic liquid on hydrolyzed cellulose material: FT-IR spectroscopy and TG-DTG-DSC analysis, *Int. J. Polym. Anal. Charact.,* **15**(7), 460–469.
4. Darie, R. N., Bodirlau, R., Teaca, C. A., Macyszyn, J., Kozlowski, M., Spiridon, I. (2013). Influence of accelerated weathering on the properties of polypropylene/polylactic acid/eucalyptus wood composites, *Int. J. Polym. Anal. Charact.,* **18**(4), 315–327.
5. Jawaid, M., Khalil, H. P. S. A., Abu Bakar, A. (2010). Mechanical performance of oil palm empty fruit bunches/jute fibres reinforced epoxy hybrid composites, *Mater. Sci. Eng. Struct. Mater. Prop. Microstruct. Process,* **527**(29–30), 7944–7949.
6. Hassan, M. M., Islam, M. R., Khan, M. A. (2005). Influence of additives on the performance of photografted jute yarn with 3-(trimethoxysilyl) propylmethacrylate, *Int. J. Polym. Anal. Charact.,* **10**(3–4), 179–194.
7. Singha, A. S., Thakur, V. K. (2009). Morphological, thermal, and physicochemical characterization of surface modified *pinus* fibers, *Int. J. Polym. Anal. Charact.,* **14**(3), 271–289.
8. Hassan, A., Ken, L. S., Jawaid, M. (2013). Flame retardancy and kinetic behavior of ammonium polyphosphate-treated unsaturated polyester/phenolic interpenetrating polymer network, *Int. J. Polym. Anal. Charact.,* **18**(2), 137–145.
9. Kommula, V. P., Reddy, K. O., Shukla, M., Marwala, T., Rajulu, A. V. (2013). Physico-chemical, tensile, and thermal characterization of Napier grass (native African) fiber strands, *Int. J. Polym. Anal. Charact.,* **18**(4), 303–314.
10. Maheswari, C. U., Reddy, K. O., Muzenda, E., Rajulu, A. V. (2012). Tensile and thermal properties of polycarbonate-coated tamarind fruit fibers, *Int. J. Polym. Anal. Charact.,* **17**(8), 578–589.
11. Jawaid, M., Alothman, O. Y., Paridah, M. T., Khalil, H. P. S. A. (2013). Effect of fiber treatment on dimensional stability and chemical resistance properties of hybrid composites, *Int. J. Polym. Anal. Charact.,* **18**(8), 608–616.
12. Singha, A. S., Thakur, V. K. (2008). Synthesis and characterization of pine needles reinforced RF matrix based biocomposites, *E. J. Chem.,* **5**, 1055–1062.

13. Thakur, V. K., Singha, A. S. (2010). Natural fibres-based polymers: Part 1-mechanical analysis of pine needles reinforced biocomposites, *Bull. Mater. Sci.*, **33**(3), 257–264.
14. Maheswari, C. U., Reddy, K. O., Muzenda, E., Shukla, M., Rajulu, A. V. (2013). Mechanical properties and chemical resistance of short tamarind fiber/unsaturated polyester composites: Influence of fiber modification and fiber content, *Int. J. Polym. Anal. Charact.*, **18**(7), 520–533.
15. Pinto, M. A., Chalivendra, V. B., Kim, Y. K., Lewis, A. F. (2014). Evaluation of surface treatment and fabrication methods for jute fiber/epoxy laminar composites, *Polym. Compos.*, **35**(2), 310–317.
16. Thakur, V., Thakur, M., Gupta, R. (2013). Eulaliopsis binata: Utilization of waste biomass in green composites. In *Green Composites from Natural Resources*, pp. 125–32. CRC Press
17. Thakur, V. K., Singha, A. S. (2011). Physicochemical and mechanical behavior of cellulosic pine needle-based biocomposites, *Int. J. Polym. Anal. Charact.*, **16**(6), 390–398.
18. Singha, A. S., Thakur, V. K., Mishra, B. N. (2009). Study of grewia optiva fiber reinforced urea-formaldehyde composites, *J. Polym. Mater.*, **26**(1), 81–90.
19. Khalil, H. P. S. A., Jawaid, M., Firoozian, P., Zainudin, E. S., Paridah, M. T. (2013). Dynamic mechanical properties of activated carbon- filled epoxy nanocomposites, *Int. J. Polym. Anal. Charact.*, **18**(4), 247–256.
20. Thakur, V. K., Singha, A. S., Misra, B. N. (2011). Graft copolymerization of methyl methacrylate onto cellulosic biofibers. *J. Appl. Polym. Sci.*, **122**(1), 532–544.
21. Pinto, M. A., Chalivendra, V. B., Kim, Y. K., Lewis, A. F. (2013). Effect of surface treatment and z-axis reinforcement on the interlaminar fracture of jute/epoxy laminated composites, *Eng. Fract. Mech.*, **114**, 104–114.
22. Atluri, R. P. V., Rao, K. M., Gupta, A. V. S. S. K. S. (2013). Experimental investigation of mechanical properties of golden cane fiber-reinforced polyester composites, *Int. J. Polym. Anal. Charact.*, **18**(1), 30–39.
23. Singha, A. S., Thakur, V. K. (2009). Synthesis and characterizations of silane treated *grewia optiva* fibers, *Int. J. Polym. Anal. Charact.*, **14**(4), 301–321.
24. Khalil, H. P. S. A., Jawaid, M., Firoozian, P., Amjad, M., Zainudin, E., Paridah, M. T. (2013). Tensile, electrical conductivity, and morphological properties of carbon black-filled epoxy composites, *Int. J. Polym. Anal. Charact.*, **18**(5), 329–338.

25. Thakur, V. K., Thakur, M. K., Gupta, R. K. (2013). Synthesis of lignocellulosic polymer with improved chemical resistance through free radical polymerization, *Int. J. Biol. Macromol.*, **61**, 121–126.

26. Thakur, V. K., Singha, A. S., Kaur, I., Nagarajarao, R. P., Liping, Y. (2010). Silane functionalization of *Saccaharum cilliare* fibers: thermal, morphological, and physicochemical study, *Int. J. Polym. Anal. Charact.*, **15**(7), 397–414.

27. Singha A. S., Thakur V. K. (2008). Synthesis and characterization of grewia optiva fiber-reinforced PF-based composites. *Int. J. Polym. Mater.*, **57**(12), 1059–1074.

28. Singha, A. S., Thakur, V. K. (2009). Synthesis, characterisation and analysis of hibiscus sabdariffa fibre reinforced polymer matrix based composites. *Polym. Polym. Compos.*, **17**(3), 189–194.

29. Singha, A. S., Thakur, V. K. (2009). Chemical resistance, mechanical and physical properties of biofibers-based polymer composites, *Polym. Plast. Technol. Eng.*, **48**(7), 736–744.

30. Thakur, V. K., Singha, A. S., Mehta, I. K. (2010). Renewable resource-based green polymer composites: Analysis and characterization, *Int. J. Polym. Anal. Charact.*, **15**(3), 137–146.

31. Singha, A. S., Thakur, V. K. (2009). Synthesis, characterization and study of pine needles reinforced polymer matrix based composites, *J. Reinf. Plast. Compos*, **29**(5), 700–709.

32. Khalil, H. P. S. A., Firdaus, M. Y. N., Jawaid, M., Anis, M., Ridzuan, R., Mohamed, A. R. (2010). Development and material properties of new hybrid medium density fibre board from empty fruit bunch and rubber wood, *Mater. Des.*, **31**(9), 4229–4236.

33. Jawaid, M., Khalil, H. P. S. A. (2011). Cellulosic/synthetic fibre reinforced polymer hybrid composites: A review, *Carbohydr. Polym.*, **86**(1), 1–18.

34. Thakur, V. K., Thakur, M. K., Gupta, R. K. (2014). Graft copolymers of natural fibers for green composites, *Carbohydr. Polym.*, **104**, 87–93.

35. Thakur, V. K., Thakur, M. K., Gupta, R. K. (2013). Development of functionalized cellulosic biopolymers by graft copolymerization, *Int. J. Biol. Macromol.*, **62**, 44–51.

Chapter 11

Development and Applications of Sustainable Polylactic Acid Parts

Khosrow Khodabakhshi and Morteza Ehsani

Plastics Department, Faculty of Processing, Iran Polymer and Petrochemical Institute, Tehran, P. O. box 115/14965, Iran

k.khodabakhshi@gmail.com

11.1 Introduction

Biosourced and biodegradable materials will gradually replace the currently existing family of oil-based polymers as they become cost- and performance-wise competitive. Polylactic acid or polylactide (PLA) is becoming the most important bio-based polymer due to its favorable properties widely investigated. Polylactic acid, also known as polylactide, is an aliphatic polyester produced from renewable sources. It has attracted huge attention due to its biodegradability and mechanical properties similar to polyethylene terephthalate (PET), polystyrene (PS) and other common petrol-based plastics. It has a wide range of applications in medical devices, food packaging, electronics, and engineering devices. Production of PLA with low molecular weight (LMW) through direct polycondensation (PC) is known since 1932. Water is the byproduct of the reaction, which

Handbook of Sustainable Polymers: Processing and Applications
Edited by Vijay Kumar Thakur and Manju Kumari Thakur

ISBN 978-981-4613-53-8 (Hardcover), 978-981-4613-54-5 (eBook)
www.panstanford.com

has to be removed from the reaction in order to shift the chemical equilibrium to the product side and enhance the polycondensation reaction to increase the polymerization degree. Reacting mixture viscosity increases during polymerization that makes the water removal very difficult, thus only LMW polymers can be produce by PC. The most effective method to manufacture high-molecular-weight PLA is through ring opening polymerization (ROP) of lactide [1]. Like ROP of ε-caprolactam [2, 3], this reaction requires strict purity of the lactide monomer, obtained by dimerization of the lactic acid monomer [4]. PLA is obtained by heating a catalyst/monomer mixture under vacuum or an inert atmosphere. Tin (II)-based catalysts, tin octoate in particular, are the most widely used catalysts for the ring-opening polymerization of lactide. They have been shown to give a controlled and living polymerization of lactide via coordination–insertion mechanism. High-molecular-weight polymer with monomer conversion above 95%, good reaction rate, and low levels of racemization can be obtained using tin (II) compounds and Sn(Oct)2 in particular [5]. The existing PLA production processes mainly employ a series of batch reactors, which are time-, energy-, and cost consuming. Since high-molecular-weight PLA can be obtained in a short time in the presence of a co-catalyst, e.g., triphenyl phosphate, through ROP, the reactive extrusion of lactide has recently attracted many researchers. This method allows the production of ready-to-use PLA parts in a shorter and cost-effective process [6].

11.2 Lactide

11.2.1 Production

Lactide was first synthesized by Pelouze in 1845 [7] by self-esterification of lactic acid to obtained a prepolymer and heating of the prepolymer to produce distillate crystals. Gruter and Pohl improved the process in 1914 [8]. The procedure was first polycondensation of lactic acid at 120–135°C with the aid of air flow to remove the water. Next, lactide was distilled off under vacuum at 200°C in the presence of zinc oxide. Tin (II)-based catalysts are the most common used catalysts in the modern industry. Since the prepolymer degradation is an equilibrium reaction, lactide must be extracted from the system in order to shift the reaction toward the product side [9]. In general, lactide synthesis involves the following steps:

11.2.1.1 Prepolymerization

Prepolymers with an average degree of polymerization (DP) of around 10 are usually obtained in a batch process at elevated temperatures of up to 190°C and vacuum pressure of 70–250 mbar after 6 h [10]. Due to the lower rate of mass transfer of water in the viscous prepolymer, thin-film and rotating-flask vacuum equipment showed faster reaction times than a stirred tank. Continuous processes have also been described in a number of patents [11–13]. Prepolymers with a DP of 7–20 are recommended as feed to the lactide synthesis. There is an optimum range for the prepolymer molecular weight. At very LMWs, feeds will contain more monomer that boils at a lower temperature than lactide, and this will contaminate the crude lactide distilled off from the reactor. The residual monomer in crude lactide leads to more acidity, which reduce the product quality. On the other hand, at higher molecular weights of the prepolymer the rate of mass transfer decreases due to the high viscosity of the mixture. It is also reported that increasing prepolymer molecular weight results in an increase in the *meso*-lactide concentration in the crude lactide [14].

11.2.1.2 Lactide synthesis

Lactide is produced by degradation reactions, mainly via intramolecular chain scission of the prepolymer. Lactide synthesis from a prepolymer with a DP in the range of 10–15 in the presence of various catalysts at 4–5 mbar and 190–245°C is reported by Noda and Okuyama [9]. The best performances were reported using 0.05–0.2 wt% tin catalysts and tin octoate (stannous 2-ethylhexanoate) in particular, which is widely available. The catalyst increases the rate of backbiting reactions from hydroxyl chain ends of prepolymers to form lactide molecules [9, 15]. The melt viscosity of the prepolymer increases because of the esterification reactions during the process, which results in decreased rate of mass transfer.

Continuous lactide synthesis in which the prepolymer is fed continuously to a reactor is also reported in the literature [11]. This procedure resulted in a product with a conversion around 70% per pass, which could be increased by recycling the residue to the prepolymer hydrolyzing section. Stirred tank reactor with different stirrer types [12], stirred reactor with a distillation section on top of the reactor [16], thin-film evaporator [17], horizontal wiped film evaporator [13], and distillation column with perforated plates [10], are reported as reaction vessels in some patents. Pressure must be kept below 10 mbar to increase the driving force for the lactide evaporation and the overall reaction rate. However, this will require detailed considerations of equipment size, vacuum systems, condensers, and so on [18].

Although increasing the catalyst concentration results in the faster overall reaction rate, enhanced rate of racemization, which is unwanted in the production of optically pure lactide may also occur. Higher synthesis temperatures, longer reaction times and presence of some metal cations such as sodium and potassium have the same effect on the stereochemical purity of the crude lactide [14, 15, 19]. Released metals through corrosion and carboxylic acid impurities formed during lactic acid fermentation are other sources of impurities [10, 14, 20–22].

11.2.1.3 Purification of lactide

A crude lactide stream produced in the lactide synthesis reactors contains lactic acid, lactic acid oligomers, water, *meso*-lactide, and further impurities. Two main separation methods, distillation and crystallization, are currently employed for lactide purification. Crystallization may be carried out either by solvent crystallization or melt crystallization. The most used method for production of ultra-pure lactide in laboratory is by repeated recrystallization of a saturated lactide solution in mixtures of toluene and ethyl acetate [15, 23, 24]. Lactide purification using C4–12 ethers [25], and an organic solvent that is immiscible with water to extract the solution with water [26] are also reported. Melt crystallization is more practical in industry for lactide purification. Several types of equipment are described in the literature for melt crystallization [17, 27–30]. This method uses the differences in the melting points of L-, D-, and *meso*-lactide for separating the different lactides from each other. In a distillation process, the crude lactide is first distilled to remove the acids and water, and then *meso*-lactide is separated from lactide [11, 31]. Different methods are reported in the literature for distillation purification of lactide [32, 33]. In general, crystallization yields highly pure lactide, whereas distillation is more affordable. Lactide with higher stereochemical purity produces PLA with higher stereochemical purity, which controls the polymer final properties such as melting point, crystallinity, crystallization rate, and mechanical strength [34–36]. In conclusion, a polymer grade lactide must be free of acids, water, and metal ions, and stereochemical pure.

11.2.2 Properties

Lactide, (3,6-dimethyl-1,4-dioxane-2,5-dione), exists in three different forms as illustrated in Fig. 11.1. In addition to the three diastereomeric structures, a racemate of D-lactide and L-lactide also exists, *rac*-lactide [37, 38]. Some physical and chemical properties of all types of lactides are listed in Table 11.1.

Table 11.1 Physical and chemical properties of the lactides

	Unit	D-Lactide	L-Lactide	*meso*-Lactide	*rac*-Lactide
Molecular weight	g/mol	144	144	144	
Melting point	°C	97 [19]	97 [19]	53 [18]	125 [19]
Boiling point	°C		255 (atm) [38]		142 (20 mbar) [18]
Heat of fusion	J/g		146 [19]	128 [18]	185 [19]
Heat of vaporization	kJ/mol		63 [19]		
Optical rotation	Degree	+260[38]	-260 [38]		
Solid density	g/mL		1.32–1.38 [19]	1.32–1.38 [19]	
Liquid viscosity	mPa.s		2.71 (110°C); 2.33 (120°C) [18]		

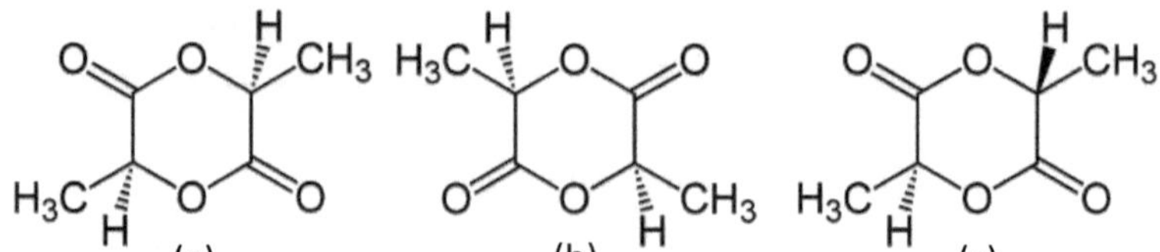

Figure 11.1 The three diastereomeric structures of lactide; (a) *S,S*-Lactide (L-lactide), (b) *R,R*-lactide (D-lactide), (c) *R,S*-lactide (*meso*-lactide).

11.3 PLA Processing and Polymerization

11.3.1 Properties Related to PLA Extrusion

Commercial PLA resins are copolymers of poly(L-lactic acid) (PLLA) and poly(D,L-lactic acid) (PDLLA), with L-isomer constituting the main fraction [39]. The final properties of the PLA depend on the proportion of the enantiomers. PLA polymers with L-enantiomer content >90% tend to be more crystalline than those with higher D-content. Moreover, melting temperature (T_m), glass transition temperature (T_g), and crystallinity decrease with decreasing L-isomer content [40, 41]. So, when high crystallinity is needed such as in heat-resistant articles and fibers, an optically pure PLA is used, whereas in applications that crystallization needs to be suppressed (e.g., thermoforming) PLA resins with higher D-isomer fraction (4–8%) are more suitable [42].

PLA has a T_g in the range of 35–60°C [43], but polymers with higher T_g are used where high mechanical properties are required. PLA melts at around 180°C in its stereochemically purest form, when this temperature can decrease to 130°C in the presence of opposite enantiomers. This may help its processing by expanding the processing window, and reducing thermal and hydrolytic degradation. PLA crystals melt over a range of temperatures, so the process temperature needs to be set to about 40–50°C above the melting point. PLA depolymerization and hydrolysis reactions occur during processing, which result in the molecular weight drop [19, 44, 45]. Residual catalysts from the polymerization process and moisture reduce the stability of the polymer during melt processing [46]. PLA resins with less than 250 ppm (0.025 wt%) moisture content are suitable for extrusion. However, in the case of longer residence times or high temperatures approaching 240°C processes, PLA with less than 50 ppm should be used [47–49]. Semicrystalline PLA pellets are dried in the range of 60–90°C, whereas the amorphous resin must be dried below the T_g (in the range of 30–40°C) in order to avoid sticking of the pellets. PLA drying is usually carried out feeding the pellets in a hopper that is purged with dry air generated by the desiccant bed. Additives and modifiers are used to improve PLA processing and final properties. Toughness, thermal stability, melt strength, polymer slippage and crystallinity can be increased using solid and liquid additives [50]. It should be noted that adding these additives at high concentrations can potentially affect the degradation and biodegradation behaviors of PLA [51, 52].

Extrusion of PLA using conventional screws consists of three sections, which are the feed section, the compression or melting section, and the metering section producing a L/D ratio of 24–30. Extruder screws for processing PET, which are design to minimize shear and degradation, are suitable for processing PLA [53]. Due to the high density (about 1.24 g/mL) of PLA, PLA extrusion requires more power than polyolefins processing [54]. Also, it is recommended to use screws with a compression ratio in the range of 2–3 [55]. PLA with a molecular weight in the range of 100,000–300,000 Da has a melt viscosity in the range of 5,000–10,000 P (500–1000 Pas) at shear rates of 10–50 s^{-1} [56].

At temperatures above 200°C, PLA undergoes degradation reactions through intra- and intermolecular ester exchange, back-biting ester interchange reactions, *cis*-elimination, and radical and concerted nonradical reactions [57, 58] in which, lactide molecules, oligomers, and acetaldehyde may form. In food packaging applications, the generated acetaldehyde can migrate into the contained food, which results in off-flavor of the product [59–61]. In addition, racemization occurs at elevated temperatures, which reduce the optical purity of the polymer and consequently the material properties of the resulting polymer [57, 62–65]. Residual lactide in PLA can result in lowering the melt viscosity and elasticity, and fouling of the processing equipment [66], in which a layer of lactide on the equipment surfaces is seen. To avoid this problem, the temperature must be increased to reduce the tendency of lactide condensation.

In cast film extrusion, the molten PLA is extruded through a lip die and quenched on highly polished chrome rollers that are cooled with circulating water as seen in Fig. 11.2. Sheet and film forming can be achieved on a horizontal three-roll stack with rollers temperatures of 25–50°C. It is recommended to position the die as close as possible to the entrance nip and slightly higher than the nip to reduce the chance of trapping air [47]. Hydraulic roll stands, capable of producing pressure around 143–161 kg/cm are required to prevent floating of the rolls that would result in uneven PLA surfaces, edge instability, and neck-in [47]. In PLA film casting, electrostatic or low-pressure air is required to eliminate streaking, and improve edge stability [67]. To overcome the low toughness of quenched PLA film, mechanical orientation is employed. Mechanically oriented PLA film exhibited higher tensile strength, modulus, and elongation at break [68]. Nipped rolls are required during drawing in order to avoid necking. Crystallinity increases during stretching and the amount of crystallinity attained through orientation depends on the mode of stretching, drawing rate, temperature, and annealing conditions [42, 69, 70]. Ou and Cakmak [69] prepared biaxially oriented PLA films by stretching cast PLA in both MD and TD to different ratios, followed by annealing the films at 110°C for 8 min. They reported that simultaneous biaxial stretching of PLA film did not cause significant crystalline order in the films. However, annealing the films maintained in the grips resulted in development of crystalline order.

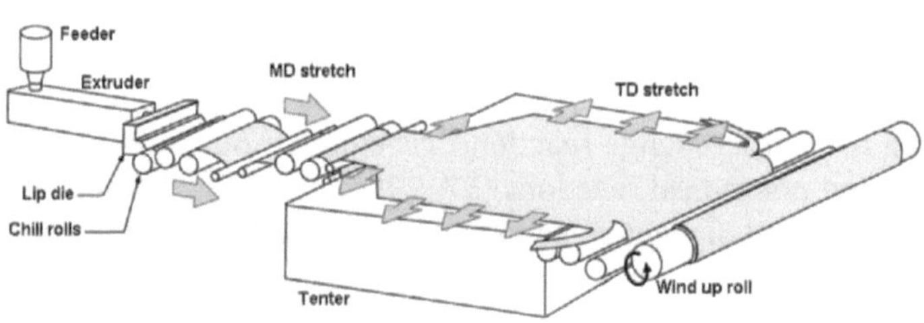

Figure 11.2 Biaxially oriented extrusion cast film machine and typical temperature conditions used during biaxial orientation film casting. Reproduced from ref. 71 with permission of Elsevier.

11.3.2 Polymerization Methods

11.3.2.1 Direct polycondensation

PLA can be produced from lactic acid by direct condensation in three steps of removal of the free water content, oligomer polycondensation, and melt polycondensation of high-molecular-weight PLA. Due to the equilibrium of lactic acid and water, low amount of oligomers can also formed in the first step. This stage can be carried out in evaporators, such as falling film evaporators. Low-molecular-weight oligomers are produced in the second step. The reaction rate is significantly controlled by temperature and the catalyst used [72]. The loss of lactic acid due to entrainment must be avoided using a reflux condenser or a rectification column. Due to the high viscosity of the melt, water removal is critical in the last step. To shift the polycondensation reaction toward the product side, the water formed in the reaction should be removed efficiently. Thus, mass and heat transfers are predominant in this stage. The catalysts used for direct polycondensation are strong acids, titanium (IV) butoxide [73], metal-based catalysts such as Ge, Sb, Zn, Fe, Al, Ti, and Sn (e.g., SnO) [74], sodium carbonate, calcium carbonate, and lanthanum oxide [75]. Depends on the polymerization conditions used, PLA polymers with molecular weights in the range of 40,000–200,000 g/mol can be obtained [73–76].

11.3.2.2 Conventional production via ring opening polymerization

PLA has a unique position in the field of bio-based products. PLA is processable with most conventional processing equipment. The main use of PLA was initially in medical applications, which is still an important one [77, 78]. Since then, there has also been an increase in its large-scale industrial production, for example in packaging [38, 79–82]. The end-of-life options of PLA have been brought into sustainability discussions. This can be seen in the form of a number of suggested approaches on how to deal with waste materials from the polymerization process, the manufacturing process of end products, or the end product after its use. Conversion of PLA into lower-molecular-weight polymers, as well as the complete hydrolysis of the polymer into lactic acid for use as new building blocks for either bio solvents or polymers, has been described [83–85].

Ring opening polymerization of L-lactide is generally the most preferred route for preparing high-molecular-weight PLA due to the possibility of an accurate control of the process. Lactide polymerization can be carried out using melt polymerization, bulk polymerization, solution polymerization, and suspension polymerization techniques, but melt polymerization is the most simple and reproducible method [86]. The reaction has been carried out in a reaction vessel with an agitator [87], a plug-flow reactor equipped with agitation blades [88], a static mixer for continuous ROP of lactide [89] and finally by reactive extrusion[90]. A large number of catalysts such as carboxylates and alkoxides of Sn [91, 92] and Al [93, 94], metal-free catalysis [95], catalysts based on Ca [96], Fe [97], Mg [98], Zn [99], Pt [100], and other metal complexes [101–103] have been employed in the ROP of lactide. Among these catalysts, stannous 2-ethylhexanoate (tinoctanoate) is the most used, which has shown to give a controlled and living polymerization via coordination–insertion mechanism (Scheme 11.1). The proposed polymerization mechanism by Stridsberg et al. [104] is by converting the stannous 2-ethylhexanoate to a stannous alkoxide in a reaction with a hydroxyl-bearing compound such as water or alcohol, followed by polymerization on the tin–oxygen bond of the alkoxide ligand.

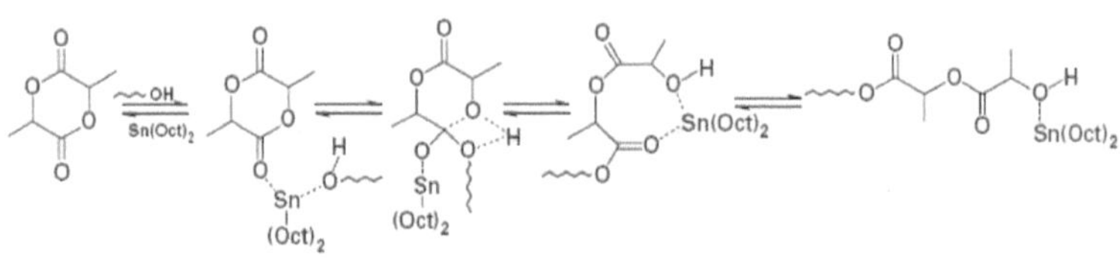

Scheme 11.1 Coordination insertion mechanism of lactide to PLA.

Due to the low melt stability of PLA, its processing is more demanding than that of commodity plastics and post-polymerization treatments are required to avoid degradations. This can be achieved by use of processing aids, additives, and catalyst deactivators. Deactivators used are phosphorous containing compounds [105], antioxidants [106], acrylic acid derivatives [107], and organic peroxides [108]. This process is generally performed in combination with a lactide removal process at low pressure and at high temperatures [109]. Residual lactide can also be reduced by applying solid-state polymerization below the melting point of PLA [110].

11.3.3 Reactive Extrusion

PLA properties and price can be changed by combining PLA with other biocompatible polymers, fillers, or reinforcements [111, 112].

Alternatively, PLA can be modified by adding plasticizers to obtain more flexible materials [113]. In order to make production of PLA economically viable, the use of a continuous single-stage process is highly desirable. Therefore a reactive extrusion process is a very attractive approach. Jacobsen et al. [90] reported a process for the ROP of L-lactide using single-stage reactive extrusion technology for the first time. They developed the processing and screw concept to polymerize L-lactide into PLA using a closely intermeshing co-rotating twin-screw extruder. They studied the influence of main processing parameters such as screw speed, throughput rate, and die resistance, resulting in a change of extruder head pressure on number average molecular weight M_n, molecular weight distribution MWD and conversion rate.

Polymerization in a mini-extruder using $Sn(Oct)_2$ as catalyst, and introducing triphenylphosphine (TPP) as a co-catalyst in an equimolar complex with $Sn(Oct)_2$ [90] is reported. It is believed that TPP enhances the polymerization rate and delays the occurrence of undesirable back-biting reactions [114, 115]. A 0.15 M solution of the equimolar $Sn(Oct)_2$/TTP has been prepared by dilution in freshly dried toluene. ULTRANOX 626 was also used as the stabilizer. The reactive extrusion polymerization was carried out using 2 kg of lactide, previously mixed with 0.5 wt% of ULTRANOX 626 stabilizer and the catalyst solution. The prepared mixture was fed to the twin-screw extruder using a constantly nitrogen purged gravimetric feeding unit. The lactide was fed into a closely intermeshing co-rotating twin-screw extruder of BERSTORFF (ZE 25), having a screw diameter of 25 mm and a L/D-ratio of 48, with 12 Sections equipped with a static mixer at the tip of the screw (Fig. 11.3). The polymerization progressed at a temperature of about 185°C. The polymer was extruded through a strand die, cooled by a constant flow of air on a take-off unit and pelletized. They also carried out comparing bulk batch polymerization at 180°C in dried and vacuum evacuated 25 mL glass ampoules using 5 g of recrystallized lactide and the catalyst solution. It has to be noted that the state of equilibrium due to the enhanced pressure level and the thermo-mechanical energy input does not necessarily have to be the same as in similar polymerization experiments conducted in glass ampoules under the same starting conditions.

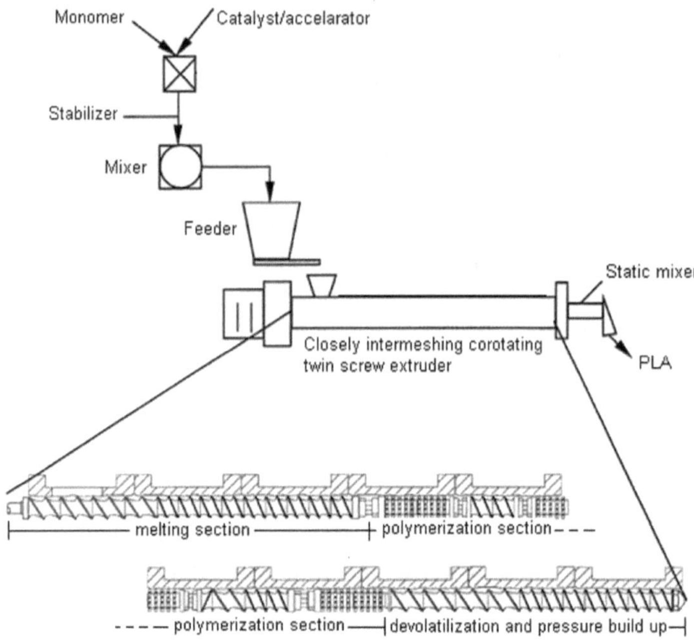

Figure 11.3 Processing and screw concept used for reactive extrusion polymerization of polylactide. Reproduced from ref. 90 with permission of Elsevier.

The extruder employed in their work contained a modular screw and consisted of several barrel parts with the screws rotation speed of 100 rpm. The solid monomer was pushed into the extruder through the first two barrel cold sections. Temperature was increased in the third barrel section to melt lactide and to disperse the catalyst and stabilizing agent within the low viscous material. Then the whole twin-screw extruder until the screw tips was heated to a temperature of 180°C for oligomerization to occur. Three combinations of shearing and kneading elements were included. Each of these combinations included a large kneading element, which enhanced the thermal polymerization conditions by energy dissipation, followed by elements working similar to counter rotating scissor blades. These elements cut the material flow in ever-new single flows, creating new neighborhoods within the melt. In the first combination a counter conveying kneading element, in the second combination a counter conveying standard screw element and finally a blister was chosen as flow restriction element. After the last pressure reduction, the melt entered a partially filled zone, using screw elements with a higher screw volume to devolatilize remaining monomer or LMW components from the polymer. Within the last two barrel-sections, screw elements with a smaller helix angle provided a pressure build-up to push the polymer through the static mixer in front of the screw tips and through the die. In this section the final polymerization occurred under increased pressure of 30–40 bar. Within the static mixer the stabilizer system ULTRANOX 626 (bis-(2,3-di-*t*-butylphenyl) pentaerythritol diphosphite) was dispersed within the polymer. The extruded high viscous polymer was finally cooled down on an air-cooled conveyer belt and pelletized afterwards.

Jacobsen et al. [90] compared the properties of PLA samples produced in a glass ampoule and the single-stage continuous reactive extrusion (Table 11.2). The resulting conversions, 98.5% and 99%, show that in both cases the polymerization reaction has

Table 11.2 Comparison of two types of poly-L-lactide polymerized in glass ampoule bulk batch polymerization technology and using a single-stage reactive extrusion polymerization process, both catalyzed with an equimolar $Sn(Oct)_2 \cdot TPP$ complex with an initial monomer to tin molar ratio of 5000 at 180°C

Time (min)	Conversion[b]	M_w/M_n[a]	$M_n \times 10^{-3}$ [a]	
40	98.5	1.9	246	Glass ampoule
7	99.0	1.8	91.1	Reactive extrusion

Source: Reproduced from ref. 90 with permission of Elsevier.

[a]Determined by GPC in $CHCl_3$ eluent at 35°C relative to polystyrene calibration.

[b]Determined by FTIR spectroscopy.

been finished. However, the time required to reach this conversion for the bulk batch polymerization in glass ampoule is approximately 40 min compared with only 7–8 min in the reactive extrusion process. This difference can be explained by the fact that as soon as a high molecular weight, i.e., a high melt viscosity is reached, the speed of the reaction is limited not by the reactivity of the chemicals but by the diffusion of the monomer or LMW compounds inside the high viscous melt to find a reactive partner. This physical movement is limited to the Brownian movement in the glass ampoule, but is supported in the twin-screw extruder by the mixing elements and by the shearing of the polymer inside the intermeshing zone [90]. The LMW distribution values smaller than 2 in both cases show that intermolecular transesterification reactions have been very limited, which is due to the use of triphenylphosphine as a cocatalyst, as well as due to the use of ULTRANOX 626 as stabilizing agent. The influence of processing parameters during reactive extrusion polymerization on the resulting polymers was also reported and is shown in Table 11.3.

Table 11.3 Influence of different processing parameters during reactive extrusion polymerization on the resulting molecular polymer parameters

Screw speed (rpm)	Mass flow rate (kg/h)	Extruder head pressure (bar)	$M_n \times 10^{-3}$	M_w/M_n	Conversion (%)
50	0.75	~30	77.0	1.80	99
50	1.00	~30	81.0	1.70	97
50	1.25	~30	66.0	2.10	94
50	1.00	~35	83.0	1.70	94
100	1.00	~35	79.0	1.70	90
200	1.00	~35	81.0	1.80	87
100	1.00	~40	79.0	1.70	94
100	1.00	~60	84.0	1.70	94
100	1.00	~85	86.0	1.80	96
100	1.00	~110	89.5	1.75	92

Source: Reproduced from ref. 90 with permission of Elsevier.

As shown in Table 11.3, the mass flow rate was varied from 0.75 to 1 kg/h and further to 1.25 kg/h, while all other processing parameters were kept constant. It has been shown that conversion decreased from 99% at a mass flow rate of 0.75 kg/h over 97% to only 94% at the highest flow rate of 1.25 kg/h. This fact can be explained by the effective residence time of the material inside the extruder. The molecular weight increased by 4000 and the molecular weight distribution decreased slightly, while increasing the mass flow rate from 0.75 to 1 kg/h, which can be explained by already starting degradation reactions at the experiment with 0.75 kg/h mass flow rate. It has been reported that raising the screw speed from 50 to 100 rpm and further up to 200 rpm reduced the degree of conversion. The reason for this behavior is the same, as the remaining residence time inside the twin-screw extruder decreases with increasing screw speed, which results in a reaction not finished at the extruder die. Increased screw speed led to the reduced molecular weight and increased polydispersity by reducing the polymerization time and increasing the rate of reaction, favoring degradation and side reactions to enhance the polydispersity. It was shown that the increase in the extruder head pressure influenced the state of reaction equilibrium, which was shifted to the polymer side for high-pressure levels. Experimental data showed an increase in number average molecular weight from 80,000 g/mol to nearly 90,000 g/mol, while increasing the extruder head pressure from 40 to 110 bar. The polydispersity showed no tendency to change with pressure, while the conversion increased slightly [90]. Figure 11.4 shows the molecular properties of PLLA samples polymerized in a reactive extrusion process as a function of mass flow rate through the extruder [116].

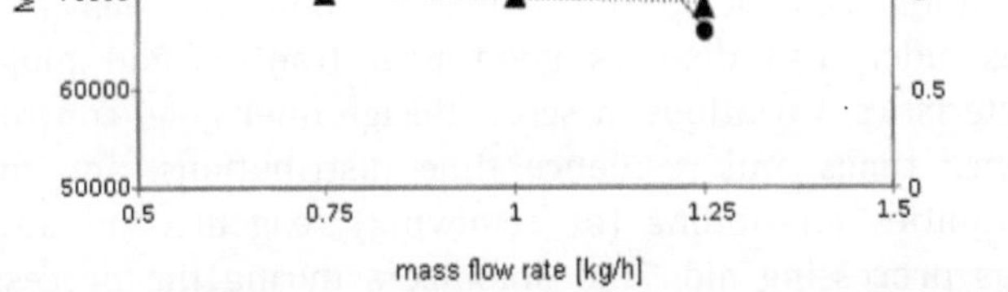

Figure 11.4 Molecular properties of PLLA samples polymerized in a reactive extrusion process, (●) number average molecular weight (M_n), (▲) monomer conversion (MC), and (■) polydispersity (PD) as a function of mass flow rate through the extruder. Reproduced from ref. 116 with permission of Elsevier.

Puaux et al. [117] studied the bulk polymerization process of L-lactide initiated by stannous octoate in a laboratory batch reactor. They conducted the polymerization reactions in a Haake Rheocord internal mixer with a R600 chamber. Their set-up allowed the monitoring of the torque and the control of the reaction temperature. They reported that the bulk polymerization process can be conducted at reactions times of the same order of magnitude as the residence times of the melts in twin-screw extruders. This permits to use these apparatus as polymerization reactors, taking the benefits of the reactive extrusion technique advantages. They suggested that the temperature has a strong effect on the monomer conversion. At a temperature of 195°C a monomer conversion level about 90% was achieved in a reaction time around 10 min. It has been shown that the polymer molecular weight was increasing faster on the first interval of reaction time, due to the relatively high propagation rates induced by the higher concentration level of monomer in the mixture. At temperatures over 180°C, the molecular weights presented a decreasing trend on the last time interval, due to the process reversibility. They reported that M_n and M_w increased with the monomer to catalyst ratio and with temperature.

A range of processes have now been explored with respect to starch modification by REX and several reviews that relate to the area have been published [118, 119]. Chemical changes to starch or starch-based materials during extrusion processing of food products have also been widely investigated and reviewed [120, 121]. Twin-screw extruders, in particular, can be used to produce modified starches in a continuous process with a more consistent product quality. Thus, with the use of REX, starch modification can be performed in a homogeneous medium. The extruder also displays good heat transfer and plug flow characteristics. Variations in screw design offer good control over residence times and residence time distributions, and provide opportunities for adding (or removing) reagents and additives such as processing aids and stabilizers during the process. For a general description of reactive extrusion processes and a detailed summary of the important features in designing reactive extrusion experiments, refer to the literature [122–124].

New plasticization ways based on low-molecular-weight plasticizers from citrates family have been investigated to improve the ductility of PLA [125]. Hassouna et al. [125] studied the grafting reactions between anhydride-grafted PLA (MAG-PLA) copolymer with hydroxyl-functionalized citrate plasticizer, i.e., tributyl citrate (TbC), carried out through reactive extrusion. All melt (reactive) blending were performed on a Leistritz co-rotating intermeshing twin-screw extruder ($L/D = 45$). The temperature profile was 160°C/180°C/180°C/180°C/180°C/180°C/180°C/180°C/180°C/180°C from the feed throat to the die, and the melt temperature was 181°C. The employed screw speed was 80 rpm. A two-hole filament die, 3 mm in diameter and equipped with a cooling sleeve, was assembled to the extruder. Maleation of PLA (MAG-PLA) was carried out by reactive extrusion.

Schmack et al. [126] spun PLA fibers through the reactive extrusion polymerization of L-lactide (92 wt%) and *meso*-lactide (8 wt%). In many potential textile technological applications (e.g., for nonwoven materials) the fiber forming process is of general importance. An effective polymer synthesis requires also an effective spinning process to reduce the still high cost of the PLA fibers compared with those of established synthetic fibers.

Takamura et al. [127] cross-linked PLLA using various types of peroxides under constant mole ratios of peroxide-derived radicals to PLLA during reactive extrusion in a single-screw extruder (D = 20 mm, $L/D = 25$) with a fixed temperature profile for the extruder (Zone 1/Zone 2/Zone 3/Die = 180 °C/185°C/190°C/190°C).

The ROP of lactide within a twin screw extruder process in Section 11.3.3 can be transferred under certain circumstances to similar polymers and copolymers with comparable monomers or comonomers. They include lactones such as ε-caprolactone, and dilactones such as glycolide. This process can be transferred to other monomers if the kinetics of reaction is fast; the equilibrium of reaction is close to complete monomer conversion, to ensure the production of high-molecular-weight polymers; side and degradation reactions can be suppressed by the use of suitable stabilizer systems. To generate copolymers from different monomers, the reaction kinetics of both monomers must be similar and can be enhanced using the same catalytic system, and both monomers must be consumed completely or nearly completely during the copolymerization reaction. These additional boundary conditions restrict the number of possible copolymers further, especially as the reaction kinetics of both monomers is different in most cases. An alternative to this problem is the possibility to incorporate pre-oligomerized, hydroxy-terminated blocks as starter molecules to a second type of monomer. Thus, the homopolymerization of this second monomer will start at both ends of the oligomers and incorporate them as blocks of predefined size into the polymer. This technology has been described by Stevels et al. (1996) [114] using large amounts of prepolymers and $Sn(Oct)_2$ as catalyst. Jacobsen et al. [116] showed that small amounts (1 and 2 wt%) of pre-oligomerized poly-ε-caprolactone blocks (M_n 4000 g mol^{-1}) have no significant influence on the propagation of the polymerization reaction of PLA during reactive extrusion. They conducted the reactive extrusion using a barrel temperature of 180°C, a throughput rate of 1 kg h^{-1} and a screw speed of 100 rpm. The $Sn(Oct)_2$/TPP catalytic system was used with a molar monomer to catalyst ratio of 5000. They reported that after exiting the extruder, all materials reached the equilibrium of reaction and the monomer conversion showed no significant indication of a negative influence of the oligomers on the reaction kinetics (Table 11.4). However, the average molecular weight was shown to decrease with increasing content of caprolactone blocks, which can be explained by the increased amount of potential starter molecules for the polymerization reaction. They also reported similar results when prepolymerized hydroxyl terminated polyethyleneglycol (PEG) was used in 2–5 wt% concentrations. The reactive extrusion was carried out using a barrel temperature of 185°C, a throughput rate of 1 kg h^{-1} and a screw speed of 50 rpm.

Table 11.4 Influence of the addition of hydroxyl terminated oligo-ε-caprolactone (OCL, M_n = 4000 g/mol, δ = 1.5) and polyethylene glycol (PEG, M_n = 1500 g/mol) on the polymerization of PLLA in a reactive extrusion process

Comonomer	Content (%)	Conversion (%)	Time (min)	Temperature (°C)	$M_n \times 10^{-3}$	PD
1 OCL	0	94	~7	180	87.0	1.7
2 OCL	1	97	~7	180	66.6	1.3
3 OCL	2	95	~7	180	43.8	1.4
4 PEG	0	97	~6	180	93.3	1.9
5 PEG	2	99	~6	180	83.3	1.6
6 PEG	5	99	~6	180	70.6	1.4

Source: Reproduced from ref. 116 with permission of Elsevier.

Gallos et al. [6] synthesized stereocomplexed polylactide using reactive extrusion process. The stereocomplexed PLA obtained in a two-step polymerization of L-lactide and D-lactide process to yield poly-L,D-lactic acid multiblocks. In the first step, DSM Micro 15 extruder was employed at 185°C with a constant nitrogen flow. It was filled with a mixture of 20 g of L-lactide, tin(II)octanoate and triphenylphosphine. The tin(II)octanoate amount was a 1/5000 molar ratio versus the L-lactide amount. The tin(II)octanoate and the triphenylphosphine were in an equimolar ratio. The polymerization conversion rate was followed by torque measurement. The polymerization process was ended when the torque stabilizes on a maximum level. In the second step, when the torque stabilizes, the half part of the PLLA formed was removed from the extruder and replaced by a mixture of 10 g of D-lactide mixed, tin(II)octanoate, and triphenylphosphine. Finally, 0.75 wt% a tropolone was added to deactivate catalyst when torque was stabilized. This reactive extrusion process permits to make PLA matrix able to form stereocomplex crystallinity and to incorporate flame-retardant additives (i.e., combination of ammonium polyphosphate, melamine, and organoclay) during synthesis at the same time. This material was molded keeping its stereocomplex crystallinity.

11.4 PLA Nanocomposites

11.4.1 General Modifications

PLA-based nanostructured materials offer unique combinations of properties including biodegradability and thermoplastic processability that may be exploited in applications such as packaging of agricultural products, and disposable materials. PLA-based nanocomposites are a very important inclusion in the family of novel biomaterials and their history is not older than 10 years. By changing the nature of the nanofillers and the processing conditions, some of the physical properties of pure PLA such as mechanical, thermal, electrical, biodegradability, and so on have been improved significantly.

Generally, PLA nanocomposites exhibit moderate improvement in modulus, strength and thermal stability, decreased gas permeability, and rate of degradation. Lee et al. [128] reported that the modulus of the PLLA nanocomposites systematically increased with increasing montmorillonite (MMT) loading. They also claimed that the crystallinity and the T_g of PLLA nanocomposites were lower than neat PLLA. In the case of PLLA/multi-walled carbon nanotube (MWCNT) nanocomposites, it was observed that Young's modulus and stiffness increased, which eventually led to brittle fracture, with MWCNT loading in the nanocomposite films compared to pure PLLA films. Sinha Ray et al. [129] reported that there was a significant increase in flexural modulus for nanocomposite prepared with 4 wt% of organoclay when compared to that of neat PLA. Bandyopadhyay et al. [130] reported that the thermal stability of PLA containing organically modified MMT nanoclay was improved. Paul et al. [131] prepared PLA layered silicate nanocomposites by melt intercalation in the presence of a stabilizer to decrease the possibility of matrix degradation. The degradation of PLA during processing (at 195°C) takes place even in the presence of antioxidant, and a 41.2% decrease in number-average molecular weight of PLA was observed compared to native PLA. An increase in the thermal stability under oxidative conditions was found, and it was suggested that a physical barrier between the polymer medium and the superficial zone of flame combustion may be generated due to the char formation.

11.4.2 Biodegradability

A major problem with the PLA matrix is the very slow rate of degradation. Sinha Ray et al. [129] studied the degradation of the PLA matrix in a compost environment. They reported that the degradability of the PLA component in PLA/qC13(OH)-mica4 (Dipolyoxyethylene alkyl(coco) methyl ammonium cation, CO-OP Chemicals, Japan) or PLA/qC16SAP4 (saponite hexadecyltributylphosphonium cation, CO-OP Chemicals, Japan) nanocomposite was enhanced significantly compared to neat PLA. On the other hand, the PLA component in PLA/C18MMT4 nanocomposite shows a slightly higher degradation rate compared to neat PLA. Any factor that affects the hydrolysis of the PLA matrix ultimately controls the degradation of PLA. The incorporation of OMLS fillers into the PLA matrix results in a small reduction in the molecular weight of the matrix. It is well known that PLA of relatively LMW may show higher rates of enzymatic degradation because of the high concentration of accessible chain end groups [129]. However, in these cases, the rate of molecular weight change of pure PLA and PLA in various nanocomposites is almost the same. So, the initial molecular weight is not a main factor here for controlling the degradability of nanocomposites. Another factor that controls degradability of PLA in nanocomposites is the different degree of dispersion of silicate layers in the polymer matrix, which actually depends on the nature of the surfactant used to modify the clay surface. Paul et al. [132] investigated the hydrolytic degradation of composites based on the same amounts of Cloisite Na, Cloisite 25A, and Cloisite 30B over a period of more than 5 months, and directly compared them to pure PLA. The results indicated that the fastest hydrolysis, leading to an increase of the crystallinity of the PLA matrix, is found for the Cloisite Na-based composite. It was concluded that the relative hydrophilicity of the clay play determining roles in the hydrolytic degradation process and incorporation of the more hydrophilic fillers result in higher degradation rates.

11.4.3 Melt Rheology

Measurements of melt rheological properties are not only important to understand the processability of nanocomposites but are also used in determining the strength of polymer–filler interfacial interactions and the structure–property relationships in nanocomposites. In the dynamic complex viscosity, η^*, master curves for pure PLA and nanocomposites, based on linear dynamic oscillatory shear measurements, at low shear rate, $\alpha_T w$, region (<10 rad/s), pure PLA exhibited almost Newtonian behavior while all the nanocomposites showed very strong shear thinning tendencies. On the other hand, M_w and PDI of pure PLA and various nanocomposites were almost the same; thus, the high viscosities of PLACNs were explained by the flow restrictions of polymer chains in the molten state due to the presence of MMT particles [133]. These observations suggest that the silicate layers are strongly oriented toward the flow direction at high shear rates and that of pure polymer dominates shear thinning behavior at high shear rates [134].

11.5 Polylactic Acid Properties

11.5.1 Chemical Structure

PLA is a chiral polymer containing asymmetric carbon atoms with a helical conformation. Polymers that have stereo centers in the repeating unit can exhibit two structures of maximum order, that is, isotactic and syndiotactic. These stereoregular polymers are typically crystalline. Isotactic polymers are used in a wide range of applications. In contrast, syndiotactic polymers have limited applications mainly due to their hard productivity and inherently alternating stereochemistry. Syndiotactic PLA is expected to be a versatile polymer with controllable stereochemistry. Syndiotactic PLA can be synthesized by syndio-specific polymerization using *meso*-lactide [135]. Since lactide have three different stereoisomers, its corresponding polymers may have different chain configurations. PLA properties are controlled by the ratio of D- to L-enantiomers used and the sequence of arrangement of the enantiomers in the polymers. PLA with the L-isomeric content higher than 90% tends to crystallize. The melting temperature (T_m), glass transition temperature (T_g), and crystallinity of PLA decrease with decreasing optical purity [39–41, 136]. The chain structure, configuration, tacticity, conformation, and the frustrated structure of PLA are parameters that govern the properties of this polymer. Francois and his co-workers [137] were the first group to report their investigations on the configuration of PLLA, PLLDA, and the stereocomplex blend using ^{1}H and ^{13}C NMR spectroscopy. Thakur et al. [138, 139] also used ^{1}H and ^{13}C NMR to determine the structure of PLA. The stereochemical microstructure of polymers can be controlled by the monomer composition or the stereochemical preference related to the initiating/catalytic system [140]. Tsuji [141] and Karst et al. [142] studied the influence of PLLA and PDLA blends on the resistance to hydrolysis of PLA. They reported that the 50/50 PLLA/PDLA blend had the greatest resistance to hydrolysis. They suggested that the 50/50 blend had the greatest resistance to hydrolysis due to its stronger hydrogen bonding and dipole–dipole interactions compared to pure PLLA or PDLA. Karst and Yang [143] reported their findings on the influence of L-lactide and D-lactide fraction and their sequences in blocks or random copolymers on the resistance to hydrolysis. It was found that PLA copolymers had less resistance to hydrolysis than PLLA/PDLA polyblends at the same percentages of lactide isomers.

11.5.2 Thermal Properties

PLA can be amorphous or semicrystalline depending on its stereochemical structure and thermal history. Amorphous PLLDA copolymer shows a faster degradation rate than its stereoregular counterparts, which makes it the preferred candidate for drug delivery vehicles and low strength scaffolding material for tissue regeneration [143, 144]. Semicrystalline PLLA consists of crystalline phases and amorphous phases, and tie molecules that appear in both phases. So, the amorphous phase consists of the rigid amorphous fraction (RAF) [145] and the mobile amorphous phase (MAP) [146]. Arnoult et al. [145] suggested that the crystalline phase and RAF do not significantly modify the fragility parameters of PLLA. Wunderlich [147] proposed that the RAF may not remain rigid until the polymer melting point and may decrease or disappear gradually above the T_g of MAP. Pyda and Wunderlich [148] reported their studies on the glass transition of amorphous and semicrystalline PLA using differential scanning calorimetry (DSC), and temperature-modulated DSC (TMDSC). The glass transition temperature shifted to higher temperatures with increasing extent of annealing. The glass transition of the RAF was determined from changes of apparent heat capacity between the MAP glass transition temperature and the polymer melting temperature. They suggested that RAF was responsible for the nonlinearity of the change in heat capacity. Picciochi et al. [149] reported a decrease in T_g correlated with the changes in the thickness of RAF. They showed that increasing the crystallization temperature resulted in decreased T_g because of the corresponding increase in the thickness of the RAF.

Like other semi-crystalline polymers, the glass transition of PLA is influenced by physical aging [150], crystallinity, morphology, and impurities [151, 152]. Other properties of PLA such as the mechanical strength also depend on the morphology, crystallinity, and orientation. It was found by Wong et al. [153] that the orientation of the crystalline phase was always higher than that of the amorphous phase.

Semicrystalline PLA has higher strength and modulus comparable to polypropylene and polystyrene. The crystalline structure of PLA has been well investigated [154–158]. PLLA can crystallize into three forms (α, β, and γ), generally referred to as polymorphism. The α-structure with a T_m of 185°C is more stable than the β-structure with a T_m of 175°C [149]. The β-form crystals are generally prepared by stretching their α-counterparts at high temperature and high draw ratio (DR). Kanamoto and coworkers

[156–158] reported that α to β transition occurred more rapidly with an increase in the DR or extrusion pressure. It was also reported that the α to β crystal transition proceeded efficiently at temperatures between 130 and 140°C. Besides, the β-form formation increased as the molecular weight increased. The detailed mechanism of formation and thermal properties of the γ-form crystals have not been investigated yet.

Pure PLLA and PDLA have the same properties, that is, a glass transition temperature (T_g) between 50 and 70°C, a melting temperature (T_m) between 170 and 190°C, and a crystallinity of around 35% [159–162]. Syndiotactic PDLLA has a lower T_g and T_m at 34 and 155°C, respectively [163]. Crystallinity of PLA depends on the polymer molecular weight, thermal history, and purity. A 50:50 blend of PLLA and PDLA results in the formation of a stereocomplex crystalline structure with a melting temperature reaching 230°C [161]. The melting enthalpy of PLLA is reported to be 93.6 J/g by Fisher et al. [164]. A value of 140 J/g is also reported by Loomis et al. [165]. Figure 11.5 shows a typical DSC analysis of amorphous PDLLA and PLLA samples [166]. In the 65–90°C heating interval, PDLLA exhibits an aging peak with an enthalpy of 8 J/g. On the other hand, the amorphous PLLA shows a crystallization peak in the range of 100–160°C centered at about 132°C, followed by a melting peak at 182°C. It has been reported that a commercial PLLA (L210 from Boehringer, Ingelheim, Germany) has a T_g and T_m at 73°C and 192°C, respectively. Table 11.5 lists T_g, T_m, heat capacity (ΔC_p), and enthalpy of fusion (ΔH_m) of various molecular weight PLLA samples, (93.6 J/g) for PLLA crystals having an infinite size [164].

Table 11.5 Glass transition temperature (T_g), melting point (T_m), enthalpy of fusion (ΔH_m), and heat capacity (ΔC_p) for different molecular weight PLLA samples

M_v (g/mol)	T_g(°C)	T_m (°C)	ΔH_m (J/g)	ΔC_p (J/g·K)
5,300	55	156	61	0.60
20,000	59	174	68	0.54
691,000	64	186	59	0.54

Note: Reproduced from ref. 169 with permission from Elsevier.

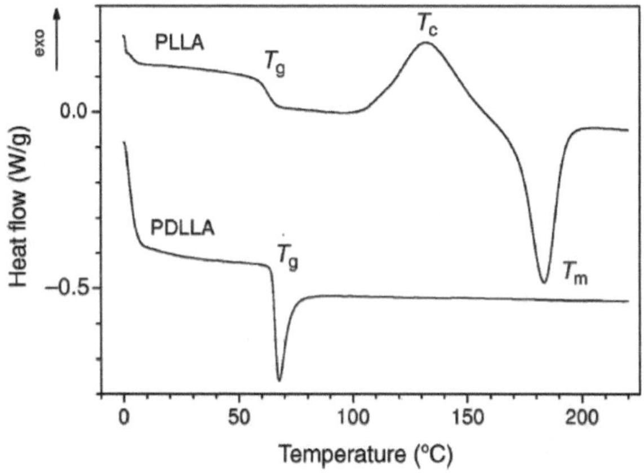

Figure 11.5 DSC thermograms of amorphous PLLA and PDLLA at a heating rate of 10°C/min.

All the products display a high crystallinity between 50% and 73%. Melting temperature is proportional to the molecular weight according to the classical Flory equation:

$$\frac{1}{T_{\mathrm{m}}} = \frac{1}{T_{\mathrm{m}}^{*}} - 2R\frac{M_0}{\Delta H_{\mathrm{mu}} M_{\mathrm{n}}},$$

where T_m is the melting temperature, T_m^* is the melting temperature at infinite molecular weight, R is the gas constant, ΔH_{mu} is the heat of fusion per mole of the repeating unit, M_0 the molecular weight of the repeating unit, and M_n is the number-average molecular weight.

As reported by Migliaresi et al. [167], thermal history has a significant influence on the thermal properties of similar molecular weight PLAs. As shown in Fig. 11.6, the crystallinity increases with annealing time and with decreasing molecular weight.

T_g values of both PLLA and PDLLA increase up to about 20 kDa molar mass and reach an almost threshold value thereinafter. PLLA displays higher T_g than PDLLA due to the hindering effect exerted on the amorphous chain mobility by the crystalline regions. While the calculated theoretical T_m values for pure PLLA are reported to be 215, 207, and 205°C by different authors, the maximum practical obtainable T_m for pure PLLA is around 190°C [168, 169]. It has been reported by Perego et al. [44] that crystallinity of PLLA of different viscometric molecular weights increased from 3–13% to 42–65%, depending on the molecular weight.

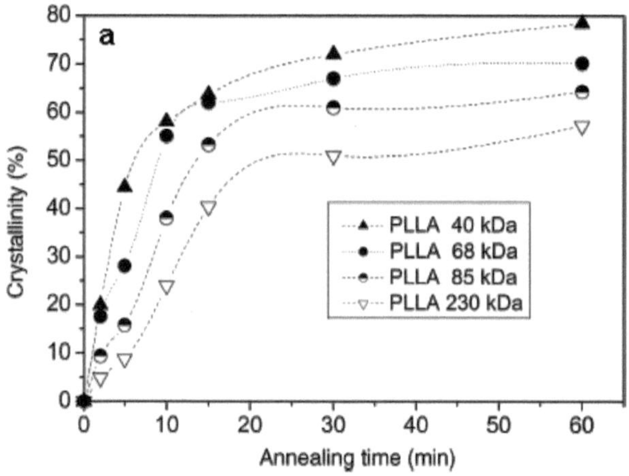

Figure 11.6 Effect of annealing time on crystallinity of compression molded PLLA at 160°C. Adapted from ref. 167 with permission from Wiley.

11.5.3 Optical Properties

The food packaging industry plays a dominant role in PLA applications; thus, this material has to provide efficient barriers against light, water vapor, atmospheric gases, and volatile organic compounds (VOCs), preventing food degradation and preserving aromas and flavors [170]. For instance, a good packaging should prevent the photo degradation of food components during storage by absorption and reflection of the incident light [171]. This can be achieved by choosing the appropriate packaging material and the addition of the right additives or stabilizers to the packaging material. The photochemical degradation of plastics is mainly due to the high-energy UV-B (315–280 nm) radiation [39]. Figure 11.7 shows that PLA transmits about 85% at 250 nm and 95% at 300 nm of the UV radiation. However, no UV radiation transmission was seen in the 190–220 nm wavelength region. PS and cellophane transmit less radiation in the UV range and PET does not transmit any light in this wavelength range. Several important parameters are used to characterize visible light transmission and color of plastics, such as transparency (ASTM D1746-03 Standard Test Method for Transparency of Plastic Sheeting) and degree of yellowness (ASTM D6290-05 Standard Test Method for Color Determination of Plastic Pellets), which are based on the absorbance or transmission of UV–Vis light [172]. The average transparency of food packaging films is around 95%. PLA, PS, and LDPE have the same degree of yellowness, while cellophane and PET have higher values [39]. The formation of a yellowish color in PLLA during processing above its melting temperature is a drawback in the production of PLLA products by melt processing techniques, such as extrusion. This is also one of the most critical problems during the melt processing of aromatic polyesters, including PET [173].

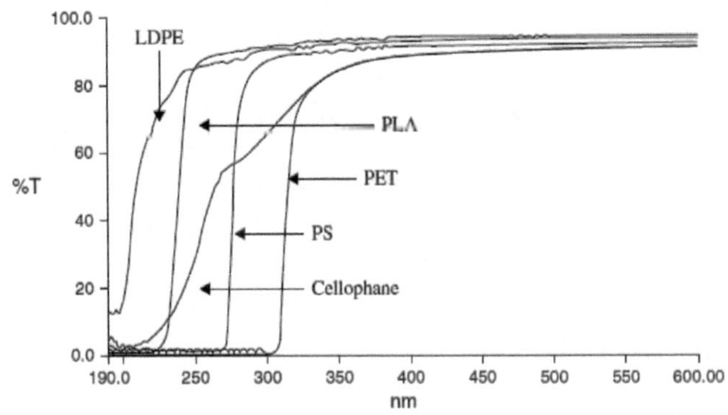

Figure 11.7 Transmission against radiation wavelength for PLLA, PS, LDPE, PET, and cellophane films. Adapted from ref. 39 with permission from Wiley.

Wang et al. [174] used UV–Vis spectroscopy to monitor the PLLA degradation during extrusion. A red shift in the sample λ_{max} due to the formation of conjugated double bonds (chromophoric groups) resulting from thermal degradation was identified [174]. It was concluded that pyrolytic elimination was responsible for the color formation and molecular weight reduction in dry PLLA; while an additional hydrolysis mechanism in moist PLLA only reduced molar mass but did not contribute to a change in UV–Vis absorption. Refractive index is a fundamental physical property of a substance, which shows how much the speed of light is reduced inside the medium. Polymeric materials have refractive indices depend on their building units. PLA contains enantiomers of both L- and D-lactic acid. Hutchinson et al. [175] did not find significant differences (1.499 to 1.448) in the refractive indices between samples of PLA with different enantiomeric compositions. The specific optical rotation of a pure material is an intrinsic property of that material at a given wavelength and temperature when dissolved in a particular solvent. The specific optical rotation $[\alpha]$ of PLLA and PDLA polymers was measured in chloroform at a concentration of 1 g/dL at 25°C using a polarimeter and wavelength of 589 nm. The values of $[\alpha]$ for PDLA and PLA were approximately 150° and –150°, respectively [176]. FTIR has been used to identify characteristic bands of PLA [177–179], its crystalline structure [180], and residual lactide in the polymer [181]. Paragkumar et al. [182] used ATR-FTIR to study the surface segregation and restructuring in PDLA and PLLA films of various thicknesses. It was reported that PDLA surface segregation and the surface restructuring of methyl side groups are influenced by the polymer film thickness. On the other hand, PLLA thin and clear films undergo surface conformational changes upon solvent treatment with organic solvents. The solvent-treated surface of PLLA becomes hazy and milky white and its hydrophobicity increases compared to untreated surfaces.

11.5.4 Degradation

Degradation of PLA proceeds via water diffusion in the free volume amorphous phase and consequent hydrolysis. The degradation kinetics of PLLA is largely affected by factors such as molecular weight, surface/volume ratio, purity, chain orientation, and crystallinity. Migliaresi et al. [183] reported that the crystallinity of PLLA samples increased during degradation, which slowed down the hydrolytic attack on the compact stable crystalline regions. This is a drawback for the biomedical use of PLLA as an implant material, which can last in the implant site for a very long time [184]. The influence of molecular weight on PLA degradation has been well-investigated [185–187]. It has been reported that the degradation rates of PLA films increased with a decrease in molecular weight and an increase in branching. PLA is stable under typical end use and storage conditions at ambient temperature. However, when it is exposed to high temperature (>60°C) and high humidity (>80% RH) conditions, such as in a composting environment, PLA disintegrates within 1 week to 1 month, followed by bacterial attack on the fragmented residues to give carbon dioxide and water. In the primary degradation phase, PLA undergoes chemical hydrolysis that is both temperature and humidity dependent, and does not involve any microorganism. As the lower molecular weights, reaching approximately 10,000 to 20,000 $gmol^{-1}$, microorganisms present in the soil begin to digest the lower-molecular-weight oligomers and lactic acid, producing carbon dioxide and water. In natural environments such as in soil or in water, the degradation of PLA proceeds slowly, which is a beneficial feature for agricultural/horticultural and geotextile applications.

11.5.5 Rheology

The melt rheology of PLA is distinct from other commodity plastics due to its chemical stability. Both hydrolysis and pyrolysis can occur at PLA processing temperatures. Therefore, the commercially available PLAs must be stabilized against thermal degradation. Tris (nonylphenyl)phosphite (TNPP), with concentration of 0.35 wt%, can be used as a degradation stabilizer [188]. Rheological measurements can provide direct evidence for the loss in molecular weight of PLA during processing. Molecular forces in polymeric materials can be classified as either bonded or nonbonded. Bonded forces are strictly intramolecular and typically consist of strong covalent bonds. Nonbonded forces are weaker interactions and may be either intramolecular or intermolecular. Examples of nonbonded forces include hydrogen bonding, dipole and multipole interactions, and van der Waals forces [189]. In good solvents, polymer chains adopt open conformations and expand. A related measure of polymer size is the radius of gyration, which is defined as the average of the distance from a segment to the chain's center of mass. The radius of gyration can also be considered as an average measure of the radius of a polymer coil. The bulk properties of PLA are greatly affected by the molecular weight of the polymer, structure, and the degree of crystallinity, influenced by the relative proportions of L- and D-lactide in the polymer backbone [190]. PLAs have physical properties useful in fibers, packaging, and other applications. A few works [191–193] have studied the rheological properties of PLA. The tensile tests on PLA showed a strong strain hardening behavior [193]. For the branched PLA samples, the zero shear viscosity and the elasticity increase with increasing branched content. The branched PLA shear thins more strongly than the linear material.

11.5.6 Mechanical Properties

Polylactic acid is a glossy, high modulus thermoplastic polymer with properties comparable to polystyrene (PS). Amorphous PDLLA is typically used for biomedical devices and in particular for slow drug release, while semicrystalline PLLA is selected for applications where higher mechanical and thermomechanical properties are required [194]. PLA usually exhibit an elastic modulus of 3000 to 4000 MPa, tensile strength of 50–70 MPa, flexural strength of 100 MPa, and flexural modulus of 4000–5000 MPa [115, 195–198]. Therefore, it can be good replacement for the conventional polymers in many applications, such as packaging, extruded and thermoformed containers, and soon. PLA has a T_g of about 60–70°C, and it is brittle at room temperature, with a low elongation at break (2–10%). The toughness can be improved by blending with other biodegradable polymers, plasticization, and rubber modification through copolymerization. Mechanical orientation upon drawing is also useful for improving the mechanical properties of PLA to convert the fragile polymer into a versatile and useful material. Due to the slow crystallization kinetic of PLA, specimens obtained by a typical injection molding process are generally almost amorphous. Tables 11.6 and 11.7 list the mechanical properties of PDLLA and PLLA injection molded specimens with different molecular weights [44]. It is obvious that the mechanical properties of PDLLA and amorphous PLLA are quite different. This difference is mainly related to the stereoregularity of the polymer chains in PLLA. Engelberg and Kohn [198] reported a 20% increase in tensile strength as a result of increase in molecular weight from 107,000 to 550,000 g/mol. In general, it seems that at high molecular weights, the variation of mechanical properties becomes less pronounced.

Extruded sheets of Nature Works amorphous PLA (grade 4060, 10% D-isomer) show tensile strength of 44 MPa, elongation at break of 7%, and elastic modulus of 2500 MPa, whereas the semicrystalline grade (grade 4032, 1.5% D-isomer) possesses a tensile strength of 54 MPa, elongation at break of 9%, and elastic modulus of 2600 MPa. Mitsui Fine Chemicals (grade Lacea H.100-E) PLLA with a molecular weight of 84,000 g/mol showed tensile strength of 66 MPa, elongation at break of 1.8%, and elastic modulus of 3300 MPa [199]. A lower value of elastic modulus of

1151 MPa is reported for PLLA manufactured by Purac, with a yield stress of 62 MPa and strain at break of 14.5% [200]. The values of notched Izod impact resistance of PLLA fall in the range of 2.0–3.0 kJ/m^2 for low-crystallinity (3–9%) and 3.0–7.0 kJ/m^2 for high-crystallinity (45–70%) samples [44]. PDLLA has lower values, ranging from 1.5 to 2.0 kJ/m^2. PLA impact resistance depends on the processing conditions (i.e., induced crystallinity and crystalline structure) and molecular weight. Charpy impact strength values of 8 and 15 kJ/m^2 have been reported for PDLLA having molecular weight of 78,000 and 108,000 g/mol, respectively [201], while PLLA is characterized by values ranging from 10 to 35 kJ/m^2 [202].

Table 11.6 Mechanical properties of PDLLA specimens with different molecular weights

	I	II	III
Viscosity average molecular weight (Da)	47500	75000	114000
Tensile properties			
Yield strength (MPa)	49	53	53
Tensile strength (MPa)	40	44	44
Yield elongation (%)	1.7	1.4	1.5
Elongation at break (%)	7.5	4.8	5.4
Elastic modulus (MPa)	3650	4050	3900
Flexural properties			
Flexural strength (MPa)	84	86	88
Maximum strain (%)	4.8	4.1	4.2
Elastic modulus (MPa)	3500	3550	3600
Impact resistance			
Izod, notched (kJ/m^2)	1.8	1.7	1.8
Izod, unnotched (kJ/m^2)	13.5	14.0	15.0
Heat resistance			
HDT (°C)	51	50	50
Vicat penetration (°C)	52	53	52
Hardness			
Rockwell hardness (scale H)	78	72	76

Source: Reproduced from ref. 44 with permission from John Wiley and Sons.

Table 11.7 Mechanical properties of PLLA specimens with different molecular weights

	I	II	III
Viscosity average molecular weight (Da)	23000	58000	67000
Melting point (°C)	178	179	181
Crystallinity (%)	9	9	3
Tensile properties			
Yield strength (MPa)	—	68	70
Tensile strength (MPa)	59	58	59
Yield elongation (%)	—	2.3	2.2
Elongation at break (%)	1.5	5.0	7.0
Elastic modulus (MPa)	3550	3750	3750
Flexural properties			
Flexural strength (MPa)	64	100	106
Maximum strain (%)	2.0	4.1	4.7
Elastic modulus (MPa)	3650	3600	3650
Impact resistance			
Izod, notched (kJ/m^2)	1.9	2.5	2.6
Izod, unnotched (kJ/m^2)	13.5	18.5	19.5
Heat resistance			
HDT (°C)	57	—	55
Vicat penetration (°C)	60	59	59
Hardness			
Rockwell hardness (scale H)	85	83	88

Source: Reproduced from ref. 44 with permission from John Wiley and Sons.

Rockwell hardness of PLA generally ranges between 70 and 90 according to scale H [44] and around 120 according to scale L. The Rockwell hardness of PLLA is affected very little by crystallinity and molecular weight variations. However, the effect of glass transition temperature (T_g) is more pronounced. Generally, PDLLA has lower hardness values, ranging from 72 to 78 H, because of its lower T_g [44].

The storage modulus of PLLA shows a slight increase before, and subsequently a rapid decrease at the temperature corresponding to α-relaxation. This behavior is attributed to the rearrangement of macromolecules, and relieving the stress generated during processing. This cold crystallization process takes place at a temperature of about 91°C [203]. Amorphous and crystalline PLLA show heat distortion temperature (HDT) values of 55 to 57°C and 60 to 66°C, respectively, and PDLLA has a HDT at around 50°C. Vicat penetration values for PDLLA and amorphous PLLA are 52 to 53°C and 59 to 60°C, respectively. On the other hand, this value for a crystalline PLLA sample is in the range of 157–165°C [44]. PLA mechanical properties at different temperatures are reported in Table 11.8 [44]. As shown, the flexural properties of types of PLA are quite different at different temperatures. Increasing temperature results in a significant drop of the flexural strength of PDLLA and amorphous PLLA, because of the lower glass transition and the lack of crystallinity. Only semicrystalline PLLA maintains its mechanical properties at this temperature.

Table 11.8 Influence of temperature on PLA flexural properties

	PDLLA	PLLA quenched	PLLA annealed
Viscosity average molecular weight (Da)	75000	58000	47000
Glass transition temperature (°C)	50	58	59
Crystallinity (%)	0	9	52
Properties at 23°C			
Flexural strength (MPa)	86	100	113
Maximum strain (%)	4.1	4.1	4.8
Elastic modulus	3550	3600	4150
Properties at 36°C			
Flexural strength (MPa)	60	77	83
Maximum strain (%)	3.3	3.9	4.3
Elastic modulus	2800	3400	3600
Properties at 56°C			
Flexural strength (MPa)	0.2	0.4	28
Maximum strain (%)	N/A	8	7.5
Elastic modulus	25	50	950

Source: Reproduced from ref. 44 with permission from John Wiley and Sons.

It has been reported [204] that the annealed PLLA samples showed cracks that propagated through spherulites and along spherulite boundaries with few or no crazes formed around the main crack. This caused the toughness of PLLA under impact loading to increase significantly by annealing. Since the crystallization temperature peak of PLLA has been estimated at about 123°C, a proper temperature for annealing is around 100°C. Also, the density and size of spherulites increase with annealing time and temperature [205]. In general, annealing of PLLA results in an increased tensile and flexural strengths, as well as impact resistance and thermomechanical properties.

A significant route to improve PLA mechanical properties is drawing and orientation. As reported by Grijpma et al. [201], PLA orientation increases tensile strength, elongation at break, Young's modulus, and impact strength. In order to avoid the poorer mechanical properties in perpendicular to the direction of orientation and minimize an isotropy, a biaxial molecular orientation must be maintained. An amorphous PLLA film forms α-crystallites when it is drawn above the T_g. A PLLA semicrystalline film with α-crystallites transforms into the oriented β-crystallites upon tensile drawing [156, 157]. Biaxial orientation of PLLA film is effective in increasing the tensile strength, elongation at break, and elastic modulus. Stretching is preferably two times or more in both directions and, typically, at a temperature between the T_g and T_m of the material [206]. Biaxial orientation of an extrusion cast film of PLA 4042D (8% D-isomer) from NatureWorks LLC results in MD elongation at break of 160% and TD elongation at break of 100% [207]. This material, when extruded without biaxial orientation, has much lower elongation at break values of only around 10%.

It is known that a 1:1 blending of pure PLLA with pure PDLA yields a stereocomplex crystalline structure [197]. While the melting temperature of α and β-crystalline forms of PLA falls in the range 170–180°C, the T_m of PLA stereocomplex is between 220 and 230°C. The mechanical properties of stereocomplex PLA are markedly improved; however, its high T_m makes it a difficult material for processing. Figure 11.8 shows the stress–strain curve for a stereocomplex film and a PLLA film, both having a molecular weight of 150,000 g/mol. Tensile strength and elongation at break of stereocomplex PLA film are almost twice as much as the nonblending film [208]. These polymers have different morphologies. The PLLA films are composed of normal spherulites, whereas for the blend films, microcrystallites are observed [208]. It is reported that the maximum achieved tensile modulus and strength, for stereocomplex PLA films with an extrusion draw ratio of 16 and prepared by solid-state coextrusion, were 9500 and 410 MPa, respectively. In addition, these films exhibited excellent thermomechanical stability showing 9500, 7000, and 3000 MPa modulus (E') values at room temperature, 100 and 200°C, respectively [209].

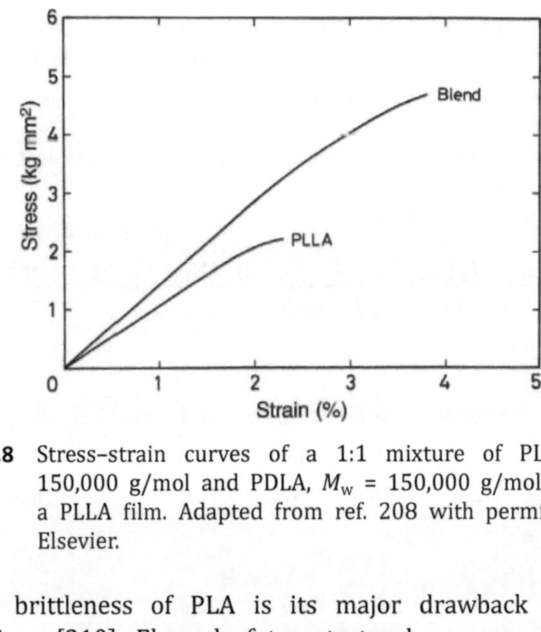

Figure 11.8 Stress–strain curves of a 1:1 mixture of PLLA, M_w = 150,000 g/mol and PDLA, M_w = 150,000 g/mol blend and a PLLA film. Adapted from ref. 208 with permission from Elsevier.

The brittleness of PLA is its major drawback for many applications [210]. Elmendorf tear test values range from 5 to 15 N/mm, which is quite low compared to conventional film grade polyolefins. This makes the extrusion casting of PLA film for packaging and many other applications relatively difficult compared to other polymers. In fact, there is no tolerance for PLA film tearing or cracking when subjected to forces during package manufacturing [203]. There are several methods for improving the flexibility of PLLA, for instance, through copolymerization with ε-caprolactone [211]. Also blending with biomaterials, such as thermoplastic starch [202], is a feasible method to prepare PLA-based materials that can be successfully used for film manufacturing. However, an interesting and practical route to modify the mechanical properties of PLLA consists of introducing biodegradable plasticizers in its composition [212]. The lactide monomer itself is an effective plasticizing agent for PLLA, but it has the disadvantage of rapid migration due to the small molecular weight of this compound [195–197]. Since PLA T_m also decreases with addition of a plasticizer, a suitable substance must plasticize this polymer without affecting its heat resistance and compostability. Also, the tensile strength decrease should be minimized as much as possible. As reported by Labrecque et al. [210] different citrate esters at up to 20–30% by weight of plasticizer concentration gave a good miscibility with PLA [210]. Tensile strength and elongation at break of PLA plasticized with citrate esters are reported in Table 11.9, and show a marked

Table 11.9 Thermal and mechanical properties of PLA plasticized with different citrate esters

	T_g (°C)	T_m (°C)	ΔH (J/g)	Tensile strength (MPA, Yield)	Elongation at break (%)
PLA	59.1	145.2	0.79	51.7	7
Triethyl citrate (wt%)					
10	42.1	134.1	0.31	28.1	21.3
20	32.6	130.9	2.86	12.6	382
30	22.0	126.8	7.57	7.2	610
Tributyl citrate (wt%)					
10	40.4	143.1	0.06	22.4	6.2
20	17.6	139.0	19.1	7.1	350
Acetyl triethyl citrate (wt%)					
10	50.8	141.7	0.91	34.5	10
20	30.0	138.1	0.91	9.6	320
30	14.2	131.6	18.34	7.6	228
Acetyl tributyl citrate (wt%)					
10	25.4	139.2	1.4	17.7	2.3
20	17.0	138.9	3.9	9.2	420

Source: Reproduced from ref. 210 with permission from Wiley.

effect of plasticizer at concentrations between 10% and 30%. As expected, the tensile strength of PLA decreases with addition of plasticizer up to 50%. On the other hand, elongation at break does not show any significant change at the lower percentages but significantly increases at higher concentrations in all cases [210]. Similar data have been published by Ljungberg et al. [213] using tributyl citrate, oligomeric tributyl citrate, obtained by transesterification with diethylene glycol and triacetine. Malonate esters, such as diethyl bishydroxymethyl malonate (DBM), have also been used as plasticizers for PLA. At a concentration of 20–25 wt%, phase separation occurs because of saturation of the amorphous phase. Aging a blend containing 15 wt% DBM for 4 months at ambient temperature also leads to phase separation and to the migration of the plasticizer to the film surface. Malonate oligomers, prepared by reacting DBM with acid chlorides, lead to more stable materials in aging tests [214]. Finally, polyethylene glycol (PEG) and oligomeric lactic acid (OLA) have been tested as the other two plasticizers for PLA [199].

11.5.7 Permeability, Sorption, and Diffusion

The ratio of L- to D-enantiomers has influence on the barrier properties of PLA. In General, purer PLLA polymers have higher barrier properties [215]. Lehermeier et al. [215] found that small changes in L:D stereochemical content (96:04 versus 98:02) had no effect on permeation properties of pure gases such as CO_2, O_2, N_2, and methane (CH_4). It has also been reported that at 0% relative humidity and temperatures in the range of 25–45°C, polymer films with the higher L-lactide content showed higher value of permeability coefficient, P, for CO_2. The values of P reported for O_2 followed the same trend. Water vapor permeability (WVP) of amorphous PLA films with different L-lactide unit contents (pure, 77:23, and 50:50 L:D content) have been reported as 2.18×10^{-14}, 1.90×10^{-14}, and 1.95×10^{-14} kg m/(m^2 s Pa), respectively, at 25°C [216]. Auras et al. [217] reported the values of P for CO_2 in PLA films range from 1.99×10^{-17} to 4.18×10^{-17} kg m/(m^2 s Pa) at 0% RH and at temperatures from 25 to 45°C, respectively. The results were obtained using semicrystalline PLA films with L-lactide contents of 94% and 98% and densities around 1250 kg/m^3. The O_2 permeability of PLA sheets has been determined by Auras et al.

[218]. The PLA permeability coefficient was (4.33×10^{-18} kg m/(m^2 s Pa)). According to Lehermeier et al. [215], the value of P for N_2 in PLA films obtained by casting is 1.93×10^{-14} kg m/(m^2 s Pa) at 30°C. N_2 permeation in PLA films has been reported to be very similar to that in PS films (3.27×10^{-14} kg m/(m^2 s Pa)) and in LDPE films (2.82×10^{-14} kg m/(m^2 s Pa)). The value of P for CH_4 of PLA cast from solution (5 wt% methylene chloride) is 1.49×10^{-14} kg m/(m^2 s Pa) at 30°C [215], which is lower than that reported for PET (5.94×10^{-17} kg m/(m^2 s Pa)) at 30°C [219] and similar to that for PS (3.42×10^{-14} kg m/(m^2 s Pa)) and LDPE (5.94×10^{-14} kg m/(m^2 s Pa)). Auras et al. [217] measured the WVP of semicrystalline PLA films containing different percentages of L-lactide (94% and 98%) and found that in the range of 40–90% RH, the values for both polymers were practically constant. P for PLA films was determined to be 1.48 to 2.20×10^{-14} kg m/(m^2 s Pa) at 10 to 37.8°C and 40–90% RH.

11.6 Applications

Polylactic acid can be formed using the conventional plastic processing methods such as extrusion molding, injection molding, blow molding, extrusion foaming, fibers, and nonwoven fabric. In the last two decades, the PLA applications was focused on biodegradable applications, such as agriculture, engineering, compostable products, and packaging materials. Since then PLA has been considered as a promising material to reduce environmental pollution and solid waste disposal problems.

11.6.1 Medical

Ring-opening polymerization of lactide is often utilized to obtain high-molecular-weight PLA for medical applications, unless the application requires fast degradation, where PLA synthesized through polycondensation of lactic acids with molecular weights <20,000 are used. The amorphous PDLLA polymer degrades faster than semicrystalline PLLA. As stated, the toughness of PLLA can be improved by a drawing process for applications such as bone fixation that requires strength and elasticity. Drawing PLLA with molecular weight over 70,000 increases its orientation and crystallinity [220]. The degradation of PLA by hydrolytic scission of ester linkages yields lactic acids. Lactic acid is a natural product associated with muscular construction in animals and humans, which can be decomposed by the body's normal metabolic pathways. Since L-lactic acid is the naturally occurring stereoisomer of lactic acid, PLLA is more commonly used for medical applications than PDLA. It has been tried to neutralize the acidic degradation products by adding agents such as calcium carbonates and/or calcium phosphates to the PLLA implants [221]. Other compounds present in the polymer are trace amounts of initiators, accelerators, and catalysts, which are mainly based on tin, and include tin(II) 2-ethylhexanoate that has been approved by the FDA.

The synthetic polymer used for sutures is mainly PGA or its copolymer with LLA (90:10) [222]. Because PLLA fibers degrade very slowly, they are not suitable for sutures. However, in applications that require long retention of the strength, PLLA fibers are the preferred material [223, 224]. Aoki et al. [225] used a PLLA felt for rotary cuff surgery as a bioabsorbable artificial tendon graft for the treatment of irreparable rotator cuff tear. PLLA fibers are also utilized in the form of biodegradable stents in cardiovascular and urological surgery. Biodegradable stents have been tested in numerous clinical studies. The degradable polymer of choice depends on the length of time required for the stent function and PLLA fibers are recommended for longer support [226]. Biodegradable materials have been replacing metal items for the fixation of fractured bones in the forms of plates, pins, screws, and wires. PLA has a large application in this field due to its high strength, similar to that of bone [227]. PLLA plates are also used mainly for the maxillofacial surgery, since they are not strong enough for load-bearing applications [228]. Recently, PLLA implants have also found applications in spine surgery [229].

One application of PLLA in the form of injectable microspheres is temporary fillings in facial reconstructive surgery. In 2004, the FDA approved PLLA for the restoration of facial fat loss in people with HIV. Microspheres and microcapsules have been widely applied in drug delivery systems (DDS). Release of drugs is based on several mechanisms that include diffusion and polymer degradation (hydrolysis or enzymatic degradation). In many cases, since fast degradation of the material is required, PLGA and a blend of PLA and other faster degrading polymers have been extensively studied. Liang et al. [230] investigated controlled release of methotrexate using a PLLA microsphere delivery system. PDLLA has been used for the slow release of recombinant human bone morphogenic protein (rhBMP-2) [231].

11.6.2 Packaging

By using extrusion casting with drawing-induced crystallization, PLA film can be manufactured with markedly improved thermal resistance, tensile strength, impact resistance, and elongation while maintaining its transparency. The PLA elongation increases greatly after drawing. Melt viscosity of PLA is temperature dependent and temperature management in molding is important. Temperature control and melt viscosity adjustment allow the use of a single PLA grade in various molding methods as shown in Table 11.10. Oriented PLA film shows better dead fold and twist retention properties than polyethylene (PE). Commercial examples of such films include vegetable bags, transparent envelope windows, laminated paper bags, and so on.

Table 11.10 Typical molding condition of PLA, M_w = 140,000

Process	Molding temperature (°C)
Injection molding	170–200
Extrusion molding	180–240
Inflation molding	170–190
Spinning	180–240
Lamination	200–240

Although PLA-oriented film is transparent and thermally resistant, some problems may occur due to its brittleness. When used as a bag, for example, PLA film tends to create a rustling sound when products are placed into the bag. Many methods have been reported to improve the flexibility of PLA films, such as addition of plasticizer [232] and blending with other flexible/biodegradable resins. Some plasticizers with enhanced performance have been developed by Daihachi Chemicals, Arakawa Chemicals, and Riken Vitamin. Lamination with transparent flexible resins such as PE or PP has been used to make flexible PLA film while maintaining transparency [233]. PLA-oriented film begins to shrink above

60°C, with a shrinkage ratio of about 70% and has a comparable shrinkage ratio and toughness to existing shrink labels. However, oriented PLA film has an insufficient shrinkage ratio around 70°C and shows whitening caused by crystallization. To overcome this problem, amorphous PLA, polymer blends with other aliphatic polyesters, and lamination with polyester film have been used to produce PLA films with shrinkage behavior similar to PVC. Unoriented PLA sheet is excellent for thermoforming shallow and deep drawn containers. Because the heat deflection temperature of PLA sheet is about 10 to 15°C, it cannot be used for microwave containers. However, PLA can be satisfactorily used in packaging for products that are not to be heated, such as single use disposable drinking cups. PLA packages have already been commercialized for eggs, vegetables, fruits, bakery, and blister pack products [233].

11.6.3 Fibers

PLA fibers are not only biodegradable but also highly functional due to their intrinsic properties such as bacteriostatic, flame-retardant, and weathering resistance [234]. PLA can be melt-spun into various types of fibers including monofilaments, multifilaments, bulked continuous filaments, staple fibers, short-cut fibers, and spun bond fabrics by conventional melt spinning machines. The fibers are then drawn and annealed to give desirable mechanical properties such as high tenacity, good toughness, and good dimensional stability. Crystallinity, crystal structure, morphology, and molecular orientation strongly affect the mechanical, thermal, and hydrolysis resistance properties of PLA fibers. The yarn properties of PLA fibers are relatively similar to those of PET. However, the density and refractive index of PLA are lower than those of PET, giving it light weight and deep silky luster without brightness. The moisture regaining and wicking properties of PLA are superior to those of PET because PLA is more hydrophilic. It is well known that lactic acid is a strong organic acid, which exhibits antibacterial and antifungal properties. It is hypothesized that small amounts of lactic acid migrating from the PLA fiber matrix may be responsible for the antibacterial/antifungal property [235, 236]. A xenon arc test of fading resistance demonstrated that PLA fibers perform better than PET fibers, with a 0% loss in elongation at break for the PLA fibers and a 30% loss in elongation at break for the PET fibers, after 100 h of xenon arc exposure [234].

The unique features of PLA fibers offer distinct benefits in the fields of geotextiles and agricultural/horticultural applications [237]. Both woven and nonwoven fabrics made from PLA fibers are being considered for automobile cabin parts applications. Toyota has already employed a new type of spare-tire box cover using Kenaf/PLA board, which is manufactured by heat compression molding of a Kenaf/PLA (weight-based blend ratio: 7/3) fiber web [238]. Another potential application of PLA fibers is filters for tea, coffee, organic wastes, and air [239]. PLA is safe and "generally recognized as safe" for food contact uses [236]. Towels made from 100% PLA or blends of PLA with cotton feel more comfortable and lead to better performance of fast drying and low shrinkage after drying when compared with 100% cotton. Fabrics of PLA with excellent hand and drape and a silk-like luster are also suitable for fashion and formalwear such as wedding dresses [239].

11.7 Conclusions

Polylactic acid is likely to have significant future potential in a variety of applications due to specific advantages compared with conventional petroleum-based polymers. The cost-performance balance of PLA has resulted in its use in many applications, including packaging, fibers, and medicals. The use of PLA as an alternative to petroleum-based polymers will increase demand for agricultural products such as corn and sugar beets, and is an advanced example of sustainable technology.

A new process has been developed for the production of PLA using reactive extrusion polymerization based on a new catalytic system, which not only enhances the ROP kinetics of L-lactide but also suppresses side and degradation reactions such as intermolecular transesterification reactions. Co-rotating closely intermeshing twin-screw extruders have often been used for polymerization reactions, but in any case, the reaction time was sufficiently smaller than the residence time in the extruder. In the case of ROP of lactide, the extraordinary effect of triphenylphosphine on the efficiency of $Sn(Oct)_2$ as a catalyst was the key point to ensure further enhancement of the polymerization reaction. Thus, it was feasible to employ a single-stage process using reactive extrusion technology to polymerize L-lactide to PLA, which can be used right away from the process. The resulting polymer exhibits thermal and mechanical properties, which allow PLA to be used in a wide range of nowadays polymer applications.

References

1. Khodabakhshi, K., Hassani, M., Ahmadi, S. (2014). *Polym. Sci. Series B*, **56**, 728–735.
2. Khodabakhshi, K., Gilbert, M., Fathi, S., Dickens, P. (2014). *J. Thermal. Anal. Calorim.*, **115**, 383–391.
3. Khodabakhshi, K., Gilbert, M., Dickens, P. (2013). *Polym. Adv. Tech.*, **24**, 503–510.
4. Ehsani, M., Khodabakhshi, K., Asgari, M., Hassani, M. (2014). *e-Polymers*, **14**, 353–361.
5. Khodabakhshi, K., Ehsani, M., Asgari, M., Ahmadi, S. Februry (2014). Polylactic acid synthesis, Project Report, Iran Polymer and Petrochemical Institute.
6. Gallos, A., Fontaine, G., Bourbigot, S. (2013). *Polym. Adv. Tech.*, **24**, 130–133.
7. Pelouze, J. (1845). *J. Chem. Pharm.*, **53**, 112–124.
8. Pohl, H. (1914) U.S. Patent 1,095,205 (to Gruter).
9. Noda, M., Okuyama, H. (1999). *Chem. Pharm. Bull.*, **47**, 467–471.
10. O'Brien, W., Cariello, L. A., Wells, T. F. (1995). WO 9606092 (to Ecological Chemical Products Company).
11. Gruber, P., Kolstad, J., et al. (1992). U.S. Patent 5,338,822, (to Cargill Inc.); U.S. Patent 5,258,488, 1993 (to Cargill Inc.).
12. Matsuo, T., et al. (2005). U.S. Patent 25222379A1 (to Hitachi/Toyota).
13. Kamikawa, M., et al. (2006). EP 1873185A1 (to Hitachi).
14. Gruber, P., Kolstad, J., et al. (1999). U.S. Patent 6277951B1 (to Cargill Inc.).
15. Hartmann, M. H. (1998). High molecular weight polylactic acid polymer, in: (Kaplan, D. L., ed.), *Biopolymers from Renewable Resources*, Springer, Berlin, Chapter 15, pp. 367–411.
16. Saitoh, S., et al. (2005). *Appl. Environ. Microbiol.*, **71**, 2789–2792.

17. Coszach, P., Bogaert, J. C., van Gansberghe, F. (2006). U.S. Patent 0014975A1 (to Galactic).
18. Groot, W., Krieken, J., Sliekersl, O., and Vos, S. (2010). Production and purification of lactic acid and lactide, in: (Aurus, R., Lim, L. T., Selke, S. E. M., and Tsuji, H., ed.), *Poly(lactic acid), Synthesis, Structure, Properties, Processing, and Application*, Wiley, New Jersey, Chapter 1, pp. 3–16.
19. Witzke, D. R. (1997). Introduction to properties, engineering, and prospects of polylactide polymers, PhD thesis, Department of Chemical Engineering, Michigan State University, East Lansing, MI.
20. Henton, D. E., Gruber, P., Lunt, J., Randall, J. (2005). Polylactic acid technology, in: (Mohanty, A. K., Misra, M., Drzal, L. T. eds.), *Natural Fibers, Biopolymers, and Biocomposites*, CRC Press, Chapter 16, pp. 527–577.
21. Martak, J., Schlosser, S. (2007). *Sep. Purif. Technol.*, **57**, 483–494.
22. Ford, T. M. (1993). U.S. Patent 5,310,599 (to Du Pont de Nemours & Co.).
23. Stolt, M., Sodergard, A. (1999). *Macromolecules*, **32**, 6412–6417.
24. Leenslag, J. W., Pennings, A. J. (1987). *Makromol. Chem.*, **188**, 1809–1814.
25. Hoessel, P., Hoffmann, G., Braun, F., Kratz, D., Brudermueller, M., Witzel, T. (1994). EP 0588222 (BASF AG).
26. de Vries, K. S. (1988). EP 0275581 (to Akzo NV).
27. Fridman, I. D., Kwok, J. (1993). U.S. Patent 5,264,592 (to Camelot Industries).
28. Scholz, R. U., van der Steen, R. P. M. (2007), WO/2007/148975, Purification of Lactide Streams.
29. Stepanski, M. (2008). 1st PLA World Congress, Munich, September 9–10.
30. O'Brian, W. G., Cariello, L. A., Wells, T. F. (1997). EP 0777664 (to Ecological Chem. Prod.).
31. Meerdink, J., Sodergard, A. (2003). WO 05056509A1 (to Hycail).
32. Bhatia, K. K. (1993). EP 0531462 (to Du Pont).
33. Meerdink, M., Sodergard, A. (2005). EP 1694626 (to Tate & Lyle Public Limited Co.).
34. Grijpma, D. W., Pennings, A. J. (1994). *Macromol. Chem. Phys.*, **196**, 1649–1663.
35. Kolstad, J. J. (1996). *J. Appl. Polym. Sci.*, **62**, 1079–1091.

36. Gruber, P., Hall, J., Kolstad, J. J., et al. (1993). WO 9509879A1 (to Cargill Inc.).
37. Benecke, H. P., Markle, R. A., Sinclair, R. G. (1994). U.S. Patent 5, 332, 839 (to BioPak Technology, Ltd.).
38. Zhu, J. J., Wang, Y. L., Quan, X. J., Xin, J., Wang, R. (2005). *Chem. World*, **3**, 173–176.
39. Auras, R., Harte, B., Selke, S. (2004). *Macromol. Biosci.*, **4**, 835–864.
40. Urayama, H., Moon, S. I., Kimura, Y. (2003). *Macromol. Mater. Eng.*, **288**, 137–143.
41. Dorgan, J. R., Jansen, J., Clayton, M. P. (2005). *J. Rheol.*, **49**, 607–619.
42. Drumright, R. E., Gruber, P. R., Henton, D. E. (2000). *Adv. Mater.*, **12**, 1841–1846.
43. Baker, G. L., Vogel, E. B., SmithIII, M. R. (2008). *Polym. Rev.*, **48**, 64–84.
44. Perego, G., Cella, G. D., Bastioli, C. (1996). *J. Appl. Polym. Sci.*, **59**, 37–43.
45. Gogolewski, S., Jovanovic, M., Perren, S. M. (1993). *Polym. Degrad. Stabil.*, **40**, 313–322.
46. Gruber, P. R., Kolstad, J. J., Ryan, C. M., Hall, E. S., Conn, R. S. E. (1996). U.S. Patent 5,484,881.
47. Sheet Extrusion Processing Guide, Nature Works LLC, Minnetonka, MN, 2005.
48. Nature Works PLA Injection Molding Guide for 3051D, Nature Works LLC, Minnetonka, MN, 2006.
49. Taubner, V., Shishoo, R. (2001). *J. Appl. Polym. Sci.*, **79**, 2128–2135.
50. http://www2.dupont.com/Biomax/en_US/index.html (accessed date: March 10, 2014).
51. Renstad, R., Karlsson, S., Sandgren, A., Albertsson, A. C. (1998). *J. Environ. Polym. Degrad.*, **6**, 201–221.
52. Fukuda, N., Tsuji, H. (2005). *J. Appl. Poly. Sci.*, **96**, 190–199.
53. *Bioplastics Magazine*, Issue May/June 03/2008, 24–26 (http://www.bioplasticsmagazine.com/en/online-archive/issues2008/200803.php).
54. Production of NatureWorks Polylactide Films on Blown Film Equipment Designed for Producing Low Density Polyethylene Film, NatureWorks LLC, Minnetonka, MN, 2003.
55. NatureWorks Polymer Technical Data Sheets, PLA Polymer 7032D, NatureWorks LLC, Minnetonka, MN, 2005.

56. Garlotta, D. (2001). *J. Polym. Environ.*, **9**, 63–84.
57. Kopinke, F. D., Remmler, M., Mackenzie, K., Moder, M., Wachsen, O. (1996). *Polym. Degrad. Stabil.*, **53**, 329–342.
58. McNeill, I. C., Leiper, H. A. (1985). *Polym. Degrad. Stabil.*, **11**, 309–326.
59. Bashir, A., Al-Uraini, A. A., Jamjoom, M., Al Khalid, A., Al Hafez, M., Ali, S. (2002). *J. Macromol. Sci.*, **A39**, 1407–1433.
60. Villain, F., Coudane, J., Vert, M. (1994). *J. Appl. Polym. Sci.*, **52**, 55–60.
61. Sugaya, N., Nakagawa, T., Sajurai, K., Morita, M., Onodera, S. (2001). *J. Health Sci.*, **47**, 21–27.
62. Tsukegi, T., Motoyama, T., Shirai, Y. (2007). *Polym. Degrad. Stabil.*, **92**, 552–559.
63. Fan, Y., Nishida, H., Shirai, Y., Endo, T. (2003). *Green Chem.*, **5**, 575–579.
64. Khabbaz, F., Karlsson, S., Albertsson, A. C. (2000). *J. Appl. Polym. Sci.*, **78**, 2369–2378.
65. Westphal, C., Perrot, C., Karlsson, S. (2001). *Polym. Degrad. Stabil.*, **73**, 281–287.
66. Kolstad, J. J., Witzke, D. R., Hartmann, M. H., Hall, E. S., Nangeroni, J. (2002). U.S. Patent 6353086B1.
67. NatureWorks PLA Processing Guide for Biaxially Oriented Film, NatureWorks LLC, Minnetonka, MN, 2006.
68. Bigg, D. M. (1996). Effect of copolymer ratio on the crystallinity and properties of polylactic acid copolymers, SPE ANTEC Technical Papers, pp. 2028–2039.
69. Ou, X., Cakmak, M. (2003). SPE ANTEC Papers, pp. 1701–1705.
70. Lee, J. K., Lee, K. H., Jin, B. S. (2001). *Eur. Polym. J.*, **37**, 907–914.
71. Lim, L. T., Auras, R., Rubino, M. (2008). *Prog. Polym. Sci.*, **33**, 820–852.
72. Yoo, D. K., Kim, D., Lee, D. S. (2005). *Macromol. Res.*, **13**, 68–72.
73. Chen, G. X., Kim, H. S., Kim, E. S., Yoon, J. S. (2006). *Eur. Polym. J.*, **42**, 468–472.
74. Moon, S., Lee, C. W., Miyamoto, M., Kimura, Y. (2000). *J. Polym. Sci. Part A*, **38**, 1673–1679.
75. Maruyama, H., Murayama, T., Yanagisawa, N., Tsuzaki, N. (1998). EP 0848026 (to Kyowa Yuka).
76. Yanagisawa, N., Nezu, T., Hotta, K., Murayama, T., Tsuzaki, N., Kodama, S., Maruyama, H., Yokomori, Y. (1998). EP 0823448 (to Kyowa Hakko Kogyo).

77. Kulkarni, R. K., Pani, K. C., Neuman, C., Leonard, F. (1966). *Arch. Surg.*, **93**, 839–843.

78. Purac Biomaterials, available at http://www.puracbiomaterials.com (access date: March 10, 2014).

79. Bogaert, J. C., Coszach, P. (2000). *Macromol. Symp.*, **153**, 287–303.

80. Crank, M., Patel, M., Marscheider-Weidemann, F., Schleich, J., Heusing, B., Angerer, G. (2005). Techno-economic Feasibility of Large-scale Production of Bio-based Polymers in Europe, in: (Wolf, O., ed.), *Technical Report Series*, European Commission, EUR 22103 EN.

81. NatureWorks LLC, available at http://www.natureworksllc.com (access date: March 10, 2014).

82. Total and Packaging, PLA (PolyLactic Acid): an ambitious research project, available at http://www.packaging.chemicals.total.com/content/documents/37_ato%20packscope%2031%20en%20-%20pla.pdf (access date: March 10, 2014).

83. Hutchinson, F. G. (1987). EP 0244114 (to ICI PLC).

84. Yamamoto, K., Tani, T., Aoki, T., Hata, Y. (2003). EP 1310517 (to Wako Pure Chem. Ind. Ltd.; Tadeka Chem. Ind. Ltd.).

85. Qu, M., Tu, H., Amarante, M., Song, Y., Sherry, S. (2014). *J. Appl. Polym. Sci.*, **131** (http://onlinelibrary.wiley.com/doi/10.1002/app.40287/abstract).

86. Nieuwenhuis, A. J. (1992). *J. Clin. Mater.*, **10**, 59–67.

87. Ebato, H., Imamura, S. (1995). EP 0661325 (to Dainippon Ink and Chem.).

88. Iiyama, T., Sato, T., Yamada, M. (1999). EP 0916684 (to Daicel Chem.).

89. Morio, T., Yasutoshi, K. (1996). U.S. Patent 5,484,882 (Dainippon Ink and Chem.).

90. Jacobsen, S., Fritz, H. G., Degée, Ph., Dubois, Ph., Jérôme, R. (2000). Polymer, **41**, 3395–3403.

91. Aubrecht, K. B., Hillmyer, M. A., Tolman, W. B. (2002). *Macromolecules*, **35**, 644–650.

92. Kricheldorf, H. R., Fechner, B. (2002). *Biomacromolecules*, **3**, 691–695.

93. Radano, C. P., Baker, G. L., Smith, M. R. (2000). *J. Am. Chem. Soc.*, **122**, 1552–1553.

94. Zhong, Z., Dijkstra, P. J., Feijen, J. (2003). *J. Am. Chem. Soc.*, **125**, 11291–11298.

95. Bourissou, D., Moebs-Sanchez, S., Martin-Vaca, B. (2007). *C. R. Chimie*, **10**, 775–794.

96. Darensbourg, D. J., Choi, W., Karroonnirun, O., Bhuvanesh, N. (2008). *Macromolecules*, **41**, 3493–3502.

97. O'Keefe, B. J., Breyfogle, L. E., Hillmyer, M. A., Tolman, W. B. (2002). *J. Am. Chem. Soc.*, **124**, 4384–4393.

98. Chamberlain, B. M., Cheng, M., Moore, D. R., Ovitt, T. M., Lobkovsky, E. B., Coates, G. W. (2001). *J. Am. Chem. Soc.*, **123**, 3229–3238.

99. Chen, H. Y., Tang, H. Y., Lin, C. C. (2006). *Macromolecules*, **39**, 3745–3752.

100. Lemmouchi, Y., Perry, M. C., Amass, A. J., Chakraborty, K., Schacht, E. (2008). *J. Polym. Sci. Part A*, **46**, 5348–5362.

101. Chisholm, M. H., Gallucci, J. C., Phomphrai, K. (2004). *Inorg. Chem.*, **43**, 6717–6725.

102. Chisholm, M. H., Gallucci, J. C., Phomphrai, K. (2005). *Inorg. Chem.*, **44**, 8004–8010.

103. Russell, S. K., Gamble, C. L., Gibbins, K. J., Juhl, K. C. S., Mitchell III, W. S., Tumas, A. J., Hofmeister, G. E. (2005). *Macromolecules*, **38**, 10336–10340.

104. Stridsberg, K. M., Ryner, M., Albertsson, A. C. (2002). *Adv. Polym. Sci.*, **157**, 41–65.

105. Higgins, N. A. (1954). U. S. Patent 2,683,136 (to Du Pont).

106. Gruber, P., Kolstad, J. J., Hall, E. S., Conn, R. S. E. (2000). EP 0615532, (to Cargill Inc.).

107. Kolstad, J. J., Witzke, D. R., Hartmann, M. H., Hall, E. S., Nageroni, J. F. (2001). EP 1070097 (to Cargill Inc.).

108. Sodergard, A., Selin, J. F., Niemi, M., Johansson, C. J., Meinander, K. (1999). EP 0737219 (to Fortum Oyj).

109. Ohara, H., Okamoto, T. (1998). U. S. Patent 5,728,847 (to Shimadzu Corp.).

110. Ohara, H., Sawa, S., Kawamoto, T. (1995). EP 0664309 (to Shimadzu Corp.).

111. Hiljanen, M., Varpomaa, P., Seppälä, J., Törmälä, P. (1996). *Macromol. Chem. Phys.*, **197**, 1503–1523.

112. Jacobsen, S., Fritz, H. G. (1996). *Polym. Eng. Sci.*, **36**, 2799–2804.

113. Jacobsen, S., Fritz, H. G. (1999). *J. Polym. Sci.: Part A: Polym. Chem.*, **37**, 2413–2420.

114. Stevels, W. M., Bernard, A., Van de Witte, P., Dijkstra, P. J., Feijen, J. (1996). *J. Appl. Polym. Sci.*, **62**, 1295.

115. Degee, Ph., Dubois, Ph., Jerome, R., Jacobsen, S., Fritz, H. G. (1999). *Polym. Eng. Sci.*, **39**, 1303–1310.

116. Jacobsen, S., Fritz, H., Degée, Ph., Dubois, Ph., Jérôme, R. (2000). *Ind. Crops. Prod.*, **11**, 265–275.

117. Puaux, J. P., Banu, I., Nagy, I., Bozga, G. (2007). *Macromol. Symp.*, **259**, 318–326.

118. Xie, F. W., Yu, L., Liu, H. S., Chen, L. (2006). *Starch Starke*, **58**, 131–139.

119. Kalambur, S., Rizvi, S. S. H. (2006). *J. Plast. Film Sheet*, **22**, 39–58.

120. Singh, S., Gamlath, S., Wakeling, L. (2007). *Int. J. Food Sci. Technol.*, **42**, 916–929.

121. Wolf, B. (2010). *Curr. Opin. Colloid Interface Sci.*, **15**, 50–54.

122. Janssen, L. P. B. M. (2004). *Reactive Extrusion Systems*, Marcel Dekker, New York.

123. Al Malaika, S. (1996). *Reactive Modifiers for Polymers*, Chapman & Hall, London.

124. Xanthos, M. (1992). *Reactive Extrusion*, Hanser, Munich.

125. Hassouna, F., Raquez, J. M., Addiego, F., Toniazzo, V., Dubois, Ph., Ruch, D. (2012). *Eur. Polym. J.*, **48**, 404–415.

126. Schmack, G., Jehnichen, D., Vogel, R., Tändler, B., Beyreuther, R., Jacobsen, S., Fritz, H. G. (2001). *J. Biotech.*, **86**, 151–160.

127. Takamura, M., Nakamura, T., Takahashi, T., Koyama, K. (2008). *Polym. Degrad. Stabil.*, **93**, 1909–1916.

128. Lee, J. H., Park, T. G., Park, H. S., Lee, D. S., Lee, Y. K., Yoon, S. C., Nam, J. D. (2003). *Biomaterials*, **24**, 2773.

129. Sinha Ray, S., Yamada, K., Okamoto, M., Ogami, A., Ueda, K. (2003). *Chem. Mater.*, **15**, 1456.

130. Bandyopadhyay, S., Chen, R., Giannelis, E. P. (1999). *Polym. Mater. Sci. Eng.*, **81**, 159.

131. Paul, M. A., Alexandre, M., Degee, P., Henrist, C., Rulmont, A., Dubois, P. (2003). *Polymer*, **44**, 443.

132. Paul, M. A., Delcourt, C., Alexandre, M., Degee, P. H., Monteverde, F., Dubois, P. H. (2005). *Polym. Degrad. Stabil.*, **87**, 535.

133. Sinha Ray, S., Okamoto, M. (2003). *Macromol. Mater. Eng.*, **288**, 936.

134. Hoffmann, B., Kressler, J., Stoppelmann, G., Friedrich, Chr., Kim, G. M. (2000). *Colloid Polym. Sci.*, **278**, 629.

135. Jiang, X., Luo, Y., Tian, X., Huang, D., Reddy, N., Yang, Y. (2010). Chemical structure of poly(lactic acid), in: (Aurus, R., Lim, L. T., Selke, S. E.

M., Tsuji, H., ed.), *Poly(lactic acid), Synthesis, Structure, Properties, Processing, and Application*, Wiley, New Jersey, Chapter 6, pp. 69–80.

136. Molinaroa, S., Romerob, M. C., Boaroc, M., Sensidonia, A., Lagaziod, C., Morrise, M., Kerry, J. (2013). *J. Food Eng.*, **117**(1), 113–123.
137. Francois, C., Michel, V., Stella, C., Pierre, G. (1983). *Polymer*, **24**, 53–59.
138. Thakur, K. A. M., Kean, R. T., Hall, E. S., Kolstad, J. J., Munson, E. J. (1998). *Int. J. Polym. Anal. Charact.*, **4**(5), 379–391.
139. Thakur, K. A. M., Kean, R. T., Hall, E. S., Kolstad, J. J., Munson, E. J. (1998). *Macromolecules*, **31**(5), 1487–1494.
140. Tang, Z., Chen, X., Yang, Y., Pang, X., Sun, J., Zhang, X., Jing, X. (2004). *J. Polym. Sci. Part A*, **42**(23), 5974–5982.
141. Tsuji, H. (2000). *Polymer*, **41**(10), 3621–3630.
142. Karst, D., Yang, Y. (2006). *Polymer*, **47**(13), 4845–4850.
143. Karst, D., Yang, Y. (2008). *Macromol. Chem. Phys.*, **209**(2), 168–174.
144. Nair, L. S., Laurencin, C. T. (2007). *Prog. Polym. Sci.*, **32**, 762–798.
145. Arnoult, M., Dargent, E., Mano, J. F. (2007). *Polymer*, **48**(4), 1012–1019.
146. Kattan, M., Dargent, E., Grenet, J. (2002). *Polymer*, **43**(4), 1399–1405.
147. Wunderlich, B. (2003). *Prog. Polym. Sci.*, **28**(3), 383–450.
148. Pyda, M., Wunderlich, B. (2005). *Macromolecules*, **38**(25), 10472–10479.
149. Picciochi, R., Wang, Y., Alves, N. M., Mano, J. F. (2007). *Colloid Polym. Sci.*, **285**(5), 575–580.
150. Pan, P. J., Zhu, B., Inoue, Y. (2007). *Macromolecules*, **40**(26), 9664–9671.
151. Reignier, J., Tatibouet, J., Gendron, R. (2009). *J. Appl. Polym. Sci.*, **112**(3), 1345–1355.
152. Bras, A. R., Viciosa, M. T., Dionisio, M., Mano, J. F. (2007). *J. Therm. Anal. Calorim.*, **88**(2), 425–429.
153. Wong, Y. S., Stachurski, Z. H., Venkatraman, S. S. (2008). *Acta Mater.*, **56**(18), 5083–5090.
154. Aou, K., Hsu, S. L., Kleiner, L. W., Tang, F. W. (2007). *J. Phys. Chem. B*, **111**(42), 12322–12327.
155. Sasaki, S., Asakura, T. (2003). *Macromolecules*, **36**(22), 8385–8390.
156. Sawai, D., Takahashi, K., Sasashige, A., Kanamoto, T., Hyon, S. H. (2003). *Macromolecules*, **36**(10), 3601–3605.
157. Takahashi, K., Sawai, D., Yokoyama, T., Kanamoto, T., Hyon, S. H. (2004). *Polymer*, **45**(14), 4969–4976.

158. Sawai, D., Yokoyama, T., Kanamoto, T., Sungil, M., Hyon, S. H., Myasnikova, L. P. (2006). *Macromol. Symp.*, **242**, 93–103.

159. Fambri, L., Migliaresi., C., Kesenci, K., Piskin, E. (2002). Biodegradable polymers, in: (Barbucci, R., ed.), *Integrated Biomaterials Science*, Kluwer Plenum Publications, New York, Chapter 4, pp. 119–187.

160. Tsusj, H. (2002). Polylactide, in: (Doi, Y., Steinbuchel, A., eds.), *Biopolymers. Polyesters III. Applications and Commercial Products*, Wiley-VCH Verlag GmbH, Weinheim, Chapter 5, pp. 129–177.

161. Ikada, Y., Jamshidi, K., Tsuji, H., Hyon, S. H. (1987). *Macromolecules*, **20**, 904–906.

162. Celli, A., Scandola, M. (1992). *Polymer*, **33**, 2699–2703.

163. Ovitt, T. M., Coates, G. W. (1999). *J. Am. Chem. Soc.*, **121**, 4072–4073.

164. Fisher, E. W., Sterzel, H. J., Wegner, G., Kolloid Z. Z. (1973). *Polymer*, **251**, 980–990.

165. Loomis, G. L., Murdoch, J. R., Gardner, K. H. (1990). *Polym. Prepr.*, **31**(2), 55.

166. Fambri, L., Migliaresi, C. (2010). Crystallization and thermal properties, in: (Aurus, R., Lim, L. T., Selke, S. E. M., Tsuji, H.ed.), *Poly(lactic acid), Synthesis, Structure, Properties, Processing, and Application*, Wiley, New Jersey, Chapter 9, pp. 113–123.

167. Migliaresi, C., Cohn, D., De Lollis, A., Fambri, L. (1991). *J. Appl. Polym. Sci.*, **43**, 83–95.

168. Tsuji, H., Ikada, Y. (1995). *Polymer*, **36**, 2709–2716.

169. Ninjenhuis, A. J., Grijpma, D. W., Pennings, A. J. (1991). *Polym. Bull.*, **26**, 71–77.

170. Oliveira, N. S., Oliveira, J., Gomes, T., Ferreira, A., Dorgan, J., Marrucho, I. M. (2004). *Fluid Phase Equilibr.*, **222**, 317–324.

171. Turhan, K. N., Sahbaz, F. (2001). *Polym. Int.*, **50**, 1138–1142.

172. Goncalves, C. M. B., Coutinho, J. A. P., Marrucho, I. M. (2010). Optical properties, in: (Aurus, R., Lim, L. T., Selke, S. E. M., Tsuji, H. ed.), *Poly(lactic acid), synthesis, structure, properties, processing, and application*, Wiley, New Jersey, Chapter 8, pp. 97–111.

173. Ciolacu, C. F. L., Choudhury, N. R., Dutta, N. K. (2006). *Polym. Degrad. Stabil.*, **91**, 875–885.

174. Wang, Y. M., Steinhoff, B., Brinkmann, C., Alig, I. (2008). *Polymer*, **49**, 1257–1265.

175. Hutchinson, M. H., Dorgan, J. R., Knauss, D. M., Hait, S. B. (2006). *J. Polym. Environ.*, **14**, 119–124.

176. Tsuji, H., Hyon, S. H., Ikada, Y. (1991). *Macromolecules*, **24**, 5651–5656.
177. Mehta, R., Kumar, V., Bhunia, H., Upadhyay, S. N. (2005). *J. Macromol. Sci. Polym. Res.*, **C45**, 325–349.
178. Gutierrez-Villarreal, M. H., Ulloa-Hinojosa, M. G., Gaona-Lozano, J. G. (2008). *J. Appl. Polym. Sci.*, **110**, 163–169.
179. Khorasani, M. T., Mirzadeh, H., Irani, S. (2008). *Radiat. Phys. Chem.*, **77**, 280–287.
180. Kister, G., Cassanas, G., Vert, M., Pauvert, B., Terol, A. (1995). *J. Raman Spectrosc.*, **26**, 307–311.
181. Jalabert, M., Fraschini, C., Prud'Homme, R. E. (2007). *J. Polym. Sci. Polym. Chem.*, **45**, 1944–1955.
182. Paragkumar, N. T., Dellacherie, E., Six, J. L. (2006). *Appl. Surf. Sci.*, **253**, 2758–2764.
183. Migliaresi, C., Cohn, D., Fambri, L. (1994). *J. Biomater. Sci. Polym. Ed.*, **5**(6), 591–606.
184. Bergsma, J. E., de Bruijn, W. C., Rozema, F. R., Bos, R. R. M., Boering, G. (1995). *Biomaterials*, **16**, 25–31.
185. Plackett, D. V., Holm, V. K., Johansen, P., Ndoni, S., Nielsen, P. V., Sipilainen-Malm, T., Sodergard, A., Verstichel, S. (2006). *Packag. Technol. Sci.*, **19**(1), 1–24.
186. Wanamaker, C. L., Tolman, W. B., Hillmyer, M. A. (2009). *Biomacromolecules*, **10**(2), 443–448.
187. Dong, C. M., Guo, Y. Z., Qiu, K. Y., Gu, Z. W., Feng, X. D. (2005). *J. Control. Release*, **107**(1), 53–64.
188. Lehermeier, H. J., Dorgan, J. R. (2001). *Polym. Eng. Sci.*, **41**, 2172–2184.
189. Dorgan, J. R. (2010). Rheology of Poly(lactic acid), in: (Aurus, R., Lim, L. T., Selke, S. E. M., Tsuji, H., ed.), *Poly(lactic acid), Synthesis, Structure, Properties, Processing, and Application*, Wiley, New Jersey, Chapter 10, pp. 125–138.
190. Lu, L., Mikos, A. G. (1999). Poly(lactic acid), in: (Mark, J. E., ed.), *Polymer Data Handbook*, Oxford University Press, New York, pp. 627–633.
191. Biresaw, G., Carriere, C. J. (2002). *J. Polym. Sci. Part B*, **40**, 2248–2258.
192. Cooper-White, J. J., Mackay, M. E. (1999). *J. Polym. Sci. Part B*, **37**, 1803–1814.
193. Palade, L. I., Lehermeier, H. J., Dorgan, J. R. (2001). *Macromolecules*, **34**, 1384–1390.

194. Perego, G., Cella, G. D. (2010). Mechanical properties, in: (Aurus, R., Lim, L. T., Selke, S. E. M., Tsuji, H., ed.), *Poly(lactic acid), Synthesis, Structure, Properties, Processing, and Application*, Wiley, New Jersey, Chapter 11, pp. 141–152.

195. Sodergard, A., Stolt, M. (2002). *Prog. Polym. Sci.*, **27**, 1123–1163.

196. Urayama, H., Kanamori, T., Kimura, Y. (2001). *Macromol. Mater. Eng.*, **286**, 705–713.

197. Garlotta, D. (2001). *J. Polym. Environ.*, **9**, 63–84.

198. Engelberg, I., Kohn, J. (1991). *Biomaterials*, **12**, 292–304.

199. Baiardo, M., Frisoni, G., Scandola, M., Rimelen, M., Lips, D., Ruffieux, K., Wintermantel, E. (2003). *J. Appl. Polym. Sci.*, **90**, 1731–1738.

200. Renouf-Glauser, A. C., Rose, J., Farrar, D. F., Cameron, R. E. (2005). *Biomaterials*, **26**, 5771–5782.

201. Grijpma, D. W., Altpeter, H., Bevis, M. J., Feijen, J. (2002). *Polym. Int.*, **51**, 845–851.

202. Martin, O., Averous, L. (2001). *Polymer*, **42**, 6209–6219.

203. Ljungberg, N., Wesslen, B. (2005). *Biomacromolecules*, **6**, 1789–1796.

204. Park, S. D., Todo, M., Arakawa, K. (2004). *J. Mater. Sci.*, **39**, 1113–1116.

205. Park, S. D., Todo, M., Arakawa, K., Koganemaru, M. (2006). *Polymer*, **47**, 1357–1363.

206. Kumar, S., Akhtar, S., Kumar, V. (2005). *Popular Plast. Packag.*, **50**, 85–89.

207. NatureWorks technical data sheet, http://forums.reprap.org/file.php?94,file=4512,filename=TechnicalDataSheet_4042D_pdf.pdf (access date: March 10, 2014).

208. Tsuji, H., Ikada, Y. (1999). *Polymer*, **40**, 6699–6708.

209. Sawai, D., Tamada, M., Kanamoto, T. (2007). *Polym. J.*, **39**, 953–960.

210. Labrecque, L. V., Kumar, R. A., Dav_e, V., Gross, R. A., McCarthy, S. P. (1997). *J. Appl. Polym. Sci.*, **66**, 1507–1513.

211. Perego, G., Vercellio, T., Balbontin, G. (1993). *Macromol. Chem.*, **194**, 2463–2469.

212. Pillin, I., Montrelay, N. Grohens, Y. (2006). *Polymer*, **47**, 4676–4682.

213. Ljungberg, N., Andersson, T., Wesslen, B. (2003). *J. Appl. Polym. Sci.*, **88**, 3239–3247.

214. Ljungberg, N., Colombini, D., Wesslen, B. (2005). *J. Appl. Polym. Sci.*, **96**, 992–1002.

215. Lehermeier, H. J., Dorgan, J. R., Way, D. (2001). *J. Membr. Sci.*, **190**, 243–251.

216. Tsuji, H., Okino, R., Daimon, H., Fujie, K. (2006). *J. Appl. Polym. Sci.*, **99**, 2245–2252.

217. Auras, R., Harte, B., Selke, S., Hernandez, R. (2003). *J. Plast. Film Sheet.*, **19**, 123–135.

218. Auras, R., Singh, S. P., Singh, J. J. (2005). *Packag. Technol. Sci.*, **18**, 207–216.

219. Michaels, A. S., Vieth, W. R., Barrie, J. A. (1963). *J. Appl. Phys.*, **34**, 13–20.

220. Ueda, H., Tabata, Y. (2003). *Adv. Drug Deliv. Rev.*, **55**, 501–518.

221. Eufinger, H., Rasche, C., Lehmbrock, J., Wehmoller, M., Weihe, S., Schmitz, I., Schiller, C., Epple, M. (2007). *Biomaterials*, **28**, 475–485.

222. Gilding, D. K. (1981). in: (Williams, D. F., ed.), *Biocompatibility of Clinical Implant Materials*, CRC Press, Boca Raton, pp. 209–232.

223. Suzuki, S., Ikada, Y. (2010). Medical applications, in: (Aurus, R., Lim, L. T., Selke, S. E. M., Tsuji, H., ed.), *Poly(lactic acid), Synthesis, Structure, Properties, Processing, and Application*, Wiley, New Jersey, Chapter 27, pp. 445–454.

224. Durselen, L., Hehl, G., Simnacher, M., Kinzl, L., Claes, L. (2001). *Clin. Biomech.*, **63**, 222–228.

225. Aoki, M., Miyamoto, S., Okamura, K., Yamashita, T., Ikada, Y., Matsuda, S. (2004). *J. Biomed. Mater. Res. Part B*, **71B**, 252–259.

226. Zilberman, M., Nelson, K. D., Eberhart, R. C. (2005). *J. Biomed. Mater. Res. Part B: Appl. Biomater.*, **74B**, 792–799.

227. Hanafusa, S., Matsusue, Y., Yasunaga, T., Yamamuro, T., Oka, M., Shikinami, Y., Ikada, Y. (1995). *Clin. Orthop. Relat. Res.*, **315**, 262–271.

228. Suzuki, T., Kawamura, H., Kasahara, T., Nagasaka, H. (2004). *J. Oral Maxillofac. Surg.*, **62**, 919–924.

229. Vaccaro, A. R., Singh, K., Haid, R., Kitchel, S., Wuisman, P., Taylor, W., Branch, C., Garfin, S. (2003). *Spine J.*, **3**, 227–237.

230. Liang, L. S., Wong, W., Burt, H. M. (2005). *J. Pharm. Sci.*, **94**, 1204–1215.

231. Schliephake, H., Weich, H. A., Schulz, J., Gruber, R. (2007). *J. Biomed. Mater. Res.*, **83A**, 455–462.

232. Japanese patent publication 08-283557, 2000-109663.

233. Obuchi, S., Ogawa, S. (2010). Packaging and other commercial applications, in: (Aurus, R., Lim, L. T., Selke, S. E. M., Tsuji, H., ed.),

Poly(lactic acid), Synthesis, Structure, Properties, Processing, and Application, Wiley, New Jersey, Chapter 28, pp. 457–467.

234. Gruber, P., O'Brien, M. (2002). Polylactide nature works_PLA, in: (Steinbuchel, A., Doi, Y., eds.), *Biopolymers, Vol. 4: Polyesters III*, Wiley-VCH Verlag GmbH, Germany, pp. 235–250.
235. Mochizuki, M., Murase, S., Matsunaga, N. (2006). *Sen'i Gakkaishi*, **62**(11), P-323–P-329.
236. Mutsuga, M. (2008). *Food Addit. Contam.*, **25**(10), 1285–1292.
237. Mochizuki, M. (2001). *Web J.*, **43**, 16–22.
238. Kawamoto, H. (2006). *Mater. Stage*, **6**(3), 12–19.
239. Mochizuki, M. (2010). Textile applications, in: (Aurus, R., Lim, L. T., Selke, S. E. M., Tsuji, H., ed.), *Poly(lactic acid), Synthesis, Structure, Properties, Processing, and Application*, Wiley, New Jersey, Chapter 29, pp. 469–476.

Chapter 12

Pectin and Its Roles in Transdermal Drug Delivery

Anuar Nor Khaizan[a,b,c] **and Tin Wui Wong**[a,b,c]

[a]*Non-Destructive Biomedical and Pharmaceutical Research Centre, Universiti Teknologi MARA, 42300 Puncak Alam, Selangor, Malaysia*
[b]*Particle Design Research Group, Faculty of Pharmacy, Universiti Teknologi MARA, 42300 Puncak Alam, Selangor, Malaysia*
[c]*CoRe Frontier Materials and Industry Application, Universiti Teknologi MARA, 40450 Shah Alam, Selangor, Malaysia*

wongtinwui@salam.uitm.edu.my

12.1 Introduction

Pectin is a linear polysaccharide that is available as the primary constituent of plant cell walls (Willats et al., 2006). It consists mainly of galacturonic acid residues joined in chains by α-(1–4) glycosidic linkages with their carboxyl groups partly esterified with methoxy moieties (Fig. 12.1). Commercial pectin is usually extracted from citrus peel and apple pomace (Wang et al., 2014). The pectin is non-toxic, water-soluble, biodegradable, and biocompatible (Ninan et al., 2013). As a natural polymer, pectin is polydisperse in molecular weight (Corredig et al., 2000). It has a complex chemical composition that can vary with its source, during plant extraction, distribution,

Handbook of Sustainable Polymers: Processing and Applications
Edited by Vijay Kumar Thakur and Manju Kumari Thakur

ISBN 978-981-4613-53-8 (Hardcover), 978-981-4613-54-5 (eBook)
www.panstanford.com

storage, formulation and processing (Sriamornsak, 2003). Pectin has excellent gelling properties and it is largely dependent on the molecular size and degree of esterification (DE) of polymer chain (Thakur et al., 1997). Pectin is broadly classified as high methoxyl (HM) (DE > 50%) and low methoxyl (LM) (DE < 50%) pectin (Itoh et al., 2008; Sriamornsak and Kennedy, 2008) (Fig. 12.1). In addition, the carboxyl groups of pectin can be further demethylated or amidated (Fig. 12.1).

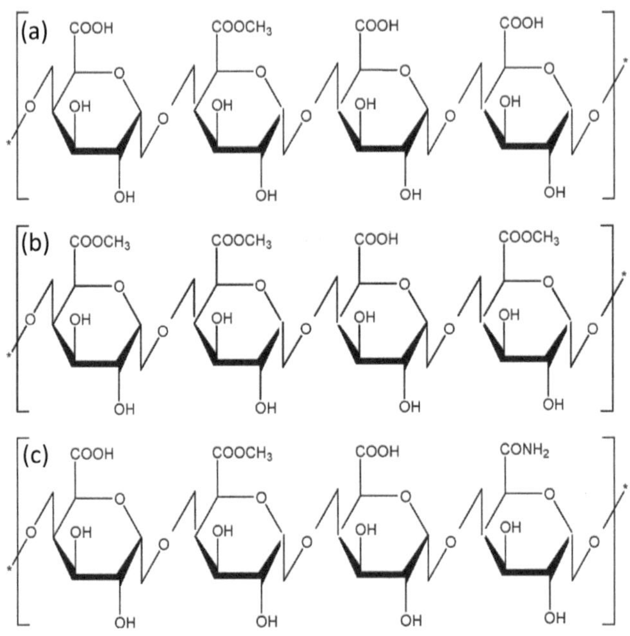

Figure 12.1 Chemical structures of (a) LM, (b) HM and (c) amidated pectin.

12.2 Pectin in Drug Delivery

Pectin shows significant potential in drug delivery. It is either used on its own or blended with other polymers to meet the desired physicochemical properties. Pectin has been used as an emulsifier to prepare nanoemulsion consisting of itraconazole, a poorly water-soluble drug (Burapapadh et al., 2010). An increase in pectin content, especially HM pectin leads to a decrease in emulsion droplet size. HM pectin, which is constituted of a high DE, provides a good emulsification property as a high fraction of hydrophobic ester moiety is available to adsorb at the oil droplet surfaces. This then enables the stabilization of oil droplets in aqueous continuous phase. The molecular interaction between itraconazole and pectin is deemed to bring about the formation of nano-sized emulsion. Under such circumstances, it is speculated that the itraconozaole may perform as a surfactant that gives a synergistic emulsifier effect to that of pectin.

Pectin has been employed as a surface coat, as well as stabilizer for liposomal drug delivery system (Nguyen et al., 2011; Smistad et al., 2012). Three different types of pectin, namely LM, HM and amidated pectin, are used to coat the positively charge liposome. Pectin coating of liposome results in a reproducible increase in liposomal particle size, change of zeta potential from positive to negative magnitude and reduction in aggregation propensity of liposome globules. However, the permeability of the liposome membrane remains unchanged with the application of pectin. It implies that the physical stability of liposome can be raised by pectin without affecting its drug release kinetics.

Pectin is extensively evaluated as an excipient for sustained- or controlled-release formulations (Carbinatto et al., 2014; Dutta and Sahu 2012a; Joseph and Venkataram, 1995; Kubo et al., 2004; Murata et al., 2004; Saravanan and Rao, 2010; Sriamornsak et al., 2007). Owing to its excellent gelling property, pectin is used as an in situ gelling vehicle for oral sustained drug delivery. The pectin has been blended with xyloglucan to increase the gel strength of xyloglucan solution (Itoh et al., 2008). The incorporation of pectin into xyloglucan solution is found to be able to minimize the gel erosion thus sustaining the release of drug. The LM pectin demonstrates a higher gelling tendency under high gastric pH in comparison to HM pectin. The LM pectin is susceptible to cross-link with di- and tri-valent ions that exist in the stomach gastric juice, hence making it potentially suitable for use as in situ gelling vehicle in oral sustained drug delivery (Itoh et al., 2007).

Pectin has been employed in magnetic field-modulated drug delivery (Da Silva et al., 2014; Dutta and Sahu, 2012a, 2012b, 2012c). The covalent TiO_2-co-pectin microspheres containing Fe_3O_4 nanoparticles are developed by ultrasound-induced crosslinking/polymerization technique with amoxicillin as the drug of interest. Sustained amoxicillin release is demonstrated by the formed nanostructured microspheres with their drug release site-specificity modulated by an external magnetic field (Da Silva et al., 2014). In another study, pectin based magnetic nanocarriers are loaded with 5-fluorouracil. These nanocarriers exhibit pH-responsive and swelling behavior that modulate the release of 5-fluorouracil in simulated gastric, intestinal and colonic fluids as a function of the magnetic field (Dutta and Sahu, 2012b).

Sol-gel conversion due to the crosslinking phenomenon between pectin and calcium ions is a widespread concept that has been used in the development of pharmaceutical innovations (Capel et al., 2006; Durand et al., 1990; Iijima et al., 2005; Lopes da Silva et al., 1996; Matia-Merino et al., 2004; Sriamornsak et al., 2005). An emulsion gel bead of calcium pectinate for use as oral floating drug delivery system has been developed using olive and peppermint oils as the floating agents (Sriamornsak et al., 2005). The floating dosage form is designed to maintain the delivery system at the upper gastrointestinal tract. This is necessary specifically for drugs that exert their therapeutic action in stomach or have poor bioavailability in post-gastric regions (Sriamornsak et al., 2007). Using metronidazole as a model drug, the floating formulation is reported to be able to sustain drug release, and float in the gastric spaces for 24 h due to matrix porosity brought about by the addition of olive and peppermint oils.

Pectin is a popular carrier for colonic drug delivery. This is attributed to its resistance to breakdown by upper gastrointestinal fluids and will, however, be degraded by the hydrolytic enzymes of bacteria in colon thus releasing the drug specifically in this regime (Elyagoby et al., 2013; Liu et al., 2003; Perera et al., 2010). Pectin/hydroxypropylmethylcellulose mixture has been used as a coating material for nisin tablet that is employed in the treatment of colonic infectious diseases (Ugurlu et al., 2007). It is found that an optimum combination of pectin and hydroxypropylmethylcellulose at the weight ratio of 4:1 preserves the integrity of nisin tablets during the 6 h dissolution study. A mere pectin coat is not adequate to maintain the integrity of nisin tablets since there is a limit to the amount of pectin that can be incorporated in an acceptable film coat for colon-specific drug delivery (Macleod et al., 1997). Further, it is learnt that the amidated pectin can create a firmer tablet and is more vulnerable to enzyme degradation in comparison to calcium pectinate (Ahrabi et al., 2000). The amidated pectin can act as a good binding agent and ferment faster in the colonic medium when compared to the calcium pectinate. With reference to colon-specific drug delivery, the latest investigation of pectin involves its use as carrier and coat materials for 5-fluorouracil in colon cancer and aberrant crypt foci treatment (Bose et al., 2014). The pectin is able to delay drug release from dosage form in upper gastrointestinal tract. Its role as a soluble binder in hydrophobic ethylcellulose coating onto pectin core can be mediated in vivo instead of conventionally prepared by fluid-bed technique in vitro.

Polyelectrolyte complexes made of chitosan and pectin have been prepared for colonic delivery of vancomycin (Bigucci et al., 2008). The pectin-chitosan complexes (molar ratio of 9:1) exhibit high mucoadhesive attribute, pH-sensitive swelling characteristics and desired colonic drug release profiles. In another study, it is demonstrated that polyelectrolyte complex of pectin and chitosan at a weight ratio of 2:1 is not able to prevent the matrix from swelling in an aqueous environment (Ghaffari et al., 2007). Incorporation of water-insoluble film-forming polymer such as Eudragit® RS can help to impede swelling and drug release during stomach and small intestine passage, thus making the composite a suitable material for sigmoidal colon drug delivery.

Pectin is used in non-viral gene delivery due to its biocompatibility, biodegradability, low immunogenicity, and minimal cytotoxicity profiles (Katav et al., 2008; Zhou et al., 2012). Pectin is developed as a DNA carrier using modified citrus pectin with three different positively charged amine groups (pectin-[NH_2], pectin-[$N^+H(CH_3)_2$] and pectin-NH_2-[$N^+(CH_3)_3$]) at physiological pHs (Katav et al., 2008). All modified pectin can produce complexes with plasmid DNA via electrostatic interaction. Nevertheless, the complexation and transfection efficacy vary with the type of amine group on pectin with pectin-NH_2-[$N^+(CH_3)_3$] showing the best performance from the perspective of a nonviral gene vehicle.

The mucoadhesive property of amidated pectin has been exploited in ocular drug delivery system design. Hydrogen bonding between the hydroxyl, carboxyl and amine groups of pectin results in its weak adhesion with mucus layer (Andrews et al., 2009). Thiolation of pectin increases polymer-mucus adhesion. The thiolated pectin is able to interact with cysteine-rich subdomains of mucus glycoprotein forming strong covalent bonds between the mucoadhesive pectin and mucus layer. The thiolated pectin nanoparticles have been fabricated as timolol maleate carrier for the treatment of ocular ailments. The ex vivo corneal permeation of drug increases with the use of thiolated pectin in nanoparticulate formulation in comparison to the conventional solution dosage form (Sharma et al., 2012).

12.3 Pectin in Transdermal Drug Delivery

Skin is the major organ of a human body and is predominantly resistant to the penetration of external molecules (Godin and Touitou, 2007). Transdermal drug delivery system is an alternative approach to the conventional drug administration methods, which attempt to utilize the skin for controlled delivery of drug molecules from the surface of the skin, through its layers, and to the systemic circulation. The drug can enter the systemic circulation without passing through the gastrointestinal wall and hepatic portal system thus avoiding the first-pass metabolism effect.

The skin is constituted of three important layers namely epidermis, dermis and subcutaneous tissue (Mezei and Gulasekharam, 1980; Richard et al., 1991) (Fig. 12.2). The epidermis is located at the outer layer of the skin and is made up of keratinocytes derived from the deepest layer of epidermis named the basal layer. The keratinocytes split and gradually migrate upwards to produce the stratum corneum at the surface of epidermis. The impermeable stratum corneum or "horny layer" contains dead and dense layer, which generally acts as a rate-limiting barrier to the drug penetration. The cells at this layer are flat and scale-like in shape. The dermis is a thick layer of fibrous and elastic tissue, which lies directly below the epidermis. The proteins in dermis, particularly collagen, elastin, and reticulin, provide strength and flexibility to the skin. Dermis contains various supporting tissues as well as blood vessels, nerves, hair roots and sweat glands. The large amount of blood volume in the skin generally acts as a "sink" for diffusing molecules reaching the capillaries, maintaining a very low concentration of penetrant in the dermis, maximizing epidermal concentration gradient, and thus promoting percutaneous absorption. Below the dermis is the subcutaneous tissue, which is a layer of fat and connective tissue that provides a mechanical cushion and thermal barrier, as well as serves as an energy storage area. The thickness of skin varies in different regions of the body as well as between individuals due to skin hydration, body temperature, ethnic differences and skin infection (Holbrook and Odland, 1974; Whitton and Everall, 1973).

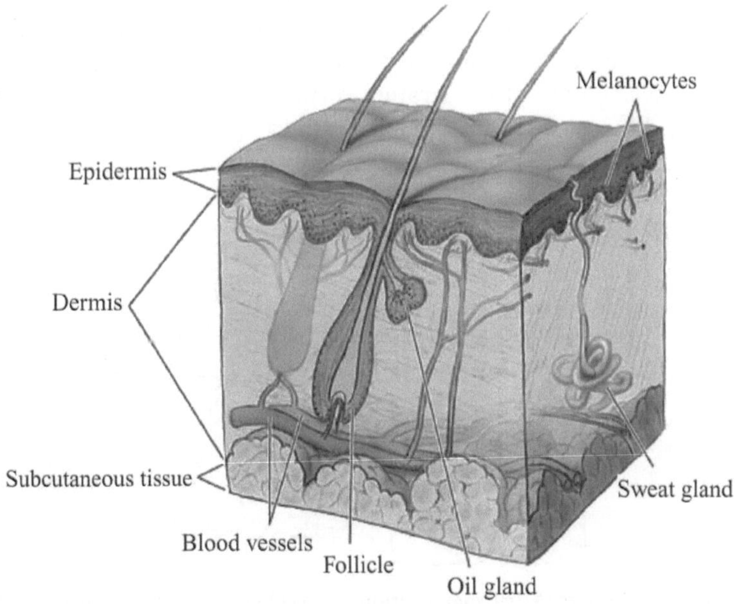

Figure 12.2 Diagram of skin layers (Bliss, 2005).

There are three potential pathways for a molecule to diffuse into the skin (Fig. 12.3). The drug molecules can penetrate through the hair follicles with their associated sebaceous glands, via the sweat ducts, or across the continuous stratum corneum between these appendages (Huang et al., 2005). The appendageal routes avoid penetration through the stratum corneum, hence known as shunt routes. Nevertheless, these routes are considered to be less significant because of their relatively small area, approximately 0.1% of the total skin area and these routes usually do not play a substantial role in steady-state drug flux. Conversely, the routes appear to be crucial for ions and large polar molecules that encounter difficulty to permeate the stratum corneum (Aulton, 2002; Reddy and Mehvar, 2004).

The structure of the stratum corneum has been illustrated as a "bricks and mortar" structure, which is similar to a wall. The "bricks" are mainly corneocytes of hydrated keratin and the "mortar" is composed of a complex lipid mixture of ceramides, fatty acids, cholesterol and cholesterol esters (Chen et al., 2013; Rim et al., 2008) (Fig. 12.3). There are two possible routes of penetration via the stratum corneum, which either through intracellular or intercellular route. Hydrophilic drugs permeate favorably into the intracellular domains though it could be challenging due to the low water content of the stratum corneum, while lipophilic drugs usually diffuse via the intercellular pathway (Moghimi et al., 1996). Most molecules traverse the stratum corneum by both routes with the lipid bilayers of the intercellular route offer the major pathway.

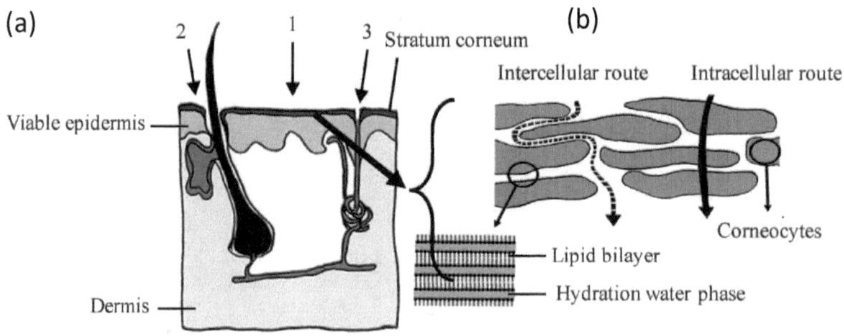

Figure 12.3 Route of drug penetration via (a) (1) stratum corneum, (2) hair follicles, and (3) sweat ducts; (b) intercellular and intracellular pathways of stratum corneum.

There is a limited investigation of pectin in transdermal drug delivery though studies indicate that it is a highly promising matrix polymer for controlled drug release and skin permeation (Table 12.1). Pectin has been used as a matrix polymer for transdermal delivery system of macromolecular drugs such as insulin (Hadebe et al., 2014; Tufts and Musabayane, 2010). The HM and LM pectins have been used to coat the vitamin C loaded liposome (Zhou et al., 2014). The pectin is mainly coated on the surfaces of liposome by means of hydrogen bonding. The coated liposome demonstrates equivalent vitamin C entrapment efficiency as the uncoated liposome. The pectin-coated liposomal formulations have their storage stability improved with reduce rates of aggregation, lipid oxidation and vitamin C leakage.

The pectin-based transdermal matrix films have been formulated to provide an effective long-term plasma drug concentration

(Basavaraj et al., 2010; Bektaş et al., 2014; Güngör et al., 2008; Kumar et al., 2010; Sharma et al., 2011; Wong and Nor Khaizan, 2013).

Table 12.1 Pectin-based transdermal drug delivery systems

Polymer	Model drug	Permeation enhancer	Dosage form	References
Pectin	Vitamin C	—	Liposome	(Zhou et al., 2014)
Pectin	Nifedipine	Nerolidol	Matrix film	(Bektaş et al., 2014)
Pectin	Sulfanilamide	• Oleic acid • Microwave	• Gel • Matrix film	(Nor Khaizan and Wong, 2013; Wong and Nor Khaizan, 2013)
Pectin	Meloxicam	—	Matrix film	(Sharma et al., 2011)
Pectin	Naproxen sodium	• Isopropyl myristate • Sodium lauryl sulfate	Gel	(Attia, 2009)
Pectin	Verapamil hydrochloride	• Nerolidol • D-limonene • Eucalyptol • Menthone • Menthol	Matrix film	(Güngör et al., 2008)
Amidated pectin	Chloroquine	—	Matrix hydrogel	(Musabayane et al., 2003; Sibiya et al., 2014)
Amidated pectin	Insulin	—	Hydrogel	(Hadebe et al., 2014; Tufts and Musabayane, 2010)
Pectin and gelatin	Testosterone	—	Hydrogel	(Mazzitelli et al., 2013)
Pectin and guar gum	Carvedilol	Polyethylene glycol 400	Matrix film	(Karemore et al., 2012)
Pectin and sodium alginate	Pioglitazone hydrochloride	Menthol	Matrix film	(Hiroji Megha et al., 2012)
Pectin and hydroxypropylmethylcellulose	Sodium nonivamide acetate	Iontophoresis	Hydrogel	(Fang et al., 1998)

Pectin films carrying model drugs such as nifedipine (Bektaş et al., 2014), sulfa drugs (Anuar et al., 2012; Wong and Nor Khaizan, 2013), meloxicam (Sharma et al., 2011), verapamil hydrochloride (Güngör et al., 2008), chloroquine (Sibiya et al., 2014), carvedilol (Karemore et al., 2012) and pioglitazone hydrochloride (Hiroji Megha et al., 2012) are developed by solvent-evaporation method. These films are examined for their in vitro release and physicochemical properties including drug-polymer compatibility, thickness, weight uniformity, tensile strength, folding endurance, flatness, swelling index and drug content uniformity, and skin irritation. Plasticizers, namely propylene glycol (Bektaş et al., 2014; Güngör et al., 2008) and glycerine (Bektaş et al., 2014; Hiroji Megha et al., 2012; Karemore et al., 2012), are deliberately added in the formulation to improve the mechanical strength, flexibility and bioadhesiveness of the matrix films that are required during transdermal administration.

Chloroquine is one of the drugs used in falciparum malaria treatment in sub-Saharan Africa (Sibiya et al., 2014). It is demonstrated that both oral and transdermal chloroquine treatment are capable of reducing the *Plasmodium berghei* parasites in the blood stream of male *Sprague dawley* rats to undetectable levels by day 5. The transdermally delivered chloroquine has successfully sustained the plasma drug concentration with practically no significant changes in the blood glucose levels, as well as urinary Na^+ and K^+ excretion of non-infected and infected rats in comparison to the oral treatment. The results suggest that the pectin-chloroquine matrix film can potentially circumvent the adverse effects related to glycemia and kidney malfunction that are associated with the oral delivery of chloroquine.

In development of transdermal drug delivery system, pectin is commonly blended with other polymers such as gelatin, hydroxypropylmethylcellulose, guar gum and sodium alginate. The blend of pectin and gelatin is employed in transdermal hydrogel patches and they are prepared using a two-step gelation procedure (Mazzitelli et al., 2013). The preparation starts off with thermal gelation of gelatin, followed by ionic gelation characterized by the typical egg-box structure formation of pectin due to the introduction of divalent cations. The other polymer blends, pectin and hydroxypropylmethylcellulose, pectin and guar gum as well as pectin and sodium alginate, are practically prepared by either solvent-evaporation method or simple gelling mechanism. Polymer blending leads to improved mechanical and physicochemical properties, along with the controlled-release transdermal drug delivery attribute. Other critical additives that are employed in pectin-based transdermal drug delivery system include plasticizer and permeation enhancer. The latter is essential to address the absorption barrier issue of stratum corneum.

12.3.1 Permeation Enhancement Mechanism

Pectin by itself can acts as a skin permeation enhancer for drug molecules (Fig. 12.4). Sulfanilamide is a polar antibiotic. It demonstrates a very low skin permeation profile with an extent of drug permeation of 13.45 ± 7.55% at 24 h of application. The formulation of sulfanilamide into pectin gel increases its skin permeation extent to 62.97 ± 1.66% (Nor Khaizan and Wong, 2013). The permeation enhancement action of pectin is possibly a resultant effect of its bioadhesivity. A mucoadhesive pectin can increase the contact between the drug matrix and skin. This denotes a closer proximity between drug and skin interface, hence increasing the drug bioavailability (Anuar et al., 2012; Zhou et al., 2014).

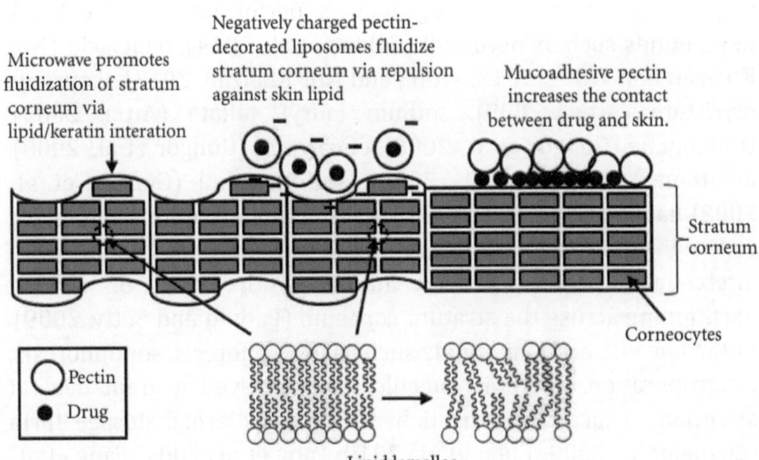

Figure 12.4 Skin permeation enhancement by pectin and microwave for transdermal drug delivery.

Similar observation is found with pectin-coated vitamin C loaded liposome. The pectin-coated liposome exhibits an increased vesicle permeability up to 2.1-fold after 24 h of application against that of the uncoated liposome (Zhou et al., 2014). The average diameter of pectin-coated liposome is larger than the uncoated liposome. The size of the liposomal vesicles does not play a significant role in the rate of vesicle permeation. In the case of liposomal system, the surface charges of vesicles can contribute to skin permeation. According to several reports (Cui and Mumper, 2001; Gillet et al., 2011; Kohli and Alpar, 2004), the negatively charged liposomes usually result in a higher drug flux and greater diffusion coefficient than the positively charged or neutral liposomes. The skin permeation of LM pectin coated-liposome with a surface charge of –35.5 mV is higher than that of the uncoated liposome, which exhibits a surface charge of –2.3 mV. The skin acts as a negatively charged membrane due to the high fractions of negatively charged lipid in the stratum corneum. The permeation enhancement effect brought about by the negative surface charges of liposomes originates from the repulsive forces between the skin and the vesicles (Gillet et al., 2011; Kohli and Alpar, 2004) (Fig. 12.4). It then results in temporary channels formed within the stratum corneum and eases molecular permeation.

Chemical permeation enhancer is commonly used with pectin to promote transdermal drug delivery performance (Table 12.1). Compounds such as nerolidol (Bektaş et al., 2014), oleic acid (Nor Khaizan and Wong, 2013; Wong and Nor Khaizan, 2013), isopropyl myristate (Attia, 2009), sodium lauryl sulfate (Attia, 2009), D-limonene (Güngör et al., 2008), eucalyptol (Güngör et al., 2008), menthone (Güngör et al., 2008), and menthol (Güngör et al., 2008) have been used. The enhancers act by disrupting the highly ordered lipid structure of stratum corneum, interacting with intercellular protein and/or improving the drug or solvent partitioning across the stratum corneum (Pathan and Setty, 2009). In the late era, active methods such as iontophoresis, sonophoresis, electroporation, and microneedle have received a great deal of attention to facilitate drug delivery of transdermal dosage form (Chang et al., 2000; Chen et al., 2010; Fang et al., 2004; Fang et al., 2002; Le et al., 2000; Lin et al., 2001; Tokumoto et al., 2006; Wang et al., 2005; Yan et al., 2010). The most recent active method refers to microwave technology that can increase the skin permeability to drug embedded in solid pectin matrix (Wong and Nor Khaizan, 2013). The microwave exerts spacing of lipid structure of stratum corneum into structureless domains (Fig. 12.4). Using solid pectin matrix enriched with chemical permeation enhancer such as oleic acid, the microwave has shown to facilitate the migration of oleic acid from skin surfaces to stratum corneum and synergistically inducing lipid/keratin fluidization and promoting drug permeation.

Microwave appears to be a less favorable active transdermal device when it is used with pectin formulated in the form of a semisolid gel and even with oleic acid (Nor Khaizan and Wong, 2013). In gel, the pectin molecules are relatively free flowing when compared to solid matrix. Through stratum corneum fluidization by microwave and/or oleic acid, a strong pectin-skin binding effect can be initiated. This hinders the penetration of drug molecules across the stratum corneum into dermis and systemic circulation. In a different study, pectin is similarly found to be unsuitable for transdermal delivery of sodium nonivamide acetate when the iontophoresis is used as the physical permeation enhancer (Fang et al., 1998). The anionic pectin could have competed with the drug to permeate the skin made porous by iontophoresis procedure.

12.4 Conclusion

Pectin has a great potential in transdermal drug delivery application with respect to its controlled release, physical stabilization, mucoadhesive, and permeation enhancement properties. Nevertheless, limited studies have been conducted in transdermal delivery research unlike oral dosage form design. Future studies are required to further probe the capacity of pectin in transdermal drug delivery as a function of dosage form, combination technique with active methods such as microwave technology, physicochemical properties of drugs, and excipients.

References

Ahrabi S. F., Madsen G., Dyrstad K., Sande S. A., Graffner C. (2000). Development of pectin matrix tablets for colonic delivery of model drug ropivacaine. *Eur. J. Pharm. Sci.*, **10**, 43–52.

Andrews G. P., Laverty T. P., Jones D. S. (2009). Mucoadhesive polymeric platforms for controlled drug delivery. *Eur. J. Pharm. Biopharm.*, **71**, 505–518.

Anuar N. K., Wong T. W., Taib M. N. (2012). Microwave modified non-crosslinked pectin films with modulated drug release. *Pharm. Dev. Technol.*, **17**, 110–117.

Attia D. A. (2009). In vitro and in vivo evaluation of transdermal absorption of naproxen sodium. *Australian J. Basic Appl. Sci.*, **3**, 2154–2165.

Aulton M. E., ed. (2002). *Pharmaceutics: The Science of Dosage Form Design*, 2nd ed.. Churchill Livingstone Press, Spain.

Basavaraj K. H., Johnsy G., Navya M. A., Rashmi R., Siddaramaiah (2010). Biopolymers as transdermal drug delivery systems in dermatology therapy. *Crit. Rev. Ther. Drug*, **27**, 155–185.

Bektaş A., Cevher E., Güngör S., Ozsoy Y. (2014). Design and evaluation of polysaccharide-based transdermal films for the controlled delivery of nifedipine. *Chem. Pharm. Bull.*, **62**, 144–152.

Bigucci F., Luppi B., Cerchiara T., Sorrenti M., Bettinetti G., Rodriguez L., Zecchi V. (2008). Chitosan/pectin polyelectrolyte complexes: Selection of suitable preparative conditions for colon-specific delivery of vancomycin. *Eur. J. Pharm. Sci.*, **35**, 435–441.

Bliss D. (2005). *Skin (Layers, Glands, Vessels)*, National Cancer Institute.

Bose A., Elyagoby A., Wong T. W. (2014). Oral 5-fluorouracil colon-specific delivery through in vivo pellet coating for colon cancer and aberrant crypt foci treatment. *Int. J. Pharm.*, **468**, 178–186.

Burapapadh K., Kumpugdee-Vollrath M., Chantasart D., and Sriamornsak P. (2010). Fabrication of pectin-based nanoemulsions loaded with itraconazole for pharmaceutical application. *Carbohyd. Polym.*, **82**, 384–393.

Capel F., Nicolai T., Durand D., Boulenguer P., Langendorff V. (2006). Calcium and acid induced gelation of (amidated) low methoxyl pectin. *Food Hydrocolloid.*, **20**, 901–907.

Carbinatto F. M., De Castro A. D., Evangelista R. C., Cury B. S. F. (2014). Insights into the swelling process and drug release mechanisms from cross-linked pectin/high amylose starch matrices. *Asian J. Pharm. Sci.*, **9**, 27–34.

Chang S.-L., Hofmann A., Zhang L., Deftos L. J., Banga A. K. (2000). The effect of electroporation on iontophoretic transdermal delivery of calcium regulating hormones. *J. Control. Release*, **66**, 127–133.

Chen L., Han L., Lian G. (2013). Recent advances in predicting skin permeability of hydrophilic solutes. *Adv. Drug Deliver. Rev.*, **65**, 295–305.

Chen B., Wei J., Iliescu C. (2010). Sonophoretic enhanced microneedles array (SEMA)—Improving the efficiency of transdermal drug delivery. *Sensor. Actuat. B Chem.*, **145**, 54–60.

Corredig M., Kerr W., Wicker L. (2000). Molecular characterization of commercial pectins by separation with linear mix gel permeation columns in-line with multi-angle light scattering detection. *Food Hydrocolloid.*, **14**, 41–47.

Cui Z., Mumper R. J. (2001). Chitosan-based nanoparticles for topical genetic immunization. *J. Control. Release*, **75**, 409–419.

Da Silva E. P., Sitta D. L. A., Fragal V. H., Cellet T. S. P., Mauricio M. R., Garcia F. P., Nakamura C. V. (2014). Covalent TiO_2/pectin microspheres with Fe_3O_4 nanoparticles for magnetic field-modulated drug delivery. *Int. J. Biol. Macromol.*, **67**, 43–52.

Durand D., Bertrand C., Clark A. H., Lips A. (1990). Calcium-induced gelation of low methoxy pectin solutions: Thermodynamic and rheological considerations. *Int. J. Biol. Macromol.*, **12**, 14–18.

Dutta R. K., Sahu S. (2012a). Development of diclofenac sodium loaded magnetic nanocarriers of pectin interacted with chitosan for targeted and sustained drug delivery. *Colloid. Surface. B*, **97**, 19–26.

Dutta R. K., Sahu S. (2012b). Development of a novel probe sonication assisted enhanced loading of 5-FU in SPION encapsulated pectin nanocarriers for magnetic targeted drug delivery system. *Eur. J. Pharm. Biopharm.*, **82**, 58–65.

Dutta R. K., Sahu S. (2012c). Development of oxaliplatin encapsulated in magnetic nanocarriers of pectin as a potential targeted drug delivery for cancer therapy. *Results Pharm. Sci.*, **2**, 38–45.

Elyagoby A., Layas N., Wong T. W. (2013). Colon-specific delivery of 5-fluorouracil from zinc pectinate pellets through in situ intra-capsular ethylcellulose-pectin plug formation. *J. Pharm. Sci.*, **102**, 604–616.

Fang J., Hung C., Fang Y., Chan T. (2004). Transdermal iontophoresis of 5-fluorouracil combined with electroporation and laser treatment. *Int. J. Pharm.*, **270**, 241–249.

Fang J.-Y., Huang Y.-B., Lin H.-H., Tsai Y.-H. (1998). Transdermal iontophoresis of sodium nonivamide acetate. IV. Effect of polymer formulations. *Int. J. Pharm.*, **173**, 127–140.

Fang J., Hwang T., Huang Y., Tsai Y. (2002). Transdermal iontophoresis of sodium nonivamide acetate V. Combined effect of physical enhancement methods. *Int. J. Pharm.*, **235**, 95–105.

Ghaffari A., Navaee K., Oskoui M., Bayati K., Rafiee-Tehrani M. (2007). Preparation and characterization of free mixed-film of pectin/chitosan/eudragit RS intended for sigmoidal drug delivery. *Eur. J. Pharm. Biopharm.*, **67**, 175–186.

Gillet A., Compère P., Lecomte F., Hubert P., Ducat E., Evrard B., Piel G. (2011). Liposome surface charge influence on skin penetration behaviour. *Int. J. Pharm.*, **411**, 223–231.

Godin B., Touitou E. (2007). Transdermal skin delivery: Predictions for humans from in vivo, ex vivo and animal models. *Adv. Drug Deliver. Rev.*, **59**, 1152–1161.

Güngör S., Bektaş A., Alp F. I., Uydeş-Doğan B. S., Özdemir O., Araman A., Özsoy Y. (2008). Matrix-type transdermal patches of verapamil hydrochloride: In vitro permeation studies through excised rat skin and pharmacodynamic evaluation in rats. *Pharm. Dev. Technol.*, **13**, 283–289.

Hadebe S. I., Ngubane P. S., Serumula M. R., Musabayane C. T. (2014). Effects of transdermally delivered insulin on some selected metabolic parameters of streptozotocin-induced diabetic male Sprague–Dawley rats. *Endocrine Abstracts* (p. 34).

Hiroji Megha B., Nagesh C., Devdatt J., Chandrashekhara S. (2012). Development and evaluation of transdermal drug delivery system using natural polysaccharides. *Res. J. Pharm. Dosage Forms Tech.*, **4**, 278–284.

Holbrook K. A., Odland G. F. (1974). Regional differences in the thickness (cell layers) of the human stratum corneum: An ultrastructural analysis. *J. Invest. Dermatol.*, **62**, 415–422.

Huang X., Tanojo H., Lenn J., Helen Deng C., Krochmal L. (2005). A novel foam vehicle for delivery of topical corticosteroids. *J. Am. Acad. Dermatol.*, **53**, S26–S38.

Iijima M., Hatakeyama T., Hatakeyama H. (2005). Swelling behaviour of calcium pectin hydrogels by thermomechanical analysis in water. *Thermochim. Acta*, **431**, 68–72.

Itoh K., Hirayama T., Takahashi A., Kubo W., Miyazaki S., Dairaku M., Togashi M. (2007). In situ gelling pectin formulations for oral drug delivery at high gastric pH. *Int. J. Pharm.*, **335**, 90–96.

Itoh K., Yahaba M., Takahashi A., Tsuruya R., Miyazaki S., Dairaku M., Togashi M. (2008). In situ gelling xyloglucan/pectin formulations for oral sustained drug delivery. *Int. J. Pharm.*, **356**, 95–101.

Joseph I., Venkataram S. (1995). Indomethacin sustained release from alginate-gelatin or pectin-gelatin coacervates. *Int. J. Pharm.*, **126**, 161–168.

Karemore, Megha N. D., Mayuri S., Belgamwar A. V., Mundhada D. R., Bhaskaran S. (2012). Design and evaluation of carvedilol transdermal patch using natural polymers. *J. Pharm. Res.*, **5**, 4947–4949.

Katav T., Liu L., Traitel T., Goldbart R., Wolfson M., Kost J. (2008). Modified pectin-based carrier for gene delivery: Cellular barriers in gene delivery course. *J. Control. Release*, **130**, 183–191.

Kohli A. K., Alpar H. O. (2004). Potential use of nanoparticles for transcutaneous vaccine delivery: Effect of particle size and charge. *Int. J. Pharm.*, **275**, 13–17.

Kubo W., Miyazaki S., Dairaku M., Togashi M., Mikami R., Attwood D. (2004). Oral sustained delivery of ambroxol from in situ-gelling pectin formulations. *Int. J. Pharm.*, **271**, 233–240.

Kumar M., Chauhan A. K., Kumar S., Kumar A., Malik S. (2010). Design and evaluation of pectin based metrics for transdermal patches of meloxicam. *J. Pharm. Res. Health Care*, **2**, 244–247.

Le L., Kost J., Mitragotri S. (2000). Combined effect of low-frequency ultrasound and iontophoresis Applications for transdermal heparin delivery. *Pharm. Res.*, **17**, 1151–1154.

Lin W., Cormier M., Samiee A., Griffin A., Johnson B., Teng C.-L., Hardee G. E. (2001). Transdermal delivery of antisense oligonucleotides with microprojection patch (Macroflux®) technology. *Pharm. Res.*, **18**, 1789–1793.

Liu L., Fishman M. L., Kost J., Hicks K. B. (2003). Pectin-based systems for colon-specific drug delivery via oral route. *Biomaterials*, **24**, 3333–3343.

Lopes da Silva J. A., Gonçalves M. P., Doublier J. L., Axelos M. A. V. (1996). Effect of galactomannans on the viscoelastic behaviour of pectin/calcium networks. *Polym. Gels Netw.*, **4**, 65–83.

Macleod G. S., Fell J. T., Collett J. H. (1997). Studies on the physical properties of mixed pectin/ethylcellulose films intended for colonic drug delivery. *Int. J. Pharm.*, **157**, 53–60.

Matia-Merino L., Lau K., Dickinson E. (2004). Effects of low-methoxyl amidated pectin and ionic calcium on rheology and microstructure of acid-induced sodium caseinate gels. *Food Hydrocolloid.*, **18**, 271–281.

Mazzitelli S., Pagano C., Giusepponi D., Nastruzzi C., Perioli L. (2013). Hydrogel blends with adjustable properties as patches for transdermal delivery. *Int. J. Pharm.*, **454**, 47–57.

Mezei M., Gulasekharam V. (1980). Liposomes: A selective drug delivery system for the topical route of administration I. Lotion dosage form. *Life Sci.*, **26**, 1473–1477.

Moghimi H. R., Williams A. C., Barry B. W. (1996). A lamellar matrix model for stratum corneum intercellular lipids. II. Effect of geometry of the stratum corneum on permeation of model drugs 5-fluororouracil and oestradiol. *Int. J. Pharm.*, **131**, 117–129.

Murata Y., Miyashita M., Kofuji K., Miyamoto E., Kawashima S. (2004). Drug release properties of a gel bead prepared with pectin and hydrolysate. *J. Control. Release*, **95**, 61–66.

Musabayane C. T., Munjeri O., Matavire T. P. (2003). Transdermal delivery of chloroquine by amidated pectin hydrogel matrix patch in the rat. *Ren. Fail.*, **25**, 525–534.

Nguyen S., Alund S. J., Hiorth M., Kjøniksen A.-L., Smistad G. (2011). Studies on pectin coating of liposomes for drug delivery. *Colloid. Surf. B*, **88**, 664–673.

Ninan N., Muthiah M., Park I.-K., Elain A., Thomas S., Grohens Y. (2013). Pectin/carboxymethyl cellulose/microfibrillated cellulose composite scaffolds for tissue engineering. *Carbohyd. Polym.*, **98**, 877–885.

Nor Khaizan A., Wong T. W. (2013). Microwave: Effects and implications in transdermal drug delivery. *Prog. Electromagn. Res.*, **141**, 619–643.

Pathan I. B., Setty C. M. (2009). Chemical penetration enhancers for transdermal drug delivery systems. *Trop. J. Pharm. Res.*, **8**, 173–179.

Perera G., Barthelmes J., Bernkop-Schnürch A. (2010). Novel pectin-4-aminothiophenole conjugate microparticles for colon-specific drug delivery. *J. Control. Release*, **145**, 240–246.

Reddy I. K., Mehvar R. (2004). *Chirality in Drug Design and Development.* CRC Press, USA.

Richard S., Querleux B., Bittoun J., Idy-Peretti I., Jolivet O., Cermakova E., Lévêque J.-L. (1991). In vivo proton relaxation times analysis of the skin layers by magnetic resonance imaging. *J. Invest. Dermatol.*, **97**, 120–125.

Rim J. E., Pinsky P. M., Van Osdol W. W. (2008). Using the method of homogenization to calculate the effective diffusivity of the stratum corneum with permeable corneocytes. *J. Biomech.*, **41**, 788–796.

Saravanan M., Rao K. P. (2010). Pectin–gelatin and alginate–gelatin complex coacervation for controlled drug delivery: Influence of anionic polysaccharides and drugs being encapsulated on physicochemical properties of microcapsules. *Carbohyd. Polym.*, **80**, 808–816.

Sharma R., Ahuja M., Kaur H. (2012). Thiolated pectin nanoparticles: Preparation, characterization and ex vivo corneal permeation study. *Carbohyd. Polym.*, **87**, 1606–1610.

Sharma K., Singh V., Arora A. (2011). Natural biodegradable polymers as matrices in transdermal drug delivery. *Int. J. Drug Dev. Res.*, **3**, 85–103.

Sibiya H., Serumula M., Musabayane C. T. (2014). Evaluation of the efficacy of transdermal delivery of chloroquine on plasmodium berghei-infected male sprague–dawley rats: Effects on blood glucose and renal electrolyte handling. *Endocrine Abstracts* (p. 34).

Smistad G., Bøyum S., Alund S. J., Samuelsen A. B. C., Hiorth M. (2012). The potential of pectin as a stabilizer for liposomal drug delivery systems. *Carbohyd. Polym.*, **90**, 1337–1344.

Sriamornsak P. (2003). Chemistry of pectin and its pharmaceutical uses: A review. *Silpakorn Univ. J. Soc. Sci. Humanit. Arts*, **3**, 206–228.

Sriamornsak P., Kennedy R. A. (2008). Swelling and diffusion studies of calcium polysaccharide gels intended for film coating. *Int. J. Pharm.*, **358**, 205–213.

Sriamornsak P., Sungthongjeen S., Puttipipatkhachorn S. (2007). Use of pectin as a carrier for intragastric floating drug delivery: Carbonate salt contained beads. *Carbohyd. Polym.*, **67**, 436–445.

Sriamornsak P., Thirawong N., Puttipipatkhachorn S. (2005). Emulsion gel beads of calcium pectinate capable of floating on the gastric fluid: Effect of some additives, hardening agent or coating on release behavior of metronidazole. *Eur. J. Pharm. Sci.*, **24**, 363–373.

Sriamornsak P., Thirawong N., Weerapol Y., Nunthanid J., Sungthongjeen S. (2007). Swelling and erosion of pectin matrix tablets and their impact on drug release behavior. *Eur. J. Pharm. Biopharm.*, **67**, 211–219.

Thakur B. R., Singh R. K., Handa A. K., Rao M. A. (1997). Chemistry and uses of pectin—A review. *Crit. Rev. Food Sci.*, **37**, 47–73.

Tokumoto S., Higo N., Sugibayashi K. (2006). Effect of electroporation and pH on the iontophoretic transdermal delivery of human insulin. *Int. J. Pharm.*, **326**, 13–19.

Tufts M., Musabayane C. (2010). Transdermal delivery of insulin using amidated pectin hydrogel patches. *Endocrine Abstracts* (p. 21).

Ugurlu T., Turkoglu M., Gurer U. S., Akarsu B. G. (2007). Colonic delivery of compression coated nisin tablets using pectin/HPMC polymer mixture. *Eur. J. Pharm. Biopharm.*, **67**, 202–210.

Wang X., Chen Q., Lü X. (2014). Pectin extracted from apple pomace and citrus peel by subcritical water. *Food Hydrocolloid.*, **38**, 129–137.

Wang Y., Thakur R., Fan Q., Michniak B. (2005). Transdermal iontophoresis: Combination strategies to improve transdermal iontophoretic drug delivery. *Eur. J. Pharm. Biopharm.*, **60**, 179–191.

Whitton J. T., Everall J. D. (1973). The thickness of the epidermis. *Brit. J. Dermatol.*, **89**, 467–476.

Willats W. G., Knox J. P., Mikkelsen J. D. (2006). Pectin: New insights into an old polymer are starting to gel. *Trends Food Sci. Tech.*, **17**, 97–104.

Wong T. W., Nor Khaizan A. (2013). Physicochemical modulation of skin barrier by microwave for transdermal drug delivery. *Pharm. Res.*, **30**, 90–103.

Yan K., Todo H., Sugibayashi K. (2010). Transdermal drug delivery by in-skin electroporation using a microneedle array. *Int. J. Pharm.*, **397**, 77–83.

Zhou W., Liu W., Zou L., Liu W., Liu C., Liang R., Chen J. (2014). Storage stability and skin permeation of vitamin C liposomes improved by pectin coating. *Colloid. Surf. B*, **117**, 330–337.

Zhou Y., Yang B., Ren X., Liu Z., Deng Z., Chen L., Deng Y. (2012). Hyperbranched cationic amylopectin derivatives for gene delivery. *Biomaterials*, **33**, 4731–4740.

Chapter 13

Development in Tunicate Cellulose

Guangzhe Piao and Dawei Zhang

Key Laboratory of Rubber-Plastics Ministry of Education,
School of Polymer Science and Engineering,
Qingdao University of Science and Technology,
Qingdao 266042, China

piao@qust.edu.cn

13.1 Lyotropic N*-LCs of Tunicate Cellulose Nanocrystals (T-CNs)

13.1.1 Introduction

Cellulose is one of the most important natural polymers, a key source of sustainable materials on industry, and an almost inexhaustible raw material. Natural cellulose based materials have been used by our society for thousands of years and their use continues today. The first example was the fabrication of regenerated cellulose filaments by spinning a solution of cellulose in a mixture of copper hydroxide and aqueous ammonia [1]. The chemical modification of cellulose on an industrial scale led to a broad range of products based on cellulose. Since then the isolation, characterization, and search for applications of novel

Handbook of Sustainable Polymers: Processing and Applications
Edited by Vijay Kumar Thakur and Manju Kumari Thakur

ISBN 978-981-4613-53-8 (Hardcover), 978-981-4613-54-5 (eBook)
www.panstanford.com

forms of cellulose, variously termed crystallites, nanocrystals, and whiskers, is generating much activity.

Cellulose is produced by plants, trees, bacteria, and some animals (tunicates) via the condensation polymerization of glucose, which is a product of the photosynthesis process in plants and trees. Long chains of anhydroglucose units, joined via β-1, 4-glycosidic linkages (C–O–C) (Fig. 13.1), are formed during this process [2, 3]. Cellulose, in its native state (termed cellulose-I) in the cell walls of plants and trees, forms a crystalline structure, where by chains aggregate via hydrogen bonding.

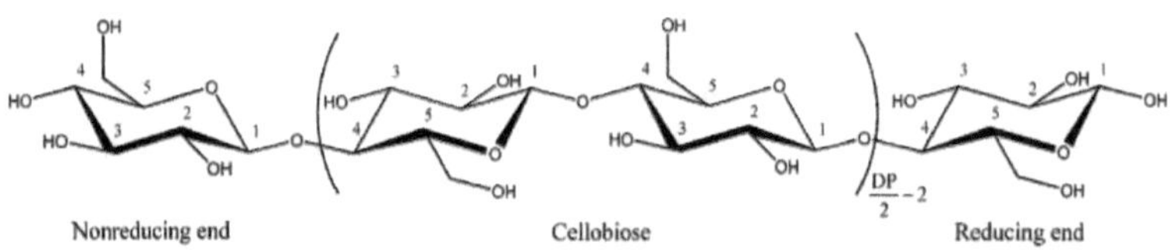

Figure 13.1 Chemical structure of cellulose [2].

There are several polymorphs of crystalline cellulose-I, II, III, and IV. Each has been extensively studied [4]. Crystalline cellulose that is naturally produced by a variety of organisms, it is sometimes referred to as "natural" cellulose. Cellulose-I has two polymorphs, a triclinic structure (I_α) and a monoclinic structure (I_β), which coexist in various proportions depending on the cellulose source. The I_α structure is the dominate polymorph for most algae and bacteria, whereas I_β is the dominant polymorph for higher plant cell wall cellulose and in tunicates [5–7]. Allomorph ratios are species specific, and this gives rise to natural structural variations in cellulose crystals. However, the mechanisms contributing to crystal formation remain unknown [8].

Tunicates, the only animals known to produce cellulose, are a family of sea animals that have a mantle consisting of cellulose microfibrils embedded in a protein matrix (Fig. 13.2). The thick leathery mantle in their mature phase is used as a new source of cellulose. Most research has used a class of tunicates that are commonly known as "sea squirts" (Ascidiacea), marine invertebrate filter feeders [9]. The structure and properties of tunicate cellulose (TC) are expected to be comparable between species, but there may be small differences in the cellulose microfibril formation process [10].

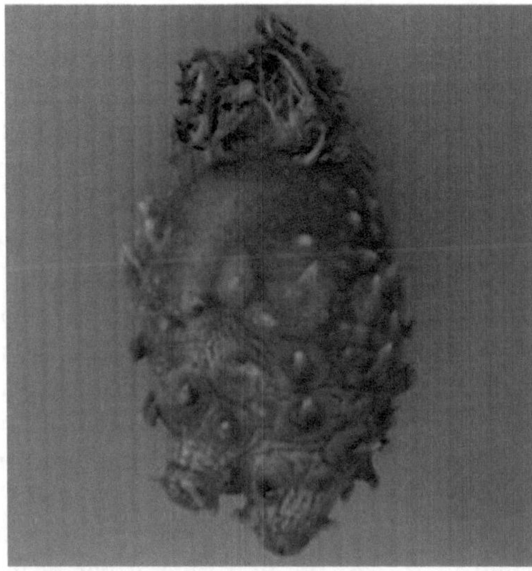

Figure 13.2 Digital photo of *Halocynthia roretzi*.

The colloidal suspension of cellulose nanocrystals (CNs) was first prepared by Ranby using sulfuric acid hydrolysis [11]. This has remained the method of choice for many studies owing to the stability of the resulting suspensions [12–19]. The acid hydrolytic cleavage is basically dependent on the acid species, acid concentration, time and temperature of the hydrolysis reaction. Under controlled conditions, CNs can be obtained using sulfuric acid that allows the grafting of the sulfate groups, randomly distributed on the cellulose surface, inducing a negative electrostatic layer covering the nanocrystals [20]. The different treatments of these charged nanocrystals, such as mechanical dispersion or ultrasonication, permit the dispersion of the aggregates and finally produce colloidal suspensions [15, 17].

At very low concentrations, these nanocrystals are randomly suspended in water and form an isotropic phase. When the concentration reaches a critical value, the nanocrystals spontaneously display ordered phases showing interesting lyotropic chiral nematic liquid crystalline (N*-LC) phase [21].

There are many parameters could affect the pitch of N*-LCs, such as ionic strength, drying temperature, suspension concentration, magnetic field [22] and sonication [23, 24]. Interestingly, the chiral nematic orders can be retained after evaporation of the solvent (generally water), leaving iridescent films of cellulose-I [20]. It is extremely difficult to align and preserve the chiral nematic structure in films of liquid crystals. However, suitably stabilized, CNs can preserve this structure upon drying to a thick film resulting in parabolic focal conic defect structures that are reminiscent of smectic and lamellar LCs of polymers and lipids [25].

In the laboratory, sonication is used as a final step to obtain a well-dispersed colloidal CNs suspension. Sonication is thought to break up side-by-side CNs aggregates in suspension [14, 15]. The effects of low power sonication on CNs suspension properties have been studied and found that brief sonication was sufficient to disperse the CNs. To our knowledge, there is no method reported in the literature which can control the pitch of N*-LCs of T-CNs by high power sonication.

In this chapter, we present mainly our group's recent research and development in TC. For the first time, it has been found that the pitch of the lyotropic N*-LCs of T-CNs can be tuned by high-power ultrasonic radiation. And after evaporation of the lyotropic N*-LCs of T-CNs, some transparent films that could selectively reflect near-infrared light were prepared. Flexible paper-like materials of T-CN nanocomposites with high electrical conductivity were also prepared via *in situ* polymerization.

13.1.2 Experimental

13.1.2.1 Preparation of T-CN suspension

Suspension of T-CNs was prepared from TC. The cellulose mantles were separated from the rest of the organs and were cut into small pieces. Two hundred gram of *Halocynthia roretzi* was soaked in a 10 wt% KOH solution for 24 h followed by washing, then bleached at 60°C with 30 mL of NaClO in 1 L of acetic buffer for 24 h and the resultant white mantles were washed till neutrality [26]. The above procedure was repeated three times. Then the pieces were homogenized into millimeter-size fragments using a multiblender mill (IKA-A11, Germany) and mixed with 64 wt% H_2SO_4 and stirred at 50°C for 12 h to obtain T-CNs. Then the acid was removed through centrifugation and prolonged dialysis with distilled water until the pH outside dialysis bag was neutral. After that the sample was poured into the dialysis bag and concentrated by a 10 wt% polyethylene glycol (PEG) solution. Then the T-CNs were dispersed by sonication and the concentration of the suspension was measured gravimetrically before and after evaporation of the water. The concentration of the sample suspension was 1.5 wt%.

13.1.2.2 Ultrasonic process

Sample of the T-CN suspension was sonicated using a Sonics vibra-cell (950 W, 25 kHz) ultrasonic processor with a 10 mm diameter probe carried out in an ice bath. T-CN suspension was submitted to ultrasonic treatment at 800 W for 8 min. After this process, the sample was concentrated by PEG solution and poured into polystyrene Petri dish. It was allowed to evaporate undisturbed at ambient conditions to obtain transparent thin film.

The composition of the TC was researched by Fourier Transform Infrared spectrum (FT-IR, BRUKER VERTEX70). Polarizing optical microscopy (POM, LEICA DM2500P) was used to observe the behavior of the N*-LCs and the change of the pitch influenced by high power sonication. Morphology of the T-CNs was characterized by using scanning electron microscope (SEM, JEOL JSM-6700F). The crystal structure of TC was studied by X-ray diffraction (XRD, D/MAX-2500/PC). The selective reflectivity of the N*-LCs film was characterized by UV-Vis-NIR reflectance spectroscopy (UV-Vis-IR, SHIMADZU UV-2400PC).

13.1.3 Results and Discussion

13.1.3.1 N*-LCs of T-CNs

The FT-IR spectrum of tunicate cellulose is shown in Fig. 13.3. The bands at 3411, 2900, 1640, 1430, 1371, 1163, and 1060 cm^{-1} in Fig. 13.3 are associated with native cellulose. A strong band at 3410 cm^{-1} arises from the stretching of hydroxyl groups. The bands at 2900 and 1642 cm^{-1} originate from the C–H stretching and the H–O–H bending of the absorbed water, respectively. The peak at 1372 cm^{-1} is attributed to the O–H bending and that at 1157 cm^{-1} corresponds to the C–O antisymmetric bridge stretching. A strong band at 1060 cm^{-1} is due to the C–O–C pyranose ring skeletal vibration. The bands at 3240/750 and 3270/710 cm^{-1} were characteristic of I_α and I_β [5], and they were used for the estimation of the ratio of these dimorphs in cellulose obtained from various organisms [6]. In this figure, we could hardly find the bands of

3240 and 750 cm^{-1}, and the tunicate cellulose consists mostly I_β phase.

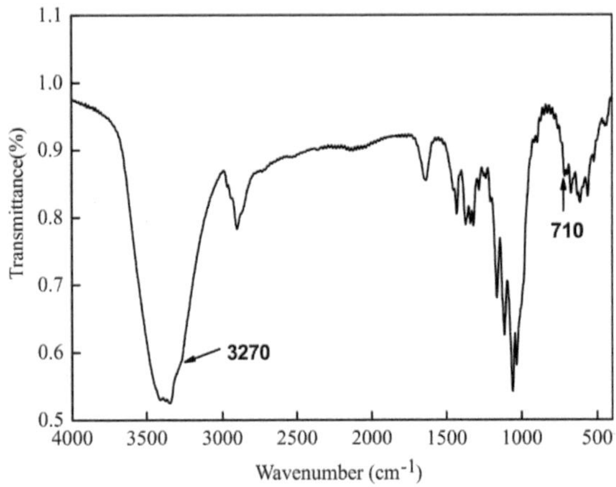

Figure 13.3 FT-IR spectrum of TC.

Figure 13.4 shows the XRD profile of the tunicate cellulose sample. The sample showed four characteristic cellulose I_β reflections at 14.7, 16.6, 22.5 and 34.5°, respectively, corresponding to the $(1\bar{1}0)$, (110), (200) and (400) crystallographic planes of the cellulose I_β lattice, this observation was correlated with other result of FT-IR [7]. The crystallinity of the tunicate cellulose is higher than 90%.

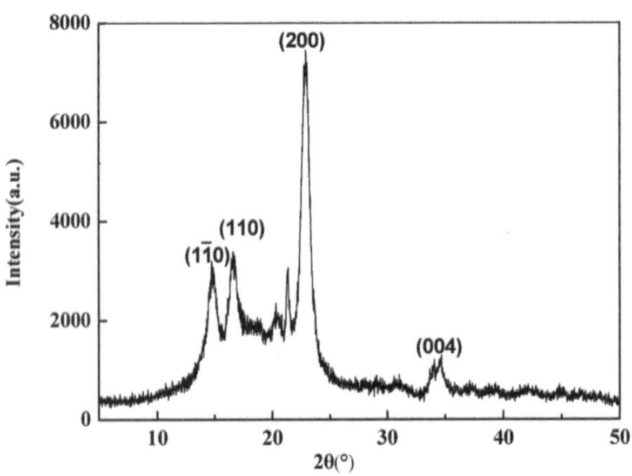

Figure 13.4 XRD pattern of tunicate cellulose.

13.1.3.2 Effects of sonication on T-CN suspension

Figure 13.5 shows textures of the T-CN suspension with different cellulose concentration observed by POM. Samples were dropped on the slides. With increasing concentration the images become colorful and exhibit different textures, (a) and (b) show typical banded texture and schlieren texture, respectively. We could also see the typically fingerprint texture for N*-LCs in (d), and the pitch is about 15 μm. However, it is more difficult to observe the fingerprint texture as other kinds of cellulose reported before. This would be due to its large aspect ratio, which hinders the arrangement of the T-CNs.

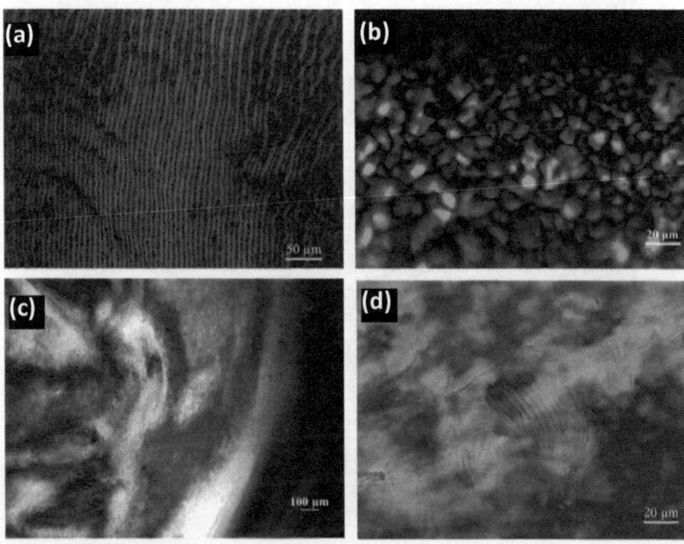

Figure 13.5 POM images of T-CNs without sonication (a) 2.5, (b) 3, (c) 3.5, and (d) 4 wt%.

Various factors such as size, shape, dispersity, charge, electrolyte, and external stimuli can affect the liquid crystallinity, pitch, domain size, ordering and other properties [9]. The sample suspension was submitted to ultrasonic treatment at 800 W for 8 min. After high power sonication, we could observe a trip-like texture (Fig. 13.6a). And "tactoid" could be observed in the following 24 h, which would aggregate with each other gradually (Figs. 13.6b–e). Then the pitch characteristic of a chiral nematic phase is clearly observed and the domain size of the chiral nematic phase is quite large as compared to the T-CN suspensions without sonication, and the pitch is about 50 μm (Fig. 13.6f). This would be due to the high power sonication change the surface charge density and the electrical double layer of the T-CNs [23].

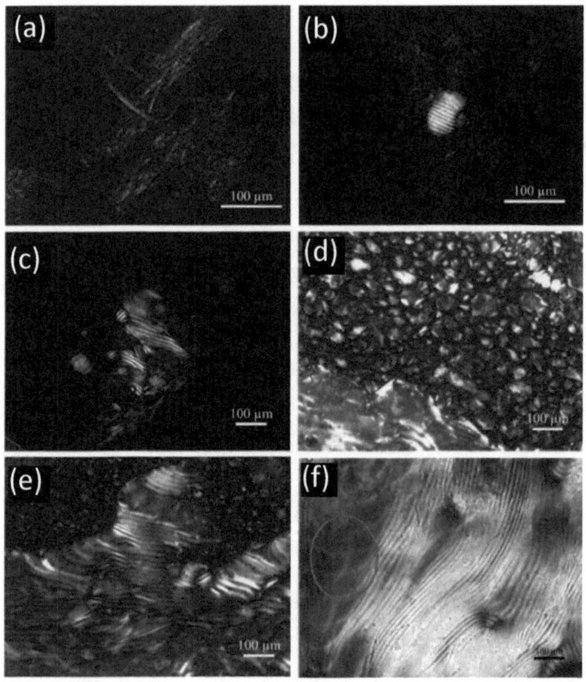

Figure 13.6 POM images of T-CNs (3 wt%) after sonication. (a–f) show the formation process of the fingerprint texture.

An interesting optical behavior was also found in the red circle of (Fig. 13.6f). It was a transformation of the fingerprint texture and the distance between the contiguous three lines was about 200 μm. That was probably because the interface was distorted by the flow of the anisotropic phase [27].

Each kind of CNs has a characteristic size, aspect ratio, morphology, crystallinity, crystal structure, and properties. Figure 13.7a shows the microstructure of the T-CNs prepared by acid hydrolyzation. T-CNs have a high-aspect-ratio (20–30 nm wide, 0.5–4 μm in length), are about 100% cellulose, highly crystalline (85–100%) and containing a high fraction of I_{β} crystal structure (~90%) [19, 28–30]. After sonication, the T-CNs became much shorter (0.1–1 μm in length). The change of the size of the T-CNs may contribute to the change of the pitch of the N*-LCs.

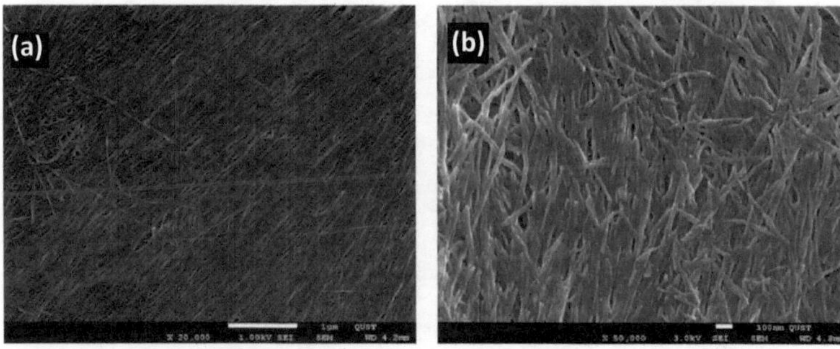

Figure 13.7 SEM images of T-CNs. (a) Without applying sonication; (b) Applying ultrasonic treatment at 800 W for 8 min.

13.1.3.3 N*-LC film of T-CNs

Films cast of CNs have excellent transparency and are being considered for use in transparent film applications, barrier film applications [31], loudspeaker membranes [32], and polymer matrix thermal and mechanical reinforcements.

Transparent film of T-CNs (Fig. 13.8) was cast from the suspensions of T-CNs at ambient conditions that retain the self-assembled helical order formed in the N*-LCs (Fig. 13.9).

Figure 13.8 Digital photo of the transparent T-CNs film.

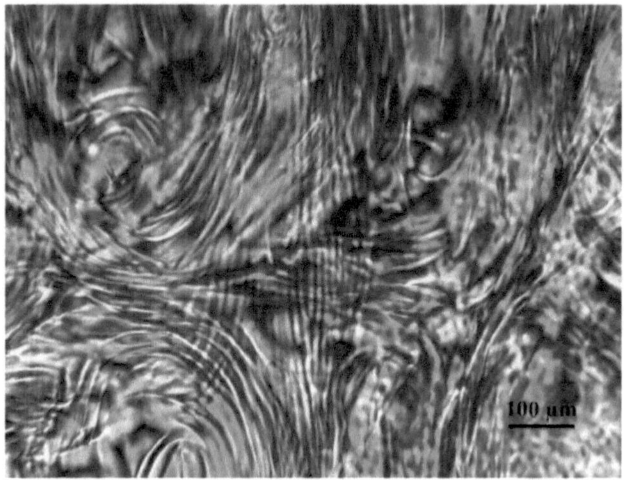

Figure 13.9 POM image of the T-CNs film.

T-CNs film has a polydomain structure in which the helical axes of different chiral nematic domains point in different directions, with the pitch of each domain giving an average pitch. Because the pitch is related to the wavelength of maximum reflection of λ, by

$$\lambda = n\,\mathrm{P}\sin\theta, \tag{13.1}$$

where n is the average refractive index and θ is the angle of incidence [33]. So it determines the reflection wavelength and the N*-LCs films can selectively reflect light with specific wavelength by control the pitch.

Selective reflectivity of the film was characterized by UV-Vis-NIR reflectance spectroscopy in the wavelength range of 200–2500 nm (Fig. 13.10), which encompasses the full range of the visible spectrum and near infrared one. The result shows that the T-CNs film could reflect 45% the near-infrared light while only reflect 5% of the visible light.

The helical structure of N*-LC could be retained in the film which results in selective reflectance that can be tuned across the entire visible spectra and into the near-infrared area. Therefore, this transparent film might be useful for near infrared shading for energy-saving in areas of smart building.

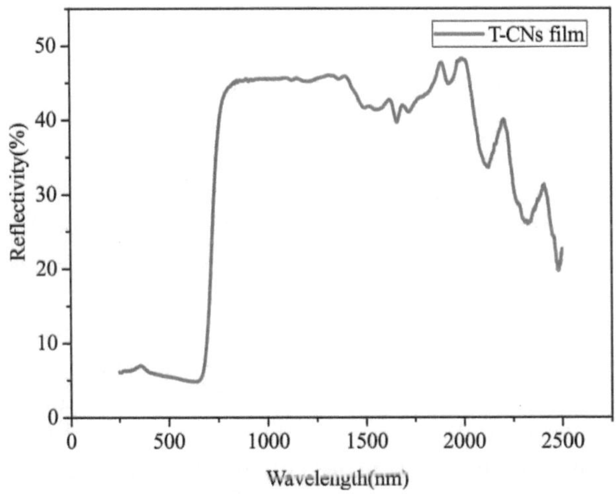

Figure 13.10 UV-Vis-NIR reflection spectrum of T-CNs film.

13.2 T-CN Nanocomposites

13.2.1 Introduction

The demand for new technologies that require high-performance materials has driven materials research toward the development of novel functional nanoscale materials with superior properties. Conducting polymers are a promising class of materials that possess unique properties that allow them to be used in a wide variety of applications [34]. The conductive properties of polyacetylene resulted in the 2000 Nobel Prize in chemistry and various analogues of it have been investigated throughout the years, including polyphenylene, polyaniline, polythiophene, and polypyrrole (PPy) [35]. However, the poor processability and inadequate mechanical properties limit its commercial applications.

In nature, fibrous biopolymers have long been used in the reinforcement of extracellular biocomposites, inspiring the reproduction of this technology using native CNs as filler in a range of host polymer matrixes. Due to the highly crystalline nature of the cellulose nanoparticles, they possess attractive mechanical properties, such as an axial Young's Modulus of around 140 GPa, which is dependent on cellulose crystallinity and axial ratio [36]. When coupled with their high axial ratio, high surface area, availability, low cost and biodegradability, CNs display great reinforcement potential to polymer matrixes of poor thermomechanical stability.

Deposition of conducting polymers on fiber surface of fabrics, such as cotton [37], bacteria cellulose (BC) [34], cellulose microcrystal (CMC) [38], cellulose derivatives [39] and alga [40] have been widely investigated in the last few years due to its importance in emerging technologies. These materials were generally obtained through *in situ* oxidative polymerization.

Pioneering research by Favier used crystallites derived from tunicates as cellulosic reinforcement in poly(styrene-co-butyl acrylate) films and concluded that hydrogen bonding between the tunicate crystallites caused their percolation through the polymer matrix, resulting in the enhanced mechanical properties observed, in the same way that the high strength of a paper sheet results from cellulose fiber percolation [13].

The bulk of research into cellulose crystallite reinforced nanocomposites uses tunicates as the source of the filler due to the high crystallinity, large aspect ratio, purity monoclinic cellulose I_β allomorph, low density and remarkable mechanical strength [42]. Although TC is not a starting material for the textile industry, it is a versatile biopolymer with a wide range of properties that can be used for several innovative applications, including biomedical devices and scaffolds for tissue engineering [41].

This work reports the production of a new functional material that exploits the inherent properties of TC with electrical conductivity of PPy. For this purpose, the T-CNs doped PPy had been prepared via the chemical oxidative polymerization of pyrrole (Py) with T-CNs as both polymeric template and matrix, and using APS as oxidant agent in aqueous solution.

13.2.2 Experimental

PPy was synthesized via an oxidative synthetic approach in this study. 5 mmol Py monomer was dissolved in 50 mL T-CNs dispersion (1.5 wt%) at room temperature for 1 h. The solution was then cooled down, which was followed by addition of APS solution (dissolved in 50 mL distilled water) with APS/Py = 1.25 mol/mol, drop by drop. The polymerization was carried out by using magnetic stirring in ice-water bath. After 6 h, the black suspensions of PPy/T-CNs (0.5 g/g) could be observed and the reaction mixtures remained suspended. Aqueous suspensions of PPy/T-CNs were then prepared by washing the reaction mixtures with distilled water and dispersed by sonication. Films were deposited onto polystyrene Petri dish by casting directly from the suspension of PPy/T-CNs and dried at room temperature for 3 days.

In this part, the composition of T-CN nanocomposites was determined by FT-IR. Morphology of the nanocomposites was characterized by using SEM and transmission scanning electron microscope (TEM, JEOL JEM-2000EX). The conductivities of the nanocomposite films were measured by the four-probe technique (GZFTL RST-8) at room temperature using a semiconductor device analyzer.

13.2.3 Results and Discussion

13.2.3.1 Mechanical property

A highly flexible paper-like material from the PPy/T-CN suspensions was prepared by casting onto the polystyrene Petri dishes (Fig. 13.11). Previous work on PPy coated on other cellulose materials employed the addition of a steric stabilizer in order to maintain a stable suspension and an insulating matrix to produce the films.

Figure 13.11 Digital photo of a flexible film of PPy/T-CNs film [28].

Electrical conductivities of the samples were determined and it was observed that the conductivity could up to 1.4 × 10^{-3} S/cm without addition of any external matrix. Additionally, T-CNs is not an electric conductor and the conductivity decreased with decreasing the amount of Py. This behavior is attributed to the formation of a continuous conducting PPy layer that homogeneously coated the surface of T-CNs. Thus, the T-CNs plays a good role of matrix as well as template. In the present work, no stabilizer was used to maintain the solution, nor was another polymer matrix required to cast the film.

The new nanocomposites might find interesting uses in chemical sensors or paint formulations for antistatic applications, as well as in the development of electrically conductive nanocomposites with better mechanical properties due to the high mechanical strength of the T-CNs.

13.2.3.2 FT-IR analysis

The FT-IR spectra of T-CNs, PPy, and PPy/T-CNs film are shown in Fig. 13.12. The peaks near 3340, 3270, 2900, 1060, and 710 cm^{-1} in Fig. 13.12a were associated with cellulose. The characteristic broad band of cellulose O–H group appears at 3340 cm^{-1}. The band at 2900 cm^{-1} is attributed to the asymmetrically stretching vibration of C–H. The broad absorption band in the range of 1060 to 1030 cm^{-1} is attributed to the contribution of various functional groups, such as C–O and C–O–C. The peaks at 3270 and 710 cm^{-1} were attributed to the I_β phase of cellulose. Figure 13.12b was the FT-IR spectrum of PPy. The peaks at 3419 cm^{-1} could be attributed to N–H stretching vibrations. The peaks at 1539 and 1450 cm^{-1} could be attributed to C–N and C–C asymmetric and symmetric ring-stretching, respectively. Additionally, the strong peaks near 1170 and 891 cm^{-1} present the doping state of PPy, the peak at 1037 cm^{-1} is attributed to C–H deformation and N–H deformation vibrations, and the broadband at 1300 cm^{-1} demonstrates the C–H and C–N in-plane deformation vibration, respectively. Moreover, these characteristic peaks of T-CNs and PPy were also found in that of the PPy/T-CN nanocomposites. Although the band at 1539 cm^{-1} for pure PPy is blue-shifted to 1551 cm^{-1} for composites and the bending modes for OH groups of T-CNs weakened, which indicates that the existence of the PPy coating layer. The observed shift of the band may be caused by chemical bonding between –N–H in the Py ring and the –OH functional group of cellulose, although the –OH absorption band of the composites is too weak to confirm that OH groups of the cellulose is intermolecular bonded to –N–H in the Py ring [35]. The results indicated that the *in situ* polymerization of Py onto the T-CNs still retained their chemical structures well.

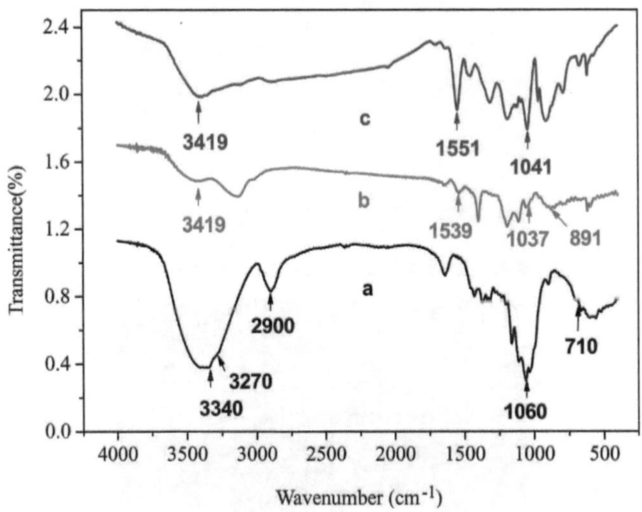

Figure 13.12 FT-IR spectra of TC (a), PPy (b) and PPy/T-CN nanocomposite film (c).

13.2.3.3 SEM analysis

Morphological analysis showed that the string-like T-CNs (Fig. 13.13a) length ranges from 1 to 2 μm and width averages 25 nm, and the aspect ratio is about 40 to 80. Moreover, such high-aspect-ratio nanocrystals should create a huge entanglements network and thus it is responsible for such highly mechanical stiffness and highly flexible paper-like materials. The influence of the amount of Py on the structure and morphologies of the PPy/T-CN nanocomposites was investigated by SEM (Fig. 13.13b). The size of PPy particles were ranged from 10 to 100 nm. The adhesion of PPy particles to the cellulose surface could attribute to the formation of hydrogen bonds between cellulose hydroxyl groups and Py NH moieties. The formed PPy particles on cellulose surface had spherical shape and controlled size, which was probably due to the charge density on T-CNs that impeded, further, disordered growth [38]. And the increase of the T-CNs could prevent the aggregation of PPy particles and acted as separant and dispersant during the *in situ* polymerization of Py, which eventually could decrease the particles size.

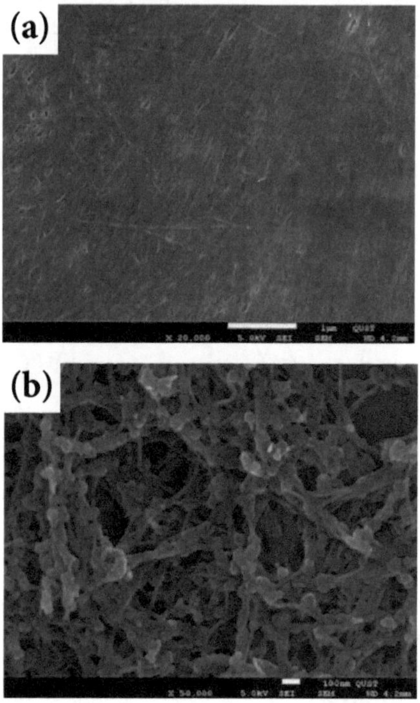

Figure 13.13 SEM images of T-CNs (a) and PPy/T-CN nanocomposites (b).

13.2.3.4 TEM analysis

In order to further prove the above speculation, the TEM images of the PPy/T-CN nanocomposites dispersed in ethanol were given in Fig. 13.14. The morphologies of the composites became smaller with the increasing T-CNs. The width of the T-CNs was about 20 nm, while that of the composites was 30–50 nm. A thickness layer of PPy about 30 nm was deposited on the T-CNs. We could find that on the surface of the nanocomposites absorbed a lot of uniform spherical nanoparticles and their diameters are about 2.5 nm. According to the previous work, they might be PPy nanoparticles [43]. And we could also find on the surface of the covered layer adhered a lot of spherical nanoparticles ranging from about 20 to 100 nm, which is consistent with the SEM results. This indicated that the T-CNs molecules acted as the templates as well as the matrix, as reported previously [44]. The micrographs revealed that PPy was constituted of particles that uniformly coated on the surface of T-CNs. The polymer coating adhered to the nanocrystals and a continuous conducting pathway was formed which was responsible for the high electrical conductivity values.

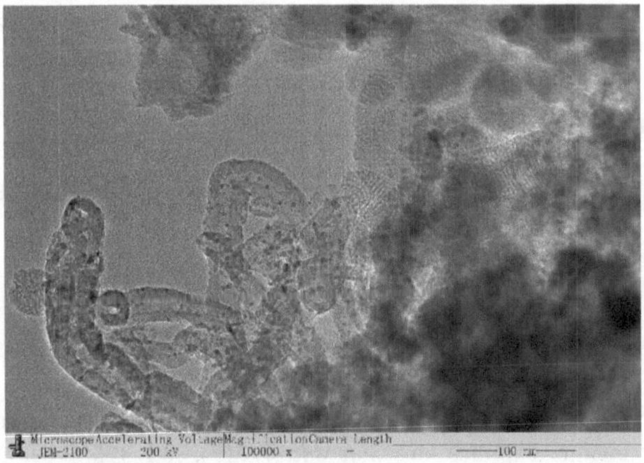

Figure 13.14 TEM image of PPy/T-CN nanocomposite [28].

13.3 Summary

The T-CNs yield stable aqueous suspensions because of the presence of negative charges on the surface of the nanocrystals during the hydrolysis process. The obtained nanocrystals are rod-shaped particles, range from 500 nm to 4 μm in length and 20 to 30 nm in diameter. At very low concentrations, T-CNs are randomly suspended in water and form an isotropic phase. When the concentration reaches a critical value, the T-CNs spontaneously display ordered phases showing interesting lyotropic N*-LCs. The chiral nematic orders can be retained after evaporation of the water, leaving transparent film with selective reflectivity, which tuned across the entire visible spectra and into the near-infrared area. The reflection wavelength can be controlled by using high power sonication. Because of their geometry, important axis ratio (L/d), and high crystallinity, these nanocrystals have been also extensively used to process conductive nanocomposites by the *in situ* oxidative chemical polymerization based on PPy, to reinforce the mechanical properties. Highly flexible paper-like materials of

PPy/T-CN nanocomposites with electrical conductivities as high as 10^{-3} S/cm were also obtained.

Acknowledgments

The authors are grateful to Prof. T. Kimura and Dr. F. Kimura (Kyoto University) for their helpful discussions and technical support.

References

1. Klemm, D., Kramer, F., and Moritz, S. (2011). Nanocelluloses: A new family of nature-based materials, *Angew. Chem. Int. Ed.*, **50**(24), 5438–5466.
2. Habibi, Y., Lucia, L. A., and Rojas, O. J. (2010). Cellulose nanocrystals: chemistry, self-assembly, and applications, *Chem. Rev.*, **110**(6), 3479–3500.
3. Eichhorn, S. J. (2011). Cellulose nanowhiskers: Promising materials for advanced applications, *Soft Matter*, **7**(2), 303–315.
4. O'SULLIVAN, A. C. (1997) Cellulose: The structure slowly unravels, *Cellulose*, **4**(3), 173–207.
5. Sugiyama, J., Persson, J., and Chanzy, H. (1991). Combined infrared and electron diffraction study of the polymorphism of native celluloses, *Macromolecules*, **24**(9), 2461–2466.
6. Imai, T., and Sugiyama, J. (1998). Nanodomains of I_α and I_β cellulose in algal microfibrils, *Macromolecules*, **31**(18), 6275–6279.
7. Sugiyama, J., Vuong, R., and Chanzy, H. (1991). Electron diffraction study on the two crystalline phases occurring in native cellulose from an algal cell wall, *Macromolecules*, **24**(14), 4168–4175.
8. Nakashima, K., Nishino, A., and Horikawa, Y. (2011). The crystalline phase of cellulose changes under developmental control in a marine chordate, *Cell. Mol. Life Sci.*, **68**(9), 1623–1631.
9. Moon, R. J., Martini, A., and Nairn, J. (2011). Cellulose nanomaterials review: Structure, properties and nanocomposites, *Chem. Soc. Rev.*, **40**(7), 3941–3994.
10. Kimura, S., and Itoh, T. (2004). Cellulose synthesizing terminal complexes in the ascidians, *Cellulose*, **11**(3), 377–383.
11. Holt, B. L., Stoyanov, S. D., and Pelan, E. (2010). Novel anisotropic materials from functionalised colloidal cellulose and cellulose derivatives, *J. Mater. Chem.*, **20**(45), 10058–10070.

12. Revol, J. F., Bradford, H., and Giasson, J. (1992). Helicoidal self-ordering of cellulose microfibrils in aqueous suspension, *Int. J. Biol. Macromol.,* **14**(3), 170–172.

13. Favier, V., Chanzy, H., Cavaille, J. Y. (1995). Polymer nanocomposites reinforced by cellulose whiskers, *Macromolecules,* **28**(18), 6365–6367.

14. Dong, X. M., Kimura, T., and Revol, J. F. (1996). Effects of ionic strength on the isotropic-chiral nematic phase transition of suspensions of cellulose crystallites, *Langmuir,* **12**(8), 2076–2082.

15. Dong, X. M., Revol, J., and Gray, D. G. (1998). Effect of microcrystallite preparation conditions on the formation of colloid crystals of cellulose, *Cellulose,* **5**(1), 19–32.

16. Roman, M., and Winter, W. T. (2004). Effect of sulfate groups from sulfuric acid hydrolysis on the thermal degradation behavior of bacterial cellulose, *Biomacromolecules,* **5**(5), 1671–1677.

17. Beck, S., Roman, M., and Gray, D. G. (2005). Effect of reaction conditions on the properties and behavior of wood cellulose nanocrystal suspensions, *Biomacromolecules,* **6**(2), 1048–1054.

18. Bondeson, D., Mathew, A., and Oksman, K. (2006). Optimization of the isolation of nanocrystals from microcrystalline cellulose by acid hydrolysis, *Cellulose,* **13**(2), 171–180.

19. Elazzouzi, S., Nishiyama, Y., and Putaux, J. L. (2007). The shape and size distribution of crystalline nanoparticles prepared by acid hydrolysis of native cellulose, *Biomacromolecules,* **9**(1), 57–65.

20. Souza, M. M., and Borsali, R. (2004). Rodlike cellulose microcrystals: Structure, properties, and applications. *Macromol. Rapid Commun.,* **25**(7), 771–787.

21. Hirai, A., Inui, O., and Horii, F. (2008). Phase separation behavior in aqueous suspensions of bacterial cellulose nanocrystals prepared by sulfuric acid treatment, *Langmuir,* **25**(1), 497–502.

22. Pan, J., Hamad, W., and Straus, S. K. (2010). Parameters affecting the chiral nematic phase of nanocrystalline cellulose films, *Macromolecules,* **43**(8), 3851–3858.

23. Beck, S., Bouchard, J., and Berry, R. (2010). Controlling the reflection wavelength of iridescent solid films of nanocrystalline cellulose, *Biomacromolecules,* **12**(1), 167–172.

24. Tischer, P., Sierakowski, M., and Westfahl, H. (2010). Nanostructural reorganization of bacterial cellulose by ultrasonic treatment, *Biomacromolecules,* **11**(5), 1217–1224.

25. Roman, M., and Gray, D. G. (2005). Parabolic focal conics in self-assembled solid films of cellulose nanocrystals, *Langmuir,* **21**(12), 5555–5561.
26. Kimura, F., Kimura, T., and Tamura, M. (2005). Magnetic alignment of the chiral nematic phase of a cellulose microfibril suspension, *Langmuir,* **21**(5), 2034–2037.
27. Revol, J. F., Godbout, L., and Dong, X. M. (1994). Chiral nematic suspensions of cellulose crystallites; phase separation and magnetic field orientation, *Liquid Cryst.,* **16**(1), 127–134.
28. Zhang, D., Zhang, Q., and Gao, X. (2013). A nanocellulose polypyrrole composite based on tunicate cellulose, *Int. J. Polym. Sci.,* **2013**, Article ID 175609, 6 pp. http://dx.doi.org/10.1155/2013/175609.
29. Bai, W., Holbery, J., and Li, K. (2009). A technique for production of nanocrystalline cellulose with a narrow size distribution, *Cellulose,* **16**(3), 455–465.
30. Habibi, Y., Hoeger, I., and Kelley, S. (2009). Development of Langmuir–Schaeffer cellulose nanocrystal monolayers and their interfacial behaviors, *Langmuir,* **26**(2), 990–1001.
31. Fukuzumi, H., Saito, T., and Iwata, T. (2008). Transparent and high gas barrier films of cellulose nanofibers prepared by TEMPO-mediated oxidation, *Biomacromolecules,* **10**(1), 162–165.
32. Henriksson, M., and Berglund, L. A. (2007). Structure and properties of cellulose nanocomposite films containing melamine formaldehyde, *J. Appl. Polym. Sci.,* **106**(4), 2817–2824.
33. Edgar, C. D., and Gray, D. G. (2001). Induced circular dichroism of chiral nematic cellulose films, *Cellulose,* **8**(1), 5–12.
34. Mihranyan, A., Nyholm, L., and Bennett, G. (2008). A novel high specific surface area conducting paper material composed of polypyrrole and Cladophora cellulose, *J. Phys. Chem. B,* **112**(39), 12249–12255.
35. Müller, D., Rambo, C. R., and Recouvreux, D. S. (2011). Chemical *in situ* polymerization of polypyrrole on bacterial cellulose nanofibers, *Synthetic Metals,* **161**(1), 106–111.
36. Eichhorn, S. J., Dufresne, A., and Aranguren, M. (2010). Review: Current international research into cellulose nanofibres and nanocomposites, *J. Mater. Sci.,* **45**(1), 1–33.
37. Zhang, D., Zhang, L., Wang, B., and Piao, G. (2013). Nanocomposites of polyaniline and cellulose nanocrystals prepared in lyotropic chiral nematic liquid crystals, *J. Mater.,* **2013**, Article ID 614507, 6 pp. http://dx.doi.org/10.1155/2013/614507.

38. Sasso, C., Zeno, E., and Petit, M. (2010). Highly conducting polypyrrole/cellulose nanocomposite films with enhanced mechanical properties, *Macromol. Mater. Eng.*, **295**(10), 934–941.

39. Beneventi, D., Alila, S., and Boufi, S. (2006). Polymerization of pyrrole on cellulose fibres using a $FeCl_3$ impregnation-pyrrole polymerization sequence, *Cellulose,* **13**(6), 725–734.

40. Mihranyan, A. (2011). Cellulose from cladophorales green algae: From environmental problem to high-tech composite materials, *J. Appl. Polym. Sci.*, **119**(4), 2449–2460.

41. Dugan, J. M., Gough, J. E., and Eichhorn, S. J. (2010). Directing the morphology and differentiation of skeletal muscle cells using oriented cellulose nanowhiskers, *Biomacromolecules,* **11**(9), 2498–2504.

42. Holt, B. L., Stoyanov, S. D., and Pelan E. (2010). Novel anisotropic materials from functionalised colloidal cellulose and cellulose derivatives, *J. Mater. Chem.*, **20**(45), 10058–10070.

43. Jang, J., and Oh, J. H. (2002). Novel crystalline supramolecular assemblies of amorphous polypyrrole nanoparticles through surfactant templating, *Chem. Commun.*, **19**, 2200–2201.

44. Bruno, F. F., Nagarajan, R., and Roy, S. (2003). Biomimetic synthesis of water soluble conductive polypyrrole and poly (3,4-Ethylenedioxythiophene). *J. Macromol. Sci. A,* **40**(12), 1327–1333.

Chapter 14

Synthesis and Characterization of Bacterial Cellulose-Based Composites and Their Applications

Muenduen Phisalaphong[a] and Suchata Kirdponpattara[a,b]

[a]*Department of Chemical Engineering, Faculty of Engineering, Chulalongkorn University, Bangkok 10330, Thailand*
[b]*Department of Chemical Engineering, King Mongkut's University of Technology North, Bangkok 10800 Thailand*

Muenduen.p@chula.ac.th

14.1 Introduction

Bacterial cellulose (BC) is a biopolysaccharide synthesized primarily by Acetobacter bacteria. In the metabolism process of Acetobacter, the consumed glucose is polymerized into cellulose and extruded as nanofibrils through tiny pores at the outer membrane of the bacterial cell. Subsequently, the nanofibrils aggregate to form cellulose ribbons in a web-like network structure. Although many bacteria genera, such as Achromobacter, Agrobacterium, Pseudomonas, and Rhizobium, can produce BC [1], Acetobacter bacteria (especially *Acetobacter xylinum*) have the ability to synthesize a large quantity of high quality cellulose organized as twisted bundles of microfibrillar ribbons [2].

Handbook of Sustainable Polymers: Processing and Applications
Edited by Vijay Kumar Thakur and Manju Kumari Thakur

ISBN 978-981-4613-53-8 (Hardcover), 978-981-4613-54-5 (eBook)
www.panstanford.com

In the literature, various carbon sources, such as monosaccharides, polysaccharides, alcohols, organic acids, and even wastes, have been examined in BC production. To obtain a 100% cellulose yield of BC, glucose has been used as the main substrate in culture media [3]. Originally, in the Philippines, BC was produced from coconut water (i.e., Nata de coco), which is also composed of glucose [4]. Coconut water, supplemented with nitrogen and monosaccharides or disaccharides, was generally used as the culture media for BC production [5–7]. For optimal culture conditions, the culture media pH should be adjusted to 4–7 using acetic acid and kept at 25–30°C [3]. Additionally, an adequate oxygen supply should be provided. All factors strongly affect the production yield and structure of BC [8].

The culture modes are broadly classified into two strategies: static and agitated cultures. In a static culture, a BC pellicle is synthesized at the air–culture medium interface. The pellicle thickens with increasing cultivation time [9]. Producing BC using a static culture results in an increased BC crystallinity when compared with that produced from an agitated culture [10]. Granulated BC is synthesized under agitated cultivation. For the production of large amounts of BC, an agitated culture is suggested because this method allows for a high cell density and high volumetric productivity due to increased oxygenation of the media [10]. The characteristics of BC are mainly affected by the culture conditions.

BC synthesis by *Acetobacter xylinum* was first described by Brown [3]. BC is an outstanding biomaterial because of its unique properties. Although BC and plant cellulose have the same chemical structure (i.e., a linear polymer of D-glucose residues joined by β-1,4-glycosidic linked as polysaccharides), their physical and chemical properties are different. The fibers of plant cellulose are much larger than those of BC, as shown in Fig. 14.1. Unlike plant cellulose, there are no lignins, hemicelluloses, or other complex carbohydrates in the BC structure. This indicates that BC has high purity. Under static conditions, the ultrafine web-network of crystalline nanofibrils tightly forms into a gel and grows downward until the media is exhausted or when cells entrapped in the pellicle are inactivated [11]. Consequently, BC displays high crystallinity and high mechanical strength. Additionally, the ultrafine web-network structure provides a large surface area and a highly porous matrix. Other unique properties of BC include a high hydrophilicity and a high water adsorption capacity (WAC).

Figure 14.1 SEM images of the surface of a cellulose film from BC (a) comparing to that from plant cellulose (b).

Numerous applications of BC in tissue engineering and biomedical fields have been reported because of its biocompatibility and non-toxicity. For instance, BC was used for an artificial blood vessel for microsurgery [12], an artificial skin for burn healing [13], nerve generation [14], a scaffold for cartilage implantation [15], vascular grafts [16], and for wound healing for human and animal subjects [17]. Other applications include membrane separation, cell immobilization, drug delivery, paper production, food production, and cosmetics. However, there are limitations to BC use. For example, dried BC exhibits poor rehydration ability. Therefore, the development of a rehydrating membrane, such as a BC–gelatin composite, was required [18]. BC–chitosan composites have also been synthesized to improve the antibacterial properties of BC [19, 20]. For additional applications, such as bone tissue engineering, BC has been modified with hydroxyapatite [21] or calcium carbonate [22]. According to the literature, BC-based composites have been used to enhance the physical and biological properties of BC to overcome its limitations and to also expand its applications.

14.2 Synthesis of BC-Based Composites

The composite synthesis of two types of materials has been used to combine their individual properties as an alternative approach to producing complex materials for various applications. The physicochemical and biological properties of the base material are modified by a supplemented material [11]. Depending on the nature of the two individual materials to be combined and the desired applications, various procedures have been utilized to synthesize the composites. Overall, the synthesis of BC-based composites is divided into three methods: biosynthesis, impregnation, and blending.

14.2.1 Biosynthesis

Biosynthesis is a simple approach that incorporates the supplement material into the BC fibril network [23]. A schematic diagram is provided in Fig. 14.2a. At the start of the BC synthesis, the supplement is directly added into the BC culture medium.

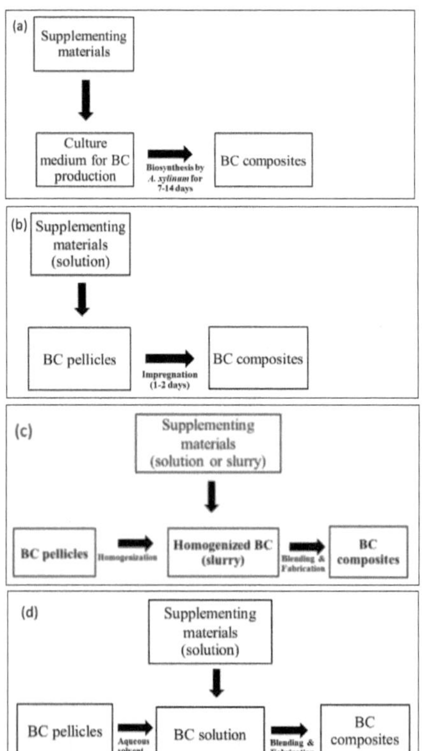

Figure 14.2 Schematic diagrams for the synthesis of BC-based composites: (a) biosynthesis, (b) impregnation, (c) slurry blending, and (d) solution blending.

During *A. xylinum* metabolism, the cells produce the composite nanofibril network by trapping the supplement within the network [24]. Consequently, the supplement becomes part of the BC fibril network. Generally, the state of supplement is liquid. However, for non-soluble solid supplements, solid powders are not entrapped within the network under static culture conditions. Because of this, an agitation culture can be used to synthesize the BC-solid supplement composite. Under BC culture agitation, multi-wall carbon nanotubes (MWNTs) were added. The subsequent presence of MWNTs in rice-like cellulose assemblies was observed [25]. Usually, there is a limited amount of supplements. For example, it was found that too much added aloe vera gel into the BC culture medium not only extremely changed the morphology of BC but also reduced cellulose production and the mechanical strength [26]. Thus, the biosynthesis approach might be restricted to a few applications. For instance, to improve the antibacterial properties of BC, antibacterial supplements cannot be directly added into the media due to their natural inhibition of bacterial growth [19, 27].

14.2.2 Impregnation

To prevent a reduction in cellulose production, the variation of morphologies, and to maintain the original nanofibril web-like network of non-composite BC, impregnation is an alternative approach for synthesizing BC composites. As shown in a schematic diagram in Fig. 14.2b, the BC pellicle is impregnated in the supplement for a period of time to form a BC-based composite. For the synthesis of a BC composite via the impregnation technique, only nano-sized molecules can easily penetrate through the nanoporous matrix of a BC and attach to the BC ribbons. In contrast, larger molecules can only coat and enwrap the surface of the BC pellicle [28–29]. However, during the impregnation step, the supplement might not be homogeneously distributed throughout the BC pellicle. Moreover, the fabrication of specific morphologies (e.g., a macroporous structure) or desired shapes of BC composites via biosynthesis and impregnation requires complicated equipment and procedures.

14.2.3 Blending

To overcome the limitations of fabricating BC composites by biosynthesis and impregnation methods, a BC gel can be transformed to a homogenized BC slurry or solution for easier mixing with a supplement at different compositions (Figs. 14.2c,d). In this way, a supplement can be homogeneously distributed in the matrix [30]. A variety of supplement types can be used with this approach [31]. Additionally, BC composites synthesized via this supplementation approach can be easily formed into desired shapes or structures by using simple techniques.

14.3 BC-Based Composites

14.3.1 BC-Alginate

14.3.1.1 Alginate

Alginate (alginic acid) is isolated from cell walls of marine brown seaweed and algae. Alginate is a linear polysaccharide which comprises primarily β-D-mannuronic acid (M block) and α-L-glucuronic acid (G block). Generally, alginate associates with monovalent salts to improve the water-soluble property of alginate, especially sodium salt (i.e., Na-alginate). Na-alginate solution can be easily formed into a hydrogel by cross-linking with a calcium chloride solution. After a sodium and calcium ion exchange, calcium ions are found in polar cavities between the G block regions of two polymer chains, as in the egg box model [32]. Consequently, the different compositions of M and G blocks affect the physical properties of the calcium alginate gel [33].

Alginate is widely used in separation, food, textile, and paper applications. Moreover, due to its biocompatibility, alginate has been extensively used in the biomedical field as a drug delivery material [34, 35], a wound dressing [36], a scaffold [37, 38] and for cell encapsulation [39]. Therefore, the BC–alginate composite has been extensively studied in many fields. Comparative characteristics of BC–alginate composites prepared from various methods are shown in Table 14.1.

Table 14.1 Comparative characteristics of BC–alginate composites prepared from various methods

	Synthesis methods	
Characteristics	**Biosynthesis**	**Blending**
Chemical structure	• Intermolecular interaction between hydroxyl group of BC and carboxyl group of alginate [23, 31, 40, 42]	
Morphology	• Well incorporation [23] • Denser structure with smaller pore size [23]	• Homogeneous structure and miscibility [31, 42] • Less homogeneous (>20% weight ratio of 3% w/v alginate) [31] • Using freeze drying technique: interconnected porous structure (≤3% w/v alginate) and multilayer structure (>3% w/v alginate) [41]
Mechanical strength	• Reduction of mechanical properties with increasing alginate content [23, 31, 40, 42]	
WAC	• Improvement of WAC with increasing alginate content [23, 31, 40, 41]	
WVTR	• Slight reduction of WVTR with increasing of alginate content [23]	• Improvement of WVTR with increasing alginate content [31]
Application	• Membrane separation [23]	• Mucosal flap in oral tissue (promotion of keratinocyte and gingival fibroblast proliferations) [40] • Yeast cells carrier for ethanol production (high ethanol productivity) [41]

14.3.1.2 Characterization

14.3.1.2.1 *Chemical structure*

The intermolecular interaction between the hydroxyl functional group of BC and the carboxyl functional group of alginate was detected using Fourier transform infrared (FT-IR) spectroscopy.

The carboxyl group spectral bands shifted to higher wavenumbers [31, 40]. A slight band shift indicative of a carbonyl group was observed in the BC–alginate composite synthesized via the biosynthesis approach [23]. Kanjanamosit et al. [23] indicated that there may be weak physical interactions between carbonyl groups of BC and alginate compounds.

14.3.1.2.2 *Morphology*

The morphology of the BC–alginate composite is dependent on the synthesis approach as shown in Fig. 14.3 and the conditions, such as the alginate content and the dryness of the composite. Because the chemical structures of BC and alginate are quite similar, the composite presented a homogeneous structure with a certain level of miscibility of the polymer blend. However, by the synthesis using dissolved BC, weight ratios greater than 20% wt. alginate might lead to less miscibility and a less homogenous composite [31]. Moreover, a higher concentration of alginate strongly affected the composite structure. Kirdponpattara and Phisalaphong [41] found that the BC–alginate composite formulated using a greater than 3% wt. alginate solution formed a multilayer structure instead of an interconnected porous structure. By the biosynthesis approach, the composite film was denser with smaller pore sizes when compared with those of the unmodified BC film, because alginate was deposited on the composite surface, incorporated into BC fibrils and filled the pores of the BC [23].

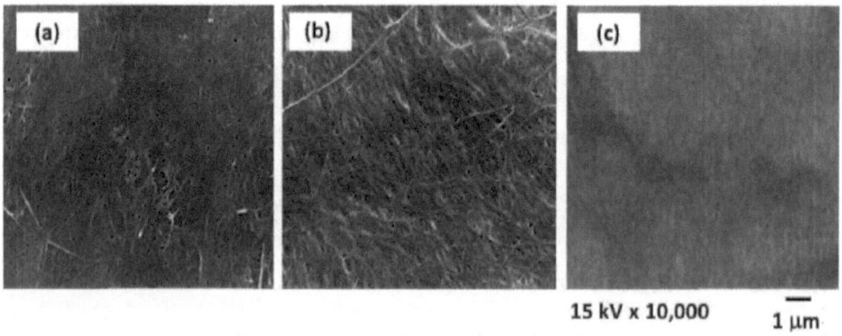

Figure 14.3 SEM images of BC–alginate films prepared by biosynthesis (a), by slurry blending (b), and by solution blending (c).

By preparing a three-dimensional BC–alginate composite sponge, the mixture of homogenized BC and alginate solution was cross-linked with a calcium chloride solution and then freeze dried at −40°C [40, 41]. BC-alginate composite formed an asymmetric structure with a nanoporous outer layer covering an interior of interconnected macropores [40, 41]. This three-dimensional structure could be utilized in several applications.

14.3.1.2.3 *Mechanical strength*

The tensile strength, elongation at break and Young's modulus decreased with increasing alginate content. These trends were observed due to the presence of alginate in the BC–alginate composite. The presence of alginate fluidized and increased the motion of cellulose in the composite, thereby perturbing the strong hydrogen bond of pure BC and resulting in the reduction of mechanical strength [23, 31, 42]. Additionally, for homogenized and dissolved BC, the naturally strong hydrogen bonding of BC was disrupted to form BC–alginate hydrogen bonds, resulting in reductions of mechanical strength and crystallinity [40].

14.3.1.2.4 *Water adsorption capacity*

Alginate is a highly hydrophilic biopolymer and easily adsorbs water molecules. Therefore, the addition of alginate caused an increase in the WAC of the BC–alginate composite. Additionally, alginate can disrupt and reduce the hydrogen bonding of BC, which also results in the easy adsorption of water [31].

14.3.1.2.5 *Water vapor transmission*

Although the pore size of the BC–alginate composite was smaller than that of BC, the water vapor transmission (WVTR) of the composite was twice as great as that of the BC. Because of the high hydrophilicity of alginate, a strong interaction formed between alginate (–COO) and water (–OH). More water molecules adhered on the composite surface, resulting in a larger concentration gradient across the membrane and a higher driving force. Therefore, water molecules easily passed through the small pores of the composite membrane [31]. However, different results were obtained from different approaches of the BC–alginate composite synthesis. The WVTR of the BC–alginate composite synthesized via biosynthesis was lower than that of the BC [23]. This may be due to the dissimilarity between structures and intermolecular interactions, leading to different degrees of pore size swelling.

14.3.1.3 Application

14.3.1.3.1 *Biomedical field and tissue engineering*

Due to the asymmetric structure of the 3D BC–alginate composite fabricated by cross-linking and freeze drying [40], the composite could be used as a wound dressing. The nanoporous outer layer could prevent bacterial infection and the macroporous inner layer could drain wound exudates. In this research, the composite was developed for use as mucosal flaps in oral tissue regeneration. Chiaoprakobkij et al. [40] used human keratinocytes (HaCat) and gingival fibroblasts (GF) to test their proliferative capabilities on the composite sponge. It was found that the composite supported the proliferations of both cell types. HaCat exhibited good attachment and spreading on the composite sponge. For GF, the cells were still in round shape. Moreover, the range of the composite pore size (100–500 μm) may be suitable for tissue engineering applications [43]. The BC–alginate composite fiber fabricated by a wet-spinning apparatus could also be used as a scaffold [42].

14.3.1.3.2 *Cell immobilization*

The BC–alginate composite sponge was also applied as a cell carrier for ethanol production [41]. The composite structure was effective for yeast cell immobilization. After 72 h (fermentation time), a large amount of yeast cells adhered on the outer layer and inside the inner layer of the composite sponge. This indicated that there were sufficient nutrients for the cells, leading to increased cell densities and ethanol productivity. The addition of homogenized BC into the alginate overcame mass transfer limitations in alginate gels.

14.3.1.3.3 *Membrane separation*

Although the permeate flux of a BC membrane was surprisingly high, the selectivity was quite poor. To improve the selectivity, alginate was supplemented to increase the hydrophilic property of BC membrane. The BC–alginate composite was synthesized using dissolved BC [31]. The resultant composite membrane displayed good characteristics, including chemical stability, high mechanical strength, high WAC and high WVTR, and could be used for membrane separation. The composite membrane synthesized by biosynthesis approach could be used in pervaporation processes [23].

14.3.2 BC–Gelatin

14.3.2.1 Gelatin

Gelatin, a polypeptide, is derived from collagen obtained from skin, bone and connective tissues of animals, such as cattle, pigs, chicken, and fish. Gelatin is produced by the partial hydrolysis of collagen. Gelatin is dissolvable in hot water and solidifies at ambient temperature. Gelatin gel exhibits transparency and flexibility. However, brittleness is observed in dried gelatin film [44].

There have been many reported attractive properties of gelatin, including good biocompatibility, excellent biodegradability, low immunogenicity [45], promotion of cell adhesion and proliferation [46], and low cost. In addition, gelatin plays a similar function as collagen in tissue engineering [47]. Numerous works have studied the use of gelatin as artificial skin [48, 49], wound dressing [50, 51], drug delivery microspheres [52, 53], and scaffolding for bone tissue engineering [54–56]. Gelatin is also extensively used in food, pharmaceutical, and cosmetic applications.

Generally, BC–gelatin composites are synthesized using impregnation with or without cross-linking to improve the rehydration of dried BC film [18, 28] and to expand the applications of BC in bone tissue engineering [29, 37]. In our research group, the BC–gelatin composite prepared using a biosynthesis approach was developed for use in wound dressings and tissue engineering [57]. Comparative characteristics of BC–gelatin composites prepared from various methods are shown in Table 14.2.

Table 14.2 Comparative characteristics of BC–alginate composites prepared from various methods

	Synthesis methods	
Characteristics	**Biosynthesis**	**Impregnation**
Chemical structure	– Interactions between the characteristic groups of BC and gelatin [57]	– Weak intermolecular hydrogen bonds [18]

(Continued)

Table 14.2 (*Continued*)

	Synthesis methods	
Characteristics	**Biosynthesis**	**Impregnation**
Crystallinity	– Reduction of crystallinity with adding gelatin [18, 29, 57]	
Morphology	– Well incorporation with denser structure and smaller pore size [57]	– Denser structure [47] and uniform gelatin coating on BC surface [29, 47, 58]
		– A honeycomb-like structure with increasing gelatin content [18] – Uneven distribution [28]
Mechanical strength	– Reductions of tensile strength and elongation at break with increasing gelatin content [57]	– Slight improvement of hardness, springiness and cohesiveness with increasing gelatin content [18]
WAC	– Improvement of WAC with increasing gelatin content (≤3% w/v gelatin) [57]	– Reduction of WAC with increasing gelatin content [18]
WVTR	– Reduction of WVTR with adding gelatin [57]	
Application	– Wound dressing and tissue engineering [57]	– Scaffold for tissue engineering (non-toxicity and promotion of NIH3T3 cell proliferation) [47] – Bone tissue engineering (promotions of chondrogenic cells [29] and mesenchymal stem cells [58]) – Rehydration membrane (high rehydration ratio) [18, 28]

14.3.2.2 Characterization

14.3.2.2.1 *Chemical structure*

For the BC–gelatin composite synthesized by impregnation with a gelatin solution and cross-linked with 1-ethyl-3-(3-dimethylamino propyl)carbodiimide (EDC; the composite is BC–gelatin/E [18]) and procyanidin (PA; the composite is BC–gelatin/PA [58]), hydroxyl group absorption bands were observed at 3200–3600 cm^{-1}, but were weakened. This implied that the intermolecular hydrogen bonds in the composite were weakened [18]. This observation was affected by gelatin covering the surface of the BC [58]. By comparing the spectra bands between gelatin and BC–gelatin/PA, Wang [58] found that the absorbance of the amide I band at 1620 cm^{-1} was shifted from 1670 cm^{-1} (i.e., the amide I of gelatin), which justified the formation of hydrogen bonds in the ordered region of the amide I band. The stable hydrogen bond interaction was formed between amide I and the PA molecule.

By using the biosynthesis approach, the shifts of the amide II and O–H stretching peaks could indicate specific interactions between the functional groups of BC and gelatin [57].

14.3.2.2.2 *Crystallinity*

During cellulose synthesis, the presence of gelatin in the culture medium increased the viscosity, resulting in a decreased bacteria movement speed, which altered the BC structure. Consequently, the crystallinity of the composite film was lower than that of the BC film [57].

14.3.2.2.3 *Morphology*

BC hydrogel was modified with gelatin and hydroxyapatite for being used in tissue engineering applications [29, 58]. From SEM images of BC–gelatin/PA composites, thin gels were formed coating the surface of BC fibers. These results were confirmed by X-ray diffusion (XRD) analyses. At 0.5% wt. gelatin of the BC–gelatin/PA, the gel coating was thicker and the gel distribution was denser than those of BC–gelatin/PA synthesized by 0.1 and 0.25% wt. gelatin [47]. To improve the adsorption of gelatin into BC fibers, the BC pellicle was firstly concentrated by centrifugation and then immersed in a gelatin solution [28]. However, similar structures were observed in the BC–gelatin composite prepared with or without concentrating.

By supplementing a gelatin solution in the culture medium, gelatin was well incorporated into the BC nanofibril structure and filled the pores, leading to a denser structure of the BC–gelatin composite film with a smaller pore size than that of the BC film [57], as shown in Fig. 14.4.

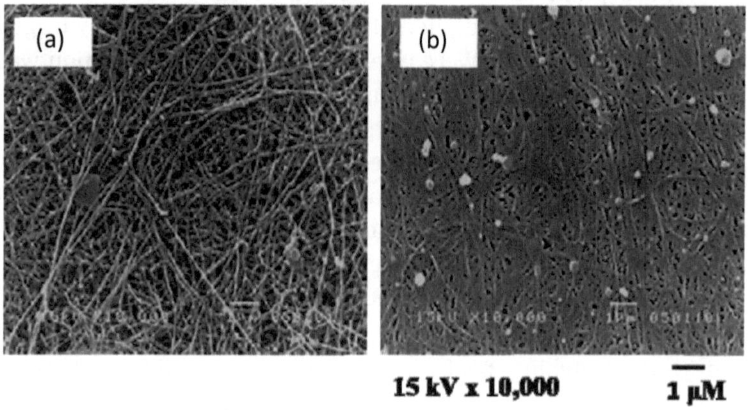

Figure 14.4 SEM images of films in re-swollen form of BC (a) and BC–gelatin prepared by addition of 10% w/v gelatin in the culture medium (b).

14.3.2.2.4 Mechanical strength

Mechanical properties including hardness, springiness, and cohesiveness of BC–gelatin/cross-linkers were characterized [18]. BC–gelatin/Es prepared with various gelatin concentrations (0.5–20%) exhibited the highest mechanical properties when compared with those of BC and BC–gelatin/other cross-linkers (i.e., transglutaminase and genipin). It was suggested that an increase of the molecule junction density of gelatin and cross-linker effectively led to the capture of water molecules of BC–gelatin/cross-linkers and enhanced the structural stability, resulting in improved mechanical properties.

For the application of these materials in tissue engineering, values of the tensile strength and elongation at break are required. Adding gelatin concentrations of 0–3% (w/v) in the culture medium slightly decreased the mechanical strengths of BC–gelatin composites due to the brittleness of gelatin. However, the presence of hydrogen bonds of BC is still possible [57]. In contrast, the structure of the BC–gelatin composite synthesized via biosynthesis with >5% (w/v) gelatin was almost completely coated with gelatin, resulting in gelatin-gelatin linkages. Due to the extremely weak and brittle properties of gelatin, the tensile strength and elongation at break of the composite film formulated from > 5% (w/v) gelatin considerably decreased [57].

14.3.2.2.5 *Water adsorption capacity*

Because of the hydrophilicity of gelatin, the WAC of the BC–gelatin composite and the BC–gelatin/cross-linkers improved with increasing concentrations of gelatin [18, 57]. Because of this, gelatin was able to disrupt the crystallization of BC by inserting between cellulose molecules [28]. Nonetheless, the randomized and weak structure of the composite would likely decrease its WAC [57].

14.3.2.2.6 *Water vapor transmission*

By comparing with BC film, the denser film with a smaller pore size was formed by the supplementation of gelatin into BC culture medium. Water molecules hardly penetrated through the small pores of the BC–gelatin composite film, resulting in the reduction of the WVTR [57].

14.3.2.3 Application

14.3.2.3.1 *Biomedical field and tissue engineering*

There was no observed toxicity in Vero cells cultured on the BC–gelatin composite synthesized via the biosynthesis approach. No significant differences were observed for cell viabilities when cultured on BC and the composite film. Due to its transparency, non-toxicity and biocompatibility, the composite film is relevant in tissue engineering and wound dressing [57]. Furthermore, cross-linked BC–gelatin/PA (produced by immersing 0.25% wt. gelatin concentration with PA) promoted the growth of NIH3T3 cells more than that of BC. The cells spread on the BC–gelatin/PA surface with normal morphology [47].

For specific applications (e.g., bone tissue engineering), BC–gelatin/PA doped with hydroxyapatite (HAp) were synthesized (BC–gelatin/PA/Hap). The cell compatibility of BC–gelatin/PA/HAp was tested with mesenchymal stem cells [58]. The results indicated that the composite supported cell growth and proliferation, over 7 days of cultivation. Studies on the effectiveness of composites in vitro and in vivo behavior should be further explored.

14.3.2.3.2 *Rehydration membrane*

Chen et al. [28] proposed a new technique to modify BC with gelatin for improving the dehydration of BC. Excluding a cross-linking step, the BC pellicle was initially centrifuged to concentrate the BC gel. Then, the concentrated BC was immersed with gelatin. Finally, the BC impregnated with gelatin was freeze dried to obtain the BC–gelatin composite film. Incredibly, the rehydration ability of the composite was approximately 125%. Consequently, the composite film could be used as a rehydrating membrane with high storage ability, a food pad with antibacterial property and wound dressing.

Morphology and physical properties of BC–gelatin composites via cross-linking were reported [18]. The mechanical strength of the composite was improved [18]. It might be implied that the cross-linking of BC–gelatin with EDC is also a potential technique by which to prepare a rehydrating membrane.

14.3.3 BC–Aloe vera

14.3.3.1 Aloe vera

Aloe vera, a succulent plant, is part of the lily family (Liliaceae). The bulk of the aloe vera leaf is filled with a clear gel-like substance (mucilage) containing approximately 99% water. For the other 1%, the main components are polysaccharide vitamins, minerals, and enzymes. The major polysaccharide, acemannan, is an acetylated β-1,4-linked glucomannan. Acemannan is used as an antiviral agent, an immunomodulator, and an agent in reducing opportunistic infections and stimulating the healing processes [59].

Aloe vera is well known for its healing properties. It has been used for centuries to heal a variety of conditions, such as burns, wounds, irritated skin, and bug bites. Regarding wound healing, many beneficial properties of aloe vera gel has been reported including the abilities to relieve pain, reduce inflammation, prevent itching, prevent bacterial, fungal and viral growth, and enhance blood flow [60–63]. Additionally, aloe vera is widely used in cosmetics, food, and dietary applications.

Although BC and aloe vera have been applied in many fields, the BC–aloe vera composite has rarely been investigated. Saibuatong and Phisalaphong [26] have characterized the biosynthesis of the BC–aloe vera composite. The Characteristics of BC–aloe vera composites are shown in Table 14.3 [26].

Table 14.3 Characteristics of BC–aloe vera composites

Characteristics	Synthesis method: Biosynthesis
Chemical structure	–Intermolecular interaction between hydroxyl group of BC and amino group of aloe vera [26]
Crystallinity	–Enhancement of crystallinity with adding aloe vera (0–30% v/v aloe vera) [26] –Reduction of crystallinity with adding aloe vera (≥ 50% v/v aloe vera) [26]
Morphology	–Well incorporation and uniform structure (0–30% v/v aloe vera) [26] –Less uniform with excessive gel on BC film (≥ 50% v/v aloe vera) [26]
Mechanical strength	–Improvement of mechanical properties with adding aloe vera (0–30% v/v aloe vera) [26]
WAC	–Improvement of WAC with adding aloe vera (0–30% v/v aloe vera) [26]
WVTR	–Improvement of WVTR with adding aloe vera (0–30% v/v aloe vera) [26]
Application	–Biomedical applications [26]

14.3.3.2 Characterization

14.3.3.2.1 *Chemical structure*

The intermolecular interaction between the carbonyl group of BC and the amino group of aloe vera was evidenced by the presence of two new peaks (1574 and 1542 cm^{-1}) in the FTIR spectra. For the supplement of 30–50% (v/v) aloe vera, the gel could adhere and form intermolecular bonds to BC fibrils [26].

14.3.3.2.2 *Morphology*

The BC–aloe vera composite synthesized with 5–30% (v/v) of aloe vera gel displayed a quite homogeneous and uniform structure in which aloe vera was well incorporated within BC fibrils (Fig. 14.5). However, a higher volume ratio of aloe vera (e.g., 50% v/v) created a less uniform network structure and led to excessive gel on the film surface. The increase of aloe vera gel content led to a composite film with small pore sizes and a narrow pore size distribution. Moreover, the supplementation of aloe vera in the culture medium directly inhibited cellulose formation. The oxygen transmission into the culture medium was also limited due to the increased culture medium viscosity, resulting in decreased *Acetobacter xylinum* metabolism [26].

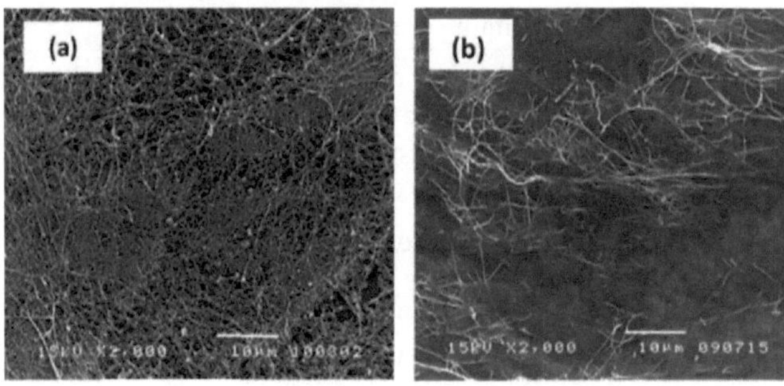

15 kV x 10,000 10 μm

Figure 14.5 SEM images of re-swollen films of BC (a) and BC-aloe vera prepared by addition of 30% v/v aloe vera gel in the culture medium (b).

14.3.3.2.3 *Mechanical strength*

The tensile strength of the BC–aloe vera composite film increased with increasing aloe vera content until 30% (v/v). Due to the randomized structure and excessive gel on the surface of the composite synthesized with more than 30% (v/v) gel, the tensile strength sharply decreased. The elongation at break and Young's modulus of the BC–aloe vera composite were analogous to that of the tensile strength. At 30% (v/v) aloe vera gel, the mechanical strength of the composite film was 1.6 times the tensile strength, 1.4 times the elongation at break and 1.2 times Young's modulus greater than those of BC film [26].

14.3.3.2.4 *Water adsorption capacity*

Because of the high hydrophilicity of aloe vera, the swelling and WAC of the BC–aloe vera composite film was increased. However, the trend of the WAC was similar to that of the mechanical strength. The optimal content of aloe vera gel was determined to be 30% (v/v) [26].

14.3.3.2.5 *Water vapor transmission*

The supplementation of aloe vera gel in the 0–30% (v/v) range resulted in enhanced WVTR. The swollen film also enlarged the pore sizes, resulting in an increase in the WVTR. Additionally, the increased hydrophilicity of the composite also increased the WVTR [26].

14.3.3.3 Application

14.3.3.3.1 *Biomedical field and tissue engineering*

The BC–aloe vera composite improved BC characteristics, such as high mechanical strength, high WAC, and high WVTR, for use in tissue engineering and other biomedical applications. The evaluation of the composite film as a material for tissue engineering is currently being studied [26].

14.3.4 BC-Chitosan

14.3.4.1 Chitosan

Chitosan is another abundant polysaccharide obtained by the deacetylation of chitin. The degree of deacetylation (DD) is an important parameter characterizing chitosan. The DD of chitosan must be higher than 50% [64]. The chemical structure of chitosan consists of β-(1-4)-linked D-glucosamin and N-acetyl-D-glucosamine. The only difference between the chemical structures of chitosan and cellulose is the presence of the amino group instead of the hydroxyl group on each glucose molecule.

Chitosan is highly soluble in dilute acids and can easily form gels at higher pH values. Chitosan has many attractive properties, such as its biocompatibility, non-toxicity, biodegradability, antimicrobial activity and its ability to easily form flat films. Chitosan has been studied by numerous groups in the biomedical, pharmaceutical and biotechnology fields [65]. Chitosan is a promising biopolymer for wound dressing applications because of its ability to infiltrate inflammatory cells, reduce scar tissue, absorb wound exudes, and as an antibacterial agent [66–68]. However, chitosan film displays weak mechanical strength and brittleness, which limit its applications [69].

Because of the structural similarities of BC and chitosan, the composites of these two biopolymers combine the physicochemical properties of chitosan with the excellent mechanical strengths of BC [70]. Comparative characteristics of BC-chitosan composites prepared from various methods are shown in Table 14.4.

Table 14.4 Comparative characteristics of BC–chitosan composites prepared from various methods.

	Synthesis methods		
Characteristics	**Biosynthesis**	**Impregnation**	**Blending**
Chemical structure	–Intermolecular interaction between hydroxyl group of BC and amino group of chitosan [19, 71, 73]		
Crystallinity	–Slight reduction of crystallinity with increasing chitosan content [19]	–Reduction of crystallinity with adding chitosan [71, 72]	–Improvement of crystallization by adding BC (a reference material: chitosan) [70]
Morphology	–Well-organized fibril networks [19] –Denser structure with smaller pore size [19, 73]	–More compact and denser structure with smaller pore size [20] –An arranged network with specially oriented pores [71] –Multilayer structure [72]	–Good dispersion [70]
Mechanical strength	–Improvements of tensile strength and Young's modulus, and reduction of elongation at break with increasing Mw and content of chitosan [19]	–Improvement of mechanical properties with increasing chitosan content [71] –Improvement of Young's modulus and slight reductions of tensile strength and elongation at break with adding chitosan [20]	–Improvement of mechanical properties with increasing chitosan content [75] –Improvements of Young's modulus and tensile strength, and reduction

Characteristics	Synthesis methods Biosynthesis	Impregnation	Blending
		–Improvement of elongation and reduction of tensile strength with adding chitosan [74]	of elongation at break with increasing BC content (a reference material: chitosan) [70]
WAC	–Improvement of WAC with increasing Mw and content of chitosan [19]	–Improvement of WAC with increasing chitosan content [71] –50% reduction of WAC with adding chitosan [20]	—
Application	–Wound dressing (non-toxicity, promotion of keratinocyte and gingival fibroblast proliferations) [73]	–Wound dressing and tissue engineering (high antibacterial ability, higher effective healing on rat when compared to those of the commercial products) [20] –Ethanol/water membrane separation (high selectivity) [74]	–Transparent biodegradable and antibacterial packaging [70] –Tissue engineering (open-interconnected microporous structure and biodegradability in vitro) [75]

14.3.4.2 Characterization

14.3.4.2.1 *Chemical structure*

Intermolecular hydrogen bonding between BC and chitosan of the BC–chitosan composite was identified from the spectra shift of amino groups (from 1556 to approximately 1559 cm^{-1}). By increasing the chitosan concentration in the culture medium, the intensity of this band gradually increased [19]. In a similar way, the hydrogen bonding interaction became stronger with increasing immersion time in the impregnation technique [71].

14.3.4.2.2 *Crystallinity*

The effect of chitosan on the crystallinity of the BC–chitosan composite from biosynthesis was not clearly observed because only a low concentration of chitosan (0–0.75% w/v) was added into the culture. At chitosan concentrations greater than 0.75% w/v, the production of cellulose was significantly affected [19].

A reduction crystallinity index (CI) of the BC–chitosan composite synthesized with the impregnation method was observed when the chitosan content was increased; the reduction also corresponded with an increased immersion time [71]. With the incorporation of chitosan in BC fibrils, BC hydrogen bonding was disturbed by the interaction between chitosan and BC [71]. Moreover, it suggested that because of this interaction, BC chains were difficult to move, resulting in the reduction of CI [72].

14.3.4.2.3 *Morphology*

With the addition of the chitosan solution into the culture medium for BC production, the BC–chitosan composite films were homogeneous with noticeably denser fibrils and smaller pore sizes compared with those of BC film (Fig. 14.6). A smaller pore size was obtained by increasing the MW of chitosan [19, 73].

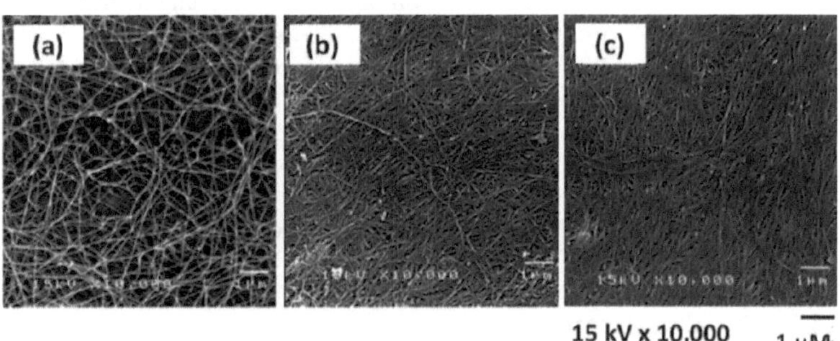

Figure 14.6 SEM images of re-swollen films of BC (a), BC–chitosan prepared by adding 0.75% chitosan of MW 30,000 (b) and of MW 80,000 (c) in the culture medium.

According to the impregnation approach, the BC pellicle was immersed in the chitosan solution for different periods of time. It was found that chitosan molecules would not only coat the BC surface but also penetrate through the BC fibrils. Chitosan was well incorporated with nanofibrils, resulting in a denser and more compact structure with a smaller pore size [20]. From SEM images of the cross-sectional area of the composite, the BC structure formed a multilayer structure with a high aspect ratio and a well-organized network [72]. Moreover, in the investigation of Ul-Islam, an arranged network with specifically oriented pores was observed [71].

As mentioned above in the composite synthesis section, using a blending technique results in a homogeneous structure. A good dispersion of both the chitosan granules and BC microfibrils without noticeable aggregation was observed by using atomic force microscopy [70].

14.3.4.2.4 *Mechanical strength*

Using the biosynthesis approach, a higher tensile strength, and Young's modulus of BC–chitosan composite were obtained by increasing the MW and concentration of chitosan. Because the fibrils became thicker, they could resist a stronger force than pure BC fibrils [19].

On the other hand, variation in the mechanical strength of the BC–chitosan composite synthesized by the impregnation technique was observed. This may be due to the differences of bacterial strain, cultivation condition and treatment procedures. It was suggested that due to the brittleness of chitosan, the mechanical strength of the composite was weaker than that of BC film [72]. Young's modulus of the composite was much greater than that of the BC [20]. For another study, a higher elongation at break of the composite was found [74]. However, in some conditions, it was reported that the composite had a higher mechanical strength than that of the BC film [71]. In this case, it was proposed that the deep penetration of the chitosan inside the BC network formed a denser fibril structure, which could share the loading force. Additionally, the thickness of the composite film increased [71].

14.3.4.2.5 *Water adsorption capacity*

The WAC of the BC–chitosan composite from biosynthesis was enhanced approximately 120% in comparison with that of the BC film. The improvement of the WAC was affected by increasing the MW and concentration of chitosan in the culture medium [73].

During the impregnation technique, chitosan incorporating with BC fibrils could also enhance the WAC of the BC–chitosan composite because of the hydrophilicity of chitosan [71]. However, with some different conditions, it was reported that the WAC of the composite decreased to approximately half that of the BC owing to the more compact and denser structure with smaller pore size of the composites [20].

14.3.4.2.6 *Antimicrobial activity*

As mentioned above, due to the small amount of chitosan added in the culture medium, the BC–chitosan composite obtained from the biosynthesis did not show inhibition effect the growth of *Escherichia coli* (*E. coli*) and *Staphylococcus aureus* (*S. aureus*). For antifungal activity, a slight inhibition was detected on the growth of *Aspergillus niger* (*A. niger*) [19].

The improved antibacterial activity of the BC–chitosan composite via the impregnation approach was reported. Firstly, the antimicrobial evaluation was performed by the direct contact of the composite with *E. coli* and *S. aureus*. The growth reductions of 99.9% were observed in both cultures. Secondly, the evaluation was performed by the disc diffusion method. Because of a tight incorporation between chitosan and BC, it resulted in no chitosan diffusion out of the composite. Therefore, no inhibitory clear zone was found [20].

14.3.4.3 Application

14.3.4.3.1 *Biomedical field and tissue engineering*

For wound healing applications, the membrane composite of BC and chitosan was synthesized by the impregnation technique. In wound healing experiments, wounds were created on the skins of Sprague Dawley rats. It was found that wounds covered with the composite membrane healed faster than those covered with BC, commercial Tegaderm™ hydrocolloid dressing or transparent film dressing (manufactured by 3M Health Care Ltd., USA). The BC–chitosan composite membrane was a strong candidate for wound dressing applications [20]. The possibility for utilizing BC–chitosan composite film for human treatment was examined by Kingkaew et al. [73]. Human-transformed skin keratinocytes and primary human skin fibroblasts were seeded on the composite film to investigate the ability for cell attachment and proliferation. It was shown that the composite film promoted the proliferation of keratinocytes and also improved cell adhesion and spreading on the surface. Typical elongated morphologies of fibroblast cells with good spreading were observed in both the BC and composite films [73]. According to these results, the BC–chitosan composite has potential to be used as a wound dressing for humans.

For scaffolds in tissue engineering, Nge et al. [75] synthesized the oxidized BC–chitosan composite scaffold by using supplementation (i.e., homogenized BC) and a lyophilization technique [75]. An open microporous structure with interconnecting pores was observed in the composite scaffold. The pore size and morphology of the composite could be controlled by the freezing temperature of lyophilization and the content of the BC. A variety of the porous nanocomposite scaffolds could be further developed in other scaffolding and regeneration applications.

14.3.4.3.2 *Membrane separation*

Dubey et al. [74] improved the BC membrane by the impregnation of chitosan (Mw 100–300 kDa). The composite membrane was dried under vacuum. The potential of the composite membrane for the pervaporative separation of the ethanol/water azeotrope was comparable to that of a polyvinyl alcohol (PVA) membrane [74]. The normalized flux, selectivity, and PSI of the composite membrane were 42.8 kg μm $m^{-2}h^{-1}$, 9.2 and 350 kg μm $m^{-2}h^{-1}$, respectively. Compared with the PVA membrane, BC membrane impregnated with chitosan had excellent dimensional stability, better mechanical strength, and improved thermal stability.

14.3.4.3.3 *Food packaging*

The transparent nanocomposite film based on BC and chitosan was proposed by Fernandes [70]. Various concentrations of microfibrillar BC with 10–200 nm width was mixed with three different chitosan solutions (i.e., low and high MW and a water-soluble quaternary ammonium derivative of high-MW chitosan). The results showed that the three nanocomposite films had significantly increased mechanical strength when compared with chitosan film [70]. Additionally, the composite films were transparent, flexible, thermally stable and had low oxygen permeability, which is promising for applications requiring transparent, biodegradable and antibacterial packaging [70].

14.3.5 BC–PVA

14.3.5.1 PVA

Polyvinyl alcohol (PVA) is produced by hydrolysis reaction of polyvinyl acetate. PVA is a water-soluble, hydrophilic, thermoplastic, biocompatible, non-toxic and biodegradable polymer. PVA can be cross-linked to form PVA hydrogel by several techniques, such as chemical and physical cross-linking, electron beam and γ-irradiation [76]. By means of physical cross-linking technique, the mechanical strength of PVA can be easily modified by adjusting parameters, such as the concentration of PVA, the number of freeze/thaw cycles, freezing temperature [77]. PVA has been widely used in the fields of medicine and food. It has been applied in reconstruction and replacements of cardiovascular [78, 79] and cartilage tissues [80–82]. An exponential stress-strain curve of PVA is similar to that of soft tissues [79].

PVA has high elasticity; however, its tensile strength is very poor. Its mechanical strength could be improved by blending with BC [83]. According to the chemical structures of PVA and BC, the presences of hydroxyl group in both polymers lead to a miscible structure of BC–PVA composite [83]. The BC–PVA composite would be an ideal composite for soft tissue application due to its sufficient tensile strength and enough elasticity. Comparative characteristics of BC–PVA composites prepared from various methods are shown in Table 14.5.

Table 14.5 Comparative characteristics of BC–PVA composites prepared from various methods.

	Synthesis methods		
Characteristics	**Biosynthesis**	**Impregnation**	**Blending**
Chemical structure	—	–Strong hydrogen bonds of intra-molecule and inter-molecule [84]	—
Crystallinity	–Improvement of crystallinity with adding BC (a reference material: PVA) [83]	–Reduction of crystallinity after cross-linking reaction [85]	—

Characteristics	Synthesis methods: Biosynthesis	Impregnation	Blending
Morphology	–Effective dispersion [83]	–Uniform structure with larger diameter of fibrils [84] –Wet state: well incorporation [85] –Dry state: heterogeneous structure (dense and porous structures) [85]	–Denser structure with increasing PVA content [87]
Mechanical strength	–Improvements of tensile strength and Young's modulus with adding BC (a reference material: PVA) [83]	–Improvement of elongation at break, and reductions of tensile strength and Young's modulus with increasing PVA content [84] –Reductions of tensile strength and Young's modulus with adding PVA [85]	–Slight difference of mechanical properties (comparison to those of BC) [87]
Application	—	–Cardiovascular application (Hemocompatibility) [88]	–Cardiovascular application (similarity of mechanical properties to those of porcine aorta and aortic heart valve) [89]

14.3.5.2 Characterization

14.3.5.2.1 *Chemical structure*

The strong hydrogen bonds of intra-molecule and inter-molecular type formed in the BC–PVA structure, resulting in the detection of the hydroxyl (O–H) stretching vibration (3200–3500 cm^{-1}) [84].

14.3.5.2.2 *Crystallinity*

In the comparison of BC and the BC–PVA composite, the crystallinity of BC in the composite was decreased by 50%. By immersing wet BC pellicles in polyvinyl alcohol (PVA) solution, PVA could penetrate the BC network structure and filled the pores. Acetal linkages were formed in the BC–PVA composites by a cross-linking reaction, which made PVA insoluble in water. However, the crystal destruction was observed after the cross-linking [84].

14.3.5.2.3 *Morphology*

By using biosynthesis technique to synthesize BC–PVA composite [83], PVA powders and glyoxal aqueous solution (as a cross-linking agent) were added into the culture medium. The homogeneous distribution of the composite ribbons was noticed throughout the matrix, which could be implied that BC, PVA and also glyoxal were compatible with each other.

PVA could penetrate into the BC structure and enwrapped the BC fibrils. From the SEM images, it could be observed that the diameter of the composite fibrils was larger than that of unmodified BC [84]. However, the overall structure of BC and BC–PVA composite, fiber thickness, fiber distribution and three-dimensional orientations were quite similar [85]. The SEM examination of BC and BC–PVA revealed that interpenetrated networks could be formed by integrating PVA fiber into the original BC pellicle [85]. The drying method and conditions also have significant effects on the composite structure. The BC–PVA composites dehydrated by freeze-drying process [84] had a structure that was different from those dried in an incubator oven [85]. When PVA fibers are heated, they contract [86].

To synthesize the BC–PVA composite by blending method, the homogenized BC was mixed with PVA solution. Then the films were prepared by casting the mixture on Teflon coated tray [87]. It was found that PVA coated the BC fibrils, leading to a denser structure and more transparency with increasing PVA content.

14.3.5.2.4 *Mechanical strength*

Using PVA as a reference, it was shown that the tensile strength and Young's modulus of the BC–PVA composites synthesized by using biosynthesis were improved with increasing BC content [83].

Although the fracture strain of PVA was very high, the fracture strain of the BC–PVA composite prepared by immersing BC pellicle in PVA was much lower than that of PVA and close to that of BC. It was suggested that bonding between BC and PVA in the composite was extremely tight, resulting in a brittle composite [84].

The mechanical behavior of the BC–PVA composite synthesized by using homogenized BC was similar to that of BC [87]. As compared to pure PVA, the presence of BC fibers in the PVA matrix enhanced the tensile strength and the elastic modulus but decreased the toughness. It was suggested that hydrogen bonding between BC itself was stronger than that between of BC and PVA [87].

14.3.5.3 Application

14.3.5.3.1 *Biomedical field and tissue engineering*

BC suspension obtained from an agitation culture system was mixed with PVA powder to fabricate PVA–BC composite by considering PVA as a matrix and BC as a reinforcement material. The composite was proposed for cardiovascular soft tissue replacement application [78]. The composite could be adjusted its mechanical properties to get close to those of specific tissue (aorta and heart valve) by changing the composition of the composite and the processing parameters.

Leitao et al. [88] characterized the hemocompatility of BC–PVA composite prepared by immersing BC pellicle in PVA solution and then using freezing/thawing method. The results indicated that the composite had good hemocompatility. In the comparison to commercial synthetic graft, expanded polytetrafluoroethylene (ePTFE), the BC–PVA composite presented a superior performance based on the results of platelet adhesion and activation profile. Therefore, the composite was suitable for cardiovascular application. According to the morphologies of the BC–PVA composite [88], a major portion of membrane consisted of large pores ranging from 500 nm to 2 μm in diameter, which might facilitate the cells migration into the composite and nutrient diffusion in the composite.

14.4 Conclusion

A wide range of BC-based composite materials have been developed to enhance the physical and biological properties of BC to overcome its limitations and to also expand its applications. Due to the specific properties of the supplemented materials, required properties such as transparency, rehydration ability, antimicrobial activity, and compatibility with specific cells of the modified composites could be developed. BC-based composites have the potential uses in a variety of applications, including biomedical materials, tissue engineering, drug delivery, immobilization matrices for cells and enzymes, food packaging, and membrane. The variability of supplemented materials, synthesis methods, and fabrication conditions considerably influence characteristics and properties of the composites. Since the requirement of materials for specific applications is constantly increasing, development in advanced materials with insight technologies is necessary.

Acknowledgments

The authors are grateful for the financial support of the Ratchadaphiseksomphot Endowment Fund of Chulalongkorn University, Project RES560530044-AM and Postdoctoral Fellowship (Ratchadaphiseksomphot Endowment Fund).

References

1. Raghunatham, D. (2013). Production of microbial cellulose from the new bacterial strain isolated from temple wash waters, *Int. J. Curr. Microbiol. App. Sci.*, **2**(12), 275–290.
2. Brown, Jr. R. M. (1996). The biosynthesis of cellulose, *J. Macromol. Sci. A*, **33**(10), 1345–1373.
3. Jonas, R., and Farah, L. F. (1998). Production and application of microbial cellulose, *Polym. Degrad Stabil.*, **59**, 101–106.
4. Halib, N., Amin, M. C. I. M., and Ahmad, I. (2012). Physicochemical properties and characterization of *Nata de Coco* from local food industries as a source of cellulose, *Sains Malays.*, **41**(2), 205–211.
5. Geyer, U., Klemm, D., and Schmauder, H. P. (1994). Kinetics of the utilization of different C sources and the cellulose formation by *Acetobacter xylinum*, *Acta Biotechnol.*, **14**, 261–266.
6. Masaoka, S., Ohe, T., and Sakota, N. (1993). Production of cellulose from glucose by *Acetobacter xylinum*, *J. Ferment. Bioeng.*, **75**, 18–22.

7. Oikawa, T., Morino, T., and Ameyanea, M. (1995). Production of cellulose from D-arabitol by *Acetobacterxylinum* KU-1, *Biosci. Biotech. Biochem.*, **59**, 1564–1565.

8. Chawla, P. R., Bajaj, I. B., Survase, S. A., and Singhal, R. S. (2009). Microbial cellulose: Fermentative production and applications, *Food Technol. Biotechnol.*, **47**(2), 107–124.

9. Borzani, W., and Souza, de S. J. (1995). Mechanism of the film thickness increasing during the bacterial production of cellulose on non-agitated liquid media, *Biotechnol. Lett.*, **17**(11), 1271–1272.

10. Czaja, W., Romanovicz, D., and Brown, Jr. R. M. (2004). Structural investigations of microbial cellulose produced in stationary and agitated culture, *Cellulose*, **11**, 403–411.

11. Shah, N., Ul-Islam, M., Khattak, W. A., and Park, J. K. (2013). Overview of bacterial cellulose composites: A multipurpose advanced material, *Carbohyd. Polym.*, **98**, 1585–1598.

12. Andrade, F. K., Costa, R., Domingues, L., Soares, R., and Gama, M. (2010). Improving bacterial cellulose for blood vessel replacement: Functionalization with a chimeric protein containing a cellulose-binding module and an adhesion peptide, *Acta Biomater.*, **6**, 4034–4041.

13. Fontana, J. D., Souza, de A. M., Fontana, C. K., Torriani, I. L., Moreschi, J. C., Gallotti, B. J., Souza, de S. J., Narcisco, G. P., Bichara, J. A., and Farah, L. F. X. (1990). Acetobacter cellulose pellicle as a temporary skin substitute, *Appl. Biochem. Biotechnol.*, **24–25**, 253–263.

14. Klemm, D., Heublein, B., Fink, H., and Bohn, A. (2005). Cellulose: Fascinating biopolymer and sustainable raw material, *Angew. Chem. Int. Ed. Engl.*, **44**, 3358–3393.

15. Svensson, A., Nicklasson, E., Harrah, T., Panilaitis, B., Kaplan, D. L., Brittberg, M., and Gatenholm, P. (2005). Bacterial cellulose as a potential scaffold for tissue engineering of cartilage, *Biomaterials*, **26**, 419–431.

16. Zahedmanesh, H., Mackle, J. N., Sellborn, A., Drotz, K., Bodin, A., Gatenholm, P., and Lally, C. (2011). Bacterial cellulose as a potential vascular graft: Mechanical characterization and constitutive model development, *J. Biomed. Mater. Res. B Appl. Biomater..*, **97**, 105–113.

17. Alvarez, O. M., Patel, M., Booker, J., and Markowitz, L. (2004). Effectiveness of a biocellulose wound dressing for the treatment of chronic venous leg ulcers: Results of a single center randomized study involving 24 patients, *Wounds*, **16**, 224–233.

18. Chang, S. T., Chen, L. C., Lin, S. B., and Chen, H. H. (2012). Nano-biomaterials application: Morphology and physical properties of bacterial cellulose/gelatin composites via crosslinking, *Food Hydrocolloids*, **27**, 137–144.

19. Phisalaphong, M., and Jatupaiboon, N. (2008). Biosynthesis and characterization of bacteria cellulose-chitosan film, *Carbohyd. Polym.*, **74**, 482–488.

20. Lin, W. C., Lien, C. C., Yeh, H. J., Yu, C. M., and Hsu, S. H. (2013). Bacterial cellulose and bacterial cellulose-chitosan membranes for wound dressing applications, *Carbohyd. Polym.*, **94**, 603–611.

21. Hong, L., Wang, Y. L., Jia, S. R., Huang, Y., Gao, C., and Wan, Y. Z. (2006). Hydroxyapatite/bacterial cellulose composites synthesized via a biomimetic route, *Mater. Lett.*, **60**, 1710–1713.

22. Stoica-Guzun, A., Stroescu, M., Jinga, S. I., Jipa, I. M., and Dobre, T. (2013). Microwave assisted synthesis of bacterial cellulose-calcium carbonate composites, *Ind. Crop. Prod.*, **50**, 414–422.

23. Kanjanamosit, N., Muangnapoh, C., and Phisalaphong, M. (2009). Biosynthesis and characterization of bacteria cellulose-alginate film, *J. Appl. Polym. Sci.*, **115**, 1581–1588.

24. Ul-Islam, M., Khan, T., and Park, J. K. (2012). Water holding and release properties of bacterial cellulose obtained by in situ and ex situ modification, *Carbohyd. Polym.*, **88**, 596–603.

25. Yan, Z., Chen, S., Wang, H., Wang, B., and Jiang, J. (2008). Biosynthesis of bacterial cellulose/multi-walled carbon nanotubes in agitated culture, *Carbohyd. Polym.*, **74**, 659–665.

26. Saibuatong, O., and Phisalaphong, M. (2010). Novo aloe vera-bacterial cellulose composite film from biosynthesis, *Carbohyd. Polym.*, **79**, 455–460.

27. Ul-Islam, M., Khattak, W. A., Ullah, M. W., Khan, S., and Park, J. K. (2014). Synthesis of regenerated bacterial cellulose-zinc oxide nanocomposite films for biomedical applications, *Cellulose*, **21**, 433–447.

28. Chen, H. H., Lin, S. B., Hsu, C. P., and Chen, L. C. (2013). Modifying bacterial cellulose with gelatin peptides for improved rehydration, *Cellulose*, **20**, 1967–1977.

29. Jing, W., Chunxi, Y., Yizao, W., Honglin, L., Fang, H., Kerong, D., Yuan, H. (2013). Laser patterning of bacterial cellulose hydrogel and its modification with gelatin and hydroxyapatite for bone tissue engineering. *Soft Mater.*, **11**, 173–180.

30. Phisalaphong, M., Suwanmajo, T., and Sangtherapitikul, P. (2008). Novel nanoporous membranes from regenerated bacterial cellulose, *J. Appl. Polym. Sci.*, **107**, 292–299.

31. Phisalaphong, M., Suwanmajo, T., and Tammarate, P. (2007). Synthesis and characterization of bacterial cellulose/alginate blend membranes, *J. Appl. Polym. Sci.*, **107**(5), 3419–3424.

32. Grierson, D. (1993). *Biosynthesis and Manipulation of Plant Products*, 1st Ed. (Blackie Academic & Professional, Chapman & Hall, UK).

33. Lee, K. Y., and Mooney, D. J. (2012). Alginate: Properties and biomedical applications, *Prog. Polym. Sci.*, **37**, 106–126.

34. Ahmad, Z., and Khuller, G. K. (2008). Alginate-based sustained release drug delivery systems for tuberculosis, *Expert Opin. Drug. Deliv.*, **5**, 1323–1334.

35. Tonnesen, H. H., and Karlsen, J. (2002). Alginate in drug delivery systems, *Drug Dev. Ind. Pharm.*, **28**, 621–630.

36. Thomas, S. (2000). Alginate dressings in surgery and wound management-Part 1, *J. Wound Care*, **9**(2), 56–60.

37. Wang, C. C., Yang, K. C., Lin, K. H., Liu, H. C., and Lin, F. H. (2011). A highly organized three-dimensional alginate scaffold for cartilage tissue engineering prepared by microfluidic technology, *Biomaterials*, **32**(29), 7118–7126.

38. Lee, S. H., Jo, A. R., Choi, G. P., Woo, C. H., Lee, S. J., Kim, B. S., You, H. K., and Cho, Y. S. (2013). Fabrication of 3D alginate scaffold with interconnected pores using wire-network molding technique, *Tissue Eng. Regen. Med.*, **2**, 53–59.

39. Shoichet, M. S., Li, R. H., White, M. L., and Winn, S. R. (1996). Stability of hydrogels used in cell encapsulation: An in vitro comparison of alginate and agarose, *Biotechnol. Bioeng.*, **50**, 374–381.

40. Chiaoprakobkij, N., Sanchavanakit, N., Subbalekha, K., Pavasant, P., and Phisalaphong, M. (2011). Characterization and biocompatibility of bacterial cellulose/alginate composite sponges with human keratinocytes and gingival fibroblasts, *Carbohyd. Polym.*, **85**, 548–553.

41. Kirdponpattara, S., and Phisalaphong, M. (2013). Bacterial cellulose-alginate composite sponge as a yeast cell carrier for ethanol production, *Biochem. Eng. J.*, **77**, 103–109.

42. Zhang, S., and Luo, J. (2011). Preparation and properties of bacterial cellulose/alginate blend bio-fibers, *J. Eng. Fiber. Fabr.*, **6**, 69–72.

43. Yang, S., Leong, K. F., Du, Z., and Chua, C. K. (2001). The design of scaffolds for use in tissue engineering. Pat I. Traditional factor, *Tissue Eng. J.*, **7**(6), 679–689.

44. Yakimets, I., Wellner, N., Smith, A. C., Wilson, R. H., Farhat, I., and Mitchell, J. (2005). Mechanical properties with respect to water content of gelatin films in glassy state, *Polymer*, **46**, 12577–12585.

45. Ramalingam, M., Tiwari, A., Ramakrishna, S., and Kobayashi, H. (2012). *Integrated Biomaterials for Biomedical Technology*, 1st Ed. (John Wiley & Sons, Inc., USA).

46. Rottensteiner, U., Sarker, B., Heusinger, D., Dafinova, D., Rath, S. N., Beier, J. P., Kneser, U., Horch, R. E., Detsch, R., Boccaccini, A. R., and Arkudas, A. (2014). In vitro and in vivo biocompatibility of alginate dialdehyde/gelatin hydrogels with and without nanoscaled bioactive glass for bone tissue engineering applications, *Materials*, **7**, 1957–1974.

47. Wang, J., Wan, Y. Z., Luo, H. L., Gao, C., and Huang, Y. (2012). Immobilization of gelatin on bacterial cellulose nanofibers surface via crosslinking technique, *Mater. Sci. Eng. C*, **32**, 536–541.

48. Choi, Y. S., Hong, S. R., Lee, Y. M., Song, K. W., Park, M. H., and Nam, Y. S. (1999). Study on gelatin-containing artificial skin: I. Preparation and characteristics of novel gelatin-alginate sponge, *Biomaterials*, **20**(5), 409–417.

49. Parvez, S., Rahman, M. M., Khan, M. A., Khan, M. A. H., Islam, J. M. M., Ahmed, M., Rahman, M. F., and Ahmed, B. (2012). Preparation and characterization of artificial skin using chitosan and gelatin composites for potential biomedical application, *Polym. Bull.*, **69**, 715–731.

50. Rujitanaroj, P., Pimpha, N., and Supaphol, P. (2008). Wound-dressing materials with antibacterial activity from electrospun gelatin fiber mats containing silver nanoparticles, *Polymer*, **49**, 4723–4732.

51. Takahashi, H., Miyoshi, T., and Boki, K. (1993). Study on hydrophilic properties of gelatin as a clinical wound dressing. I. Hydrophilic properties of gelatin as a wound dressing, *Tokushima J. Exp. Med.*, **40**, 159–167.

52. Cortesi, R., Esposito, E., Osti, M., Squarzoni, G., Menegatti, E., Davis, S. S., Nastruzzi, C. (1999). Dextran cross-linked gelatin microspheres as a drug delivery system, *Eur. J. Pham. Biopharm.*, **47**, 153–160.

53. Adhirajan, N., Shanmugasundaram, N., and Babu, M. (2007). Gelatin microspheres cross-linked with EDC as a drug delivery system for doxycyline: Development and characterization, *J. Microencapsul.*, **24**, 647–659.

54. He, C., Zhang, F., Cao, L., Feng, W., Qiu, K., Zhang, Y., Wang, H., Mo, X., and Wang, J. (2012). Rapid mineralization of porous gelatin scaffolds by electrodeposition for bone tissue engineering, *J. Mater. Chem.*, **22**, 2111–2119.

55. Liu, X., Smith, L. A., Hu, J., and Ma, P. X. (2009). Biomimetic nanofibrous gelatin/apatite composite scaffolds for bone tissue engineering, *Biomaterials*, **30**, 2252–2258.

56. Narbat, M. K., Orang, F., Hashtjin, M. S., and Goudarzi, A. (2006). Fabrication of porous hydroxyapatite-gelatin composite scaffolds for bone tissue engineering, *Iran. Biomed. J.*, **10**(4), 215–223.

57. Taokaew, S., Seetabhawang, S., Siripong, P., and Phisalaphong, M. (2013). Biosynthesis and characterization of nanocellulose-gelatin films, *Materials*, **6**, 782–794.

58. Wang, J., Wan, Y., Han, J., Lei, X., Yan, T., and Gao, C. (2011). Nanocomposite prepared by immobilising gelatin and hydroxyapatite on bacterial cellulose nanofibres, *Micro Nano Lett.*, **6**(3), 133–136.

59. Moore, E. D., and Mac Analley, B. H. (1995). A drink containing mucilaginous polysaccharides and its preparation, U.S Patent 5, 443–530.

60. Heggers, J. P., Kucukcelebi, A., Listengarten, D., Stabenau, J., Ko, F., Broemeling, L. D., Robson, M. C., and Winters, W. D. (1996). Beneficial effect of Aloe on wound healing in an excisional wound model, *J. Altern. Complement. Med.*, **2**, 271–277.

61. Maenthaisong, R., Chaiyakunapruk, N., Niruntraporn, S., and Kongkaew, C. (2007). The efficacy of aloe vera used for burn wound healing: A systematic review, *Burns*, **33**, 713–718.

62. Reynolds, T., and Dweck, A. C. (1999). Aloe vera leaf gel: A review update, *J. Ethnopharmacol.*, **68**, 3–37.

63. Yao, H., Chen, Y., Li, S., Huang, L., Chen, W., and Lin, X. (2009). Promotion proliferation effect of a polysaccharide from Aloe barbadensis Miller on human fibroblasts in vitro, *Int. J. Bio. Macromol.*, **45**, 152–156.

64. Krajewska, B. (2004). Application of chitin-and chitosan-based material for enzyme immobilizations: A review, *Enzyme Microb. Tech.*, **35**, 126–139.

65. Shigemasa, Y., and Minami, S. (2013). Applications of chitin and chitosan for biomaterials, *Biotechnol. Genet. Eng.*, **13**, 383–420.

66. Barbosa, J. N., Amaral, I. F., Aguas, A. P., and Barbosa, M. A. (2010). Evaluation of the effect of the degree of acetylation on the inflammatory response to 3D porous chitosan scaffolds, *J. Biomed. Mater. Res. A*, **93**, 20–28.

67. Dai, T., Tanaka, M., Huang, Y. Y., Hamblin, M. R. (2011). Chitosan preparations for wounds and burns: Antimicrobial and wound-healing effects, *Expert Rev. Anti Infect. Ther.*, **9**, 857–879.

68. Matsunaga, T., Yanagiguchi, K., Yamada, S., Ohara, N., Ikeda, T., and Hayashi, Y. (2006). Chitosan monomer promotes tissue regeneration on dental pulp wounds, *J. Biomed. Mater. Res. A*, **76**, 711–720.

69. Elsabee, M. Z., and Abdou, E. S. (2013). Chitosan based edible films and coatings: A review, *Mater. Sci. Eng. C Mater. Biol. Appl.*, **33**, 1819–1841.

70. Fernandes, S. C. M., Oliveira, L., Freire, C. S. R., Silvestre, A. J. D, Neto, C. P., Gandini, A., and Desbrieres, J. (2009). Novel transparent nanocomposite films based on chitosan and bacterial cellulose, *Green Chem.*, **11**, 2023–2029.

71. Ul-Islam, M., Shah, N., Ha, J. H., and Park, J. K. (2011). Effect of chitosan penetration on physico-chemical and mechanical properties of bacterial cellulose, *Korean J. Chem. Eng.*, **28**(8), 1736–1743.

72. Kim, J., Cai, Z., Lee, H. S., Choi, G. S., Lee, D. H., and Jo, C. (2011). Preparation and characterization of a Bacterial cellulose/Chitosan composite for potential biomedical application, *J. Polym. Res.*, **18,** 739–744.

73. Kingkaew, J., Jatupaiboon, N., Sanchavanakit, N., Pavasant, P., and Phisalaphong, M. (2010). Biocompatibility and growth of human keratinocytes and fibroblasts on biosynthesized cellulose-chitosan film, *J. Biomater. Sci.*, **21**, 1009–1021.

74. Dubey, V., Pandey, L. K., and Saxena, C. (2005). Pervaporative separation of ethanol/water azeotrope using a novel chitosan-impregnated bacterial cellulose membrane and chitosan-poly(vinyl alcohol) blends, *J. Membr. Sci.*, **251**, 131–136.

75. Nge, T. T., Nogi, M., Yano, H., and Sugiyama, J. (2010). Microstructure and mechanical properties of bacterial cellulose/chitosan porous scaffold, *Cellulose*, **17**, 349–363.

76. Hwang, M. R., Kim, J. O., Lee, J. H., Kim, Y. I., Kim, J. H., Chang, S. W., Jin, S. G., Kim, J. E., Lyoo, W. S., Han, S. S., Ku, S. K., Yong, C. S., and Choi, H. G. (2010). Gentamicin-loaded wound dressing with polyvinyl alcohol/dextran hydrogel: Gel characterization and in vivo healing evaluation, *Pharm. Sci. Tech.*, **11**, 1092–1103.

77. Millon, L. E., Guhados, G., and Wan, W. (2008). Anisotropic polyvinyl alcohol-Bacterial cellulose nanocomposite for biomedical applications, *J. Biomed. Mater. Res. B Appl. Biomater.*, **86**, 444–452.

78. Millon, L. E., Mohammadi, H., and Wan, W. K. (2006). Anisotropic polyvinyl alcohol hydrogel for cardiovascular applications, *J. Biomed. Mater. Res. B Appl. Biomater.*, **79**, 305–311.

79. Millon, L. E., and Wan, W. K. (2006). The polyvinyl alcohol-bacterial cellulose system as a new nanocomposite for biomedical applications, *J. Biomed. Mater. Res. B Appl. Biomater.*, **79**, 45–253.

80. Grant, C., Twigg, P., Egan, A., Moody, A., Smith, A., Eagland, D., Crowther, N., Britland, S. (2006). Poly(vinyl alcohol) hydrogel as a biocompatible viscoelastic mimetic for articular cartilage, *Biotechnol. Prog.*, **22**, 1400–1406.

81. Kobayashi, M., and Hyu, H. S. (2010). Development and evaluation of polyvinyl alcohol-hydrogels as an artificial atrticular cartilage for orthopedic implants, *Materials.*, **3**, 2753–2771.

82. Noguchi, T., Yamamuro, T., Oka, M., Kumar, P., Kotoura, Y., Hyon, S., and Ikada, Y. (1991). Poly(vinyl alcohol) hydrogel as an artificial articular cartilage: Evaluation of biocompatibility, *J. Appl. Biomater*, **2**, 101–107.

83. Castro, C., Vesterinen, A., Zuluaga, R., Caro, G., Filpponen, I., Rojas, O. J., Kortaberria, G., and Ganan, P. (2014). In situ production of nanocomposites of poly(vinyl alcohol) and cellulose nanofibrils from Gluconacetobacter bacteria: Effect of chemical crosslinking, *Cellulose*, in Press.

84. Qiu, K., and Netravali, A. N. (2012). Bacterial cellulose-based membrane-like biodegradable composites using cross-linked and noncross-linked polyvinyl alcohol, *J. Mater. Sci.*, **47**, 6066–6075.

85. Leitao, A. F., Gupta, S., Silva, J. P., Reviakine, I., and Gama, M. (2013). Hemocompatibility study of a bacterial cellulose/polyvinyl alcohol nanocomposite, *Colloids Surf., B*, **11**, 493–502.

86. Kudo, S., Otsuka, E., and Suzuki, A. (2010). Swelling behavior of chemically crosslinked PVA gels in mixed solvents, *J. Polym. Sci. Pol. Phys.*, **48**, 1978–1986.

87. Indriyati, L., Yudianti, R., and Karina, M. (2012). Development of nanocomposites from bacterial cellulose and poly(vinyl alcohol) using casting-drying method, *Procedia Chem.*, **4**, 73–79.

88. Leitao, A. F., Silva, J. P., Dourado, F., and Gama, M. (2013). Production and characterization of a new bacterial cellulose/poly(vinyl alcohol) nanocomposite, *Materials*, **6**, 1956–1966.

Chapter 15

Chitosan Polymer as Novel Drug Carrier

Aakanchha Jain,[a,b] Sourabh Jain,[b,c] Vikas Jain,[d] and Dharmveer Kohli[a]

[a]*Department of Pharmaceutical Sciences, Dr. H. S. Gour University, Sagar, Madhya Pradesh 470003, India*
[b]*Bhagyoday Tirth Pharmacy College, Sagar, Madhya Pradesh 470002, India*
[c]*School of Pharmaceutical Sciences, Shobhit University, Meerut, India*
[d]*Mahakal Institute of Pharmacy, Ujjain, Madhya Pradesh, India*

pharmaaj@hotmail.com, drdvkohali@gmail.com

15.1 Chitosan as Polymer

Chitosan is a second abundant heteropolysaccharide composed of β-(1–4)-2-deoxy-2-amino and -2-acetamino D-glucopyranose units [31]. Chitosan is the partial deacetylated (involving alkaline hydrolysis) product of chitin. Chitin is the main component of outer skeleton of small insects, cuticles, and the shells of shrimp, crab, lobster, etc. Chitosan has been expansively investigated in the pharmaceutical industries for the development of various sustained and controlled release drug delivery systems due to its versatile biological activity (as food and non-food items) and excellent biocompatibility. Hirano [22] conducted an experiment with both healthy and infected skin and found the LD50 in case of orally administered mouse to be (>16 g/kg). Wu et al. [76] worked on

Handbook of Sustainable Polymers: Processing and Applications
Edited by Vijay Kumar Thakur and Manju Kumari Thakur

ISBN 978-981-4613-53-8 (Hardcover), 978-981-4613-54-5 (eBook)
www.panstanford.com

acylated chitosan derivatives and reported their ability to increase membrane permeability in vitro and in vivo. It is a safe and friendly substance for humans; therefore, medical and pharmaceutical applications can be easily worked out with joint efforts from specialist in various fields [57]. Biodegradability of chitosan is of paramount importance because it is easily degraded by ubiquitous enzymes in the human body, and the oligomers so formed can activate macrophages too.

Chitosan as a polymer is so well known that it has found place in many official books, encyclopedias, monographs, American Standard Testing Materials, and in Pharmacopoeias of various countries.

15.1.1 Grades of Chitosan and Their Solubilities

The average molecular weight of various types of chitosan available ranges between 3800 and 500,000 g/mol and vary in degree of deacetylation (66% to 99.8%) and viscosity. Great ranges of grades based on viscosity are also available in the market. The viscous nature is probably due to its unbranched nature and high molecular weight, for which it has been used as an excellent viscosity-enhancing agent, especially in acidic environments (due to its solubility in acidic medium) [65]. It acts as a pseudoplastic material where a decrease in viscosity is inversely related to the rate of shear and thus exhibits thixotropy. As the concentration of chitosan is increased, there is an increase in its viscosity due to its gelling tendency. The same phenomenon is observed when there is a decrease in the temperature or increase in the degree of deacetylation [65]. Precisely, chitosan is chemically similar in nature to plant fiber, cellulose, in terms of its insolubility in water and organic solvents, its biodegradable nature, and that it can be broken down into smaller units of glucose on treatment with concentrated acids.

The three main active functional groups of chitosan include an amino/acetamido group and both primary and secondary hydroxyl groups at the C2, C3, and C6 positions, respectively. Easy generation of both inter- and intra-molecular hydrogen bonds, which causes variation in their structures and physicochemical properties is mainly associated with the amount of amino contents

[87]. Applications of chitosan are summarized in Figure 15.1 (http://www.webmd.com/accessed on 3/4/14) [26].

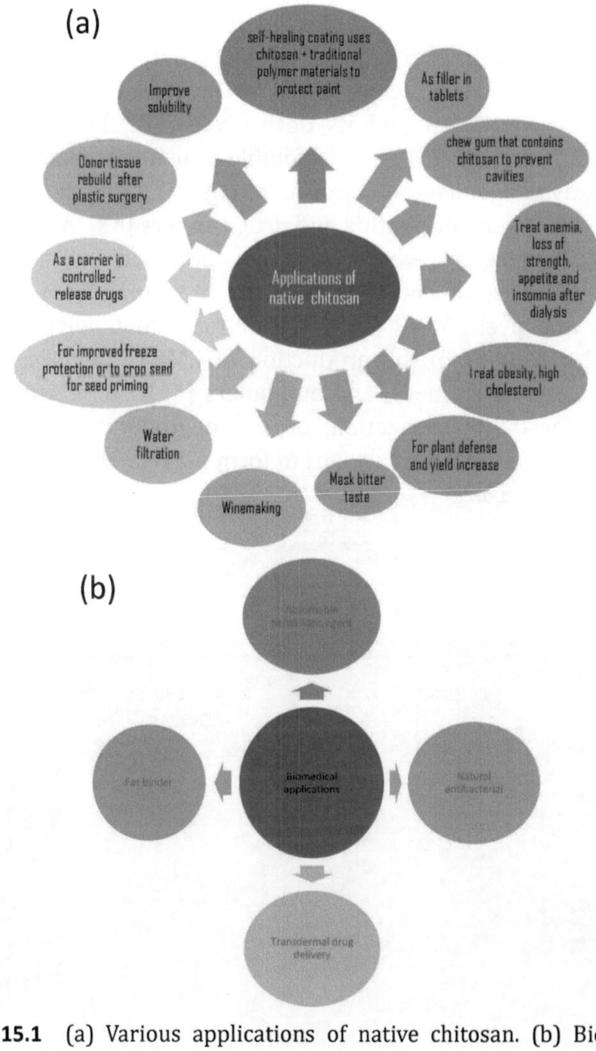

Figure 15.1 (a) Various applications of native chitosan. (b) Biomedical applications of chitosan.

15.1.2 Low-Molecular-Weight Chitosan/Chito-Oligomers

The high molecular weight leading to the very high viscosity of chitosan poses its pitfalls in several biological applications. Due to the crystalline structure of chitosan and its high molecular weight in contrast to its superiority as a biomaterial, the application of chitosan is limited. The applicability of chitosan as a bioactive agent in the form of drug delivery carriers, polypeptide carrier, and gene delivery carrier is limited due to its lipophillicity or lesser aqueous solubility and its solubility only in acidic solution [37]. Molecular weight contributes to nearly all the functional properties of chitosan and its derivatives [56]. Low-molecular-weight chitosan (LMWC) is fairly soluble in neutral aqueous media and it is advantageous over chitosan in that it is easily modified and can be used as gene or peptide and drug carriers [43]. Also, smaller particles (diameter wise) can be obtained with LMWC. Depolymerization of chitosan leads to the breakdown of large molecule and gives LMWC. Subjecting chitosan to depolymerization (chitonolysis) leads to the production of LMWC oligosaccharides (named as chitooligomers) and monomers [31]. There are various methods for depolymerization, or say degradation, of chitosan polymer of high molecular weight to form LMWC. The methods are depicted in Fig. 15.2.

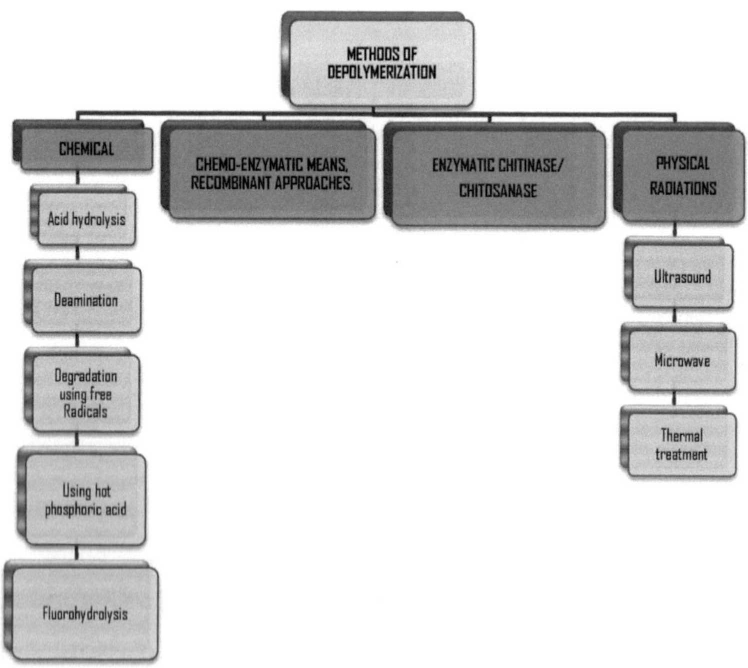

Figure 15.2 Flow chart of various methods of depolymerization of chitosan.

15.1.2.1 Applications

Following are important examples of LMWC that currently have found prominent place in research and pharmaceuticals [31]:

- antibacterial
- modulating plant resistance to diseases
- stimulating murine peritoneal macrophages
- antitumor
- functional food formulations
- induction of phytoalexins
- hemostatic effects
- immunopotentiating in plants
- increasing the mitogenic response and chemotactic activities in animals
- good lipid binding
- hypoglycaemic
- antioxidants
- wound reparation

15.1.3 Structural Unit of Chitosan: D-Glucosamine

Acid hydrolysis of chitosan produces structural unit of chitosan, D-glucosamine. The oxidized product of D-glucosamine is D-glucosaminic acid, a carbohydrate unit used to manufacture various biotics. It has various pharmaceutical functions. It has been recently identified as a promising sweetener and condiment [75]. Also, various amino acids leading to the formation of proteins and several glycosidase inhibitors can be synthesized using this as the starting material [87].

The most remarkable medical application of D-glucosaminic acid is its biocompatible and non-toxic ligand chelating ability with many metals. As ferric-D-glucosaminic acid complexes are not hydrolyzed in living systems they have potentials in pharmaceuticals and veterinary iron therapy. Because of its unhydrolyzable nature, D-glucosaminic acid has been used for cancer targeting too. A new drug using chromium with D-glucosaminic acid has been designed having anti-diabetic property. Two different complexes of chromium:glucosaminate were synthesized using the ratios 1:1 and 2:3, by neutralization and exchange reaction [87]. It caused a decrease in the blood sugar level comparable to that of widely used the chromium-picolinate complex (Cr (pic) 3).

15.2 Chitosan Conjugates

Polymer (CH)–drug conjugates, broadly referred to as macromolecular therapeutics, have therapeutic and pharmaceutical advantages in comparison to the polymer alone. The discovery of drug–polymer conjugation dates back to 1955, when Jatzkewitz [32] conjugated mescaline, a psychedelic alkaloid, to poly(N-vinylpyrrolidone) using glycyl-L-leucine as a linker. The formulation thus developed was a depot alkaloidal preparation. However, the concept of use of polymer–drug conjugates can be credited to Ringsdorf [58]. Since then numerous drug have been conjugated with polymers and evaluated for their therapeutic efficacy.

There are various advantages of polymeric-drug conjugates, such as decreased non-specific toxicity, improved hydrophillicity/water solubility, effective organ targeting through cell surface ligands/receptors, slower rate of elimination, preferential accumulation in solid tumors, and ability to overcome multidrug resistance. In addition to these advantages, there are many disadvantages of polymeric-drug conjugates; for instance, higher substitution may lead to aggregation and thus micelles. The micelles formation may affect the biodistribution of the drug. Moreover, the drug may act as haptens and evoke unnecessary immune responses in body. However, targeting may be achieved by use of ligands such as folic acid, transferring, etc. Few examples of chitosan–polymer drug conjugates are given in Table 15.1.

Table 15.1 Examples of some polymer–CH composites

Polymer name	Purpose	Ref.
Poly(e-caprolactone)-CH* Composite	bone regeneration	[29]
Methylpyrrolidinone-CH*	prolonged vascularisation with heparin	[6]
Albumin-CH*	increase in vascularization	[15]
Polylacticacid-CH*	osteogenesis	[41]
Glycol-CH*	improve the intracellular delivery of hydrophilic macromolecules	[14]
CH*-prednisolon	Colon specific delivery	[53]

*CH, chitosan.

15.3 Derivatives of Chitosan

The reactive hydroxy and amino groups of chitosan have been used for its derivatization or chemical modification. The amino group being nucleophilic in nature allows easy formation of imine bond with aldehyde or corresponding amide derivatives. However, in acidic solution, the cationic property of chitosan is revealed and the amino groups act like alkali, receive protons, and generate salts [87]. Some of the derivatives with their method of syntheses have been summarized in Table 15.2.

Table 15.2 Combined applications of various chitosan derivatives and reagents used for their preparation

Chitosan derivatives	Reagents used		Ref.	Combined applications
Ch-4-Thiobutylamidine	1% acetic acid + 1 M NaOH + 2-imminothiolane HCl		[59]	Organ specific drug delivery
Quaternized Aromatic Derivatives of Ch [Methylated (aminobenzyl) Ch; Methylated (pyridynyl) Ch; Methylated (benzyl) Ch]	1% acetic acid + 1 N NaOH + corresponding aldehydes [4-*N*,*N*-dimethylbenzal-dehyde; 4-Pyridin carbaldehyde; 4-Benzaldehyde]		[48]	Inhibitor of hemagglutination of human erythrocytes by influenza virus Antibacterial activity against
Thiolated Ch	HCl (1 mol/L) + NaOH (5 mol/L) + reduced glutathione + EDC and NHS (200 mmol/L)		[61]	Gram-negative bacteria Improved intracellular
N-Trimethyl Ch	Methyl iodide		[60]; [69]	delivery of hydrophilic
Ch sulfates	Chlorosulfonic acid-sulfuric acid	+ Dimethyl Formamide/ DMF-dichloroacetic acid/ tetra-hydrofuran/ formic acid	[22]; [24]	macromolecules
	Sulfurtrioxide		[25]	Osteogenesis/ bone regeneration
	Sulfurtrioxide/ trimethylamine		[33]	Cancer targeting
	Concentrated sulfuric acid		[45]	Antifungal activity
	Oleum		[51]	Antihypercholest-erolemic effect
	Chlorosulfonic acid		[20]	

(*Continued*)

Table 15.2 *(Continued)*

Chitosan derivatives	Reagents used		Ref.	Combined Applications
Methoxyphenyl Ch	Vanillin, *O*-vanillin, syringaldehyde, veratraldehyde		[57]	Hair and skin Cosmetics
Tyrosine glucan	4-hydroxy phenyl pyruvic acid + tyrosinases			Carrier for corneal endothelial cell transplantation
Highly Cationic Chitosans e.g. *N*-(2-hydroxyl) propyl-3-trimethyl-ammonium chitosan chloride	Dialkylaminoalkyl chloride in alkaline condition glycidyl-trimethyl-ammonium chloride		[84]	Thermosensitive hydrogel
Phosphorylated Ch e.g. *N*-mono- and di-phosphonic-methylene Ch *N*-Lauryl-*N*-ethelyne-phosphonic Ch	Phosphorous acid + formaldehyde reductive amination 2-chloro-1,3,2-dioxaphosphos-pholane		[31]	Prolonging drug release as well as augmenting drug adsorption properties Increased Mucoadhesive
Fatty acid derivatives	Stearic acid, 450–451	+ acetic acid + methanol	[27]	Development of porous scaffold structure guaranteeing cell anchorage, proliferation, and tissue formation in three dimensions
	Linoleic acid, 452–453	+ water + sodium	[12]; [13]	
	Oleic acid 454	cyanoboro-hydrate	[86]	
Azidated Ch	4-azidobenzoic acid + O-butyryl Ch		[88]	
Thiourea Ch	Glutaraldehyde 379		[19]	
	Ammonium thiocyanate 393		[9]	Inhibition of zinc dependent proteasease and carboxy-peptidases
EDTA Ch e.g. Diethylene-tri-aminepentaacetic acid Ch	Acid + EDC (1-ethyl-3-(3-dimethylaminop ropyl)carbodiimide)/ acid anhydride		[17]; [47]	
Cyclodextrin linked Ch	*N*-succinyl Ch + aminated β cyclodextrin		[4]	Improved wound dressings with more absorbability
	1,6-hexamethylene diisocyanate		[66]	

Carboxyalkyl Ch N-carboxyalkyl O-carboxyalkyl N-carboxybenzyl chitosan	Carboxyaldehydes in a reductive amination sequence. Chloroacetic acid + sodium hydroxide 2-carboxy benzaldehyde + glutaraldehyde	[31]
Crown ether bound Ch	Schiff's base type/reduced form ($NaBH_4$)	[63]
Sugar modified Ch	Sodium cyanoborohydrate + unmodified sugar	[21]
	Sodium cyanoborohydrate + sugar-aldehyde derivative	
e.g. Lactosaminated *N*-succinyl Ch/Sialic acid–Ch	Sodium cyanoborohydrate + *N*-succinyl Ch/ sialic acid	[36]
e.g. 6-*O*-linked carboxtmethyl Ch	Sodium borotetrahydrate	[34]
N-Acyl Ch	Aqueous acetic acid/ ethanol, pyridine, pyridine/chloroform, trichloroacetic acid/ dichloroethane, ethanol/ methanolmixture, methanol/formamide	[90]
e.g. Hydroxyacyl Ch	Lactones	[31]
e.g. *N*-Carboxyacyl Ch (uccinic, maleic, glutaric, itaconic, phthalic)	Cyclic anhydrides	[23]
O-Acyl Chitosan	Methanesulfonic acid	[64]
Ch-Dendrimer Hybrid	*N*-methoxycarbonyl ethyl chitosan + PAMAM dendrimer	[74]

15.4 Chitosan-Based Novel Drug Delivery Systems

Nanoparticles can be targeted to different organs such as stomach, brain, colon, liver, kidney, heart, blood, etc., and induce lethal effects [40]. Nanoparticles being spherical in shape possess unique properties of a large surface: volume ratio, thus, possessing high surface reactivity, which could be oppressed in many biomedical applications. The drug stability, duration of effect can be improved through nanoparticulate delivery system. Also, novel drug delivery systems allow administration through other non-parental routes too [18].

Nanoparticles, due to their unique properties and surface characteristics, can protect the drug from P-gp, cytochrome P-450, and the destructive factors in the GI tract and can increase the permeability of drugs through the gastrointestinal barrier [62]. These reports confirm that nanoparticulate systems with unique properties can increase the transport of many compounds across the GI barrier. Numerous examples of nanoparticles being used to deliver antibodies, cytokines, other proteins, and drug substances into specific cells or tissues with good safety and trustworthiness are reported in the literature [49, 71, 85].

15.4.1 Chitosan as Novel Drug Carrier

A wide range of natural materials such as gelatin, cellulose, sodium alginate, synthetic and semi-synthetic polymers, dendrimers, lipoids, triglycerides, Tweens, and Spans have been employed to formulate drug carriers [55]. Among these, polysaccharides being biodegradable in nature have established escalating consideration because of their stupendous physical and biological properties [45]. The chemical modification of chitosan imparts amphiphilicity to the molecule. This is an important characteristic for the formation of self-assembled drug delivery systems, viz. nanoparticles. The electrostatic interaction/absorption of antigens onto chitosan microparticles are common and relatively slow phenomenon usually best applicable for mucosal vaccines development. An important aspect of chitosan is that being a natural polymer, it enhances the immunogenicity of haptens, i.e., poor immune response generating antigens either in the form of solution or as micro/nanoparticles [3, 82]. Chitosan particles can be prepared by several methodologies [1, 7, 52] including the following:

- ionotropic gelation
- emulsification
- reverse micelles formation
- solvent evaporation
- spray drying
- coacervation
- sieving methods

Li et al. [44] reported that alginate coated chitosan microparticles could be an effective means for mucosal immunization and oral intake of antigens. Mourya and Inamdar [50] prepared chitosan capsules encapsulating 5-aminosalicylic acid for colon-specific delivery.

Trickler et al. [70] reported that chitosan/glycerylmonooleate nanoparticles are mucoadhesive and showed fourfold increased cellular uptake and a 1000-fold reduction in the IC50 of paclitaxel. The prospective use of chitosan nanoparticles as antibacterial and enhancer of intracellular delivery of ceftriaxone to macrophages and enterocytes of *Salmonella* typhimurium has also been reported by Zaki and Hafez [83]. The potential advantages of such a system for antibiotic nanoparticles include enhancement of oral absorption and promotion of intracellular delivery. Zhu et al. [89] prepared galactosylated chitosan oligosaccharide nanoparticles encapsulating adenosine triphosphate for liver carcinoma cell specific uptake. Their results indicate that the prepared nanoparticles could be an attractive delivery system for intracellular drug delivery to hepatocellular carcinoma cell targeting; however, the results need to be checked in vivo.

The data presented in the study done by Yang et al. [80] prepared nanoparticles incorporated with chitosan and polylactic-co-glycolic acid. Their results demonstrate that nanoparticles formulated could potentially be used as a safe prolonged release delivery system for low-molecular-weight heparin. The property of very high positive zeta potential enables chitosan to be cytotoxic (in vitro) against various human cancer cell lines. Talking about PLGA based drug delivery systems due to scarcity of functional groups on its surface it is quite difficult to achieve either passive or active targeting. Thus, in addition to the chitosan PLGA incorporated nanoparticles, PLGA nanoparticles were modified with chitosan coating by Wang et al. [73] for improving the efficacy of PLGA for mucosal targeted systems. Oral Delivery of docetaxel using thiolated chitosan nanoparticles prepared showed promising results in terms of pharmacokinetic study and cell line study, for the development of an oral drug delivery system for anticancer drugs [62]. According to Azizi et al. [5] pH-dependent swelling properties of chitosan and alginates could be used to control drug release. They reported that nanoparticles could protect EGFR coding DNA for nearly 6 h. The topical delivery for chitosan nanoparticles has also been determined by Tan et al. [68]. They compared quercetin solution and quercetin-loaded Ch nanoparticles. The nanoparticles significantly increased accumulation of quercetin in the skin layers with higher permeation, especially in the epidermal layer. Lee et al. [42] suggested that the side effects of hair dye can be alleviated using poly-γ-glutamic acid/glycol chitosan microparticles as the delivery system, thus proving its capability in cosmetic field. In spite of all the above-said applications and advantages of chitosan nanoparticles, the native antifungal property of chitosan could not be proved to be effective when administered in the form of trimethyl chitosan nanoparticles. However, the antifungal activity against *C. albicans* and *F. solani* was observed for all the nanoparticles prepared from different molecular weight chitosan [30].

15.4.2 Chitosan DNA Complex

One of the newer applications of chitosan is its potential as a mucosal gene transporter in the form of chitosan gene nanoparticles. These nanoparticles have proved to be effective in inducing mucosal (Immunoglobulin A) as well as systemic immune responses in several experiments [72, 78, 81]. DNA being negatively charged has great potentials to interact with positively charged chitosan and form complexes, used for the preparation of nanoparticulates to improve the delivery of DNA [77]. Xu et al. [77] also reported that chitosan nanoparticles enhance the immunogenicity of the psaA DNA vaccine. Previous studies have suggested that chitosan enhanced the mucosal immunity of a nasal vaccine by providing longer residence times in the nasal cavity and by opening momentarily the firm junctions all through the mucosal cell lining, thus, allowing better access of vaccine to the nasal-associated lymphoid tissue [2].

15.4.3 Chitosan as Biomaterial

Chitosan has been widely used in tissue engineering, wound healing, drug delivery, and other biomedical applications. It can be formulated in a variety of forms, such as powder, film, sphere, gel, and fiber. These features make chitosan an almost ideal biomaterial. Recently adding nanoparticles in cell cultures for treatments and therapeutic uses is among the newest developments other than probing and imaging in nanotechnology and biotechnology [18].

15.4.3.1 Beads of chitosan/microspheres:

Solvent evaporation, coacervation and emulsification are the main reported methods of preparation of Chitosan beads. Blended beads of chitosan have been widely researched. The applicability of chitosan/alginate microparticles, chitosan/xanthan microspheres [8] and chitosan/gelatin microspheres have also been reported [16].

15.4.3.2 Chitosan sponge

The elasticity of sponges provides logical strength and allows a chance to utilize chitosan for the construction of wound dressings and absorption of wound fluid [54]. As chitosan solution is unaccompanied non-foamable, it is frequently combined with many other good foaming polymers, such as gelatin. Chitosan and sodium alginate blended sponge incorporating silver sulfadiazine were prepared by Kim and Raha [38] and were applicable as wound dressings. Yang et al. [79] enhanced the growth of liver tissue through hepatocyte loaded alginate/galactosylated chitosan sponges. Strobin et al. [67] prepared the sponge of chitosan-fibroin blend, which was found to be suitable for a construction of dressing materials.

15.4.3.3 Fibers

The first form of chitosan as fiber was reported by Kunike [39]. A significant advantage of chitosan fiber includes its good mechanical properties as compared to the same weight of chitosan. However, the biggest disadvantage of such fibers is its high production cost, and thus, it is necessary to search for blends or composites of chitosan with other yarns. Nanofiber of chitosan/polyethylene oxide blends has been prepared by high electrospinning to be used in air and water filtration [11]; fibers of chistosan blends with collagen [10] starch, poly(ethylene oxide), PVA, silk fibroin and alginate have also been reported [16].

15.5 Toxicity of Chitosan Nanoparticles

Although reports regarding the toxicity evaluation of many types of lipidic nanoparticles are available but still the knowledge about the toxicity of biodegradable nanomaterials, generally considered safe, is still limited. Hu et al. [28] evaluated the toxicity of chitosan nanoparticles using zebrafish as the model. They incubated different concentrations of nanoparticles with embryos of zebrafish. ZnO nanoparticles were used as positive control during the experiment. Embryos when treated with chitosan nanoparticles showed an increased in apoptotic and necrotic cells, i.e., dying cells with high generation of reactive oxygen species, as well as overexpression of heat shock protein 70, indicating that chitosan nanoparticles can lead to the death of zebrafish by physiological stress.

15.6 Electrokinetic and Stability Behavior of Chitosan

Physical properties (electrokinetic properties and colloidal stability) of nanoparticles formed by chitosan ionically cross-linked by sodium tripolyphosphate have been studied by López-León et al. [46] as a function of pH and ionic strength of the medium. They observed that due to glucosamine groups, change of pH from acid to basic values leads to de-swelling. This reduction in swelling could be ascribed to reduction in the intramolecular electric repulsions present inside the particle mesh. Also, at the pH values near the isoelectric point of the chitosan–TPP particles, the system became colloidally unstable.

15.7 Aspects of Chitosan Uncovered in This Chapter

Nanoparticles can enter and affect different organs and tissues such as brain, liver, kidney, heart, and blood and induce cytotoxic effects [40]. These particles may alter and inhibit cell growth leading to various patho-physiological states in humans and animals. Consequently, nanotoxicology research is now gaining much more attention and its importance is gradually being recognized.

15.8 Conclusion

As a biocompatible and biodegradable material, chitosan has found many and diverse medical, pharmaceutical, and therapeutic applications. Nevertheless, there is a lot of scope for improvisation in mechanical properties and biocompatibility of chitosan before it becomes an ideal biomaterial for tissue engineering applications. For instance, the fat binding ability of chitosan has wide scope. However, the mechanism of interaction between chitosan and fat is not well understood and has not been proved clinically. Research is still under way for several potential dietary or clinical applications for chitosan to be used as a soluble dietary fiber and as a means to retard emptying of the stomach and create a sense of satiety. Potentials such as alteration of bile acid composition are still unknown, leaving a wide gap in many important fields where chitosan is to be utilized.

Acknowledgment

Aakanchha Jain is thankful to INSPIRE DST for PhD Fellowship.

Reference

1. Agnihotri, S. A., Mallikarjuna, N. N., and Aminabhavi, T. M. (2004). Recent advances on chitosan-based micro- and nanoparticles in drug delivery, *J. Control. Release.*, **100**, 5–28.
2. Alpar, H. O., Somavarapu, S., Atuah, K. N., and Bramwell, V. W. (2005). Biodegradable mucoadhesive particulates for nasal and pulmonary antigen and DNA delivery, *Adv. Drug Deliv. Rev.*, **57**, 411–430.
3. Amidi, M., Romeijn, S. G., Coos Verhoef, J., Junginger, H. E., Bungener, L., Huckriede, A., Crommelin, D. J. A., and Jiskoot, W. (2007). N-Trimethyl chitosan (TMC) nanoparticles loaded with influenza subunit antigen for intranasal vaccination: Biological properties and immunogenicity in a mouse model, *Vaccine*, **25**, 144–153.
4. Aoki, N., Nishikawa, M., and Hattori, K. (2003). Synthesis of chitosan derivatives bearing cyclodextrin and adsorption of p-nonylphenol and isphenol A, *Carbohydr. Polym.*, **52**, 219–223.
5. Azizi, E., Namazi, A., Haririan, I., Fouladdel, S., Khoshayand, M. R., Shotorbani, Parisa, Y., Nomani, A., and Gazori, T. (2010). Release profile and stability evaluation of optimized chitosan/alginate nanoparticles as EGFR antisense vector, *Int. J. Nanomed.*, **5**, 455–461.
6. Berscht, P. C., Nies, B., Liebendorfer, A., and Kreuter, J. (1994). Incorporation of basic fibroblast growth factor into methylpyrrolidinone chitosan fleeces and determination of the in vitro release characteristics, *Biomaterials*, **15**, 593–600.
7. Calvo, P., Remunan-Lopez, C., Vila-Jato, J. L., and Alonso, M. J. (1997). Chitosan and chitosan/ethylene oxide propylene oxide block copolymer nanoparticles as novel carriers for proteins and vaccines, *Pharm. Res.*, **14**, 1431–1436.
8. Chellat, F., Tabrizian, M., Dumitriu, S., Chornet, E., Rivard, C. H., and Yahia, L. J. (2000). *Inc J Biomed. Mater. Res.*, **53**(5), 592–599.
9. Chen, S., Wu, G., and Zeng, H. (2005). Preparation of high antimicrobial activity thiourea chitosan–Ag+ complex, *Carbohydr. Polym.*, **60**, 33–38.
10. Chen, Z., Mo, X., He, C., and Wang, H. (2008). Intermolecular interactions in electrospun collagen-chitosan complex nanofibers, *Carbohydr. Polym.*, **72**, 410–418.
11. Desai, K., Kit, K., Li, J., and Zivanovic, S. (2008). Morphological and surface properties of electrospun chitosan nanofibers, *Biomacromol.*, **9**(3), 1000–1006.

12. Du, Y. Z., and Wang, L. (2009). Preparation and characteristics of linoleic acid-grafted chitosan oligosaccharide micelles as a carrier for doxorubicin, *Colloids Surf B.*, **69**, 257–263.

13. Du, Y. Z., and Wang, L. (2011). Linoleic acid-grafted chitosan oligosaccharide micelles for intracellular drug delivery and reverse drug resistance of tumor cells, *Int. J. Biol. Macromol.*, **48**, 215–222.

14. Dufes, C., Schatzlein, A. G., Tetley, L., Gray, A. I., Watson, D. G., Olivier, J. C., Couet, W., and Uchegbu, I. F. (2000). Niosomes and polymeric chitosan based vesicles bearing transferrin and glucose ligands for drug targeting, *Pharm. Res.*, **17**, 1250–1258.

15. Elcin, Y. M., Dixit, V., and Gitnick, G. (1996). Controlled release of endothelial cell growth factor from Chitosan-albumin microspheres for localized angiogenesis: In vitro and in vivo studies, *Artif. Cells Blood Substit. Immobil. Biotechnol.*, **24**, 257–271.

16. El-hefian, E. A., Nasef, M. M., and Yahaya, A. H. (2011). Chitosan physical forms: A short review, *Aust. J. Basic Appl. Sci.*, **5**(5), 670–677.

17. El-Sharif, A. A., and Hussain, M. H. (2011). Chitosan-EDTA new combination is a promising candidate for treatment of bacterial and fungal infections, *Curr. Microbiol.*, **62**, 739–745.

18. Gao, W., Lai, J. C. K., and Leung, S. W. (2012). Functional enhancement of chitosan and nanoparticles in cell culture, tissue engineering, and pharmaceutical applications, *Front. Physiol.*, **3**, 1–13.

19. Guibal, E., Vincent, T., and Navarro, M. R. (2000). Synthesis and characterization of a thiourea derivative of chitosan for platinum recovery, *J. Appl. Polym. Sci.*, **75**, 119–134.

20. Hagiwara, K., Kuribayashi, Y., Iwai, H., Azuma, I., Tokura, S., Ikuta, K., and Ishihara, C. (1999). A sulfated chitin inhibits hemagglutination by Theileria sergenti merozoites, *Carbohydr. Polym.*, **39**, 245–248.

21. Hall, L. D., and Yalpani, M. (1980). Formation of branched-chain, soluble polysaccharides from chitosan, *J. Chem. Soc. Chem. Commun.*, **23**, 1153–1154.

22. Hirano, S. (1999). Chitin and chitosan as novel biotechnological materials, *Polym. Int.*, **48**, 732–734

23. Hirano, S., and Moriyasu, T. (2004). Some novel N-(carboxyacyl) chitosan filaments, *Carbohydr. Polym*, **55**, 245–248.

24. Hirano, S., Tanaka, Y., Hasegawa, M., Tobetto, K., and Nishioka, A. (1985). Effect of sulfated derivatives of chitosan on some blood coagulant factors, *Carbohydr. Res.*, **137**, 205–215.

25. Holme, K. R., and Perlin, A. S. (1997). Chitosan N-sulfate A water-soluble polyelectrolyte, *Carbohydr. Res.*, **302,** 7–12.
26. http://www.webmd.com/accessed on 3/4/14.
27. Hu, F. Q., and Wu, X. (2008). Cellular uptake and cytotoxicity of shell cross-linked stearic acid-grafted chitosan oligosaccharide micelles encapsulating doxorubicin, *Eur. J. Pharm. Biopharm.*, **69,** 117–125.
28. Hu, Y.-L., Wang, Q., Han, F., Shao, J.-Z., and Gao, J.-Q. (2011). Toxicity evaluation of biodegradable chitosan nanoparticles using a zebrafish embryo model, *Int. J. Nanomed.*, **6**, 3351–3359.
29. Im, S. Y., Cho, S. H., Hwang, J. H., and Lee, S. J. (2003). Growth factor releasing porous poly(epsiloncaprolactone)-chitosan matrices for enhanced bone regenerative therapy, *Arch. Pharmacol. Res.*, **26**, 76–82.
30. Ing, L. Y., Zin, N. M., Sarwar, A., and Katas, H. (2012). Antifungal activity of chitosan nanoparticles and correlation with their physical properties, *Int. J. Biomater.*, 1–9, 2012;2012:632698. doi: 10.1155/2012/632698.
31. Jain, A., Gulbake, A., Shilpi, S., Jain, A., Hurkat, P., and Jain, S. K. (2013). A new horizon in modifications of chitosan: Syntheses and applications, *Crit. Rev. Ther. Drug Carrier Syst.*, **30**(2), 91–181.
32. Jatzkewitz, H. (1955). Peptamin (glycyl-L-leucyl-mescaline) bound to blood plasma expander (polyvinylpyrrolidone) as a new depot form of a biologically active primary amine (mescaline), *Z. Naturforsch.*, **10b**, 27–31.
33. Je, J. Y., Park, P. J., and Kim, S. K. (2005). Prolyl endopeptidase inhibitory activity of chitosan sulfates with different degree of deacetylation, *Carbohydr. Polym.*, **60**, 553–556.
34. Jia, Z., Shen, D., and Xu, X. (2001). Synthesis and antibacterial activities of quaternary ammonium salt of chitosan, *Carbohydr. Res.*, **333**(1), 1–6.
35. Jiao, L.-Z., Wang, Y., Zhou, C., and Zhang, Z. (2008). Polysaccharides-based nanoparticles as drug delivery systems, *Adv. Drug Deliv. Rev.*, **60** (15), 1650–1662.
36. Kato, Y., Onishi, H., and Machida, Y. (2001). Biological characteristics of lactosaminated N-succinyl-chitosan as a liver-specific drug carrier in mice, *J. Control. Release*, **70,** 295–307.
37. Kim, I.-Y., Seo, S.-J., Moon, H.-S., Yoo, M.-K., Park, I.-Y., Kim, B.-C., and Cho, C.-S. (2008). Chitosan and its derivatives for tissue engineering applications, *Biotechnol. Adv.*, **26,** 1–21.

38. Kim, S. K., and Rha, C. (1989). Transdermal permeation of proteins in chitosan capsules; Eds.; *Chitin and Chitosan-Sources; Chemistry, Physical Properties and Application*, Elsevier: London, 365.

39. Kunike, G., (1926). Chitin und chitinseide, *Kunsideide* (*Chemiefasern*), **8**, 182–188.

40. Lai, J. C. K., Lai, M. B., Jandhyam, S., Dukhande, V. V., Bhushan, A., Daniels, C. K., and Leung, S. W. (2008). Exposure to titanium dioxide and other metallic oxide nanoparticles induces cytotoxicity on human neural cells and fibroblasts, *Int. J. Nanomed.*, **3**, 533–545.

41. Lee, H.-W., Park, Y.-S., Jung, J.-S., and Shin, W.-S. (2002). Chitosan oligosaccharides, dp 2–8, have prebiotic effect on the Bifidobacterium bifidium and Lactobacillus sp, *Anaerobe*, **8**, 319–324.

42. Lee, H.-Y., Jeong, Y.-I., and Choi, K.-C. (2011). Hair dye-incorporated poly-γ-glutamic acid/glycol chitosan nanoparticles based on ion-complex formation, *Int. J. Nanomed.*, **6**, 2879–2888.

43. Lee, J. Y., Nam, S. H., Im, S. Y., Park, Y. J., Lee, Y. M., Seol, Y. J., and Lee, S. J. (2002). Enhanced bone formation by controlled growth factor delivery from chitosan-based materials, *J. Control. Release*, **78**, 187–197.

44. Li, X.-Y., Kong, X.-Y., Shi, S., Zheng, X.-L., Guo, G., Wei, Y. Q., and Qian, Z. Y. (2008). Preparation of alginate coated chitosan microparticles for vaccine delivery, *BMC Biotechnol.*, **8**(89), 1–11.

45. Liu, Y., Yu, Z. L., Zhang, Y. M., Guo, D. S., and Liu, Y. P. (2008). Supramolecular architectures of beta-cyclodextrin-modified chitosan and pyrene derivatives mediated by carbon nanotubes and their DNA condensation, *J. Am. Chem. Soc.*, **130**, 10431–10439.

46. López-León, T., Carvalho, E. L. S., Seijo, B., Ortega-Vinuesa, J. L., and Bastos-González, D. (2005). Physicochemical characterization of chitosan nanoparticles: Electrokinetic and stability behavior, *J. Colloid Interface Sci.*, **283**, 344–351.

47. Loretz, B., and Bernkop-Schnurch, A. (2006). In vitro evaluation of chitosan-EDTA conjugate polyplexes as a nanoparticulate gene delivery system, *AAPS J.*, **8**, 756–764.

48. Mahjub, R., Dorkoosh, F. A., Amini, M., Khoshayand, M. R., and Rafiee-Tehrani, M. (2011). Preparation, statistical optimization, and in vitro characterization of insulin nanoparticles composed of quaternized aromatic derivatives of chitosan, *AAPS PharmSciTech*, **12** (4), 1407–1419.

49. Mi, F. L., Wu, Y. Y., Chiu, Y. L., Chen, M. C., Sung, H. W., Yu, S. H., Shyu, S. S., and Huang, M. F. (2007). Synthesis of a novel glycoconjugated

chitosan and preparation of its derived nanoparticles for targeting HepG2 cells, *Biomacromolecules.*, **8**, 892–898.

50. Mourya, V. K., and Inamdar, N. N. (2008). Chitosan-modi. cations and applications: Opportunities galore, *React. Funct. Polym.*, **68,** 1013–1051.
51. Nagasawa, K., Tohira, Y., Inoue, Y., and Tanoura, N. (1971). Reaction between carbohydrates and sulfuric acid: Part I Depolymerization and sulfation of polysaccharides by sulfuric acid, *Carbohydr. Res.*, **18,** 95–102.
52. Ohya, Y., Shiratani, M., Kobayashi, H., and Ouchi, T. (1994). Release behavior of 5-fluorouracil from chitosan-gel nanospheres immobilizing 5-fluorouracil coated with polysaccharides and their cell specific cytotoxicity, *Pure Appl. Chem.*, **A31**, 629–642.
53. Oosegi, T., Onishi, H., and Machida, Y. (2008). Novel preparation of enteric-coated chitosan–prednisolone conjugate microspheres and in vitro evaluation of their potential as a colonic delivery system, *Eur. J. Pharm. Biopharm.*, **68**, 260–266.
54. Oungbho, K., and Muller, B. W. (1997). Chitosan sponges as sustained release drug carriers, *Int. J. Pharm.*, **156**, 229–237.
55. Patel, M. P., Patel, R. R., and Patel, J. K. (2010). Chitosan mediated targeted drug delivery system: A review, *J. Pharm. Pharm. Sci.*, **13**(3), 536–557.
56. Prashanth, K. V. H., and Tharanathan, R. N. (2007). Chitin/chitosan: Modifications and their unlimited application potential an overview, *Trends Food Sci. Technol.*, **18,** 117–131.
57. Ravi Kumar, M. N. V., Muzzarelli, R. A. A., Muzzarelli, C., Sashiwa, H., and Domb, A. J. (2004). Chitosan chemistry and pharmaceutical perspectives, *Chem. Rev.*, **104**(12), 6018–6084.
58. Ringsdorf, H. (1975) Structure and properties of pharmacologically active polymers, *J. Polym. Sci. Polym. Symp.*, **51**(1), 135–153.
59. Sakloetsakun, D., Perera, G., Hombach, J., Millotti, G., and Bernkop-Schnürch, A. (2010). The impact of vehicles on the mucoadhesive properties of orally administrated nanoparticles: A case study with chitosan-4-thiobutylamidine conjugate, *AAPS PharmSciTech*, **11**(3), 1185–1192.
60. Sandri, G., Bonferoni, M. C., Rossi, S., Ferrari, F., Boselli, C., and Caramella, C. (2010). Insulin-loaded nanoparticles based on N-trimethyl chitosan: In vitro (caco-2 model) and ex vivo (excised rat jejunum, duodenum, and ileum) evaluation, *AAPS PharmSciTech*, **11**(1), 362–371.

61. Saremi, S., Atyabi, F., Akhlaghi, S. P., Ostad, S. N., and Dinarvand, R. (2011). Thiolated chitosan nanoparticles for enhancing oral absorption of docetaxel: Preparation, in vitro and ex vivo evaluation of penetration enhancement properties, *Int. J. Nanomed.*, **6**, 119–128.

62. Saremi, S., Dinarvand, R., Kebriaeezadeh, A., Ostad, S. N., and Atyabi, F. (2013). Enhanced oral delivery of docetaxel using thiolated chitosan nanoparticles: Preparation, in vitro and in vivo studies, *BioMed Res. Int.*, 1–8, Volume 2013 (2013), Article ID 150478, 8 pp. http://dx.doi.org/10.1155/2013/150478.

63. Sashiwa, H., and Sei-ichi, A. (2004). Chemically modified chitin and chitosan as biomaterials, *Prog. Polym. Sci.*,**29**(9), 887–908.

64. Sashiwa, H., Norioki, K., Atsuyoshi, N., Einosuke, M., Noboru, Y., and Sei-ichi, A. (2002). Chemical modification of chitosan: Synthesis of water-soluble chitosan derivatives by simple acetylation, *Biomacromolecules*, **3**(5), 1126–1128.

65. Shaji, J., Jain, V., and Lodha, S. (2010). Chitosan: A novel pharmaceutical excipient, *Int. J. Pharm. Appl. Sci.*, **1**(1), 11–28.

66. Sreenivasan, K. (1998). Synthesis and preliminary studies on a β-cyclodextrin-coupled chitosan as a novel adsorbent matrix, *J. Appl. Polym. Sci.*, **69**, 1051–1055.

67. Strobin, G., Kucharska, M., Ciechaiska, M., Wawro, D., Stiplewski, W., Jóiwicka, J., and Sobczak, S. (2006). *Pol. Chitin Soc.*, Monograph XI, 61.

68. Tan, Q., Liu, W., Guo, C., and Zhai, G. (2011). Preparation and evaluation of quercetin-loaded lecithin-chitosan nanoparticles for topical delivery, *Int. J. Nanomed.*, **6**, 1621–1630.

69. Thanou, M., Florea, B. I., Geldof, M., Junginger, H. E., and Borchard, G. (2002). Quaternized chitosan oligomers as novel gene delivery vectors in epithelial cell lines, *Biomaterials*, **23**(1), 153–159.

70. Trickler, W. J., Nagvekar, A. A., and Dash, A. K. (2008). A novel nanoparticle formulation for sustained paclitaxel delivery, *AAPS PharmSciTech*, **9**(2), 486–493.

71. Wang, Q., Zhang, L., Hu, W., Hu, Z. H., Bei, Y. Y., Xu, J. Y., Wang, W. J., Zhang, X. N., and Zhang, Q. (2010). Norcantharidin-associated galactosylated chitosan nanoparticles for hepatocyte-targeted delivery, *Nanomedicine*, **6**, 371–381.

72. Wang, X., Zhang, X., Kang, Y., Jin, H., Du, X., Zhao, G., Yu, Y., Li, J., Su, B., Huang, C., and Wang, B. (2008). Interleukin-15 enhance DNA vaccine elicited mucosal and systemic immunity against foot and mouth disease virus, *Vaccine*, **26**, 5135–5144.

73. Wang, Y., Li, P., and Kong, L. (2013). Chitosan-modified PLGA nanoparticles with versatile surface for improved drug delivery, *AAPS PharmSciTech*, **14**(2), 585–592.

74. Wen, Y., Tan, Z., Sun, F., Sheng, L., Zhang, X., and Yao, F. (2012). Synthesis and characterization of quaternized carboxymethyl chitosan/poly(amidoamine) dendrimer core-shell nanoparticles, *Mater. Sci. Eng. C*, **32**, 2026–2036.

75. Wipff, G., Weiner, P., and Kollman, P. (1982). A molecular–mechanics study of 18–crown–6 and its alkali complexes: An analysis of structural flexibility, ligand specificity, and the macrocyclic effect, *J. Am. Chem. Soc.*, **104** (12), 3249–3258.

76. Wu, Y., Seo, T., Maeda, S., Sasaki, T., Irie, S., and Sakurai, K. (2005). Circular dichroism induced by the helical conformations of acylated chitosan derivatives bearing cinnamate chromophores, *J. Polym. Sci. Polym. Phys.*, **43**, 1354–1364.

77. Xu, J., Dai, W., Wang, Z., Chen, B., Li, Z., and Fan, X. (2011). Intranasal vaccination with chitosan-DNA nanoparticles expressing pneumococcal surface antigen a protects mice against nasopharyngeal colonization by streptococcus pneumonia, *Clin. Vaccine Immunol.*, **18**(1), 75–81.

78. Xu, W., Shen, Y., Jiang, Z., Wang, Y., Chu, Y., and Xiong, S. (2004). Intranasal delivery of chitosan-DNA vaccine generates mucosal SIgA and anti-CVB3 protection, *Vaccine*, **22**, 3603–3612.

79. Yang, J., Chung, T. W., Nagaoka, M., Goto, M., Cho, C. S., and Akaike, T. (2001). Hepatocyte-specificporous polymer-scaffolds of alginate/galactosylated chitosan sponge for liver-tissue engineering, *Biotech. Lett.*, **23**, 1385–1389.

80. Yang, T., Nyiawung, D., Silber, A., Hao, J., Lai, L., and Bai, S. (2012). Comparative studies on chitosan and polylactic-co-glycolic acid incorporated nanoparticles of low molecular weight heparin, *AAPS PharmSciTech*, **13**(4), 1309–1318.

81. Yuan, X., Yang, X., Cai, D., Mao, D., Wu, J., Zong, L., and Liu, J. (2008). Intranasal immunization with chitosan/pCETP nanoparticles inhibits atherosclerosis in a rabbit model of atherosclerosis, *Vaccine*, **26**, 3727–3734.

82. Zaharoff, D. A., Rogers, C. J., Hance, K. W., Schlom, J., and Greiner, J. W. (2007). Chitosan solution enhances both humoral and cell-mediated immune responses to subcutaneous vaccination, *Vaccine*, **25**, 2085–2094.

83. Zaki, N. M., and Hafez, M. M. (2012). Enhanced antibacterial effect of ceftriaxone sodium-loaded chitosan nanoparticles against

intracellular salmonella typhimurium, *AAPS PharmSciTech*, **13**(2), 411–421.

84. Zambito, Y., Zaino, C., Uccello-Barretta, G., Balzano, F., and Di Colo, G. (2008). Improved synthesis of quaternary ammonium-chitosan conjugates (N^+–Ch) for enhanced intestinal drug permeation, *Eur. J. Pharm. Sci.*, **33**, 343–350

85. Zhang, J., Chen, X. G., Peng, W. B., and Liu, C. S. (2008). Uptake of oleoyl-chitosan nanoparticles by A549 cells, *Nanomedicine*, **4**, 208–214.

86. Zhang, J., and Chen, X. G. (2010). Effect of molecular weight on the oleoyl-chitosan nanoparticles as carriers for doxorubicin, *Colloids Surf. B*, **77**, 125–130.

87. Zhang, J., Xia, W., Liu, P., Cheng, Q., Tahirou, T., Guvand Bo Li, W. (2010). Chitosan modification and pharmaceutical/biomedical applications, *Mar. Drugs*, **8**, 1962–1987.

88. Zhu, A., Zhang, M., and Shen, J. (2002). Cell adhesion behavior of chitosan surface modified by bonding 2-methacryloyloxyethyl phosphorylcholine, *J. Biomater. Sci. Polym. Ed.*, **13**(5), 501–510.

89. Zhu, X. L., Du, Y. Z., Yu, R. S., Liu, P., Shi, D., Chen, Y., Wang, Y., and Huang, F. F. (2013). Galactosylated chitosan oligosaccharide nanoparticles for hepatocellular carcinoma cell-targeted delivery of adenosine triphosphate, *Int. J. Mol. Sci.*, **14**, 15755–15766.

90. Zong, Z., Kimura, Y., Takahashi, M., and Yamane, H. (2000). Characterization of chemical and solid state structures of acylated chitosans, *Polymer*, **41**, 899–906.

Chapter 16

Chitosan and Its Roles in Transdermal Drug Delivery

Asif Nawaz,[a,b] Nauman Rahim Khan,[a,b] and Tin Wui Wong[a,b,c]

[a]*Non-Destructive Biomedical and Pharmaceutical Research Centre Universiti Teknologi MARA, 42300, Puncak Alam, Selangor, Malaysia*
[b]*Particle Design Research Group, Faculty of Pharmacy, Universiti Teknologi MARA, 42300, Puncak Alam, Selangor, Malaysia*
[c]*CoRe Frontier Materials and Industry Application, Universiti Teknologi MARA, 40450, Shah Alam, Selangor, Malaysia*

wongtinwui@salam.uitm.edu.my

16.1 Introduction

Human skin represents the largest organ, which is characterized by a surface area of 18,000 cm^2. It constitutes 4% of the total body weight. With a thickness of approximately 2 mm, the skin is about 10^2 to 10^4 times less permeable than blood capillary wall (Banga, 1998). Skin functions to protect the host from physical as well as chemical assault; prevent excessive transepidermal water loss and dehydration; and perform thermoregulation, sensing, and endocrine activity (Menon, 2002; Fartasch, 1996; Labrie et al., 2001).

Skin is composed of different layers: epidermis (the uppermost), dermis, and hypodermis (Fig. 16.1) (Labrie et al., 2001).

Handbook of Sustainable Polymers: Processing and Applications
Edited by Vijay Kumar Thakur and Manju Kumari Thakur

ISBN 978-981-4613-53-8 (Hardcover), 978-981-4613-54-5 (eBook)
www.panstanford.com

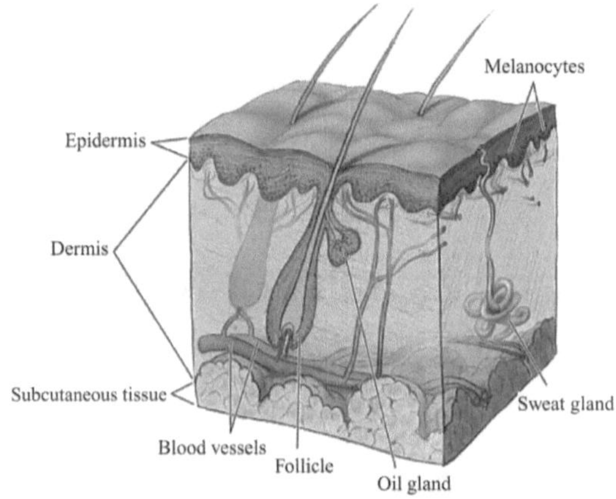

Figure 16.1 Schematic diagram of skin tissue (Bliss, (2005)).

The epidermis is an epithelium consisting of inner viable epidermis, a living hydrophilic layer, and outer nonviable epidermis, a hydrophobic layer made from dead cells. It is differentiated into stratum corneum, stratum lucidum, stratum granulosum, stratum spinosum, and stratum basale in the direction of dermis (Fig. 16.2). The viable epidermal layer has a thickness of about 0.02 to 0.2 mm. It is composed of many layers of keratinocytes, a widespread distribution of melanocytes, Langerhans cells, dendritic T cells, epidermotropic lymphocytes and Merkel cells, and a number of catabolic enzymes such as esterases, proteases, phosphatases, nucleotidases and lipases (Walters and Roberts, 2002; Barry, 2001). The outer and nonviable epidermis, namely, stratum corneum, is about 10 to 20 μm thick (Gregor and Ulrich, 2010). The stratum corneum is deemed to be the major obstacle of drug permeation.

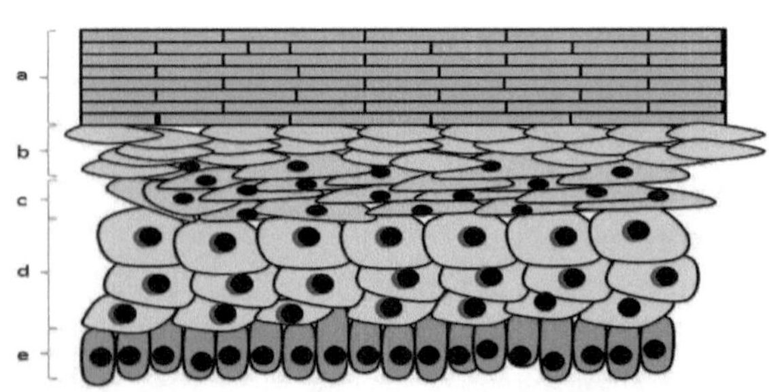

Figure 16.2 Epidermal differentiation.

The dermis is a hydrophilic layer supplied with rich blood circulation and nervous system. It is made of a dense network of collagen and elastin embedded in mucopolysaccharides. The dermis is thicker than epidermis and makes up the bulk of the skin. It provides skin with elasticity and mechanical strength and contains immune cells (Buck, 2004). In human, dermis is divided into thin lower surface of epidermis and thick underlying reticular layer. The dermis plays a major role in temperature, pressure, and pain regulation (Walters and Roberts, 2002). The subcutaneous fat or hypodermis is a fibrofatty layer that is loosely connected to dermis. The hypodermis participates in thermoregulation through fat acting as a heat insulator. It is a protective cushion over its underlying structure (Freinkel and Woodley, 2001).

16.2 Stratum Corneum

The stratum corneum is the outer most layer of nonviable epidermis. It has a thickness of about 10 to 12 μm. The stratum corneum consists of 15 to 25 layers of flattened, stacked, hexagonal, and cornified cells known as corneocytes. Each cell is approximately 40 μm in diameter and 0.5 μm in thickness (Bouwstra, 1997). The thickness of stratum corneum varies with the site of human body. The body extremities such as palms and soles have a thicker stratum corneum (Walters and Roberts, 2002). The stratum corneum is characterized by an array of keratin-rich corneocytes surrounded by lipid lamella made of cholesterol, free fatty acids, and ceramides (Bouwstra, 1997). The corneocytes are arranged in brick and mortar structure. Such structural arrangement creates a tortuous intercellular diffusion pathway for water or any other molecules that transverse the stratum corneum. The hydrophobic lipids that surround these diffusion paths or water pores are organized in tight lamellar structure. The summative effects translate to the formation of a tight permeation barrier (Menon, 2002).

The skin permeation capacity of a drug is a function of the physicochemical properties of the permeating agent. Broadly, three possible routes of skin transport, namely, intercellular, transcellular, and appendageal routes, are advocated (Roberts et al., 2002) (Fig. 16.3).

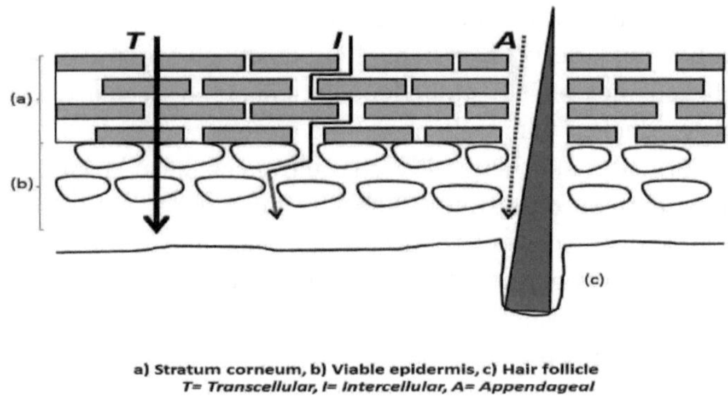

Figure 16.3 Schematic diagram of the routes of drug permeation through stratum corneum. I: intercellular, A: appendageal, T: transcellular.

16.3 Chitosan

Chitosan is a linear copolymer composed of N-acetyl-D-glucosamine and D-glucosamine units (Fig. 16.4). It is derived from alkaline deacetylation of chitin that is the second most abundant polysaccharide next to cellulose (Fig. 16.4) (Sloan, 1992). Chitin is the main component in the shells of crustaceans such as shrimp, crab, and lobster. It is also found in the exoskeleton of mollusks and insects, and the cell walls of some fungi (Rinaudo, 2008; Raabe et al., 2007; Vincent and Wegst, 2000).

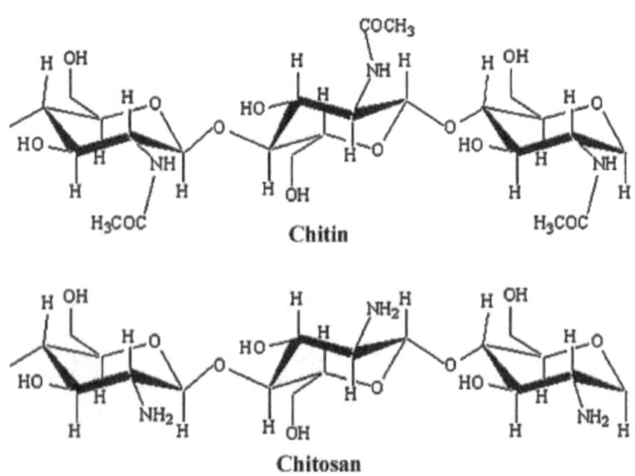

Figure 16.4 Chemical structures of chitin and chitosan.

The chitosan is first brought to attention as deacetylated chitin by Rouget in 1859 (Dodane and Vilivalam, 1998). During the past 20 to 25 years, it has been extensively used in various bioapplications (Sloan, 1992; Zohuriaan-Mehr, 2005; Hirano, 1996; Kurita, 1995). The chitosan receives a widespread utilization in pharmaceutical, medical, biotechnology, cosmetics, food, agriculture, paper, and textile industries due to its biodegradability, biocompatibility, antibacterial activity, nontoxicity, and versatile chemical and physical properties (Rinaudo, 2008; Dodane and Vilivalam, 1998; Mourya and Inamdar, 2008; Jayakumar et al., 2007; Hirano, 1999; Dash et al., 2011; Prabaharan, 2008; Illum, 1998).

The chitosan is characterized by a degree of N-deacetylation between 40% and 98%, and molecular weight between 50,000 and 2,000,000 Da. The chitosan of a molecular weight lower than 50 KDa is known as oligochitosan (Casettari et al., 2012). Chitosan possesses functional O–H and NH_2 moieties that can give rise to hydrogen bonding or be exploited for covalent derivatization and electrostatic complexation. Both molecular weight and degree of deacetylation properties have a major impact on the physicochemical and biological properties of chitosan such as aqueous solubility, viscosity, biodegradability, biocompatibility, mucoadhesion, mucopermeation enhancement, antimicrobial and wound healing activity (Table 16.1) (Dash et al., 2011; Hejazi and Amiji, 2003). On the basis of molecular weight, chitosan is divided into low molecular weight of less than 150 KDa, high-molecular-weight chitosan of 700 to 1000 KDa and medium-molecular-weight chitosan (Jon et al., 2007).

Chitosan is a weak base with a pka value of about 6.2 to 7. It is therefore insoluble at neutral and alkaline pH values. It does however soluble in acidic milieu through the salt formation with inorganic or organic acids such as hydrochloric acid, acetic acid, lactic acid, and glutamic acid. In acidic medium, the amino moiety of chitosan is protonated and has it transformed into a soluble and positively charged polysaccharide that has a high charge density. Being a cationic polyelectrolyte, the chitosan can form gels by interacting with different types of divalent and polyvalent anions (Hejazi and Amiji, 2003; Errington et al., 1993; Amiji and Patel, 1996). In the case of monovalent salt, its addition can likewise interfere with the aqueous solubility of chitosan. A high ionic strength ambience is envisaged to lower the solubility of chitosan via charge shielding (Filion et al., 2007).

Table 16.1 Physicochemical and biological properties of chitosan as a function of its molecular weight and degree of deacetylation

Physicochemical/ biological property	Structural characteristics	References
Aqueous solubility↑	DD↑	(Yi et al., 2005; Schiffman and Schauer, 2007; Zhang and Neau 2001; Chatelet et al., 2001; Khan et al., 2002)
Viscosity↑	DD↑	(Kofuji et al., 2005)
Biodegradability↑	DD↓, MW↓	(Zhang and Neau 2001; Kofuji et al., 2005; Huang et al., 2004)
Biocompatibility↑	DD↑	(Chatelet et al., 2001; Schipper et al., 1996)
Mucoadhesion↑	DD↑, MW↑	(Huang et al., 2004; El-Kamel et al., 2007; Roldo et al., 2004)
Mucopermeation enhancement↑	DD↑, MW↓	(Schipper et al., 1996; Schipper et al., 1997; Soane et al., 2001)
Antimicrobial activity↑	DD↑, MW↑	(Helander et al., 2001; Liu et al., 2001)
Wound healing↑	DD↑, MW↑	(Minagawa et al., 2007; Alsarra, 2009)

DD, degree of deacetylation; MW, molecular weight.

The polycationic nature of chitosan contributes to its mucoadhesive property where an electrostatic interaction can be elicited between the chitosan and the negatively charged sialic acid at the mucosa (Lehr et al., 1992). The chitosan is able to promote mucoadhesion of a dosage form. It facilitates drug absorption through enhancing drug availability to the mucosal epithelium (Lehr et al., 1992). The polycationic chitosan has also been reported to be able to transiently open the tight junctions between adjacent epithelial cells with reference to oral, nasal, pulmonary and skin delivery (Junginger and Verhoef, 1998; Smith et al., 2004; Artursson et al., 1994; Ranaldi et al., 2002; Yeh et al., 2011). It can act as a permeation enhancer and increase transmucosal transport of drugs (Merkel et al., 2012; Artursson et al., 1994). The chitosan has been considered for pharmaceutical formulation and drug delivery applications pertaining to its mucoadhesive, controlled release and permeation enhancement properties. In conventional pharmaceutical dosage forms, chitosan is employed as binder, disintegrant, coating and gelling agents (Sinha et al., 2004; Issa et al., 2005). In the late era, the chitosan has been deemed to be a promising excipient, and it is presently being designed into novel carriers of small-molecule drugs (Li et al., 2011; Rejinold et al., 2011; Rajan et al., 2013; Zhu et al., 2009), genes (Gerrit, 2001; Jayakumar et al., 2010; Anitha et al., 2009; Calvo et al., 1997; Huang et al., 2004; Janes and Alonso, 2003; Van Der Lubben et al., 2003; Xu and Du, 2003), proteins and peptides (Kim et al., 2013; Chua et al., 2011; Chu et al., 2012; Amidi et al., 2010) for oral (Lehr et al., 1992; Deshpande et al., 1996; Sabnis et al., 1997; Tozaki et al., 1997; Wong et al., 2002), parenteral (Thanoo et al., 1992; Patashnik et al., 1997; Kamiyama et al., 1999), ocular (Genta et al., 1997; Felt et al., 1999; Calvo et al., 1997) and nasal (Illum et al., 1994; Fernández-Urrusuno et al., 1999; Tengamnuay and Mitra, 1997; Costantino et al., 2005; Vyas et al., 2005) delivery, and tissue engineering (Peter et al., 2010; Madhumathi et al., 2009; Shalumon et al., 2009; Nagahama et al., 2008) as indicated by a large number of studies published over the last few years. In drug delivery, the chitosan is commonly fabricated into film or membrane (Li et al., 2002), hydrogel (Bhattarai et al., 2010) and particulate systems (Sunil et al., 2004; Wong and Nurulaini, 2012; Wong and Nurjaya, 2008).

The transdermal drug delivery has generated much interest among scientists and industrialists during the past decade. Transdermal delivery technology can provide local drug therapy at a specific skin site. It warrants an opportunity to reduce adverse side effects in association with extensive systemic drug absorption. Transdermal delivery of drugs bypasses hepatic first pass metabolism and pH-induced physicochemical degradation occurring in gastrointestinal tract. The transdermal drug delivery can be executed without causing any pain unlike injection. Despite these facts, the barrier function of stratum corneum remains an impediment to succeed transdermal drug delivery. The key approach nowadays is active disruption and fluidization of stratum corneum to make paths for drug diffusion (Kumar et al., 2004; Wong and Khaizan, 2013).

16.4 Transdermal Drug Delivery

Numerous dosage forms have been designed by means of synthetic and natural matrix substances for transdermal drug administration. Among the polymers, chitosan emerges as a unique constituent due to its mucoadhesive and permeation enhancement attributes that facilitate transmucosal drug absorption (Junginger and Verhoef, 1998; Di Colo et al., 2004; Sinswat and Tengamnuay, 2003; Jonker et al., 2002; Hamman et al., 2002; Thanou et al., 2001; Thanou et al., 2000; Khalil et al., 2012).

16.4.1 Chitosan as Transdermal Permeation Enhancer

Tight junction is intercellular connections that hold epithelial cells together at their epical end (Fig. 16.5). The intercellular tight junction is one of the major barriers to the paracellular transport of macromolecules and polar compounds. Tight junction structure and permeability are regulated by physiological factors such as concentration of cyclic AMP, intercellular calcium concentration and transient mucosal osmotic loads (Junginger and Verhoef, 1998). Several studies have shown that one of the possible mechanisms of permeation enhancer is to loosen the tight junctions of epithelial membranes (Murakami et al., 1982; Murakami et al., 1984). Chitosan enhances the paracellular permeability of mucosal epithelia in intestine, nasal and buccal cavities by transiently opening the tight junctions through the change in the cytoskeletal protein F-actin from a filamentous to a globular structure. This then increases the absorption of hydrophilic and macromolecular drugs via the intercellular route (Junginger and Verhoef, 1998). In the past years, it has been shown that the skin is similarly characterized by tight junctions. The stratum corneum is constituted of dead keratinized cells and fibrous protein. Although its composition differs from that of the epithelial cells, the stratum corneum demonstrates fixed negative charges in tight junctions between cells as in the cases of epithelial cells in distant anatomies (Hamman et al., 2002). As such, it is speculated that the chitosan can act as a permeation enhancer effective for transdermal drug delivery.

The chitosan has appeared to be able to interact with negative charges of lipids in the skin (Taveira et al., 2009). In a study by Wen He et al. (2009), chitosan and its derivatives are found to significantly change the secondary structure of keratin in stratum corneum, increase the water content in stratum corneum, decrease HaCaT cell membrane potential, and enhance cell membrane fluidity to various degrees. The chitosan compounds are deemed to interact with both lipids and proteins of stratum corneum. Such association could lead to disorganization of lipid lamella, formation of larger aqueous pores and fluidized membrane that result in an increased permeation extent of drugs by intercellular and/or transcellular routes.

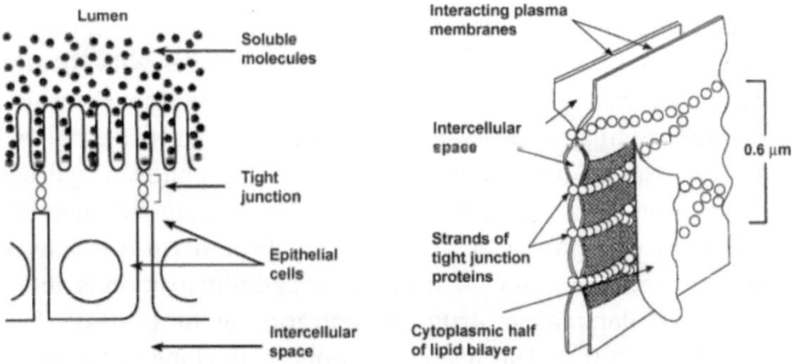

Figure 16.5 Tight junction as intercellular connection that holds epithelial cells together at their apical end (Junginger and Verhoef, 1998).

16.4.2 Chitosan-Based Transdermal Drug Delivery Systems

For the purpose of transdermal delivery, chitosan has been largely processed into hydrogel, membrane, microparticle, nanoparticle, and nanocapsule. In addition, it is also used in the processing of liposome and polymersome.

16.4.2.1 Hydrogel

Hydrogel offers attractive features as a transdermal drug delivery system owing to its high water content, which provides a comfortable feeling on the patient's skin and results in a higher compliance level for the therapy (Thomas and Finnin, 2004). Glimepride is a third-generation oral antidiabetic sulfonylurea drug. Its oral administration is met with low bioavailability. The formulation of glimepride into chitosan hydrogel is reported to translate to effective release and consistent therapeutic efficacy over 48 h in mice (Ammar et al., 2008). The chitosan hydrogel has also been employed to deliver berberine alkaloid (Tsai et al., 1999) and active S-enantiomer of racemic propranolol (Suedee et al., 2008) transdermally. The latter uses a composite membrane made of chitosan hydrogel reservoir with poloxamer as the drug release modifier. In another study, the chitosan hydrogel is physically cross-linked with lauric, myristic, palmitic, or stearic acid and is harvested by means of freeze drying process. This chitosan hydrogel provides a higher transcutaneous permeation propensity of propranolol hydrochloride than the corresponding drug solution. The improved delivery is ascribed to the interaction of hydrogel with stratum corneum, which increases the drug solubility in skin (Cerchiara et al., 2002). Similar observations are obtained in capsaicin delivery across the excised nude mouse skin where the chitosan hydrogel is found to be more effective as a drug vehicle than commercially available creams (Wang et al., 2001). In bromocriptine delivery across the rabbit skin in vivo, transdermal administration is able to provide a similar plasma drug concentration as oral dosage form (Degim et al., 2003). The use of iontophoresis along with dermal application of chitosan hydrogel results in a greater drug flux across the skin barrier when compared to either method individually (Kumar et al., 2004).

16.4.2.2 Film/membrane

Chitosan is well known for its film forming property. A number of studies have been performed with chitosan fabricated into membrane as transdermal device. Cross-linked chitosan membrane and chitosan gel have been concurrently used in the development of transdermal drug delivery system. The gel functions as a drug reservoir. The membrane acts as a rate-controlling device with its cross-link density dictating the drug permeability (Thacharodi and Panduranga Rao, 1996a). The rate-controlling property of the membrane can also be modulated via formulating chitosan membrane with collagen (Thacharodi and Panduranga Rao, 1996b). In a singly membrane system, the chitosan has been plasticized with polyethylene glycol to accommodate captopril (Kumar et al., 2004). The chitosan film improves drug bioavailability with no significant irritant reaction in male albino rats when compared against an oral dosing at 4.3 mg/kg.

Asymmetric chitosan membrane has been developed by Mi et al. (2003) to compartmentalize drug load based on its intended release kinetics by means of a two-step (dry/wet) phase separation technique. In this membrane, burst-release sulfadiazine and slow-release silver are installed at sponge-like porous layer and dense layer respectively for bacterial control in skin burn wound (Fig. 16.6). The differences in drug release kinetics are attributed to chemical equilibrium of the ionic interactions between sulfadiazine, silver ions, chloride ions and chitosan membrane, as well as the pore structures of asymmetric chitosan membranes. The silver sulfadiazine loaded asymmetric chitosan membrane is less cytotoxic than the traditional cream owing to controlled release of actives. It is very effective for long-term bacterial growth inhibition on infected wound. The lower phase of membrane, in addition to acting as a drug reservoir, is an essential anchor that provides spaces for wound tissue growth and regeneration (Mi et al., 2003). The same membrane system has also been developed for guided bone tissue regeneration in treatment of periodontal diseases (Ho et al., 2010).

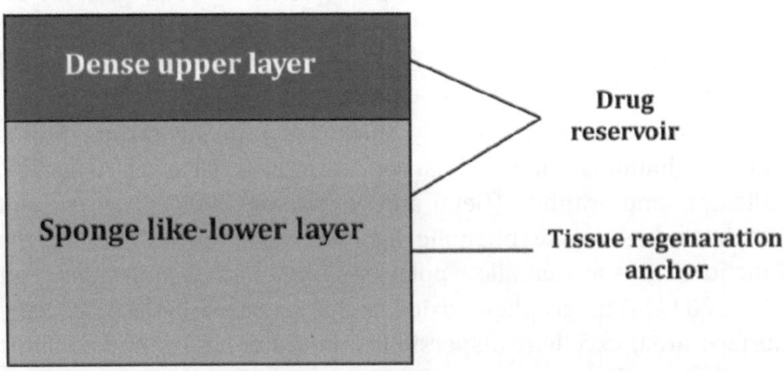

Figure 16.6 Asymmetric chitosan membrane as skin wound dressing.

16.4.2.3 Nanoparticle

Nanoparticles are defined as particles with physical dimensions between 1 nm and 1000 nm (Lövestam et al., 2010). They have been used to restrict the access of drug at the target sites and to deliver the drug at a controlled rate at the site of action (Sagar et al., 2011). Khalil et al. formulate warfarin-β-cyclodextrin loaded chitosan nanoparticles for transdermal delivery. Chitosan is a hydrophilic carrier. The hydrophobic warfarin is first complexed with β-cyclodextrin prior to incorporating into chitosan nanoparticles. The nanoparticles are formed through ionotropic gelation of chitosan matrix with tripolyphosphate counterions. The release profile of warfarin-β-cyclodextrin complex from chitosan nanoparticles is characterized by an initial burst release followed by a slow and continuous release phase. The nanoparticulate formulation enables an increase in the permeation of warfarin through excised rat skin in a constant and continuous setting (Khalil et al., 2012).

Tan et al. investigate lecithin–chitosan nanoparticles as a delivery system for quercetin. In comparison to quercetin solution, the quercetin loaded nanoparticles exhibit a higher skin permeation ability and have significantly increased the cumulative amount of quercetin in the skin especially in the epidermis. The positive charge of chitosan has rendered a close contact between the nanoparticles and the skin. The released lecithin mixes and fuses with the lipids in skin. It loosens the skin microstructure and disturbs the lamellar arrangement of lipids thereby resulting in increased accumulation and permeation of drug (Tan et al., 2011). Similar findings have also been reported by Senyigit et al. (2010), where clobetasol was used as the drug of interest instead of quercetin.

Graphene is an atomically thick sheet of sp^2 carbon atoms and the building block of carbon structures such as nanotube, fullerene, and graphite (Geim and Novoselov, 2007). The oxide of graphene has ample phenolic hydroxyl, epoxide and carboxylic functional groups that allow polar bonding to other molecules (Pan et al., 2011). The graphene oxide is characterized by large specific surface area, excellent dispersibility in water and other aqueous mediums, and low nanotoxicity (Liu et al., 2011). It is a highly interesting material for use in the biomedical field including drug delivery. Recently, Justin and Chen (2014) developed a biodegradable chitosan–graphene oxide nanocomposite with the aim to improve the mechanical property and drug delivery performance over chitosan. The chitosan–graphene oxide nanocomposite undergoes a slower biodegradation rate than chitosan. This nanocomposite, however, offers a faster and a more substantial release of drug than system containing the chitosan alone, owing to the abundant oxygenated functional groups, hydrophilic character and large specific surface area provided by graphene oxide sheets. In the presence of amino-rich chitosan, the nanocomposite can have controllable, pH-sensitive drug delivery profiles and is deemed applicable for transdermal drug delivery.

16.4.2.4 Microparticle

Microparticles refer to particulate systems of sizes between 1 and 1000 μm (Kim and Pack, 2006). Unlike nanoparticles, the microparticles are larger in size and skin penetration using intercellular route is deemed impossible. Their accumulation in skin is largely mediated via transfollicular route. With reference to transdermal drug delivery, the late studies examine skin penetration profiles of microparticles and its relationship with drug bioavailability and effectiveness. Gelfuso et al. (2011) fabricated porous chitosan microparticles of 3 μm in size as a carrier of minoxidil sulfate at different drug/chitosan weight ratios for alopecia treatment. These microparticles were able to sustain the minoxidil release. The chitosan microparticles are concluded to be potentially useful to deliver minoxidil for alopecia therapy.

Hydrophilic sunscreen agent, phenylbenzimidazole sulfonic acid, has been formulated in microparticle system. The incorporation of the sunscreen in chitosan microparticles improves its in vitro ultra violet screening effect through sustained release and particle penetration (Gomaa et al., 2010). Retinoic acid is an analogue of vitamin A that is used in acne treatment. Unfortunately, local irritation reactions strongly limit its topical use (Lehman et al., 1988). Lira et al. (2009) investigated the efficacy of chitosan–alginate coacervate microparticles in acne treatment. The chitosan–alginate microparticles are able to sustain the release of retinoic acid and be accumulated in deeper layers of the skin to release the retinoic acid thereby negating its topical adverse effects. Catechins are major antioxidants in green tea. They are known to have poor skin permeation. Wisuitiprot et al. (2011) assessed the influence of chitosan microparticles on cutaneous absorption of catechins. The chitosan microparticles are reported to improve the skin permeation of catechins significantly via particle penetration and also prevent the catechin from enzymatic changes. Yenilmez et al. (2011) prepared vitamin E-chitosan microparticles as skin cosmetics. In vivo study conducted on human cheeks and crow's feet skin demonstrates that the vitamin E-chitosan microparticles significantly increase the skin moisture and elasticity, and reduce the skin wrinkle and roughness. This is ascribed to chitosan moisture retention and controlled release of active ingredients.

16.4.2.5 Nanocapsule/microcapsule

Unlike nanoparticles or microparticles, the capsulated particles consist of matrices coated or layered with membrane. Shin et al. (2012) fabricated chitosan-based polymer nanocapsules. The nanocapsules are equipped with positive charges on their periphery to disturb the lamellar layer of stratum corneum into a fluidized membrane. This eventually enables a higher diffusion propensity of encapsulated drug, riboflavin-5-monophosphate. Transdermal delivery of drugs using chitosan microcapsules as carriers has also been investigated. Lam et al. (2012) demonstrated the topical delivery of 5-fluorouracil using chitosan-based microcapsules. The chitosan-based microcapsules of 5-fluorouracil give rise to a stronger growth inhibition on human keratinocytes when compared to the free drug. The untreated skin cells or those treated by drug-free microcapsules show a high integrity of cellular morphology. On the other hand, the skin cells subjected to 5-fluorouracil microcapsules and free 5-fluorouracil display a substantial degree of cellular damage, with 5-fluorouracil microcapsules exhibit a stronger cell growth inhibitory effect than the free drug. The latter may be associated with sustained-release characteristics of 5-fluorouracil from microcapsules, which provide a continuous therapeutic effect against the growth of skin keratinocytes (Lam et al., 2012).

Yuen et al. (2012) reported the potential use of chitosan-based microcapsules for topical delivery of two antifungal agents, miconazole nitrate and clotrimazole in tinea pedis treatment. The release rates of two drugs are investigated under different compression pressures from 0.028 to 5 kg and different pH values between 5.5 and 8.5, mimicking the conditions of human skin and actual application environment. The drug release is found to increase with an increase in compression pressure. The release kinetics of drugs is however not affected by the pH of the surrounding medium. Chitosan microparticles help in the continuous release of antifungal drugs in a controlled manner.

16.4.2.6 Liposome

Liposome is primarily composed of phospholipids, which are nontoxic, nonimmunogenic, biocompatible, and biodegradable. The liposome can store the drug load in its hydrophobic shell or the hydrophilic interior depending on the physicochemical nature of the drug. Lingling et al. (2011) developed chitosan-coated elastic liposome of local anesthetic lidocaine hydrochloride. The elastic liposome is composed of soya lecithin and is prepared by rotary evaporation-sonication method. The introduction of a chitosan coat onto the surfaces of elastic liposome modulates its surface charge towards positivity, improve the physicochemical stability, and prolong the drug release profile. With reference to both in vitro and in vivo studies, the chitosan-coated elastic liposome is able to give a statistically significant improvement in skin permeation of lidocaine. Accordingly, it can reduce the pain sensation significantly in comparison to the corresponding elastic liposome and chitosan solution. Similar results are obtained by Wang et al. (2013). In this study, the researchers used lycin-modified chitosan to form polymeric liposome with cholesterol as the additive for transdermal delivery of lidocaine hydrochloride. In vitro and in vivo skin permeation studies indicate that the polymeric liposome can control and enhance transdermal penetration of lidocaine hydrochloride into deeper layers of skin.

16.4.2.7 Polymersome

In the past years, polymersome has attracted tremendous attention as a versatile drug carrier with respect to their colloidal stability, tunable membrane properties and ability in encapsulating or integrating a broad range of drugs and molecules (Lee and Feijen, 2012). Coan et al. (2014) developed polystyrene-b-poly(acrylic acid) polymersome decorated with chitosan for topical delivery of finasteride. The polymersome decorated with chitosan appears to be a more appropriate drug carrier option since it provides a greater extent of drug retention in the skin and a better control over the drug release profile. Chitosan-decorated polymersome interacts more strongly with the skin components than non-decorated samples. This is probably due to the availability of positive surface charge, which increases mucoadhesion of polymersome and drug retention time.

16.4.3 Chitosan-Based Transdermal Gene Delivery

Table 16.2 summarizes drug types that have been formulated in various transdermal dosage forms. Macromolecular drugs, namely, insulin and oligonucleotide, have been investigated for their possibility to administer via skin. Gene therapy treats and prevents diseases via gene transfer. Broadly, there are three main classes of gene delivery vectors: viral, physical (needle-free injectors or electroporation) and nonviral (naked DNA, DNA complex with cationic lipids and polymers). The viral vector represents the most successful gene delivery system with respect to gene expression. Nonetheless, there are several disadvantages and risks such as marked immunogenicity that activates inflammatory system leading to the degeneration of transfected tissue, toxin production including mortality, insertional mutagenesis, and limited transgenic capacity. This renders the development of efficient nonviral systems being critical (Kim, 2013).

Table 16.2 Drugs encapsulated in dosage forms of chitosan or its derivatives for transdermal delivery

Dosage form	Drug	Reference
Hydrogel	Terbinafine hydrochloride	(Özcan et al., 2009)
	Glimepride	(Ammar et al., 2008)
	Berberine	(Tsai et al., 1999)
	bromocriptine	(Degim et al., 2003)
Membrane/film	Captopril	(Kumar et al., 2004)
	5-fluorouracil	(Rana et al., 2005)
	Etoricoxib	(Wahid et al., 2008)
	Silver compounds	(Hye Kim et al., 2013)
Nanoparticle	Warfarin	(Khalil et al., 2012)
	Quercetin	(Tan et al., 2011)
	Clobetasol	(Şenyiğit et al., 2010)
	Aceclofenac	(Jana et al., 2014)
	Melatonin	(Hafner et al., 2011)
	Oligonucleotides	(Ozbas-Turan et al., 2010)
	Insulin	(Wong, 2009)
Microparticle	Minoxidil	(Gelfuso et al., 2011)
	phenylbenzimidazole sulfonic acid	(Gomaa et al., 2010)
	Catechins	(Wisuitiprot et al., 2011)

The ability of chitosan to condense DNA to form complexes is the basis for its use in nonviral gene delivery. Lee et al. (2008) developed biodegradable polymeric nanoparticles composed of chitosan and poly-γ-glutamic acid by ionotropic gelation method for transdermal DNA delivery using a low-pressure gene gun. The formed nanoparticles penetrate into the mouse skin and enhance gene expression. Their biological performance is promoted by compact and dense structure of nanoparticles, which is brought about by strong chitosan-DNA and chitosan-poly-γ-glutamic acid interaction. The nanoparticulate carrier performance of chitosan for gene delivery is also examined by Lee et al. (2010). The nanoparticles that consist of hydrophobic poly(lactic-co-glycolic acid) core and positively charged glycol chitosan shell are designed for transdermal DNA delivery and epidermal Langerhans cell tracking. The glycol chitosan acts as a vehicle for DNA loading while poly(lactic-co-glycolic acid) core of nanoparticles is used to load quantum dots for ultrasensitive detection of Langerhans cells' migration from skin to lymph nodes, following the bombardment of nanoparticles into the skin.

16.4.4 Chitosan Skin Toxicity

Almost all functional properties of chitosan depend on its chain length, charge density, and charge distribution. Numerous studies have demonstrated that the salt form, molecular weight, and degree of deacetylation of chitosan, as well as the pH at which the chitosan is used, influence the functional properties of this polymer (Kean and Thanou, 2010).

The chitosan and its derivatives show no acute toxicity and are not absorbed via transdermal route. The European pharmacopoeia contains a single monograph on chitosan hydrochloride. In the United States, chitosan is currently being included in the US Pharmacopoeia (Sarmento and das Neves, 2012). The chitosan and its derivatives are deemed safe for use as permeation enhancer for transmucosal delivery of hydrophilic drugs and offer promising prospects for novel pharmaceutical applications (Junginger and Verhoef, 1998). Despite the chitosan and its derivatives interact with lipids and proteins of the membranes of stratum corneum, they may not penetrate into deeper layers of the skin. This can be inferred from the absence of skin irritation by chitosan and its derivatives in Draize test (Aoyagi et al., 1991).

16.5 Conclusion

Chitosan demonstrates its usefulness in transdermal drug delivery via its mucoadhesive, permeation enhancement, and controlled-release properties. They have been formulated in both solid and semisolid states, from macro to nanoscales. The unique feature of chitosan lies in its polycationic character that has a strong affinity for negatively charged lipids of skin. The chitosan can raise skin permeability. With skin acting as a low metabolic diffusion path and posing convenient administration, transdermal delivery of macromolecular drugs with the aid of chitosan remains a great interest for further future investigation.

References

Alsarra, I. A. (2009). Chitosan topical gel formulation in the management of burn wounds, *Int. J. Biol. Macromol.*, **45**(1), 16–21.

Agnihotri, S. A., Mallikarjuna, N. N., Aminabhavi, T. M. (2004). Recent advances on chitosan-based micro- and nanoparticles in drug delivery, *J. Control. Release*, **100**, 5–28.

Amidi, M., Mastrobattista, E., Jiskoot, W., Hennink, W. E. (2010). Chitosan-based delivery systems for protein therapeutics and antigens, *Adv. Drug Deliv. Rev.*, **62**(1), 59–82.

Amiji, M. M., Patel, V. R. (1996). Preparation and characterization of freeze-dried chitosan-poly(ethylene oxide) hydrogels for site specific antibiotic delivery in the stomach, *Pharm. Res.*, **13**, 588–593.

Ammar, H. O., Salama, H. A., El-Nahhas, S. A., Elmotasem, H. (2008). Design and evaluation of chitosan films for transdermal delivery of glimepiride, *Curr. Drug Deliv.*, **5**, 290–298.

Anitha, A., Divya Rani, V. V., Krishna, R., Sreeja, V., Selvamurugan, N., Nair, S. V. (2009). Synthesis, characterization, cytotoxicity and antibacterial studies of chitosan, O-carboxymethyl and N, O-carboxymethyl chitosan nanoparticles, *Carbohydr. Polym.*, **78**, 672–677.

Aoyagi, T., Terashima, O., Nagase, Y., Matsui, K. (1991). Preparation of a polymer containing hexadecylpyridinium bromide groups and its utilization as a transdermal drug penetration enhancer, *Polymer*, **32**, 2106–2111.

Artursson, P., Lindmark, T., Davis, S. S., Illum, L. (1994). Effect of chitosan on the permeability of monolayers of intestinal epithelial cells (Caco-2), *Pharm. Res.*, **11**(9), 1358–1361.

Banga, A. K. (1998). *Electrically Assisted Transdermal and Topical Drug Delivery*, Bristol, PA: Taylor and Francis, Inc.

Barry, B. W. (2001). Novel mechanisms and devices to enable successful transdermal drug delivery, *Eur. J. Pharm. Sci.*, **14**, 101–114.

Bhattarai, N., Gunn, J., Zhang, M. (2010). Chitosan-based hydrogels for controlled, localized drug delivery, *Adv. Drug Deliv. Rev.*, **62**(1), 83–99.

Bliss D. (2005). *Skin (Layers, Glands, Vessels)*, National Cancer Institute.

Bouwstra, J. A. (1997). The skin barrier, a well-organised membrane. *Colloids Surf. A Physicochem. Eng. Aspects*, **123**, 403–413.

Buck, P. (2004). Skin barrier function: Effect or age, race and inflammatory disease, *Int. J. Aromatherapy*, **14**, 70–76.

Calvo, P., Remunan-Lopez, C., Vila-Jato, C. L., Alonso, M. J. (1997). Chitosan and chitosan/ethylene oxide-propylene oxide block copolymer nanoparticles as novel carriers for proteins and vaccines, *Pharm. Res.*, **14**, 1431–1436.

Calvo, P., Vila-Jato, J. L., Alonso, M. J. (1997). Evaluation of cationic polymer-coated nanocapsules as ocular drug carriers, *Int. J. Pharm.*, **153**, 41–50.

Caon, T., Porto, L. C., Granada, A., Tagliari, M. P., Silva, M. A. S., Simões C. M. O., Borsali, R., Soldi, V. (2014). Chitosan-decorated polystyrene-b-poly (acrylic acid) polymersomes as novel carriers for topical delivery of finasteride, *Eur. J. Pharm. Sci.*, **52**, 165–172.

Casettari, L., Vllasaliu, D., Castagnino, E., Stolnik, S., Howdle, S., Illum, L. (2012). PEGylated chitosan derivatives: Synthesis, characterizations and pharmaceutical applications, *Prog. Polym. Sci.*, **37**(5), 659–685.

Cerchiara, T., Luppi, B., Bigucci, F., Orienti, I., Zecchi, V. (2002). Physically cross-linked chitosan hydrogels as topical vehicles for hydrophilic drugs, *J. Pharm. Pharmacol.*, **54**, 1453–1459.

Chatelet, C., Damour, O., Domard, A. (2001). Influence of the degree of acetylation on some biological properties of chitosan films, *Biomaterials*, **22**, 261–268.

Chu, B. Y., Kobaisi, M. A., Zeng, W., Mainwaring, D., Jackson, D. C. (2011). Chitosan microparticles and nanoparticles as biocompatible delivery vehicles for peptide and protein-based immunocontraceptive vaccines, *Mol. Pharm.*, **9**(1), 81–90.

Chu, B. Y., Kobiasi, M. A., Zeng, W., Mainwaring, D., Jackson, D. C. (2012). Chitosan-based particles as biocompatible delivery vehicles for peptide and protein-based vaccines, *Procedia Vaccinol.*, **6**, 74–79.

Costantino, H. R., Sileno, A. P., Johnson, P. H. (2005). Pharmacokinetic attributes of intranasal delivery: Case studies and new opportunities, *Drug Deliv.*, **3**, 8–11.

Dash, M., Chiellini, F., Ottenbrite, R. M., Chiellini, E. (2011). Chitosan—a versatile semi-synthetic polymer in biomedical applications, *Prog. Polym. Sci.*, **36**, 981–1014.

Degim, I. T., Acartürk, F., Erdogan, D., Demirez, L. N. (2003). Transdermal administration of bromocriptine, *Biol. Pharm. Bull.*, **26**(4), 501–505.

Deshpande, A. A., Rhodes, C. T., Shah, N. H., Malick, A. W. (1996). Controlled-release drug delivery systems for prolonged gastric residence: An overview, *Drug Dev. Ind. Pharm.*, **22**, 531–539.

Di Colo, G., Burgalassi, S., Zambito, Y., Nardini, I., Saettone, M. F. (2004). Effect of chitosan and of N carboxymethylchitosan on intraocular penetration of topically applied ofloxacin, *Int. J. Pharm.*, **273**, 37–44.

Dodane, V., Vilivalam, V. D. (1998). Pharmaceutical applications of chitosan, *Pharm. Sci. Technol. Today*, **1**(6), 246–253.

El-Kamel, A. H., Ashri, L. Y., Alsarra, I. A. (2007). Micromatricial metronidazole benzoate film as a local mucoadhesive delivery system for treatment of periodontal diseases, *AAPS Pharm. Sci. Tech.*, **8**, E75/1–E75/11.

Errington, N., Harding, S. E., Varum, K. M., Illum L. (1993). Hydro dynamic characterization of chitosans varying in degree of acetylation, *Int. J. Biol. Macromol.*, **15**, 113–117.

Fartasch M. (1996). The nature of the epidermal barrier: Structural aspects, *Adv. Drug Deliv. Rev.*, **18**, 273–282.

Felt, O., Furrer, P., Mayer, J. M., Plazzonet, B., Bun, P., Gurny, R. (1999). Topical use of chitosan in ophthalmology: Tolerance assessment and evaluation of precorneal retention, *Int. J. Pharm.*, **180**, 185–193.

Fernández-Urrusuno, R., Calvo, P., Remuñán-López, C., Vila-Jato, J. L., Alonso, M. J. (1999). Enhancement of nasal absorption of insulin using chitosan nanoparticles, *Pharm. Res.*, **16**(10), 1576–1581.

Filion, D., Lavertu, M., Buschmann, M. D. (2007). Ionization and solubility of chitosan solutions related to thermosensitive chitosan/glycerol-phosphate systems, *Biomacromolecules*, **8**(10), 3224–3234.

Freinkel, R. K., Woodley, D. T. (eds.). (2001). *The Biology of the Skin*, Parthenon Publishing, New York.

Geim, A. K., Novoselov, K. S. (2007). The rise of grapheme, *Nat. Mater.*, **6**(3), 183–191.

Gelfuso, G. M., Gratieri, T., Simao, P. S., de Freitas, L. A. P, Lopez, R. F. V. (2011). Chitosan microparticles for sustaining the topical delivery of minoxidil sulphate, *J. Microencapsul.*, **28**(7), 650–658.

Genta, I., Conti, B., Perugini, P., Pavanetto, F., Spadaro, A., Puglisi, G. (1997). Bioadhesive microspheres for ophthalmic administration of acyclovir, *J. Pharm. Pharmacol.*, **49**, 737–742.

Gerrit, B. (2001). Chitosans for gene delivery, *Adv. Drug Deliv. Rev.*, **2**, 145–150.

Gomaa, Y. A., El-Khordagui, L. K., Boraei, N. A., Darwish, I. A. (2010). Chitosan microparticles incorporating a hydrophilic sunscreen agent, *Carbohydr. Polym.*, **81**, 234–242.

Gregor, C., Ulrich, V. (2010). Nanotechnology and the transdermal route: A state of the art review and critical appraisal, *J. Control. Release*, **141**(3), 277–299.

Hafner, A., Lovric, J., Pepic, I., Filipovic-Grcic, J. (2011). Lecithin/chitosan nanoparticles for transdermal delivery of melatonin, *J. Microencapsul.*, **28**(8), 807–815.

Hamman, J. H., Stander, M., Kotze, A. F., Monti, D., Chetoni, P. J. (2002). Effect of degree of quaternization of N-trimethyl chitosan chloride on absorption enhancement: In vivo evaluation in rat nasal epithelia, *Int. J. Pharm.*, **232**, 235–242.

Hejazi, R., Amiji, M. (2003). Chitosan-based gastrointestinal delivery systems, *J. Control. Release*, **89**, 151–165.

Helander, I. M., Nurmiaho-Lassila, E. L., Ahvenainen, R., Rhoades, J., Roller, S. (2001). Chitosan disrupts the barrier properties of the outer membrane of Gram-negative bacteria, *Int. J. Food Microbiol.*, **71**, 235–244.

Hirano, S. (1996). Chitin biotechnology applications, *Biotechnol. Annu. Rev.*, **2**, 237–258.

Hirano, S. (1999). Chitin and chitosan as novel biotechnological materials, *Polym. Int.*, **48**, 732–734.

Ho, M. H., Hsieh, C. C., Hsiao, S. W., Van Hong Thien, D. (2010). Fabrication of asymmetric chitosan GTR membranes for the treatment of periodontal disease, *Carbohydr. Polym.*, **79**, 955–963.

Huang, M., Khor, E., Lim, L. Y. (2004). Uptake and cytotoxicity of chitosan molecules and nanoparticles: Effects of molecular weight and degree of deacetylation, *Pharm. Res.*, **21**, 344–353.

Hye Kim, J., Kim, S., Kwon, I. B., Hyun Kim, M., I. K., Lim, J. (2013). Simple fabrication of silver hybridized porous chitosan-based patch for transdermal drug-delivery system, *Materials Letters*, **95**, 48–51.

Illum, L. (1998). Chitosan and its use as a pharmaceutical excipient, *Pharm. Res.*, **15**, 1326–1331.

Illum, L., Farraj, N. F., Davis, S. S. (1994). Chitosan as a novel nasal delivery system for peptide drugs, *Pharm. Res.*, **11**, 1186–1189.

Issa, M. M., Köping-Höggård, M., Artursson, P. (2005). Chitosan and the mucosal delivery of biotechnology drugs, *Drug Discov. Today*, **2**(1), 1–6.

Jana, S., Manna, S., Nayak, A. K., Sen, K. K., Basu, S. K. (2014). Carbopol gel containing chitosan-egg albumin nanoparticles for transdermal aceclofenac delivery, *Colloids Surf. B*, **114**, 36–44.

Janes, K. A., Alonso, M. J. (2003). Depolymerized chitosan nanoparticles for protein delivery: Preparation and characterization, *J. Appl. Polym. Sci.*, **88**, 2766–2779.

Jayakumar, R., Chennazhi, K. P., Muzzarelli, R. A. A., Tamura, H., Nair, S. V., Selvamurugan, N. (2010). Chitosan conjugated DNA nanoparticles in gene therapy, *Carbohydr. Polym.*, **79**, 1–8.

Jayakumar, R., New, N., Tokura, S., Tamura, H. (2007). Sulfated chitin and chitosan as novel biomaterials, *Int. J. Biol. Macromol.*, **40**, 175–181.

Jon, S., Eunhye, L. E. E., Lee, J. J., Lee, I.-H. (2007). Transmucosal delivery of pharmaceutical active substances, U.S. Patent Application 11/847, 237, filed August 29.

Jonker, C., Hamman, J. H., Kotze, A. F. (2002). Intestinal paracellular permeation enhancement with quaternized chitosan: In situ and in vitro evaluation, *Int. J. Pharm.*, **238**, 205–213.

Junginger, H. E., Verhoef, J. C. (1998). Macromolecules as safe penetration enhancers for hydrophilic drugs—a fiction? *PSTT*, **1**, 370–376.

Justin, R., Chen, B. (2014). Characterisation and drug release performance of biodegradable chitosan–graphene oxide nanocomposites, *Carbohydr. Polym.*, **103**, 70–80.

Kamiyama, K., Onishi, H., Machida, Y. (1999). Biodisposition characteristics of N-succinyl-chitosan and glycol-chitosan in normal and tumor-bearing mice, *Biol. Pharm. Bull.*, **22**(2), 179–186.

Kean, T., Thanou, M. (2010). Biodegradation, biodistribution and toxicity of chitosan, *Adv. Drug Deliv. Rev.*, **62**(1), 3–11.

Khalil, S. K., El-Feky, G. S., El-Banna, S. T., Khalil, W. A. (2012). Preparation and evaluation of warfarin-β-cyclodextrin loaded chitosan nanoparticles for transdermal delivery, *Carbohydr. Polym.*, **90**(3), 1244–1253.

Khan, T. A., Peh, K. K., Ch'ng, H. S. (2002). Reporting degree of deacetylation values of chitosan: The influence of analytical methods, *J. Pharm. Sci.*, **5**, 205–212.

Kim, J. Y., Choi, W. I., Kim, Y. H., Tae, G. (2013). Brain-targeted delivery of protein using chitosan-and RVG peptide-conjugated, pluronic-based nano-carrier, *Biomaterials*, **34**(4), 1170–1178.

Kim, K. K., Pack, D. W. (2006). Microspheres for drug delivery. In *BioMEMS and Biomedical Nanotechnology* (pp. 19–50). Springer US.

Kim, S. K., ed. (2013). *Chitin and Chitosan Derivatives: Advances in Drug Discovery and Developments*. CRC Press.

Kofuji, K., Qian, C. J., Nishimura, M., Sugiyama, I., Murata, Y., Kawashima, S. (2005). Relationship between physicochemical characteristics and functional properties of chitosan, *Eur. Polym. J.*, **41**, 2784–2791.

Kumar, M. R., Muzzarelli, R., Muzzarelli, C., Sashiwa, H., Domb, A. J. (2004). Chitosan chemistry and pharmaceutical perspectives, *Chem. Rev.*, **104**(12), 6017–6084.

Kurita, K. (1995). Chemistry and application of chitin and chitosan, *Polym. Degrad. Stab.*, **59**, 117–120.

Labrie, F., Labrie, C., Pelletier, G., El-Alfy, M. (2001). Intracrinology and the skin, *Horm. Res. Paediatr.*, **54**(5–6), 218–229.

Lam, P. L., Lee, K. K. H., Wong, R. S. M., Cheng, G. Y. M., Cheng, S. Y., Yuen, M. C. W., Lam, K. H., Gambari, R., Kok, S. H. L., Chui, C. H. (2012). Development of hydrocortisone succinic acid/and 5-fluorouracil/chitosan microcapsules for oral and topical drug deliveries, *Bioorg. Med. Chem. Lett.*, **22**, 3213–3218.

Lee, J. S., Feijen, J. (2012). Polymersomes for drug delivery: Design, formation and characterization, *J. Control. Release*, **161**(2), 473–483.

Lee, P. W., Hsu, S. H., Tsai, J. S., Chen, F. R., Huang, P. J., Ke, C. J., Liao, Z. X., Hsiao, C. W., Lin, H. J., Sung, H. W. (2010). Multifunctional core-shell polymeric nanoparticles for transdermal DNA delivery and epidermal Langerhans cells tracking, *Biomaterials*, **31**(8), 2425–2434.

Lee, P. W., Peng, S. F., Su, C. J., Mi, F. L., Chen, H. L., Wei, M. C., Lin, H. J., Sung, H. W. (2008). The use of biodegradable polymeric nanoparticles in combination with a low-pressure gene gun for transdermal DNA delivery, *Biomaterials*, **29**(6), 742–751.

Lehman, P. A., Slattery, J. T., Franz, T. J. (1988). Percutaneous absorption of retinoids: Influence of vehicle, light exposure, and dose, *J. Invest. Dermatol.*, **91**, 56–61.

Lehr, C. M., Bowstra, J. A., Schacht, E. H., Juginger, H. E. (1992). In vitro evaluation of mucoahesive properties of chitosan and some others natural polymers, *Int. J. Pharm.*, **78**, 43–48.

Li, F., Liu, W. G., Yao, K. D. (2002). Preparation of oxidized glucose-crosslinked Nalkylated chitosan membrane and in vitro studies of pH-sensitive drug delivery behavior, *Biomaterials*, **23**, 343–347.

Li, P., Wang, Y., Peng, Z., She, F., Kong, L. (2011). Development of chitosan nanoparticles as drug delivery systems for 5-fluorouracil and leucovorin blends, *Carbohydr. Polym.*, **85**(3), 698–704.

Lingling, L., Zhang, Y., Han, S., Qu, Z., Zhao, J., Chen, Y., Chen, Z., Duan, J., Pan, Y., Tang, X. (2011). Penetration enhancement of lidocaine hydrochlorid by a novel chitosan coated elastic liposome for transdermal drug delivery, *J. Biomed. Nanotechnol.*, **7**(50), 704–713.

Lira, A. A. M., Rossetti, F. C., Nanclares, D. M., Neto, A. F., Bentley, M. V. L., Marchetti, J. M. (2009). Preparation and characterization of chitosan-treated alginate microparticles incorporating all-trans retinoic acid, *J. Microencapsul.*, **26**(3), 243–250.

Liu, X. F., Guan, Y. L., Yang, D. Z., Li, Z., Yao, K. D. (2001). Antibacterial action of chitosan and carboxymethylated chitosan, *J. Appl. Polym. Sci. Symp.*, **79**, 1324–1335.

Liu, Z., Robinson, J. T., Tabakman, S. M., Yang, K., Dai, H. (2011). Carbon materials for drug delivery & cancer therapy, *Mater. Today*, **14**(7), 316–323.

Lövestam, G., Rauscher, H., Roebben, G., Klüttgen, S., Gibson, N., Putaud, J. P., Stamm, H. (2010). Considerations on a definition of nanomaterial for regulatory purposes. In (Centre, J. R., ed.), *JCR Reference Reports, Luxembourg*, 40.

Madhumathi, K., Sudheesh Kumar, P. T., Kavya, K. C., Furuike, T., Tamura, H., Nair, S. V. (2009). Novel chitin/nanosilica composite scaffolds for bone tissue engineering applications, *Int. J. Biol. Macromol.*, **45**, 289–292.

Menon, G. K. (2002). New insights into skin structure: Scratching the surface, *Adv. Drug Deliv. Rev.*, **54** (1), S3–S17.

Merkel, O. M., Zheng, M., Debus, H., Kissel, T. (2012). Pulmonary gene delivery using polymeric nonviral vectors, *Bioconjug. Chem.*, **23**, 3–20.

Mi, F. L., Wu, Y. B., Shyu, S. S., Chao, A. C., Lai, J. Y., Su, C. C. (2003). Asymmetric chitosan membranes prepared by dry/wet phase separation: A new type of wound dressing for controlled antibacterial release, *J. Membr. Sci.*, **212**(1), 237–254.

Minagawa, T., Okamura, Y., Shigemasa, Y., Minami, S., Okamoto, Y. (2007). Effects of molecular weight and deacetylation degree of chitin/chitosan on wound healing, *Carbohydr. Polym.*, **67**(4), 640–644.

Mourya, V. K., Inamdar, N. N. (2008). Chitosan-modifications and applications: Opportunities galore, *React. Funct. Polym.*, **68**, 1013–1051.

Mudshinge, S. R., Deore, A. B., Chetan, S. P., Bhalgat, C. M. (2011). Nanoparticles: Emerging carriers for drug delivery, *Saudi Pharm. J.*, **19**, 129–141.

Murakami, T., Sasaki, Y., Yamajo, R., Yata, N. (1984). Effect of bile salts on the rectal absorption of sodium ampicillin in rats, *Chem. Pharm. Bull.*, **32**(5), 1948–1955.

Murakami, T., Yata, N., Tamauchi, H., Kamada, A. (1982). Studies of absorption promoters for rectal delivery preparations. II. A possible mechanism of promoting efficacy of enamine derivatives in rectal absorption, *Chem. Pharm. Bull.*, **30**(2), 659–665.

Nagahama, H., New, N., Jayakumar, R., Koiwa, S., Furuike, T., Tamura, H. (2008). Novel biodegradable chitin membranes for tissue engineering applications, *Carbohydr. Polym.*, **73**, 295–302.

Ozbas-Turan, S., Akbuga, J., Sezer, A. D. (2010). Topical application of antisense oligonucleotide-loaded chitosan nanoparticles to rats, *Oligonucleotides*, **20**, 147–53.

Özcan, Ý., Abacý, Ö., Uztan, A. H., Aksu, B., Boyacýoðlu, H., Güneri, T., Özer, Ö. (2009). Enhanced topical delivery of terbinafine hydrochloride with chitosan hydrogels, *AAPS PharmSciTech*, **10**(3), 1024–1031.

Pan, Y., Wu, T., Bao, H., Li, L. (2011). Green fabrication of chitosan films reinforced with parallel aligned graphene oxide, *Carbohydr. Polym.*, **83**(4), 1908–1915.

Patashnik, S., Rabinovich, L., Golomb, G. (1997). Preparation and evaluation of chitosan microspheres containing bisphosphonates, *J. Drug Target.*, **4**(6), 371–380.

Peter, M., Binulal, N. S., Nair, S. V., Selvamurugan, N., Tamura, H., Jayakumar, R. (2010). Novel biodegradable chitosan-gelatin/nano-bioactive glass ceramic composite scaffolds for alveolar bone tissue engineering, *Chem. Eng. J.*, **158**, 353–361.

Pillai, C. K. S., Paul, W., Sharma, C. P. (2009). Chitin and chitosan polymers: Chemistry, solubility and fiber formation, *Prog. Polym. Sci.*, **34**(7), 641–678.

Prabaharan, M. (2008). Review paper: Chitosan derivatives as promising materials for controlled drug delivery, *J. Biomater. Appl.*, **23**, 5–36.

Raabe, D., Al-Sawalmih, A., Yi, S. B., Fabritius, H. (2007). Preferred crystallographic texture of chitin as a microscopic and macroscopic design principle of the exoskeleton of the lobsterHomarus americanus, *Acta Biomater.*, **3**, 882–895.

Rajan, M., Raj, V., Al-Arfaj, A. A., Murugan, A. M. (2013). Hyaluronidase enzyme core-5-fluorouracil-loaded chitosan-PEG-gelatin polymer nanocomposites as targeted and controlled drug delivery vehicles, *Int. J. Pharm.*, **453**(2), 514–522.

Rana, V., Babita, K., Goyal, D., Tiwary, A. (2005). Sodium citrate cross-linked chitosan films: Optimization as substitute for human/rat/rabbit epidermal sheets, *J. Pharm. Pharm. Sci.*, **8**, 10–17.

Ranaldi, G., Marigliano, I., Vespignani, I., Perozzi, G., Sambuy, Y. (2002). The effect of chitosan and other polycations on tight junction permeability in the human intestinal Caco-2 cell line, *J. Nutr. Biochem.*, **13**(3), 157–167.

Rejinold, N. S., Chennazhi, K. P., Nair, S. V., Tamura, H., Jayakumar, R. (2011). Biodegradable and thermo-sensitive chitosan-g-poly (N-vinylcaprolactam) nanoparticles as a 5-fluorouracil carrier, *Carbohydr. Polym.*, **83**(2), 776–786.

Rinaudo, M. (2008). Main properties and current applications of some polysaccharides as biomaterials, *Polym. Int.*, **57**, 397–430.

Roberts, M. S., Cross, S. E., Pellett, M. A. (2002). Skin transport. In *Dermatological and Transdermal Formulations* (Walters, K. A., ed.), Marcel Dekker Inc. New York, USA. 89–195.

Roldo, M., Hornof, M., Caliceti, P., Bernkop-Schnurch, A. (2004). Mucoadhesive thiolated chitosans as platforms for oral controlled drug delivery: Synthesis and in vitro evaluation, *Eur. J. Pharm. Biopharm.*, **57**, 115–121.

Sabnis, S., Rege, P., Block, L. H. (1997). Use of chitosan in compressed tablets of diclofenac sodium: Inhibition of drug release in an acidic environment, *Pharm. Dev. Technol.*, **2**(3), 243–255.

Sarmento, B., das Neves, J. (eds.). (2012). *Chitosan-Based Systems for Biopharmaceuticals: Delivery, Targeting and Polymer Therapeutics.* John Wiley & Sons.

Schiffman, J. D., Schauer, C. L. (2007). Cross-linking chitosan nanofibers, *Biomacromolecules*, **8**, 594–601.

Schipper, N. G., Olsson, S., Hoogstraate, J. A., deBoer, A. G., Varum, K. M., Artursson, P. (1997). Chitosans as absorption enhancers for poorly absorbable drugs 2: Mechanismof absorption enhancement, *Pharm. Res.*, **14**, 923–929.

Schipper, N. G., Varum, K. M., Artursson, P. (1996). Chitosans as absorption enhancers for poorly absorbable drugs. 1: Influence of molecular weight and degree of acetylation on drug transport across human intestinal epithelial (Caco-2) cells, *Pharm. Res.*, **13**, 1686–1692.

Senyiğit, T., Sonvico, F., Barbieri, S., Özer, Ö., Santi, P., Colombo, P. (2010). Lecithin/chitosan nanoparticles of clobetasol-17-propionate capable of accumulation in pig skin, *J. Control. Release*, **142**(3), 368–373.

Shalumon, K. T., Binulal, N. S., Selvamurugan, N., Nair, S. V., Deepthy, M., Furuike, T. (2009). Electrospinning of carboxymethyl chitin/poly (vinyl alcohol) nanofibrous scaffolds for tissue engineering applications, *Carbohydr. Polym.*, **77**, 863–869.

Shin, J., Shin, K., Lee, H., Suh, K. D., Kim, H., Kim, J. W. (2012). Enhanced transdermal delivery by using electrostatically interactive chitosan nanocapsules, *Colloid Polym. Sci.*, **290**(6), 553–559.

Sinha, V. R., Singla, A. K., Wadhawan, S., Kaushik, R., Kumria, R., Bansal, K., Dhawan, S. (2004). Chitosan microspheres as a potential carrier for drugs, *Int. J. Pharm.*, **274**(1), 1–33.

Sinswat, P., Tengamnuay, P. (2003). Enhancing effect of chitosan on nasal absorption of salmon calcitonin in rats: Comparison with hydroxypropyl- and dimethyl—cyclodextrins, *Int. J. Pharm.*, **257**, 15–22.

Smith, J., Wood, E., Dornish, M. (2004). Effect of chitosan on epithelial cell tight junctions, *Pharm. Res.*, **21**(1), 43–49.

Soane, R. J., Hinchcliffe, M., Davis, S. S., Illum, L. (2001). Clearance characteristics of chitosan based formulations in the sheep nasal cavity, *Int. J. Pharm.*, **217**, 183–191.

Suedee, R., Bodhibukkana, C., Tangthong, N., Amnuaikit, C., Kaewnopparat, S., Srichana, T. (2008). Development of a reservoir-type transdermal enantioselective-controlled delivery system for racemic propranolol using a molecularly imprinted polymer composite membrane, *J. Control. Release*, **129**, 170–178.

Tan, Q., Liu, W., Guo, C., Zhai, G. (2011). Preparation and evaluation of quercetin-loaded lecithin-chitosan nanoparticles for topical delivery, *Int. J. Nanomed.*, **6**, 1621.

Taveira, S. F., Nomizo, A., Lopez, R. F. V. (2009). Effect of the iontophoresis of a chitosan gel on doxorubicin skin penetration and cytotoxicity, *J. Control. Release*, **134**, 35–40.

Tengamnuay, S. P., Mitra, A. K. (1997). Chitosans as nasal absorption enhancers of salmon calcitonin: Comparison with hydroxypropyl-and dimethyl-b-cyclodextrins, *Pharm. Res.*, **14**, S129.

Thacharodi, D., Rao, K. P. (1996a). Rate-controlling biopolymer membranes as transdermal delivery systems for nifedipine: Development and in vitro evaluations, *Biomaterials*, **17**(13), 1307–1311.

Thacharodi, D., Rao, K. P. (1996b). Collagen-chitosan composite membranes controlled transdermal delivery of nifedipine and propranolol hydrochloride, *Int. J. Pharm.*, **134**(1), 239–241.

Thanoo, B. C., Sunny, M. C., Jayakrishnan, A. (1992). Cross-linked chitosan microspheres: Preparation and evaluation as a matrix for the controlled release of pharmaceuticals, *J. Pharm. Pharmacol.*, **44**, 283–286.

Thanou, M. M., Florea, B. I., Langemeyer, M. W. E., Verhoef, J. C., Junginger, H. E. (2000). N-trimethylated chitosan chloride (TMC) improves the intestinal permeation of the peptide drug buserelin in vitro (Caco-2) and in vivo (rats), *Pharm. Res.*, **17**, 27–31.

Thanou, M. M., Verhoef, J. C., Junginger, H. E. (2001). Oral drug absorption enhancement by chitosan and its derivatives, *Adv. Drug Deliv. Res.*, **52**, 117–126.

Thomas, B. J., Finnin, B. C. (2004). The transdermal revolution, *Drug Discov. Today*, **9**, 697–703.

Tozaki, M., Emi, Y., Horisaka, E., Fujita, T., Yamamoto, A., Muranishi, S. (1997). Degradation of insulin and calcitonin and their protection by various protease inhibitors in rat ceacal contents: Implications in peptide delivery to the colon, *J. Pharm. Pharmacol.*, **49**, 164–168.

Tsai, C. J., Hsu, L. R., Fang, J. Y., Lin, H. H. (1999). Chitosan hydrogel as a base for transdermal delivery of berberine and its evaluation in rat skin, *Biol. Pharm. Bull.*, **22**, 397–401.

Van Der Lubben, I. M., Kersten, G., Fretz, M. M., Beuvery, C., Verhoef, J. C., Junginger, H. E. (2003). Chitosan microparticles for mucosal vaccination against diphtheria: Oral and nasal efficacy studies in mice, *Vaccine*, **21**, 1400–1408.

Vincent, J. F. V., Wegst, U. G. K. (2000). Design and mechanical properties of insect cuticle. *Arthropod. Struct. Dev.*, **46**, 187–199.

Vyas, T. K., Shahiwala, A., Marathe, S., Misra, A. (2005). Intranasal drug delivery for brain targeting, *Curr. Drug Deliv.*, **2**(2), 165–175.

Wahid, A., Sridhar, B. K., Shivakumar, S. (2008). Preparation and evaluation of transdermal drug delivery system of etoricoxib using modified chitosan, *Indian J. Pharm. Sci.*, **70**(4), 455–460.

Walters, K. A., Roberts, M. S. (2002). The structure and function of skin. In *Dermatological and Transdermal Formulations*, Marcel Dekker Inc., New York, USA. 1–39.

Wang, Y. Y., Hong, C. T., Chiu, W. T., Fang, J. Y. (2001). In vitro and in vivo evaluations of topically applied capsaicin and nonivamide from hydrogels, *Int. J. Pharm.*, **224**(1), 89–104.

Wang, Y., Su, W., Li, Q., Li, C., Wang, H., Li, Y., Cao, Y., Chang, J., Zhang, L. (2013). Preparation and evaluation of lidocaine hydrochloride-loaded TAT-conjugated polymeric liposomes for transdermal delivery, *Int. J. Pharm.*, **441**(1), 748–756.

Wen, H., Xianxi, G., Lihai, X., Feng, M. (2009). Study on the mechanisms of chitosan and its derivatives used as transdermal penetration enhancers, *Int. J. Pharm.*, **382**, 234–243.

Wisuitiprot, W., Somsiri, A., Ingkaninan, K., Waranuch, N. (2011). In vitro human skin permeation and cutaneous metabolism of catechins from green tea extract and green tea extract-loaded chitosan microparticles, *Int. J. Cosmetic. Sci.*, **33**(6), 572–579.

www.Healthfavo.com/detailed-skin-diagram.html

Wong, T. W. (2009). Chitosan and its use in design of insulin delivery system, *Recent Patents Drug Deliv. Formulation*, **3**(1), 8–25.

Wong, T. W., Chan, L. W., Kho, S. B., Heng, P. W. S. (2002). Design of controlledrelease solid dosage forms of alginate and chitosan using microwave, *J. Control. Release*, **84**(3), 99–114.

Wong, T. W., Khaizan, A. N. (2013). Physicochemical modulation of skin barrier by microwave for transdermal drug delivery, *Pharm. Res.*, **30**(1), 90–103.

Wong, T. W., Nurjaya, S. (2008). Drug release property of chitosan–pectinate beads and its changes under the influence of microwave, *Eur. J. Pharm. Biopharm.*, **69**(1), 176–188.

Wong, T. W., Nurulaini, H. (2012). Sustained-release alginate-chitosan pellets prepared by melt pelletization technique, *Drug Dev. Ind. Pharm.*, **38**(12), 1417–1427.

Xu, Y., Du, Y. (2003). Effect of molecular structure of chitosan on protein delivery, *Int. J. Pharm.*, **250**, 215–226.

Yeh, T. H., Hsu, L. W., Tseng, M. T., Lee, P. L., Sonjae, K., Ho, Y. C., Sung, H. W. (2011). Mechanism and consequence of chitosan-mediated reversible epithelial tight junction opening, *Biomaterials*, **32**(26), 6164–6173.

Yenilmez, E., Başaran, E., Yazan, Y. (2011). Release characteristics of vitamin E incorporated chitosan microspheres and in vitro–in vivo evaluation for topical application, *Carbohydr. Polym.*, **84**, 807–811.

Yi, H., Wu, L. Q., Bentley, W. E., Ghodssi, R., Rubloff, G. W., Culver, J. N., Payne, G. F. (2005). Biofabrication with chitosan, *Biomacromolecules*, **6**, 2881–2894.

Yuen, C. W. M., Yip, J., Liu, L., Cheuk, K., Kan, C. W., Cheung, H. C., Cheng, S. Y. (2012). Chitosan microcapsules loaded with either miconazole nitrate or clotrimazole, prepared via emulsion technique, *Carbohydr. Polym.*, **89**, 795–801.

Zhang, H., Neau, S. H. (2001). In vitro degradation of chitosan by a commercial enzyme preparation: Effect of molecular weight and degree of deacetylation, *Biomaterials*, **22**, 1653–1658.

Zhu, L., Ma, J., Jia, N., Zhao, Y., Shen, H. (2009). Chitosan-coated magnetic nanoparticles as carriers of 5-fluorouracil: Preparation, characterization and cytotoxicity studies, *Colloids Surf. B*, **68**(1), 1–6.

Zohuriaan-Mehr, M. J. (2005). Advances in chitin and chitosan modification through graft copolymerization: A comprehensive review, *Iran Polym. J.*, **14**, 235–265.

Chapter 17

Chiral Electrochemical Sensors Based on Molecularly Imprinted Polymers with Pharmaceutical Applications

Bogdan-Cezar Iacob, Ede Bodoki, and Radu Oprean

Analytical Chemistry Department, "Iuliu Hatieganu" University of Medicine and Pharmacy, 4 Louis Pasteur St., 400349, Cluj-Napoca, Romania

bodokie@umfcluj.ro

17.1 Introduction

There is a noticeable increase in the use of biopolymers and synthetic polymers in medicine and pharmaceutical analysis as drug delivery systems, therapeutic systems, macromolecular prodrugs, as biomimetic receptors, and as recognition element and as stationary phase in chromatographic and chiral separations [1–3]. Because of their appealing extraction capacity, electrochemical properties, large modifiable surfaces and multiple active sites, it is easily understandable the focus toward developing sensing devices based on nanostructured polymeric preconcentration and recognition elements. A highly selective interface is essential in creating electrochemical sensors capable of detecting the target molecule in a complex matrix, even in the presence of close-related structural analogues of the analyte. Generally, the interface

Handbook of Sustainable Polymers: Processing and Applications
Edited by Vijay Kumar Thakur and Manju Kumari Thakur

ISBN 978-981-4613-53-8 (Hardcover), 978-981-4613-54-5 (eBook)
www.panstanford.com

is represented by synthetic chemical compounds able to interact by nonselective chemical bonds with the analyte of interest but also with chemical compounds that possess same chemical functionalities as the analyte. An alternative increasingly employed in electrochemistry is represented by a class of materials whose selectivity can be directed in the fabrication process toward the target molecule. These materials, molecularly imprinted polymers (MIPs), although they were exploited in electrochemistry for quite some time, only recently have begun to be used as the recognition element in the electrochemical sensors for the chiral discrimination of racemic compounds.

In this chapter, we present the evolution in the development of MIP-based chiral electrochemical sensors and also discuss the impediments encountered in the imprinting process of one of the enantiomers in the polymer. The nature of the polymers and their electrical properties are also addressed. An important part of the chapter is dedicated to the analytical applications of these sensors.

17.2 Molecular Imprinting Approach

Molecular recognition is fundamental to almost every biological process, such as ligand–receptor binding, substrate-enzyme reactions, translation and transcription of the genetic code, being therefore of universal interest for life sciences. Over the past four decades, researchers have been inspired by nature to produce biomimetic materials with molecular recognition properties using various techniques and procedures, following a careful and rational design rather than by evolution. Although the technique of molecular imprinting originated since 1949 [4], it gained a widespread attention only in the last 20 years, becoming shortly a very versatile and valuable tool in the hands of analytical specialists. This higher degree of selectivity compared to conventional materials offers to MIPs an advantage in analytical chemistry.

One of the most important features of molecular imprinting is to provide a means for creating robust polymeric recognition materials with predetermined ligand selectivity [5–7]. Thus, the main application of MIPs turned out to be in the field of separation sciences, as stationary phases in (electro)chromatographic analysis and chiral separations and as selective sorbents in solid phase extractions. Other important areas in which MIPs have found their use are: chemical sensors [8], catalysis [9], drug delivery [10], alternatives to biological antibodies and receptor systems [11, 12].

The process of MIP creation is generally depicted as shown in Fig. 17.1. MIPs are obtained by adding to a functional monomer's solution a template (target) molecule, which in case of chiral analysis is represented by one of the enantiomers. After polymerization and extraction of the template, a cavity is left behind in the structure of the rigid polymer, which is complementary by shape and functionality to the template molecule. The key to the manufacture of MIPs with good binding properties resides in the optimization of synthesis parameters. The nature of functional monomer possessing appropriate substituents is very important because their non-covalent molecular interactions with the template are crucial for the imprinting process. The resulting MIPs can selectively recognize and rebind the imprint molecule and in some cases even bind with other closely related molecules. MIPs present obvious analytical advantages in comparison with macromolecules of biological origin in terms of their availability, ease of preparation and fine-tuning of their properties, rigidity of matrix and low cost.

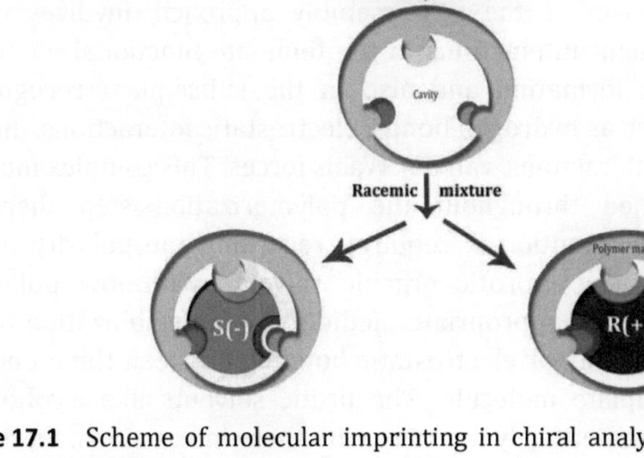

Figure 17.1 Scheme of molecular imprinting in chiral analysis using the noncovalent approach.

The versatility of molecular imprinting resides in the fact that the template may be any substance, ranging from small molecules such as amino acids, pharmaceutical compounds, metal ions to larger molecules (proteins) or even cells and viruses. However, along with the increase in the size of the template, its removal from the formed three-dimensional polymeric structure is becoming more and more difficult, with the risk of being permanently stuck inside the polymer. This steric hindrance also prevents the access of the target molecule in the imprinted cavities for the rebinding process [13].

Based on the occurring interactions during the functional monomer–template complex formation and template rebinding, three approaches regarding molecular imprinting are reported:

(a) *Covalent Imprinting*

The covalent approach or the pre-organized approach implies the formation of a template-functional monomer complex through reversible covalent bonds prior to polymerization. After synthesis and removal of the template, in the subsequent rebinding step, the initial covalent linkage is reconstituted between the polymer and template. Therefore, only a low number of non-selective binding sites are expected to be formed because of the well-defined stoichiometry taking place between the functional monomer and template. Unfortunately, this approach is only applicable to a limited number of template molecules.

(b) *Noncovalent Imprinting*

Also known as the self-assembly approach, involves weak, noncovalent interactions in the template-functional monomer complex formation and also in the subsequent recognition step, such as hydrogen bonds, electrostatic interactions, dipole-dipole interactions, van der Waals forces. This complex must be maintained throughout the polymerization step therefore a special attention is required regarding the polarity of the solvent used. Aprotic organic solvents with low polarities represent the appropriate medium for the stabilization of the hydrogen and/or electrostatic bonding between the monomer and template molecule. The protic solvents like alcohols or water competitively interfere with these interactions; therefore, they should be avoided. Usually, during the polymerization, a large excess of functional monomer compared to the template is required in order to favor the formation of template-monomer assemblies. In fact, the association between the monomer and the template is governed by Le Chatelier's principle; therefore increasing the concentration of components shifts the equilibrium toward complex formation. However, this excess leads also to the formation of non-specific binding sites responsible for the heterogeneous binding site distribution observed in these MIPs.

(c) *Semi-covalent Imprinting*

The semi-covalent approach combines the advantages of the previous two methods, employing reversible covalent bonds in the imprinting step and noncovalent interactions in the recognition process, after the cleavage of the template from the polymer.

In the fabrication process of MIPs, the first step represents the choice of proper imprinting approach. This selection is performed depending on the nature of the used template molecule and the potential linkages that can be established between the functional groups on the target molecule and on the functional monomer. The noncovalent approach has the advantages of being applicable to almost any type of template and from the synthesis point of view, it is far simpler than the covalent protocol, being the most widely adopted technique in the MIP preparation. Due to the weak interactions on which this approach is relying, it allows rapid and reversible bindings, most needed in chromatographic separations. Even though the polymers obtained by this approach present a higher degree of heterogeneity in the binding sites and therefore higher degree of nonspecific interactions, the latter being considered a limiting factor in chiral analysis, nonetheless, because of its simple and straightforward preparation protocol it was imported also in the MIP-based sensor development.

Although the practicalities of MIP fabrication are simple (directly mixing of template with all the other polymerization mixture components), there are numerous factors influencing the process of molecular imprinting that need rigorous control. A typical polymerization mixture is made up in a solvent in which four types of substances are dissolved: one or several functional monomers, template(s), initiator, and cross-linking agent(s). The nature of these components, as well as their ratio, plays a crucial role in the imprinting process, along with the initiation type, temperature, and duration of polymerization. Unfortunately, the optimization of all these variables most often was achieved empirically following tedious and frequently inefficient protocols by keeping one component constant at a time. However, nowadays having many chemometric optimization tools (design of experiments) at disposal, they are starting to be employed in the selection of suitable monomers, as well as for the fine-tuning of appropriate ratios between the polymerization mixture's constituents [14, 15].

Based on their chemical composition, two major types of polymeric matrices can be distinguished: organic and inorganic based-MIPs. Although the first system on which the molecular imprinting effect was discovered was based on inorganic silica [4], the polymers of organic nature experienced the largest spread and popularity: styrene [16] and acrylate [17–22] based-MIPs. The vast majority of publications deal with the acrylate based-polymers, employing mostly the same functional monomer, methacrylic acid (MAA) and the same cross-linker agent, ethylene glycol dimethacrylate (EDMA) [23–27]. This combination of monomers is the most popular nowadays in applications involving separation techniques (capillary electrochromatography, liquid chromatography), leading to a good imprinting effect, even though other monomers have been proven their superiority in the imprinting process. This particular type of monomer combination was transferred also to the fabrication of MIP-based electrochemical sensors, although acrylate polymers present no electrical properties. Thus, it appears that researchers, as an act of convenience, prefer to broaden the range of applications of a previously described functional polymer composition, rather than to test other (novel) structural and functional monomers.

On the other hand, to obtain and maintain the memory of the template in the polymeric network, a high degree of cross-linking is needed, resulting in a rigid organic matrix. Unfortunately, only 15% of the cavities can rebind the template and the remaining 85% are collapsing after the elimination of template, never returning to their initial state [28]. In order to favor a facile elution and rebinding of the template in the cavities, a high level of polymer flexibility is required. Thus, for the synthesis of MIPs with high imprinting factor and maximum efficiency a well-chosen balance between polymer rigidity–flexibility has to be achieved.

17.3 Chiral Electrochemical Sensors

Chirality is a major concern in the modern pharmaceutical industry. The human body is highly chiral selective, in which optically active endogenous (bio)molecules are usually present in only one of their chiral forms, which represents also the basis of one theory regarding the origin of life [29]. For example, human proteins are built exclusively by L-amino acids, whereas glucose, having five asymmetric carbon atoms, occurs biologically in only one of its chiral isomer. This quintessential nature of life is a consequence of the fact that the enzymes that synthesize these chiral biomolecules are also chiral. Furthermore, enantiomers of pharmaceutically active compounds, entering the body are metabolized by separate pathways to produce different pharmacological activities. This is largely attributed to the current awareness that enantiomers of a racemic drug placed in a chiral environment such as a living organism act as completely different molecules, frequently manifesting different pharmacological activities, as well as different pharmacokinetic and pharmacodynamic effects. It is also worth mentioning their economic impact, since nowadays chiral drugs comprise more than half the drugs approved for therapy worldwide, including many of the top-selling ones. In 2008, from the total drugs approved by FDA, 63% were single enantiomers, 32% achiral drugs and only 5% racemates [30]. Single enantiomer drugs had sales of $225 billion in 2005, representing 37% of the total pharmaceutical market of $602 billion based on estimates from Technology Catalysts International and IMS Health [31]. As a result of advances in chemical technologies associated with the synthesis, separation and analysis of the individual enantiomers of a racemate, seconded by recent international regulatory requirements in the pharmaceutical field, the number of chiral drugs marketed as single enantiomers is continuously growing, leading to an entire process called the "chiral switch" [32, 33].

Thus, chiral analysis is progressively becoming an organic part of the pharmaceutical analysis, where MIPs may play an important part.

One of the alternatives to chromatographic methods in chiral analysis, due to the simpler design, manufacturing, and maintenance, is the use of chiral electrochemical sensors. The enantioselectivity of these sensors can be based on various principles of molecular recognition of the target analytes on a chiral sensing interface. They may include the use of quasi-chiral multiwalled carbon nanotubes (MWCNT)/ionic liquids composites [34], chiral ionophores [35], natural biochemical recognition mechanisms (e.g., chiral enzymatic biosensors or immunosensors) [36], or they can employ biomimicking mechanisms with synthetic selectors (e.g., chiral sensors based on MIPs) [37]. Generally, the electrochemical discrimination between enantiomers is based on the difference in stability constants of diastereoisomeric complexes formed with the chiral sensing interface. The basic difference and in the meantime the biggest challenge in chiral sensing in comparison with chiral separations comes from the fact that in case of the direct chiral sensing devices there is only one single unitary process of separation that corresponds to one theoretical plate in chromatographic separations [38]. Nevertheless, electrochemical sensors became increasingly employed for detecting different ionic and/or molecular species because of the possibility of their reproducible and low cost mass production, with great potential for single use analysis, avoiding cross-contamination especially in case of biological samples.

17.4 Chiral Electrochemical MIP-Based Sensors

Nature always amazed researchers related to its power of molecular recognition, so they endeavored to mimic it. Thus, molecular imprinting emerged as a technique that it is capable of creating synthetic receptors with comparable recognition properties to the biological systems [39], but with an excellent chemical and thermal stability. MIPs present also obvious advantages over recognition biomolecules, such as antibodies, natural receptors and enzymes, in terms of their availability, ease of preparation and fine-tuning of their properties, rigidity of matrix and low cost.

Although molecular imprinting has more than six decades of history [4], MIPs were introduced in electrochemistry only in

1990s by Mosbach and coworkers [40, 41]. Since then, important progress was made in the field of MIP-based electrochemical sensors. However, even today, the use of MIPs in the electroanalytical methods is still very limited, and only a very small number of publications deal with the preparation of enantioselective electrochemical sensors, leaving plenty of room for novelty and innovation in a fairly immature high-tech field. Moreover, amongst the small number of papers reporting MIP-based enantioselective electrochemical sensors, the vast majority of them deals with the analysis of chiral amino-acids [42–45] or monosaccharides [46]. Yet, the situation is different in most of the cases of chiral sensors developed for pharmaceutically active compounds in which the chiral center is located more inwards of the molecule and the functional groups that are involved in the functional monomer-template complexes are not directly bound to the asymmetric carbon. In such electrochemical applications, a certain degree of selectivity toward the template enantiomer was observed, but absolute enantioselectivity was never achieved [47, 48]. Using MIPs as (pseudo)stationary phases in separation techniques (like high performance liquid chromatography, capillary electrochromatography), where up to several hundred thousand theoretical plates are achieved, enantioselectivity is hardly an issue, and the chiral separation of such molecules is easily accomplished most often with very high resolutions.

17.4.1 MIP Anchoring Onto the Transducer

In order to be incorporated into an electrochemical sensor as the chiral recognition element, MIPs must be intimately integrated with a physicochemical transducer (i.e., the surface of the electrode). In this case, the role of the MIP is to specifically bind the target molecule from the sample, whereas the signal generated by the oxidation or reduction of analytes at the MIP-modified electrode is converted by the transducer into a quantifiable signal. Probably the most important obstacle in the MIP expansion in electrochemical field is the unestablished methodology for transducing specific binding events into electric signals [49].

One convenient MIP integration strategy into an electrochemical sensor is by incorporating small MIP particles into a conductive material (e.g., carbon paste). MIP particles are prepared in advance, mostly by bulk polymerization, resulting in blocks of polymers that are grounded and sieved into particles of desired size. This method gives particles of irregular size and shape and leads to an excessive loss of material. In order to obtain small and spherical particles, other methods were developed, like suspension polymerization, core–shell emulsion polymerization and dispersion/precipitation polymerization techniques. This type of MIP-based electrochemical sensors employed acrylate-based polymers exclusively for the non-chiral recognition and determination of different analytes [21, 22, 24, 50]. Although these sensors present the advantage of a facile renewal of the active surface just by polishing it on a white clean paper, they suffer from a poor homogeneity of MIP distribution in the conductive matrix and thus a low inter-assay reproducibility. Moreover, the majority of the MIP particles are found embedded in the matrix and only a small number are exposed at the surface of the electrode, rendering available only a few specific recognition sites. Furthermore, problems such as bleeding or strong nonspecific adsorption interactions were observed when employing MIP particles as chemical receptors.

A more reproducible sensor modification that offers a larger coverage and a more intimate contact of the polymer with the electroactive area is achieved by grafting the MIP onto the surface of the solid electrode. By introducing a previously silanized indium-tin oxide (ITO) glass in the polymerization mixture, followed by a thermal initiation (60°C), a polymer block surrounding the ITO glass was obtained. After breaking the resulting polymer and the ultrasonication of the modified glass, a nanometer range thin layer [51] of acrylate-based MIP with chiral discrimination properties was obtained [49]. A more intimate contact between the polymeric recognition element and the electrode's electroactive surface is achieved by immobilizing the iniferter (initiator-transfer agent-terminator) onto the sensor's surface, also known as the "grafting from" method. The polymerization mixture is then uniformly dispersed over the initiator-modified surface by spin coating, followed by an UV [52] or temperature (60°C) [45] initiated free-radical polymerization. This approach allows a polymer film growth of nanometer-range controlled thickness due to the fact that polymerization takes place only at the surface of the electrode.

Compared to the "grafting to" approach, the "grafting from" method presents an evident superiority, offering a more homogenous MIP coverage of the sensor's surface with a larger number of cavities and a 100-fold faster diffusion of analyte [45]. However, the initiator's immobilization directly onto the surface is usually complex and chemically not tenable [53].

The low degree of attachment because of the steric hindrance offered by the "grafting to" technique was overcame by covalently bonding the chiral MIP film onto the sensor's surface through a vinyl functionalized ormosil (organically modified silica) monolayer [53]. A silver working electrode was used on the surface of which a silane self-assembled monolayer (SAM) was first attached through a thiol functional group, followed by functionalization of SAM with a vinyl silane. Three MIP layers were thermally cross-linked onto the sensor's modified surface after the spin coating of the prepolymerization mixture. The electrode-to-electrode reproducibility is questionable because of the numerous modification steps involved in the chiral sensor's fabrication, even though authors claim a RSD of 1.2% for three different electrodes used for the analysis of thyroxine in blood serum samples.

The problem related to MIP-transducer integration is eliminated preparing a chiral MIP monolithic composite by the polymerization of a MIP pre-polymerization mixture in which carbon particles were added, resulting a conductive MIP fiber used as a working electrode [37]. This approach involves a straight-forward sensor fabrication methodology, in which all components are mixed together in a solvent and the resulting mixture is poured into glass capillaries followed by thermal polymerization. The rather odd thing is that the majority of carbon particles were arranged as a strip along the entire length of the fiber and yet the composite showed very good electro-conductive properties. An alternative to the carbon powder addition, in order to induce electrical features to the MIP composite, is to employ conducting MIP polymers. Thus, a cylindrical electrode based on conductive stationary phase was fabricated using overoxidized polypyrrole (PPy) [54]. The polymerization and overoxidation of PPy were realized in the presence of the template, allowing the creation of complementary cavities to the template. The conductive poly(aniline-co-m-aminophenol) copolymer was also tested as a stationary phase filled into a porous ceramic column for enantioselective recognition [55].

Nevertheless, probably the most direct way of binding the chiral MIP to the electrode's surface is through electropolymerization [56]. Electropolymerization implies polymer formation on the electroactive surface upon oxidation or reduction of certain organic compounds. This procedure is also the most often used method in the fabrication of chiral MIP-based sensors due to the facile preparation of the MIP-film and the possibility of controlling polymer thickness and deposition density by easily changing the polymerization conditions (e.g., applied range of voltage and scan rate, number of voltammetric cycles, concentration of monomer, etc.). The most common type of electrochemical MIP formation is by means of cyclic voltammetry, which allows the growth of ultra-thin polymer films on the surface of the electrode; thus, a good adherence to the transducer surface is achieved in a simple and quick manner.

A L-histidine (His)-imprinted sol solution, consisting of tetraethylorthosilicate (TEOS), phenyltrimethoxysilane (PTMOS), and methyltrimethoxysilane (MTMOS), was electrodeposited onto a MWCNT-modified ITO glass by using cyclic voltammetry in the range of −0.8 V to +0.9 V [57]. This technique allowed the formation of a thin, uniform MIP film distributed regularly on the MWCNT-modified electrode surface. Electropolymerization was also employed in the preparation of chiral imprinted organic membranes onto gold electrodes or gold-coated quartz crystal electrodes [58]. They were achieved by cycling the potential between −0.4 and −1.4 V in an aqueous solution containing acrylamide, *N,N′*-methylenebisacrylamide, $NaNO_3$, $Na_2S_2O_8$, and L- or D-histidine. The electropolymerization of acrylamide in solution is possible only if electrolytes such as tartaric acid, zinc salts, perchlorate, and nitrate [59] are used.

Another type of chiral MIP sensor using conducting imprinted polymer, poly(indole-3-acetic acid) (PI3AA), was electrogenerated in phosphate buffer, in the presence of L- or D-aspartic acid, acting as template, on a MWCNT-coated pencil graphite electrode (PGE) by cycling the potential 12 times on a large interval, from −1.6 V to +1.6 V vs. Ag/AgCl (Fig. 17.2) [60]. A nano-MIP film of controlled thickness was achieved on the modified pencil tip, which was subsequently over-oxidized using a more anodic potential, up to

+2.0 V. During over-oxidation, dopant molecules (L- or D-aspartic acid) were de-doped from the polymeric network. Benzidine was

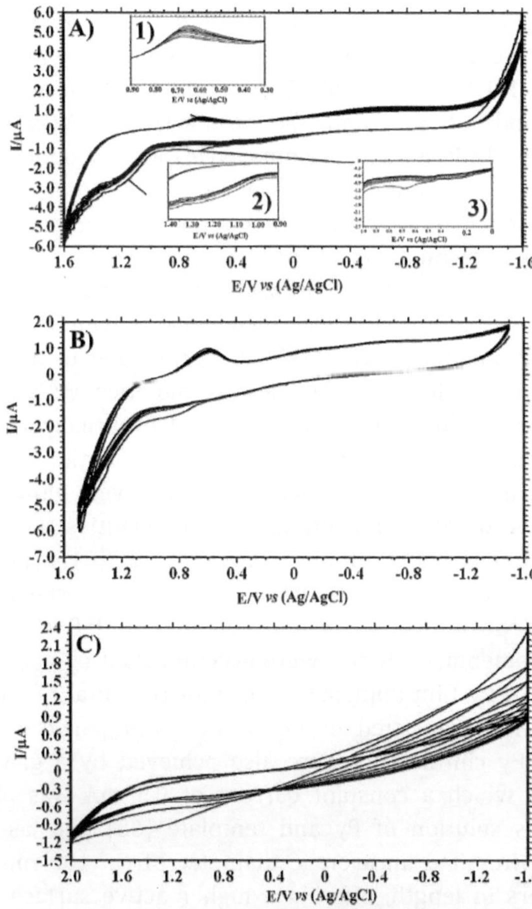

Figure 17.2 MIP deposition by electropolymerization using cyclic voltammetry: (a) MI-PI3AA-asp-adduct/MWCNTs-PGE [inset (1) reduction peak of monomer, inset (2) oxidation peak of aspartic acid, and inset (3) oxidation peak of monomer], (b) NI-PI3AA/MWCNTs-PGE, and (c) over-oxidized MI-PI3AA/MWCNTs-PGE [electropolymerization conditions; 0.05 mM l-asp, 0.1 mM PI3AA, 0.01 M phosphate buffer (supporting electrolyte pH 5.0), no. of scan cycles 12, potential range −1.6 to +1.6 V vs. Ag/AgCl, over-oxidation potential range −1.6 to +2.0 V, scan rate 100 mV s^{-1})]. Reproduced with permission from [60].

also used as an electro-active monomer in the electrochemical fabrication of an ultra-thin nanofilm of MIP for the chiral discrimination of methionine (MET) [61]. The electropolymerization was effectuated in an aqueous solution (pH = 2.3) over three consecutive cyclic voltammetric scans in the potential range of −1.0 and +1.0 V vs. Ag/AgCl. The resulted polymeric membrane showed good adhesion onto the carboxylated MWCNT- modified PGE surface. The PGE was also employed in the electropolymerization of copper(II)–5-methyl-thiophene–2-carboxylic acid complex, in the presence of D- or L-pyroglutamic acid, giving forth of a molecularly imprinted metallo-polymer [62].

One of the most utilized conducting polymers in electrochemistry for direct electrochemical polymerization is PPy. It was also adopted in the MIP fabrication due to its particular electrical properties and the various moieties with which the pyrrole ring could be functionalized and eventually could bring certain favorable features. With the over-oxidation of PPy, the positive charges from its structure are removed, along with the introduction of oxygen containing functionalities, i.e., carbonyl group [63]. Thin films of PPy were created in the presence of L-tryptophan (Trp) (template molecule) on a Pt working electrode by electrodeposition at a constant potential of +0.8 V vs. SCE [64]. The complementary cavities were accomplished by over-oxidizing the formed PPy film applying a constant potential of +1.0 V. The polymeric film consisted of uniformly sized particles of about 1–2 μm. PPy chiral MIPs were also achieved by a galvanostatic method in which a constant current of 0.3 mA was applied in an aqueous solution of Py and template [65]. The as-prepared PPy nanowires had an average diameter of ca. 100 nm and few micrometers in length, offering a higher active surface area and thus more enantiospecific sites compared to PPy films.

The acrylate polymers are well known for their electrical insulating properties, which limit the efficiency of the signal transduction, fact that restricted their use as active components of electrochemical sensors. However, there is a wide selection of acrylic monomers containing a variety of functional groups that eventually could improve the resulting polymer's electric properties. Moreover, these polymers were intensely studied and the underlying imprinting process is well known. On the other hand, conductive MIP polymers offer a good signal transduction process, reproducibility and a more controllable growth of the polymer [66]. As a downside, they have a more compact and rigid structure [67], features that may interfere with the imprinting effect. The advantages of these two approaches can be combined by assembling the conducting polymers with the conventional imprinted polymers into one MIP sensor. Thus, a chiral hybrid MIP sensor for the analysis of L-Phe was prepared by the electropolymerization of a polymer precursor, poly [2-(*N*-carbazolyl)ethyl-methacrylate-co-meth-acrylic acid] [66]. The insulating polymethacrylic acid played the role of functional monomer while the conducting polycarbazole was used as "cross-linker agent" offering a better grafting on the electrode surface.

Another hybrid MIP example is represented by the copolymerization of phenol with a covalently constructed complex between 3-hydroxyphenyl boronic acid (functional monomer) and monosaccharide (template). The copolymer was deposited on a gold electrode by cyclic voltammetry using two successive scans [46]. The obtained polymer presented an irregular morphology with an average thickness of 36 nm.

17.4.2 Chiral MIP Applications

Although the use of polymers in electrochemistry attracted a remarkable interest, scientists experienced some reluctance in employing imprinted polymers as sensing element in sensor development. Several drawbacks seem to be the limiting factor in the MIP expanding, like the low density of imprinted cavities, the electro-insulating properties of traditional MIPs and the slow diffusion of analyte. In case of chiral MIP-based sensors, the situation is even more austere. Because of the identical physicochemical properties of the two enantiomers, differing only in the three-dimensional geometry, a more specific receptor has to be developed in order to achieve chiral discrimination.

17.4.2.1 Amino acids

Various publications have already described MIP-modified sensors, which showed good chiral recognition toward the enantiomers of different amino acids. The feasibility of achieving electrosynthesized polyacrylamide membranes imprinted with L- or D-His was explored by using AC impedance spectroscopy and piezoelectric quartz crystal technique [58]. It was observed that the electron-transfer between the electrode and the $K_3[Fe(CN)_6]$/$K_4[Fe(CN)_6]$ (1:1) redox probe was facilitated after the interaction of MIP with both enantiomers of His due to the swelling of the polyacrylamide membrane. However, in the case of template enantiomer, the electron-transfer resistance decreased in a larger extent compared to the other enantiomer, with 72% and 40%, respectively. A more considerable difference was noticed in the case of Au-coated quartz crystals modified with L- or D-His. The modified crystals were also tested for the interference of other amino acids, such as tyrosine, arginine and phenylalanine, showing only a limited affinity toward these molecules. An inorganic MIP-based sensor was also constructed for the chiral analysis of His and was evaluated by amperometric measurements [57]. When a solution of L-His was added, the sharpest increase in the current was observed for the MIP/sol/MWCNTs/Si-ITO electrode, followed by the MIP/sol/Si-ITO electrode and the lowest current value was for the non-imprinted polymer/sol/MWCNTs/Si-ITO electrode. However, by this protocol, the authors fail to evaluate the effect of D-His on the pristine MIP systematically presenting only the addition of L-His followed by the addition of D-His. With the addition of D-His in the case of all three electrodes, the current slightly decreases, proving a saturation of the binding sites previously occupied with L-His. Two MIP-based chiral sensors were proposed by Prasad et al. [52, 68] for trace analysis of His, claiming an absolute enantiodiscrimination. First sensor was based on the modification of hanging mercury drop electrode with the copolymerized poly (ethylenediamine tetraacetic acid-co-chloranil) imprinted polymer. Interestingly, the chiral sensor exhibited two different linear relationships between the differential pulse cathodic stripping voltammetry (DPCSV) peak currents and target enantiomer concentration: from 0.0005 to 0.03946 μg mL^{-1} and from 4.95 to 90.16 μg mL^{-1}, respectively. A remarkable limit of detection of 0.128 ng mL^{-1} was obtained for the template enantiomer and a negligible cross-reactivity for the twelve studied interferents was reported. The proposed MIP-sensor was applied for the L-His determination in human blood serum samples, and due to their high dilution, no distortion of peak current or sorption of proteins was observed. The second sensor was developed by "metal-ion mediated imprinting" approach, in which Cu^{2+} ions are able to form a complex with template (L-His) and acryl phosphate functional monomer, as shown in Fig. 17.3. The presence of Cu^{2+} offered an electroconductive feature of the film during rebinding. The MIP-modified PGE was tested using differential pulse anodic stripping voltammetry (DPASV) for L-His detection in blood serum samples and pharmaceuticals with no matrix effect and cross-reactivity. Both chiral MIP-based sensors showed an excellent recoveries and reproducibility.

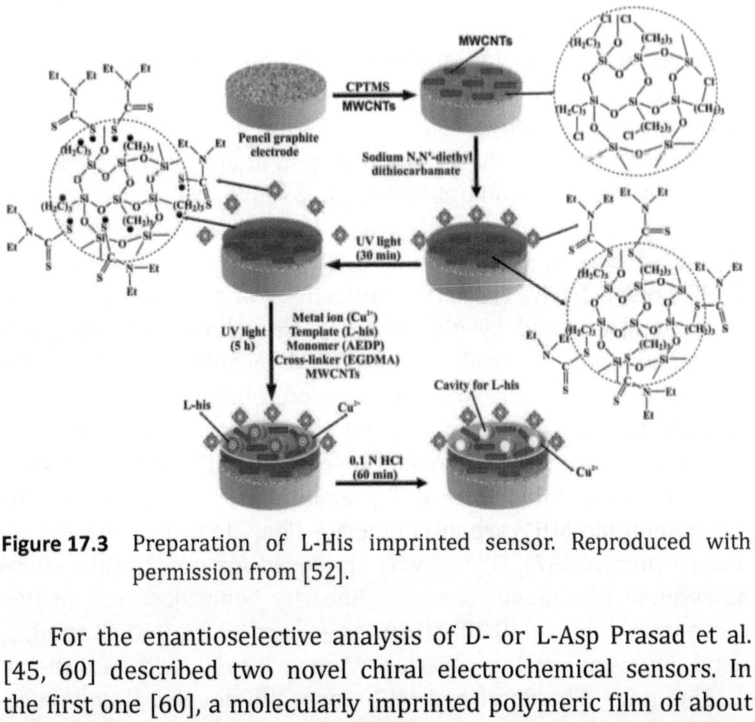

Figure 17.3 Preparation of L-His imprinted sensor. Reproduced with permission from [52].

For the enantioselective analysis of D- or L-Asp Prasad et al. [45, 60] described two novel chiral electrochemical sensors. In the first one [60], a molecularly imprinted polymeric film of about 2 nm thickness was electro-generated onto a MWCNT-modified PGE surface. During the polymerization, a polymeric backbone with positively charged indole rings was formed, on which Asp anions as dopant were bound. The resulting electrodes were characterized by DPASV on the linear range of 0.15–8.9 μM, with a LOD of 0.016 μM. Recently, a second MIP-based sensor was developed via surface-initiated activators regenerated by electron transfer for atom transfer radical polymerization approach over the matrix of TiO_2 nanoparticles/MWCNTs [45]. TiO_2 nanoparticles and MWCNTs were introduced for the enhancement of electron conductivity and sensitivity. Interestingly, both Asp imprinted chiral sensors showed noteworthy detection ability without any cross-reactivity and an extraordinary reproducibility and stability. Moreover, they were tested in aqueous, pharmaceutical and biological samples, without any false-positive response.

A thin film of overoxidized PPy was imprinted with L-Trp and its enantioselectivity was characterized by stripping voltammetry and quartz crystal microbalance (QCM) [64]. The imprinting effect was greatly affected by the synthesis parameters, such as overoxidation potential and time and the thickness of PPy film. However, only a partial enantiodiscrimination was obtained due to the strong MIP surface adsorption of the interfering enantiomer. PPy was also employed in the fabrication of L-Trp imprinted cylindrical electrode tested by the potential-induced technique in which the ejection–rebinding steps are controlled by the applied potential on the electrode [54]. The resulted polymer presented a re-binding ratio of L-Trp and D-Trp of 52.6% and 17.3%, respectively. The enantioselectivity of the L-Trp imprinted PPy was also tested by differential pulse voltammetry on glassy carbon working electrode and by QCM technique on Au-coated quartz crystals. Similar selectivities were obtained in both cases, proving the formation of imprinted cavities for the template enantiomer during the overoxidation process of PPy. Absolute chiral discrimination ability toward Trp enantiomers was presented by the monolithic MIP-carbon composite fiber developed by Prasad and coworkers [37]. DPASV was employed for quantitative chiral assessment of aqueous samples, linearity being observed in two concentration ranges: 0.90–18.60 ng mL^{-1} and 24.23–840.22 ng mL^{-1}. Good precision and recoveries were achieved with a LOD of 0.24 ng mL^{-1}. The described MIP-sensor did not reveal any cross-reactivity with none of the eight studied common interferents of Trp and was successfully applied for the analysis of biological fluids (human serum and cerebrospinal fluid) without any preliminary treatment.

The enantiomers of phenylalanine anilide were analyzed on an imprinted poly(EDMA-co-MAA) polymer grafted onto an ITO glass electrode based on the chiral discriminative gate effect of the MIP in organic solvents [49]. The chiral differentiation implied the evaluation of the effect of phenylalanine anilide enantiomers on the faradic current of redox species (ferrocene), effect caused by a morphological change in the polymer's matrix. In the presence of template enantiomer, an important decrease in ferrocene's anodic current was observed in comparison with the same effect in the presence of the other enantiomer. Because of the very narrow linear dynamic range (0.5–1 μM), the obtained sensor can only be employed for qualitative analysis and exclusively in non-polar solvents.

The chiral recognition of another amino acid, phenylalanine, was achieved by electro-generating PPy nanowires using a pseudo-template, D- or L-camphorsulfonic acid (CSA) [65]. The acidic surfactant, CSA, in aqueous solutions formed micelles with the monomer, acting as a protecting agent for the prevention of radial overgrowth of PPy nanowires. By the electrostatic interactions between its $-SO_3^-$ group and the positively charged PPy backbone as well as by hydrogen bonding between its own carbonyl oxygen and the N atom of PPy also ensured an efficient imprinting process. The de-doped PPy nanowires coated Pt electrodes were able not only to recognize the template enantiomers, but also enantiomers of molecules with similar molecular size and conformation, i.e., phenylalanine. The enantioselective capabilities of the MIP nanowires were investigated by electrochemical impedance spectroscopy and circular dichroism. A new hybrid MIP chiral sensor was also developed using L-phenylalanine as template and its chiral performance was investigated by open circuit potential-time technique [66]. The proposed sensor showed a relatively good ability to discriminate between the enantiomers with a selectivity coefficient of $K_D^L = 5.75 \times 10^{-4}$ and a limited interference of amino acids with similar configuration. A linear range of $2.5 \times 10^{-6} - 2.5 \times 10^{-2}$ M and a detection limit of 1.37 μM were achieved.

Electropolymerization of benzidine in the presence of L- or D-MET was employed in order to generate enantiospecific cavities for MET enantiomers [61]. During the MIP film electrogeneration as well as during the template enantiomer's re-binding process MET is converted to its oxidized form, methionine sulfone. The obtained sensor presented a rapid and cost-effective preparation and was characterized by DPCSV showing absolute discriminative properties between MET enantiomers. It proved to have a good intermediate precision (RSD < 4%) and a linearity in the concentration range of 11.7–206.3 ng mL^{-1} (aqueous solutions),

11.7–197.4 ng mL^{-1} (pharmaceutical matrix), and 11.8–152.3 ng mL^{-1} (blood serum matrix). The lowest LOD (2.4 ng mL^{-1}) compared to other sensors developed for Met analysis, along with the striking electrode reproducibility and stability, in the absence of any nonspecific binding of potential interferents, makes it the best option for trace analysis of Met in different clinical applications.

A new poly(aniline-co-m-aminophenol) chiral conductive stationary phase was developed for L-glutamic acid analysis [55]. The resulted column was tested by applying different potentials on the copolymer for the release (negative potential for reduction of polymer) and then for the re-attachment (positive potential) of template. The enantioselective recognition was evaluated by monitoring the fluorescence of L- and D-glutamic acid re-bounded to the packed electrode column. The results were also confirmed by QCM analysis. The re-binding ratio of L- to D-glutamic acid was found to be 2.38. Chiral analysis of one of glutamic acid's derivative, pyroglutamic acid, was achieved with the aid of Cu^{2+} incorporated MIP modified PGE [62]. The highly conductive imprinted film was evaluated by DPASV showing a linear response in the concentration interval of 2.8–170 ng mL^{-1}, with a detection limit of 0.77 ng mL^{-1}. The obtained sensor was applied in the pyroglutamic acid's chiral determination from different real samples (urine, cerebrospinal fluid, plasma) without any cross-reactivity and false-positive results.

Self-assembled monolayer and layer-by-layer approaches were combined by thermal cross-linking on the surface of a silver electrode in order to develop a MIP-based sensor for the chiral analysis of thyroxine, a thyroid hormone with an amino-acid side chain [53]. The DPASV response peak current was linear to the concentration of thyroxine in the range from 0.010 to 17.2 ng mL^{-1} and the achieved detection limit was 0.0060 ng mL^{-1}. The molecularly imprinted film displayed absolute enantioselectivity to the template enantiomer and was applied for the determination of thyroxine in aqueous, blood and pharmaceutical samples with no interference from related compounds.

17.4.2.2 Monosaccharides

A well-known shortcoming of MIPs based on non-covalent interactions is the relatively low number of specific imprinted cavities. However, this drawback may be overcome by employing covalent interactions between the template and the functional monomer. A popular approach in saccharide analysis is based on their complexation with boronic acid [69]. Based on these interactions, Granot et al. [46] prepared imprinted polymeric films with D-glucose or D-mannose, as shown in Fig. 17.4. The electrochemical detection of monosaccharides is achieved by competitive interactions with the polymer between the target monosaccharide and a ferrocene-labeled monosaccharide. For

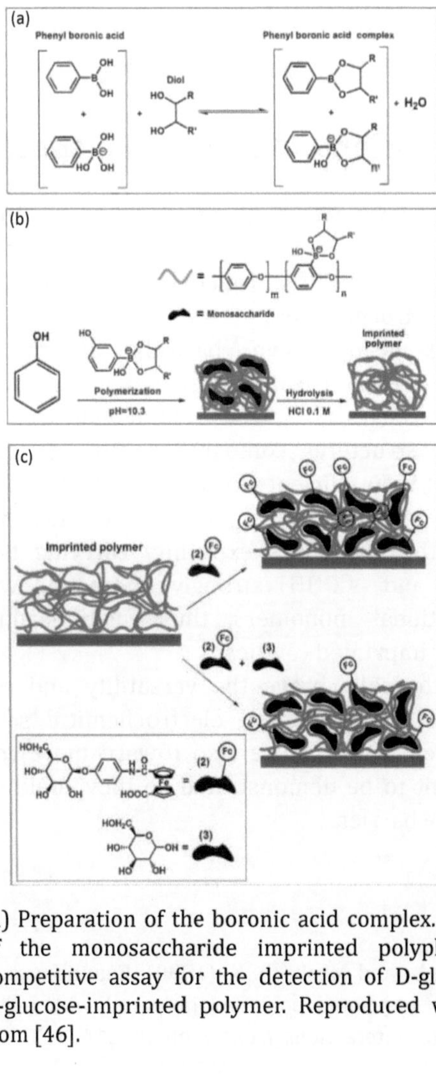

Figure 17.4 (a) Preparation of the boronic acid complex. (b) Preparation of the monosaccharide imprinted polyphenol. (c) The competitive assay for the detection of D-glucose using the D-glucose-imprinted polymer. Reproduced with permission from [46].

For example, in the analysis of D-glucose using the D-glucose imprinted polymer, the amperometric response of ferrocene-functionalized D-glucose was monitored. With the increasing concentration of D-glucose, the current response of the redox-labeled monosaccharide decreases due to the competitive interactions toward the D-glucose imprinted cavities of the polymer (D-glucose removes the ferrocene-derivative from the polymer). The obtained sensor showed good selectivity toward other monosaccharides and a slight discrimination between glucose enantiomers. Nonetheless, this method may be applicable for the development of sensors designed for different diol-containing molecules.

17.5 Conclusions and Future Perspectives

Molecularly imprinting technology seems to expand every year being more and more frequently applied in various fields of science. Only recently, MIPs started to be employed as chiral sensing interfaces in electrochemistry.

As it has been shown, basically all chiral MIP-based electrochemical sensors were developed up till now for amino acids or monosaccharides. However, chiral pharmaceuticals present more complex structures compared to the already mentioned molecules; thus, their efficient molecular imprint is considered to be more difficult. Moreover, in the case of amino acids, the asymmetric carbon is at the molecule's extremity carrying two functional groups ($-NH_2$ and $-COOH$) strongly interacting with the used common functional monomers, thus easily leading to highly enantiospecific imprinted cavities.

In order to really prove the versatility and real analytical potentials of MIP-based chiral electrochemical sensors, in the near future their performance also toward more complex chiral molecules ought to be demonstrated so they could overcome the laboratory scale barrier.

References

1. Andersson, L., and Mosbach, K. (1990). Enantiomeric resolution on molecularly imprinted polymers prepared with only non-covalent and non-ionic interactions, *J. Chromatogr.*, **516**, 313–322.

2. Schweitz, L., Andersson, L. I., and Nilsson, S. (1997). Capillary electrochromatography with predetermined selectivity obtained through molecular imprinting, *Anal. Chem.*, **69**, 1179–1183.
3. Duan, A., Xie, S. M., and Yuan, L. M. (2011). Nanoparticles as stationary and pseudo-stationary phases in chromatographic and electrochromatographic separations, *Trends Anal. Chem.*, **30**(3), 484–491.
4. Dickey, F. (1949). The preparation of specific adsorbents, *Proc. Natl. Acad. Sci.*, **35**(5), 227–229.
5. Wulff, G. (1984). Molecular imprinting, *Ann. N. Y. Acad. Sci.*, **434**, 327–333,.
6. Mosbach, K., and Ramstrom, O. (1996). The emerging technique of molecular imprinting and its future impact on biotechnology, *Nat. Biotechnol.*, **14**, 163–170.
7. Sellergren, B. (2001). *Molecularly Imprinted Polymers: Man-Made Mimics of Antibodies and Their Applications in Analytical Chemistry*, Elsevier, Amsterdam, The Netherlands.
8. Piletsky, S. A., Turner, N. W., and Laitenberger, P. (2006). Molecularly imprinted polymers in clinical diagnostics-future potential and existing problems, *Med. Eng. Phys.*, **28**(10), 971–977.
9. Li, W., and Li, S. (2007). In *Oligomers—Polymer Composites: Molecular Imprinting*, Springer Berlin Heidelberg, vol. 206, pp. 191–210.
10. Puoci, F., Iemma, F., and Picci, N. (2008). Stimuli-responsive molecularly imprinted polymers for drug delivery: A review, *Curr. Drug Deliv.*, **5**(2), 85–96.
11. Longo, L., and Vasapollo, G. (2008). Phthalocyanine-based molecularly imprinted polymers as nucleoside receptors, *Met. Based Drugs.*, **2008**, 1–5.
12. Ge, Y., and Turner, A. P. (2009). Molecularly imprinted sorbent assays: Recent developments and applications, *Chemistry* (*Weinheim an der Bergstrasse, Germany)*, **15**(33), 8100–8107.
13. Yan, M., and Ramström, O. (2005). *Molecularly Imprinted Materials. Science and Technology*, Marcel Dekker, New York.
14. Lanza, F., and Sellergren, B. (2004). Molecularly imprinted polymers via high-throughput and combinatorial techniques, *Macromol. Rapid Commun.*, **25**, 59–68.
15. Piletsky, S., Karim, K., Piletska, E. V., Day, C. J., Freebairn, D. W., Legge, C., and Turner, A. P. F. (2001). Recognition of ephedrine enantiomers by molecularly imprinted polymers designed using a computational approach, *Analyst*, **126**, 1826–1830.

16. Huang, B.-Y., Chen, Y.-C., and Liu, C.-Y. (2011). An insight into the mechanism of CEC separation of template analogues on a norepinephrine-imprinted monolith, *J. Sep. Sci.*, **34**, 2293–2300.
17. Qu, P., Zhang, L., Sheng, J., Lei, J., and Ju, H. (2011). Convenient enantio-separation by monolithic imprinted capillary clamped in a chip with electrochemical detection, *Electrophoresis*, **32**, 1522–1529.
18. Wei, Z.-H., Mu, L.-N., Pang, Q.-Q., Huang, Y.-P., and Liu, Z.-S. (2012). Preparation and characterization of grafted imprinted monolith for capillary electrochromatography, *Electrophoresis*, **33**, 1–7.
19. Zaidi, S. (2013). Dual-templates molecularly imprinted monolithic columns for the evaluation of serotonin and histamine in CEC, *Electrophoresis*, **34**, 1375–1382.
20. Gholivand, M., and Torkashvand, M. (2011). A novel high selective and sensitive metronidazole voltammetric sensor based on a molecularly imprinted polymer-carbon paste electrode, *Talanta*, **84**, 905–912.
21. Alizadeh, T., Ganjali, M. R., Norouzi, P., Zare, M., and Zeraatkar, A. (2009). A novel high selective and sensitive para-nitrophenol voltammetric sensor, based on a molecularly imprinted polymer-carbon paste electrode, *Talanta*, **79**, 1197–1203.
22. Sadeghi, S., Motaharian, A., and Moghaddam, A. Z. (2012). Electroanalytical determination of sulfasalazine in pharmaceutical and biological samples using molecularly imprinted polymer modified carbon paste electrode, *Sens. Actuators B*, **168**, 336–344.
23. Zhao, L., Zhao, F., and Zeng, B. (2013). Electrochemical determination of methyl parathion using a molecularly imprinted polymer–ionic liquid–graphene composite film coated electrode, *Sens. Actuators B*, **176**, 818–824.
24. Li, J., Chen, Z., and Li, Y. (2011). A strategy for constructing sensitive and renewable molecularly imprinted electrochemical sensors for melamine detection, *Anal. Chim. Acta*, **706**, 255–260.
25. Tan, Y., Zhou, Z., Wang, P., Nie, L., and Yao, S. (2001). A study of a bio-mimetic recognition material for the BAW sensor by molecular imprinting and its application for the determination of paracetamol in the human serum and urine, *Talanta*, **55**, 337–347.
26. Liu, X., Wei, Z.-H., Huang, Y.-P., Yang, J.-R., and Liu, Z.-S. (2012). Molecularly imprinted nanoparticles with nontailing peaks in capillary electrochromatography, *J. Chromatogr. A*, **1264**, 137–142.
27. Shi, X.-X., Xu, L., Duan, H.-Q., Huang, Y.-P., and Liu, Z.-S. (2011). CEC separation of ofloxacin enantiomers using imprinted microparticles prepared in molecular crowding conditions, *Electrophoresis*, **32**, 1348–1356.

28. Wulff, G. (2002). Enzyme-like catalysis by molecularly imprinted polymers, *Chem. Rev.*, **102**, 1–27.

29. Rubenstein, E., Bonner, W. A., Noyes, H. P., and Brown, G. S. (1983). Supernovae and life, *Nature*, **306**, 118.

30. Lin, G., You, Q. D., and Cheng, J. F. (2011). *Chiral Drugs: Chemistry and Biological Action.*, John Wiley & Sons, New Jersey.

31. Erb, S. (2006). Single-enantiomer drugs poised for further market growth, *Pharmaceut. Technol.*, **30**, 14–18.

32. Hutt, A., and Valentová, J. (2003). The chiral switch: The development of single enantiomers drugs from racemates, *Acta Facult. Pharm. Univ. Comenianae.*, **50**, 7–23.

33. Agranat, I., Caner, H., and Caldwell, J. (2002). Putting chirality to work: The strategy of chiral switches, *Nat. Rev. Drug Discov.*, **1**(10), 753–768.

34. Chen, L., Li, K., Zhu, H., Meng, L., Chen, J., Li, M., and Zhu, Z. (2013). A chiral electrochemical sensor for propranolol based on multi-walled carbon nanotubes/ionic liquids nanocomposite, *Talanta*, **105**, 250–254.

35. Xu, L., Yang, Y., Wang, Y., and Gao, J. (2009). Chiral salen Mn(III) complex-based enantioselective potentiometric sensor for l-mandelic acid, *Anal. Chim. Acta*, **653**(2), 217–221.

36. Stefan, R.-I., Bala, C., and Aboul-Enein, H. Y. (2003). Biosensor for the enantioselective analysis of S-captopril, *Sens. Actuators B*, **92**(1–2), 228–231.

37. Prasad, B., Madhuri, R., Tiwari, M. P., and Sharma, P. S. (2010). Enantioselective recognition of D- and L-tryptophan by imprinted polymer-carbon composite fiber sensor, *Talanta*, **81**, 187–196.

38. Trojanowicz, M., and Kaniewska, M. (2009). Electrochemical chiral sensors and biosensors, *Electroanalysis*, **21**(3–5), 229–238.

39. Vlatakis, G., Andersson, L. I., Muller, R., and Mosbach, K. (1993). Drug assay using antibody mimics made by molecular imprinting, *Nature*, **361**, 645–647.

40. Andersson, L., Miyabayashi, A., O'Shannessy, D. J., and Mosbach, K. (1990). Enantiomeric resolution of amino acid derivatives on molecularly imprinted polymers as monitored by potentiometric measurements, *J. Chromatogr.*, **516**, 323–331.

41. Hedborg, E., Winquist, F., Lundström, I., Andersson, L. I., and Mosbach, K. (1993). Some studies of molecularly imprinted polymer membranes in combination with field-effect devices, *Sens. Actuators. A.*, **37–38**, 796–799.

42. Deore, B., Chen, Z. D., and Nagaoka, T. (2000). Potential-induced enantioselective uptake of amino acid into molecularly imprinted overoxidized polypyrrole, *Anal. Chem.*, **72**, 3989–3994.

43. Deore, B., Yakabe, H., Shiigi, H., and Nagaoka, T. (2002). Enantioselective uptake of amino acids using an electromodulated column packed with carbon fibres modified with overoxidised polypyrrole, *Analyst*, **127**, 935–939.

44. Okuno, H., Kitano, T., Yakabe, H., Kishimoto, M., Deore, B. A., Siigi, H., and Nagaoka, T. (2002). Characterization of overoxidized polypyrrole colloids imprinted with l-lactate and their application to enantioseparation of amino acids, *Anal. Chem.*, **74**, 4184–4190.

45. Prasad, B., Srivastava, A., and Tiwari, M. P. (2013). Molecularly imprinted polymer-matrix nanocomposite for enantioselective electrochemical sensing of D- and L-aspartic acid, *Mater. Sci. Eng. C*, **33**, 4071–4080.

46. Granot, E., Tel-Vered, R., Lioubashevski, O., and Willner, I. (2008). Stereoselective and enantioselective electrochemical sensing of monosaccharides using imprinted boronic acid-functionalized polyphenol films, *Adv. Funct. Mater.*, **18**(3), 478–484.

47. Haupt, K., Noworyta, K., and Kutner, W. (1999). Imprinted polymer-based enantioselective acoustic sensor using a quartz crystal microbalance, *Anal. Commun.*, **36**, 391–393.

48. Piacham, T., Josell, A., Arwin, H., Prachayasittikul, V., and Ye, L. (2005). Molecularly imprinted polymer thin films on quartz crystal microbalance using a surface bound photo-radical initiator, *Anal. Chim. Acta*, **536**, 191–196.

49. Sekine, S., Watanabe, Y., Yoshimi, Y., Hattori, K., and Sakai, K. (2007). Influence of solvents on chiral discriminative gate effect of molecularly imprinted poly(ethylene glycol dimethacrylate-co-methacrylic acid), *Sens. Actuators B*, **127**, 512–517.

50. Gholivand, M., Malekzadeh, G., and Torkashvand, M. (2013). Determination of lamotrigine by using molecularly imprinted polymer–carbon paste electrode, *J. Electroanal. Chem.*, **692**, 9–16.

51. Yoshimi, Y., Ohdaira, R., Iiyama, C., and Sakai, K. (2001). Gate effect of thin layer of molecularly-imprinted poly (methacrylic acid-co-ethyleneglycol dimethacrylate), *Sens. Actuators B*, **73**, 49–53.

52. Prasad, B., Kumar, D., Madhuri, R., and Tiwari, M. P. (2011). Metal ion mediated imprinting for electrochemical enantioselective sensing of l-histidine at trace level, *Biosens. Bioelectron*, **28**, 117–126.

53. Prasad, B., Madhuri, R., Tiwari, M. P., and Sharma, P. S. (2010). Layer-by-layer assembled molecularly imprinted polymer modified silver electrode for enantioselective detection of d- and l-thyroxine, *Anal. Chim. Acta*, **681**, 16–26.

54. Kong, Y., Wei, J., Wang, W., and Chen, Z. (2011). Separation of tryptophan enantiomers with polypyrrole electrode column by potential-induced technique, *Electrochim. Acta*, **56**, 4770–4770.

55. Kong, Y., Li, X., Ni, J., Yao, C., and Chen, Z. (2012). Enantioselective recognition of glutamic acid enantiomers based on poly (aniline-co-m-aminophenol) electrode column, *Electrochem. Commun.*, **14**, 17–20.

56. Piletsky, S., and Turner, A. (2006). *Molecular Imprinting of Polymers*, Landes Bioscience, Georgetown.

57. Zhang, Z., Hu, Y., Luo, L., and Yao, S. (2010). Layer-by-layer assembly sensitive electrochemical sensor for selectively probing l-histidine based on molecular imprinting sol–gel at functionalized indium tin oxide electrode, *Biosens. Bioelectron.*, **26**, 696–702.

58. Liao, H., Zhang, Z., Nie, L., and Yao, S. (2004). Electrosynthesis of imprinted polyacrylamide membranes for the stereospecific L-histidine sensor and its characterization by AC impedance spectroscopy and piezoelectric quartz crystal technique, *J. Biochem. Biophys. Methods*, **59**, 75–87.

59. Ersoz, A., Gavalas, V. G., and Bachas, L. G. (2002). Potentiometric behavior of electrodes based on overoxidized polypyrrole films, *Anal. Bioanal. Chem.*, **372**, 786–790.

60. Prasad, B., and Pandey, I. (2013). Electrochemically imprinted molecular recognition sites on multiwalled carbon-nanotubes/pencil graphite electrode surface for enantioselective detection of D- and L-aspartic acid, *Electrochim. Acta*, **88**, 24–34.

61. Prasad, B., Pandey, I., Srivastava, A., Kumar, D., and Tiwari, M. P. (2013). Multiwalled carbon nanotubes-based pencil graphite electrode modified with an electrosynthesized molecularly imprinted nanofilm for electrochemical sensing of methionine enantiomers, *Sens. Actuators B*, **176**, 863–874.

62. Prasad, B. B., and Pandey, I. (2013). Metal incorporated molecularly imprinted polymer-based electrochemical sensor for enantioselective analysis of pyroglutamic acid isomers, *Sens. Actuators B*, **186**, 407–416.

63. Ramanavicius, A., Ramanaviciene, A., and Malinauskas, A. (2006). Electrochemical sensors based on conducting polymer-polypyrrole, *Electrochim. Acta*, **51**(27), 6025–6037.

64. Kong, Y., Zhao, W., Yao, S., Xu, J., Wang, W., and Chen, Z. (2010). Molecularly imprinted polypyrrole prepared by electrodeposition for the selective recognition of tryptophan enantiomers, *J. Appl. Polym. Sci.*, **115**, 1952–1957.
65. Huang, J., Wei, Z., and Chen, J. (2008). Molecular imprinted polypyrrole nanowires for chiral amino acid recognition, *Sens. Actuators B*, **134**, 573–578.
66. Chen, Y., Chen, L., Bi, R., Xu, L., and Liu, Y. (2012). A potentiometric chiral sensor for L-Phenylalanine based on crosslinked polymethylacrylic acid–polycarbazole hybrid molecularly imprinted polymer, *Anal. Chim. Acta*, **754**, 83–90.
67. Xiang, C., Xie, Q., Hu, J., and Yao, S. (2006). Studies on electrochemical copolymerization of aniline with o-phenylenediamine and degradation of the resultant copolymers via electrochemical quartz crystal microbalance and scanning electrochemical microscope, *Synthetic Metals*, **156**(5–6), 444–453.
68. Prasad, B., Srivastava, S., Tiwari, K., and Sharma, P. S. (2009). A new zwitterionic imprinted polymer sensor using ethylenediamine tetraacetic acid and chloranil precursors for the trace analysis of L-histidine, *Mater. Sci. Eng. C*, **29**, 1781–1789.
69. Saito, S., Massie, T. L., Maeda, T., Nakazumi, H., and Colyer, C. L. (2012). A long-wavelength fluorescent squarylium cyanine dye possessing boronic acid for sensing monosaccharides and glycoproteins with high enhancement in aqueous solution, *Sensors*, **12**, 5420–5431.

Chapter 18

Molecularly Imprinted Polymers: A Versatile Tool in Pharmaceutical Applications

Sushma Pardeshi,[a] Rita Dhodapkar,[b] and Anupama Kumar[a]

[a]*Department of Chemistry,*
Visvesvaraya National Institute of Technology, Nagpur 440010, India
[b]*Waste Water Technology Division,*
CSIR—National Environmental Engineering Research Institute, Nagpur 440020, India

drkumaranupama@rediffmail.com, anupamakumar@chm.vnit.ac.in,
drkumaranupama21@gmail.com, rs_dhodapkar@neeri.res.in

18.1 Introduction

The field of sustainable polymers is budding and growing at an unrivaled rate. Molecularly imprinted polymers (MIPs) are the polymer networks endowed with the ability to recognize specific molecules and have an enormous potential for the variety of applications. A perspective is offered in this chapter for the computational modeling-based rational design of MIPs for pharmaceuticals and preparation of different MIPs formats using various polymerization methods. The versatility of application of MIPs in pharmaceutical industries is also emphasized.

Handbook of Sustainable Polymers: Processing and Applications
Edited by Vijay Kumar Thakur and Manju Kumari Thakur

ISBN 978-981-4613-53-8 (Hardcover), 978-981-4613-54-5 (eBook)
www.panstanford.com

18.1.1 Molecular Imprinting and Molecularly Imprinted Polymers (MIPs)

Molecular recognition is defined as the specific chemical interaction of one molecule with another. It is an ability to exclusively pick out own partner molecule from a number of molecules in the system, completely ignoring all molecules other than the partner (Komiyama et al., 2003).

Molecular imprinting is an attractive approach for the preparation of synthetic receptors molecules. Molecular imprinting involves preparation of template selective recognition sites in a matrix using a molecular template in a casting procedure. The resulting material is known as MIPs (Ratner, 2004).

The terminology of molecular imprinting consists of the following:

- The molecules are used to create the marks or imprints within a network polymer. These molecules are denoted by a variety of expressions such as "templates," "target molecules," "analytes," and "imprint or print molecules."
- The monomers used to create the polymer are known as functional monomers or monomers.
- The co-monomers used to fix the template–monomer complexes in the MIPs are known as "cross-linkers."
- The solvent used in the imprinting procedure is denoted by the term "porogen."

The process of molecular imprinting is represented schematically in Fig. 18.1.

The template interacts with the polymerizable functional monomer in presence of a porogen to form a template–monomer complex. The mechanisms of interactions involved in the template–monomer complexes are either, covalent, semi-covalent or non-covalent. This template–monomer complex is then polymerized in the presence of a cross-linking monomer resulting in the formation of a highly cross-linked polymeric matrix. The cross-linking is performed to fix the template within a rigid polymer matrix. The template is then removed from this polymeric matrix by a solvent that is capable of disrupting the interactions between the template and the monomer. This results in a vacant recognition/binding site in the polymer matrix.

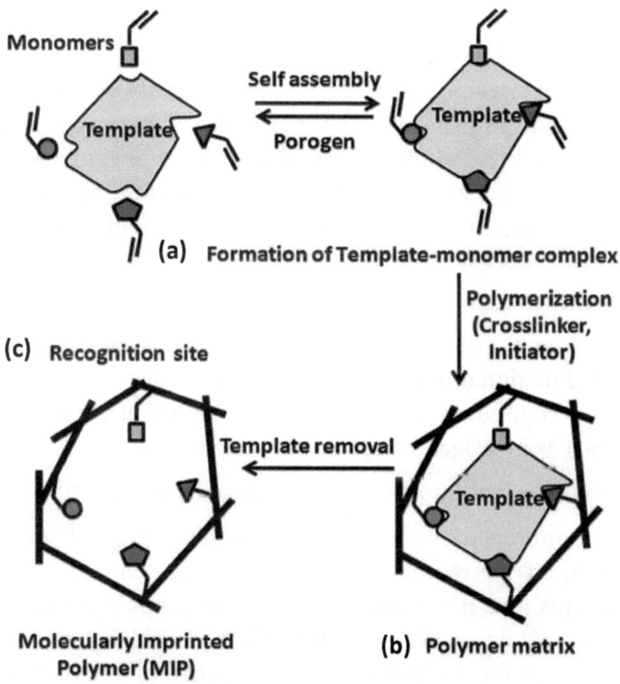

Figure 18.1 Schematic representation of molecular imprinting: (a) Formation of a template–monomer complex, (b) formation of polymer matrix after polymerization step, and (c) formation of recognition site after template removal from the polymeric matrix.

The shape and size of the recognition site and the spatial arrangement of the functional groups are complementary to the structure of the template molecule. Upon re-introduction of the template, preferential binding within the cavity occurs. Thus, recognition site has high affinity for the target molecule and can re-bind the template during the application of MIPs.

18.1.1.1 Molecular imprinting strategies

The molecular imprinting is classified into three types according to the type of interactions between functional monomer and target molecules in the pre-polymerization complex and during the re-binding of the template (Komiyama et al., 2003; Sellergren and Hall, 2001).

- Covalent imprinting (pre-organized approach)

  The template and monomers are covalently linked to form a template–monomer pre-polymerization complex in the covalent imprinting. The template is removed from the polymeric matrix by chemical cleavage after polymerization. The functionality remaining in the binding site is capable of binding the target molecule by re-establishment of the covalent bond.

- Semi-covalent imprinting

  The template is covalently bound to a polymerizable group and the functionality is recovered after cleavage of the templates. The re-binding takes place via non-covalent interactions such as hydrogen bond or ionic interactions.

- Non-covalent imprinting (self-assembly approach)

  The approach relies on the formation of a pre-polymerization complex between monomers carrying suitable functional groups and the template. A cross-linker is then added and the polymerization initiated. Then, the highly cross-linked polymer forms around the template–monomer complexes. The template is then removed from the polymer via extraction with a solvent, which disrupts the non-covalent interactions present in the pre-polymerization complex. Subsequent template re-binding takes place through the formation of the same non-covalent interactions.

The advantages of non-covalent imprinting over the preceding approaches include the following:

- It is open to a wide range of templates.
- Simple processes are used to extract the template from the MIP matrix.
- Synthesis of monomer–template adducts is unnecessary (which is essential in covalent imprinting).
- The need to chemically cleave and reform covalent bonds is not required in the non-covalent approach. Thus, template-binding and template-release are fast.

Given its versatility and closest resemblance of naturally occurring recognition mechanisms, the non-covalent imprinting is most widely used in the molecular imprinting.

18.1.1.2 Characteristic of MIPs

The characteristic features and advantages of MIPs include ease of preparation and cost effectiveness. MIPs are considered a cost-effective alternative to biomolecule-based recognition. They have high physical robustness and high mechanical strength and remain stable in presence of extremes of acids and bases. MIPs show high resistance to elevated temperature and pressure and can be stored for long time without loss of affinity for target molecule. They can be reused for several binding cycles without decreasing the binding capacity. MIPs can be prepared in different formats such as blocks, beads, thin films following the need of the application (Cormack and Elorza, 2004; Ensing and Boer, 1999; Mahony et al., 2005; Piletsky et al., 2001; Mayes and Whitcombe, 2005).

18.1.2 Progress in MIP Technology

Molecular imprinting was first introduced in the early 1930s by M. V. Polykov, a Soviet chemist who performed a series of investigations on silica for use in chromatography (Polyakov, 1931). The molecular imprinting was applied to organic polymers in 1972 by the groups of Gunter Wulff (Wulff and Sarhan, 1972) and Klotz independently (Takagishi and Klotz, 1972). The most important development in the area of molecular imprinting was the introduction of non-covalent approach by the group of K. Mosbach in the early 1980s (Mosbach et al., 1984) and was a newer period of intensive development and continues until today.

The past few years have witnessed an almost exponential increase in interest in the area of molecular imprinting, as illustrated by the number of informative reviews and research articles appearing in the area. Concerning their commercialization, there has been great progress during the past decade, in particular in the analytical chemistry and biochemistry fields. Number of companies such as MIP technologies AB, Lund, Sweden; POLYIntell, Rouen, France, and Semorex Inc., North Brunswick, NJ, USA, sells tailor-made imprints for various templates.

There are number of potential application areas that have been identified for MIPs; all based on their capability to specifically recognize molecular targets: affinity separation (Wei and Mizaikoff, 2007), chemical sensors and assays (Qiu et al., 2012; Tokonami et al.,

2009), directed synthesis and enzyme-like catalysis (Chai et al., 2012), and biomedical applications (Yañez et al., 2011).

As MIPs made gradual transition from the academic research laboratory to pharmaceutical application areas, it became essential that the MIPs are designed, both chemically and structurally, in order to optimize its performance for these application. There have been several attempts on the development of generic procedure for MIPs such as combinatorial synthesis/ screening (Batra and Shea, 2003) and Chemometrics (Gholivand et al., 2012); however, the in silico–based computational design has been in prime focus in the past decade.

18.2 Computational Modeling for Rational Design of MIPs for Pharmaceutical Templates

It is essential to have overview of computational methods in mind before one proceeds to computational modeling of MIPs for pharmaceuticals.

18.2.1 An Overview of Computational Methods Used in Rational Design of MIPs

Computational chemistry simulates chemical structures and reactions numerically, based on the fundamental laws of physics, and allows studying the chemical reactions by running calculations on computer rather than performing the experiments (Young, 2001). The broad areas of computational chemistry, i.e., molecular mechanics (MM), molecular dynamics (MD), and quantum chemical methods, perform the same basic types of calculations such as the following:

(a) performing geometry optimizations (to locate the lowest energy molecular structure and spatial arrangement of atoms, nuclei and electrons)
(b) computations of energy and thermodynamic properties
(c) computation of vibrational frequency of molecule resulting from the interatomic motion within the molecule

The mathematical background and theory of the computational methods is beyond the scope of this chapter and can be found elsewhere (Young, 2001; Ramachandran et al., 2008).

18.2.1.1 Molecular mechanics and molecular dynamics simulations

Molecular mechanics and MD simulations use the law of classical physics to predict the structure and properties of molecule. These calculations perform computations based on the interactions among the nuclei and do not explicitly treat the electrons in a molecular system. This approximation makes them quite inexpensive and allows its use for very large system containing thousands of electrons. However the limitations associated with these methods are, they achieve good results for a limited class of molecules. Due to the negligence of electrons, these methods cannot treat chemical problems where electronic effects predominate (Young, 2001; Ramachandran et al., 2008; Foresman, 2004).

18.2.1.2 Quantum chemical methods

Quantum chemical methods, also known as the electronic structure methods, use the laws of quantum mechanics. Quantum mechanics states that the energy and other related properties of a molecule may be obtained by solving the Schrodinger equation (Eq. 18.1).

$$H\Psi = E\Psi \tag{18.1}$$

However, the exact solution to the Schrodinger equation is not computationally feasible. Quantum chemical methods are characterized by their various mathematical approximations to its solution (Foresman, 2004).

The semi-empirical method, the ab initio method, and the density functional theory (DFT) are the three classes of quantum chemical methods. The semi-empirical method uses parameter derived from experimental data to simplify the computations. Ab initio methods are solely based on the laws of quantum mechanics. DFT is similar to the ab initio method in many ways except that it includes one attractive feature of electron correlation. Electron correlation can be described as "the electrons in a molecular system reacts to one another's motion and attempts to keep out of one another's way" (Young, 2001; Foresman, 2004).

Different model chemistries, i.e., combination of levels of theory and basis sets are used in quantum chemical methods (Fig. 18.2). Levels of theory denote a hierarchy of procedures corresponding to different mathematical transformations and approximations. A basis set is a mathematical representation of the molecular orbitals within a molecule.

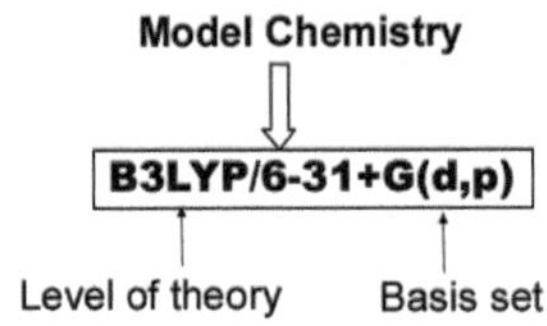

Figure 18.2 Description of model chemistry by taking example of B3LYP/6-31+G(d,p) with its different components.

More information on model chemistries, levels of theories, and the basis sets can be found elsewhere (Young, 2001, 2008; Foresman, 2004).

Properties of molecules differ considerably in the vacuum and in the presence of solution in the model during the simulation. The effect of solvent is included in the model by numerical representation of the solvent reaction field termed self-consistent reaction field (SCRF) methods (Foresman, 2004). The polarized continuum model (PCM) is one of such numerical approximations developed by Tomasi (Tomasi and Persico, 1994) and is the most widely used SCRF model to study the effect of solvent in quantum chemical simulations.

18.2.2 Computational Modeling for the Rational MIP Design

The "rational MIP design" involves better insights to the molecular imprinting process. An in-depth understanding of the molecular level events occurring in the MIPs is modeled mathematically. The earliest attempts to describe aspects of the molecular imprinting involved physical or mathematical formalism based on thermodynamic models (Pande et al., 1997; Sergeyeva et al., 1999; Nicholls et al., 1995).

The rapid growth in computer power over the past decade and a concurrent development in establishing new and improved software have opened the new possibilities for simulating aspects of the complex molecular imprinting process (Nicholls et al., 2009). The use of computational modeling in the field of molecular imprinting began around in the year 2000. During that decade (2000–2010), computational modeling flourished and was widely used, as evident from the several articles and reviews that appeared in the area (Nicholls et al., 2009; Nicholls et al., 2013; Nicholls et al., 2011), wherein comprehensive information can be found on the topic.

The computational methods are used to investigate various aspects of the molecular imprinting and polymer performance as presented in Fig. 18.3.

- The MIP performance can be optimized through the selection of suitable functional monomers and porogens.
- The characterization of molecular complexes formed between templates and monomers gives a clearer picture of the interactions between the template and monomer, which is the basis of the molecular recognition.
- Computational modeling has also been used in the study of thermodynamic and spectroscopic properties of a system.

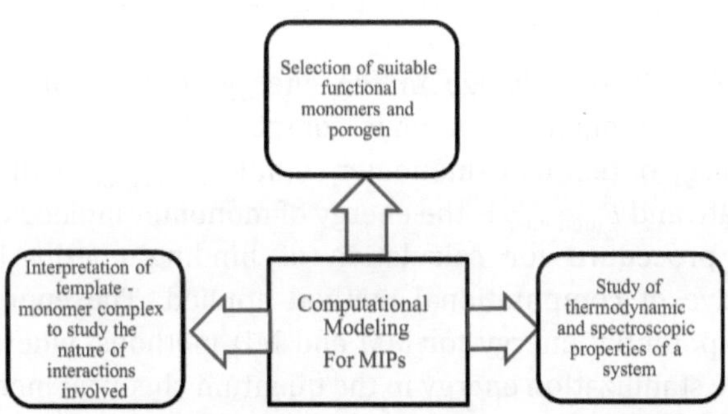

Figure 18.3 Role of computational modeling in optimization of MIP performance.

18.2.2.1 Rational selection of monomers and porogen and characterization of template–monomer prepolymerization complex

An important step in the polymer synthesis is the formation of prepolymerization complex between the template and a functional monomer or a number of functional monomers. The more stable or stronger this complex is, the more selective the MIP will be (Nicholls et al., 2001).The broad range of functional monomers (approximately 4000 different polymerizable compounds) available makes it possible to synthesize MIP for any type of stable chemical compound (Karim et al., 2005). Nevertheless, selection of a functional monomer with the high interaction energy with the template molecule is significant for the better performance of MIPs.

Computational modeling makes it possible to simulate the template–monomer prepolymerization complex. Binding energy (also termed interaction energy or binding affinity) is the change in the energy (ΔE) due the formation of template–monomer complex and is the measure of interaction between them. The monomer that gives the highest value of binding energy due to the formation of most stable template–monomer complex is considered an ideal candidate for MIP preparation. The energies (E) of the most stable conformers of template, monomer, and template–monomer prepolymerization complex are computed and the change in the energy (ΔE) due to the complex formation is calculated using Eq. 18.2.

$$\Delta E = E_{\text{template–monomer complex}} - [E_{\text{template}} + E_{\text{monomer}}], \tag{18.2}$$

where, (ΔE) is the change in the energy (k cal mol^{-1}) on the formation of template–monomer complex, $E_{\text{template–monomer complex}}$ is the energy of template–monomer complex, E_{template} is the energy of template and E_{monomer} is the energy of monomer molecules.

The procedure for calculation of binding energy is same irrespective of computational method applied. The energy is in terms of potential energy for MM and MD methods, whereas it is electronic stabilization energy in the quantum chemical method.

The selection of an optimal porogen is another important criterion for the development of MIP with high binding capacity. The porogen plays a crucial role by affecting the physical properties of the resulting polymer, including its porosity, specific surface area, and the swelling behavior. The porogen also governs the stability of the prepolymerization complex, which in turn significantly determines the recognition properties of the resulting MIPs (Andersson and Nicholls, 2001). The effect of porogen can be included explicitly by including the porogen molecules in the model in MM and MD methods. The effect of porogen can be included explicitly by including the porogen molecules in the model or implicitly by considering its presence using SCRF methods such as PCM in the model in quantum chemical methods (Foresman, 2004). In general, the presence of porogen may decrease the stability of the complex, which results in increasing the energy. The porogen with the least interference on the template–monomer complex is considered the best porogen for MIP synthesis. The rational selection of porogen for MIPs synthesis has been reported in several studies (Dineiro et al., 2005; Wu et al., 2005).

The nature of interactions involved between the template and the monomers (hydrogen bonding, electrostatic and Π–Π staking) can be studied by the characterization of template–monomer complex model. The functional groups taking part in the complex formation and site of attachment can be visualized by the clear picture of molecular complexes. Moreover, the mole ratio of template: monomer can also be optimized using computational methods.

MIPs with high binding capacity and selectivity have been obtained for different templates following computational methods (Nicholls et al., 2009; Nicholls et al., 2013; Nicholls et al., 2011). MIPs for pharmaceuticals have been rationally designed using the computational modeling and are described in detail in the following sections.

18.2.2.2 Optimization of MIP performance through molecular mechanics and molecular dynamics simulation

Ibuprofen is a member of the class of drugs termed non-steroidal anti-inflammatory drugs (NSAIDS). A MIP capable of recognizing Ibuprofen in the aqueous media was developed (Farrington and Regan, 2007). The nature of the interaction between monomer and Ibuprofen was investigated by Amber, using MM method in conjugation with the NMR spectroscopy. The nature of the interaction between Ibuprofen and monomer 2-vinylpyridine was a Π–Π stacking interaction between their aromatic rings. The MIP could be selectively extracted from aqueous conditions by molecularly imprinted solid phase extraction (MISPE) with good selectivity and recoveries up to >80%.

Similar kind of studies using MM method and Hyperchem 7.5 software package was performed on caffeine-imprinted polymers (Farrington et al., 2006). The nature of interaction between template and monomer and the active sites of molecules taking part in complex formation were studied. The relative ratios of template to functional monomer and the choice of solvent (porogen) were also optimized.

Abacavir, which is a HIV-1 reverse transcriptase inhibitor, was chosen as the target analyte (Chianella et al., 2006). MD based LEAPFROG algorithm (Sybyl) was used to screen virtual library of functional monomers to select the best monomers leading to polymers with high binding capacity for the target compound. The MIP prepared using the best monomer acrylamide showed a high binding capacity, up to 157 mg of drug/g of adsorbent in MISPE of abacavir in several different conditions (buffered/non-buffered solutions and in the presence/absence of organic solvents).

Yanez et al. employed MD simulations for the development of MIPs capable of acting as traps for bile salts (cholate salts) in the aqueous medium (Yanez et al., 2010). *N*-(3-aminopropyl)-methacrylate hydrochloride (APMA•HCl), *N,N*-diethylamino ethyl methacrylate (DEAEM) and ethyleneglycol methacrylate phosphate (EGMP) were found as suitable functional monomers from a virtual library of 18 monomers.

The MIPs were synthesized using these monomers and their binding performance was evaluated. These MIPs showed a high capability to remove sodium cholate from aqueous medium. However, the non-imprinted polymers (NIPs) prepared with APMA•HCl (monomer with the highest interaction energy with the template) have shown equal binding capacity as the MIP, for the template. This is due to the high affinity of monomer for the template leading to unspecific interactions that made the imprinting effect (binding by the recognition sites) less important.

18.2.2.3 Optimization of MIP performance through quantum chemical methods

The reports of MIP-design protocols utilizing quantum chemical methods, i.e., semi-empirical, ab initio and DFT are becoming increasingly common in the design and evaluation of MIPs, in particular for studies of the MIP prepolymerization stage (Nicholls et al., 2009).

MIPs were computationally designed to achieve high selectivity toward methocarbamol, an active muscle relaxant (Gholivand and Khodadadian, 2011). A virtual library consisting of 18 functional monomers was built and possible interactions between the template and monomers were investigated using a semi-empirical approach. The monomers with the highest binding scores were then considered for additional calculations using the DFT at B3LYP/6-31G(d,p) level. Acrylic acid (AA) and tetrahydrofuran (THF) were found to be the best choices of functional monomer and porogen, respectively.

Ahmadi et al. used MIPs as a selective sorbent for extraction of 3,4-methylene dioxymethamphetamine (MDMA), a strong central nervous system stimulant from plasma samples (Ahmadi et al., 2011). Ab initio calculations were used to measure ΔE of template–monomer complexes on Gaussian 03 software utilizing B3LYP/6-31-G(d,p) basis set. The MDMA interacted more strongly with methacrylic acid (MAA) in comparison to other functional monomers and the MAA-based MIPs, was used for the cleanup of MDMA from human plasma with the sufficient accuracy and precision.

Computational designing of MIPs was performed with an objective to develop MISPE protocol for selective extraction of methadone from plasma and saliva samples using the DFT at B3LYP/6-31G+(d,p) level (Ahmadi et al., 2012). The MAA gave the largest ΔE in acetonitrile as a porogen. Two MIP were synthesized for methadone using MAA as functional monomer in acetonitrile and methanol, respectively. As expected, the best results were obtained for the MIP prepared in acetonitrile. The extraction recovery of the methadone from plasma and saliva samples using the optimized MIP was 85.0% and 90%, respectively.

Ab initio method was used for synthesis of MIPs for furosemide, which is a potent diuretic drug (Bagher et al., 2010). The ΔE values were calculated through HF/6-31G(d). The furosemide interacted more strongly with acrylamide (AAm) in comparison with other functional monomers. Using AAM-based MIPs, a MISPE procedure was developed for the cleanup of furosemide with sufficient accuracy and precision for trace analysis in plasma samples.

We have used methodology based on DFT calculations for the selection of a suitable functional monomer for a rational design of MIP with a high binding capacity for the important antioxidant gallic acid (GA) (Pardeshi, et al., 2012). A virtual library of 18 functional monomers was created and screened for the template GA at B3LYP/6-31G(d) and the changes in the Gibbs free energy

(ΔG) due to complex formation were determined. Acrylic acid and acrylamide (AAm) gave the highest value of ΔG due to complex formation with GA. 4-vinyl pyridine (4-Vp) had intermediate value of ΔG while, methyl methacrylate (MMA) gave least value of ΔG due to complex formation with GA. The optimized structure of template–monomer complexes is shown in the Fig. 18.4.

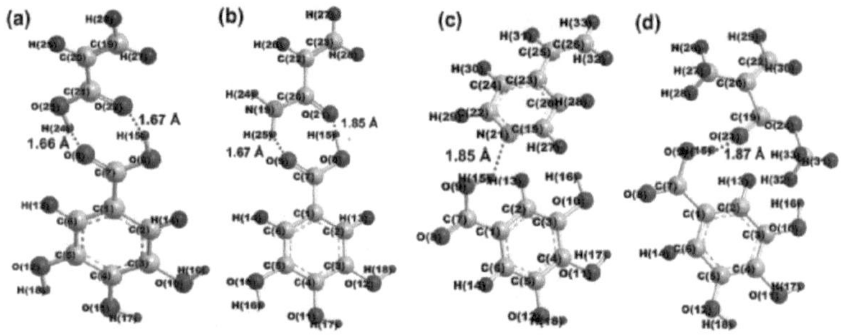

Figure 18.4 The optimized structures of (a) GA-AA, (b) GA-AAm, (c) GA-4Vp, and (d) GA-MMA complex.

Based on this study, the MIPs were synthesized and binding performance was evaluated. The imprinting factor for AA and AAm based MIPs were 5.28 and 4.80, respectively, 4-Vp based MIP had imprinting factor of 2.59 while MMA based MIP exhibited an imprinting factor of 1.95. The experimental results were in good agreement with the computational predictions.

An enantioselective-imprinted polymer for S-warfarin, an anticoagulant (Ahmadi et al., 2014) was designed computationally by using the DFT at B3LYP/631G+(d,p) level. The computational method showed that the MAA and acetonitrile had the highest ΔE for the pre-polymerization complex. Additionally, the mole ratio of 1:3 gave the highest ΔE. The polymer was synthesized with the theoretical optimized conditions. The results of the enantioselective extraction of MIP for R- and S-warfarin revealed about 90% extraction recovery of the S-warfarin from plasma sample.

18.2.2.4 Computational modeling: A tool for study of thermodynamic and spectroscopic properties of MIP

Computational methods are capable of predicting many properties of molecules and reactions and other important data in vacuum and in various solvents, in addition to computing the energy of a particular molecular structure, such as

- thermodynamic and thermochemical properties
- molecular orbitals
- dipole moments
- atomic charges and electrostatic potentials
- polarizability and hyperpolarizabilities
- vibrational frequencies (IR and Raman spectra)
- NMR properties

Additionally, computational chemistry provides data on structural arrangement of species involved in bonding and detailed characterization of existing interactions including interaction energies, bonding distances and charge transfer information. Theoretically calculated IR spectrum includes assignments of vibrational frequencies of different functional groups as well as potential energy distribution (PED) of normal modes. Molecular simulation software packages are employed to depict a three-dimensional vision from molecules, which help us understand and imagine the molecule configurations (Michalska and Bien, 2006).

An understanding of the physical rules governing the formation of monomer–template complexation is fundamental in being able to rationally design polymerization systems with high selectivity and affinity. There is emergence of several reports in MIP literature in the past five years that give a detailed insight into the thermodynamic and spectroscopic properties of MIP system in addition to the selection of optimized precursors. A few examples are discussed in the following paragraphs.

Karlsson et al. simulated a model of prepolymerization system using MD simulations with the aim of exploring the binding site heterogeneity observed in the MIPs (Karlsson et al., 2009). The model comprised total 1199 discrete molecules of template bupivacaine, monomer MAA, explicit solvent molecules of chloroform, cross-linker EGDMA and initiator 2,2′-azobis-isobutyronitrile. The influence of cross-linker, porogen-template interactions, and template conformational variability was studied together in the model. Simulation results were investigated in conjunction with NMR spectroscopic data. Collectively, through a series of statistical analyses of molecular trajectories in conjunction with spectroscopic data it was demonstrated that an ensemble of complex structures is present in the prepolymerization mixture, which is the basis for the binding site heterogeneity observed in the MIPs prepared using the non-covalent strategy.

Computational methods were employed in efforts to elucidate physical mechanisms underlying the ligand selectivity of polymeric sorbents produced through the molecular imprinting of 17-β-estradiol (Olsson et al., 2013). A series of comprehensive MD simulations was performed on the prepolymerization systems. The results from the studies highlight the role of the cross-linker and the importance of the interplay between functionalities of the various monomers employed in template complexation.

A computational approach was proposed to study of MIP selectivity for template chlorphenamine (CPA) over the non-template diphenhydramine (DHA) (Riahi et al., 2009). The complexes were optimized using the density-functional tight-binding, which is an approximate version of the DFT method at the B3LYP/6-311+(d,p). The effects of porogen on template–monomer interactions were also investigated using PCM and changes in the interaction energies, atomic charges, vibrational spectra, dipole moment, and polarizability were studied. The theoretical results indicated that CPA was more strongly bonded to the monomer MAA in comparison with its analogue DHA, which is responsible for the selective behavior of MIPs.

An investigation on the effects of the electrostatic force and the Mulliken charge distribution on the selectivity of MIPs was carried out based on DFT for hydroxyzine- and cetirizine-imprinted polymers (Azimi et al., 2014). The results showed a correlation between the selectivity coefficients and the theoretical charge distributions and also showed that charge distribution based model was able to predict the selectivity coefficients of MIP based potentiometric sensors.

Prasad et al. performed detailed computational modeling on two important pharmaceutical templates. The binding energies of the tryptophan, an amino acid template complexed with different monomers were computed using second order Moller Plesset theory (MP2) at 6–311++g** level in vacuum (Prasad et al., 2012). The monomer, p-nitrophenyl acrylate (2 mol), was found to be the optimized monomer for the MIP development. The binding energies of Hydroxyurea, an antineoplastic agent (Prasad et al., 2013), complexed with different monomers were computed using (MP2) at 6-311++G(d,p) level both in the vacuum and in porogen. This was followed by a detailed theoretical and experimental IR spectroscopic analysis.

Our group reported the quantum chemical computations based on DFT on the template GA, monomer AA and GA–AA complex to study the nature of interactions involved in the GA–AA complex (Pardeshi et al., 2013). B3LYP/6-31+G(2d,2p) model chemistry was used to optimize their structures and frequency calculations. The effect of porogen ACN on complex formation was included by using PCM. The results demonstrated the formation of a stable GA–AA complex through the intermolecular hydrogen bonding between carboxylic acid groups of GA and AA. The Mulliken atomic charge analysis and simulated vibrational spectra (Fig. 18.5) also supported the stable hydrogen bonding interaction between the carboxylic acid groups of GA and AA with minimal interference of porogen ACN.

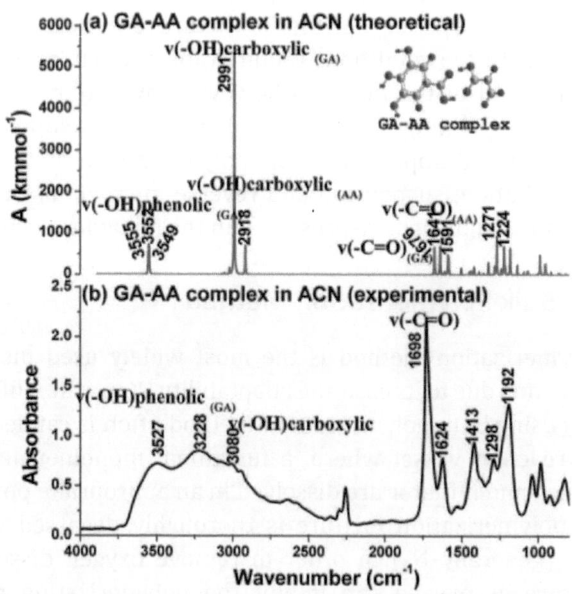

Figure 18.5 (a) Simulated and (b) experimental IR spectra of GA–AA complex in ACN.

These approaches to the rational design of MIPs have shown that in silico screening of suitable monomers for each specific template can significantly shorten the process and improve the success rate. Applying proposed theoretical approach one can evaluate variety of interacting monomer–template systems without laborious synthesis. The next step of MIP development after rational selection of precursors is polymerization. Different polymerization methods are used for synthesis of MIPs.

18.3 Polymerization Methods for Synthesis of MIPs for Pharmaceutical Applications

As the number of technological applications of MIPs continues to increase, there is a growing demand for MIPs tailored to specific physical forms. Polymerization methods to yield the MIPs in the form of size-defined spherical particles at the micro- or nanoscale, monoliths, thick and thin films and molded microstructures have all been published in the last few years (Biffis et al., 2012).

It should be noted that the synthetic conditions required to form a specific type of polymer structure are not necessarily compatible with the conditions required for optimal imprinting of the template. All of the methods however, have utility in at least one area of imprinting and each of them was developed to suit specific target and application. Few polymerization methods (i.e., bulk, precipitation, suspension and reverse microemulsion) along with their pros and cons are presented in the following sections.

18.3.1 Bulk Polymerization Method

Bulk polymerization method is the most widely used method of polymerization due to its ease and adaptability (Kan et al., 2008). MIP synthesis using bulk polymerization method often is carried out in a single reaction vessel where, a functional monomer, template, cross-linker, and initiator are dissolved in an appropriate porogen.

The polymerization mixture is thoroughly degassed with an inert gas (generally N_2) in order to remove oxygen dissolved in this mixture as oxygen can inhibit the polymerization process. The degassing step is normally carried out in an ice bath in order to minimize the possible evaporation of any of the components present, especially those at the lowest concentrations.

After degassing, the flask is closed quickly and tightly and the mixture is ready to polymerize. Thermally initiated polymerization is the most commonly used protocol and is normally carried out at 60°C in an oil or water bath.

UV radiation for polymerization is normally used when any of the compounds involved in the polymerization process is thermally unstable or low temperatures are required for whatever reason.

During polymerization, a minimum amount of porogen is used. Thus, the growing polymer chains precipitate in bulk forming a single monolith solid. The resulting polymer monolith is ground and sieved in order to obtain the desired particle size fraction for the intended application once the polymerization reaction is complete.

Table 18.1 MIPs prepared by bulk polymerization method for the pharmaceutical templates

Template	Sample matrix	Application	References
Tributylammonium cefadroxil	Milk samples	MISPE	Claude et al., 2012
3-Nitro-L-tyrosine	Urine sample	MISPE	Margola et al., 2013
(E)-Resveratrol and related phenolic Compounds	Fruit juice	MISPE	Crescenzi et al., 2012
Penicillin G	Fermentation broth	MISPE	Javanbakht et al., 2012
Enrofloxacin	Water sample	MISPE	Moreno-Bondi et al., 2008
Sitagliptin	Rat plasma and urine	MISPE	Rao et al., 2011
Methidathion	Olive oil	MISPE	Piletska et al., 2011
Tramadol	Plasma and urine samples	MISPE	Javanbakht et al., 2010
Florfenicol	Chicken meat, fish, and honey samples	MISPE	Sadeghi et al., 2013
Esculetin	Ash bark of Chinese traditional medicine	MISPE	He et al., 2005

This has been the only method for MIP synthesis for a very long time. Most of the MIPs prepared using the bulk polymerization with pharmaceuticals as templates has performed well in the SPE and chromatography applications (Table 18.1).

The drawbacks of bulk polymerization are related to issues of postprocessing, amount of material lost, heterogeneous binding sites, and irregular shaped particles (Fig. 18.6) creating the back-pressure in columns (Kan et al., 2008; Wang et al., 2012).

Figure 18.6 Typical particles produced after processing the monolithic polymer obtained by bulk polymerization.

- This method, although simple, convenient, and reliable, does not allow control of the process and produces highly irregular particles with a loss of up to 80% of the material.
- Some (or most) of the imprinted sites may be located inside the bulk of the polymeric matrix and therefore be inaccessible to the desired analyte molecules.

The chromatographic columns or SPE cartridges require specific morphologies of MIPs, i.e., as micron size beads for more efficient packing. Various polymerization methods for MIP preparation had evolved due to the apparent limitations of the bulk polymerization and demand of MIP microspheres in applications.

18.3.2 Precipitation Polymerization

Precipitation polymerization is a simple and attractive method for producing high-quality MIPs as spherical particulates (Kan et al.,

2008; Ye et al., 2001). It is similar to the bulk polymerization, except that higher amount of porogen (about 10 times) is employed in precipitation polymerization. It is a surfactant-free polymerization method that involves polymerization of monomers in dilute solution (typically <5% w/v) in near theta solvents (Sambea et al., 2006).

All the components involved in the synthesis of the polymer are dissolved in a porogenic solvent. This solvent must be able to solubilize all the components involved in the polymerization process at the temperature of polymerization. At the same time, the solvent must be able to solubilize the growing polymeric chains up to certain extent.

Once the polymer reaches a certain critical mass, this solvent is no longer able to hold the polymeric chain in solution, thus precipitating the particles off the solution. The polymerization temperature in which the solvent is able to hold the polymerization process up to a certain critical mass is known as theta (θ) temperature. The polymer precipitates off the solution after this point. Solvents with this particularity for a given polymer are known as θ solvents. Therefore, the selection of a proper θ solvent is essential to obtain discrete particles in the precipitation polymerization method.

The desired product cannot be obtained if the solvent either solubilizes the polymer at the polymerization temperature selected or does not solubilize the oligomers formed in the first steps of the polymerization process.

The advantages of this method include quick delivery of high-quality imprinted polymers with minimal optimization in a one-step process. Furthermore, near monodisperse, spherical particles can be prepared in good yields as presented in Fig. 18.7.

The uniqueness of the precipitation polymerization is development of high level of cross-linking in individual particles. The coagulation between the particles is prevented due to the hard and resilient nature of each particle (Yang et al., 2004).

The precipitation polymerization was invented by Goh and Stover, (2002) and this method was first introduced by Ye et al. (2001) in molecular imprinting, since then it is applied to a large variety of target molecules and successfully applied for the synthesis of MIPs for various pharmaceutical templates (Table 18.2).

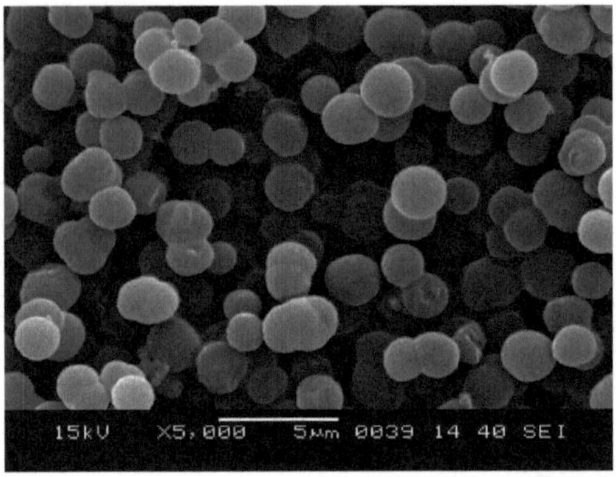

Figure 18.7 Typical microspheres produced by precipitation polymerization for MIP using Gallic acid as a template molecule.

Table 18.2 MIPs synthesized via precipitation polymerization for the different pharmaceutical compounds

Template	Application	Reference
Carbamazepine and oxcarbazepine	HPLC application	Cormack et al., 2009
Matrine and oxymatrine	MISPE of herb sophora flavescens	Haginaka et al., 2011
Andrographolide	MISPE of drug sample	Luo et al., 2011
Diclofenac	MISPE of water sample	Zhang et al., 2011
Cephalosporins	MISPE of Milk sample	Claude et al., 2012
Dipyridamole	Controlled release in serum	Javanbakht et al., 2011
Quercetin	Chemiluminescence sensor	Luo et al., 2012
Barbital	MISPE of barbiturates from human urine	Borrull et al., 2011
Fluoroquinolone	HPLC application	A. Beltran et al., 2012
Caffeic acid	HPLC application	E. Benito-Peña, et al., 2011
Nicotine	MISPE from in cigarette smoke extracts	Haginaka et al., 2006
Morphine	Amperometric detection of morphine based on poly (3,4 ethylenedioxythiophene)	Ho et al., 2005

Achieving a suitable particle size with better yield is important in precipitation polymerization as many parameters affect its mechanism. We have prepared GA based MIPs by the precipitation polymerization and observed the effect of porogen on particle size and specific molecular recognition properties (Pardeshi et al., 2014). MIP, M-100 prepared in the porogen acetonitrile and MIP, M-75 prepared in a mixture of acetonitrile-toluene (75:25 v/v), resulted in the formation of microspheres with approximately 4 μm particle size and surface area of 96.73 m^2g^{-1} and nanoparticles (0.8–1000 nm) and a surface area of 345.9 m^2g^{-1}, respectively. The results have shown that effect of toluene on the particle size of MIPs depends on the type of cross-linker used and its solubility parameter. Matching the solubility parameter of solvent mixture and cross-linkers is important to obtain the desired particle size in MIPs. The MIPs selectively recognized GA in presence of its structural analogues. Pure GA with percent recovery of 75 (±1.6) and 83.4 (±2.2) was obtained from the aqueous extract of herb *Emblica officinalis* by M-100 and M-75, respectively.

The influence of the various parameters on the resulting MIP morphology with the pharmaceutical templates nicotinamide (Sole et al., 2009) and caffeine (Wang et al., 2003) is also reported.

The drawbacks of this method are: Many solvent systems can not be used for this method because their properties do not suit the requirements of method. Another notable limitation of this method is that it is not robust as bulk polymerization as many parameters affect the particle size of resulting MIPs. Sometimes the particle size range produced below the submicron range and the MIP particles cannot be used in desired chromatography applications.

18.3.3 Suspension Polymerization

Suspension Polymerization is another synthetic polymerization method designed to deliver spherical particles with an appropriate size and shape in a single preparative step.

A dispersing solvent with a suitable stabilizing agent is mixed thoroughly in the first step of this polymerization method. Afterwards, an immiscible solution of the dispersing agent, containing template molecule, functional monomer, cross-linking agent, porogen, and initiator is added carefully with vigorous stirring. Immiscibility of the dispersing phase in the porogen and high-speed stirring during the process, leads to the formation of discrete droplets in the micron size range. Once the droplets of the polymerization mixture have formed within the dispersing phase, the polymerization reaction is induced within any discrete droplet. In this way, every droplet acts as a unique reactor. The typical size for the particles obtained under this procedure range from 5 to 100 μm.

The most widely used dispersing agent is water. However, the water can interrupt template–monomer bonding more especially if the non-covalent imprinting method is used. Perfluorocarbon liquid as a dispersing medium was used to produce MIPs with good recognition properties (Mayes and Mosbach, 1996; Hantash et al., 2006).

Mineral oil (liquid paraffin) was also used for the formation of pre-polymerization droplets (Kempe and Kempe, 2006). Being chemically inert in nature, these liquids do not affect the non-covalent interactions in template–monomer complex when used as a dispersing medium. There are some recent reports utilizing suspension polymerization for MIP synthesis for *Staphylococcus* aureus protein A-imprinted polyacrylamide (Pan et al., 2009), Stigmasterol MIP microspheres (Han et al., 2008), MIP hydrogels for the peptide hepcidin (Abbate et al., 2010) and Promethazine based MIPs (Alizadeh et al., 2012).

The major drawback of suspension polymerization method is formation of polydispersed MIP particles. Also, porogenic solvents such as chloroform and toluene cannot be used as these are miscible with mineral oil (Haginaka et al., 2008).

18.3.4 Reverse Micro-Emulsion Polymerization

Reverse microemulsion polymerization method is used in a variety of applications such as to produce the nanotubes (Jang and Yoon, 2013) and quantum dots (Zhang and Fan, 2012). However, it is an emerging method of MIP synthesis.

This technique consists of an aqueous solution of monomer which is emulsified in organic solvent and the polymerization is initiated with initiator which is soluble in organic solvent (Thomas and Zaikov, 2008). A water-in-oil-in-water (W/O/W) double emulsion is formed by adding oil soluble surfactants to the discrete monomer (oil) phase. Water is absorbed from the continuous water phase by the monomer droplets as a result of the stabilizing effect of the oil soluble surfactants.

The use of water in polymerization is environmentally friendly and also allows excellent heat dissipation during the course of the polymerization.

The reverse microemulsion polymerization for MIP preparation has two main advantages.

- First, the polymeric particles obtained are ready to be used immediately without following the tedious postprocessing of polymers such as grinding and sieving and the final product consists of spherical morphology, which is suitable for chromatographic applications (Fig. 18.8).
- Second, since the non-covalent interactions are formed between template and monomer in aqueous phase and large quantity of water is present during the synthesis, MIPs prepared using this protocol is expected to perform well in the aqueous environment.

This can overcome the challenge of MIP to reveal recognition properties in aqueous environment and enable imprinting of the pharmaceutical compounds that are present in aqueous matrices or biological fluids.

Figure 18.8 SEM micrograph of GA-imprinted polymers synthesized via reverse-microemulsion polymerization.

However, there are only few reports on synthesis of MIPs using the reverse microemulsion polymerization method, i.e., antioxidant GA (Pardeshi et al., 2013) and an ophthalmic drug gatifloxacin (Zhang et al., 2012) imprinted polymers. Rhodamine-B (Liu et al., 2009) and 2,4-dichlorophenol (Li et al., 2009) were also imprinted using this method. Molecular imprinting of template was the focus of these reports and none of these mentioned a detailed investigation on the mechanism of the polymerization method and parameters affecting it. Detailed study of these aspects is required to investigate the suitability of method in the area of molecular imprinting of pharmaceuticals.

18.4 Application of MIPs in Pharmaceutical Industries

Use of MIPs as the tailor made sorbent for solid phase extraction (SPE) in selective enrichment, separation or pre-concentration of pharmaceuticals from the complex natural matrices is one of the important applications of the MIPs. MIPs are widely used as the stationary phase for analytical racemic separations of drugs. MIPs also find the use as in vitro controlled/sustained drug delivery devices, although its application in this field is just at a budding stage.

18.4.1 Molecularly Imprinted Solid Phase Extraction of Pharmaceuticals

Molecularly imprinted solid-phase extraction (MISPE) is widely used for selective extraction of pharmaceuticals in variety of matrices (Rao et al., 2011; Javanbakht et al., 2010). The potential value of MISPE lies in the ability to selectively isolate specific compounds or their structural analogues from a complex matrix. The application of MIPs as sorbents allows not only preconcentration and cleaning of the sample but also selective extraction of the target analyte, which is important, particularly when the sample is complex and impurities can interfere with the quantification.

Some advantages of use of MIPs in SPE are the higher sample load capacity and higher stability. The MIPs potentially offer a high level of sample cleanup than that obtained using conventional extraction materials (C18 or silica based material) as demonstrated in various reports.

18.4.1.1 MISPE procedure

The principle of MISPE is similar to classical SPE. The different steps in the MISPE method are described in Fig. 18.9 and each step can be optimized for the best performance of the MISPE method.

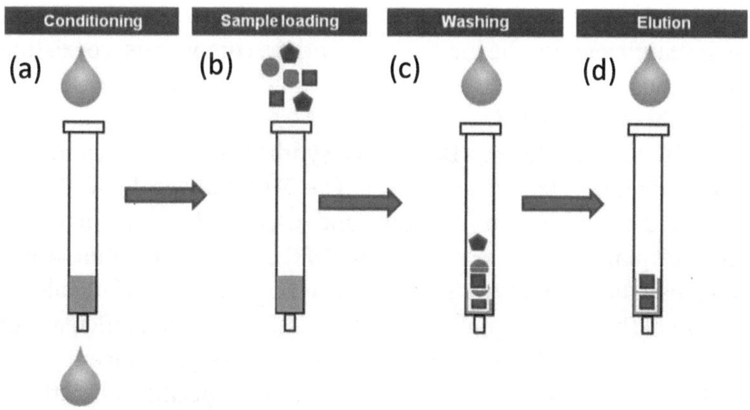

Figure 18.9 MISPE procedures: (a) conditioning, (b) sample loading, (c) washing, and (d) elution step.

(a) **Conditioning**: The conditioning is achieved by passing a suitable solvent through the MIP sorbent. Proper conditioning ensures consistent interaction of the analytes with sorbent and maximizes the retention of analytes in the later steps.

(b) **Sample loading**: A certain volume of the sample is introduced into the MISPE cartridge. The analytes of interest are bound on the stationary phase of the MIPs during this step.

(c) **Washing**: Washing involves use of a suitable solvent to selectively remove unwanted matrix interferences that were co-extracted with the analyte. The solvent should be chosen so that the analytes are not extracted during this stage.

(d) **Elution**: The analytes of interest are extracted from the MIPs sorbents using an appropriate solvent. It is best to choose a solvent that can allow direct injection of elute into an analytical technique. However, if required evaporation and reconstitution of the eluent with mobile phase can be done prior to analysis.

Some recent development in the field is discussed in the subsequent paragraphs.

A MIP was prepared using levonorgestrel (LEV) as template using MAA as functional monomer and EGDMA as cross-linking monomer via a bulk polymerization and MIP particles were packed into a stainless steel column (150 mm × 4.6 mm i.d.). The LEV-imprinted polymer was applied for MISPE of LEV from human serum. The binding ability of the prepared MIP for LEV was found to be sufficient in the presence of other compounds coexisting in serum sample with the recoveries value above 80% (Khorrami and Mehrseresht, 2013).

MIP for diclofenac (DFC) was synthesized by precipitation polymerization (Dai et al., 2011). The MIP showed high affinity toward DFC in aqueous solution and was used as SPE material for the quantitative enrichment of DFC in environmental water samples. The parameters such as washing solvent, elution solvent and breakthrough volume, which affects the extraction efficiency of the MIPs were evaluated to achieve the selective pre-concentration of DFC from water samples and to reduce non-specific interactions. Recoveries of DFC extracted from tap water, river water and wastewater samples were higher than 95%. The MIPs were found to be highly reusable as the stability of the MIP did not vary even after 30 adsorption and desorption cycles.

MIPs were used as sorbent material in MISPE for metoclopramide (MCP) in biological fluids (Javanbakht et al., 2009). The binding capability and selectivity study of the MIPs revealed the high affinity of MIP for MCP in the aqueous environment. The optimized MISPE procedure was used to selectively extract the MCP from human serum and urine with an extraction recovery of more than 90%. The MISPE procedure resulted in efficient cleanup of trace amounts of MCP from serum and urine and the limits of detection of MCP in human serum and urine were 3 and 1.2 μgL^{-1}, respectively.

Lamivudine (an antiviral agent with potent activity against AIDS and hepatitis B virus) based MIPs were prepared and applied as sorbent in MISPE of Lamivudine in the biological fluids (human serum and urine) (Shekarchi et al., 2013). The recoveries of Lamivudine in serum and urine samples were found to be 84.2–93.5% and 82.5–90.8%, respectively. The limit of detection and quantization of lamivudine was 19.34 and 58.6 μg/L in serum and 7.95 and 24.05 μg/L in urine, respectively.

18.4.2 MIPs as the Stationary Phase for Analytical Racemic Separations

Many pharmaceutical industries started to market the pure enantiomer of some drugs, since biological recognition systems are based on optically active molecules and two enantiomers of a racemic drug may interact by different mechanisms with these systems. One of the enantiomers may exert pharmacologically different or unwanted side effects (Kempe, 2001). Most important drugs of whose optically active pure form is demanding are anti-inflammatory, analgesics, antiviral, anticancer, cardiovascular, dermatological, gastrointestinal, ophthalmic, respiratory diseases and other pharmaceuticals used for the central nervous system (Ali et al., 2006). The importance of enantiomerically pure compounds is growing with the increasing structural complexity of new drugs.

The economic interest is very important in the development of the drugs. The separation methods are more economical than the asymmetric synthetic methods, and hence the development of rapid, reproducible, and inexpensive resolution methods is a demanding area. Conventionally, enantiomeric separation is performed by fractional crystallization, microbiological methods, kinetic enzymatic resolution, or by chromatographies (Ilisz et al., 2006), which have the problems of robustness, reproducibility, ease, and cost effectiveness.

MIP-based enantiomeric separation is attractive due to their high enantioselectivity and stability. When an enantiomer is used as the template, then the resulting MIP can show capability of enantiomeric recognition between the pair of enantiomers.

The reviews provide a more comprehensive coverage on topic (Ansell, 2005; Jo et al., 2013). The recent applications in the area are presented in the Table 18.3.

Table 18.3 The enantiomeric separation procedures for various pharmaceuticals using MIPs

Template	Process	Reference
(S)-citalopram	The MIP coated stir-bars for chiral chromatographic separation	Gomez-Caballero et al. 2011
L-tryptophan	Anodic stripping cyclic voltammetry and chronocoulometry.	Prasad et al. 2011
S-Warfarine	Chiral separation chromatography and polarimetry methods.	Ahmadi et al. 2014
bis[1-phenylethyl] amine	Frontal chromatography experiments on the (MIPs) prepared using the chiral functional monomer (S)-2-(2-methyl-acryloylamino)-3-phenyl propionic acid (MAPP)	Torres et al., 2012
L-aspartic acid	L-aspartic acid-imprinted chitosan evaluation by batch rebinding experiments	Monier et al., 2010
aromatic amino acids	Chiral resolution of dl-tryptophan (dl-Trp), dl-phenylalanine (dl-Phe), dl-tyrosine (dl-Tyr) from phosphate buffer solution was achieved by surface functionalized magnetic nanoparticles (MNPs).	Ghosh et al., 2013
β-blockers	The separation of propranolol, metoprolol, atenolol, and pindolol on four different types of chiral stationary phases (CSPs): Chiralpak AD-H, Lux Cellulose-1, Chirobiotic T, and Sumichiral OA-4900 were compared using polar organic (PO) elution mode and normal phase (NP) elution mode.	Morante-Zarcero et al., 2012

18.4.3 MIPs as Drug Delivery Devices

Drug delivery systems (DDS) refer to systems for transporting a pharmaceutical compound in the body as needed to safely achieve its desired therapeutic effect. A number of significant advances are made in the development of new technologies for DDS in the last few years.

An efficient DDS must be capable of regulating the rate of release (delayed or extended-release systems) and/or targeting the drug to a specific site. An ideal DDS should provide a desired rate of delivery of the therapeutic dose, at the most appropriate place in the body, in order to prolong the duration of pharmacological action and reduce the adverse effects, minimize the dosing frequency and enhance patient compliance (Gong et al., 2008).

MIPs have the ability to recognize specific bioactive molecules and have a sorption capacity dependent on the template concentration of the surrounding medium. Thus, it has an enormous potential for creating satisfactory DDS, although imprinted DDS have not reached clinical application yet. However, there is the progressive increase in the number of papers devoted to the application of MIPs in the design of new DDS (Bajpai et al., 2000).

A MIP-based DDS can act as a hydrogels, i.e., as carriers for controlled release of therapeutic compounds. Hydrogels are used as prime carriers for pharmaceutical applications, predominantly as carriers for delivery of drugs, peptides, or proteins (Gong et al., 2008). Another example of MIP hydrogels is imprinted soft contact lenses able to load and release drugs in a controlled way for the treatment of ocular pathologies.

MIPs based DDS can also act as trap systems for removal of undesired substances from the body, e.g., sugars (Parmpi et al., 2004), cholesterol (Tong et al., 2011) and some amino acids (Gong et al., 2008). The systems able to bind such substances in the gastrointestinal tract, hindering their absorption by the body and may be used instead of or as a complement of the pharmacological therapy for an efficient treatment of disease.

The following aspects are of prime importance for the development of MIP based DDS (Gong et al., 2008):

- *Compromise between rigidity and flexibility*:

  The structure of the imprinted cavities should be stable enough to maintain the conformation in the absence of the template, but somehow flexible enough to facilitate the attainment of a fast equilibrium between the release and re-uptake of the template in the cavity.

- *High chemical stability*:

  MIPs based DDS should be stable enough to resist enzymatic and chemical attack and mechanical stress. The system should be stable on contact with biological fluids of complex composition and different pH, in which the enzymatic activity is intense.

- *Safety and toxicological concerns*:

  The system is going to enter into contact with sensitive tissues; therefore, it should not be toxic, nor should be its components, residual monomers, impurities or possible products of degradation.

Recently, specific binding and controlled release of thiamine hydrochloride (THC) to and from a MIP matrix namely, THC-templated-silylated montmorillonite-itaconic acid polymer was investigated. The swelling study of MIP in distilled water showed the maximum swelling capacity is 329.0 g/g at about 8 h. In vitro release experiments showed that MIP releases drug in intestinal fluid in controlled manner (Anirudhan et al., 2013).

The MIPs were prepared with nicotine as the template to evaluate the feasibility of MIPs as excipients for the controlled transdermal delivery of nicotine (Luís et al., 2014). The transdermal delivery of nicotine has been suggested as an aid to smoking cessation therapy because a constant plasma level of nicotine reduces the craving for nicotine and therefore, aids in the abstinence from smoking. The MIPs were prepared and included in transdermal systems with vehicles of different polarities. The drug release kinetics revealed that the MIPs can be considered as suitable candidates for the controlled transdermal administration of nicotine.

18.5 Conclusions

An exciting recent trend is going towards the use of MIPs for pharmaceutical applications.

MIPs are rationally designed using various computational methods for pharmaceutical applications. Theoretical predictions can work hand-in-hand with the experiments to design the MIPs, saving time and money and limiting environmentally hazardous wastes. Application of theoretical calculations can significantly limit the number of experiments performed for the selection of the best compound for a particular experiment. These calculations extract the best precursors out of the pool of possible systems quickly and cost efficiently, thereby accelerating the experiment itself.

The various polymerization methods are used due to demand of specific morphology of MIPs in chromatography and SPE applications. Presently MIPs for pharmaceutical templates are prepared mostly using bulk and precipitation methods. Application of a variety of polymerization methods in mixed organic solvent systems possibly even aqueous polymerizations may allow for better tailoring of the imprinting environment, which may result in increase water compatibility of MIP for extraction of the target pharmaceutical compound from aqueous environment.

MIPs find applications for many purposes in the pharmaceutical industries. A number of highly selective pre-concentrations of pharmaceuticals from biological and environmental samples have been reported. Where, several studies have made direct comparisons with conventional extraction materials and demonstrated superior cleanup using the MIP material. MIPs for the recognition of single enantiomers have advanced considerably during the last few years. The drug delivery application of MIP is in its budding stage however, it is certainly not inconceivable that fully validated comprehensive studies could turn the current situation on its head.

References

Abbate, V., Frascione, N., Bansal, S. S. (2010). Preparation, characterization, and binding profile of molecularly imprinted hydrogels for the peptide hepcidin, *J. Polym. Sci. Part A: Polym. Chem.*, **48**, 1721–1731.

Ahmadi, F., Ahmadi, J., Rahimi-nasrabadi, M. (2011). Computational approaches to design a molecular imprinted polymer for high selective extraction of 3,4-methylenedioxymethamphetamine from plasma, *J. Chromatogr. A*, **1218**, 7739–7747.

Ahmadi, F., Rezaei, H., Tahvilian, R. (2012). Computational-aided design of molecularly imprinted polymer for selective extraction of methadone from plasma and saliva and determination by gas chromatography, *J. Chromatogr. A*, **1270**, 9–19.

Ahmadi, F., Yawari, E., Nikbakht, M. (2014). Computational design of an enantioselective molecular imprinted polymer for the solid phase extraction of S-warfarin from plasma, *J. Chromatogr. A*, **1338**, 9–16.

Ali, I., Kumerer, K., Aboul-Enein, H. Y. (2006). Mechanistic principles in chiral separations using liquid chromatography and capillary electrophoresis, *Chromatographia*, **63**, 295–307.

Alizadeh, T., Ganjali, M. R., Akhoundian, M. (2012). Synthesis and application of different nano-sized imprinted polymers for the preparation of promethazine membrane electrodes and comparison of their efficiencies. *Int. J. Electrochem. Sci.*, **7**, 7655–7674.

Alvarez-lorenzo, C., Guney, O., Oya, T., Sakai, Y., Kobayashi, M., Enoki, T., Takeoka, Y., Ishibashi, T., Kuroda, K., Tanaka, K., Wang, G., Grosberg, A. Y., Masamune, S., Tanaka, T. (2000). Polymer gels that memorize elements of molecular conformation, *Macromolecules*, **33**, 8693–8697.

Andersson, H. S., Nicholls, I. A. (2001). A historical perspective of the development of molecular imprinting. In *Molecularly Imprinted Polymers: Man-made Mimics of Antibodies and Their Applications in Analytical Chemistry* (Sellergren, B., ed.), Elsevier, Amsterdam.

Anirudhan, T. S., Divya, P. L., Nima, J. (2013). Silylated montmorillonite based molecularly imprinted polymer for the selective binding and controlled release of thiamine hydrochloride, *React. Funct. Polym.*, **73**, 1144–1155.

Ansell, R. J. (2005). Molecularly imprinted polymers for the enantioseparation of chiral drugs, *Adv. Drug Deliv. Rev.*, **57**, 1809–1835.

Azimi, A., Javanbakht, M. (2014). Computational prediction and experimental selectivity coefficients for hydroxyzine and cetirizine molecularly imprinted polymer based potentiometric sensors, *Anal. Chim. Acta*, **812**, 184–190.

Bagher, M., Khodadadian, M., Ahmadi, F. (2010). Computer aided-molecular design and synthesis of a high selective molecularly imprinted polymer for solid-phase extraction of furosemide from human plasma, *Anal. Chim. Acta*, **658**, 225–232.

Bajpai, A. K., Shukla, S. K., Bhanu, S., Kankane, S. (2008). Responsive polymers in controlled drug delivery, *Prog. Polym. Sci.*, **33**, 1088–1118.

Batra, D., Shea, K. J. (2003). Combinatorial methods in molecular imprinting, *Curr. Opin. Chem. Biol.*, **7**, 434–442.

Beltran, A., Borrull, F., Cormack, P. A., Marcé, R. M. (2011). Molecularly imprinted polymer with high-fidelity binding sites for the selective extraction of barbiturates from human urine, *J. Chromatogr. A*, **1218**, 4612–4618.

Beltran, A., Marcé, R. M., Cormack, P. A., Borrull, F. (2009) Synthesis by precipitation polymerisation of molecularly imprinted polymer microspheres for the selective extraction of carbamazepine and oxcarbazepine from human urine, *J. Chromatogr. A*, **1216**, 2248–2253.

Benito-Peña, E., Martins, S., Orellana, G., Moreno-Bondi, M. (2009). Water-compatible molecularly imprinted polymer for the selective recognition of fluoroquinolone antibiotics in biological samples, *Anal. Bioanal. Chem.*, **393**, 235–245.

Biffis, A., Dvorakova, G., Falcimaigne-cordin, A. (2012). Physical forms of MIPs. In *Topics in Current Chemistry, Molecular Imprinting* (Haupt, K., ed.), Springer, New York.

Chai, S. N., Zhao, G. H., Zhang, Y. N., Wang, Y. J., Nong, F. Q., Li, M. F., Li, D. M. (2012). Selective photoelectrocatalytic degradation of recalcitrant contaminant driven by an n-p heterojunction nanoelectrode with molecular recognition ability, *Environ. Sci. Technol.*, **46**, 10182–10190.

Chianella, I., Karim, K., Piletska, E. V., Preston, C., Piletsky, S. A. (2006). Computational design and synthesis of molecularly imprinted polymers with high binding capacity for pharmaceutical applications-model case: Adsorbent for abacavir, *Anal. Chim. Acta*, **559**, 73–78.

Cormack, P. A., Elorza, A. Z. (2004). Molecularly imprinted polymers: Synthesis and characterization, *J. Chromatogr. B*, **804**, 173–182.

Dai, C., Zhou, X., Zhang, Y., Liu, S., Zhang, J. (2011). Synthesis by precipitation polymerization of molecularly imprinted polymer for the selective extraction of diclofenac from water samples, *J. Hazard. Mater.*, **198**, 175–181.

David Young, C. (2001). *Computational Chemistry: A Practical Guide for Applying Techniques to Real-World Problems*, John Wiley & Sons, Inc. 42–46.

Del Sole, R., Lazzoi, M. R., Arnone, M., Della Sala, F., Cannoletta, D., Vasapollo, G. (2009). Experimental and computational studies on non-covalent imprinted microspheres as recognition system for nicotinamide molecules, *Molecules*, **14**, 2632–2649.

Dineiro, Y., Menendez, M. I., Blanco-Lopez, M. C., Lobo-Castanon, M. J., Miranda-Ordieres, A. J., Tunon-Blanco, P. (2005). Computational approach to the rational design of molecularly imprinted polymers for voltammetric sensing of homovanillic acid, *Anal. Chem.*, **77**, 6741–6746.

Ensing, K., Boer, T. D. (1999). Tailor-made materials for tailor-made applications: Application of molecular imprints in chemical analysis, *TrAC, Trends Anal. Chem.*, **18**, 138–145.

Esfandyari-manesh, M., Javanbakht, M., Atyabi, F., Mohammadi, A. (2011), Dipyridamole recognition and controlled release by uniformly sized molecularly imprinted nanospheres, *Mater. Sci. Eng., C*, **31**, 1692–1699.

Euterpio, M. A., Pagano, I., Piccinelli, A. L., Rastrelli, L., Crescenzi, C. (2013). Development and validation of a method for the determination of trans-resveratrol and related phenolic compounds in beverages using molecularly imprinted solid phase extraction, *J. Agric. Food. Chem.*, **61**, 1640–1645.

Farrington, K., Magner, E., Regan, F. (2006). Predicting the performance of molecularly imprinted polymers: Selective extraction of caffeine by molecularly imprinted solid phase extraction, *Anal. Chim. Acta*, **566**, 60–68.

Farrington, K., Regan, F. (2007). Investigation of the nature of MIP recognition: The development and characterisation of a MIP for Ibuprofen, *Biosens. Bioelectron.*, **22**, 1138–1146.

Foresman, J. B., Frisch, A. (2004). *Exploring Chemistry with Electronic Structure Method*, Gaussian, Inc Wallingford CT.

Funaya, N., Haginaka, J. (2012). Matrine- and oxymatrine-imprinted monodisperse polymers prepared by precipitation polymerization and their applications for the selective extraction of matrine-type alkaloids from Sophora flavescens Aiton, *J. Chromatogr. A*, **1248**, 18–23.

Gholivand, M. B., Khodadadian, M. (2011). Rationally designed molecularly imprinted polymers for selective extraction of methocarbamol from human plasma, *Talanta*, **85**, 1680–1688.

Gholivand, M. B., Shariati-Rad, M., Karimian, N., Torkashvand, M. (2012), A chemometrics approach for simultaneous determination of cyanazine and propazine based on a carbon paste electrode modified by a molecularly imprinted polymer, *Analyst*, **137**, 1190–1198.

Ghosh, S., Hui, T., Uddin, M. S., Hidajat, K. (2013). Enantioselective separation of chiral aromatic amino acids with surface functionalized magnetic nanoparticles, *Colloids Surf. B*, **105**, 267–277.

Goh, E. C., Stover, H. D. (2002). Cross-linked poly (methacrylic acid-co-poly(ethylene oxide) methyl ether methacrylate) microspheres and microgels prepared by precipitation polymerization: A morphology study, *Macromolecules*, **35**, 9983–9989.

Gomez-caballero, A., Guerreiro, A., Karim, K., Piletsky, S., Goicolea, M. A., Barrio, R. J. (2011). Chiral imprinted polymers as enantiospecific coatings of stir bar sorptive extraction devices, *Biosens. Bioelectron.*, **28**, 25–32.

Gong, C., Wong, K., Lam, M. H. (2008). Photoresponsive molecularly imprinted hydrogels for the photoregulated release and uptake of pharmaceuticals in the aqueous media, *Chem. Mater.*, **20**, 1353–1358.

Haginaka J. (2008). Monodispersed, molecularly imprinted polymers as affinity-based chromatography media, *J. Chromatogr. B*, **866**, 3–13.

Han, Y. P., Lin, Q., Wang, C. (2008). Preparation of stigmasterol molecular imprinting polymeric microspheres by suspension polymerization. *Chin. J. Appl. Chem.*, **26**, 556–559.

Hantash, J., Bartlett, A., Oldfield, P., Dénès, G., O'Rielly, R., Roudiere, D., Menduni, S. (2006). Use of an on-line imprinted polymer pre-column, for the liquid chromatographic-UV absorbance determination of carbaryl and its metabolite in complex matrices, *J. Chromatogr. A*, **1125**, 104–111,

Hu, S., Li, L., He, X. (2005). Solid-phase extraction of esculetin from the ash bark of Chinese traditional medicine by using molecularly imprinted polymers, *J. Chromatogr. A*, **1062**, 31–37.

Ilisz, I., Aranyi, A., Pataj, Z., Péter, A. (2012). Enantiomeric separation of nonproteinogenic amino acids by high-performance liquid chromatography, *J. Chromatogr. A*, **1269**, 94–121.

Jang, J., Yoon, H. (2003). Facile fabrication of polypyrrole nanotubes using reverse microemulsion polymerization, *Chem. Commun.*, **720**, 720–721.

Javanbakht, M., Ali, K., Heidari, A., Akbari-adergani, B. (2012). Extraction and purification of penicillin G from fermentation broth by water-compatible molecularly imprinted polymers, *Mater. Sci. Eng. C*, **32**, 2367–2373.

Javanbakht, M., Mohammad, A., Hadi, M. (2010). Solid-phase extraction of tramadol from plasma and urine samples using a novel water-compatible molecularly imprinted polymer, *J. Chromatogr. B*, **878**, 1700–1706.

Javanbakht, M., Shaabani, N., Akbari-adergani, B. (2009). Novel molecularly imprinted polymers for the selective extraction and determination of metoclopramide in human serum and urine samples using high-performance liquid chromatography, *J. Chromatogr. B*, **877**, 2537–2544.

Jo, W., Ali, F., Choi, J. H., Lee, J. O., Sung, K. Y. (2013). Recent applications of molecular imprinted polymers for enantio-selective recognition, *Talanta*, **106**, 45–59.

Karim, K., Breton, F., Rouillon, R., Piletska, E. V., Guerreiro, A., Chianella, I., and Piletsky, S. A. (2005). How to find effective functional monomers for effective molecularly imprinted polymers? *Adv. Drug Deliv. Rev.*, **57**, 1795–1808.

Kan, X., Zhao, Q., Zhang, Z., Wang, Z., Zhu, J. (2008). Molecularly imprinted polymers microsphere prepared by precipitation polymerization for hydroquinone recognition, *Talanta*, **75**, 22–26.

Karlsson, C. G., Mahony, J. O., Karlsson, J. G., Bengtsson, H., Eriksson, L. A., Nicholls, I. A. (2009). Structure and dynamics of monomer-template complexation: An explanation for molecularly imprinted, *J. Am. Chem. Soc.*, **9**, 13297–13304.

Kempe, M. (2001). Molecularly imprinted polymers in enantiomer separations. In *Molecularly Imprinted Polymers: Man-Made Mimics of Antibodies and Their Applications in Analytical Chemistry* (Sellergren, B., ed.). Elsevier, Amsterdam.

Kempe, H., Kempe, M. (2006). Development and evaluation of spherical molecularly imprinted polymer beads, *Anal. Chem.*, **78**, 3659–3666.

Khorrami, A. R., Mehrseresht, S. (2008). Synthesis and evaluation of a selective molecularly imprinted polymer for the contraceptive drug levonorgestrel, *J. Chromatogr. B*, **867**, 264–269.

Komiyama, M., Takeuchi, T., Mukawa, T. (2003). *Molecular Imprinting from Fundamentals to Applications*, Weinheim, Germany: WILEY-VCH Verlag GmbH & Co. KGaA, Chapter 1.

Li, Y., Li, X., Li, Y., Qi, J., Bian, J., Yuan, Y. (2009). Selective removal of 2,4-dichlorophenol from contaminated water using non-covalent imprinted microspheres, *Environ. Pollut.*, **157**, 1879–1885.

Liu, R., Li, X., Li, Y., Jin, P., Qin, W., Qi, J. (2009). Effective removal of rhodamine B from contaminated water using non-covalent imprinted microspheres designed by computational approach, *Biosens Bioelectron.*, **25**, 629–634.

Luís, A., Ruela, M., Figueiredo, E. C., Pereira, G. R. (2014). Molecularly imprinted polymers as nicotine transdermal delivery systems, *Chem. Eng. J.*, **248**, 1–8.

Mahony, O., Nolan, K., Smyth, M. R., Mizaikoff, B. (2005). Molecularly imprinted polymers—potential and challenges in analytical chemistry, *Anal. Chim. Acta*, **534**, 31–39.

Mayes, A. G., Mosbach, K. (1996). molecularly imprinted polymer beads: Suspension polymerization using a liquid perfluorocarbon as the dispersing phase, *Anal. Chem.*, **68**, 3769–3774.

Mayes, A. G., Whitcombe, M. J. (2005). Synthetic strategies for the generation of molecularly imprinted organic polymers, *Adv. Drug Deliv. Rev.*, **57**, 1742–1778.

Mergola, L., Scorrano, S., Sole, R. D., Lazzoi, M. R., Vasapollo, G. (2013). Developments in the synthesis of a water compatible molecularly imprinted polymer as artificial receptor for detection of 3-nitro-L-tyrosine in neurological diseases, *Biosens. Bioelectron.*, **40**, 336–341.

Wysokińskia, R., Michalskaa, D., Bieńkoa, D. C., Ilakiamanib, S., Sundaraganesanb, N., Ramalingamc, K. (2006). Density functional study on the molecular structure, infrared and Raman spectra, and vibrational assignment for 4-thiocarbamoylpyridine, *J. Mol. Struct.*, **791**, 70–76.

Monier, M., Ayad, D. M., Wei, Y., Sarhan, A. A. (2010). Preparation of cross-linked chitosan/glyoxal molecularly imprinted resin for efficient chiral resolution of aspartic acid isomers, *Biochem. Eng. J.*, **51**, 140–146.

Morante-zarcero, S., Sierra, I. (2012). Comparative HPLC methods for β-blockers separation using different types of chiral stationary phases in normal phase and polar organic phase elution modes, Analysis of propranolol enantiomers in natural waters, *J. Pharm. Biomed. Anal.*, **62**, 33–41.

Nicholls, I. A., Adbo, K., Andersson, H. S., Andersson, P., Ankarloo, J., Hedin-Dahlström, J., Jokela, P., Karlsson, J. G., Olofssona, L., Rosengrena, J., Shoravi, S., Svenson, J., Wikman, S. (2001). Can we rationally design molecularly imprinted polymers? *Anal. Chim. Acta*, **435**, 9–18.

Nicholls, I. A. (1995). Thermodynamic considerations for the design of and ligand recognition by molecularly imprinted polymers, *Chem. Lett.*, **24**, 1035–1036.

Nicholls, I. A., Andersson, H. S., Charlton, C., Henschel, H., Karlsson, B. C., Karlsson, J. G., Mahony, J. O., Rosengren, A. M., Rosengren, K. J., Wikman, S. (2009). Theoretical and computational strategies for rational molecularly imprinted polymer design, *Biosens. Bioelectron.*, **25**, 543–552.

Nicholls, I. A., Andersson, H. S., Golker, K., Henschel, H., Karlsson, B. C. G., Olsson, G. D., Rosengren, A., Shoravi, S., Suriyanarayanan, S., Wiklander, J. G., Wikman, S. (2011). Rational design of biomimetic molecularly imprinted materials: Theoretical and computational strategies for guiding nanoscale structured polymer development, *Anal. Bioanal. Chem.*, **400**, 1771–1786.

Nicholls, I. A., Olsson, G. D., Rosengren, A. M. (2013). Computational strategies for the design and study of molecularly imprinted materials, *Ind. Eng. Chem. Res.*, **52**, 13900–13909.

Norrlöw, O., Glad, M., Mosbach, K. (1984). Acrylic polymer preparations containing recognition sites obtained by imprinting with substrates, *J. Chromatogr. A*, **299**, 29–41.

Olsson, G. D., Karlsson, B. C., Schillinger, E., Sellergren, B., Nicholls, I. A. (2013). Theoretical studies of 17-β-estradiol-imprinted prepolymerization mixtures: Insights concerning the roles of cross-linking and functional monomers in template complexation and polymerization, *Ind. Eng. Chem.*, **52**, 13965–13970.

Pan, J., Xue, X. H., Wang, J. H., Xie, H. M., Wu, Z. Y. (2009). Recognition property and preparation of Staphylococcus aureus protein A-imprinted polyacrylamide polymers by inverse-phase suspension and bulk polymerization, *Polymer*, **50**, 2365–2372.

Pande, V. S., Grosberg, A., Tanaka, T. (1997). How to create polymers with protein-like capabilities: A theoretical suggestion, *Phys. D*, **107**, 316–321.

Pardeshi, S., Dhodapkar, R., Kumar, A. (2013). Water compatible molecularly imprinted microspheres for recognition of gallic acid in aqueous environment, *Proceedings of National Conference on Green Chemistry & Engineering*, 22 March 2013, Nagpur, pp. 13–20.

Pardeshi, S., Dhodapkar, R., Kumar, A. (2013). Quantum chemical density functional theory studies on the molecular structure and vibrational spectra of Gallic acid imprinted polymers, *Spectrochim. Acta Part A*, **116**, 562–573.

Pardeshi, S., Dhodapkar, R., Kumar, A. (2014). Molecularly imprinted microspheres and nanoparticles prepared using precipitation polymerisation method for selective extraction of gallic acid from *Emblica officinalis*, *Food Chem.*, **146**, 385–393.

Pardeshi, S., Patrikar, R., Dhodapkar, R., Kumar, A. (2012). Validation of computational approach to study monomer selectivity toward the template Gallic acid for rational molecularly imprinted polymer design, *J. Mol. Model.*, **18**, 4797–4810.

Parmpi, P., Kofinas, P. (2004). Biomimetic glucose recognition using molecularly imprinted polymer hydrogels, *Biomaterials*, **25**, 1969–1973.

Pérez-Moral, N., Mayes, A. G. (2006). Direct rapid synthesis of MIP beads in SPE cartridges, *Biosens. Bioelectron.*, **21**, 1798–1803.

Piletsky, S. A., Alcock, S., Turner, P. F. (2001). Molecular imprinting: At the edge of the third millennium, *Trends Biotechnol.*, **19**, 9–12.

Polyakov, M. V. (1931). Adsorption properties and structure of silica gel, *Zhurnal Fizicheskoi Khimii*, **2**, 799–805.

Prasad, B. B., Madhuri, R., Tiwari, M. P., Sharma, P. S. (2010). Enantioselective recognition of d- and l-tryptophan by imprinted polymer-carbon composite fiber sensor, *Talanta*, **81**, 187–196.

Prasad, B. B., Rai, G. (2012). Study on monomer suitability toward the template in molecularly imprinted polymer: An ab initio approach, *Spectrochim. Acta, Part A*, **88**, 82–89.

Prasad, B. B., Rai, G. (2013). Molecular structure, vibrational spectra and quantum chemical MP2/DFT studies toward the rational design of hydroxyurea imprinted polymer, *Spectrochim. Acta, Part A*, **105**, 400–411.

Qiu, H., Luo, C., Sun, M., Lu, F., Fan, L., Li, X. (2012). A novel chemiluminescence sensor for determination of quercetin based on molecularly imprinted polymeric microspheres, *Food Chem.*, **134**, 469–473.

Quesada-Molina, C., Claude, B., García-Campaña, A. M., Olmo-Iruela, M., del, Morin, P. (2012). Convenient solid phase extraction of cephalosporins in milk using a molecularly imprinted polymer, *Food Chem.*, **135**, 775–779.

Ramachandran, K. I., Deepa, G., Namboori, K. (2008). *Computational Chemistry and Molecular Modeling, Principles and Applications*, Springer-Verlag Berlin, Heidelberg.

Rao, R. N., Maurya, P. K., Khalid, S. (2011). Development of a molecularly imprinted polymer for selective extraction followed by liquid chromatographic determination of sitagliptin in rat plasma and urine, *Talanta*, **85**, 950–957.

Ratner, B. D. (2004). Diagnostics and biomaterials. In *Biomaterials Science: An Introduction to Materials in Medicine* (Ratner, B. D., ed.), Elsevier Academic Press, China, Chapter 7.18, p. 691.

Riahi, S., Edris-tabrizi, F. (2009). A computational approach to studying monomer selectivity towards the template in an imprinted polymer, *J. Mol. Model.*, **15**, 829–836.

Sadeghi, S., Jahani, M. (2013). Selective solid-phase extraction using molecular imprinted polymer sorbent for the analysis of Florfenicol in food samples, *Food Chem.*, **141**, 1242–1252.

Sambea, H., Hoshina, K., Moaddel, R., Wainer, I. W., Haginaka, J. (2006). Uniformly-sized, molecularly imprinted polymers for nicotine by precipitation polymerization, *J. Chromatogr. A*, **1134**, 88–94.

Sellergren, B., Hall, A. J. (2001). Fundamental aspects on the synthesis and characterisation of imprinted network polymers. In *Molecularly Imprinted Polymers: Man-made Mimics of Antibodies and Their Applications in Analytical Chemistry* (Sellergren, B., ed.), Elsevier, Amsterdam.

Sergeyeva, T. A., Piletsky, S. A., Brovko, A. A., Slinchenko, E. A., Sergeeva, L. M., El'skaya, A. V. (1999). Conductimetric sensor for atrazine detection based on molecularly imprinted polymer membranes, *Analyst*, **124**, 331–334.

Shekarchi, M., Pourfarzib, M., Akbari-adergani, B., Mehramizi, A., Javanbakht, M., Dinarvand, R. (2013). Selective extraction of lamivudine in human serum and urine using molecularly imprinted polymer technique, *J. Chromatogr. B*, **931**, 50–55.

Takagishi, T., Klotz, I. M. (1972). Macromolecule–small molecule interactions; introduction of additional binding sites in polyetyleneimine by disulfide cross-linkages, *Biopolymers*, **11**, 483–491.

Tamayo, F. G., Casillas, J. L. (2003). Highly selective fenuron-imprinted polymer with a homogeneous binding site distribution prepared by precipitation polymerization and its application to the clean-up of fenuron in plant samples, *Anal. Chim. Acta*, **482**, 165–173.

Thomas, S., Zaikov, G. E. (2008). *Polymer Nanocomposites Research Advances, U. S. A.*, Nova Science Publishers, New york.

Tokonami, S., Shiigi, H., Nagaoka, T. (2009). Review: Micro- and nanosized molecularly imprinted polymers for high-throughput analytical applications, *Anal. Chim. Acta*, **641**, 7–13.

Tomasi, J., Persico, M. (1994). Molecular interactions in solution: An overview of methods based on continuous distributions of the solvent, *Chem. Rev.*, **94**, 2027–2094.

Tong, Y. J., Guan, H. M., Wang, S. F., Xu, J. W., He, C. B. (2011). Syntheses of chitin-based imprinting polymers and their binding properties for cholesterol, *Carbohydr. Res.*, **346**, 495–500.

Torres, J. J., Gsponer, N., Ramírez, C. L., Vera, D. M., Montejano, H. A., Chesta, C. A. (2012). Experimental and theoretical studies on the enantioselectivity of molecularly imprinted polymers prepared with a chiral functional monomer, *J. Chromatogr. A*, **1266**, 24–33.

Turiel, E., Martin-Esteban, A., Tadeo, J. L. (2007). Molecular imprinting-based separation methods for selective analysis of fluoroquinolones in soils, *J. Chromatogr. A*, **1172**, 97–104.

Piletsky, S., Turner, A. (2006). *Molecular Imprinting of Polymers*, Landes Bioscience, U.S.A.

Valero-Navarro, A., Gómez-Romero, M., Fernández-Sánchez, J. F., Cormack, P. A. G., Segura-Carretero, A., Fernández-Gutiérrez, A. (2011). Synthesis of caffeic acid molecularly imprinted polymer microspheres and high-performance liquid chromatography evaluation of their sorption properties, *J. Chromatogr. A*, **1218**, 7289–7296.

Wang, D., Hong, S. P., Yang, G., Row, K. H. (2003). Caffeine molecular imprinted microgel spheres by precipitation polymerization, *Korean, J. Chem. Eng.*, **20**, 1073–1076.

Wang, Y., Ding, Y., Rong, F., Fu, D. (2012). A study of the precipitation polymerization of bisphenol A-imprinted polymer microspheres and their application in solid-phase extraction, *Polym. Bull.,* **68**, 1255–1270.

Wei, S., Mizaikoff, B. J. (2007). Recent advances on noncovalent molecular imprints for affinity separations, *Sep. Sci.*, **30**, 1794–1805.

Wu, L. Q., Zhu, K. C., Zhao, W. P., Li, Y. Z. (2005). Theoretical and experimental study of nicotinamide molecularly imprinted polymers with different porogens, *Anal. Chim. Acta*, **549**, 39–44.

Wulff, G., Sarhan, A. (1972). Use of polymers with enzyme-analogous structures for the resolution of racemates, *Angew. Chem. Int. Ed.*, **11**, 341–344.

Yanez, F., Chianella, I., Piletsky, S. A., Concheiro, C., Alvarez-Lorenzo, C. (2010). Computational modeling and molecular imprinting for the development of acrylic polymers with high affinity for bile salts, *Anal. Chim. Acta*, **659**, 178–185.

Yañez, F., Martikainen, L., Braga, M. E. M., Alvarez-Lorenzo, C., Concheiro, A., Duarte, C. M. M., Gil, M. H., de Sousa, H. C. (2011). Supercritical fluid-assisted preparation of imprinted contact lenses for drug delivery, *Acta Biomater.*, **7**, 1019–1030.

Yang, S., Shim, S. E., Lee, H., Kim, G. P., Choe, S. (2004). Size and uniformity variation of poly(MMA-Co-DVB) particles upon precipitation polymerization, *Macromol. Res.*, **12**, 519–527.

Ye, L., Cormack, P. A., Mosbach, K. (2001). Molecular imprinting on microgel spheres, *Anal. Chim. Acta*, **435**, 187–196.

Yin, X. Y., Liu, Q. S., Jiang, Y. F., Luo, Y. M. (2011). Development of andrographolide molecularly imprinted polymer for solid-phase extraction, *Spectrochim. Acta Part A*, **79**, 191–196.

Zhang, H., Dramou, P., He, H., Tan, S. H., Pham-Huy, C., Pan, H. J. (2012). Molecularly imprinted stationary phase prepared by reverse micro-emulsion polymerization for selective recognition of gatifloxacin in aqueous media, *J. Chromatogr. Sci.*, **50**, 499–508.

Zhang, X., Fan, Y. (2012). Preparation of spherical silica particles in reverse micro emulsions using silicon tetrachloride as precursor, *J. Non-Cryst. Solids*, **358**, 337–341.

Chapter 19

Chitosan and Nano Chitosan: Properties and Application to Textiles

D. P. Chattopadhyay[a] and M. S. Inamdar[b]

[a]*Department of Textile Chemistry, Faculty of Technology and Engineering, The Maharaja Sayajirao University of Baroda, Vadodara, India*
[b]*Faculty of Textile Processing, Sarvajanik College of Engineering & Technology, Surat, India*

19.1 Introduction

Prof. Henri Braconot, director of Botanical Garden in Nancy (France) in 1811, first isolated a fibrous substance from the cell walls of mushroom, which he called FUNGINE. Later in 1823, Odier discovered that this compound is also one of the major constituents of the exoskeleton of insects and then he renamed as CHITIN (from Greek *khitōn* meaning tunic or envelope). Prof. C. Rouget in 1859 prepared a compound from chitin by treatment with concentrated caustic solution and observed that, unlike chitin, the resulting substance dissolved in acids. This compound was then named CHITOSAN (pronounced as kite-o-san) by Hoppe-Seiler in 1894 [1, 2].

Ecological considerations, now days, are becoming important factors in the selection of consumer goods, all over the world. The

Handbook of Sustainable Polymers: Processing and Applications
Edited by Vijay Kumar Thakur and Manju Kumari Thakur

ISBN 978-981-4613-53-8 (Hardcover), 978-981-4613-54-5 (eBook)
www.panstanford.com

consumers demand not only the right quality product, at right time, at a reasonable price, but also with no harm to ecology during the manufacture as well as in the use. Hence, there has been a constant urge to scientists and industrialists to explore and adopt the substitutes that are non-hazardous and eco-friendly. The use of natural dyes on textiles has been one of the consequences of increased environmental awareness. Enzymes are commercially available for the processes such as desizing, scouring, bleaching, and finishing of cotton fabric treatments. Eyes are today focused on biopolymers to minimize the use of hazardous synthetic polymers in textile processing. Natural polymers are becoming more and more attractive as raw material for the manufacture of novelty products with wider applications textiles, medicine, and agriculture. Cellulose, the most common natural polymer, is a well-established source for standard textiles. Others such as polyaminosaccharides, alginates, starch, collagen, gelatin, etc., are promising naturally occurring starting materials for specialty products [3]. Chitosan, another promising example, is biodegradable, biocompatible with animal and plant tissues, and non-toxic [4]. Chitin, the precursor of chitosan, is a nitrogen-containing polysaccharide and is the second most abundant biopolymer after cellulose and is renewable. It is widely distributed in the shells of crustaceans such as crabs, shrimps, and lobsters, as well as in the exoskeleton of marine zoo-plankton, including coral, jellyfish, and squid pens [1, 2, 4].

The structure of chitosan is very much close to that of cellulose except the hydroxyl group in C (2) of cellulose is being replaced by amino group in chitosan (Fig. 19.1).

Chitosan, due to its polycationic nature, is gaining enormous attention for its unique properties such as biodegradability, biocompatibility with animal and plant tissues, non-toxic nature, anti-bacterial, anti-fungal, anti-viral, anti-acid, anti-ulcer, non-toxic, non-allergenic, etc., as well as film formation, fiber formation, bead formation, hydrogel formation, etc. By virtue of these properties, chitosan has prospective applications in many fields such as medical, water treatment, cosmetics, dentifrices, food, and textile industries. In textiles, it can be used in the primary production of fibers, textile auxiliary chemicals and finishing agents [1, 4–6].

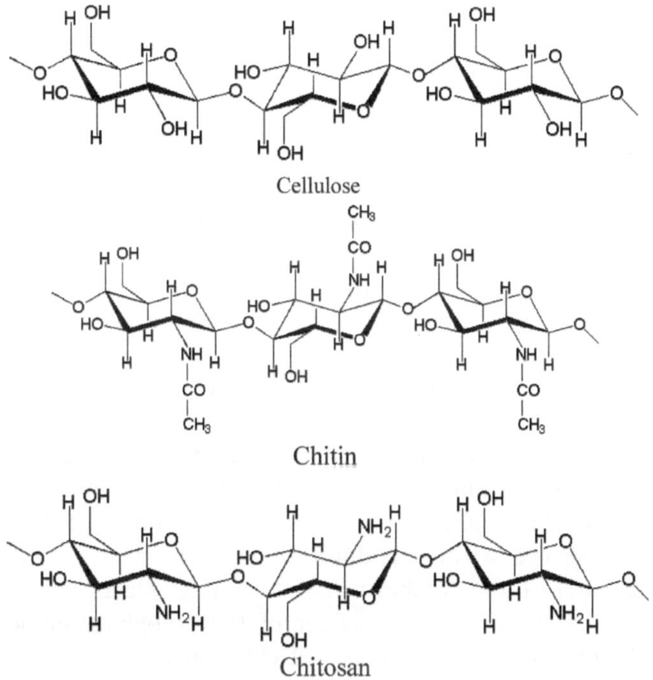

Figure 19.1 Structures of chitin, chitosan and cellulose.

19.2 Chemistry of Chitin and Chitosan

Chitosan [$(C_6H_{11}O_4N)_n$] is a polysaccharide composed of a linear (1–4) linked 2-amino-2-deoxy-β-d-glucan (i.e., β-d-glucosamine) in the chair $^4C_1$ conformation. It is derived by alkaline deacetylation of chitin, i.e., (1–4) linked 2-acetaamido-2-deoxy-β-d-glucan (i.e., *N*-acetyl-β-d-glucosamine) (Fig. 19.2). When chitin is boiled with highly concentrated sodium hydroxide solution, deacetylation takes place producing chitosan having free amino groups and sodium acetate as the byproduct. The mechanism of deacetylation is similar to alkali hydrolysis acid amides [7]. In highly alkaline medium, the hydroxyl ions of sodium hydroxide (OH^-) attack the electron deficient carbonyl carbon of acetamide group on chitin to form an intermediate anion. Chain scission follows and release free amino group on main chain with liberation of acetic acid. The liberated acid then gets neutralized with sodium hydroxide.

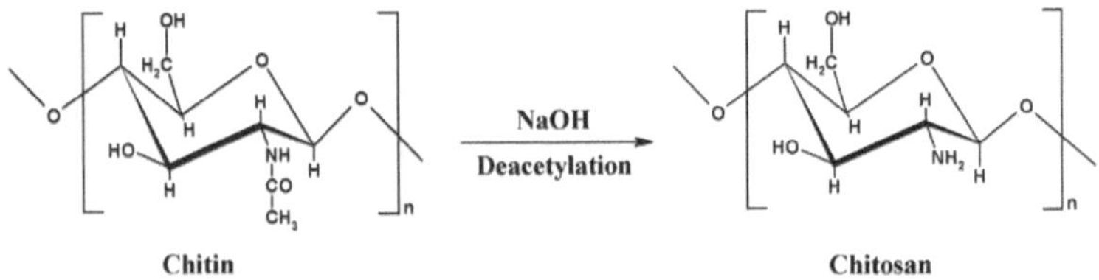

Figure 19.2 Synthesis of chitosan by deacetylation of chitin.

Between chitin and chitosan, a number of derivatives of partially *N*-acetylated chitosans exist, characterized by the degree of deacetylation (DAC). The degree of deacetylation has a direct impact on the secondary structure of the polymeric chain and can also influence the solubility of the polymer in organic or aqueous solvents. As an evolved nomenclature, chitineous substances that do not dissolve in dilute organic acids, e.g., 1–2% acetic acid, are collectively called "chitin." On the other hand, chitineous substances that can dissolve in aqueous dilute acids are referred as chitosan, approximately 60% or above deacetylated products fall in this category. Indeed, it is a copolymer of *N*-acetyl-glucosamine and glucosamine units. The structure of chitosan is very much close to that of cellulose except the hydroxyl group in C(2) of cellulose is being replaced by amino group in chitosan (Fig. 19.1) [8–12].

The structural characterization of chitosan is investigated using FTIR spectra (Fig. 19.3). The region above 1500 cm^{-1} wave number is commonly known as "functional group region" to identify functional groups such as primary amine, alcohols, methylene groups, etc. Another region that is below 1500 cm^{-1} wave number is known as the "finger print region" and is known to characterize the saccharide backbone [13]. A broad band observed at wave number around 3355 cm^{-1} is mainly attributed to O–H, NH, and NH_2 stretch [14–16] and the absorption peak in region 2864 to 2874 cm^{-1} is due to the C–H symmetrical and unsymmetrical stretch. The absorption band at 1651 cm^{-1} is assigned to C=O (carbonyl) stretching of secondary (amide I) amide group, which is characteristic of *N*-acetyl group and the medium peak at 1585 cm^{-1} is due to bending vibrations of N–H of amide II bond (*N*-acetyl residue) and the primary amine [14, 15]. Another medium absorption peak at 1374 to 1376 cm^{-1} characterizes the N–H of amide III bonds [15]. Absorption peaks at wave numbers 1151, 1066, 984, and 895 cm^{-1} are attributed, mainly, to secondary hydroxyl groups (-CH–OH) due to C–O bending in cyclic alcohols or saccharide structure [14–16]. A strong absorption peak at 1025 cm^{-1} is due to 1° (primary) hydroxyl group, characteristic peak of $-CH_2OH$ in primary alcohols, arising from C–H bending [15]. Also, various bending vibrations of C–O at different peaks like 894 and 1151 cm^{-1} in fingerprint region characterize the saccharide structure.

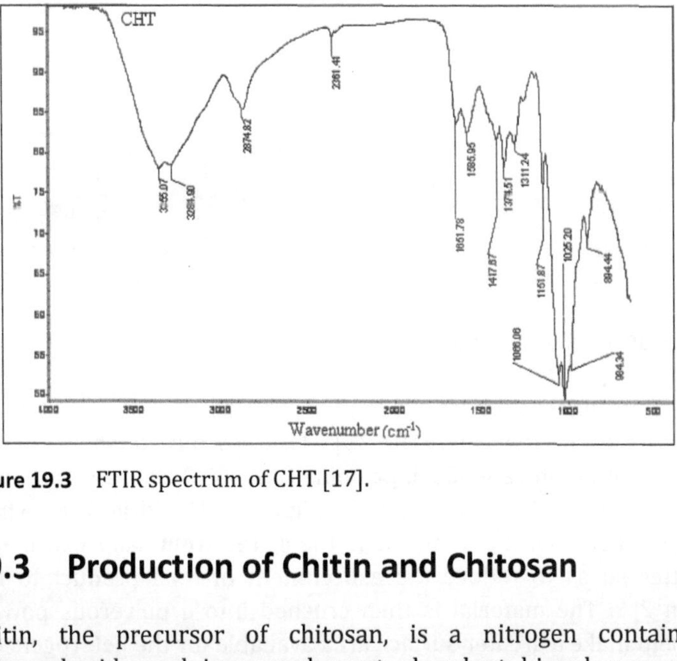

Figure 19.3 FTIR spectrum of CHT [17].

19.3 Production of Chitin and Chitosan

Chitin, the precursor of chitosan, is a nitrogen containing polysaccharide and is second most abundant biopolymer after cellulose. It is widely distributed in the shells of crustaceans such as crabs, shrimps, lobsters, prawns, squilla, etc., as well as in the exoskeleton of marine zoo-plankton, including coral, jellyfish, and squid pens. About 20–40% chitin is present the exoskeleton of these animals. It is also present in smaller quantities in insects such as butter flies ladybugs, and the cell walls of yeast, mushrooms, and other fungi (Fig. 19.4). However, since the crustacean shells (crabs, shrimps, lobsters, etc.) are waste products of food industry, these are commercially employed for the production of chitin and chitosan [1, 4, 18]. It is believed that at least 10 gigaton of chitin is synthesized and degraded and it is also estimated that over 150,000 tons of chitin is available for commercial use annually.

Figure 19.4 Sources of chitin.

Chitosan is commercially produced by the alkaline deacetylation of crustaceans' chitins [19, 20]. The crustacean shells mainly consist of chitin (20–30%), proteins (30–40%), calcium carbonate (30–50%), lipids, and traces of pigment. The dried/wet shells are washed thoroughly to make them free from sand extraneous matter so as to reduce the ash content of final product to less than 2%. The material is then crushed into a pulverous powder to help make a greater surface area available for the heterogeneous processes to follow. The proteins are removed by treatment with sodium hydroxide solution (5%) at about 85–100°C for 30 min in a stainless steel vessel. The boiled raw material is allowed to cool and is then washed to with water to remove all the traces of alkali. The deproteinized shells are transferred to a mild steel vessel lined with fiber glass and is treated with hydrochloric acid (3%) at room temperature to remove calcium carbonate. The excess acid is decanted and the residue is washed until the pH is neutral. Excess water is removed using a screw press and the dried. Lipids are extracted by soaking in organic solvents such as acetone or ethanol. An oxidative bleaching treatment with hydrogen peroxide or sodium hypochlorite is also given to obtain a white chitin powder.

Conversion of chitin into chitosan involves the deacetylation process, which is a harsh treatment usually performed with concentrated sodium hydroxide solution. Chitin flakes are treated in suspension with aqueous 40–50% caustic solution at 80–120°C with constant stirring for 4–6 h and this treatment is repeated once or more than once for obtaining high-amino-content product. To avoid depolymerization due to oxidation, sodium borohydrate is added. Excess alkali is drained off and the mixture is washed with water several times until it is free from alkali. Most of the alkali is then used either in deproteinization or in deacetylation. Excess water is removed in screw press and the wet chitosan cake is either sun dries or in drier at 60°C. Chitosan thus obtained is in the form of flakes and can be pulverized to powder. The flowchart for the manufacture of chitosan from the starting material (crustacean shells) is shown in Fig. 19.5.

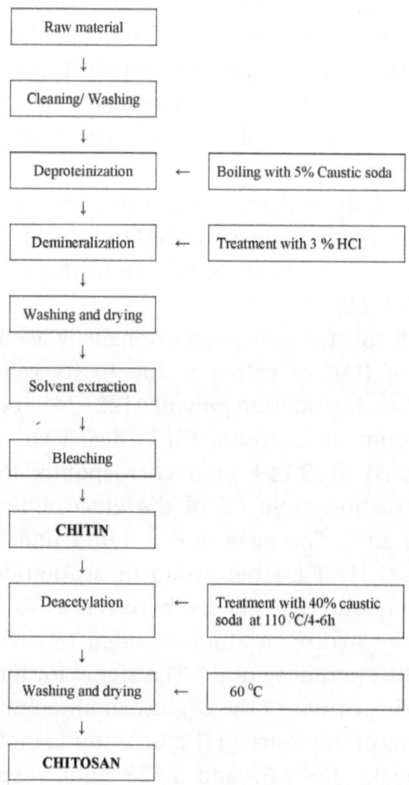

Figure 19.5 Flowchart for the manufacture of chitin and chitosan.

19.4 Physical Properties of Chitosan

Chitosan is a white fibrous material produced in different forms such as fibers, beads and membranes. The physical properties of chitosan are governed, at large, by two factors: the DAC and the molecular weight. Various factors such as source of chitin and processing parameters deacetylation determine the grades of chitosan. Chitosan being a primary aliphatic amine, it can be protonated by selected acids (pKa of chitosan = 6.3). It is insoluble in water, organic solvents, and alkalis but is soluble in organic acid solutions.

19.4.1 Degree of Deacetylation

The deacetylation process involves the removal of acetyl groups from chitin molecules. The DAC is defined as the average number of glucosamine units per 100 monomers expressed as a percentage. It determines the content of free amino groups ($-NH_2$) in the chitosan and is one of the most important chemical characteristics that influence the physicochemical properties, biological properties, antibacterial activity, and applications of chitosan. In other words, DAC value determines the functionality, reactivity, polarity, and water solubility of the polymer. Chitin does not dissolve in dilute acetic acid. When chitin is deacetylated to a certain degree (~60% deacetylation) where it becomes soluble in the acid, it is referred to as chitosan [18, 21].

The ^{1}H-NMR spectra have been extensively used in quantitative determination of DAC of chitosan due to its reliability and are shown to cause no degradation polymer [22–24]. It can be observed from the spectrum of chitosan, CHT (Fig. 19.6) that the signal (chemical shift, δ) at 3.114 ppm corresponds to the hydrogen bonded to the carbon atom C2 of the glucosamine ring and the integral (I_{H2}) is 3.66. The peak at δ = 1.964 ppm corresponds to acetyl group (i.e., H of methyl group of acetamido moiety) with intensity 1.01. While the signals between 3.30 and 4.00 ppm correspond to the hydrogen atoms bonded to carbons C3, C4, C5 and C6 of the glucopyranose units. The signal for anomeric protons H1 (the anomeric protons of the D-glucosamine units) and H1′ (the anomeric protons of the *N*-acetyl D-glucosamine units), respectively, are traced at peaks δ = 4.67 and 5.423 ppm, respectively having the intensities 1.03 and 0.73, respectively.

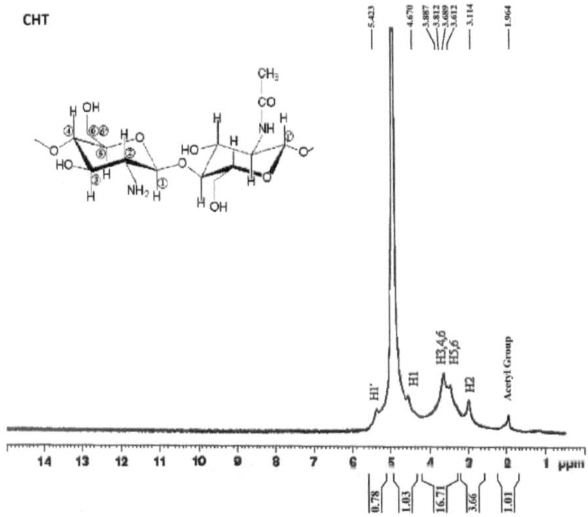

Figure 19.6 ^{1}H NMR spectrum of CHT.

From the signal integrals of ^{1}H-NMR spectrum, the (DAC, %) can be evaluated by using Eq. (19.1):

$$\mathrm{DAC}(\%)=\left(1-\frac{\frac{1}{3}(\text{Signal Intensity due to acetyl group}, I_{\mathrm{NAc}})}{(\text{Signal Intensity due to } H \text{ of C2}, I_{\mathrm{H2}})}\right)\times 100 \quad (19.1)$$

$I_{\mathrm{NAc}} = \delta$ at 1.964 ppm = 1.01
$I_{\mathrm{H2}} = \delta$ at 3.114 ppm = 3.66

$$\mathrm{DAC}=\left(1-\frac{1/3(1.01)}{(3.66)}\right)\times 100 = 90.8\%$$

19.4.2 Solubility

The solubility of chitosan is very important for its utilization, such as for chemical modification and film or fiber formation. Although the polymer backbone consists of hydrophilic functional groups chitosan is normally insoluble in water and most common organic solvents (e.g., DMSO, DMF, NMP, organic alcohols, pyridine). The insolubility of chitosan in aqueous and organic solvents is a result of its crystalline structure, which is attributed to extensive intramolecular and intermolecular hydrogen bonding between the chains and sheets, respectively (Fig. 19.7) [25, 26].

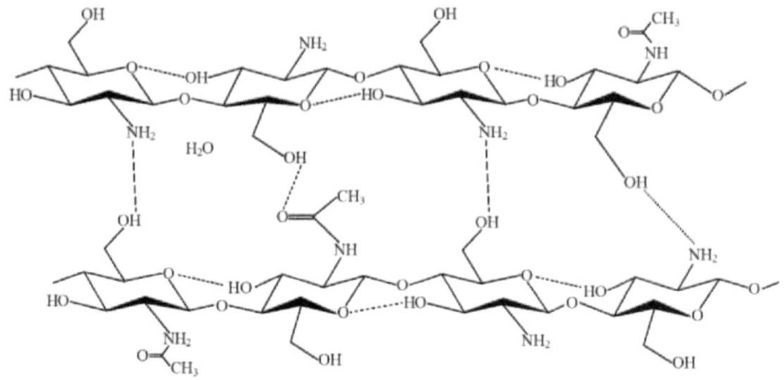

Figure 19.7 Crystalline structure of chitosan.

The pKa value of chitosan is 6.3 and is therefore soluble in dilute mineral or organic acids by protonation of free amino groups at pH below about 6.5. This cationic nature is the basis of a number of applications of chitosan. Acetic and formic acids are most widely used for the solubilization of chitosan [27]. The solubility of chitosan is determined by molecular weight, DAC value and pH and ionic valency of acid used. Generally, the solubility decreases with an increase in molecular weight. Oligomers of chitosan with a degree of polymerization (DP) of 8 or less are water-soluble regardless of pH [2, 18].

19.4.3 Molecular Weight

The molecular weight of chitosan is another important property that determines its suitability for a particular application. It determines the viscosity of its solution and the strength of chitosan fiber and film. The molecular weight of chitin and chitosan depends on its source and deacetylation conditions (time, temperature, and concentration of NaOH), respectively [1, 4]. Molecular weight of chitosan determined by various methods such as viscometry [28], gel permeation chromatography [29] and light scattering spectrophotometry [30].

The viscosity of polymer solution, at the molecular level, is a direct measure of the hydrodynamic volume of the polymer molecules which in turn is governed by the molecular size or chain length and hence the molecular weight [31, 32]. Therefore, the viscosity, measured by capillary viscometer, is widely employed to determine the average molecular weight of a polymer by using the Mark–Houwink equation, i.e., Eq. (19.2).

$$[\eta] = k[M_V]^{\alpha}, \quad (19.2)$$

where M_V is the viscosity average molecular weight of polymer, α and k are constants (α = 0.83 and $k = 1.4 \times 10^{-4}$ for 0.25 M acetic acid and 0.25 M sodium acetate solvent system) [28] and $[\eta]$ is the limiting viscosity number or intrinsic viscosity and can be determined from the Huggins equation (Eq. 19.3) [32].

$$[\eta] = \lim_{c \to 0} (\eta - \eta_s)/\eta_s C \quad (19.3)$$

where η is the viscosity of solution and η_s is the viscosity of solvent and C is the concentration. As indicated in Eq. (19.3), when $(\eta-\eta_s)/\eta_s C$, i.e., reduced viscosity (η_{red}) is plotted against concentration (C), the intercept to the Y-axis corresponds to intrinsic viscosity (η).

In general, the viscosity average molecular weight (M_v) of chitosan is determined using capillary viscometer, e.g., Ubbelohde viscometer (Fig. 19.8) by measuring the flow time (T) of chitosan solutions at varying concentrations (C) through the determination of reduced viscosity and intrinsic viscosity as shown in Fig. 19.11 and using Mark–Houwink equation, i.e., Eq. (19.2).

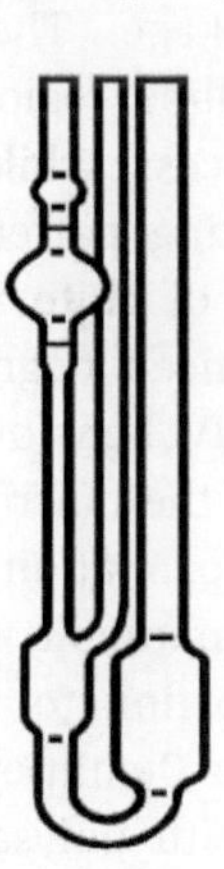

Figure 19.8 Ubbelohde capillary viscometer.

Chitosan obtained from deacetylation of crustacean chitin may have a molecular weight over 10^6 da. Consequently, it is necessary to reduce the molecular weight by chemical methods to much lower molecular weight for easy application as a textile finish. Depolymerization is a process of preparation of low-molecular-weight chitosan from high-molecular-weight chitosan. Chitosans of desired molecular weight can be obtained by controlled depolymerization by chemical, physical and enzymatic methods such as acid hydrolysis (HCl, HNO_2, etc.) [33–37], free radicals (H_2O_2, $K_2S_2O_8$) [38], enzymatic [36, 39], radiations (UV, γ rays) [40, 41], ultrasound [42], microwave and thermal treatments [43]. In physical methods, Baxter et al. [42] studied the effect of ultrasonication on the degradation of chitosan. They found that intrinsic viscosity of samples decreased exponentially with increasing sonication time and rates of intrinsic viscosity decreased linearly with ultrasonic intensity. Choi et al. [41] investigated the depolymerization of chitosan using irradiation with different doses (2~200 KGy) of gamma rays. Their results showed that viscosity of irradiated chitosan rapidly decreased with increasing irradiation dose.

Feng Tian et al. [38] reported the method for depolymerization of chitosan by hydrogen peroxide and their IR and ^{1}HNMR studies confirmed the depolymerization leading to breakage of 1–4-β-D-glucoside bonds of chitosan. They proposed the attack of per hydroxyl radical (HO^-), generated from the H_2O_2 decomposition, at amino group leading to hydrogen abstraction. The resulting unstable glucosamine radical in presence of water causes rupture of adjacent glucosidic linkage. They also proposed that the depolymerization in crystalline region mainly takes place at the surface by peeling-off process while the amorphous region is depolymerized by penetrating pattern. Trzcinski [40] employed the combined degradation of chitosan using hydrogen peroxide and UV light for the enhanced degradation rate. Photochemical degradation by means of UV light also undergoes as a result of free radical reactions along the macromolecules. Kabal'nova et al., 2001 [44] employed the ozonation of chitosan in dilute acid solution and showed a remarkable decrease of molecular mass of polysaccharide in proportion to reaction time or amount of applied ozone. According to Demin et al. [45], the initial stage of the interaction of ozone with polysaccharide is its electrophilic attack on C(1)-H bond with the formation of labile hydrotrioxides, destruction of which results in depolymerization of polysaccharide.

Enzymatic degradation of chitosan polymer is possible to get low-molecular-weight chitosan. Enzymatic process has the advantage of its specificity and the ease in the fractionation of product [46]. Lee et al. [39] used commercial bromelain for the partial depolymerization of chitosan. The enzymatic reactions were determined by the liberation of reducing sugar and the reduction in viscosity. They proposed/recommended 7% w/w enzyme treatment for 2 h at pH 5 for the satisfactory results. Ilyina et al. [47] reported chitinase and chitosanase as most effective enzymes, which are found in fungus, bacteria, and plants, to hydrolyze chitosan. Other enzymes such as glucosanase, lipase, tannase, and protease were found to have chitosanolitic activities, as reported by Yalpani and Pantaleone [46].

Nitrous acid method for the oxidative depolymerization of chitosan has been well documented [35, 37, 48, 49]. When chitosan solution is treated with nitrous acid, generated from acidic sodium nitrite, undergoes deamination reaction with subsequent cleavage of β-glycosidic linkages. A series of chitosans with desired molecular weights could be obtained by changing chitosan/$NaNO_2$ molar ratio, chitosan initial concentration and reaction time [49, 50]. The molecular weight of the depolymerized chitosan was linear with chitosan/$NaNO_2$ ratio. The reproducibility of this method was fairly good. The overall stoichiometry of the reaction between chitosan and nitrous acid (HONO) is shown Fig. 19.9. Nitrosating species originating from HONO attack the amino groups and subsequently cleave the β-glycosidic linkages to produce low-molecular-weight product and 2,5-anhydro-D-mannose is formed as the reducing end group of the cleaved polymer [35, 37, 48].

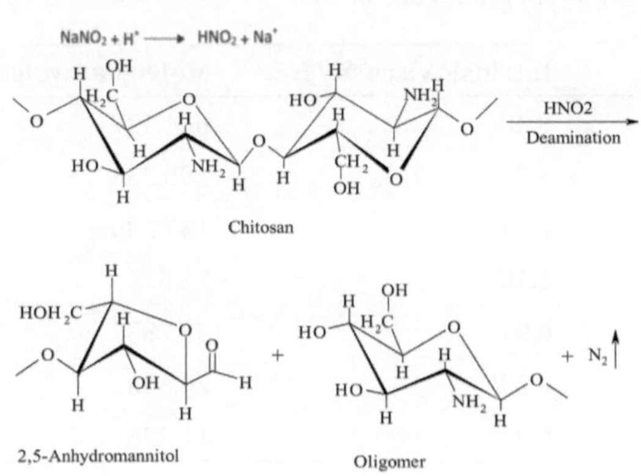

Figure 19.9 Depolymerization of chitosan by nitrous acid.

This method has the advantage of ease of operation with better control over depolymerization level besides reproducibility. It is envisaged from Fig. 19.10 that the IR spectra of chitosan and depolymerized chitosan are almost similar which indicates that the process of depolymerization caused no chemical change in the structure of the polymer except reduction in molecular weight, which is evident from the change in viscosity as presented in Table 19.1. Similar observations are reported by Hebeish et al. [35], and Mao et al. [49].

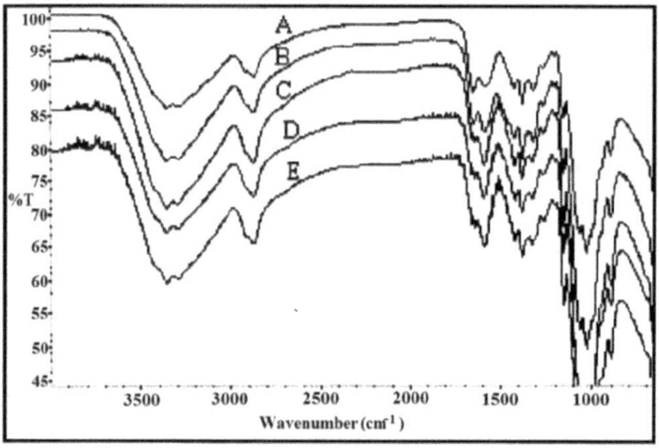

Figure 19.10 FTIR spectra of parent chitosan and its low-molecular-weight derivatives (A = CHT, B = CHT-D2, C = CHT-D3, D = CHT-D4 and E = CHT-D5) [50].

Table 19.1 Intrinsic viscosity and viscosity average molecular weights of different grades of chitosan

Chitosan	Intrinsic viscosity (η)	Molecular weight, M_v
CHT-MC	9.40	654,127
CHT-D1	4.72	285,231
CHT	2.55	135,839
CHT-D2	1.50	71,676
CHT-D3	0.91	38,733
CHT-D4	0.535	20,698
CHT-D5	0.34	11,986

19.4.4 Viscosity Behavior

Viscosity of chitosan solution is an important property that determines its suitability for commercial applications and is determined by various factors such as DAC, molecular weight, concentration, ionic strength, pH, and temperature [51]. The viscosity of chitosan solution increases with increase in molecular weight and concentration [52]. Several studies have shown that physical and chemical treatments affect its viscosity. Viscosity of chitosan decreased with increasing treatment time of grinding, heating, autoclaving, ultrasonication and ozonation [53], and decreased from 248 to 32 cP with increasing deproteinization time from 0 to 30 min [54].

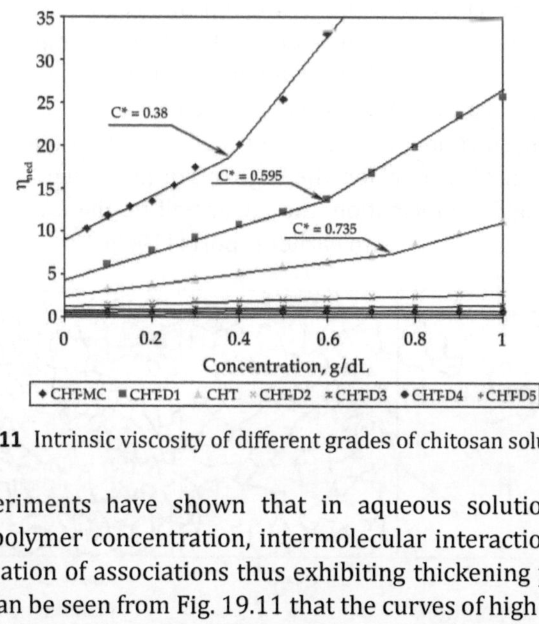

Figure 19.11 Intrinsic viscosity of different grades of chitosan solutions [50].

Experiments have shown that in aqueous solutions, above certain polymer concentration, intermolecular interactions lead to the formation of associations thus exhibiting thickening properties [50]. It can be seen from Fig. 19.11 that the curves of high molecular weight of chitosan, namely CHT-MC, CHT and CHT-D1, do not strictly follow the linearity of the Huggins Eq. (19.3) in the selected range of concentration. These curves show that the inflection at a certain critical concentration (C^*) and then after the curves bended upward. Further, the value of C^* increases with decrease in molecular weight and ultimately the curves flatten for low-molecular-weight chitosan, used in this research work. This can be explained on the fact that, when chitosan or any other polymer is added into a solvent, the solvent gradually diffuse into the polymer aggregates resulting into the swelling of the polymer. As swelling continues, the segments of polymer are solvated and loosened out. Since the molecules in a solid polymer are entangled with neighboring ones, polymer molecules during dissolution diffuse out as bunches of entangled molecules. Even though all chain segments of a polymer molecule in solution are unfolded and fully solvated, the molecules does not assume the shape of an extended straight chain but present in a coil form. These coils or aggregates offer resistance for the mobility or flow of molecules and hence impart viscosity [31, 55]. When the molecular size and concentration is increased, as in our case, the extent of entanglement is increased as shown in Fig. 19.12. In other words, the critical concentration (C^*) is, indeed, the "overlap" concentration. Thus, when $C > C^*$ the intermolecular entanglements or aggregation predominate and precludes the overall molecular motion of polymers, while individual polymers molecules are statistically separated from other molecules at $C < C^*$. Thus, the critical concentration (C^*) is a measure of molecular size and conformation of a polymer, the higher the molecular weight and the more rigid conformation, the lower will be the C^*. The results are in close agreement with earlier reports [19, 56].

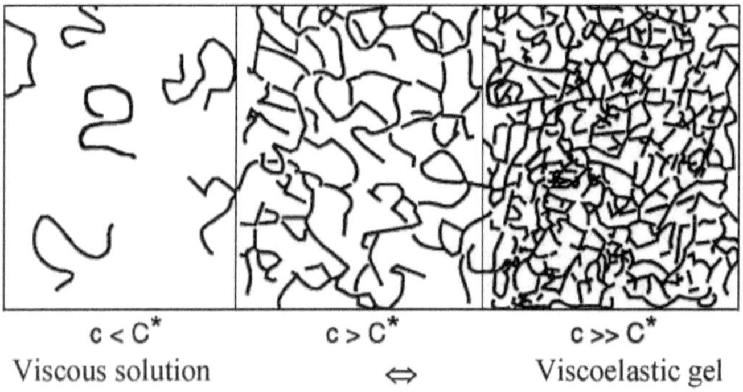

Figure 19.12 Aggregation of chitosan molecules as a function of molecular weight and concentration.

Polymeric chemicals are generally applied to textiles by padding technique where it is required to prepare a standing bath. Thus, the chemical remains in contact with water for a longer period. As biodegradability of chitosan is a well-known phenomenon, the effect of storage time on the stability of its solution was studied in terms of change in solution viscosity [50]. Figure 19.13 shows that the viscosity of chitosan solution is adversely affected with increase in storage time. It seems that the initial molecular weight and the concentration of polymer influenced the stability of the solution as found from the shift of *C**. Figures 19.13 and 19.14 reveal that the drop in viscosity in first 24 h is much faster than the latter and is more significant at higher concentration, i.e., above *C**. The loss in viscosity may be attributed to the biodegradation of chitosan molecules and/or hydrolysis of polymer molecules and slow detachment of polymer segments from the entanglements which are present above *C**. Initially the large molecules, especially at higher concentration, occupy large "hydrodynamic" volume due to aggregation as result of intermolecular cross-linkages, which leads to higher viscosity. The hydrolytic degradation of the polymer leads to the production of smaller molecular entities, which in turn causes a drop in hydrodynamic volume of the polymer molecules resulting in higher molecular mobility and as a result reduces the viscosity of the solution [28, 31, 55, 56]. It means chitosan solutions made from low-molecular-weight samples (e.g., CHT-D5) at lower concentrations are more stable in terms of viscosity behavior as illustrated in Fig. 19.15. Hence, lower-molecular-weight chitosan solutions particularly at low concentration may be more preferred for textile applications as their standing baths have more consistent viscosity.

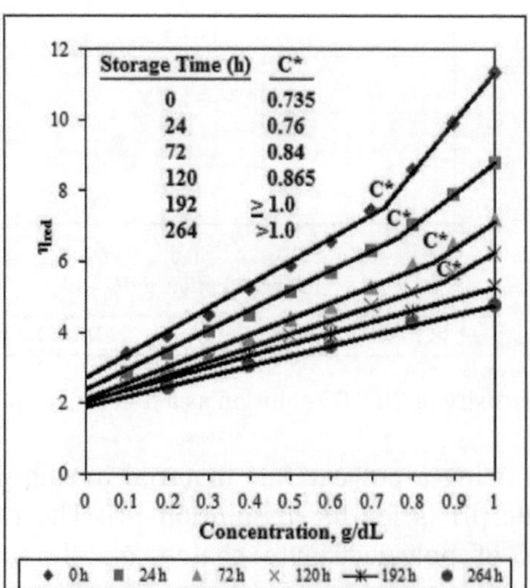

Figure 19.13 Viscosity of chitosan (CHT) solution as a function of storage time [50].

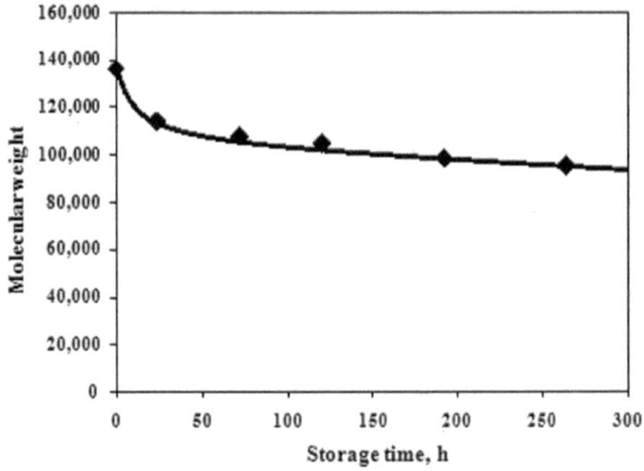

Figure 19.14 Effect of storage time of chitosan (CHT) solution on molecular weight.

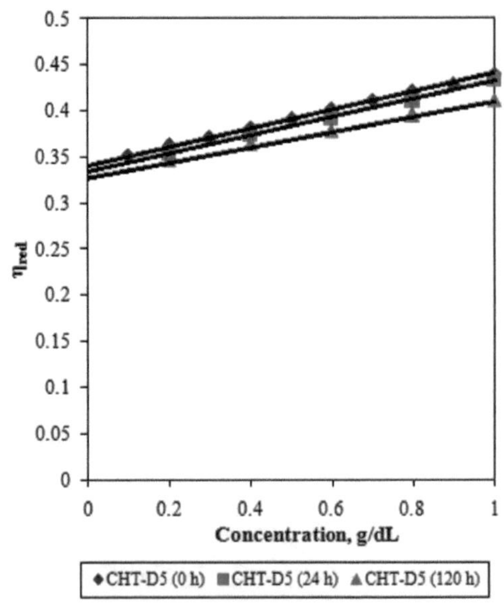

Figure 19.15 Viscosity of CHT-D5 solution as a function of storage time.

Chitosan, being a polycationic material having pKa value 6.3, requires acidic pH is for its dissolution [4]. The conformational arrangements of polyelectrolyte chains, mainly responsible for viscosity behavior, are influenced by various factors such as pH, concentration of polyelectrolyte, molecular weight, nature of counter ion and added electrolyte, etc. [57–59]. Chitosan salts such as acetate, formate, lactate, citrate, chloride, and nitrate are soluble in water [4] while sulfate is insoluble [60, 61]. Poly anions such as sodium tripolyphosphate (TPP) lead to ionotropic gelation of chitosan, which is one of the fundamental principles of nano chitosan synthesis [62]. Velásquez et al. [63] reported the effect of sodium chloride on the behavior of two chitosan salts namely nitrate and chloride. The η_{red} values for chitosan nitrate were found to be markedly more affected than chitosan chloride by the increasing added ionic strength of the medium. According to them, nitrate counter ions cause a smaller stiffness than chloride counter ions, which has been attributed to higher screening caused by bigger nitrate ions.

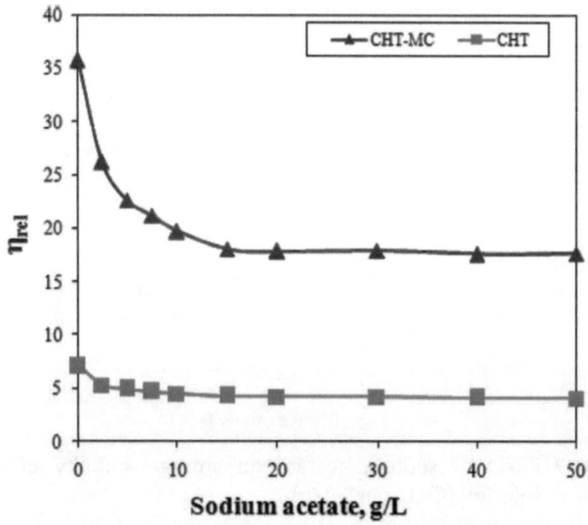

Figure 19.16 Effect of sodium acetate concentration on viscosity of chitosan solutions [50] (Chitosan 5 g/L, Acetic acid 15 g/L).

Incorporation of sodium acetate has been found to drop the viscosity of chitosan solution (Fig. 19.16) [50]. The effect of electrolyte was observed to be more pronounced on high-molecular-weight chitosan, CHT-MC. With decrease in molecular weight of chitosan, the amount of sodium acetate required to attain a minimal viscosity was decreased. The decrease in viscosity with increase in electrolyte concentration can be attributed to the shielding effect of counter ions [4]. Due to ion dipole forces, the acetate ions form a cascade negative charge over each chitosan molecule establishing repulsive forces between them. This offers low resistance to the flow or mobility of polymer molecules. As the number of amino groups content is large in high-molecular-weight chitosan, higher charge screening is effected due to added electrolytes resulting in to intensive lowering in viscosity of the solution [63]. The drop in viscosity of chitosan solution due to sodium acetate was found to be almost uniform during storage period as shown in Fig. 19.17. The viscosity of chitosan solution, thus, is influenced by the storage time and also by the presence of electrolyte.

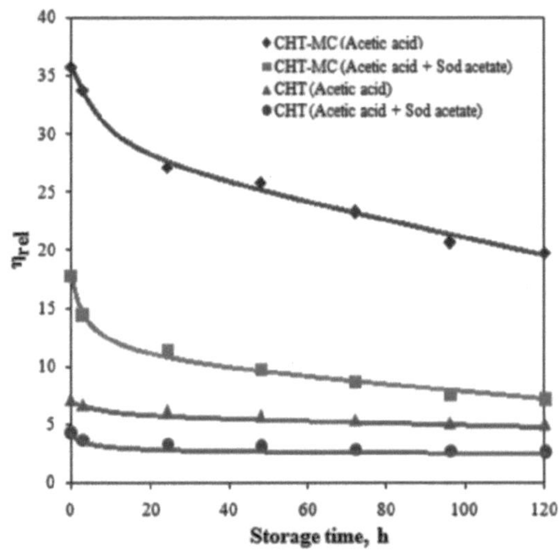

Figure 19.17 Effect of sodium acetate on storage stability of chitosan solution (Chitosan 5 g/L).

19.4.5 Chitosan Hydrogel

Hydrogels are the cross-linked polymer networks that hold a large amount of water. The polymers used to prepare hydrogels normally consist of a large portion of hydrophilic groups and the formed networks are prevented from dissolving due to the chemical or physical bonds between the polymer chains. Water can penetrate into the networks, resulting in the swelling of the hydrogels. Depending on the methods fabrication, the dimensions of hydrogels can vary from nanometer to centimeters in width and in different shapes such as films, capsules, sponges, microparticles, composites, beads, etc.

Chitosan-based gels may be broadly divided into thermally reversible gels and non-reversible gels. Reversible gels are normally obtained by physical cross-linking via ionic gelation with anionic molecules while non-reversible gels are produced by *N*-acylation and Schiff's base (aldemide) formation. One of the simplest ways to prepare chitosan gel is to treat chitosan acetate solution with carbodiimide. Chitosan hydrogel beads by physically cross-linking with TPP were developed as a pH-sensitive drug release system to encapsulate glipizide, an insulin stimulating drug. The hydrogel beads were proposed to release the encapsulated drug slowly upon subcutaneous injection, with good tolerability and prolonged half life [64]. In chemical cross-linking, the Schiff's base formation system with dialdehydes such as glyoxal and glutarldehyde is the most widely accepted one. This involves the covalent cross-linkages formation between chains [4]. Further, for the food industry, hydrogel beads may be more feasible to achieve and industrialize, due to the low cost and controllable preparation parameters, and scalable procedures, compared with nano/microparticles delivery systems in food industry. Hydrogels are also used as a media for affinity chromatography for enzymes, media for gel permeation chromatography, for the isolation of bovine serum albumin, and a wound-dressing material [1, 4, 65–67].

19.4.6 Chitosan Membrane

Chitosan, particularly high molecular weight, exhibits good film-forming properties as a result of intra- and intermolecular hydrogen bonding. The films are mostly flexible, tough, transparent, and colorless with smooth and shiny surface. Chitosan films, in general, are produced from acidic solutions containing volatile acid such as formic or acetic acid. The viscous solution is spread on glass plate and dried on water bath to evaporate the acid and moisture. The film, i.e., chitosan formate or acetate, which is soluble in water, is detached from glass sheet and the treated with 1M NaOH solution and washed thoroughly to neutral pH and then dried at 60°C [19, 68].

19.5 Chemical Properties of Chitosan

Chitosan possesses three reactive groups, viz., primary (C-6) and secondary (C-3) hydroxyl groups on each repeat unit, and the amino group at the C-2 position on each deacetylated unit. These reactive groups can be chemically modified to alter the mechanical and physical properties, and solubility of chitosan. The typical reactions involving the hydroxyl groups of chitosan are etherification and esterification. The nucleophilic amino group in chitosan allows reactions such as *N*-alkylation, *N*-acylation and reductive alkylation by reacting chitosan with alkyl halides, acid chlorides and aldehydes or ketones, respectively. Selective *O*-substitution can be carried out by protecting the amino group during the reaction. Cross-linking or graft copolymerization can also be carried out to modify chitosan [1, 43, 69].

Chitin and chitosan are biodegradable in the biosphere, in the agriculture soil, and in the hydrosphere to produce oligosaccharides [1]. These are mostly attacked, in vivo, by several non-specific proteases such as lysozyme, papain, pepsin, etc., present in all mammalian tissues. Their biodegradation leads to the release of non-toxic oligosaccharides of variable length, which can be subsequently incorporated to glycosaminoglycans and glycoproteins, to metabolic pathways or be excreted. The rate of degradation is governed by molecular weight, DAC and the distribution of acetyl groups. Degradation has been shown to increase as DAC decreases [70–72].

Chitosan is totally non-toxic and its degradation products namely glucosamines are biocompatible. Although the gastrointestinal enzymes can partially degrade both chitin and chitosan, when both polymers are orally administered, they are not absorbed. Chitosan shows a LD_{50} of around 16 g/kg, very similar to the salt and glucose values in assays carried out on mice [1, 73]. Toxicity of chitosan is reported to depend on DAC, which decreases with increase in DAC. On the other hand, Mw of chitosan did not influence toxicity [73].

19.6 Applications of Chitosan

The major driving force in the development of new applications for chitosan lies in the fact that the polymer is biodegradable, biocompatible, and possesses specific adsorption properties.

Attributing to the polycationic nature and several inherent properties, chitosan is currently receiving a great deal of interest as regards its applications in several diversified areas such as medical, food, textiles, water processing, cosmetics, agriculture, and paper [2].

19.6.1 Chitosan in Agriculture

Chitosan has many potential applications in agriculture for its essentially biodegradable characteristics. It is used primarily as a plant growth enhancer, and as an eco-friendly biopesticide substance that boosts the ability of plants to defend against fungal infections such as *Fusarium solani*. Chitosan triggers the defensive mechanisms in plants, acting much like a vaccine in humans [74].

In addition to growth stimulation properties, chitosan is used for seed-coating on cotton, corn, seed potatoes, soybeans, sugar beets, tomatoes, wheat, etc, frost protection, protective coating for fruits and vegetables, controlled release of fertilizers, nutrients, etc., into the soil. Chitosan increases photosynthesis, promotes and enhances plant growth, stimulates nutrient uptake, increases germination and sprouting, and boosts plant vigor. Due to its ability to form a semipermeable coating, chitosan extends the shelf life of treated fruit and vegetables by minimizing the rate of respiration and reducing water loss. Horticultural applications of chitosan increases blooms and extend the life of cut flowers and Christmas trees [6, 75].

19.6.2 Chitosan in Food Technology

Chitosan has been already been used as a food ingredient in Japan, Europe and in United States as a lipid trap, an important dietetic breakthrough. Since, chitosan is not digested by the human body; it acts as a fiber, a crucial diet component. It binds the lipids arriving in the intestine, thereby reducing the cholesterol absorption by about 20 to 30% by the human body [70, 76]. Chitosan can be used in meat preservation by inhibiting some meat spoilage bacteria such as *Pseudomonas fragi, Bacillus subtilis, Escherichia coli,* and *Staphylococcus aureus* and some meat starter cultures namely *Lactobacillus plantarium, Pediococcus pentosaceus,* and *Micrococcus varians.* Besides, chitosan films can be used as a packaging material for the quality preservation of a variety of food [70, 77]. Other important applications of chitosan in food sector are as follows: as a thickening and stabilizing agent in sauces and other culinary dishes, as flocculating agent it is used to clarify beverages, etc. Use of chitosan for fruit juice clarification is associated with the elimination of pectin and carbohydrates suspended particles from beverages [6, 78].

19.6.3 Chitosan in Cosmetics

Chitosan is a natural cationic gum that has been used for various cosmetic applications, it is used to maintain skin moisture, treat acne, tone skin, protect epidermis, reduce static electricity in hair, fight dandruff, etc. Incorporation of chitosan salt in shampoos confers shine and strength to hair due to the ionic interactions between chitosan and hair proteins. When applied to the surface of the skin, chitosan forms a protective and moisturizing elastic film. This makes chitosan useful in the formulation of moisturizing agents (e.g., lotions, sunscreens). Bath lotion containing chitosan lactate, chitosan succinate and chitosan alkyl phosphate increases skin softness. It may be noted that substances absorbing the harmful UV radiation or different dyes can be easily covalently linked to chitosan amino groups. Compositions based on chitosan and other hydrocollids containing antioxidants, anti-allergic and anti-inflammatory substance of vegetable origin, new types of depilatory and means for curling and doing the hair are being work out [6, 79].

19.6.4 Chitosan in Biomedicine

It is in the field of health that the many properties of chitosan (bacteriostatic, immunologic, antitumoral, cicatrizant, hemostatic, and anticoagulant) are of interest. For example, because of its biocompatibility with human tissue, chitosan's cicatrizant properties have proved its effectiveness as a component, notably, in all types of dressings (artificial skin, corneal dressings, etc.), surgical sutures, dental implants, and in rebuilding bones and gums. Due to the high *N*-amino content, chitosan acts as a powerful natural magnetic attraction for lipids, fats, and bile in the digestive tract, and actually binds with them to prevent their absorption into the bloodstream. The attracting ability of chitosan can possibly reduce cholesterol and triglycerides blood plasma levels, which contribute to obesity and cardiovascular disease. Various hypolipemic formulations including particles, powders, solutions, and injections containing chitosan, were prepared for oral administration. Some of the important medical applications of chitosan involve; wound healing and wound dressing [6, 69, 80], burn treatment [69], surgical adhesion [6, 81], ophthalmology [69, 82], drug delivery systems [43, 69, 70, 83], tissue engineering [6, 69, 84, 85] and bone tissue engineering [6, 43].

19.6.5 Chitosan in Paper Industry

Chitosan can be used to strengthen the recycled paper and increase the environmental friendliness of packaging and other products. Chitosan is already involved in the manufacture of paper because chitosan molecules greatly resemble those of cellulose the main constituent of plant walls [6, 69].

Chitosan can be a useful component in paper printing due its film forming property. Coating of paper with chitosan/gelatine coatings improves dye fixation on the paper. In ink jet printing, pigment based inks containing a polystyrene maleic anhydride polymer system that reacts with a fixer fluid containing chitosan salt such as chitosan acetate. A gel precipitate is formed on the substrate by the interaction of the chitosan in the fixer fluid and the reactive polymer system in the ink. The gel forms a protective film on the substrate providing benefits such as increased drying time, smear fastness, smudge fastness, and water fastness [86].

19.6.6 Chitosan in Chromatography

Chitosans find a wide variety of applications in chromatographic separations. The presence of free amino and hydroxyl groups in chitosan makes it a useful chromatographic support. Chitosan can be used in thin layer chromatography for separation of nucleic acid and solid phase extraction of phenols and chlorophenols [69].

19.6.7 Chitosan in Solid-State Batteries

Chitosan dissolved acetic acid solution can be employed for the fabrication of solid-state proton conducting batteries. The conductivity is due to the transport of protons through microvoids in polymer. Small dielectric constants from piezoelectric studies attributed the presence of many microvoids in polymer structure. The choice of a more stable electrode material may produce better battery system [69].

19.6.8 Chitosan in Biocatalysis

Chitosan is widely used as supports for enzyme and cell immobilization due to its appropriate characteristics. Immobilization is the process in which the enzyme, cells or organelles is confined in a definite position thus rendering an insoluble form that retains the catalytic activity and can be reused several times. The enzymes or cells are bound to the carrier material via reversible surface interactions. The forces involved are van der Waals forces and ionic and H-bonding interaction as well as hydrophobic forces. Chitosan support being a positively charged polymer binds negatively charged proteins bind easily [43, 70].

19.6.9 Chitosan in Textiles

Chitosan is polycationic biopolymer that exhibits several valuable inherent properties such as antibacterial, antifungal, antiviral, antacid, non-toxic, total biodegradable as well as film formation, fiber formation, and hydrogel formation properties. By virtue of these properties, chitosan has prospective applications in many fields such as biomedical, wastewater treatment, cosmetics, dentifrices, food, agriculture, pulp, and paper, which are discussed in the previous chapter. The unique properties of chitosan make it suitable for a number of textile applications. The application of chitosan in textiles can be categorized into two main areas namely in the production of fibers and in textile wet processing that includes dyeing and finishing [2].

19.6.9.1 Chitosan fibers

The linear structure of chitosan is mainly responsible for the fiber forming property. Chitosan fibers are produced by wet spinning method. In general, chitosan is dissolved in acetic acid to give a solution of 3% w/v and then extruded through the spinneret into a caustic coagulation bath (0.5%) to obtain a regenerated fiber. To obtain the good fiber out of chitosan a homogeneous solution with maximum polymer to solvent ratio is required apart from non-gelling tendency. By extruding the chitosan dope into a coagulating bath containing sodium acetate, sodium hydroxide, sodium dodecyl sulfate in aqueous medium at 70°C, the strength was found to be improved to 3.8 gpd. The fibers obtained were tough and flexible. Fuji Spinning employed urea along with acetic acid for dissolving the chitosan. The dope was extruded through ethanolic sodium hydroxide coagulating bath. The strength of fibers was 3.8 gpd. Sakurai and Takahasi used formic acid instead of acetic acid for the preparation of chitosan dope. This was then extruded through methanolic sodium hydroxide coagulating bath. The tenacity of fiber was 3.8 gpd [87].

Chitosan fibers find use in the production of textiles having antimicrobial, antithrombogenic, haemostatic, deodorizing, moisture controlling, and non-allergenic properties which are in turn used as bandages for wound-dressing, as sutures, as perfume releasing fabrics, carriers to for active drugs and artificial limbs [1, 3, 87]. Hemostatic activity of chitosan is high when used as a high-molecular-weight solid. Therefore, this has been explored as a modulator of wound healing. It has been found that chitosan has the ability to form coagulum on contact with erythrocytes, defibrinated blood and washed red blood cells and therefore the bandages for wound dressing were found to be effective in regenerating skin tissue of wound area [88]. Recently, chitosan has been proposed to serve as a non-protein matrix for three-dimensional tissue growths. Chitosan provides the biological primer for cell tissue proliferation. The glucosaminoglycan constitutes the wound tissue play an important role in giving structure and strength to newly formed collagen in the granulating tissue of the healing wound. Chitosan provides amino sugars, which in turn can be made available to the fibroblasts that proliferate under the action of interleukin-1 for incorporation into hyaluronate and glycoaminoglycans, thus guiding ordered deposition of collagen leading to wound healing [80, 88, 89]. Introduction of carboxymethyl groups on to chitosan fibers imparts typical gelling property that can absorb large quantity of fluid. These fibers were found to be more effective the normal chitosan fibers as a wound dressing material. By holding water inside the fiber structure, such wound dressings can limit the lateral spreading of wound exudates, reduce wound maceration, and generally improve the quality of healing process [90]. Chitosan fibers can be converted back to chitin fibers by acetylation reaction with acetic anhydride. These fibers find application as surgical suture, which has enough strength for clinical uses. The suture is digestible in the tissues by lysozyme and chitinase [1].

A composite material of chitin/chitosan and cellulose are produced by mixing powder chitin/chitosan with viscose pulp and then wet spun, known as crabyon (Omikenshi). These fibers have high moisture keeping property than cellulosic fibers and have dyeability toward direct and reactive dyes. These fibers are used as textile materials for under wears, socks, etc., as these keep skin from drying. At the same time, these give velvet touch and no irritation to skin. Therefore, clothes made up of these fibers are excellent for babies and elderly people, who have weak and sensitive skin [91]. Similar type of fiber exhibiting deodorant property was produced by a Kokai Tokkyo Koh (Japan), which is useful for clothing, beddings, interior materials, medical care materials, curtains, and carpets [92].

19.6.9.2 Application of chitosan on cotton fabric

Chitosan can be applied to cotton fabric by exhaustion method or by padding method. The viscous nature of chitosan solution, however, restricts its application by exhaust method due the possibility of uneven application. The second method, which is prevalent in continuous dyeing or finishing processes of textiles, produces uniform effects and is most suitable for polymeric applications. Chitosan solution of desired concentration is prepared using acetic acid and sodium acetate and then can be applied on cotton fabric by pad-dry cure method and then washed thoroughly [93]. The surface morphology of the treated and untreated cotton was studied under scanning electron microscope, which is presented in Fig. 19.18 [94]. Chitosan exhibits an inherent property of film formation, which is clearly seen as gloss on fiber surface as shown in Figs. 19.18b,c. Further the film deposition on fiber surface can be confirmed by prolong boiling of treated sample in distilled water so that the broken appearance of film can be viewed under SEM, as presented in Fig. 19.18d. The molecular weight of the treated chitosan seems to play some role on the surface appearance in micrograph, the higher gloss noticed in high-molecular-weight chitosan (CHT-MC)–treated fibers. The micrograph of low-molecular-weight chitosan (CHT-D4)–treated cotton fibers, however, was found to be somewhat different with non-glossy surface (Figs. 19.18e,f) indicating the penetration of polymeric material in to the fiber structure. This may be attributed to the lower viscosity of low-molecular-weight chitosans that facilitates better penetration and high-molecular-weight chitosan having high viscosity confined to surface depositions only.

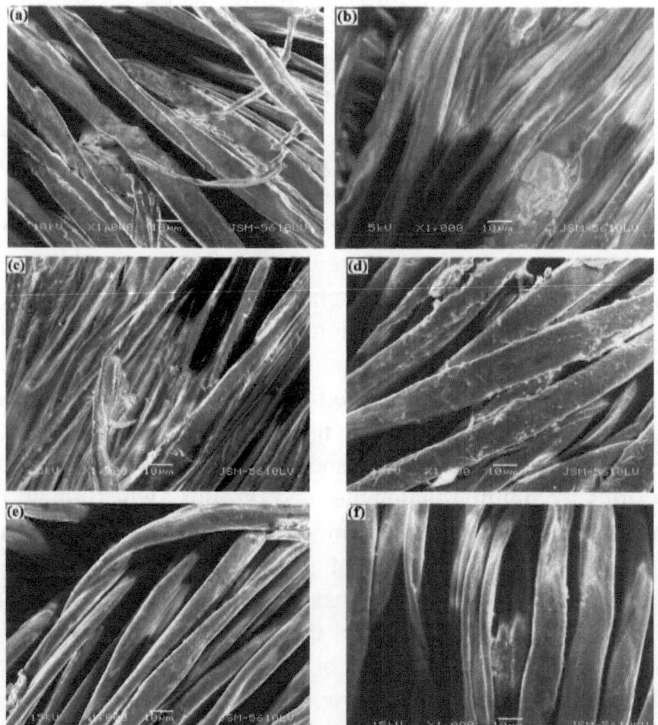

Figure 19.18 Scanning electron microphotographs (×1000) of (a) cotton fiber (control), (b) CHT-MC-treated fibers, (c) CHT-treated fibers (d) CHT-treated and then prolong-boiled cotton fibers, (e) CHT-D4-treated fibers and (f) CHT-D5-treated fibers.

19.6.9.2.1 *Appearance and feel of chitosan-treated cotton fabric*

The appeal of the fabric is manifested by its appearance and the feel. Various factors such as color and transparency of the polymeric film (chitosan), presence of functional groups, penetration level of chitosan in to fiber/fabric structure, etc., influence the appearance of treated cotton fabric. The add-on of polymer film on fibrous material is determined by the viscosity of its solution (padding liquor) which in turn is governed by its molecular weight and concentration in pad liquor [31, 32]. These two factors also influence the feel of the treated fabric. The effect of chitosan treatment on appearance and feel of the cotton is reported, which are presented in Figs. 19.19 and 19.20 [93]. The whiteness of parent chitosan, i.e., CHT-MC and CHT-treated fabrics are shown to be satisfactory and only slightly decreased with increase in concentration of chitosan nevertheless they were in tolerable limits. The appearance of depolymerized chitosan-treated fabrics, however, is adversely affected. The whiteness and brightness depolymerized chitosan-treated samples decreased severely with decrease in the molecular weight and increase in the concentration. The loss in whiteness in depolymerized chitosan-treated fabric may be due the possible liberation of nitric oxide gas from $NaNO_2$ in acidic medium that gets adsorbed on various functional groups of chitosan imparting yellowness [95]. The higher extent of (increased) discoloration of treated fabric due depolymerized chitosan was thus governed by the amount of sodium nitrite used. It seems further that the excessive depolymerization by sodium nitrite produces undesired impurities containing aldehyde end groups, which may react with free amino groups to form –N=C bond causing yellowness. The reduction in brightness of depolymerized chitosan-treated fabric may be ascribed to the loss in gloss of fiber surface due to morphological changes occurred due to diffusion of oligomer into the fiber structure, which can be observed as matty surface as shown in scanning electron microphotographs in Figs. 19.18d,e.

In same study, the fabric stiffness is progressively increased with increase in concentration and the molecular weight of chitosan as shown in Fig. 19.20. The fabric stiffness and feel treated with low-molecular-weight chitosan, namely CHT-D4 and CHT-D5, were found to be satisfactory and well in acceptable limits. The fabric surface was also found to be excessively harsh in case of high-molecular-weight chitosan and the inherent appeal of cotton or "cotton feel" was almost lost. The rigid conformation of chitosan structure and due to the formation of large number of intra and intermolecular cross-linkages due to amino and hydroxyl groups as shown in Fig. 19.7, chitosan produces stiff films [25]. High-viscosity solutions of large chitosan molecules confine the rigid film deposition onto fabric surface only thus imparting stiffness to fabric. The existence of chitosan films of high-molecular-weight chitosans (CHT-MC and CHT) can be clearly seen from Figs. 19.18b,c. This property is undesired in pretreated fabrics but may be beneficial when applied during finishing process, which impart firmness and body to the fabric. Bhuvana et al. [96], in benefit to the uniform film formation of chitosan on fiber surface, have reported the low frictional values offering itself a better candidate as stiffening agent. The high viscosity solutions, however, cause difficulties during application in padding mangle.

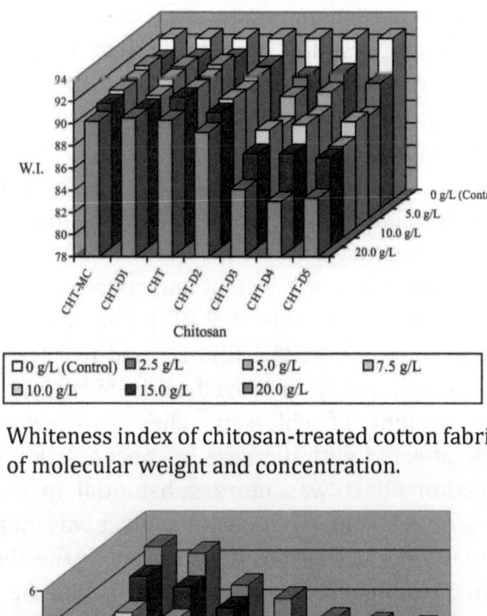

Figure 19.19 Whiteness index of chitosan-treated cotton fabric as a function of molecular weight and concentration.

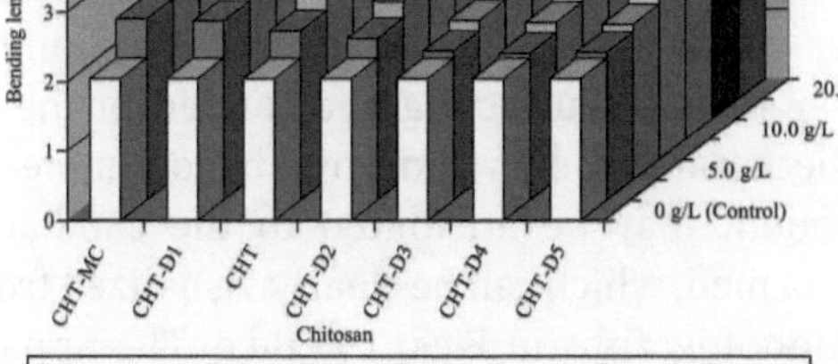

Figure 19.20 Warp way bending length of chitosan-treated cotton fabric as a function of molecular weight and concentration.

19.6.9.2.2 *Chitosan in cotton preparatory processes*

Over last few years, the usage of chlorine containing bleaching compounds has become limited due to the formation of highly toxic chlorinated organic byproducts (AOX). Therefore, chorine-free bleaching agents are gained interest as an alternative such as peracetic acid, an eco-friendly bleaching agent [97]. The bleaching carried out by the reaction of tetraacetylethylenediamine (TAED) with hydrogen peroxide, in which in situ peracetic acid (PAA) formation takes place. Hashem et al. [98] used chitosan-Mo and chitosan-Co complexes for the activation of peracetic acid and were able to obtain satisfactory results at low temperature of 70°C. Replacement of hydrogen peroxide with other oxidizing agents, namely ammonium persulfate and potassium bromate was not successful. In contrast, the perborate did succeed but with lower but with lower efficiency when compared with hydrogen peroxide.

Rapid and uniform absorbency for any pretreated fabric is indispensable for the better penetration of dyes and chemicals during the subsequent unit operations like dyeing, printing and finishing. The absorbency of chitosan-treated fabric was found to be influenced by the molecular weight and concentration of applied chitosan, as observed from Table 19.2 [93]. The absorbency of high molecular chitosan (e.g., CHT-MC, CHT-D1 and CHT)–treated fabric was high and progressively improved with lowering of (decrease in) molecular weight of chitosan. The absorbency was also observed to be affected with increase in concentration of chitosan. The concentration effect was more substantial in case of high-molecular-weight chitosan, where as this effect was meager in case of low-molecular-weight chitosan. In other words, the absorbency of chitosan-treated samples at low concentrations for the wide range of molecular weight chitosan grades and at all concentrations of chosen range for low-molecular-weight chitosans was satisfactory, nevertheless slightly affected. The decreased absorbency in case of high molecular chitosan may be due to the formation of rigid film of chitosan over surface thus acting as a barrier for the penetration of water. The better absorbency conferred by low-molecular-weight chitosan treatment may be attributed to the capillary action of matty surface formed, which can be clearly visualized from scanning electron microphotographs in Figs. 19.18d,e. The apparent loss of absorbency of cotton fabric due chitosan pretreatment, however, is not anticipated to cause any deleterious effects on subsequent processes such as dyeing, printing or finishing. Since the poor absorbency of the fabric is due to glossy rigid film formation on surface and not due to the hydrophobicity as is normally observed in oily fabrics. Such chitosan films have accessible sites such as –OH and $-NH_2$; nevertheless, they are heavily cross-linked by hydrogen bonding.

Table 19.2 Absorbency of chitosan-treated cotton fabric as a function of molecular weight and concentration [93]

Conc. (g/L)	Absorbency: Time taken for water drop to penetrate, seconds					
	CHT-MC	CHT-D1	CHT	CHT-D2	CHT-D4	CHT-D5
2.5	5.1	5.2	4.5	5.0	4.2	4.2
5.0	5.3	5.3	4.6	5.0	4.5	4.3
7.5	8.4	7.2	5.9	6.4	5.3	5.4
10.0	12.7	8.1	7.2	5.5	5.3	5.4
15.0	46.0	26.4	7.1	7.1	6.0	7.0
20.0	72.3	44.6	7.2	7.1	6.5	6.9

Note: Absorbency of control sample: 4.02 sec.

19.6.9.2.3 *Chitosan in cotton dyeing*

Cotton and other cellulosic fibers, conventionally, are dyed with direct, reactive, vat, reactive, azoic, etc., dyes, which are anionic in nature. Cotton also acquires negative surface charge when immersed in dye baths of above dyes leading to repelling action to them. To dissipate this positive surface charge and to facilitate the dyeing, large amount of electrolytes such as common salt or Glauber's salt are added into the dye bath. These electrolytes and unexhausted dyes add to the pollution load when discharged through effluents. Thus, attempts have been made to adopt salt-free and alkali-free dyeing by cationization of cotton by treatment with glycidyl trimethyl ammonium chloride, *N*,*N*-dimethyl azetidinium chloride (DMAAC), *N*-methylol acrylamide (NMA), chloropropionyl chloride (CPC), etc. However, the question of ecological aspects arises here too [99]. Pretreatment of chitosan, due to its polycationic nature, has found to improve the direct dye uptake significantly and also reduce the salt consumption [80, 100]. Bandyopadhyay et al. [101] examined the effect of chitosan treatment on reactive dyeing of cotton. They reported decrease in salt requirement by 50% to produce a comparable shade to that of untreated fabric.

The chitosan-treated fabric also showed improvement in fixation of reactive dyes. This result was explained by the increased exhaustion of negatively charged reactive dyes to the cotton, whose negative potential at the fiber surface was suppressed by the cationic chitosan treatment. Consequently, when alkali was added to the dye bath, a substantial quantity of dye was available for the reaction with cotton. It was also suggested that the amino groups of chitosan reacted with the reactive group in the dye and the fixation was further improved. The chitosan-treated fabric showed comparable color fastness properties to the untreated fabric. Kavitha et al. [102] studied the effect of chitosan treatment on natural dyeing of cotton namely turmeric. They reported increased tensile strength, flexural rigidity and shear strength. Cotton yarn coated with chitosan was found to be darker compared to uncoated yarn while dyeing for the same shade percentage. They reported the dyed yarn coated exhibited excellent activity against bacteria.

Often cotton fibers show small lightly colored or white spots due to the presence of immature cotton fibers known as neps. Metha and Combs [103] evaluated nep coverage in the direct dyeing of cotton by the pretreatment with chitosan. The chitosan pretreatment was done by exhaust method. They reported by addition of sodium sulfate during the exhaust process improved the absorption of chitosan on cotton and fully covered the neps. The neps coverage was only partial when chitosan treatment was done in absence of salt. Further, the pretreatment was more effective for neps coverage in direct dyeing than reactive dyeing of cotton. Rippon [104] applied chitosan by three different methods namely pad-batch, pad-dry and exhaustion methods to evaluate the application of chitosan to improve the dyeability of immature cotton fibers. The author postulated that the affinity of chitosan to cotton would be by van der Waals forces between them because of the similar structures of chitosan and cotton. Another possibility mentioned for the binding chitosan to cellulose was cross-linking by formation of Schiff base between cellulose's reducing end (–CHO) and the amino group of chitosan. In addition to the two possible bindings suggested by the authors, hydrogen bonding should also play an important role. Although the application of chitosan by a pad-dry method was the most effective in terms of the amount of chitosan on cotton after rinsing, this method caused uneven dyeing on the fabric because of migration of the chitosan during drying.

Further, the structure of chitosan is very much analogous to cellulose (except that the –OH group in cellulose at C2 is being replaced by $-NH_2$ in chitosan), which contributes in enhancing the dyeability. Effect of pre and post dyeing chitosan treatment of cotton fabric using direct dye is reported in [93]. Accordingly, two direct dyes namely C. I. Direct Red 81 (Mol. weight 675.6) and C. I. Direct Blue 71 (Mol. weight 1029.9), as shown by Figs. 19. 21 and 19.22, were chosen. Attributing to the presence of cationic amino groups, the effect of chitosan treatment on the dyeability toward acid dyes (C. I. Acid Blue 158; Fig. 19.23) was also investigated. The dye strength on fabric was measured in terms of k/s values using computer color matching system. The uptake was increased with increase in concentration of chitosan for a particular molecular weight grade and also was found to be influenced by the molecular weight, which increased with increase in molecular weight especially at higher concentration of chitosan applications as reported/demonstrated in Figs. 19.24 and 19.25. The dye uptake of low-molecular-weight chitosan-treated samples from CHT-D2 to CHT-D5 was almost similar at any particular concentration of chitosan. The dyeing behavior of chitosan-treated fabric, however, was somewhat different for two dyes selected for the experiment. Blue dye showed high substantivity toward chitosan-treated fabric than the red.

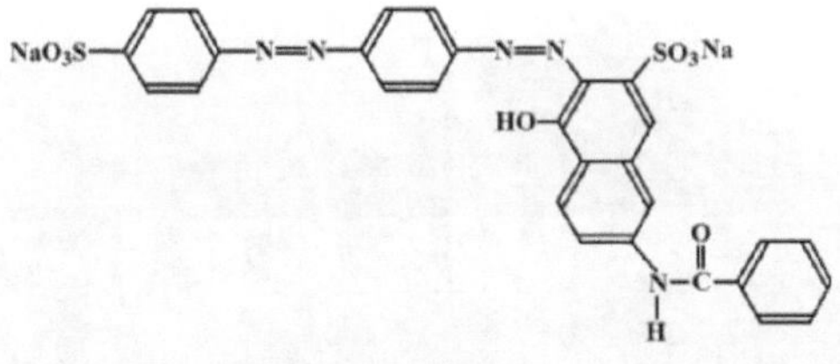

Figure 19.21 Structure of C. I. Direct Red 81.

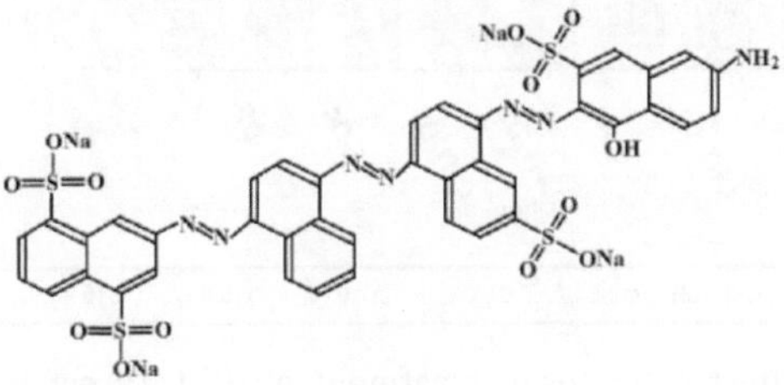

Figure 19.22 Structure of C. I. Direct Blue 71.

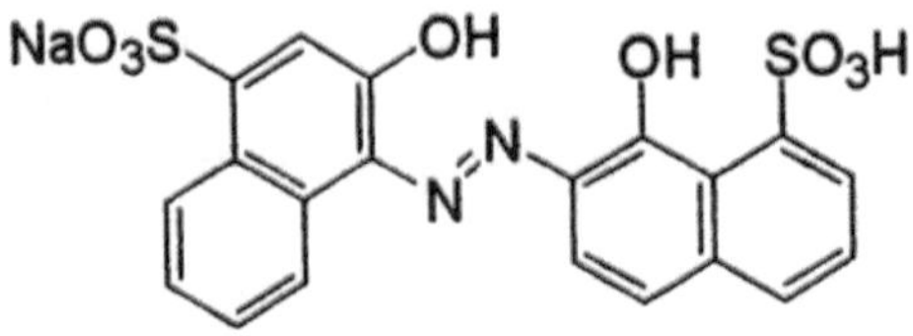

Figure 19.23 Structure of C. I. Acid Blue 158.

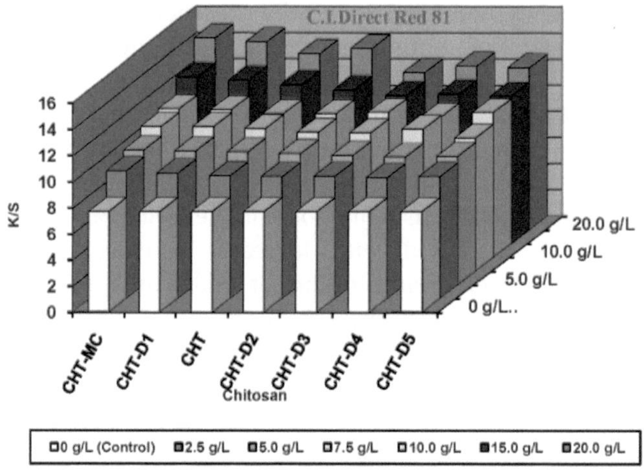

Figure 19.24 Effect of chitosan treatment on C. I. Direct Red 81 uptake (Dye: 1% o.w.m.).

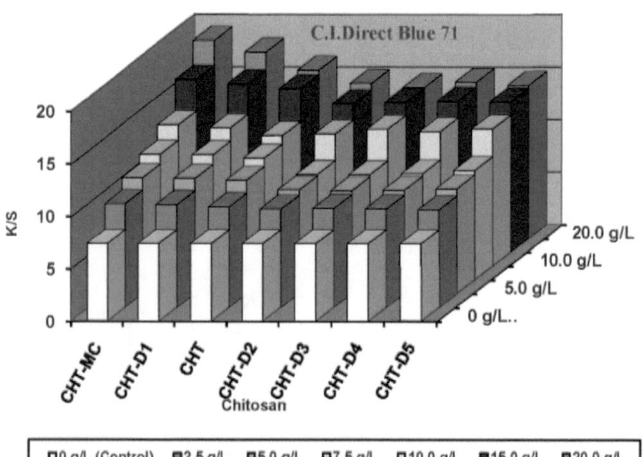

Figure 19.25 Effect of chitosan treatment on C. I. Direct Blue 71 uptake (Dye: 1% o.w.m.).

The enhanced dye uptake due to chitosan may be attributed to the cationic amino groups forming dye sites, which interact with direct dye through hydrogen bonding as well as ionic linkages. high-molecular-weight chitosan, due to highly viscous nature of their solutions, are confined mostly on the fiber surface compelling greater accumulation of dye on the surface. Also, the number of amino groups is increased with increase in molecular weight and with increase in concentration.

It can be observed from the structures of C. I. Direct Red 81 and C. I. Direct Blue 71 namely Figs. 19.21 and 19.22; the molecular weight of blue dye is high and is characterized by greater degree of planarity due conjugation system possessing high substantivity and moderate to good wet fastness properties. The structure also shows four anionic ($-SO_3^-$) groups that can form more strong linkages with amino groups. Attributing to these two features, C. I. Direct Blue 71 showed greater substantivity toward chitosan-treated fabric.

The polycationic chitosan can also dissipate the negative surface charge on cotton and drives dye molecules to the fiber. Such effects can be studied by comparing the dyeing results of chitosan-treated samples conducted in electrolyte-free dye bath with that in presence of electrolyte. The progressive decrease in importance of sodium sulfate taken as electrolyte with increase in chitosan concentration, as shown in Figs. 19.26 and 19.27, clearly elucidates the role of chitosan in the improvement of dye uptake even at reduced electrolyte concentration.

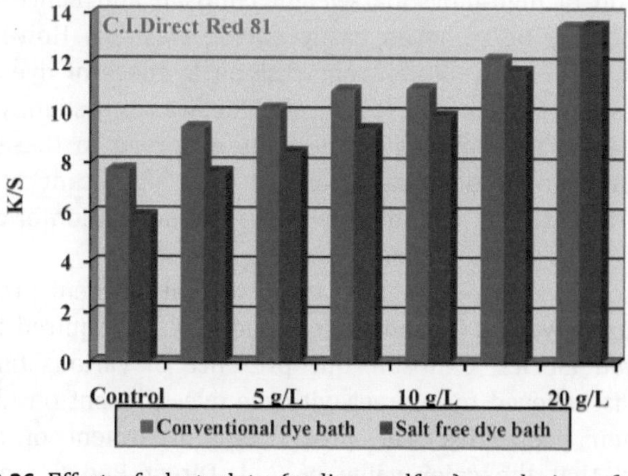

Figure 19.26 Effect of electrolyte (sodium sulfate) on dyeing of chitosan (CHT)-treated cotton fabric with C. I. Direct Red 81.

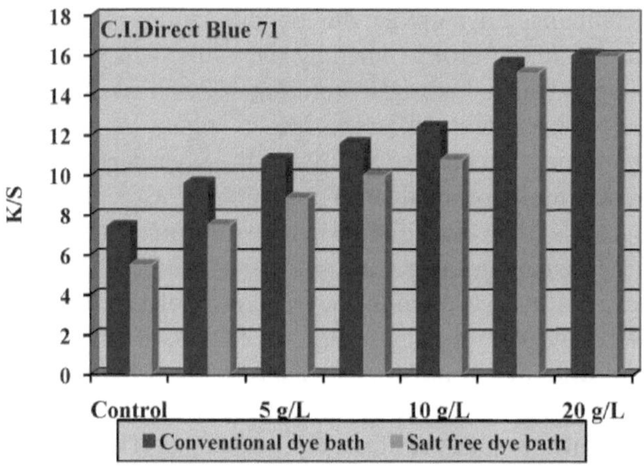

Figure 19.27 Effect of electrolyte (sodium sulfate) on dyeing of chitosan (CHT)-treated cotton fabric with C. I. Direct Blue 71.

The same paper reported the poor washing fastness ratings for red dyed samples. C. I. Direct Red 81 dye, being low-molecular-weight dye, is characterized by poor washing fastness. The fastness of this dye on chitosan-treated fabric was reported to be improved slightly with increase in concentration of chitosan, particularly in case of relatively low-molecular-weight chitosan-treated samples (Table 19.3). On the other hand, there was slight decline in ratings with increase in concentration of high-molecular-weight chitosan. This may be attributed to the preferential surface deposition of high-molecular-weight chitosan and hence the dye also that may be removed easily during washing. However, the effect of chitosan treatment on washing fastness of dyes having good washing fastness, i.e., C. I. Direct Blue, was only slightly altered (Table 19.4). The rubbing fastness, as observed in these tables, was found to be deprecated and it was significant at higher concentrations. This may be due to surface deposition of chitosan than can be easily rubbed off.

Chitosan, with due regards to its several inherent properties, can be employed as finishing agent and may be required to apply onto dyed fabrics. Chitosan, due presence of various functional groups, is believed to interact with the dyes present on fiber and alter their properties. The post dyeing treatment of chitosan reported that the color value of C. I. Direct Red 81 improved whereas that of C. I. Direct Blue 71 decreased by the chitosan after treatment, nevertheless to very small extent (Table 19.5). Any regular trend on the color depth, however, was not noticed due to the molecular weight of treated chitosan. In all cases, the bloom was somewhat enhanced at higher concentration treatment. In case of red dyed samples, blooming was on higher side on high-molecular-weight chitosan-treated samples, whereas low molecular weight showed somewhat lesser but almost similar level of blooming. In blue dyed samples, the color change was negligible when treated with high-molecular-weight chitosan and the loss in color value was observed in low-molecular-weight chitosan-treated samples [93].

Table 19.3 Effect of chitosan pretreatment on fastness of C. I. Direct Red 81 dyed fabric

	Fastness ratings of C. I. Direct Red 81 dyed samples pretreated with:							
	CHT-MC		CHT		CHT-D3		CHT-D5	
Conc. (g/L)	Washing fastness	Rubbing fastness	Washing fastness	Rubbing fastness	Washing fastness	Rubbing fastness	Washing fastness	Rubbing fastness
5.0	2–3	2	2–3	2	3	2–3	3	2–3
10.0	3	2	3	2	2–3	2–3	3–4	2–3
15.0	2–3	1–2	3–4	2	3–4	2	3–4	2
20.0	2	1–2	3	1–2	3–4	2	3–4	2

Note: Dye 1% o.w.m.; fastness ratings of control sample: washing 3 and rubbing 2–3.

Table 19.4 Effect of chitosan pretreatment on fastness of C. I. Direct Blue 71 dyed fabric

	Fastness ratings of C. I. Direct Blue 71 dyed samples pretreated with:							
	CHT-MC		CHT		CHT-D3		CHT-D5	
Conc. (g/L)	Washing fastness	Rubbing fastness	Washing fastness	Rubbing fastness	Washing fastness	Rubbing fastness	Washing fastness	Rubbing fastness
5.0	4–5	2	4–5	2	4	2–3	4–5	2–3
10.0	4	2	4–5	2	4–5	2–3	4	2–3
15.0	4	1–2	4–5	2	4–5	2	4–5	2
20.0	4–5	1–2	4–5	1–2	4–5	2	4–5	2

Note: Dye 1% o.w.m.; fastness ratings of control sample: washing 4–5 and rubbing 2–3.

Table 19.5 Effect of chitosan treatment on color depth and washing fastness of direct dyed cotton fabric [93]

		K/S values		Washing fastness rating			
				C. I. Direct Red 81		C. I. Direct Blue 71	
Sample	Conc. (g/L)	C. I. Direct Red 81	C. I. Direct Blue 71	Change in color	Staining	Change in color	Staining
Control	—	6.59	7.29	3	3	4–5	3
CHT-MC Treated	5	6.79 (3.03)	7.24(−0.69)	3	2–3	4	4
	10	7.12 (8.04)	7.49 (2.74)	3–4	2–3	4–5	4–5
CHT-D1 Treated	5	7.06 (7.13)	7.19 (−0.10)	3–4	2–3	4–5	4–5
	10	7.28 (10.47)	7.33 (0.55)	3–4	2–3	4–5	4–5
CHT Treated	5	7.25 (10.02)	6.83 (−6.3)	3–4	2–3	4–5	4
	10	7.54 (14.42)	7.31 (0.27)	3–4	3	4	4–5
CHT-D2 Treated	5	6.98 (5.92)	6.75 (−7.41)	3–4	3	4–5	4
	10	7.21 (9.41)	7.02 (−3.70)	3–4	3	4–5	4–5
CHT-D3 Treated	5	7.17 (8.80)	6.59 (−9.60)	3–4	2–3	4–5	4–5
	10	6.99 (6.07)	6.62 (−9.19)	4–5	3	4–5	4–5
CHT-D4 Treated	5	7.21 (9.41)	6.61 (−9.33)	4	3	4–5	4–5
	10	7.39(12.14)	6.81 (−6.58)	4	3	5	4–5
CHT-D5 Treated	5	7.19 (9.11)	7.01 (−3.84)	4	3	4–5	4–5
	10	7.44(12.90)	7.24 (−0.69)	4–5	3	5	4–5

Note: Dye 1% o.w.m. Values inside the parentheses indicate the percentage change in color strength dyed fabric due to chitosan treatment. The negative sign indicates the decrease/drop in color strength.

The apparent changes in shade may be attributed to the migration of dye from the fiber phase to the chitosan phase during padding and subsequently during drying operations due to the interaction of the anionic sulfonate group of dye with cationic groups of chitosan. The higher dye migration of C. I. Direct Red 81 may be attributed to its low molecular weight and poor washing fastness. The washing fastness of post dyeing chitosan derivative treatment was improved to some extent. This may be attributed to the complex formation between dye and the chitosan. Rubbing fastness, however, was not significantly altered.

Chitosan possesses one amino group in its glucosamine unit, which forms positive charge in presence of acid. This positively charged amino group can form salt linkage with anions. To characterize the chitosan-treated fabric, the work was extended further to investigate its dyeability toward acid dye (C. I. Acid Blue158; Fig. 19.23), which is non-dyeable toward normal cotton.

It was revealed from Fig. 19.28 that the chitosan-treated cotton fabrics dyed substantially with acid blue158 as against only a tint on control and remained fast to hard soaping. A progressive increase in dye uptake was observed with increase in concentration of respective chitosan. At a particular concentration the dye uptake, however, was almost identical for samples treated with varying molecular weight chitosans. This kind of dyeability can be purely attributed to the binding of acid dye to chitosan by salt linkages as shown in Fig. 19.29. Thus, the dyeability toward acid dye can be taken as one of the characterization tests for the retention of chitosan on cotton fabric and the progressive increase in dye uptake corresponds to the number of protonated amino groups of chitosan present on treated cotton fabric forming ionic linkage with stoichiometric amount of anionic acid dye.

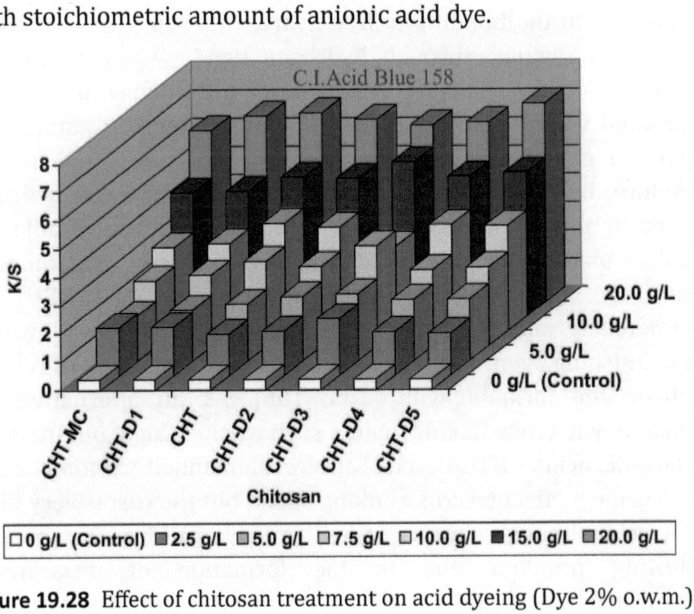

Figure 19.28 Effect of chitosan treatment on acid dyeing (Dye 2% o.w.m.).

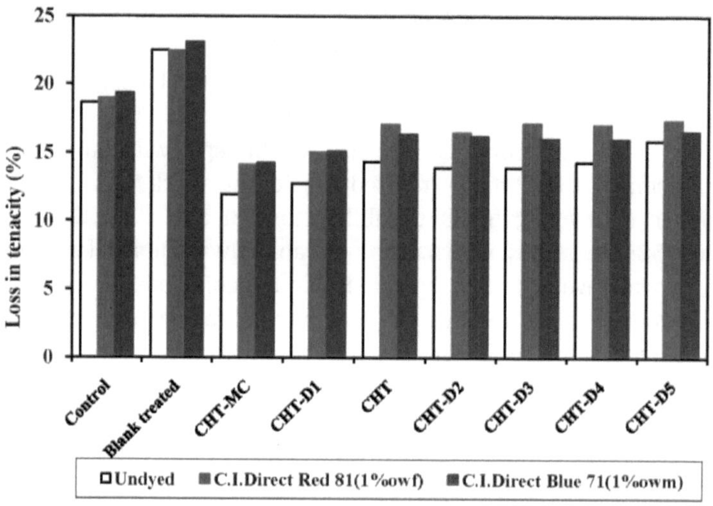

Figure 19.29 Reaction of chitosan with acid dye.

19.6.9.2.4 *Chitosan in anticrease finishing*

The aesthetic appeal of cotton cloth or garments is severely affected due to their creasing tendency. Creasing in cotton fabrics occurs due to the bonding of free hydroxyl groups, present in the amorphous regions, through hydrogen bonds when pressed or folded. Thus, the creasing behavior of cotton may be directly associated with the ability of free hydroxyl groups in amorphous region to get bound to each other. The creasing problem can, therefore, be minimized by blocking or masking these hydroxyl groups by means of cross-linking of hydroxyl groups of adjacent cellulose macromolecules. Cross-linking agent based on aminoplast resins, e.g., dimethylol dihydroxy ethylene urea (DMDHEU) are commercially employed easy care finishing of cotton. However, these finishing agents suffer from one serious drawback of release of toxic-free formaldehyde [105, 106]. As an alternative, on formaldehyde cross-linking agents such as citric acid, butane tetra carboxylic acids (BTCA), etc., are recommended. Among these, BTCA is most effective cross-linking agent; but the cost is very high. Citric acid, although cheaper, is less effective and has drawback of yellowing problem due to the formation of unsaturated polycarboxylic acid (due to dehydration of citric acid). To counteract this yellowness, additives such as hydroxyethyl amines, borates or polyethylene glycols to citric acid bath are recommended. However, the process is still less effective [107].

The crease recovery power of citric acid can be improved satisfactorily by the incorporation of chitosan in presence of sodium hypophosphite monohydrate (SHP) to the citric acid bath as reported by Waly and Okeil [35]. Here, the esterification reaction not only occurs between citric acid and cellulose but also between citric acid and hydroxyl groups of chitosan and the free carboxylic groups can also react with amino groups of chitosan by salt linkages. In a similar study, Chung et al. [108] applied chitosan and citric acid by means of conventional pad-dry-cure process. They reported the durable press appearance ratings (DP ratings) of 3.5–4. This DP performance was retained through 20 washings and tumble drying cycles. Authors proposed citric acid reacts with amino and hydroxyl groups of chitosan and hydroxyl groups and cotton through to form ester cross-linking or an inter-ionic attraction. It is well known that the amino groups of chitosan readily react with aldehyde groups to form Schiff's base. This property of chitosan was employed by Bhattacharyya et al. [109] to scavenge free formaldehyde released from DMDHEU-finished cotton fabrics. They reported that the use of chitosan as an additive in DMDHEU finishing after dyeing was more effective in reduction of formaldehyde release as compared to the fabric which is chitosan-pretreated, dyed, and then DMDHEU-finished. The authors indicated that this result might be due to the blocking of the amino groups of chitosan by dye molecules on subsequent dyeing in the case of chitosan-pretreated cotton, and as a result, reduced reactions between chitosan and formaldehyde. Huang et al. [110] treated cotton fabric with low molecular weight along with anti-crease agent, i.e., DMDHEU. The results showed that the crease resistance of processed fabric sustained for 20 washes, the softness of the fabric was improved, and the strength of the fabric decreased slightly after wash treatment. The performance of various grades of chitosan and DMDHEU in terms of crease recovery angle (CRA) in easy care finishing of cotton are illustrated in Table 19.6 and 19.7, respectively [93].

Compared to commercial cross-linking agent DMDHEU treatment, the wrinkle recovery of chitosan-treated cotton fabrics was not satisfactory. There was tremendous decline in CRA with increase in concentration of chitosan. However, there is slight improvement in wrinkle recovery by lowering the molecular weight especially for chitosan applications at low concentrations. The high- molecular-weight chitosan is believed to form mostly a surface coating thus ignoring the possibility of cross-linking. Therefore, stiff and brittle film formed may deform almost permanently when stressed. This may be the fundamental reason for low CRA values for any stiff finishes. In order to minimize the loss in resiliency, addition of commercial cross-linking agents to the pad bath formulation is recommended (Table 19.8).

Table 19.6 Wrinkle recovery property of chitosan-treated cotton fabric

Conc. (g/L)	CRA° of cotton fabrics pretreated with:						
	CHT-MC	CHT-D1	CHT	CHT-D2	CHT-D3	CHT-D4	CHT-D5
2.5	137	140	140	140	176	167	170
5.0	143	140	140	160	164	163	164
7.5	129	128	119	152	141	151	152
10.0	94	108	125	138	128	140	144
15.0	96	98	110	127	134	141	140
20.0	90	91	98	116	121	138	139

Note: CRA of control sample= 161°.

Table 19.7 Wrinkle recovery property of DMDHEU-treated cotton fabric

DMDHEU (g/L)	CRA°
Control	161
20	180
40	207
60	215
80	226
100	233

Note: Pad liquor: $MgCl_2$ 10 g/L; acetic acid 15 ml/L; curing temperature 150°C; time 4 min.

Table 19.8 Effect of DMDHEU on wrinkle recovery property of chitosan

DMDHEU concentration (g/L)	Crease Recovery Angle°		
	DMDHEU treated	DMDHEU + CHT (10 g/L) treated	DMDHEU + CHT (20 g/L) treated
Control	161	125	98
20	180	153	134
40	207	163	139
60	215	177	158
80	226	191	170
100	233	198	179

Note: Pad liquor: $MgCl_2$ 10 g/L; acetic acid 15 ml/L; curing temperature 150°C; time 4 min; CRA of control 161, CHT (10g/L) treated 125°; CHT (20 g/L) treated 98°.

19.6.9.2.5 *Chitosan in antimicrobial finishing*

Textile products made out of natural fibers provide favorable environment for the growth of microorganisms (algae, fungi, bacteria, etc), due to moisture and warmth. These organisms are mainly responsible for discoloration, stains, strength loss, etc., of fabric and skin allergies, infection diseases, etc., to human body [111–114]. The unpleasant odor is produced when bacteria that are present on the skin get to work on sweat and decompose it. The decomposition products that are responsible for odor are ammonia, methyl amine, hydrogen sulfide, low-molecular-weight fatty acids, urea, etc. [9, 115, 116]. Antimicrobial finishes can give rise to hygienic freshness and also can be used to fight against pathogenic and parasitic microorganism. Antimicrobial agents either inhibit the growth (-static) or kill (-cidal) the microorganism. Using number of chemicals such as organo-metallics, phenols, thiophenols, formaldehyde derivatives and several quaternary ammonium compounds, microbial growth can be inhibited. These, however, chemicals are non-biodegradable and toxic. Some of the commercial antimicrobial textile products recently/currently marketed/available are: Biogaurd produced by Aegis Environments (formerly Dow Corning) is quaternary ammonium compound, namely, 3-trimethoxy silyl propyldimethyl octadecyl ammonium chloride (AEM), Reputex 20 (Arch Chemicals) is polyhexamethylene biguanide (PHMB), Triclosan (2,4,4′-trichloron-2′-hydroxydiphenyl ether), etc. Environmental issues of these products are still of concern [106, 112]. Cotton fabric with good antimicrobial activity is obtained by using chitosan, which is attributed to the amino groups that are present on chitosan macromolecule. Thus, the DAC, molecular weight and concentration of chitosan in fluence the antimicrobial activity [87]. Seong et al. [37] reported chito-oligosaccharide (DP = 3 and 10) found to exhibit good antimicrobial activity and durability to washing without the need of cross-linking agent. Their report reveal that the antimicrobial activity after 15 wash cycle was able to maintain 95% (for a DP of 3) and 100% (for a DP of 10) bacterial reductions.

The polycationic nature of chitosan is mainly responsible the inhibition of bacterial and fungal growth. Fang et al. [117] reported that the chitosan inhibited growth of *Aspergillus niger* and induced considerable leakage of UV-absorbing and proteinaceous materials from it at pH 4.8, which was not induced at pH 7.6. Similarly, Tsai and Su [118] observed the chitosan-induced leakage of glucose and lactate dehydrogenase from *E. coli* cells and suggested that the death of cells resulted from the interaction between chitosan and the *E. coli* cell. The maximum antimicrobial activity exhibited by chitosan at acidic pH is also reported by other workers [119, 120].

The early work indicated that the antimicrobial effect was potent against a range of microbes, but the finishing was not durable. To improve durability, chitosan has been cross-linked to cotton using chemicals such as DMDHEU, 1,2,3,4-butane tetra carboxylic acid (BTCA), citric acid or glutaric dialdehyde [111]. These chemicals, some of which are used in cotton durable press, cross-link chitosan to cotton through hydroxyl groups. Chung et al. [108] showed, besides improved wrinkle recovery and D.P. rating, a high antimicrobial property level, which was retained through 20 washing and tumble drying cycles. Ye et al. [121, 122] synthesized nanoscale core–shell particles of poly(*n*-butyl acrylate) core and chitosan shells and applied them to cotton fabric in a pad-dry cure process. The antibacterial activity was maintained at over 90% reduction levels after 50 washes.

The composted soil bed composed of variety of microbes (e.g., bacteria and fungi) can be employed in soil burial test. The microbial attack of cellulolytic microflora in a composted soil bed is considered to be the most rigorous and practical means for the evaluation of anti-deterioration treatments. The treated and untreated fabric strips are buried and exposed to the microbial attack (cellulolytic microflora) for a stipulated period. The change in fiber strength (tenacity) of the sample during incubation is taken as a measure of the effectiveness of the biocide compound [123]. The effect of different molecular weight grades of chitosan treatment on undyed and dyed cotton fabrics for resistance against microbial attack demonstrated in Fig. 19.30.

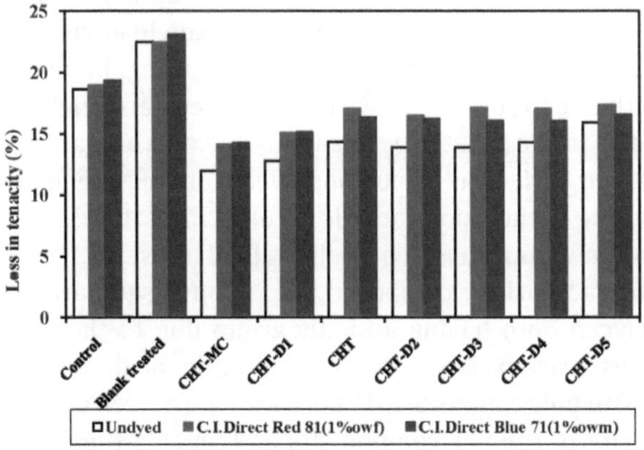

Figure 19.30 Effect of chitosan treatment on resistance against microbial attack of cotton fabric. Dye 1% (o.w.m., Concentration of chitosan derivatives in pad liquor 10 g/L. Blank treatment was given with acetic acid 15 ml/L, sodium acetate 10 g/L by pad-dry cure method.

The undyed and dyed control cotton fabrics, as revealed from Fig. 19.30, were more prone to microbial attack of cellulolytic microflora in a composted soil bed. The less loss in tenacity of fiber means better resistance against microbes. The blank treated samples, both undyed and dyed, showed maximum strength loss followed by control sample. Samples treated with chitosan were found to be more resistant to microbial attack as manifested by the lesser drop in strength. In case of undyed fabric, with increase in the molecular weight of chitosan, the strength loss was found to be progressively decreased. The rigidity of high-molecular-weight chitosan and interaction of aldehyde end groups of anhydromannitol on depolymerized chitosan with cellulose may probably be the contributing factor in strength determination [37]. Further, the dyed fabrics treated with chitosan were slightly more susceptible for microbial attack in soil burial test. The elongation capacity of fiber was also found to be affected due to rotting;

nevertheless, the extent was almost of same level for all chosen grades of chitosan.

The susceptibility of blank treated samples to microbial attack may be probably due to the acid hydrolytic degradation occurred during thermo curing. The mode of interaction between chitosan, its derivatives, and the microorganism may be considered to be the combination of different mechanisms. Chitosan believed to form a rigid film over the fiber surface that share the load and also act as a protective layer against microbes during rotting. Increased losses in strength due to rotting in case of low-molecular-weight chitosans may be due to the greater permeability for microorganisms due to more opened surface as can be observed from SEM (Figs. 19.18e,f). Secondly, due to their polycationic nature, chitosan exhibit inherent antimicrobial properties. The cell wall of most of the microbes is a polysaccharide composed of lipopolysaccharide and/or peptidoglycan both having an ionic group due to the presence of phosphates, carboxylates, *N*-acetylmuramic acid, etc., that can interact with poly cations of CHT derivatives. This cell wall maintains the integrity of cellular components and shields the cell from the extracellular environments. Immediately beneath the cell wall is a semi-permeable membrane that encloses intracellular organelles and a myriad of enzymes and nucleic acid. The enzymes are responsible for the chemical reactions that take place within the cell, and the nucleic acids store all the genetic information of the organism. The survival or growth of microorganisms depends on the integrity of the cell and the concerted action and proper state of all these components. It is believed that the polycationic nature of chitosan initiates binding with the cell membrane by means of electrostatic attraction with negatively charged microbial cell membrane. Once bound to the cell surface, chitosan is thought to affect membrane permeability, which results in the leakage of proteinaceous material and other intracellular constituents of the microbial cell causing death due to the loss of essential fluids [112, 118, 124]. Chitosan is also observed to bind DNA and inhibit mRNA and protein synthesis. Low-molecular-weight chitosan is more effective as it penetrates deeper into the cell of microorganisms [125, 126]. Due to chelation property, chitosan also binds trace of essential metal ions present in the intracellular fluid. Deficiency of such metal ions may inhibit production of toxins, enzymes and the microbial growth [127].

The effect of chitosan on the resistance against specific bacteria namely Gram-positive *S. aureus* (ATCC6538P), which is commonly found on human skin and Gram-negative *E. coli* (ATCC10229) was also evaluated by agar diffusion method. The method is applicable testing hygienic finishes of hydrophilic, air permeable materials. Effect of chitosan treatment on antimicrobial activity is presented/demonstrated in Fig. 19.31. The results envisaged that the antibacterial activity of tested fabrics against *S. aureus* was greater than *E. coli*. Further, the resistance to bacteria was improved with decrease in molecular weight of treated chitosan. The resistance of *E. coli* may be attributed to their very complex cell walls. On the other hand, Gram-positive cocci/bacterium (*S. aureus*) have a simple cell wall structure in which the cytoplasm membrane has rigid peptidoglycan layer composed of network with plenty of pores, which allow foreign molecules to enter the cell without any difficulty [128]. A comparative better resistance of low-molecular-weight chitosan may be due to more solubility, which can diffuse easily into agar domain [129].

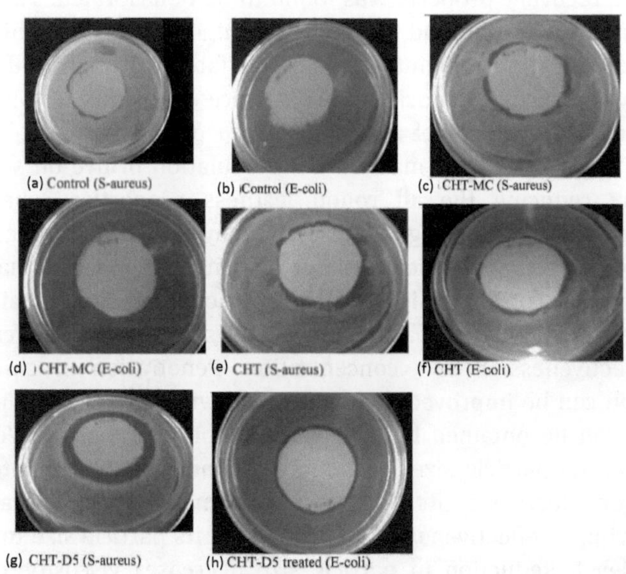

Figure 19.31 Determination of antimicrobial activity of chitosan-treated cotton fabric by agar diffusion method: (a) Control (*S. aureus*), (b) Control (*E. coli*), (c) CHT-MC (*S. aureus*), (d) CHT-MC (*E. coli*), (e) CHT (*S. aureus*), (f) CHT (*E. coli*), (g) CHT-D5 (*S. aureus*), (h) CHT-D5 (*E. coli*).

19.7 Nano Chitosan Applications to Textiles

19.7.1 Nano Concept

Preliminary experiments reported above shown that various properties of the cotton fabric improved on application of chitosan on it. The dyeability of cotton fabric toward direct dyes was enhanced significantly due to chitosan pretreatment and the degree of improvement was found to be a function of molecular weight and concentration of chitosan. The fastness to washing of direct dye on chitosan-pretreated fabric, however, was only slightly improved especially for the low-molecular-weight chitosan applications. However, the post-dyeing chitosan treatment, in general, has improved the washing fastness of direct dyed cotton fabric. The moisture related properties were in tolerable limits. Chitosan treatment was found to impart resistance to microbial attack. The appearance and handle of the treated fabric, however, was severely affected and lost its natural "cotton feel." The wrinkle recovery property was found to be deteriorated. The very large molecular size and consequently high viscosity of chitosan restricts its penetration into the fiber and fabric structure and leads to only the surface deposition. The surface deposition of this high polymer affects the feel and appearance of the treated textiles. This may also leads to maximum accumulation of dye on surface thereby reducing the all round fastness properties especially washing, rubbing and light fastness. Today's need, however, is to improve above properties without altering the inherent natural qualities of cotton. It is possible by achieving the maximum penetration of polymer particles into fiber structure and increasing its effectiveness at low concentration. Penetration of chitosan solution can be improved by lowering the viscosity of its solution, which can be obtained by lowering the concentration and/or by reducing its particle size. Reduction in concentration of chitosan, however, decreases its effectiveness. An alternative way of improving its effectiveness is to drop down its particle size toward nano level. Reduction in particle size decreases viscosity, offers greater surface area and hence enhances the effectiveness of chitosan. This is the basic of "nano technology."

The concept of "nanotechnology" lies in the fact that the properties of substances dramatically change when their size is scaled down to nanometer range. The first use of this concepts (but predating use of that name) was in "There's Plenty of Room at the Bottom," a talk given by a Nobel laureate physicist Richard P. Feynman at an American Physical Society meeting at Caltech on December 29, 1959. In general, "Nanotechnology is the engineering and fabrication of objects with size less than 100 nm. Below 100 nm the properties like melting point, hardness, catalytic activity and magnetic properties vary with size, otherwise these properties in other material are considered to be constant" or "Nanotechnology is concerned with developing the tools for characterizing and manipulating materials on nanoscale (1–100 nm) and exploiting these tools for the development of new products and processes" [130–132]. A nanometer is one billionth of a meter, roughly the width of three or four atoms. The average human hair is about 25,000 nanometers wide.

Nanotechnology basically deals with the individually arranging atoms or molecules in desired places to obtain a hybrid product with desired and diverse properties [133]. Fabrication of nano matrices can be broadly done by two approaches:

- **Top-down approach:** The top-down approach involves the fabrication of components from larger materials
- **Bottom-up approach:** An approach to building things by combining smaller components, as opposed to carving them out of larger ones (top down)

Nature follows the "bottom-up approach" rather than the "top-down approach" usually followed by humans to produce materials. The shapes are "grown" rather than cut. All the living being can trace their origin to a single cell. Reduction in particle size to nanoscale can lead to changes in properties related to specific surface area, reactivity, quantum effects, strength, electrical characteristics, optical properties, magnetic behavior, etc. As the particle size decreases, greater proportion of atoms are found at the surface compared to those in the "body" [130, 133].

It has been demonstrated in recent years that the nanotechnology can be applied to textiles to enhance various properties, especially in functional coatings, such as fabric softness, durability and breathability, water repellency, fire retardency, antimicrobial properties, anticrease properties, UV protection, self-cleaning properties and like in fiber, yarn and fabrics [131, 134–136]. Undyable polypropylene fiber can be made dyeable by dispersing nanoclay, modified with quaternary ammonium salts, into polypropylene melt before spinning. After fiber formation, the infused quaternary ammonium groups act as efficient dye sites [137]. A combined effect of wrinkle-free and stain repellency can be obtained by treating fabric with 10 nm tiny particles with both polar and non-polar moieties imparting extremely low free surface energy. These nanoparticles cross-link with cellulose to give the desired amount of wrinkle resistance. Such hybrid nanoparticles when incorporated with highly fluorinated silanes impart stain repellency. Nanotechnology can also be made applicable in the production of smart textiles. Application of carbon nano tubes can give rise to textile materials that have thermal and electrical conductivity sensible to touch and feel. Garments of such fabric can understand any abnormality in heartbeat and then send a signal to family doctor or spouse [134, 138, 139]. The inorganic UV blockers are preferable to organic blockers as they are non-toxic and chemically stable under exposure to both high temperature and UV. Usually certain semiconductor oxides such as TiO_2, SiO_2, ZnO, and Al_2O_3 are used as UV blockers. Rayleigh's scattering is dependent upon the wavelength where the scattering is inversely proportional to the wavelength to the fourth power. This theory predicts that in order to scatter UV radiation between 200 and 400 nm, the optimum particle size will be between 20 and 40 nm. UV blocking treatment for cotton fabric is given by using sol-gel method. A thin layer of titanium dioxide is formed on the surface of the treated fabric, which provides excellent UV protection fast to washing [134, 140]. Fabrics treated with nano TiO_2 can provide effective protection against bacteria and discoloration of stain due to the photo catalysis effect of this agent. Nano ZnO provides effective photo catalytic properties once it is illuminated by light and it is employed to impart antibacterial properties to textiles [141–144]. Nanosilver is very reactive to protein when contacting with bacteria and fungus; it will adversely affect the cellular metabolism and inhibit the cell growth [145].

Chitosan is a biopolymer that has received much attention and has been extensively studied for micro- and nanoparticles preparation. It is possible, for a given molecular size chitosan, to reduce the particle size to nano level by "bottom-up" approach as a result of a self-assembling or cross-linking processes in which the molecules arrange themselves in to ordered nano scale structure either by physical or covalent inter- or intramolecular interactions. One of the trends in synthesis process is to pursue a nano scale emulsion, through which finishes can be applied to textile material in a more thorough, even and precise manner. Finishes can be emulsified into nano-micelles, made into nano-sols or wrapped in nano-capsules that can adhere to textile substrates more evenly. One popular method of nano fabrication of chitosan is gel ionization technique by reaction with polyanions such as TPP [146, 147]. A comprehensive study on the synthesis of nano chitosan by gel ionization technique comprising the effect of various parameters on the particle size of nano chitosan [50, 148] and the performance evaluation of nano chitosan in context to various properties on cotton fabric is discussed [94].

19.7.2 Synthesis and Characterization Nano Chitosan

Chitosan has fairly long linear structure with rigid conformation. These long molecules in solid state are, mostly, in the form of tightly folded random coils. Individual molecular coils are also not discrete and separate but are interpenetrating and entangled with each other. In solution, the solvent gradually diffuse into the polymer aggregates resulting into the swelling of the polymer. As swelling continues, the segments of the polymer are solvated and loosened out. The loosened polymer molecule then diffuses slowly out of the polymer phase and disperses in solvent phase, forming the solution. Since the molecules in a solid polymer remains entangled with neighboring ones, polymer molecules during dissolution diffuse out as bunches of entangled molecules. Even when all chain segments of a polymer molecule in solution are unfolded and fully solvated, the molecules does not assume the shape of an extended straight chain but present in a coil form with the "bound" solvent in the empty space between the unfolded segments. Such polymer coils along with "bound," known as "hydrodynamic" sphere or ellipsoid and the apparent volume is referred to as "hydrodynamic volume" as shown in Fig. 19.32 [31, 32, 55].

The characteristic size of CHT hydrodynamic sphere, in our case, at 1 g/L concentration was determined to be 4014 nm. Such higher particle size offer higher viscosity to the solution confining the surface deposition of polymer film. The particle size of chitosan molecule can be scaled down by to nano level by "bottom-up" approach as a result of self-assembling or cross-linking processes [146]. Various methods of the synthesis of nano chitosan are described in literature [149, 150], which include precipitation or coagulation or desolvation method, covalent cross-linking, ionic cross-linking, emulsion droplet coalescence and reverse micellar method.

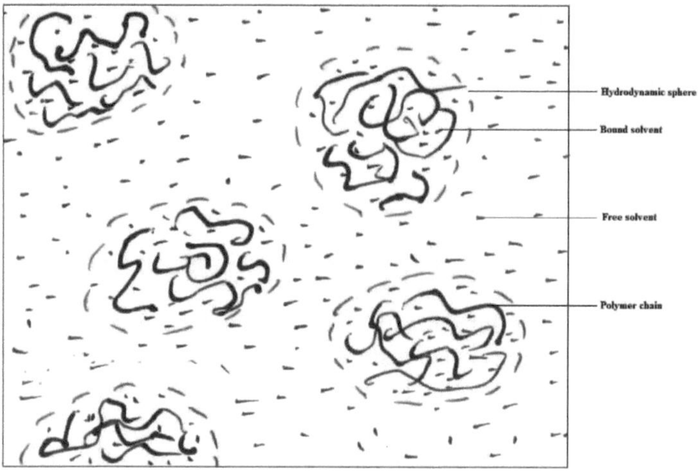

Figure 19.32 Hydrodynamic spheres of chitosan molecules in solution.

Drop-wise addition of sodium sulfate into a solution of chitosan and polysorbate 80 (used as a stabilizer for the suspension) under both stirring and ultrasonication, desolvated chitosan in a particulate form. Although the investigators called the resulting suspensions micro spheres, the precipitated particles were at micro/nano interface (900 ± 200 nm). Comparatively larger particle was attributed to their higher porosity observed by higher swelling ability [60, 151]. Dambies et al. [152] prepared chitosan particles using molybdate. It was observed a double layer structure corresponding to a very compact 100 μm-thick external layer and an internal structure of small pores. Emulsion-droplet coalescence method, introduced by Tokumitsu et al. [153], utilizes the principles of both emulsion cross-linking and precipitation. In this method, instead of cross-linking the stable droplets, precipitation is induced by allowing coalescence of chitosan droplets with sodium hydroxide droplets. A stable emulsion containing aqueous solution of chitosan is produced in liquid paraffin. Microscopically, they are homogenous and isotropic structures consisting of water-in-oil droplets separated by surfactant-rich films. Nanoparticles prepared by conventional emulsion polymerization methods are not only large (200 nm), but also possess a broad size range. Preparation of ultrafine polymeric nanoparticles with narrow size distribution could be achieved by using reverse micellar medium [154]. In this method, the surfactant is dissolved in an organic solvent to prepare reverse micelles. To this, an aqueous solution of chitosan is added gradually with constant vortexing to avoid any turbidity. To this transparent solution, a cross-linking agent is added with constant stirring overnight. Gong et al. [155] reported a facile nonaqueous electrochemical approach to synthesizing different single-crystal chitosan nanostructures on a stainless steel substrate, without using a template, catalyst, or surfactant. Chitosan was dispersed in propylene carbonate (PC) under mild ultrasonication and $LiClO_4$ was used as the supporting electrolyte during the electrochemical process.

By virtue of primary amino groups, chitosan undergoes Schiff's base formation with aldehydes and ionic interactions with anionic compounds [4]. With the former property, chemically cross-linked leading to quite stable matrixes of nano chitosan are obtained. Dialdehydes such as glutaraldehyde and salicylaldehyde are broadly used as cross-linking the molecule in covalent formulations [146, 150, 156]. In the latter, chitosan hydrogels can be obtained by ionic gelation, where nanoparticles are formed by means of electrostatic interactions with polyanions such as pentasodium tripolyphosphate (TPP), ethylene diamine tetra acetic acid (EDTA), etc. [149, 157–160]. Owing to faster ionic reactions between chitosan and TPP, non-toxic nature of these components [146] and ease of operation, the gel ionization technique for the synthesis of nano chitosan particles has been most accepted method. From the physicochemical standpoint, the interaction of chitosan with TPP is accepted to be mediated by the intramolecular cross-linking of tripolyphosphoric, ($P_3O_{10}^{-5}$) ionic species, product of the dissociation of TPP in aqueous solution, with $-NH_3$ groups in chitosan. The intramolecular cross-linking in chitosan molecule by gel ionization is schematically illustrated in Figs. 19.33 and 19.34.

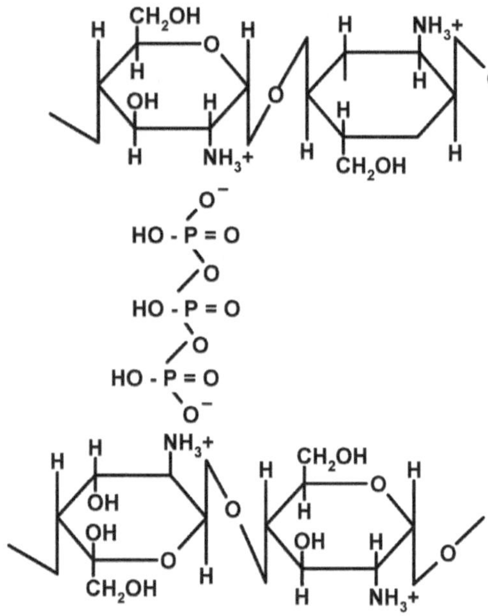

Figure 19.33 Chitosan-TPP complex formed as a result of ionic gelation [50].

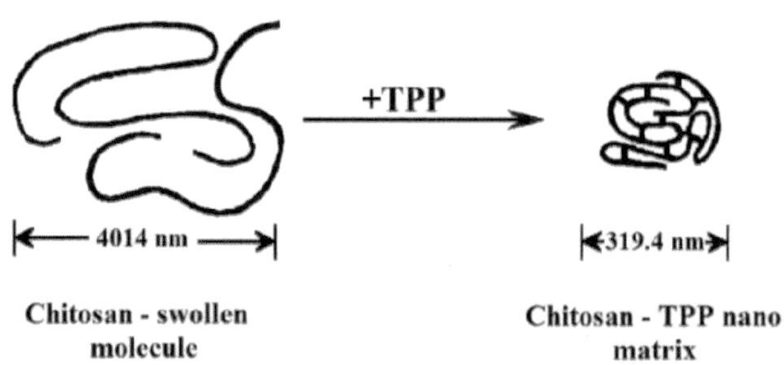

Figure 19.34 Schematic presentation of ionic gelation of chitosan with TPP [148].

Stabilization of nano chitosan dispersion can be explained on the principle of Coulombic or electrostatic repulsion. Particles in a colloid, due to smaller size, are often pushed around by the molecular collisions of the surrounding media, an effect called Brownian motion. The Brownian motion is a rather random, causing the particles to collide with each other frequently and aggregate to form larger particles, which settle down due to their weight. A prerequisite to utilization of colloids for nanotechnology is that they remain colloidally stable, i.e., they remain in suspension and resist settling down. This stability of a colloid can be achieved by means of electrostatic stabilization due to polycationic nature of chitosan in acidic medium and/or involving the creation of an electrical double layer arising from ions adsorbed on the surface of the particle and associated counter ions that surround the particle. Thus, if the electric potential associated with the double layer is sufficiently high, the Coulombic repulsion between the particles will prevent their agglomeration (Fig. 19.35) [161]. The diffused solvent in nano gel exerts pressure on polymeric chain of loop, known as osmotic pressure. The osmotic pressure and intramolecular ionic repulsion tend to swell or enlarge the particle while the elastic contribution and the degree of cross-linking act to shrink the gel. Thus, from thermodynamics point of view, the stability or equilibrium of nano gel is attained when the forces responsible for swelling are balanced with the forces causing shrinkage [146].

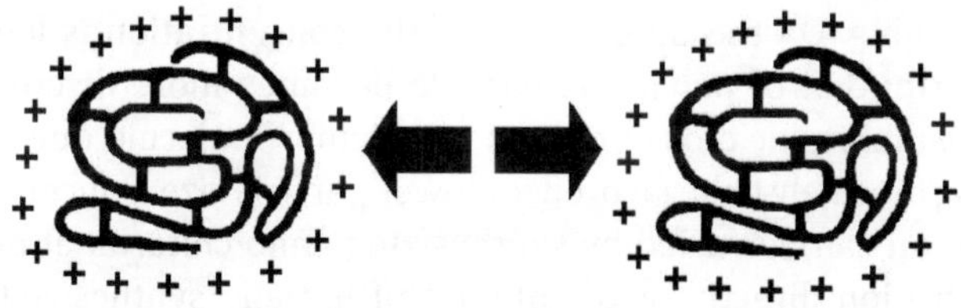

Figure 19.35 Stability of nanoparticles due to electrostatic repulsion between the same ionic charges.

A simple experimental setup designed for the synthesis of nano chitosan sols is shown in Fig. 19.36 [148]. Chitosan solution is taken in a glass beaker and subjected to rapid stirring on a magnetic stirrer at ambient temperature (30°C). Sodium tripolyphosphate solution is then added drop wise. The particle size distribution of above nanochitosan dispersions derived from varying the molecular weight of chitosans is shown in Fig. 19.37 and their particle size is given in Table 19.9, which was determined using particle size analyzer commonly known as zeta sizer. It was reported that, under a given condition of concentrations of CHT and TPP, with decrease in molecular weight, the particle size also decreased progressively and obeys a curvilinear relation as shown in Fig. 19.38. The tendency to form "loop" is expected to be more favored in larger chitosan chains than in shorter ones and can be expected to accommodate greater amount of solvent to produce higher hydrodynamic volume [31, 157] and hence packed into relatively larger gelled nanoparticles and vice versa. The large amount of bound solvent in nano gel–derived high-molecular-weight chitosan exerts higher osmotic pressure and the intramolecular ionic repulsion in acidic pH also contributes to larger size [146]. This relation of particle size with molecular weight in a definite condition of parameters may be useful for the preparation of nano chitosan of desired particle size. The regularity in particle size is determined by polydispersity index (pdi). Higher pdi value indicates the distribution in larger band width, multi-population and varying degree of size. The above study [148] also revealed that chitosan concentration in solution determines the particle size of synthesized nano chitosan, which decreases with decrease in concentration. The larger size of nanoparticles at higher concentration may be due to the aggregation of polymer molecules as a result of overlapping and also to the intermolecular cross-linking through TPP bridging. On the other hand, as the concentration is lowered, the distribution of polymeric particle becomes more discrete and the intramolecular cross-linkages in polymer molecule due to TPP bridging are likely to be favored for lower particle size. Concentration of chitosan can be varied by synthesizing nano chitosan at desired concentration directly or by dilution of initially synthesized nano chitosan at high concentration with rapid stirring. The former method was reported to be more effective.

Figure 19.36 Experimental setup for the preparation of nano chitosan by ionic gelation method.

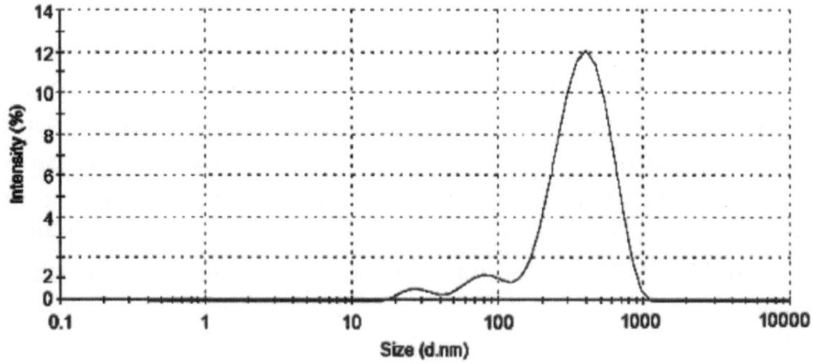

Figure 19.37 Size distribution of nano chitosan (CHTN) by intensity (particle size 319.4 nm, pdi 0.422) [148].

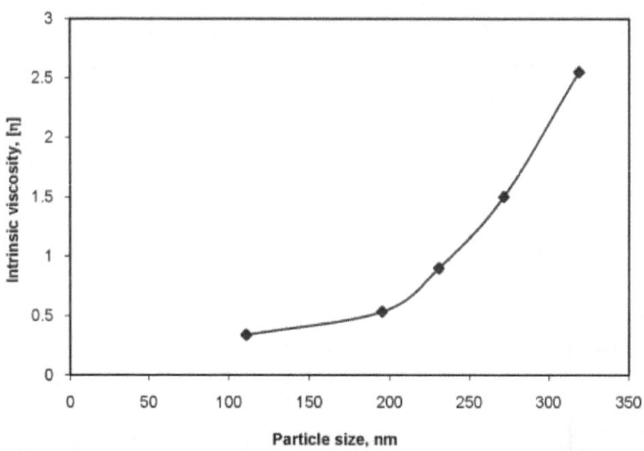

Figure 19.38 Particle size of chitosan as a function of intrinsic viscosity [148].

Table 19.9 Effect of molecular weight on particle size [148]

Parent chitosan			Synthesized nano chitosan		
Sample code	**Intrinsic viscosity (η)**	**Molecular weight (M_v)**	**Sample code**	**Particle size (nm)**	**Polydispersity index (pdi)**
CHT	2.55	135,839	CHTN	319.4	0.422
CHT-D2	1.5	71,676	CHT-D2N	271.6	0.564
CHT-D3	0.9	38,733	CHT-D3N	231.1	0.466
CHT-D4	0.535	20,698	CHT-D4N	195.2	0.278
CHT-D5	0.34	11,986	CHT-D5N	110.74	0.467

Note: Concentration of chitosan 1 g/L, CHT:TPP::1:0.15.

In ionic gelation reaction, TPP a major ingredient for cross-linking has a pronounced effect on the properties of CHTN dispersion and therefore the concentration of TPP is believed to influence/determine the particle size of nano chitosan. It was reported that [148] the particle size of chitosan reduces with increase in concentration of TPP, reaches to minimum at certain concentration and then increases with further increase in concentration of TPP as shown in Fig. 19.39. At low concentration of TPP, very few phosphate ions are present to produce effective ionic linkages with chitosan amino groups. As the concentration of TPP is increased gradually, the solution becomes opalescent indicating the formation of nano chitosan. Higher concentrations of TPP result in precipitation, which may be attributed to the aggregation of chitosan molecules due to excessive cross-linking through TPP bridging. Similar trend in terms of viscosity change with change in concentration of was also reported in same paper.

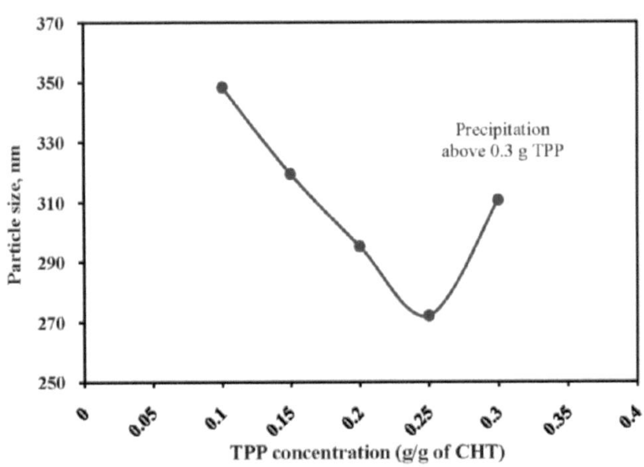

Figure 19.39 Effect of TPP concentration on particle size of CHTN [148]. Concentration of chitosan 1g/L; size of CHT molecular sphere in absence of TPP was 4014 nm.

The viscosity of polymer solution, at the molecular level, is a direct measure of the hydrodynamic volume of the polymer molecules, which in turn is governed by the molecular size or the chain length and hence the molecular weight [32]. The viscosity of nano chitosan dispersion decreases with reduction in the particle size, as reported in the literature [148]. It was also reported that the percentage drop in viscosity from molecular solution of chitosan (CHT) to corresponding nano chitosan (CHTN) dispersion follows a straight line. It means higher molecular weight chitosan scales down to nano size to greater extent than the lower-molecular-weight chitosan. This can be explained as follows; large size chitosan molecules in solution accommodate comparatively more amount of solvent and occupy large "hydrodynamic" volume. These swollen molecules compress to greater extent by ionotropic intramolecular cross-linking with TPP by displacing the solvent, as demonstrated in Fig. 19.34. On the other hand, the extent of swelling of low-molecular-weight chitosan is comparatively less [55] and hence lesser will be its tendency to compress.

The biodegradability of chitosan is anticipated to be influenced by its particle size. Therefore, the stability behavior of standing baths of nano chitosan dispersion should be taken into consideration during its applications particularly to textile fabrics. The loss in viscosity of high-molecular-weight parent chitosan solution was somewhat higher than that of corresponding synthesized nano chitosan dispersions. The integrity of nano chitosan, i.e., small particle size was found to lose which were obtained from low-molecular-weight chitosans. The viscosity changes for most of the nano chitosan dispersions in 24 h storage were tolerable and can be safely employed for applications. Complete biodegradation of nano chitosan dispersion, in general, was resulted in 3–4 days showing the formation of white globules as shown in Fig. 19.40 [148].

Figure 19.40 Stability study: white globular residue formed by microbial attack on CHTN [148].

19.7.3 Effect of Nano Chitosan Treatment on Cotton Fabric

A detailed experimental study on the performance evaluation of nano chitosan on cotton fabric has been reported earlier [94]. The surface morphology of nano chitosan-treated cotton fibers studied under scanning electron microscope shows non-glossy and somewhat swollen appearance (Fig. 19.41). This effect is more significant when the particle size is reduced indicating the greater extent of penetration of chitosan particles into the fiber structure. The appearance of treated fabric in terms of whiteness index and stiffness are reported to be satisfactory. The paper also shown that nano chitosan treatment improved the fiber strength, which further increased with the reduction in particle size as against the reduction in fiber strength of normal chitosan-treated fabrics (Table 19.10). The elongation property, however, was slightly affected with the scaling down of particle size. Normal chitosan mostly confines its film deposition on fiber surface only and thus contribute to very small extent in load bearing phenomenon rather may affect its symmetrical distribution of load. The improvement in fiber strength may be attributed to greater penetration of small particles and cross-link the adjacent fiber molecules by various forces between amino ($-NH_2$) and hydroxyl ($-OH$) groups of chitosan and hydroxyl ($-OH$) groups of cellulose molecules. The smaller the particle size, higher will be the surface area and hence the higher will be the cross-links. The formation of in situ three-dimensional networks probably resists the adjacent fiber molecules to slip and lowers the elongation at break.

The absorbency was found to be decreased with the reduction in particle size (Fig. 19.42). This may be elucidated by the example of lotus leaf effect. Distribution of nano chitosan particles as a thin layer over and beneath the surface may roll out the water droplets. Nevertheless, the absorbency of nano chitosan-treated samples is still within the tolerable limits of conventional wet processing conditions since this rise in water drop penetration time is due to the initial resistance offered by nano chitosan particles and not due to the hydrophobicity.

Table 19.10 Effect of nano chitosan treatment on tensile properties of cotton fabric [94]

Parent chitosan-treated fibers			Nano chitosan-treated fiber		
Sample code	Tenacity (g/tex)	Elongation at break (%)	Sample code	Tenacity (g/tex)	Elongation at break (%)
Untreated cotton fabric	23.33	5.25	Untreated cotton fabric	23.33	5.25
Control	20.87	4.75	Control	20.87	4.75
CHT	20.48	4.75	CHTN	25.17	4
CHT-D2	21.01	4.5	CHT-D2N	25.62	4
CHT-D3	21.45	4.5	CHT-D3N	25.56	4
CHT-D4	22.19	4.25	CHT-D4N	25.71	3.75
CHT-D5	21.81	4.5	CHT-D5N	25.72	3.5

Note: Concentration of chitosan derivatives in pad liquor 1 g/L. Control: Blank treatment was given with acetic acid (10 g/L) by pad-dry cure method.

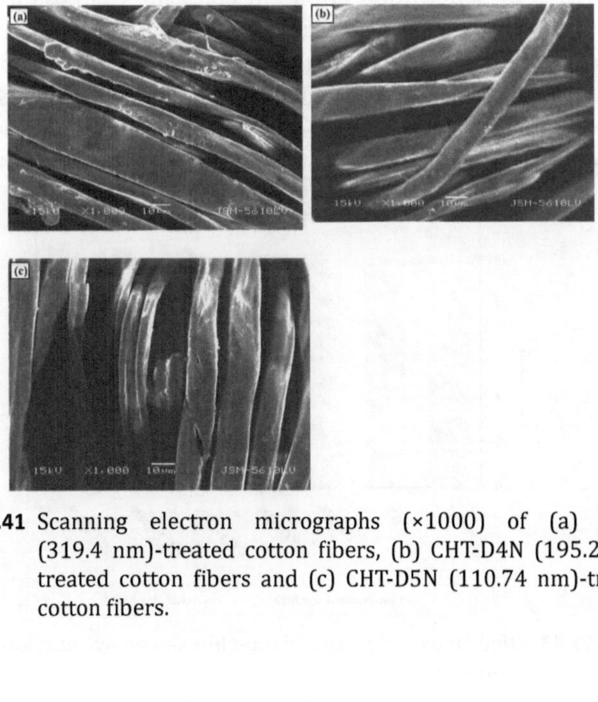

Figure 19.41 Scanning electron micrographs (×1000) of (a) CHTN (319.4 nm)-treated cotton fibers, (b) CHT-D4N (195.2 nm)-treated cotton fibers and (c) CHT-D5N (110.74 nm)-treated cotton fibers.

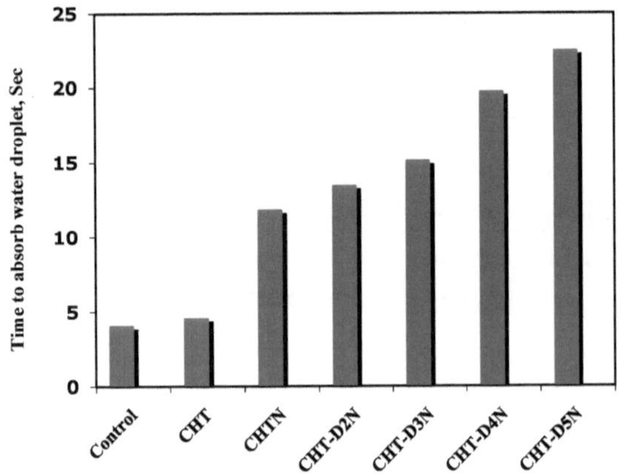

Figure 19.42 Effect of particle size of nano chitosan on absorbency of treated cotton fabric [94]. Concentration of chitosan derivative in pad liquor 1 g/L.

Chitosan treatment has shown improved dye uptake on cotton fiber both in presence and absence of electrolyte as discussed earlier. However, the dye was mostly confined to surface of fabric due to lack of penetration of parent chitosan. Fastness properties were also found to be deprecated. A study on the effect of nano chitosan treatment on the direct dyeing of cotton is reported in this paper [94]. Further, the dye bath was made slightly acidic with acetic acid (0.5 g/L) and treated for 15 min after the conventional dyeing was over (i.e., after the 60 min) (Figs. 19.43 and 19.44).

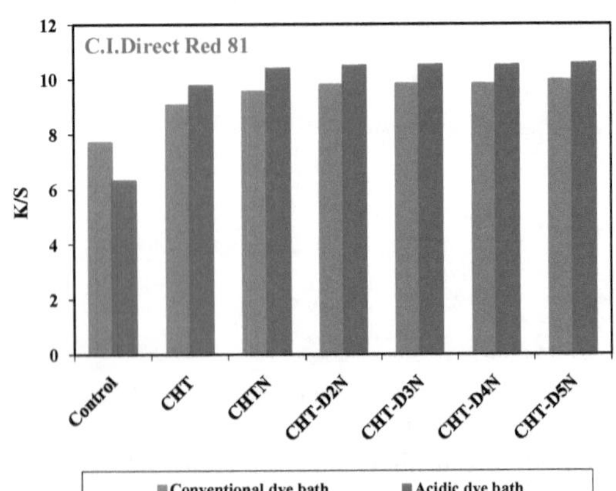

Figure 19.43 Effect of particle size of nano-chitosan on dye uptake of treated cotton fabric.

Results revealed that the dye uptake by cotton fabric, in conventional process, increased marginally with normal chitosan treatment and further improved with the reduction in particle size of nano chitosan. On acidification, the dye uptake still increased compared to conventionally dyed samples. The increased dye uptake due to chitosan treatment may be attributed to the presence of primary amino groups on chitosan. These cat ions dissipate the negative surface charge on cotton and drives dye molecules to the fiber. Further, the dye uptake may also been enhanced due to the dyeability of chitosan itself with direct dyes. The nano chitosan due to increased surface area and hence higher accessibility for dye sites put much added value. The primary amino groups on chitosan get protonated in acidic medium having enhanced positive charge can now form salt linkages with anionic (sulfonate) groups of residual dye present in the dye bath. This leads to almost complete exhaustion of dye bath. The fastness to washing was found to be improved with reduction in particle size. This may be regarded to the formation of CHTN-Dye complex in situ. The fastness to rubbing was also slightly improved with decrease in particle size. This property is mainly associated with the rubbing fastness of dyes chitosan molecules on cotton fiber.

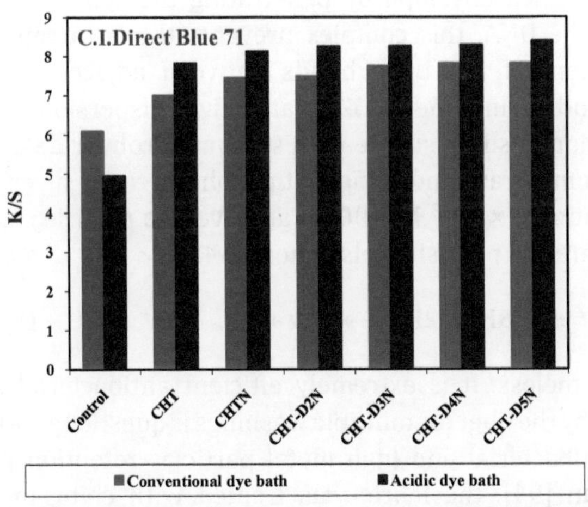

Figure 19.44 Effect of particle size of nano-chitosan on dye uptake of treated cotton fabric.

It has been earlier shown that the chitosan treatment adversely affected the wrinkle recovery property of cotton fabric.

The molecular size of polymer was the major reason for this drawback. Reduction of the particle size of chitosan has found to enhance various properties at minimal concentration and is expected the possibility to restore the aesthetics. It was reported that the treatment of cotton fabric with chitosan of lower particle size improved its crease recovery property. Yet it could not gain the rating of commercially used cross-linking agent DMDHEU. Conventional chitosan is believed to form a surface coating that lowers the possibility of cross-linking and therefore cannot contribute to the load sharing phenomenon. The improved wrinkle recovery property in case of nano chitosan treatment may be attributed to the greater penetration into fabric structure. These polycationic nanoparticles, due to better penetration, may bound the fiber molecules and resist creasing to some extent. However, a final touch up with conventional easy care finish is desired [94].

The antibacterial property of nano chitosan can further be enhanced by loading it with other antibacterial agents such as silver nanoparticle. Silver nanoparticles are very effective antimicrobial and antifungal agents at lower concentrations and are much more effective than other metals such as mercury, copper, lead, chromium, and tin. At lower concentration, silver nanoparticles directly damage the cell envelope by penetrating the cell and then silver binds to the DNA, this complex prevents the DNA replication by displacement of hydrogen bonds between adjacent nitrogen of purines and pyrimidines [162]. Nano silver dispersion is produced by reduction of silver sulfate with sodium borohydride in presence of trisodium citrate under inert atmosphere. Nano silver colloid of concentration 1×10^{-3} M/100 ml and average particle size 110 nm was prepared as published elsewhere [145].

$$4Ag^{+2} + NaBH_4 + 2H_2O \rightarrow 4Ag + BH_4 + Na^+ + 4H^+ + O_2$$

Nevertheless, it is extremely efficient antibacterial agent; its retention by the fiber for multiple washings is questioned. Attributing to the antibacterial and high metal particles retention properties of chitosan [94], the fabric was treated with chitosan (or nano chitosan) and then with nano silver colloid (two bath process). The resistance against bacterial attack of untreated and treated samples of cotton was determined by determining their loss in strength due to soil burial test. The results are presented in Fig. 19.45. It reported that the chitosan can be employed as an efficient antibacterial agent.

The effect is enhanced with the reduction in particle size of nano chitosan and coupling with nano silver.

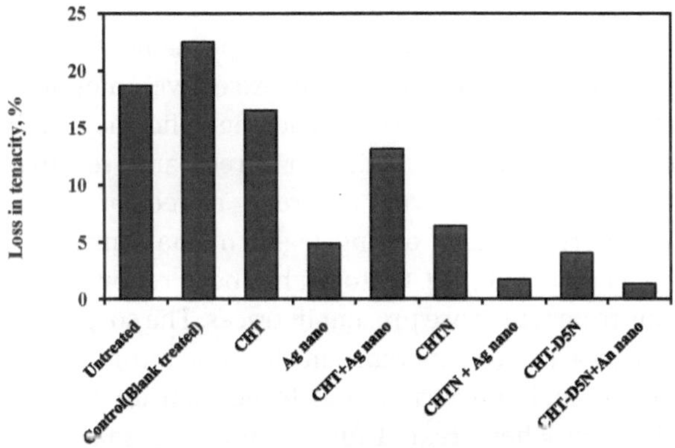

Figure 19.45 Effect of nano chitosan on resistance toward microbial attack. Concentration of chitosan derivative 1 g/L.

This effect can also be realized for specific bacteria namely Gram-positive *S. aureus* and Gram-negative *E. coli* in agar diffusion test [128] as demonstrated in Fig. 19.46.

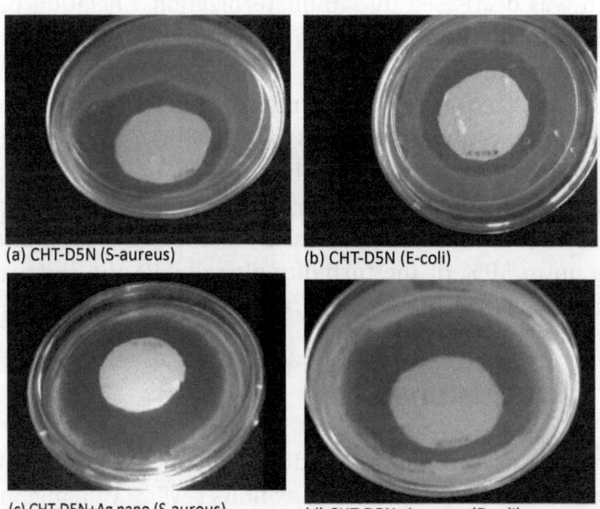

Figure 19.46 Determination of antimicrobial activity of chitosan and nano-chitosan-treated cotton fabric by agar diffusion method: (a) CHT-D5N (*S. aureus*), (b) CHT-D5N (*E. coli*), (c) CHT-D5N +Ag nano (*S. aureus*) and (d) CHT-D5N+Ag nano (*E. coli*).

19.8 Chitosan in Water Processing Used in Textile Industry

Supply of good quality water is indispensable for textile wet processing industry. The presence of excessive amount of copper ions was found to reduce the bleaching efficiency of hydrogen peroxide and caused off shades in direct and reactive dyeing. Application of chitosans of different molecular weights and quaternized derivatives of varying degree of quaternization in water processing is reported [17]. Chitosan has been reported to scavenge copper ions from which are present in traces. The sorption of copper ions was increased with increase in the concentration of chitosan. The extent of chelation was found to be high for low-molecular-weight chitosan when treated for shorter time and reduced with increase in the molecular weight, i.e., it followed the following trend, whereas for higher sorption time a reverse trend was noticed. Highly acidic pH was not found to be suitable for chelation of metal ions. A milder acidic condition (pH 5.5) showed better results. Reduction in particle size of chitosan enhanced both the rate and amount of scavenging of metal ions. The chelation efficiency of chitosan was decreased due to quaternization. Chelation property of chitosan for calcium ions is also reported [50].

Chitosan can also be used in water processing engineering as a part of a filtration process. Chitosan causes the fine sediment particles to bind together and is subsequently removed with the sediment during sand filtration. Chitosan also removes phosphorus, heavy minerals, and oils from the water. Chitosan is an important additive in the filtration process. Sand filtration apparently can remove up to 50% of the turbidity alone while the chitosan with sand filtration removes up to 99% turbidity [6, 69, 163].

19.9 Conclusions

Chitosan is a versatile polycationic biopolymer derived from alkaline deacetylation of chitin. It exhibits several valuable inherent properties such as antibacterial, antifungal, antiviral, antacidity, chelation, non-toxicity, and biodegradability as well as properties such as film formation. Further, due to its possession of hydroxyl and amino functional groups, chitosan can be fabricated to tailor products with desired functional properties. The molecular weight of parent chitosan is generally determined viscometrically. Structural features of chitosan are characterized by FTIR spectroscopy ^{1}NMR spectroscopy. Low-molecular-weight chitosans are synthesized by nitrous acid depolymerization method.

The viscosity of chitosan solution was found to be greatly influenced by its molecular weight. The initial molecular weight and the concentration of chitosan seem to influence the stability of its solution. The drop in viscosity in first 24 h was very fast and the critical concentration (C^*) point shifted toward right when the storage time was prolonged. The stability of low-molecular-weight chitosan was not significantly affected within the chosen concentration range. The viscosity of chitosan solution was dropped due to the incorporation of electrolyte (sodium acetate); the effect was more prominent for high-molecular-weight chitosan solution. Therefore, low-molecular-weight chitosans in presence of suitable electrolyte can be preferably used in textile applications.

The SEM of high-molecular-weight chitosan-treated fibers was appeared to be glossy indicating the surface deposition of chitosan, while low-molecular-weight chitosan-treated fiber surface was matty indicating non-filmed surface. The stiffness of the fabric was increased due to chitosan treatment, which increased with increase in molecular weight and concentration. The absorbency of the fabric was found to be reduced after chitosan treatment, which was deprecated with increase in molecular weight and concentration of chitosan.

Substantial enhancement in dye uptake of cotton fabric was noticed due to chitosan pretreatment. The dye uptake was increased with increase in concentration and molecular weight of treated chitosan. The extent of improvement was more for the dyes having high molecular weight and containing more number of anionic (sulfonate) groups (e.g., C. I. Direct Blue 71). Almost a salt-free dyeing was possible by chitosan pretreatments. However, the fastness properties of chitosan pretreated and dyed samples were not satisfactory. A stoichiometric amount of acid dye was taken up by the chitosan-treated cotton fabric. Acid dye adsorption test can be employed as a tool for the characterization of chitosan and also the amino groups on cotton fiber.

Chitosan treatment did not, however, prove quite suitable for easy care finishing. Chitosan treatment showed improved resistance to rotting in soil burial test. The resistance to microbial attack was due to concerted action of chitosan as polycationic nature, to form a rigid protective coating and as a chelating agent to scavengethe essential metal ions from cytoplasm of microorganism.

The very large molecular size and consequently high viscosity of chitosan restricts its penetration into the fiber and fabric structure and leads to only surface deposition resulting into deterioration in appeal of the fabric. Scaling down the particle size of chitosan to nano level increases the extent of penetration into fiber structure and sustains the inherent properties of cotton fiber. Ionotropic gelation technique using TPP as cross-linker can be the most convenient method for the synthesis of nano chitosan. The concentration of chitosan in the formulation bath was found to influence the size of particle. Particle size was reduced with reduction in concentration. Direct preparation method produced particles with comparatively lower size than that found in dilution method. Reduction in molecular weight and in turn drop in intrinsic viscosity decreased the particle size and showed a curvilinear dependence. This relation may be useful in preparation of nano chitosan dispersion of desired particle size. TPP concentration was found to play an important role in controlling the particle size. With increase in concentration of TPP, the particle size was first reduced, reached to minimum and again increased. Excessive TPP led to precipitation. Reduction in particle size reduced the viscosity of chitosan solution significantly, but the storage stability was affected adversely. Use of freshly prepared nano-chitosan dispersions prior to applications may be the remedy.

The appearance and handle of nano chitosan-treated cotton fabric was much better than the parent chitosan-treated one. Nano chitosan treatment showed improvement in fiber strength that increased with the reduction in particle size. Nano chitosan treatment reduced the water absorbency to some extent. The dyeability of both chitosan and nano chitosan-treated cotton fabric toward direct dyes was improved reasonably. Acidification of dye bath further improved the dye adsorption and wash fastness of dyed fabric.

Nano chitosan together with nano silver treatment showed enhanced resistance to microbial attack.

Chitosan has found to scavenge copper ions from which are present in traces. The sorption of copper ions was increased with increase in the concentration of chitosan. The extent of chelation was found to be high for low-molecular-weight chitosan when treated for shorter time and reduced with increase in the molecular weight, i.e., it followed the following trend whereas for higher sorption time a reverse trend was noticed. Highly acidic pH was not found to be suitable for chelation of metal ions. A milder acidic condition (pH 5.5) showed better results. Reduction in particle size of chitosan enhanced both the rate and amount of scavenging of metal ions.

References

1. Hirano, S. "Chitin and Chitosan", *Ullmann's Encyclopedia of Industrial Chemistry*, Wiely-VCH, **6** (2003), 679–691.
2. Inamdar, M. S., and Chattopadhyaya, D. P. Chitosan and its versatile applications in textile processing, *Man Made Text. India*, **XLIX**(6) (2006), 211–216.
3. Struszczyk, H., Kivekas, O., Niekraszewicz, A., and Urbanonowski, A. Chitosan-new forms and uses, *Text. Asia*, **24**(7) (July 1993), 80–83.
4. Terbojevich, M., and Muzzarelli, R. A. A., Chitosan *Handbook of Hydrocolloids*, Phillips, G. O., and Williams, P. A. (eds.), Woodhead Publishing Ltd., Cambridge, England (2000), 367–378.
5. http://dalwoo.com/chitosan/applications.html.
6. Gavhane Yogeshkumar N., Gurav Atul S., and Yadav Adhikrao V., Chitosan and its applications: A review of literature, *Int. J. Res. Pharm. Biomed. Sci.*, **4**(1) (2013), 312–331.
7. Bahl, B. S., and Tuli, A., *Advanced Organic Chemistry*, 2nd ed., Chand, S., and Co Ltd., N. Delhi, India (1983).
8. Inoue, Y., NMR determination of degree of acetylation in: *Chitin Handbook*, Muzzarelli, R. A. A., and Peter, M. G. (eds.), Atec, Grottammare, Itali (1997), 133–136.
9. Singh, M. K., 21st century with deodorant fabrics, *Man Made Text. India*, **14**(7) (2002), 279–286.
10. Singh, V. L., Biopolymers, *Asian Text. J.*, **14**(1–2) (2005), 65–68.
11. Sekar, N., Chitosan in textile processing-an update, *Colourage*, **XLVII**(7) (2000), 33–34.

12. El-Tahlawy, K. F., Utilization of citric acid—chitosan, sodium hypophosphite system for effecting concurrent dyeing and finishing, *Colourage,* **XLVI**(5) (1999), 21–26.
13. Kalsi, P. S., Chapter 3, Infrared spectroscopy, in *Spectroscopy of Organic Compounds,* 6th ed., New Age International Publisher, N. Dehli, India (2004), 59–164.
14. Bayat, A., Sadeghi, A. M. M., Avadi, M. R., Amini, M., Rafiee-Tehrani, M., Shafiee, A., Junginger, H. E., Synthesis of *N,N*-dimethyl *N*-ethyl chitosan as a carrier for oral delivery of peptide drugs, *J. Bioact. Compat. Polym.,* **21** (2006), 433–444.
15. Li, P., Dai, Y.-N., Zhang, J.-P., Wang, A.-Q., and Wei, Q., Chitosan–alginate nanoparticles as a novel drug delivery system for nifedipine, *Int. J. Biomed. Sci.,* **4**(3) (2008), 221–228.
16. Avadi, M. R., Amini, G. M., Sadegi, A. M., Irfan, M., Amini, M., Tehrani, M. R., and Shafiee, A., Synthesis and characterization of *N*-diethyl methyl Chitosan, *Iranian Polym. J.,* **13**(5) (2004), 431–436.
17. Chattopadhyay, D. P., and Inamdar, M. S. Application of chitosan and its derivatives in removal of Cu(II) ions from water used in textile wet processing, *Text. Res. J.,* Published online on February 21, 2014, doi: 10.1177/0040517514523176. http://trj.sagepub.com/content/early/2014/02/20/0040517514523176.
18. H. K. No and Meyers, S. P., Preparation and characterization of chitin and chitosan—A Review, *J. Aquat. Food Prod. Technol.,* **49**(2) (1995), 27–52.
19. Shephard, R., Reader, S., and Falshaw, A., Chitosan functional properties, *Glycoconj. J.,* **14** (1997), 535–542.
20. Meyers, S. P., No, H. K., and Lee, K. S., Isolation and characterization of chitin from crawfish shell waste, *J. Agric. Food Chem.,* **37**(3) (1989), 575–579.
21. Cho, Y. W., Jang, J. H., Park, C. R., and Ko, S. W., Preparation and solubility in acid and water of partially deacetylated chitins, *Biomacromolecules,* **1** (2000), 609–614.
22. Curtia, E., and Campana-Filho, S. P., Viscosity behavior of chitosan and *N,N,N*-trimethylchitosan chloride salts in acid-free aqueous solution, *J. Macromol. Sci. Part A: Pure Appl. Chem.,* **43** (2006), 555–572.
23. Bobu, E., Nicu, R., Lupei, M., Ciolacu, F., and Desbrières, J., Synthesis and characterization of *N*-alkyl chitosan for papermaking applications, *Cell. Chem. Technol.,* **45**(9–10) (2011), 619–625.

24. Hirai, A., Odani, H., and Nakajima, A., Determination of degree of deacetylation of chitosan by ^{1}H NMR spectroscopy, *Polym. Bull.*, **26**(1) (1991), 87–94.

25. Yui, T., Imada, K., Okuyama, K., Obata, Y., Suzuki, K., and Ogawa, K., Molecular and crystal structure of the anhydrous form of CS, *Macromolecules,* **27**(26) (1994), 7601–7605.

26. Kurita, K., Kamiya, M., and Nishimura, S., Solubilization of a rigid polysaccharide: controlled partial *N*-acetylation of chitosan to develop solubility, *Carbohydr. Polym.*, **16** (1991), 83–92.

27. Rathke, T., and Hudson, S., Review of chitin and chitosan as fiber and film formers, *J. Macromol. Sci. R. M. C.*, **C34**(3) (1994), 375–437.

28. Knaul, J. Z., Kassai, M. R., Bui, V. T., and Greber, K. A. M., Characterization of deacetylated chitosan and chitosan molecular weight—review, *Can. J. Chem.*, **76**(11) (1998), 1699–1706.

29. Ottoy, M. H., Varum, K. M., Christensen, B. E., Anthonsen, M. W., and Smidsrod, O., Preparative and analytical size-exclusion chromatography of chitosans, *Carbohydr. Polym.*, **31**(4) (1996), 253–261.

30. Domard, A., and Rinaudo, M., Preparation and characterization of fully deacetylated chitosan, *Int. J. Biol. Macromol.*, **5**(1) (1983), 49–52.

31. Gowariker, V. R., Viswanathan, N. V., and Sreedhar, Y., Polymer solutions, in *Polymer Science,* New Age International Publisher, N. Dehli, India (1986), 332–362.

32. Shaw, D. J., Rheology, in *Introduction to Colloid and Surface Chemistry,* 4th ed., Butterworth-Heinemann, Oxford (UK) (1992), 244–260.

33. Nagasawa, K., Tohira, Y., Inoue, Y., and Tanoura, N., Reaction between carbohydrates and sulfuric acid: Part I. Depolymerization and sulfation of polysaccharides by sulfuric acid, *Carbohydr. Res.*, **18**(1) (1971), 95–102.

34. Vårum, K. M., Antohonsen, M. W., Grasdalen, H., and Smidsrød, O., Determination of the degree of *N*-acetylation and the distribution of *N*-acetyl groups in partially *N*-deacetylated chitins (chitosans) by high-field n. m. r. spectroscopy, *Carbohydr. Res.*, **211**(1) April (1991), 17–23.

35. Hebeish, A., Waly, A., and Aou-Okeil, A., The effect of molecular weight of chitosan on cotton fabric treated with citric acid and its impact on dyeing with some acid dyes, *J. Text. Assoc.*, **65**(5), Jan-Feb (2005), 219–227.

36. Knittel, D., Materne, G., and Schollmeyer, E., Degradation of chitosan sizes, *Melliand Engl.,* **87**(9) (2006), E142–E144.

37. Seaong, H.-S., Kim, J.-P., and Ko, S.- W., Preparing chito-oligosaccharide as antimicrobial agents for cotton, *Text. Res. J.,* **69**(7) July (1999), 483–488.

38. Tian, F., Liu, Y., Hu, K., Zhao, B., The depolymerization mechanism of chitosan by hydrogen peroxide, *J. Mater. Sci.*, **38** (2003), 4709–4712.

39. Lee, F., Lee, W. K., Maskat, M. Y., Illias, R. M., Aziz, S. A., Kamarulzaman, K., and Osman, H., Partial depolymerization of chitosan with the aid of bromelain, *Pak. J. Biol. Sci.*, **8**(01) (2005), 73–77.

40. Trzciński, S., Combined degradation of chitosans, *Polish Chitin Soc. Monogr.*, **XI** (2006), 103–111.

41. Choi, W. S., Ahn, K. J., Lee, D. W., Byun, M. W., and Park, H. J., Preparation of chitosan oligomers by irradiation, *Polym. Degradation Stab.*, **78** (2002), 533–538.

42. Baxter, S., Zivanovic, S., and Weiss, J., Molecular weight and degree of acetylation of high-intensity ultrasonicated chitosan, *Food Hydrocolloids*, **19** (2005), 821–830.

43. Prashant, K. V. H., and Tharanathan, R. N., Chitin/chitosan: Modifications and their unlimited application potential–an overview, *Trends food Sci. Technol.,* **18** (2007), 117–131.

44. Kabal'nova, N. N., Murinov, K. Y., Mullagaliev, R., Krasnogorskaya, N. N., Shereshovets, V. V., Monakov, Y. B., and Zaikov, G. E., Oxidative destruction of chitosan under effect of ozone and hydrogen peroxide, *J. Appl. Poly. Sci.*, **81** (2001), 875–881.

45. Demin, V. A., Kabal'nova, N. N., Osipova, G. I., and Shereshovetz, V. V., Depolymerization of cellulose upon ozonation, *Russ. J. Appl. Chem.*, **66** (1993), 2562.

46. Yalpani, M., and Pantaleone, D., An examination of the usual susceptibilities of aminoglycans to enyamatic hydrolysis, *Carbohydr. Res.*, **256** (1994), 159–175.

47. Ilyina, A. V., Tikhonov, V. E., Albulov, A. I., Varlamov, V. P., Enzymatic preparation of acid-free-water soluble chitosan, *Process Biochem.*, **35** (2000), 563–568.

48. Abou-Okeil, A., and Hakeim, O. A., Effect of metal ion binding of chitosan on the printability of pretreated wool fabric, *Coloration Technol.*, **121** (2005) 41–44

49. Mao, S., Shuai, X., Unger, F., Simon, M., Bi, D., and Kissel, T., The depolymerization of chitosan: Effects on physicochemical and biological properties, *Int. J. Pharm.*, **281** (2004), 45–54.

50. Chattopadhyay, D. P., and Inamdar, M. S., Aqueous behaviour of chitosan, *Int. J. Polym. Sci.*, **2010** (2010), 1–7.

51. Li, Q., Dunn, E. T., Grandmaison, E. W., and Goosen, M. F., Applications and properties of chitosan, *J. Bioact. Compat. Polym.*, **7** (1992), 370–397.

52. Cho, Y. I., No, H. K., and Meyers, S. P., Physicochemical characteristics and functional properties of various commercial chitin and chitosan products, *J. Agric. Food Chem.*, **46** (1998), 3839–3843.

53. No, H. K., Kim, S. D., Kim, D. S., Kim, S. K., and Meyers, S. P., Effect of physical and chemical treatment on chitosan viscosity, *J. Chitin Chitosan*, **4**(4) (1999), 177–183.

54. No, H. K., Lee, S. H., Park, N. Y., and Meyers, S. P., Comparison of physicochemical, binding, and antibacterial properties of chitosans prepared without and with deproteinization process, *J. Agric. Food Chem.*, **51** (2003), 7659–7663.

55. Tager, A., *Physical Chemistry of Polymers*, 2nd ed., MIR Publishers, Moscow (1978), pp. 241–272.

56. Hwang, J. K., and Shin, H. H., Rheological properties of chitosan solutions, *Korea-Aust. Rheol. J.*, **12**(3/4), December (2000), 175–179.

57. Tsaih, M. L., and Chen, R. H., Effects of ionic strength and pH on the diffusion coefficients and conformation of chitosans molecule in solution, *J. Appl. Polym. Sci.*, **73**(10) September (1999), 2041–2050.

58. Cho, J. Y., Heuzey, M. C., Begin, A., and Carreau, P. J., Viscoelastic properties of chitosan solutions: Effect of concentration and ionic strength, *J. Food Eng.*, **74**(4) (2006), 500–515.

59. Boucard, N., David, L., Rochas, C., Montembault, A., Viton, C., and Domard, A., polyelectrolyte microstructure in chitosan aqueous and alcohol solutions, *Biomacromolecules*, **8**(4) (April 2007), 1209–1217.

60. Al-Remawi, M. M. A., Properties of chitosan nanoparticles formed using sulfate anions as crosslinking bridges, *Am. J. Appl. Sci.*, **9**(7) (2012), 1091–1100.

61. Ling, X., Zu-yu, Y., Chao, Y., Hua-yue, Z., and Yu-min, D., Swelling studies of chitosan-gelatin films cross-linked by sulfate, *Wuhan Univ. J. Nat. Sci.*, **9**(2) (2004), 247–251.

62. Bhumkar, D. R., and Pokharkar, V. B., Studies on effect of pH on cross linking of chitosan with sodium tripolyphosphate: A technical note, *AAPS Pharm. Sci. Tech*, **7**(2) (2006), E1–E6.

63. Lárez Velásquez, C., Albornoz, J. S., and Barrios, E. M., Viscometric stidies of chitosan nitrate and chitosan chlorhydrate in acid free NaCl aq solution, e-*Polymers*, No. 014 (2008), 1–8.

64. Sun, P., Li, P., Li, Y. M., Wei, Q., and Tian, L. H., A pH-sensitive chitosan-tripolyphosphate hydrogel beads for controlled glipizide delivery, *J. Biomed. Mater. Res. Part B: Appl. Biomater.*, **97**(1) (2011), 175–183.

65. Krajewska, B., Application of chitin-and chitosan-based materials for enzyme immobilizations: A review, *Enzyme Microb. Technol.*, **35**(2) (2004), 126–139.

66. Elzatahry, A., Eldin, M., Soliman, E., and Hassan, E., Evaluation of alginate–chitosan bioadhesive beads as a drug delivery system for the controlled release of theophylline, *J. Appl. Polym. Sci.*, **111**(5) (2009), 2452–2459.

67. Yang, J., Chen, J., Pan, D., Wan, Y., and Wang, Z., pH-sensitive interpenetrating network hydrogels based on chitosan derivatives and alginate for oral drug delivery, *Carbohydr. Polym.*, **92** (01)(2013), 719–725.

68. Wiles, J. L., Vergano, P. J., Barron, F. H., Bunn, J. M., and Testin, R. F., Water vapor transmission rates and sorption behaviour of chitosan films, *J. Food Sci.*, **65**(7) (2000), 1175–1179.

69. Dutta, P. K., Dutta, J., and Tripati, V. S., Chitin and chitosan: Chemistry, properties and applications, *J. Sci. Ind. Res.*, **63** (2004), 20–31.

70. Inmaculada, A., Marian, M., Ruth, H., Inés, P., Beatriz, M., Niuris, A., Gemma, G., and Ángeles, H., Functional characterization of chitin and chitosan, *Curr. Chem. Biol.*, **3** (2009), 203–230.

71. Hirano, S., Tsuchida, H., and Nagao, N., *N*-acetylation in chitosan and the rate of its enzymic hydrolysis, *Biomaterials*, **10**(8) (1989), 574–576.

72. Sashiwa, H., Saimoto, H., Shigemasa, Y., Ogawa, R., and Tokura, S., Distribution of the acetamide group in partially deacetylated chitins, *Carbohydr. Polym.*, **16**(3) (1991), 291–296.

73. Schipper, N. G. M., Varum, K., Artursson, P., Chitosans as absorption enhancers for poorly absorbable drugs. 1: Influence of molecular weight and degree of acetylation on drug transport across human intestinal epithelial (Caco-2) cells, *Pharm. Res.*, **13**(11) (1996), 1686–1692.

74. Naeem, M., Hassan, A., Ahmed, M., and El- Sayed, A., Radiation induced degradation of chitosan for possible use as a growth promoter in agricultural purposes, *Carbohydr. Polym.*, **79** (2010), 555–562.

75. http://www.epa.gov/pesticides/biopesticides.

76. Fereidoon, S., Arachchi, J. K. V., and Jeon, Y. J., Food applications of chitin and chitosans, *Trends Food Sci. Technol.*, **10** (1999), 37–51.

77. Dutta, P. K., Tripathi, S., Mehrotra, G. K., Dutta, J., Perspectives for chitosan based antimicrobial films in food application, *Food Chem.*, **114** (2009), 1173–1182.

78. Vilai, R., Nijarin, W., Nilada, K., and Pachara, C., Application of fungal chitosan for clarification of apple juice, *Process Biochem.*, **41** (2006), 589–593.

79. http://www.plasticstrends.net/index.php.

80. Giridev, V. R., Neelkandan, R., Sudha, N., Shamugasundaram, O. L., and Nadaraj, R. N., Chitosan–A polymer with wider applications, *Text. Mag.*, July (2005), 83–86.

81. http://www.seprafilm.com/medprof/efficacy.asp.

82. Jing, H., Su, W., Caracci, S., Bunning, T. G., Cooper, T., and Adams, W., Optical waveguiding and morphology of chitosan thin films, *J. Appl. Polym. Sci.*, **61** (1996), 1163.

83. Dutta, P. K., Vishwanathan, P., Mimrot, L., and Ravikumar, M. N. V., Use of chitosan-amine-oxide gel as drug carrier, *J. Polymer. Mater.*, **14** (1997), 531.

84. Giri, T. K., Amrita, T., Amit, A., Hemant, A. B., and Tripathi, D. K., Modified chitosan hydrogels as drug delivery and tissue engineering systems: Present status and applications, *Acta Pharm. Sinica B*, **2** (2012), 439–449.

85. Jaykumar, R., Prabhaharan, M., Reis, R. L., and Mano, J. F., Graft copolymerization of chitosan-present status and applications, *Carbohydr. Polym.*, **62** (2005), 142–158.

86. Peniche, C., Monal, W. A., Peniche, H., and Acosta, N., Chitosan: An attractive biocompatible polymer for microencapsulation, *Macromol. Biosci.*, **3**(10) (2003), 511–520.

87. Sudha, S., Giridev, V. R., Neelkandan, R., and Kumar, M. S., Chitosan. A versatile polymer for textile applications, *J. Text. Assoc.*, **64**(4) (2006), 165–166.

88. Ghosh, S., and Jassal, M., Use of polysaccharide fibres for modern wound dressing, *Indian J. Fibre Text. Res.*, **27**, Dec (2002), 434–450.

89. Tuzlakoglu, K., Alves, C. M., Mano, J. F., and Reis, R. L., Production and characterization of chitosan fibers and 3D fiber mesh scaffolds for tissue engineering applications, *Macromol. Biosci.*, **4** (2004), 811–819.

90. Qin, Y., Gelling fibres from cellulose, chitosan and alginate, *Chem. Fibres Int.*, **58**(1)March (2008), 30–32.
91. http://www.swicofil.com/products/055chitosan.
92. Jajpura, L., Harad, A., and Maitra, S., Chitin and chitosan in antimicrobial composite fibres, *Asian Text. J.*, **15**(2), Feb(2006), 55–58.
93. Chattopadhyay, D. P., and Inamdar, M. S., Studies on the properties of chitosan treated cotton fabric, *Asian Dyer*, **6**(5) (2009), 43–53.
94. Chattopadhyay, D., and Inamdar, M. S., Improvement in the properties of cotton fabric through synthesized nano chitosan application, *Indian J. Fibre Text. Res.*, **38** (2013), 14–21.
95. Cooper, H. R., Yellowing of textiles due to atmospheric pollution, *Text. Prog.* (A special issue on Update on Yellowing), **15**(4) (1987), 1–6.
96. Giri Dev, B., Subramaniam, R., Studies on frictional behaviour of chitosan coated fabrics, *AUTEXRes. J.*, **6** (4) December (2006), 216–222.
97. Streiner, N., Evaluation of peracetic acid as an ecofriendly safe alternative for hypochlorite, *Text. Chem. Colorists*, **27** (1995), 29–32.
98. Hashem, M., Catalytic activation of per acetic acid using chitosan-metal complex for low temperature bleaching of cotton, *Indian J. Fibre Text. Res.*, **28**, Dec (2003), 444–449.
99. Chattopadhyay, D. P., Cationization of cotton for low salt and salt free dyeing, *Indian J. Fibre Text. Res.*, **24**, March (2001), 108–115.
100. Giridev, V. R., Kumar, M. S., Neelkandan, R., and Murugesan, M., Effect of chitosan treatment on cotton fabric dyeing, *Indian Text. J.*, March (2004), 29–31.
101. Bandyopadhyay, B. N., Sheth, G. N., and Moni, M. M., Chitosan can cut salt use in reactive dyeing, *Int. Dyer*, **183**(11) (1998), 39–42.
102. Kavitha, T., Padmashwini, R., Swarna, A., Giri Dev, V. R., Neelkandan, R., and Kumar, M. S., Effect of chitosan treatment on the properties of turmeric dyed cotton yarn, *Indian J. Fibre Text. Res.*, **32**, March (2007), 53–56.
103. Metha, R. D., and Combs, R. N., An improved process for nep coverage in dyeing cotton, *Am. Dyestuffs Rep.*, **80**(9) (1991), 74–79.
104. Rippon, J. A., Improving the dye coverage of immature cotton fibres by treatment with chitosan, *J. Soc. Dyers Colourists*, **100** (1984), 298–303.
105. Shenai, V. A., and Saraf, N. M., *Technology of Finishing*, vol. X, Sevak Publications, Mumbai, India (1987).

106. Schindler, W. D., and Hauser, P. J., *Chemical Finishing of Textiles*, Woodhead Publishing Limited, Cambridge, England (2004).

107. Blanchard, E. J., Reinhardt, R. M., and Andrews, B. A. K., Finishing with modified polycarboxylic acid systems for dyeable durable press cottons, *Text. Chem. Colorists*, **23**(5) (1991), 25–28.

108. Chung, Y. S., Lee, K. K., and Kim, J. W., Durable press and antibacterial finishing of cotton fabrics with citric acid and chitosan treatment, *Text. Res. J.*, **68**(10) (1998), 772–775.

109. Bhattacharya, N., Doshi, B. A., Sahasrabudhe, A. S., and Mistry, P. R., Use of chitosan in dyeing and finishing of cotton fabric, in *Resume of Papers, 34th Joint Technological Conference of ATIRA, BTRA, SITRA, and NITRA* (1993), 115–121.

110. Huang, K. S., Wu, W. J., Chen, J. B., and Lian, H. S., Application of low-molecular-weight chitosan in durable press finishing, *Carbohydr. Polym.*, **73** (2008), 254–260.

111. Enescu, D., Use of chitosan in surface modification of textile materials, *Roum. Biotechnol. Lett.*, **13**(6) (2008), 4037–4048.

112. Gao, Y., and Cranston, R., Recent advances in antimicrobial treatments of textiles, *Text. Res. J.*, **78**(1) (2008), 60–72.

113. Bajaj, P., Ecofriendly finishes for textiles, *Indian J. Fibre Text. Res.*, **26**, March–June (2001), 162–186.

114. Vigo, T. L., and Benjaminson, M. A., Antimicrobial fiber treatments and disinfection, *Text. Res. J.*, **51**(7) (1981), 454–462.

115. Holme, I., Antimicrobial Imparts durable freshness, *Int. Dyer*, **187**(01) (2002), 9–11.

116. Hasebe, Y., Kuwahara, K., and Tokunaga, S., Chitosan hybrid deodorant agent for finishing textiles, *AATCC Rev.*, **1**(11) (2001), 23–27.

117. Fang, S. W., Li, C. F., and Shih, D. Y. C., Antifungal activity of chitosan and its preservative effect on low-sugar candied kumquat, *J. Food Prot.*, **56**(2) (1994), 136–140.

118. Tsai, G. J., and Su, W. H., Antibacterial activity of shrimp chitosan against *Escherichia coli*, *J. Food Prot.*, **62**(3) (1999), 239–243.

119. Sudardshan, N. R., Hoover, D. G., and Knorr, D., Antibacterial action of chitosan, *Food Biotechnol.*, **6**(3) (1992), 257–272.

120. Wang, G.-H., Inhibition and inactivation of five species of food borne pathogens by chitosan, *J. Food Prot.*, **55**(11) (1992), 916–919.

121. Ye, W. J., Leung, M. F., Xin, J., Kwong, T. L., Lee, D. K. L., and Li, P., Novel core-shell particles with poly(*n*-butyl acrylate) cores and

chitosan shells as an antibacterial coating for textiles, *Polymers*, **46** (2005), 10538–10543.

122. Ye, W., Xin, J. H., Li, P., Lee, K.-L. D., Kwong, T.-L., Durable antibacterial finish on cotton fabric by using chitosan-based polymeric core-shell particles, *J. Appl. Polym. Sci.*, **102**(2) October (2006), 1787–1793.

123. Schmidt, E. L., and Ruschmeyer, O. R., Cellulose decomposition in soil burial beds: I. Soil properties in relation to cellulose degradation, *Appl. Microbiol.*, **6**(2) Mar (1958), 108–114.

124. Kim, C. H., and Choi, K. S., Synthesis and antibacterial activity of quaternized derivatives having different methylene spacers, *J. Ind. Eng. Chem.*, **8**(1) (2002), 71–76.

125. Sudarshan, N. R., Hoover, D. G., and Knorr, D., Antibacterial action of chitosan, *Food Biotechnol.*, **6** (1992), 257–272.

126. Liu, X. F., Guan, Y. L., Yang, D. Z., Li, Z., and Yao, K. D., Antibacterial action of chitosan and carboxymethylated chitosan, *J. Appl. Polym. Sci.*, **79** (2001), 1324–1335.

127. Cuero, R. G., Osuji, G., and Washington, A., *N*-carboxymethyl chitosan inhibition of aflatoxin production: Role of zinc, *Biotechnol. Lett.*, **13** (1991), 441–444.

128. Kimiran Erdem, A., and Sanli Yurudu, N. O., The evaluation of antibacterial activity of fabrics impregnated with dimethyl tetradecyl (3-trimethoxysilyl propyl) ammonium chloride, *IUFS J. Biol.*, **67**(2) (2008), 115–122.

129. Sekar, N., Reactive cellulose fibres for no salt dyeing—a conceptual development towards redressal of environmental issues in reactive dyeing: An update, *Colourage*, **XLVI**(6) May (1999), 29–31.

130. Poole (Jr), C. P., and Owens, F. J., *Introduction To Nanotechnology*, John Wiley & Sons, Inc., New Jersey (2003).

131. Sawhney, A. P. S., Condon, B., Singh, K. V., Pang, S. S., Li, G., and Hui, D., Modern applications of nanotechnology in textiles, *Text. Res. J.*, **78**(8) (2008), 731–739.

132. Feyman, R., There's plenty of room at the bottom, *Eng. Sci.*, **23**(5) (1960), 22–36.

133. Sugunan, A., and Dutta, J., Nanoparticles for nanotechnology, *J. Phys. Sci. Idea*, **4**(1&2) (2004), 50–57.

134. Butola, B. S., and Mishra, S., Nanotechnology in textiles, *Asian Dyer*, **4**(01) (2007), 70–76.

135. Gowri, S., Almeida, L., Amorim, T., Carneiro, N., Souto, A. P., and Esteves, M. F., Polymer nanocomposites for multifunctional finishing of textiles—a review, *Text. Res. J.*, **80**(13) (2010), 1290–1306.

136. Amberg-Schwab, S., Functional coating using nanotechnology, *Int. Text. Bull.*, **1** (2004), 14–18.

137. Inamdar, M. S., Khan, J. S., Halbe, A. V., and Khoja, A. K., Prospects for innovative technologies in textile processing industry, *J. Text. Assoc.*, **66**(2) (2005), 73–75.

138. Schondelmair, D. S., Cramm, R., Klingeler, R., Morenz, J., Zilken, C., and Eberhardt, W., Orientation and self assembly of hydrophobic fluoroalkyl silanes, *Langmuir*, **18** (2002), 6242–6245.

139. Qian, L., and Hinestroza, J. P., Application of nanotechnology for high performance textiles, *J. Text. Appar. Technol. Manag.*, **4**(1) (2004), 1–7.

140. Yang, H. Y., Zhu, S. K., and Pan, N., Studying the mechanism of TiO_2 as UV blocking additive for fibers and fabrics by an improved scheme, *J. Appl. Polym. Sci.*, **92** (2003), 3201–3210.

141. Chen, R. Q., Nanometric materials and health care textiles, *Dyestuff Ind.*, **39**(2) (2002), 4–28.

142. Wang, R. H., Xin, J. H., Yang, Y., Liu, H. F., Xu, L. M., and Hu, J. H., The characteristic and photo catalytic activities of silver doped ZnO, *Nano Cryst.*, **227** (2004), 312–317.

143. Bozzi, A., Yuranova, T., and Kimi, J., Self cleaning of wool-polyamide and polyester textiles by TiO_2 rutile modifications under day light irradiation at ambient temperature, *J. Photochem. Photobiol. A: Chem.*, **172** (2005), 27–34.

144. Cui, S. Y., Zu, Y. D., Hui, H. Q., and Zhang, J. Y., Study on antibacterial properties of nano ceramics, *J. Hebei Univ. Sci. Technol.*, **24** (2003), 19–22.

145. Chattopadhyay, D. P., and Patel, B. H., Improvement in the physical and dyeing properties of natural fibres through pre-treatment with silver nanoparticles, *Indian J. Fibre Text. Res.*, **34** December (2009), 368–373.

146. Lopez-Leon, T., Carvalho, E. L. S., Seijo, B., Ortega-Vinuesa, J. L., Bastos-Gonzalez, D., Physicochemical characterization of chitosan nanoparticles: Electrokinetic and stability behavior, *J. Colloid Interface Sci.*, **283** (2005), 344–351.

147. Zhang, H.-L., Wu, S.-H., Tao, Y., Zang, L.-Q., and Su, Z.-Q., Preparation and characterization of water-soluble chitosan nanoparticles as protein delivery system, *J. Nanomater.*, **2010** (2010), 1–5.

148. Chattopadhyay, D. P., and Inamdar, M. S., Studies on synthesis, characterization and viscosity behaviour of nano chitosan, *Res. J. Eng. Sci.*, **1**(4) (2012), 9–15.

149. Huang, K.-S., Sheu, Y.-R., and Chao, I.-C., Preparation and properties of nanochitosan, *Polym.–Plast. Technol. Eng.*, **48** (2009), 1239–1243.

150. Patel, J. K., and Jivani, N. P., Chitosan based nanoparticles in drug delivery, *Int. J. Pharm. Sci. Nanotechnol.*, **2**(2) July -September(2009), 517–522.

151. Berthold, A., Cremer, K., Kreuter, J., Preparation and characterization of chitosan microspheres as drug carrier for prednisolone sodium phosphate as model for anti-inflammatory drugs, *J. Control. Release*, **39** (1996), 17–25.

152. Dambies, L., Vincent, T., Domard, A., and Guibal, E., preparation of chitosangelbeadsbyionotropicmolybdategelation,*Biomacromolecules*, **2** (2001), 1198–1205.

153. Tokumitsu, H., Ichikawa, H., Fukumori, Y., Chitosan–gadopentetic acid complex nanoparticles for gadolinium neutron capture therapy of cancer: Preparation by novel emulsion-droplet coalescence technique and characterization. *Pharm. Res.*, **16** (1999), 1830–1835.

154. Leong, Y. S., Candau, F., Inverse microemulsion polymerization, *J. Phys. Chem.*, **86** (1982), 2269–2271.

155. Gong, J., Hu, X., Wong, K., Zheng, Z., Yang, L., Lau, W., and Du, R., Chitosan nanostructures with controllable morphology produced by a nonaqueous electrochemical approach, *Adv. Mater.*, **20** (2008), 2111–2115.

156. Wang, R.-M., He, N.-P., Song, P.-F., He, Y.-F., Ding, L., and Lei, Z.-Q., Preparation of nano-chitosan Schiff-base copper complexes and their anticancer activity, *Polym. Adv. Technol.*, **20** (2009), 959–964.

157. Worawan Boonyo, Hans E. Junginger, Neti Waranuch, Assadang Polnok and Tasana Pitaksuteepong, "Preparation and characterization of particles from chitosan with different molecular weights and their trimethyl chitosan derivatives for nasal immunization, *J. Metals Mater. Minerals*, **18**(2) (2008), 59–65.

158. Racoviţă, S., and Vasiliu, S., Popa, M., and Luca, C., Polysaccharides based on micro- and nanoparticles obtained by ionic gelation and their applications as drug delivery systems, *Rev. Roum. Chim.*, **54**(9) (2009), 709–718.

159. Ge, H., and Huang, S., Microwave preparation and adsorption properties of EDTA-modified cross-linked chitosan, *J. Appl. Polym. Sci.*, **115** (2010), 514–519.

160. Loretz, B., and Bernkop–Schnürch, A., In vitro evaluation of chitosan–EDTA conjugate polyplexes as a nanoparticulate gene delivery system, *AAPS J.*, **8**(4) (2006), article no. 85.

161. Dutta, J., and Hofmann, H., Self-organization of colloidal nanoparticles, *Encyclopedia Nanosci. Nanotechnol.*, **9** (2004), 617–640.

162. Velmurugan, N., Kumar, G. G., Han, S. S., Nahm, K. S., and Lee, Y. S., Synthesis and characterization of potential fungicidal silver nano-sized particles and chitosan membrane containing silver particles, *Iranian Polym. J.*, **18**(5) (2009), 383–392.

163. No, H. K., Prinyawiwatkul, W., and Meyers, S. P., Chapter 19, Treatment of wastewaters with the biopolymer chitosan, in *Handbook of Carbohydrate Engineering*, Taylor & Francis Group, LLC (2005), 535–562.

Chapter 20

Silver Nanoparticle–Incorporated Hydrogels: Synthesis and Antibacterial Applications

P. Souda and Lisa Sreejith

Soft Materials Research Laboratory, Department of Chemistry, National Institute of Technology, Calicut, India

lisa@nitc.ac.in

20.1 Introduction

Stimuli responsive nanocomposite hydrogels have gained much attention worldwide due to its application in biomedicine, environment, industry, etc. The targeted functional group in the hydrogel can act as reactive site for the metal nanoparticles Hydrogels (HG), the three-dimensional polymeric network, have enough space in the swollen state that can act as nanoreactor for nucleation and growth of nanoparticles (Murali et al., 2007). The metal nanoparticles have high specific surface area and high surface-to-volume ratio, which have been studied extensively due to their unique properties such as magnetic, optical, catalytic, and antimicrobial properties. Among the widely studied nanoparticles, silver is superior due to its high electrical conductivity, antimicrobial

Handbook of Sustainable Polymers: Processing and Applications
Edited by Vijay Kumar Thakur and Manju Kumari Thakur

ISBN 978-981-4613-53-8 (Hardcover), 978-981-4613-54-5 (eBook)
www.panstanford.com

effect, optical properties, oxidative catalysis, etc. (Zhongping et al., 2001). These properties were not shown in molecules or in bulk but could be observed only in nanometric range. Silver nanoparticles are being considered non-toxic eco-friendly antibacterial material (Murali et al., 2007).

Silver nanoparticle–impregnated hydrogels (AgHG) show both the properties of nanomaterials and polymeric hydrogels. There are different approaches used to prepare hydrogel/metal nanoparticles. Several synthetic routes toward preparation of silver nanocomposite are detailed in this chapter. The particle size distribution, agglomeration, etc., are found to affect eventually the preparation and properties of AgHG. In conventional methods for the synthesis of the hydrogel/composites the metal nanoparticles and the polymerization of the monomers were done separately and then mixed them mechanically. However, due to the non-homogeneous dispersion of nanoparticles in to the hydrogel matrix, this method of preparation is not successful for the production of metal nanocomposites. Recent studies are exploiting the in situ synthesis of metal nanoparticles within the polymeric hydrogels. The mechanism of the action of silver nanocomposites on the microbial action is explained here. The characterization technique for the identification of silver nanoparticles on hydrogel matrix, its distribution, and particle size are also explained.

20.2 Ex situ Approach for the Synthesis of Silver/Hydrogel Composite

In 2005, Wen-Fu et al. synthesized hydrogel nanocomposite containing silver nanoparticles by ex situ polymerization. They prepared a series of hydrogel nanocomposites of poly(acrylic acid), poly(ethylene glycol) methyl ethyl acrylate (PEGMEA) containing silver. For this, they prepared silver nanoparticle by adding ascorbic acid solution into aqueous silver nitrate solution at room temperature. The formed silver nanoparticles were then incorporated into acrylic acid and PEGMEA monomer solution and allowed to polymerize. The resultant hydrogel/nanocomposites obtained were characterized and confirmed for its formation by Wen-Fu and Kai-Tai (2006). They found that these materials showed excellent *E. coli* resistance. However, the ex situ preparation method for AgHG is very rare in the literature as its difficulty for the production of non-agglomerated silver nanoparticles.

20.3 In situ Approach for the Synthesis of Silver/Hydrogel Composite

Poly acrylonitrile/silver nanoparticles were synthesized by Zhang et al. (2001) by the UV irradiation technique for polymerization and reduction of Ag^+ ions of $AgNO_3$, simultaneously. Here the silver nitrate directly dissolves in acrylonitrile monomer, form silver–olefin complex.

The complex formed of silver ions and acrylonitrile monomers provides an excellent precursor for the synthesis of polymer/silver nanocomposite (Wen-Fu and Kai-Tai, 2010; Carotenuto et al., 2005). When this solution is irradiated with UV, the acrylonitrile polymerized simultaneously Ag^+ gel reduced Ag^0 (Zhongping et al., 2001).

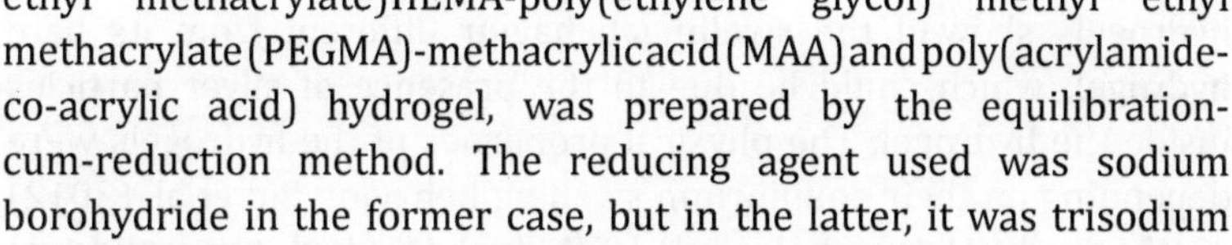

Hydrogel/silver nanocomposite, based on poly(2-hydroxyl ethyl methacrylate)HEMA-poly(ethylene glycol) methyl ethyl methacrylate(PEGMA)-methacrylicacid(MAA)andpoly(acrylamide-co-acrylic acid) hydrogel, was prepared by the equilibration-cum-reduction method. The reducing agent used was sodium borohydride in the former case, but in the latter, it was trisodium citrate solution. These hydrogel-nanocomposites showed swelling and deswelling properties—the properties of hydrogel and electrical conductivity, antimicrobial properties, etc., the properties of silver nanoparticles. For the finer distribution of metal nanoparticles the interaction of hydrogel matrix and the metal precursor is important. Since the interaction between $-COO^-$ group with Ag^+ is higher than that of the –COOH with Ag, the $H_3BO_3/CaCl_2$ buffer solution was used to deprotonize carboxylic acid (Yuanqing and Dajun, 2007). During reduction using borohydride, the borohydride and borate in water are absorbed by the silver nanoparticles, which can inhibit the aggregation of silver colloid (Ung et al., 1999; Van Hyning and Zukoski, 1998). The formation of a brownish color indicates the formation of silver nanoparticle on the hydrogel. The most advanced feature is that the formed silver nanoparticle provided excellent stability over a long period (Thomas et al., 2007).

In the γ irradiated polymerization-reduction technique (Co radiation), the polymer samples were immersed in silver nitrate and in 2-propanol solution until the complete reduction of Ag^+ ions. The mechanism during the process can be

$$H_2O \rightarrow \overset{\circ}{e}(2.7),\ \ O\dot{H}(2.7)\ \dot{H}(0.6),\ \ H_2(0.45),\ \ H_2O_2(0.7)$$

The number in the parenthesis represents the respective G values. The Ag^+ is then reacted by strongly reducing hydrated electrons and alcohol radicals (Kacarevic-Popovic et al., 2007).

$$nA\overset{+}{g} + n\bar{e}_{aq}/CH_3\dot{C}\,OHCH_3 \longrightarrow (Ag)_n$$

20.4 Swelling Studies of the Gel

In order to understand the cross-linked structures of the gel, the hydrogel was immersed in a thermodynamically compatible solvent, which is found to penetrate in to the polymeric network. Here the unsolvated glassy phase is separated from rubbery hydrogel region with a moving boundary. The nanocomposite hydrogels showed the swelling behavior different from its bare hydrogel, which could be due to the presence of silver particles inside the hydrogel. The physical properties of the hydrogels were depending on their equilibrium swelling behavior. Eid et al. (2012) synthesized poly(vinyl alcohol) (PVA)/poly(N-vinyl-2-pyrrolidone) (PVP) based hydrogel containing silver nanoparticles. They studied the swelling properties of the synthesized hydrogels and observed that the swelling of PVA/PVP nearly tends to increase (from 0 to 60%) by merely increasing PVP concentration having more hydrophilic amide group. When comparing the swelling ratios of hydrogels and composites, swelling ratio, SR was found to follow the order, hydrogel < hydrogel–Ag^+ < hydrogel–Ag nanocomposite. They explained the observed behavior based on the fact that the polar head groups of the polymeric chains should have high affinity for salts so the silver ions loaded through the gel network cause repulsion of network leading to improved swelling behavior. The further swelling observed by the addition of the reducing agent could be due to the increased overall porosity of the system (Eid et al., 2012). Ravindra et al. (2012) also observed that the swelling ratio of silver-loaded hydrogel nanocomposite in water was higher than that its bare hydrogel. However, Wen-Fee-Lu observed an opposite behavior for the swelling. Here SR of silver nanocomposite hydrogel was lower than bare hydrogel. Moreover, the swelling ratio was found to decrease with the increase of silver content in the composite. They interpreted their observation as by follows:

(1) Anionic carboxylate charges ($-COO^-$) were neutralized by the unchelated silver ions. This resulted decrease in repulsion among anionic carboxylate groups.
(2) The silver nanoparticles in the gel reduced from silver ions would occupy the mesh space or pore volume of the gel network that resulted in the decrease of its swelling ratio (Wen-Fu and Kai-Tai, 2006).

20.5 Action of Silver Nanoparticle/Hydrogel on Microbes

The exact mechanism of silver nanoparticles interact with bacteria is still unexplored. In contact with water and dissolved oxygen silver nanoparticles release small amount of silver ions according to the following equation (Kacarevic et al., 2007; Bajpai et al., 2013)

$$O_{2aq} + 4H_3\overset{+}{O} + 4Ag(S) \longrightarrow 4A\overset{+}{g}_{aq} + 6H_2O$$

It was interpreted that the silver (I) ions, released from silver nanoparticles offer a physical barrier to its diffusion of the bacterial cells might be the reason for killing bacteria. One of the proposed mechanisms is that ionic silver strongly interact with thiol groups of vital enzymes and inactivates them (Jay and Tanju, 2002; Yoshinobu et al., 2003).

20.6 Identification of Silver Particles in the Hydrogel/Metal Nanocomposite

UV-Vis absorption spectra are a quite sensitive technique for the determination of formation of silver nanoparticles. It exhibits an intense absorption peak due to the surface plasmon resonance (SPR) excitation. SPR explains the collective excitation of conduction electrons in a metal. The UV absorption at 417 nm is attributed to the Mie scattering by silver metal. Aggregates of silver formation were confirmed from the absence of peak at 335 and 560 nm in the spectrum (Amit et al., 1998). As the hydrophilic group on the hydrogel increases, the UV absorption maximum value decreases. This might be due to the reduction of free space in the hydrogel network, which could reduce the number of silver nanoparticles formed inside the gel network. In poly(acrylamide-co-acrylic acid) hydrogel/silver nanocomposite, a strong characteristic absorption peak around 406 nm was obtained. However, no such peak was obtained for plain hydrogel. Y. Xiang prepared a hydrogel/silver nanocomposite that shows a strong absorption peak at 425 nm (Yuanqing and Dajun, 2007). Corresponding to the swelling ratio of 20, but swelling ratio 10 showed peak at 430 nm—the red shift. This may be attributed to the aggregation of silver nanoparticles. The sharp absorption pattern indicates that the particle size distribution is somewhat quite narrow.

The crystallinity of hydrogel/silver nanocomposite could be established by diffraction studies. The formation of silver nanoparticle confirm by International Centre of Diffraction Data, ICDD, library of silver. The diffraction peaks for hydrogel/silver nanocomposites found at 2θ value of 38.1°, 44.26°, 64.50°, and 77.42° are attributed due to (111), (200), (220), and (311) planes of FCC, respectively (Bajpai et al., 2013; Daniel et al., 1996; Murali Mohan et al., 2010). Pure hydrogel failed to show any peak in the XRD indicating its amorphous nature. In order to check whether the silver nanoparticles were oxidized to silver oxide, the XRD pattern of composite after storing it for three months was compared with original one. If it has the same pattern as that of fresh hydrogel/silver nanocomposite, this indicates its stability at ambient condition.

The interaction between the hydrogel and silver particles can be determined by FTIR spectral studies. For hydrogels containing

–OH and –COOH groups, the silver clusters get associated with these functional groups and thus leads to broadening and shifting of –OH stretching from 3273 cm^{-1} (free gel) to 3258 cm^{-1}(silver loaded) (Juby et al., 2012).

Thermal stability of silver loaded hydrogel can be determined by thermogravmetric analysis. The degradation temperature of nanocomposite might be higher than that of bare hydrogel. The weight difference between the bare and silver loaded hydrogel is due to the presence of silver nanoparticles present in the hydrogel. Since silver nanoparticles have higher thermal stability than polymeric hydrogel, silver nanoparticle loaded hydrogel show higher thermal stability. Also, the silver particles interact with the functional groups of hydrogel resulting in weak intermolecular cross-links between the polymeric chains thereby restricting their mobility. This also leads to the increased thermal stability of silver nanocomposite hydrogel (Juby et al., 2012).

Surface morphology of the hydrogel nanocomposite can be studied by scanning electron microscopy (SEM) and Tunneling electron microscopy (TEM). The SEM image of poly(acrylamide-co-acrylic acid) hydrogel and its silver loaded nanocomposite reveals that the white spot on the surface of the nanocomposite was the silver nanoparticle formation (Thomas et al., 2007). The cross-sectional view clearly shows the presence of the silver nanoparticles inside the hydrogel matrix. The particle size was calculated from TEM image where the uniform distribution of silver particles as three diffraction rings could be clearly seen. It was thus indexed to the FCC structure of silver. It was interpreted that the ring closest to the centre is probably due to the combination of (111) and (200) reflections, second is due to (222) reflection and third one is either (420) or and/or the (422) reflections (Ramesh Babu et al., 2010; Andersson et al., 2002). The silver nanoparticle distribution, whether it is dispersed or aggregated can also identify from SEM and TEM analysis.

References

Amit, G., Maria, M., Simon, S. (1998). Effects of halides on plasmid-mediated silver resistance in *Escherichia coli*. *Appl. Environ. Microbiol.*, **64**, 5042–5045.

Andersson, M., Alfredsson, V., Kjellin, P., Palmqvist, A. E. C. (2002). Macroscopic alignment of silver nanoparticles in reverse hexagonal liquid crystalline templates. *Nano Lett.*, **2**(12), 1403–1407.

Bajpai, S. K., Navin, C., Manika, M. (2013). In situ formation of silver nanoparticles in poly(methacrylic acid) hydrogel for antibacterial applications. *Polym. Eng. Sci.*, DOI 10.1002/pen.

Carotenuto, G., Nicolais, L., Martorana, B., Perlo, P. (2005) In: Nicolais, L., Carotenuto, G. (eds.), *Metal-Polymer Nanocomposites: Metal-Polymer Nanocomposite Synthesis: Novel ex situ and in situ Approaches*, Chapter 5, Wiley Interscience, New York, p. 155.

Eid, M., El-Armoury, M. B., Salah, M., Soliman, E.-S., Hegazy, E.-S. A. (2012). Radiation synthesis and characterization of poly(vinyl alcohol)/ poly(N-vinyl-2-pyrrolidone) based hydrogels containing silver nanoparticles, *J. Polym. Res.*, **19**, 9835.

Jay, S. H., Tanju, K. (2002). Removal and sequestration of iodide using silver-impregnated activated carbon. *Environ. Sci. Technol.*, **36**, 784–789.

Juby, K. A., Charu, D., Manmohan, K., Swathi, K., Misra, H. S., Bajaj, P. N. (2012). Silver nanoparticle-loaded PVA/gum acacia hydrogel: Synthesis, characterization and antibacterial study, *Carbohydr. Polym.*, **89**, 906–913.

Kacarevic-Popovic, Z., Tomic, S., Krkljes, A., Micica, M., Suljovrujic, E. (2007). Radiolytic synthesis of Ag-poly(BIS-co-HEMA-co-IA) nanocomposites. *Radiat. Phys. Chem.*, **76**, 1333–1336.

Leff, D. V., Brandt, L., James, R. H. (1996). Synthesis and characterization of hydrophobic, organically-soluble gold nanocrystals functionalized with primary amines. *Langmuir*, **12**, 4723–4730.

Murali Mohan, Y., Lee, K., Premkumar, T., Geckeler, K. E. (2007). Hydrogel networks as nanoreactors: A novel approach to silver nanoparticles for antibacterial applications, *Polymer*, **48**, 158–164.

Murali Mohan, Y., Vimala, K., Varsha, T., Varaprasad, K. A., Sreedhar, B., Bajpai, S. K., Mohana Raju, K. (2010). Controlling of silver nanoparticles structure by hydrogel networks. *J. Colloid Interface Sci.*, **342**, 73–82.

Palapparambil, S. G., Debajyoti, R., Prafulla Kumar, S. (2010). Designing of silver nanoparticles in gum arabic based semi-IPN hydrogel. *Int. J. Biol. Macromol.*, **46**, 237–244.

Ramesh Babu, V., Changda, K., Sangsu, K., Chuljin, A., Yong-Ill L. (2010). Development of semi-interpenetrating carbohydrate polymeric hydrogels embedded silver nanoparticles and its facile studies on *E. coli*. *Carbohydr. Polym.*, **81**, 196–202.

Ravindra, S., Antoine, F. M., Rajinikanth, V., Varaprasad, K., Narayana Reddy, N., Mohana Raju, K. (2012). Development and characterization of curcumin loaded silver nanoparticle hydrogels for antibacterial and drug delivery applications. *J. Inorg. Organomet. Polym.*, **22**, 1254–1262.

Thomas, V., Murali Mohan, Y., Sreedhar, B., Bajpai, S. K. (2007). A versatile strategy to fabricate hydrogel–silver nanocomposites and investigation of their antimicrobial activity. *J. Colloid Interface Sci.*, **315**, 389–395.

Ung, T., Liz-Marzán, L. M., Mulvaney, P. (1999). Redox catalysis using Ag@ SiO_2 colloids. *J. Phys. Chem. B*, **103**(32), 6770.

Van Hyning, D. L., Zukoski, C. F. (1998). Characterization of colloidal stability during precipitation reactions. *Langmuir*, **14**, 7034.

Wen Fu, L., Kai-Tai, T. (2006). Preparation and properties of nanocomposite hydrogels containing silver nanoparticles by ex situ polymerization. *J. Appl. Polym. Sci.*, **100**, 3653–3661.

Wen-Fu, L., Kai-Tai, T. (2010). Effect of silver nanoparticles content on the various properties of nanocomposite hydrogels by in situ polymerization. *J. Mater. Sci.*, **45**, 89–97.

Yoshinobu, M., Kuniaki, Y., Shin-ichi, K., and Tetsuaki, T. (2003) Mode of bactericidal action of silver zeolite and its comparison with that of silver nitrate. *Appl. Environ. Microbiol.*, **69**(7), 4278.

Yuanqing, X., Dajun, C. (2007). Preparation of a novel pH-responsive silver nanoparticle/poly(HEMA–PEGMA–MAA) composite hydrogel. *Eur. Polym. J.*, **43**, 4178–4187.

Zhang, Z., Lide, Z., Shixing, W., Wei, C. Yong, L. (2001). A convenient route to polyacrylonitrile/silver nanoparticle composite by simultaneous polymerisation-reduction approach. *Polymer*, **42**, 8315–8318.

Chapter 21

Biodegradable Polyesters: From Monomer to Application

Geert-Jan Graulus,[a,b] Thomas Billiet,[a] Sandra Van Vlierberghe,[a,b] Hugo Thienpont,[a,b] Heidi Ottevaere,[b] and Peter Dubruel[a]

[a]*Polymer Chemistry and Biomaterials Research Group, Department of Organic and Macromolecular Chemistry, Ghent University, Krijgslaan 281 (building S4), B-9000 Ghent, Belgium*

[b]*Brussels Photonics Team, Department of Applied Physics and Photonics, Vrije Universiteit Brussel, Pleinlaan 2, B-1050 Elsene, Belgium*

Peter.Dubruel@UGent.be

21.1 Introduction

Over the past years, plastics have gathered a rather bad reputation as they are able to resist degradation and therefore accumulate in the environment [1]. The fact that most of these plastic materials are derived from fossil natural sources has contributed significantly to these environmental issues since plastics are generally labelled "unsustainable". Additionally, their low cost of production renders them easily disposable, resulting in high amounts of plastic waste being discarded annually.

Despite these shortcomings, it is hard to imagine our modern society without polymers. To overcome the negative view on

Handbook of Sustainable Polymers: Processing and Applications
Edited by Vijay Kumar Thakur and Manju Kumari Thakur

ISBN 978-981-4613-53-8 (Hardcover), 978-981-4613-54-5 (eBook)
www.panstanford.com

plastics, two strategies can be distinguished. In a first strategy, the research focuses on sustainability, which can be realised by selecting naturally occurring polymers, by polymerising monomers derived from natural sources (i.e., biomass) or by applying environmentally benign synthesis routes (e.g., biotechnological production of certain polymers).

A second strategy focuses on minimising the environmental impact of plastics. A first way to achieve this goal is by applying more efficient sorting and recycling methodologies [2]. Another approach acts on the materials applied and aims at engineering (bio)degradable polymers. Polymers such as polyethylene and polypropylene can be blended with catalysts that induce oxidation [3]. However, there is concern that the catalysts, which are often based on heavy metals, may induce an environmental problem of their own. To circumvent this issue, polymers with the ability to degrade can be engineered by applying the appropriate monomers. The latter strategy will be the focus of the present contribution. Literature often refers to "*biopolymers*" or "*bioplastics*" for both bio-based as well as (bio)degradable materials. In this chapter, however, the distinction will be made between bio-based and biodegradable polymers as bio-based polymers are not necessarily degradable and vice versa [4, 5]. Biodegradation has been defined by the International Union of Pure and Applied Chemistry (IUPAC) as the degradability of substances by biological activity by lowering of the molar masses of the macromolecules that form the substance [6]. This IUPAC definition clearly excludes degradation through enzymatic activity as livings cells have to be present to be able to use the term biodegradation. Herein, however, an extended definition will be handled since enzymatic degradation and certain abiotic reactions (e.g., photo-degradation, hydrolysis and oxidation) will be included. Biodegradable should therefore be understood as environmentally degradable [7].

Biodegradable and bio-based polymers make up about 300,000 metric tons of the total plastics market, which only accounts for 1% of the global market of synthetic plastics [7, 8]. However, the market for degradable and sustainable plastics grows about 20–30% each year, which is considerably higher than the total market growth. Polyesters, i.e., poly(lactic acid) and poly(glycolic acid), make up a large fraction of the biodegradable plastics market. Polyesters have gained increasing interest in this regard as by simply selecting the appropriate monomers in a straightforward manner, the (bio)degradation and other material characteristics can be effectively tuned towards the final application [9]. Degradability should indeed not come at the expense of other properties, since the object should still be able to fulfil its function [10]. After a short introduction on polyesters, their synthesis, their degradation behaviour and potential applications will be discussed in depth.

21.2 Polyesters

As their name suggests, polyesters are a class of polymers in which the monomer units are linked via ester bonds. Since these bonds can be hydrolysed, polyesters play an important role within the biodegradable materials. Polyesters can be generally subdivided into two groups including the aliphatic polyesters and the aromatic polyesters of which poly(ethylene terephtalate) (PET), which is used in everyday objects like food or beverage containers, is the best known representative. PET as such is not considered (bio)degradable since its decomposition rate is very low [11, 12]. However, attempts have been made to come up with degradable PET derivatives by including other monomers which introduce less stable ester linkages [11, 13]. PET owes its slow degradation to its hydrophobic nature (cfr. the presence of aromatic rings in the polymer backbone). In addition, PET is semi-crystalline and the crystalline domains effectively reinforce the material rendering hydrolysis less likely to occur.

In contrast, degradable polyesters are generally derived from aliphatic monomers including lactic acid, glycolic acid, ε-caprolactone, etc. In these materials, no aromatic structures are present in the main chain, enabling hydrolysis to occur more easily. Although produced synthetically, these polymers are often preferred over natural polymers, as their properties can be tailored and the resulting materials have more predictable lot-to-lot uniformity than their natural counterparts [9].

Polyesters can be synthesised via various synthesis routes, which will be discussed in the following sections.

21.2.1 Polycondensation

A first synthesis route is the polycondensation of dicarboxylic acids with diols or of hydroxyacids. Since this procedure occurs according to a step-wise addition of monomer units, this method is associated with long reaction times as polymer is only obtained at high conversions. In some cases, the addition of a suitable catalyst can, however, decrease the required reaction time [14].

Additionally, the control over this reaction is troublesome as the slightest imbalance in stoichiometry between the diol and the dicarboxylic acid will result in the termination of the growing chains. Furthermore, any monofunctional acid or alcohol present in the reaction mixture will result in a dead chain-end. As a result, molecular weight distributions are generally quite broad for this polymerisation method. Hydroxyacids are therefore favoured for these reactions, since these monomers do not pose the problem related to an unbalanced stoichiometry. During the condensation reaction, water is generally formed as a side-product. When applying (di)esters as starting products a transesterification mechanism is followed and alcohols make up the by-product [15, 16].

When performing direct polycondensation polymerisations in the melt state, the obtained molecular weights are generally limited, since the increasing viscosity of the reaction mixture will hinder the efficient removal of the water [17]. This disadvantage limits the use of the direct polycondensation method [18].

To compensate for this, chain-coupling reagents are sometimes applied. These dicarboxylic acids or diols are able to react with the alcohol or carboxylic acid end-groups respectively, effectively coupling two shorter polymer strands and thus increasing the overall molecular weight.

In the case of semi-crystalline polymers, reports can be found on a two-step process in which short chains are crystallised after which the polymerisation is continued in the solid state to obtain, e.g., high-molecular-weight PLLA [19]. In this process, an inert gas stream is applied to remove by-products from the surface. The coupling mainly takes place in the amorphous regions of the materials, where the reactive end groups reside.

To drive the condensation reaction towards completion and thus to increase the molecular weight of the obtained polymers, the water by-product can be removed more effectively via an azeotropic distillation. The reagents are dissolved in a suitable solvent (e.g., toluene or mixed xylenes) after which the water fraction is collected in a Dean–Stark trap, while the dried solvent is recycled back to the flask. This method can lead to high-molecular-weight polymers in a one-step reaction, although such reactions can still be time-consuming [20].

Despite the inherent drawbacks of polycondensation reactions with respect to the molecular weight and the dispersity, some applications do not require high molecular weights rendering step-growth polymerisations interesting because of their simplicity. Zhu et al. produced hyperbranched polyesters with relatively low molecular weights and introduced cinnamate pendant groups to obtain a photocross-linkable material [21]. The range of polyesters that can be synthesised via polycondensation reactions is only limited by the number of available monomers. A lot of effort is therefore placed in the synthesis of new polyester building blocks to further broaden the scope [22].

21.2.2 Ring-Opening Polymerisations of Lactones

An alternative for the above-mentioned polycondensation mechanism can be found in the ring-opening polymerisation of cyclic (di)esters, which follows a chain growth mechanism. The latter implies rapid polymer propagation resulting in high molecular weights in shorter reaction times as compared to the polycondensation. In addition, narrow molecular weight distributions can be achieved by using appropriate catalysts increasing the reproducibility of the obtained polymers. Ring opening polymerisation is, however, limited by the number of available monomers. In contrast to the vast number of dicarboxylic acids and diols, the number of available cyclic (di)esters is far lower [23, 24]. However, new lactones and di(esters) are being synthesised nowadays to advance the field beyond the current state-of-the-art [25–28].

It is beyond the scope of the present contribution to summarise all catalyst systems applied in the ROP of lactones, as the interested reader can find more information in an excellent review by Dechy-Cabaret et al. [29]. Four classes of ROP can be distinguished, including anionic, cationic, metal-insertion, and organo-catalytic ROP.

21.2.2.1 Anionic ring-opening polymerisation

In anionic ROP, nucleophilic metal alkoxides initiate the polymerisation reaction. Some bulky alkoxides, however, will rather act as a base thereby introducing a new anionic centre on a monomer unit, which can subsequently act as the anionic initiator [30, 31]. Two different propagation mechanisms can be distinguished as depicted in Fig. 21.1.

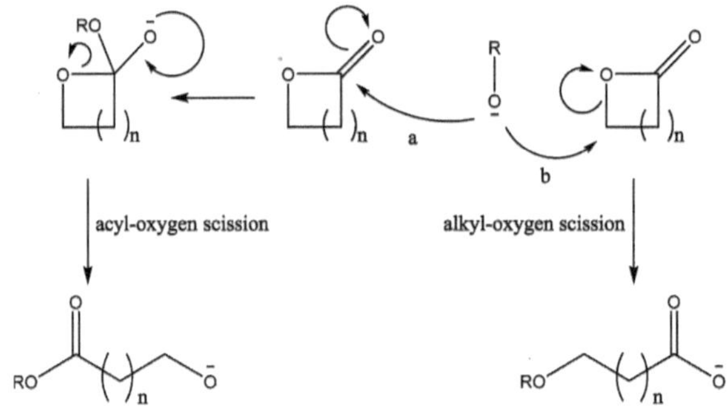

Figure 21.1 Possible mechanisms for anionic ring-opening polymerisations of non-substituted lactones showing both acyl-oxygen scission (a) as well as alkyl-oxygen scission (b).

In the first scheme (Fig. 21.1, reaction a), the alkoxide attacks the ester group after which an acyl-oxygen bond is broken. In the other case (Fig. 21.1, reaction b), the alkoxide attacks the carbon atom adjacent to the alcohol residue of the ester, after which an alkyl-oxygen scission takes place. This last mechanism is of particular importance for the polymerisation of four-membered lactones (β-lactones) in which the first mechanism is disfavoured by stereo-electronic effects [32].

In order to maintain control over the polymerisation reaction and to achieve low dispersities, transfer and termination reactions should be limited to the greatest extent possible. If transfer and termination reactions are completely absent, the polymerisation reaction can be considered as "living" [33]. In anionic ROP, the growing chains remain active. In addition, the number of active species will thus remain constant once the initiation phase has passed. As long as monomer is added, the chains will thus be able to grow. When no protic compounds are present, termination by quenching of the anionic centre will generally be negligible. The loss of control is thus mainly determined by the occurrence of transfer reactions. As the monomer gets depleted, the anionic centre can react with ester groups in the polymer chain as the reactivity of the ester groups in the polymer differs only slightly from the lactone ester group. This so-called "back-biting" reaction will thus start to occur resulting in the formation of cyclic products. In addition, intermolecular transesterification reactions will alter the obtained molecular weight.

Control over the polymerisation reaction can be increased using less reactive initiators, thereby disfavouring transesterification. One strategy applies sterically hindered initiators to oppose reaction occurring with growing polymer chains. Another strategy involves the use of certain less electropositive metals. The latter will be further referred to as metal-insertion ROP.

21.2.2.2 Metal-insertion ring-opening polymerisation

As already mentioned earlier, it is possible to increase control over the ring-opening polymerisation by applying organometallic derivatives showing d-orbitals of favourable energy. Aluminium alkoxides are one possibility to increase control over the reaction. Such alkoxides are easily synthesised by reacting alcohols with triethyl aluminium or aluminium isopropoxide. Interestingly, the applied alcohols can be substituted with reactive groups including bromides, olefins and tertiary amines to obtain end-group functionalised polymers [34, 35].

The metal-insertion ROP owes its name to the propagation mechanism (Fig. 21.2). After coordination of the metal-alkoxide with the carbonyl group of the lactone, the addition of the nucleophilic alkoxides takes place onto the electrophilic ester bond. Subsequently, an elimination reaction occurs via acyl-oxygen scission. The novel alkoxide will act as the newly generated propagating species.

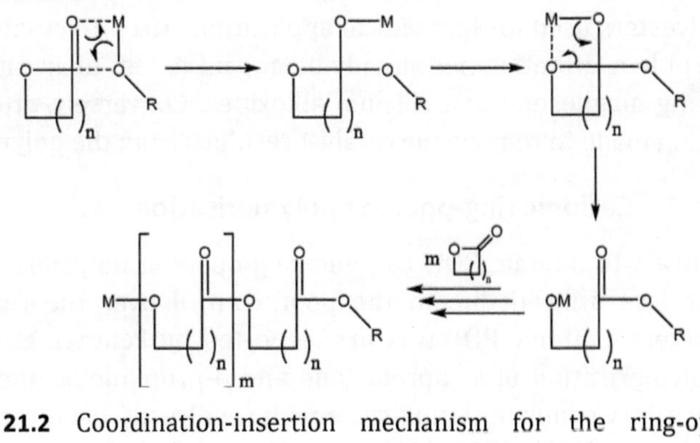

Figure 21.2 Coordination-insertion mechanism for the ring-opening polymerisation of lactones. Adapted from [30].

After the initial work on aluminium alkoxides, the metal-insertion ROP was extended to other metal ions including titanium [36–38], tin (II) [39–41], tin (IV) [42–44], zirconium [45, 46] and, more recently, rare-earth metals (Y, Er, Sm, Dy, Nd and La) [47–49].

Currently, tin(II) octoate, also known as tin(II) bis(2-ethylhexanoate), is the most frequently used catalyst for metal-insertion ROP. It owes its privileged position to the fact that it has been approved by the American Food and Drug Administration (FDA) for the formulation of polymer coatings which come in contact with food. The latter is of relevance in case the resulting materials are to be applied for food packaging applications. In addition, tin octoate is less sensitive towards water and other protic impurities, rendering it even more interesting for laboratory and industrial purposes. Tin octoate contains two carboxylates which have already been shown to be poor initiators for ROP. However, Penczek et al. discovered that the actual initiating species is formed in situ by the reaction of tin octoate with water or alcohols [39]. The formed tin alkoxide will thus be the main initiator and a similar mechanism as already discussed earlier is applicable (Fig. 21.2).

Despite the fact that the metal ions enable to control the polymerisation, they also pose an important downside as the formed polymers will always contain metal residues. These metal ions might have detrimental health effects. Aluminium derivatives have already been linked to Alzheimer's disease, although this hypothesis is controversial [50], while tin octoate, despite being FDA approved, has already been shown to be cytotoxic. As tin levels often exceed 1000 ppm, the FDA has set a limit of 20 ppm for polyesters used for biomedical applications [51]. To circumvent this problem, attention has already been paid to less toxic catalysts including magnesium and calcium alkoxides. Conversely, attempts are being made to remove the catalyst residues from the polymer.

21.2.2.3 Cationic ring-opening polymerisation

In contrast to anionic ROP, cationic ring-opening polymerisations are far less utilised due to the poor control over the reaction parameters. Cationic ROP was first reported by Penczek et al. for the polymerisation of ε-caprolactone and β-propionolactone [52]. Afterwards, various initiating systems have already been reported on which can be divided in four classes including alkylating agents, acylating agents, Lewis acids and protonic acids. Depending on the type of initiating reagent, different mechanisms can be distinguished. When alkylating agents are applied, the main reaction pathway will involve the exocyclic oxygen (Fig. 21.3).

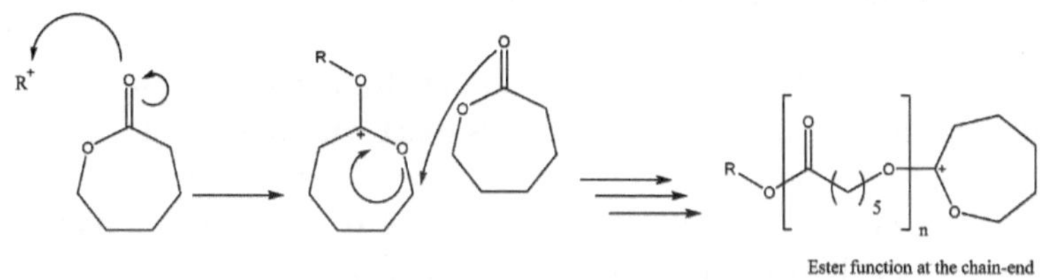

Figure 21.3 Proposed mechanism for cationic ring-opening polymerisations initiated through alkylating agents. Adapted from [30].

The mechanism becomes more complex in the case of acylating reagents as both the endo- as well as the exocylic oxygen are able to react [53]. When the endoclyclic oxygen is involved, a six-membered intermediate can be formed. The reaction of this intermediate with additional monomer units can result in both ester as well as anhydride functionalities present at the chain-end as shown in Fig. 21.4.

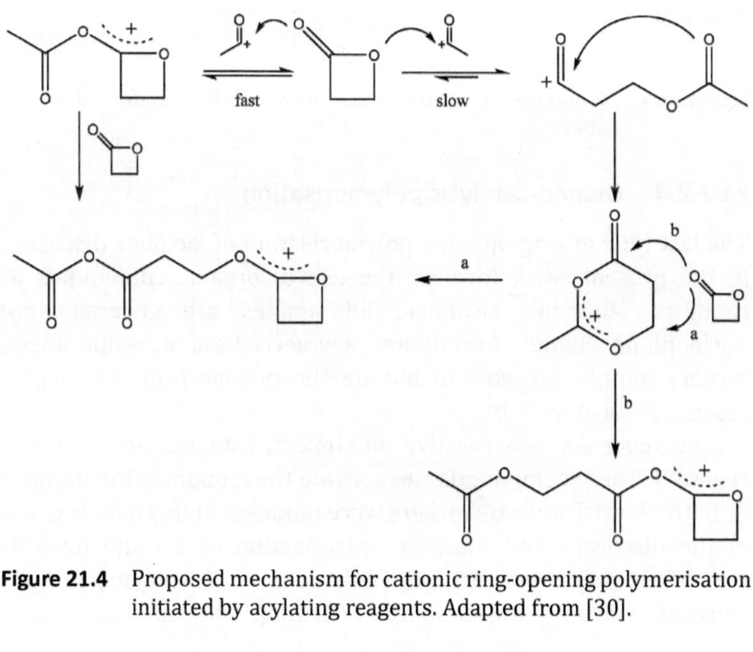

Figure 21.4 Proposed mechanism for cationic ring-opening polymerisation initiated by acylating reagents. Adapted from [30].

In case of Lewis acid initiation, a cationic ROP will only occur if the counterion is not too nucleophilic [54]. When nucleophilic counterions are present, the metal-insertion mechanism will be followed instead, as already reported for several catalytic systems including $ZnCl_2$ [55], $TiCl_4$ and $AlCl_3$ [56].

The final possibility includes the application of amines and alcohols as nucleophilic initiators under Bronsted acid catalysis. In this mechanism, activated monomers react with the initiator or the nucleophilic chain-end. After protonation of the exocyclic oxygen and the nucleophilic attack, ring-opening occurs via acyl-oxygen scission as shown in Fig. 21.5. This mechanism is more favourable than polymerisations applying alkylating agents. However, the molecular weights obtained are generally limited, unless great care is spent on the selection of suitable catalysts. Besides inorganic acids also organic compounds including citric acid [57], fumaric acid [58, 59], lactic acid [57, 60] and amino acids [57] are capable of initiating ring-opening polymerisations.

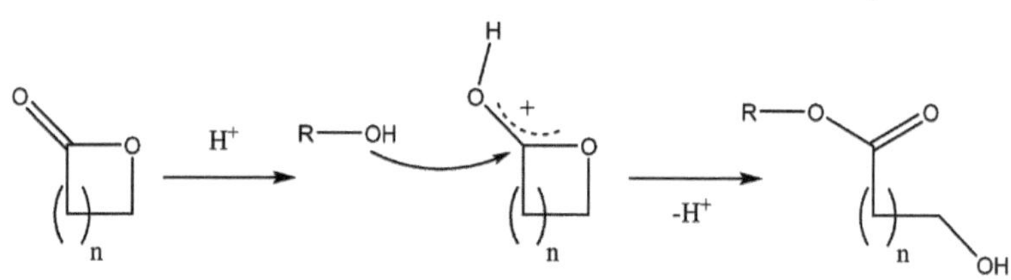

Figure 21.5 Monomers reacting with nucleophiles under Bronsted catalysis.

21.2.2.4 Organo-catalytic polymerisation

The last type of ring-opening polymerisation of lactones discussed in the present work involves the use of organic compounds as catalysts. Although alcohols and amines are generally not nucleophilic enough to initiate polymerisation, in some cases, tertiary amines are able to initiate the polymerisation of highly reactive β-lactones [30].

However, for less reactive monomers, catalysts are generally required. These compounds can activate the monomer, the initiator or both. Hedrick and co-workers were pioneers in this field because of the discovery and thus, the introduction of a route towards metal-free polyesters [61–65]. Organo-catalytic polymerisations generally involve nucleophiles activating the monomer ring followed by the release of the catalyst by the nucleophilic attack of the initiator as shown in Fig. 21.6. In order to acquire more information, a review by Waymouth, Hedrick and co-workers is highly recommended [66].

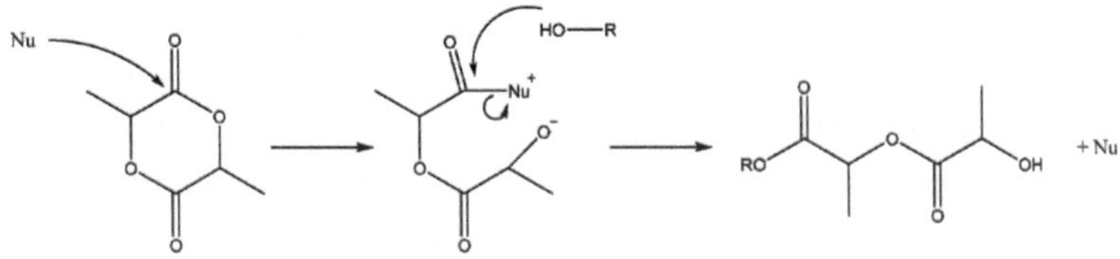

Figure 21.6 Nuceleophilic catalysis of lactide polymerisation.

The most widely applied catalyst systems in organocatalytic ROP include N-heterocyclic carbenes [62, 64, 67], guanidine and amidine bases [51, 65] and phosphorane-based compounds [68, 69]. Strong organic acids are excluded from this section as their mechanism follows the one described for cationic ring-opening polymerisations [70, 71].

In some cases, the polymerisation is performed in the presence of a thiourea compound as co-catalyst to extend the application of organo-catalysis towards less reactive monomers [63, 65, 69]. These compounds activate the carbonyl group via hydrogen bonding and thus render the carbonyl carbon more electrophilic as shown in Fig. 21.7.

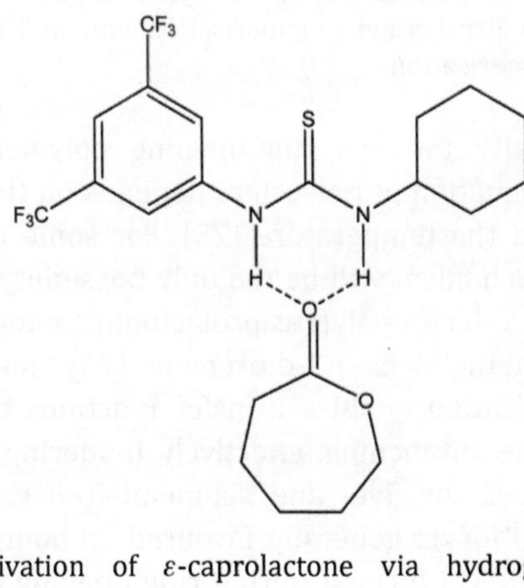

Figure 21.7 Activation of ε-caprolactone via hydrogen bonding to thiourea.

21.2.3 Ring-Opening Polymerisation of Cyclic Ketene Acetals

A less familiar route towards polyesters applies the ring-opening of cyclic acetal ketenes [72–74]. After the initiation phase, two possible propagation mechanisms can be distinguished (Fig. 21.8) including a direct vinyl polymerisation and an indirect ring-opening polymerisation. The latter can occur due to the formation of a more stable C=O double bond (ca. 50 kcal/mol more stable than a C=C double bond) and the release of ring strain [72].

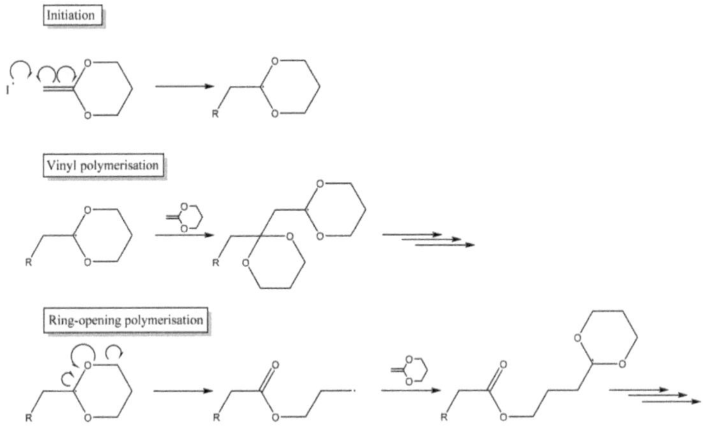

Figure 21.8 Radical polymerisation of cyclic ketene acetals showing both direct vinyl polymerisation and indirect ring-opening polymerisation.

The selectivity towards ring-opening polymerisation which results in the formation of polyesters depends on the ring size, the substituents and the temperature [75]. For some monomers, the ring-opening mechanism will be the only possibility. Jin et al. used this method to produce a poly(ε-caprolactone) analogue by the ring-opening of 2-methylidene-1,3-dioxepane [76]. Interestingly, the radical polymerisation enables transfer reactions to occur which introduce polymer branching, effectively hindering crystallisation [76, 77]. However, for five- and six-membered rings, the direct vinyl polymerisation is generally favoured, although substituents stabilising the radical formed during ring-opening could shift the propagation towards the ring-opening mechanism.

Since roughly half of the total production of plastics is generated using radical polymerisation, the radical ring-opening polymerisation shows great promise to extend these techniques towards polymers containing heteroatoms in their backbone or carrying reactive pendant groups [77]. Despite these advantages, this polymerisation method has been focussed on to a lesser extent than the previously described routes (i.e., polycondensation and ring-opening of lactones) leaving room for further improvement [74]. Recently, a controlled radical polymerisation technique has been applied by Delplace et al. to gain control over the molecular weight and the dispersity [78].

21.2.4 Enzymatic Polyester Synthesis

Finally, polyesters can also be synthesised via enzymatic processes. This type of polymerisations possesses some advantages over the earlier described chemical synthesis routes, including mild reaction conditions, a higher selectivity, a high tolerance of functional groups and the synthesis of pure reaction products which are metal free [79, 80]. However, reaction times are longer and yields lower compared to chain growth polymerisations.

Lipase-catalysed synthesis of poly(butylene succinate) (PBS) is an example of this polymerisation route. In 2006, Azim et al. used *Candida antarctica* lipase B for the production of PBS from diethyl succinate and 1,4-butanediol in a monophasic system. The molecular weights that were obtained, were quite low as the polymer precipitated after a few hours. The PBS synthesised via lipase activity showed narrower molecular weight distributions than can be achieved via polycondensation methods. Lipase or esterase enzymes can also be applied in ring-opening polymerisation of cyclic esters [81]. Although the absence of metal catalysts is a clear advantage, the residual lipase may cause problems during processing.

Poly(β-hydroxyalkanoates) (PHAs) are a class of poly(esters) which are synthesised by many bacteria as energy reserve and carbon source when exposed to an excess of carbon under unbalanced growth conditions [82]. The produced PHAs can account for up to 80% of the bacterium's dry weight. This has led to the biotechnological production of PHAs. Depending on the substrates provided to the bacteria, different monomers can be produced which results in different PHA (co)polymers [83].

Poly(3-hydroxybutyrate) was the first PHA to be produced on an industrial scale, but its brittle nature, its poor mechanical properties and its high production cost limited its application potential. In the early 1990s, Imperial Chemical Industries (ICI) started the production of poly(3-hydroxybutyrate-co-3-hydroxyvalerate) (P3HB3HV) under the trade name Biopol®. This material showed lower degrees of crystallinity and superior mechanical properties. Later on, the production of Biopol® was continued by Monsanto and subsequently followed up by Metabolix. PHAs were originally intended as bio-based alternatives for polyolefins used in plastic containers, films and bottles. Despite the large interest in PHAs, their application remains, however, limited due to their narrow processing window [84, 85].

In addition to the research being done on the incorporation of a larger monomer variety in β-hydroxyalkanoates, a mutant form of the PHA synthase enzyme was observed to enable the polymerisation of 2-hydroxyacids and the incorporation of lactic acid in the polymer backbone [80]. The properties of these materials were found to be distinctly different from those of the PHAs synthesised starting from 3-hydroxyacids.

21.2.5 Other Synthesis Routes

The synthesis routes described in the previous section are certainly not the only potential strategies towards the development of degradable polyesters. In the upcoming paragraphs, some alternative research efforts will be described which were beyond the focus of the previous sections.

In addition to the polymerisation of pure lactones, these cyclic (di)esters have already been included as co-monomer to improve the properties of closely related polymers like polyamide obtained starting from lactams [86].

Sanhao et al. have proposed to utilise a Baylis–Hillman type reaction between dialdehydes and diacrylates to obtain polymers with a polyester backbone and alcohol and double bond side-groups [87]. The authors reported this material to be a very interesting intermediate towards side-chain modified polymers.

As the field progresses, an increasing amount of research is performed on novel polyester architectures and properties. These include among others block copolymers [88, 89], dendrimers [90,

91], cyclic poly(esters) [92], star-shaped polymers [58, 59, 93], responsive materials [94, 95], stereocontrolled polymerisations [96], etc.

Finally, the polyesters, regardless of the synthesis route applied, can be modified after polymerisation to introduce reactive end-functionalities or alter the physical properties of the material [97, 98].

21.3 Biodegradation

Over time, the stability and the durability of plastics have been improved in order to resist environmental influences. Over the years, the substantial and the continuous increase in the production of commercial polymers—especially plastics used in packaging, industry and agriculture—pinpointed the potentially huge environmental accumulation and pollution problem that could persist (i.e., the plastic seas) [99–101]. Therefore, two approaches are currently being explored in order to reduce the impact on the environment: (i) designing polymers for "long" duration applications, preferentially stemming from renewable resources and (ii) the design of polymers susceptible towards degradation in a microbial active environment. In the end, innovation aims at the production of polymers exhibiting predesigned life spans. At present, biodegradable polymers are still of minor importance compared to the conventional petrochemical-based polymer materials.

The degradation of polymers generally encompasses the scission of the main and/or the side-chains of the polymer, resulting in any physical or chemical change. These scissions can be induced by thermal activation, oxidation, photolysis, radiolysis, biological activity, ozone or hydrolysis [99, 102, 103]. Singh et al. presented an extensive overview on the mechanistic implications of photo-oxidative, thermal, ozone-induced, mechano-chemical, and catalytic degradation [104]. In the following section, biodegradation will be discussed in depth. Depending on the author, biodegradation has been defined as the changes occurring in surface properties or the loss of mechanical strength, the assimilation by microorganisms (MOs) [101], degradation in the presence of enzymes [105], backbone chain scission and subsequent reduction in the average molecular weight [106, 107]. Conversely, the American Society for Testing and Materials (ASTM) define biodegradation as "the process which is capable of decomposition of materials into carbon dioxide, methane, water, inorganic compounds or biomass in which the predominant mechanism is the enzymatic action of MOs, that can be measured by standard tests, in a specified period of time, reflecting available disposal conditions". Biodegradation, therefore, can be described as the degradation of polymers due to biological environments (e.g., living cells, microorganisms) either by enzymatic or non-enzymatic hydrolysis, and thus indicates the predominance of biological activity. However, biotic and abiotic factors act synergistically and several studies have already reported that the abiotic degradation precedes microbial assimilation and can therefore not be neglected [100]. Abiotic factors include mechanical forces (e.g., shear or compressive forces), the action of light irradiation (e.g., photo-ionisation), thermal effects and chemical degradation (e.g., oxygen, hydrolysis).

Aerobic or anaerobic degradation pathways can occur. When oxygen is present, carbon dioxide and water are produced during the aerobic biodegradation processes, while carbon dioxide, water and methane are produced during anaerobic biodegradation. Biodegradable materials can be converted into gases, water, salts, minerals and residual biomass (i.e., mineralisation). Mineralisation is complete when all the biodegradable materials are or biomass is consumed and all the carbon is converted into carbon dioxide. The aerobic biodegradation of polyesters can be described according to the following mechanism [101, 108]:

$$\text{Polymer} + H_2O \xrightarrow{K1} \text{Monomer} + O_2 \xrightarrow{K2} \text{Biomass} + H_2O + CO_2 \tag{21.1}$$

The first reaction (K1) is governed by the hydrolysis of the ester bonds, generally catalysed by extracellular enzymes, into oligomers and monomers (i.e., depolymerisation). However, polymers with ester bonds may undergo hydrolysis at room temperature and neutral pH in the presence of moisture, even in the absence of enzymes of MOs. This primary depolymerisation is an erosion process, leading to water-soluble intermediates. If in the next step

(K2), these low-molecular-weight fragments migrate through the cell membrane and are mineralised by MOs, the degradation is called biomineralisation. From equation 1, it can be deduced that a temporary build-up of monomers can arise in case K1 >> K2. The biodegradable polyester family is presented in Fig. 21.9.

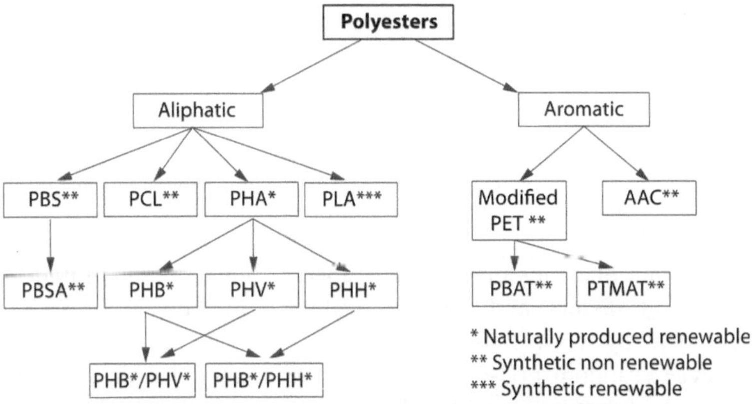

Figure 21.9 The biodegradable polyester family: poly(hydroxyalkanoates) (PHA), poly(hydroxybutyrate) (PHB), poly(hydroxyhexanoate) (PHH), poly(hydroxyvalerate) (PHV), poly(lactic acid) (PLA), poly(caprolactone) (PCL), poly(butylene succinate) (PBS), poly(butylene succinate adipate) (PBSA), aliphatic-aromatic copolyesters (AAC), poly(ethylene terephthalate) (PET), poly(butylene adipate/terephthalate) (PBAT), poly(methylene adipate/terephthalate) (PTMAT). Adapted from [103].

In general, two modes of erosion for biodegradable polymers can be distinguished including bulk and surface erosion [103]. Both modes are schematically illustrated in Fig. 21.10. In surface erosion, MOs consume the polymer matrix from the surface inwards. This is governed by a slow reduction of the molecular weight. Conversely, during bulk erosion, a polymer is degraded throughout its cross section, causing a rapid decrease in molecular weight. The low-molecular-weight compounds subsequently diffuse out of the matrix. Bulk erosion is limited to polymers containing hydrolysable functional groups in the backbone.

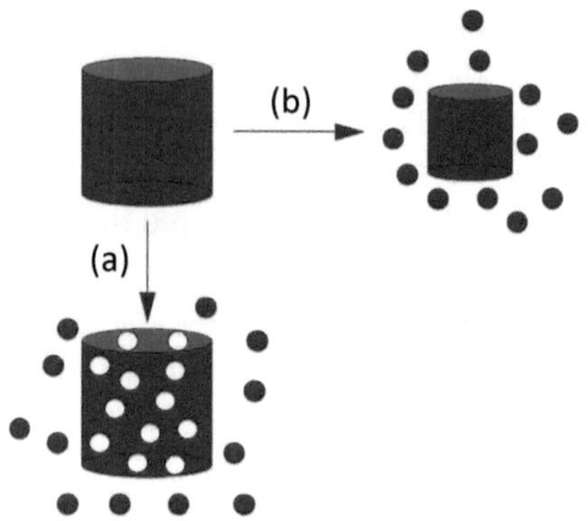

Figure 21.10 Erosion mechanisms of biodegradable polymers including (a) bulk and (b) surface erosion.

Certain enzymes may degrade specific bonds as illustrated in Fig. 21.11. Esters and ester derivatives can, however, also be degraded by conventional hydrolysis. The initial degradation involves a random, non-enzymatic hydrolytic ester cleavage, leading to rapid molecular weight reductions (i.e., bulk erosion). In this process, both the molecular mass as well as the crystallinity play a pivotal role [109].

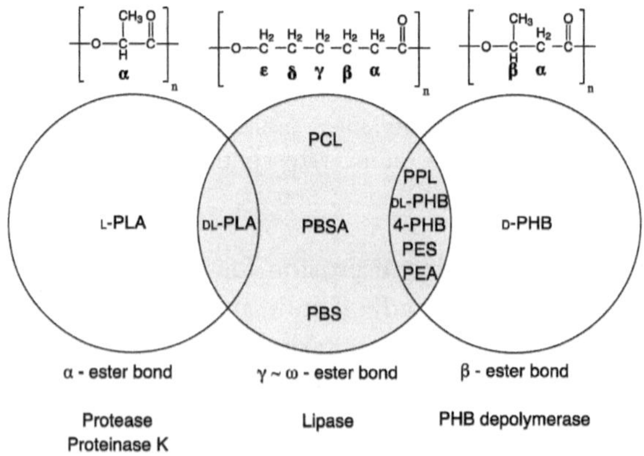

Figure 21.11 Degradation of ester bonds by different enzymes. Poly(L-lactic acid) (L-PLA), poly(D,L-lactic acid) (DL-PLA), poly(caprolactone) (PCL), poly(butylene succinate adipate) (PBSA), poly(butylene succinate) (PBS), poly(propiolactone) (PPL), poly(3-hydroxybutyrate) (DL-PHB), poly(4-hydroxybutyrate) (4-PHB), poly(ethylene succinate) (PES), poly(ethylene adipate) (PEA) and poly(3R-3-hydroxybutyrate) (D-PHB). Adapted from [110].

For most polymers, degradation in nature is a slow process. The mechanisms involved in the biodegradation are complex due to the interaction of different oxidative processes which are caused by the oxygen present in air, by the MOs or by the combination of both [104]. The intrinsic chemical structure of the polymer will determine its biodegradability. Several mechanisms act during the biodegradation of polymers [104]:

- solubilisation: hydration based on the hydrophilicity of the polymer rendering it soluble and prone to chemical or enzyme-catalysed degradation;
- charge formation followed by dissolution: ionisation or protonation of a pendant group leading to increased solubility;
- hydrolysis: hydrolytically unstable bonds, which are reasonably hydrophilic for water access leading to chain scission and low-molecular-weight by-products;
- enzyme-catalysed degradation: enzymes selectively catalyse specific reactions (i.e., oxidation, reduction, hydrolysis, ...);
- microbial degradation: MOs such as bacteria, fungi, algae act in the polymer degradation.

Factors affecting the biodegradation process (Fig. 21.12) include polymer characteristics (i.e., mobility, tacticity, crystallinity, molecular weight, type of functional groups, substituents, plasticisers, additives), type of MOs, physical properties (i.e., surface area, mechanical damage as a result of heating/cooling, freezing/thawing, wetting/drying) and other environmental conditions such as moisture, temperature, pH and oxygen [99, 101, 104, 111, 112]. For example, increasing molecular weights generally result in a decline of the degradability induced by MOs. However, monomers, dimers and oligomers are more easily degraded and mineralised [103]. Furthermore, generally, a small fraction of the polymeric material will be incorporated into microbial biomass, humus, and/or other natural products. As such, the biodegradation rate rarely reaches 100%.

Abiotic hydrolysis, photo-oxidation and physical disintegration of polymers may enhance the biodegradation rate of polymers by increasing their surface contact area for microbial colonisation or by reducing the molecular weight [113]. In the case of lactic acid homo- and heteropolymers, abiotic hydrolysis is the most important reaction for initiating the environmental degradation

[114, 115]. In order to better predict the degradation and, thus, the performance of degradable materials in various applications, a good working model is desired. To achieve a better insight in this regard, Vieira et al. have attempted to construct a mathematical model to assess the impact of hydrolysis on the material properties [116].

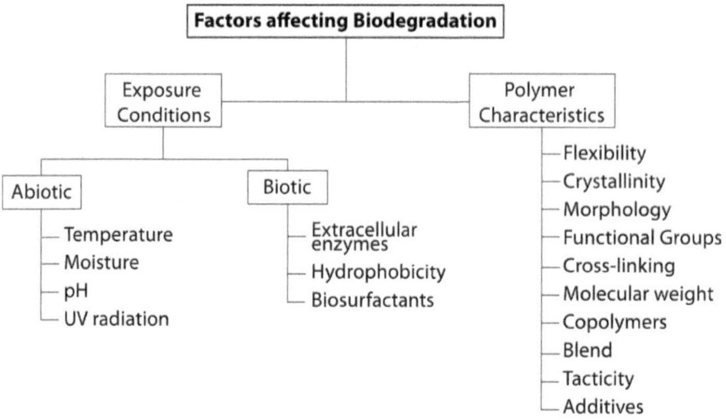

Figure 21.12 Factors affecting the biodegradation. Reprinted from [112] with the permission of Wiley.

Monitoring the (bio)degradation of polymers can, for instance, be performed by the quantification of the molecular weight reduction. However, following this strategy, information regarding biodegradation into CO_2 and H_2O of the monomers and oligomers is excluded. Therefore, biodegradation is generally measured as the conversion of organic carbon into inorganic carbon (i.e., conversion into CO_2). Standardised methodologies and procedures for testing polymer biodegradation have been introduced by the (ASTM), the Japanese Industrial Standards (JIS), the European Normative Reference (ECN) and the Organization for Economic Co-operation and Development (OECD). For an elaborate description on validated and accepted test methods, the readers can consult a review by Eubeler et al. [110].

21.4 Processing of Polyesters

(Bio)degradable polymers can be processed using similar techniques as applied for other thermoplasts. After melting, they can be processed into fibres, rods, films or moulded parts via extrusion, injection moulding, compression moulding, solvent spinning or solvent casting [9]. Moreover, especially for biomedical applications, 3D-printing is currently being explored to obtain patient-specific implants.

Regardless of the processing technique, special attention has to be given to the moisture content present in the polymer melt. As polyesters are prone to hydrolysis, the presence of water can lead to a decrease in molecular weight and a change in the properties of the obtained parts. Therefore, the polymer should be properly dried prior to its processing and contact with moisture during processing should be avoided [9, 117]. The first prerequisite can be achieved by drying the polymer material *in vacuo*. When the drying is carried out at temperatures exceeding the glass transition temperature, amorphous polymer granules may fuse together, thereby hindering the efficient removal of moisture. As a result, amorphous polymers are generally dried at room temperature.

In addition to applying excessively high processing temperatures, the presence of residual catalysts can induce depolymerisation. When monomer is formed during processing, it can act as a plasticiser, effectively altering the materials' thermal and mechanical properties.

In the upcoming sections, the principles of polyester processing will be briefly touched upon. This overview is not exhaustive but should enable the interested reader to navigate through the various options.

21.4.1 Production of Polymer Fibres

As non-degradable polyesters are quite common as textile materials, it comes as no surprise that their degradable counterparts are also readily processable into fibres. Polymer fibres are particularly interesting for biomedical applications, including wound dressings, controlled-release formulations and tissue engineering. Several spinning techniques result in the formation of polymer fibres.

In wet spinning, a viscous polymer solution is passed through a fine needle into a bath filled with non-solvent (Fig. 21.13a). As the polymer solution comes into contact with this non-solvent, the solution solidifies or precipitates as a fibre. By collecting the fibre on a rotating spool, the fibres are wound up in a continuous process.

Dry spinning follows a similar approach (Fig. 21.13b). A polymer solution is passed through a needle or spinneret, but the fibres are not passed through a non-solvent bath. Instead, a hot air stream is passed through the set-up in the opposite direction to dry the fibres.

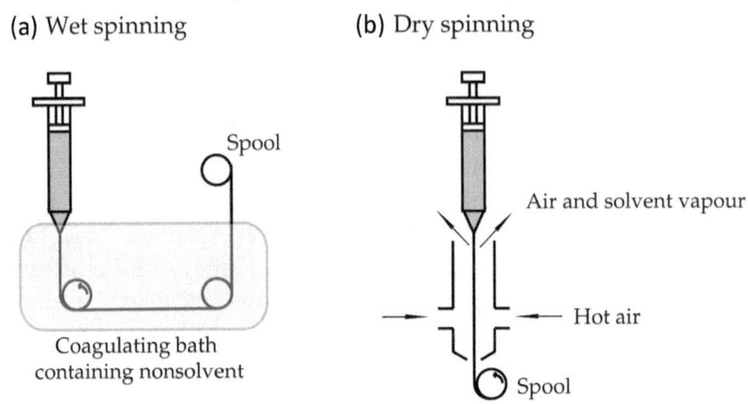

Figure 21.13 Representation of the wet (a) and dry (b) spinning process starting from a polymer solution.

Unlike the above-mentioned techniques, melt spinning starts form a polymer melt. Polymer granules are melted directly followed by passing the melted polymer through a spinneret. As the fibres emerge they are cooled down by passing a cold air stream through the collecting area. Melt spinning usually results in rapid formation of uniform fibres. However, this technique is only applicable for polymers that have a high thermal stability.

Electrospinning uses a high-voltage electrical field (typically 10–20 kV) to form micro- and even nanoscale fibres from a suspended droplet of polymer melt or solution [118]. When the repulsive electrostatic interactions overcome the droplet's surface tension, a Taylor cone is formed and a polymer jet is ejected from the tip of this Taylor cone [119]. The polymer jet is then accelerated towards a grounded collector screen. As the jet moves through the air, a stretching process occurs and the solvent evaporates which results in a non-woven polymer fabric or polymer mat [120]. Electrospinning has already been applied for both synthetic as well as natural polymers [121–124]. The set-up of the electrospinning process is shown in Fig. 21.14.

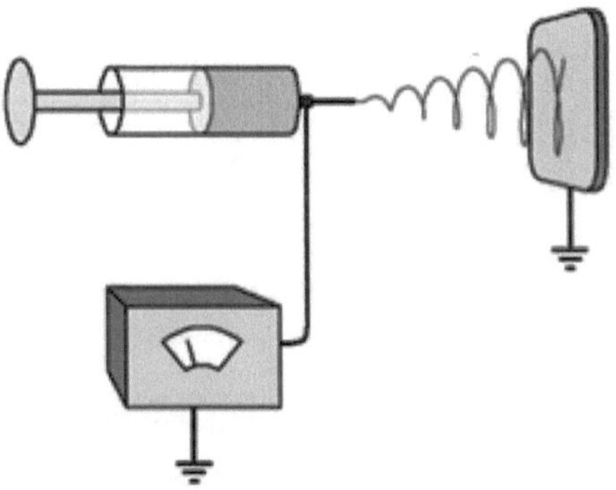

Figure 21.14 Schematic overview of the electrospinning process.

21.4.2 Production of Films and Sheets

Polymer foils and sheets possess unique properties including a high aspect ratio, a great flexibility and physical adhesion to ubiquitous surfaces [125]. When such films are produced from biodegradable or biocompatible materials, they can serve as coating material for biomedical implants.

Melt pressing is a first technique which results in thin polymer films. Polymer granules are heated between electrically heated plates, which can be forced against each other. Since this is a discontinuous process, it has little industrial application. However, if the mould plates are micro-structured, structured polymer sheets can be developed. In this case, the term hot-embossing is generally applied.

Melt extrusion is a continuous process, rendering it more suitable in an industrial setting (Fig. 21.15a). Polymer powder or pellets are fed to a screw-extruder which melts the polymer while transporting the material towards the nozzle. At the end, the melted material is forced through a flat die. The formed polymer sheet is then collected by rotating drums, which also assist in the cooling and the orientation of the polymer. Interestingly, this method is not limited to polymer sheets. By selecting the appropriate die, it is also feasible to produce rods, tubes and filaments.

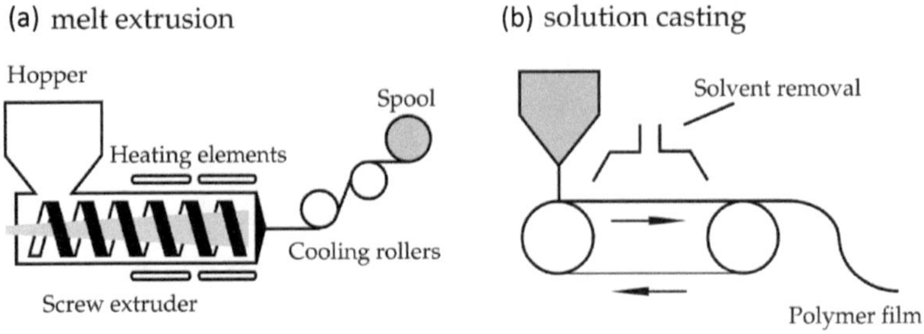

Figure 21.15 Illustration showing the production of polymer sheets via melt extrusion (a) and solution casting (b).

Bubble blowing can be seen as an alternative for melt extrusion. In this technique, a screw extruder forces the material through an annular die. At the same time, an inert gas stream passes through the inner tube to blow the extruded material into a cylindrical bubble. This bubble is then collected by a series of rollers to produce a continuous film.

In case of thermally sensitive polymers, solution casting (Fig. 21.15b) can be applied. In this technique, a viscous polymer solution is poured onto a non-adhesive surface. When the solvent is allowed to evaporate, a thin polymer film is formed which can be peeled of the mould. For optimal results, the molecular weight of the applied polymers should be high enough to form a tough film. In case of low-molecular-weight polymers, the lower amount of chain entanglements would result in weak conglomerates. Additionally, the boiling point of the solvent should ideally lie between 60 and 100°C. This prevents bubbling of the materials at ambient temperature while still allowing efficient solvent evaporation.

21.4.3 Production of Three-Dimensional Objects

In the previous sections, polymer processing techniques for one (fibres) or two dimensions (foils, sheets) have been discussed. The upcoming processing methods result in three-dimensional objects.

In compression moulding, both thermoplasts as well as thermosets can be processed. Polymer granules or resin materials are placed between two complementary moulds as indicated in Fig. 21.16a.

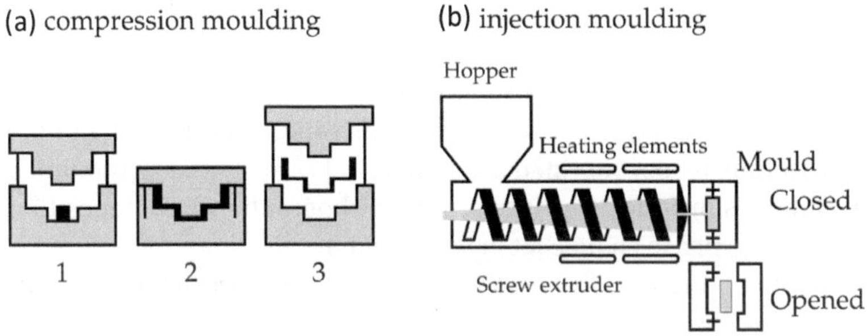

Figure 21.16 Illustration showing the production of polymer objects. On the left side compression moulding (a) starts with a preheated polymer pellet (1), which is subsequently moulded (2). After cooling (3) the object is released from the mould. On the right side, injection moulding (b) involves the injection of melted polymer in a mould by means of a plunger-type screw extruder.

Generally, the lower halve contains a cavity, while the upper part has a projection. After preheating the material, the mould is closed. As the pressure is increased, the polymer material will deform and fills up the mould cavity. In case of thermosets, the pressure and the temperature has to be maintained until the material is fully cured. For thermoplasts, it suffices to cool the mould, thereby fixating the polymer's shape. When the mould is cooled down, the object is ejected from the mould and the process can be repeated.

Injection moulding is a high-throughput method to produce three-dimensional objects. Polymer granules are fed to a plunger-type screw extruder which mixes and melts the material. As the screw moves back, a cavity will form near the nozzle which is filled up with melted polymer. Next, the plunger moves forward to inject the hot polymer mixture into a closed mould. To prevent shrinkage during cooling, an excess of polymer is usually injected. By cooling down the mould, the plastic material is fixated in its new shape after which the newly formed object is ejected from the device. Interestingly, the entire cycle only lasts 10–30 s per part. This technique is also applicable to thermosets. In this case, the pressure and the temperature should be maintained until the resin is fully cured.

21.4.4 Additive Manufacturing

Additive manufacturing, also known as solid freeform fabrication (SFF), offers many possibilities towards rapid prototyping and the fabrication of porous medical implants [126]. Rapid prototyping techniques include 3D printing, stereolithography, fused deposition modelling, phase change jet printing and 3D plotting.

Interestingly, polyesters can be readily processed via 3D printing. Polymer granules are melted inside the device. As the dispensing head moves, a polymer strand is ejected from a needle onto a collecting plate where the material sets. The deposition of subsequent layers enables the formation of complex three-dimensional structures. In some cases, the same technology can be extended towards polymer solutions.

Although it is very promising, the technique also shows some very important shortcomings [120, 127]. A first drawback is its limited resolution, which is mainly influenced by the needle diameter and the rate at which the printing head operates [128]. The Bioscaffolder^{TM} technology can also be very time-consuming since the processing parameters have to be optimised for each new polymer. Another issue is the necessity of a well-defined viscoelastic response allowing the material to quickly set [128]. In addition, the deposited material should contain a high colloid-volume ratio to prevent shrinkage by solvent evaporation.

Despite these restrictions, a variety of polymers including poly(L-lactic acid) (PLLA), poly(ε-caprolactone) (PCL), agarose, gelatin, chitosan and polyelectrolytes have already been applied as starting materials to develop porous scaffolds using the Bioscaffolder^{TM} technology [120, 129].

21.5 Applications

After discussing the synthesis and the processing of biodegradable polyesters in the previous sections, herein, their applications will be reviewed. Biomedical applications and their use in packaging materials constitute the major application areas. However, the use of biodegradable polyesters is not limited to these research fields as will become clear in the upcoming section.

21.5.1 The Role of Biodegradability in the Biomedical Field

Like soils, seas, and rivers, the human body can be regarded as a biological environment. Often, confusion arises when using the term "biodegradable". For instance, poly(glycolide) hydrolyses into water-soluble monomers and oligomers, leaving no trace of its remnants. In the biomedical sciences, this is referred to as absorbable, resorbable, or bioabsorbable as well as biodegradable. Some authors stipulate that biodegradable should be reserved for polymers designed for ecological purposes, while polymers for implantation should be termed resorbable or absorbable instead of biodegradable. However, in the field of biomaterials, biodegradable is a widely accepted synonym. Two major advantages can be correlated to the use of biodegradable biomaterials: (i) due to the gradual absorption by the human body, the materials will not evoke any chronic foreign-body reactions, and; (ii) in the field of tissue engineering (TE) these materials can act as temporary scaffolds for guided tissue regeneration, through the interaction with immunologic cells like macrophages. The concept of TE is schematically presented in Fig. 21.17.

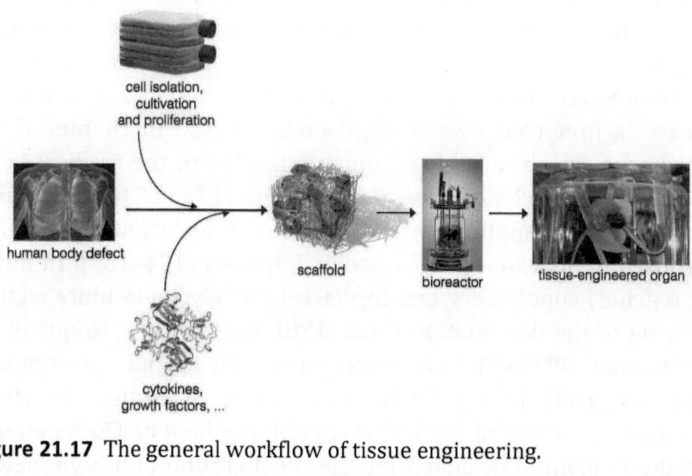

Figure 21.17 The general workflow of tissue engineering.

Recently, a paradigm shift from synthetic, non-resorbable polymers towards predesigned biodegradable polymers has taken place in the field of biomedical sciences. Especially in tissue engineering, degradable porous scaffolds integrated with biological cells or molecules are receiving increasing attention. For biomaterials, biocompatibility is of utmost importance. In the case of implanted biodegradable polymers, cytotoxic effects are induced when cell disturbances caused by the release of the low-molecular-weight compounds are large enough to result in irreversible cell damage [102].

The most commonly applied biodegradable polyesters used in the biomedical field are the linear aliphatic polyesters such as PGA, PLA, PCL, PHB and their copolymers. PLA, PGA, and PLGA are bioresorbable polyesters belonging to the group of the poly α-hydroxy acids. Lactic acid, formed during the hydrolysis of PLA is a normal by-product of the anaerobic metabolism in the human body and is incorporated in the tricarboxylic acid (TCA) cycle. Finally, it is excreted by the body as carbon dioxide and water. PGA is degraded in the body by the combination of hydrolytic scissions and enzymatic action resulting in glycolic acid. Glycolic acid will either be secreted through the urine or enters the TCA cycle. For PLGA, the degradation time can be controlled or designed from weeks up to a year or longer. Unfortunately, glycolic or lactic-acid rich degradation products can significantly lower the local pH in closed and/or less fluid buffered regions leading to complications. For example, increased acidity tends to cause bone resorption and/or demineralisation [130]. Another side effect is the potentially accelerated acid-catalysed hydrolysis [131]. Therefore, a controlled release is preferred over a rapid burst release to counter these drawbacks. PCL has attracted much attention in the medical field, mainly as controlled-release drug carrier [132, 133]. It has also been frequently applied as starting material for rapid prototyping purposes to develop patient-specific implants [134–137]. In other approaches, copolymers bearing lactide or glycolide units enable tailoring of the degradation rates [138]. For example, Hoque et al. synthesised PEG-PCL-PLA copolymers for rapid prototyping purposes [139]. In Fig. 21.18, an example illustrates the effect and the importance of biodegradation in the field of TE. A patient-specific implant is designed for the regeneration of a bone defect and the rate of degradation ideally overlaps with the rate of bone regeneration.

PLLA block copolymers with poly(amino acids) are a class of copolymers which contain both ester and amide linkages, referred to as poly(depsipeptides) or poly(esters-amides). These copolymers have been reported as interesting materials for controlled drug delivery systems [141, 142]. Table 21.1 lists a range of representative synthetic biodegradable polyesters used in the biomedical field.

Table 21.1 Representative polyesters used in the biomedical field

Polyester	M_n (kDa)	Degradation rate	Medical application
Poly(glycolide)	—	100% in 2–3 months	Sutures, soft tissue anaplerosis
Poly(glycolide-co-lactide)	40–100	100% in 50–100 days	Sutures, fraction fixation, oral implant, drug delivery
Poly(L-lactic acid)	100–300	50% in 1–2 years	Fracture fixation, ligament augmentation
Poly(L-lactic acid-co-ε-caprolactone)	100–500	100% in 3–12 months	Sutures, dural substitute
Poly(ε-caprolactone)	40–80	50% in 4 years	Contraceptive delivery implant, drug delivery
Poly(orthoesters)	100–150	60% in 50 weeks	Contraceptive delivery implant
Poly(p-dioxane)	—	100% in 30 weeks	Sutures, fracture fixation
Poly(depsipeptides)	—	—	Drug delivery

Source: Adapted from [102].

One of the major limitations in the study of biodegradable biomaterials is the discrepancy which can and mostly will also occur between in vitro and in vivo research. A systematic theoretical modelling study using computational chemistry to theoretically model the effects of steric hindrance and the inductive effect on the degradation properties of glycolide/lactide based polyesters has already been performed in order to predict structure–property relationships [143]. It was found that the rate of ester hydrolysis is greatly affected by both alkyl and halogen substituents (due to either steric hindrance or charge delocalisation). Furthermore, free radicals can interact in the degradation process [144, 145]. Inflammatory cells, for example macrophages, can produce reactive oxygen species such as superoxide and hydrogen peroxide [146, 147].

21.5.2 Biodegradable Polymers and Photonics

Photonics can be defined as the science and technology which uses light's properties for novel or better performing products. As the name suggests, photonics forms the bridge between optics and electronics. It thus comes as no surprise that materials which are well known in the electronics industry (silicon, various semiconductors and silicon oxide) are also regularly applied in photonics. However, in order to enable the widespread application of novel technologies currently being developed, the possibility to realise cheap mass manufacturing is becoming increasingly important. Unlike their inorganic counterparts, polymers are readily processable into a variety of shapes at a relatively low cost. In addition, polymers can be engineered to exhibit the desired optical properties by selecting the appropriate (co)monomers.

In the upcoming sections, the synergy between polymer chemistry and photonics will be highlighted.

21.5.2.1 Photonics in polymer chemistry

Light has been an important resource in polymer chemistry since the introduction of photo-initiated polymerisations. As this is, however, not the topic of this section, we will focus on advanced applications of light in polymer science including two-photon polymerisation and light-responsive materials.

An interesting processing technique applying light in material science is two-photon polymerisation (2PP) [127, 148]. Using 2PP, a photo-curable material is subjected to spatially well-defined laser irradiation (Fig. 21.19).

Since two photons are required to initiate the polymerisation, curing will only take place in the focal point of the laser (i.e. where the concentration of photons is the highest). The microscope objective used to focus the laser beam can be moved relative to the sample. After removal of the uncured material, complex three-dimensional structures can be achieved via direct laser writing [127, 149]. Interestingly, biodegradable polyester oligomers have already been functionalised with photo-curable moieties to extend 2PP to polyester-based materials [150].

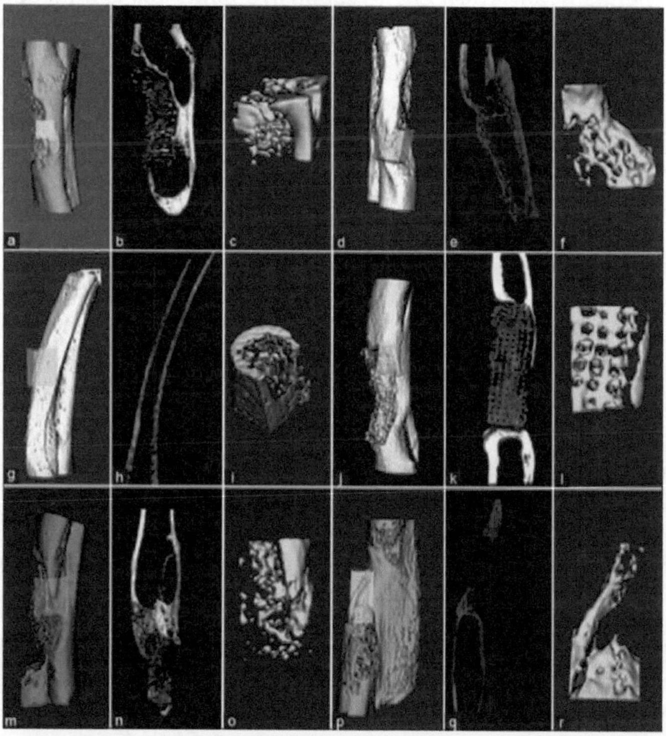

Figure 21.18 Micro-CT images: 1. Bone marrow mesenchymal stem cells (BMSCs)-polylactic-co-glycolic acid (PLGA)/β-tricalciumphosphate (β-TCP)-collagen I/apatite composites (a-i): 12 weeks postoperation showing new bone along lateral edge of residual scaffolds (a–c); 24 weeks postoperation showing nearly recanalised marrow cavity (d–f); 36 weeks postoperation showing remodelling (g–i); 2. BMSCs-PLGA/β-TCP composites (j–r): 12 weeks postoperation (j–l); 24 weeks postoperation (m-o); 36 weeks postoperation (p–r). 3D reconstruction image (a, d, g, j, m, p); reconstruction image in the coronal plane (b, e, h, k, n, q); magnified view (c, f, i, l, o, r) of yellow cubic area (a, d, g, j, m, p) showing no new bone formation. Reprinted from [140].

Light-responsive materials make up a second example of using light in polymer science to introduce novel functionalities. These materials change their characteristics upon irradiation with a specific wavelength. The best studied examples of this phenomenon are polymers containing azo-benzene moieties [151–153] and/or spirobenzopyran [154]. Light-responsive materials can be applied for controlled drug delivery [155], holography [156], sensors [157] and smart textiles [158].

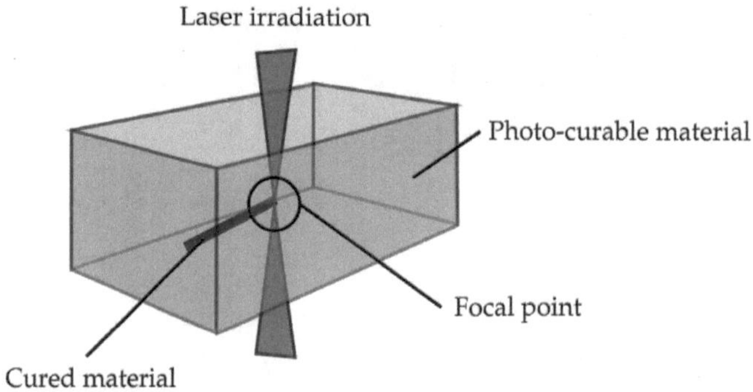

Figure 21.19 Two-photon polymerisation set-up curing a photo-curable resin at the focal point.

21.5.2.2 Degradable polymers in photonic applications

The application of degradable polymers in general and polyesters in particular in photonic applications is still in its infancy, although (bio)degradability can certainly bring an added value to a number of applications.

A first example was proposed by Ibrišimović et al., who used poly(lactic acid-co-glycolic acid) (PLGA) as a detection platform for pathogens [158]. By coating a mirror layer made from a nickel-chromium alloy with PLGA, Ibrišimović et al. were able to detect lytic enzymes secreted by certain micro-organisms. As the enzymes degraded the PLGA coating, the optical properties of the metal layer changed resulting in a wavelength or colour change.

Song et al. elegantly combined therapeutics with diagnostics via the surface initiated ROP of lactide from gold nanoparticles [159]. These polymer functionalised gold particles were able to self-assemble into plasmonic vesicles. When cancer targeting moieties were included on the surface, the authors could use the gold nanoparticles to demonstrate the affinity for cancer cells via surface-enhanced Raman scattering (SERS). In addition, irradiating the nanoparticles with infrared radiation was shown to effectively raise their temperature. This photothermal effect could be applied to locally heating a tumour and killing the cancerous cells in the process. Interestingly, drug loaded vesicles would release their chemotherapeutics as the lactide corona degrades resulting in a combined chemo-photothermal therapy for certain cancer types.

Cell visualisation and controlled delivery applications were also proposed by Chong et al. who synthesised a degradable, fluorescent polyester [160]. Although the proposed material consists of a highly hydrophobic backbone, the authors demonstrated degradability at neutral pH although this process occurs at a much lower rate than when subjected to a high pH. Interestingly, both the polymer as its degradation product showed lower cytotoxicity as other studied components including quantum dots and organic dyes.

Polyesters have also been applied in wave guide materials [161, 162]. In this regard, fluorine atoms are often introduced to modify the materials refractive index: as fluorine is bulkier than hydrogen the material becomes less dense effectively lowering its refractive index. However, due to these modifications the polymer becomes more hydrophobic which inhibits hydrolysis of these materials. These materials are thus no longer considered biodegradable.

21.5.3 Biodegradation as Emerging Trend in Packaging

In the previous sections, some high-end applications of biodegradable poly(esters) have been discussed as the added value posed by the use of these materials compensates for their higher cost compared to commodity polymers. However, due to the dropping price of production of degradable plastics and legislative pressure, these materials are also being introduced in less high-tech applications (e.g., as packaging materials) to bring down the amounts of plastics accumulating on landfills.

With the introduction of oil-based polymers such as poly(ethylene), poly(propylene), poly(styrene), poly(ethylene terephthalate) and poly(vinyl chloride) the packaging industry has been truly revolutionised. Despite the advances in polymer synthesis and processing, their end of life still poses a problem [159]. The first choice for processing plastic waste is reuse, but only some plastics can be reused after proper processing, and many of those are difficult to recycle [102]. In these cases, the waste is either incinerated with energy recuperation or dumped on landfills. Recent figures have shown that only 26.3% of plastic waste gets recycled [8]. Considering the 57 Mtonne European plastics market, huge amounts of plastics are thus disposed without further purpose. In packaging, three distinct categories can be observed: (i) primary packaging, i.e., in direct contact with the goods and taken home by consumers; (ii) secondary packaging covering assemblies of the primary packaged goods, and; (iii) tertiary packaging such as wooden pallets and plastic wrapping [3]. Especially, the materials used for the primary packaging applications pose problems in recycling or reuse. In fact, this kind of speciality packaging is potentially the largest use of biodegradable polymers to date [159]. Especially, biodegradable polymers that have undergone toxicological and cytocompatibility testing show great potential towards specialty packaging for, among others, food and pharmaceutical products.

Already in the 1970s, some requirements were formulated for biodegradable materials [159]. If the materials are to be degraded in landfills, they should ideally:

(1) be non-toxic and generate non-toxic degradation products that do not affect drainage water from landfill sites;
(2) possess suitable properties for specific applications;
(3) be economically viable;
(4) show tunable degradation kinetics via polymer modification;
(5) be processable via standard techniques.

21.5.3.1 Commodity packaging

When Wella marketed its biodegradable shampoos in the 1990s, they decided to make the packaging of this product biodegradable as well. These shampoo bottles were the first commercial product made from PHAs. Since then, PHAs have found their way to other packaging applications including motor oil bottles and disposable razors.

Bionolle is a copolymer of 1,4-butanediol and succinic acid (i.e., poly(butylene succinate), PBS) or 1,4-butanediol and both succinic and adipic acid (i.e., poly(butylene succinate/adipate), PBSA) developed by Showa Denko. It can be considered to be one of the frontrunners in packaging applications [160]. Bionolle is available in various grades depending on its molecular weight and the ratio between succinic and adipic acid and is used in plastic grocery bags.

21.5.3.2 Food packaging

For food packaging, gas permeability is an important property to take into account, as oxygen and water vapour are responsible for changes in the quality of food products [161]. Compared to glass, metal or ceramic containers, plastic packaging generally shows a high gas permeability. PLA was found to have poor gas barrier properties, which could limit its use in food packaging [7]. However, by adding inorganic components a composite material with superior characteristics can be obtained [162–164].

Another important parameter when food packaging is concerned, is the materials' resistance towards chemicals (e.g., acids or vegetable oils) present in its contents. Auras et al. have compared oriented PLA with oriented PS and PET. Interestingly; PLA only showed a minimal strength degradation after exposure to both strong and weak acids [165].

To date, PHAs have been applied extensively for food packaging applications. Poly(3HB-3HV) shows suitable gas barrier properties to enable its application in beverage bottles. In addition PHA lattices can be used to coat paper and cardboard used in paper milk cartons [82]. Currently, drink cartons are often coated with polyethylene and/or aluminium foils to improve the barrier properties. As the resulting composite material is hard to recycle, PHA-coated cartons offer an interesting alternative.

21.5.4 Agricultural and Veterinary Applications

Non-degradable materials being used in horticulture and agriculture result in an environmental concern due to the vast amounts being applied. One example can be found in mulch films which are used to conserve water and to suppress weeds. If these films are made from biodegradable materials, farmers no longer need to remove them from their fields before the next planting cycle. Bionolle has been successfully marketed as mulch film material in Japan by Showa Denko [160, 166]. The low tear resistance of Bionolle was leveraged by blending it with starch. The same materials have also been marketed as composting bags.

Degradable materials also offer new perspectives in agriculture (e.g., the controlled release of fertilisers, insecticides and herbicides) [167]. The latter could have been anticipated as the first PHAs were synthesised and degraded by soil bacteria. Chang et al. evaluated the impact of PLA on the growth and the yield of soybeans and observed that PLA increased the leaf area, the pod number, the bean number and the bean and plant dry weight [168]. Applying PLA as herbicide matrix thus not only contributes to improved weed control but also improves crop yield as PLA and lactic acid act as growth stimulants.

Another use of degradable polyesters in agriculture lies in bacterial inoculants which can enhance nitrogen fixation in plants. The bacterial cultures used in this application have to withstand harsh conditions prior to their application, making the use of a suitable carrier material crucial. To improve the survival rates within these carriers, nutrients can be added, although the type of bacteria applied may play a more important role in this regard.

In addition, the rumen of cows has been reported to be a suitable environment for PHA degradation. Therefore, it is possible to use degradable polyesters as controlled-release vesicles for certain medicines.

In horticulture, seedlings can be planted in biodegradable flower pots. When the re-potting of the plants is required, it suffices to put the original pot into the new soil. As the material degrades, the plant's roots can further expand.

21.5.5 Other Applications

In this section, some alternative application areas are summarised, which were beyond the scope of the previous sections.

The first example can be found in disposable consumer products. In products which are designed for single-use, the shorter shelf-life of degradable polymers is not an issue. Especially in nappies, feminine hygiene products and wet wipes, biodegradable materials present interesting alternatives [169].

She et al. have proposed the use of PS-PLA block copolymers to form structured thin films via self-assembly [170]. By this method, porous PS membranes could be obtained after the degradation of the PLA blocks. This example illustrates that progress in the field of biodegradable materials could also generate added value in alternative applications using non-degradable materials.

21.6 General Conclusions and Future Prospects

Despite their inherent higher production cost, biodegradable polymers are catching up with their non-degradable counterparts. At present, biodegradable polyester are emerging towards applications in which their degradability offers an added value, including biomedical implants and devices. In addition, packaging applications offer an interesting market for biodegradable polyesters in response to legislative pressure and growing ecological awareness. We expect this research field to further mature as more specialty monomers are currently being synthesised and processing issues are solved. If so, disposable products will no longer be associated with plastic waste accumulating in the environment.

Acknowledgements

The authors would like to thank the Research Foundation Flanders (FWO) for financial support (Project 3G008413) and the CHANGE2Micro Project supported by Enterprise Flanders (Agentschap Ondernemen) and Flanders In Action (ViA). S. Van Vlierberghe would like to acknowledge the Research Foundation Flanders (FWO-Flanders) for financial support under the form of a Research Grant ("Development of the ideal tissue engineering scaffold by merging state-of-the-art processing techniques", FWO Krediet aan Navorsers, 1.5.194.12N).

References

1. Jin, H.-J., Lee, B.-Y., Kim, M.-N., Yoon, J.-S. Thermal and mechanical properties of mandelic acid-copolymerized poly(butylene succinate) and poly(ethylene adipate). *J. Polym. Sci. Part B Polym. Phys.*, 2000, **38**, 1504–1511.
2. Slomkowski, S., Penczek, S., Duda, A. Polylactides-an overview. *Polym. Adv. Technol.*, 2014, **25**, 436–447.
3. Endres, H. J., Siebert-Raths, A. *Engineering Biopolymers: Markets, Manufacturing, Properties, and Applications.* Hanser Gardner Publications; 2010.

4. Siegenthaler, K. O., Künkel, A., Skupin, G., Yamamoto, M. Ecoflex® and Ecovio®: biodegradable, performance-enabling plastics. *Adv. Polym. Sci.*, 2012, **245**, 91–136.
5. Scott, G., Wiles, D. M. Programmed-life plastics from polyolefins: a new look at sustainability. *Biomacromolecules*, 2001, **2**, 615–622.
6. Vert, M., Doi, Y., Hellwich, K.-H., Hess, M., Hodge, P., Kubisa, P., et al. Terminology for biorelated polymers and applications (IUPAC Recommendations 2012). *Pure Appl. Chem.*, 2012, **84**, 377–410.
7. Madhavan Nampoothiri, K., Nair, N. R., John, R. P. An overview of the recent developments in polylactide (PLA) research. *Bioresour. Technol.*, 2010, **101**, 8493–8501.
8. Plastics Europe. Plastics—the Facts 2013. An analysis of European latest plastics production, demand and waste data. Brussels: 2013.
9. Middleton, J. C., Tipton, A. J. Synthetic biodegradable polymers as orthopedic devices. *Biomaterials*, 2000, **21**, 2335–2346.
10. Vert, M. Aliphatic polyesters: Great degradable polymers that cannot. *Biomacromolecules*, 2005, **6**, 538–546.
11. Müller, R. J., Kleeberg, I., Deckwer, W. D. Biodegradation of polyesters containing aromatic constituents. *J. Biotechnol.*, 2001, **86**, 87–95.
12. Wang, C.-H., Tsai, P.-H., Kan, L.-S., Chen, C. W. Synthesis and characterization of copolymeric aliphatic-aromatic esters derived from terephthalic acid, 1,4-butanediol, and ε-caprolactone by physical, thermal, and mechanical properties and NMR measurements. *J. Appl. Polym. Sci.*, 2013, **127**, 4385–4394.
13. Kint, D. A. review on the potential biodegradability of poly(ethylene terephthalate). *Polym. Int.*, 1999, **48**, 346–352.
14. Zeng, J., Huang, C., Jiao, L., Lu, X., Wang, Y., Wang, X. Synthesis and properties of biodegradable poly (butylene succinate-co-diethylene glycol succinate) Copolymers. *Ind. Eng. Chem. Res.*, 2012, **51**, 12258–12265.
15. Jacquel, N., Freyermouth, F., Fenouillot, F., Rousseau, A., Pascault, J. P., Fuertes, P., et al. Synthesis and properties of poly(butylene succinate): Efficiency of different transesterification catalysts. *J. Polym. Sci. Part A Polym. Chem.*, 2011, **49**, 5301–5312.
16. Marques, D. S., Gil, M. H., Baptista, C. M. S. G. Bulk polytransesterification of L-lactic acid esters: An alternative route to synthesize poly (lactic acid). *J. Appl. Polym. Sci.*, 2012, n/a, 10.1002/app.36825.
17. Luo, S.-H., Wang, Z.-Y., Mao, C.-X., Huo, J.-P. Synthesis of biodegradable material poly(lactic acid-co-glycerol) via direct melt polycondensation and its reaction mechanism. *J. Polym. Res.*, 2011, **18**, 2093–2102.

18. Gupta, A. P., Kumar, V. New emerging trends in synthetic biodegradable polymers–Polylactide: A critique. *Eur. Polym. J.*, 2007, **43**, 4053–4074.

19. Moon, S. I., Lee, C. W., Taniguchi, I., Miyamoto, M., Kimura, Y. Melt/solid polycondensation of L-lactic acid: An alternative route to poly(L-lactic acid) with high molecular weight. *Polym. Commun.*, 2001, **42**, 5059–5062.

20. Ajioka, M., Enomoto, K., Suzuki, K., Yamaguchi, A. The basic properties of poly(lactic acid) produced by the direct condensation polymerization of lactic acid. *J. Environ. Polym. Degrad.*, 1995, **3**, 225–234.

21. Zhu, C., Kustra, S. R., Bettinger, C. J. Photocrosslinkable biodegradable elastomers based on cinnamate-functionalized polyesters. *Acta Biomater.*, 2013, **9**, 7362–7370.

22. Dusselier, M., Van Wouwe, P., De Smet, S., De Clercq, R., Verbelen, L., Van Puyvelde, P., et al. Toward functional polyester building blocks from renewable glycolaldehyde with Sn cascade catalysis. *ACS Catal.*, 2013, **3**, 1786–800.

23. Robert, C., de Montigny, F., Thomas, C. M. Tandem synthesis of alternating polyesters from renewable resources. *Nat. Commun.*, 2011, **2**, 586.

24. Marubayashi, H., Asai, S., Hikima, T., Takata, M., Iwata, T. Biobased copolymers composed of L-lactic acid and side-chain-substituted lactic acids: Synthesis, properties, and solid-state structure. *Macromol. Chem. Phys.*, 2013, **214**, 2546–2561.

25. Liu, T., Simmons, T. L., Bohnsack, D. A., Mackay, M. E., Smith, M. R., Baker, G. L. Synthesis of polymandelide: A degradable polylactide derivative with polystyrene-like properties. *Macromolecules*, 2007, **40**, 6040–6047.

26. Jing, F., Smith, M. R., Baker, G. L. Cyclohexyl-substituted polyglycolides with high glass transition temperatures. *Macromolecules*, 2007, **40**, 9304–9312.

27. Jiang, X., Smith, M. R., Baker, G. L. Water-soluble thermoresponsive polylactides. *Macromolecules*, 2007, **41**, 318–324.

28. Jiang, X., Vogel, E. B., Smith, M. R., Baker, G. L. "Clickable" polyglycolides: Tunable synthons for thermoresponsive, degradable polymers. *Macromolecules*, 2008, **41**, 1937–1944.

29. Dechy-Cabaret, O., Martin-Vaca, B., Bourissou, D. Controlled ring-opening polymerization of lactide and glycolide. *Chem. Rev.*, 2004, **104**, 6147–6176.

30. Lecomte, P., Je, C. Recent developments in ring-opening polymerization of lactones. *Adv. Polym. Sci.*, 2012, **245**, 173–218.

31. Penczek, S., Cypryk, M., Duda, A., Kubisa, P., Slomkowski, S. Living ring-opening polymerizations of heterocyclic monomers. *Prog. Polym. Sci.*, 2007, **32**, 247–282.

32. Coulembier, O., Degée, P., Hedrick, J. L., Dubois, P. From controlled ring-opening polymerization to biodegradable aliphatic polyester: Especially poly(β-malic acid) derivatives. *Prog. Polym. Sci.*, 2006, **31**, 723–747.

33. Szwarc, M. "Living" polymers. *Nature*, 1956, **178**, 1168–1169.

34. Dubois, P., Jerome, R., Teyssie, P. Macromolecular engineering of polylactones and polylactides. *Polym. Bull.*, 1989, **22**, 475–482.

35. Ropson, N., Dubois, P., Jerome, R., Teyssie, P. Macromolecular engineering of polylactones and polylactides. 20. Effect of monomer, solvent, and initiator on the ring-opening polymerization as initiated with aluminum alkoxides. *Macromolecules*, 1995, **28**, 7589–7598.

36. Okuda, J., Rushkin, I. L. Mono(cyclopentadienyl)titanium complexes as initiators for the living ring-opening polymerization of ε-caprolactone. *Macromolecules*, 1993, **26**, 5530–5532.

37. Asandei, A. D., Saha, G. Living ring-opening polymerization of cyclic esters with epoxide-derived titanium alkoxides. *Macromol. Rapid Commun.*, 2005, **26**, 626–631.

38. Zeng, J.-B., Srinivansan, M., Li, Y.-D., Narayan, R., Wang, Y.-Z. Rapid ring-opening polymerization of 1,4-dioxan-2-one initiated by titanium alkoxides. *J. Polym. Sci. Part A Polym. Chem.*, 2010, **48**, 5885–5890.

39. Kowalski, A., Duda, A., Penczek, S. Kinetics and mechanism of cyclic esters polymerization initiated with tin(II) octoate, 1. Polymerization of ε-caprolactone. *Macromol. Rapid Commun.*, 1998, **19**, 567–572.

40. Kowalski, A., Duda, A., Penczek, S. Mechanism of cyclic ester polymerization initiated with Tin(II) octoate, 2. Macromolecules fitted with Tin(II) alkoxide species observed directly in MALDI–TOF spectra. *Macromolecules*, 2000, **33**, 689–695.

41. Möller, M., Kånge, R., Hedrick, J. L. Sn(OTf)2 and Sc(OTf)3: Efficient and versatile catalysts for the controlled polymerization of lactones. *J. Polym. Sci. Part A Polym. Chem.*, 2000, **38**, 2067–2074.

42. Kricheldorf, H. R., Eggerstedt, S. Macrocycles 2. Living macrocyclic polymerization of ε-caprolactone with 2,2-dibutyl-2-stanna-1,3-dioxepane as initiator. *Macromol. Chem. Phys.*, 1998, **199**, 283–290.

43. Kricheldorf, H. R., Stricker, A., Langanke, D. Polylactones, 50. The reactivity of cyclic and noncyclic dibutyltin bisalkoxides as initiators in the polymerization of lactones. *Macromol. Chem. Phys.*, 2001, **202**, 2525–2534.

44. Kricheldorf, H. R. Biodegradable polymers with variable architectures via ring-expansion polymerization. *J. Polym. Sci. Part A Polym. Chem.*, 2004, **42**, 4723–4742.

45. Chmura, A. J., Cousins, D. M., Davidson, M. G., Jones, M. D., Lunn, M. D., Mahon, M. F. Robust chiral zirconium alkoxide initiators for the room-temperature stereoselective ring-opening polymerisation of rac-lactide. *Dalt Trans.*, 2008, **21**, 1437–1443.

46. Dobrzynski, P. Synthesis of biodegradable copolymers with low-toxicity zirconium compounds. II. Copolymerization of glycolide with ε-caprolactone initiated by zirconium(IV) acetylacetonate and zirconium(IV) chloride. *J. Polym. Sci. Part A Polym. Chem.*, 2002, **40**, 1379–1394.

47. Yamashita, M., Takemoto, Y., Ihara, E., Yasuda, H. Organolanthanide-initiated living polymerizations of ε-caprolactone, δ-valerolactone, and β-propiolactone. *Macromolecules*, 1996, **29**, 1798–1806.

48. Stevels, W. M., Ankone, M. J. K., Dijkstra, P. J., Feijen, J. A versatile and highly efficient catalyst system for the preparation of polyesters based on lanthanide tris(2,6-di-tert-butylphenolate)s and various alcohols. *Macromolecules*, 1996, **29**, 3332–3333.

49. Shen, Y., Shen, Z., Zhang, Y., Yao, K. Novel rare earth catalysts for the living polymerization and block copolymerization of caprolactone. *Macromolecules*, 1996, **29**, 8289–8295.

50. Armstrong R. A. Review article What causes Alzheimer's disease? *Folia Neuropathol.*, 2013, **3**, 169–188.

51. Appel, E. A., Lee, V. Y., Nguyen, T. T., McNeil, M., Nederberg, F., Hedrick, J. L., et al. Toward biodegradable nanogel star polymers via organocatalytic ROP. *Chem. Commun. (Camb.)*, 2012, **48**, 6163–6165.

52. Hofman, A., Szymański, R., Słomkowski, S., Penczek, S. Structure of active species in the cationic polymerization of β-propiolactone and ε-caprolactone. *Die. Makromol. Chem.*, 1984, **185**, 655–667.

53. Slomkowski, S., Szymański, R., Hofman, A. Formation of the intermediate cyclic six-membered oxonium ion in the cationic polymerization of β-propiolactone initiated with $CH_3CO\oplus SbF\ominus 6$. *Die. Makromol. Chem.*, 1985, **186**, 2283–2290.

54. Albertsson, A.-C., Palmgren, R. Cationic polymerization of 1,5-dioxepan-2-one with lewis acids in bulk and solution. *J. Macromol. Sci. Part A*, 1996, **33**, 747–758.

55. Abraham, G. A., Gallardo, A., Lozano, A. E., San Roman, J. ε-caprolactone/ZnCl2 complex formation: Characterization and ring-opening polymerization mechanism. *J. Polym. Sci. Part A Polym. Chem.*, 2000, **38**, 1355–1365.

56. Kricheldorf, H. R., Sumbél, M.-V. Polylactones, 15. Reactions of δ-valerolactone and ε-caprolactone with acidic metal bromides. *Die. Makromol. Chem.*, 1988, **189**, 317–331.

57. Casas, J., Persson, P. V., Iversen, T., Córdova, A. Direct organocatalytic ring-opening polymerizations of lactones. *Adv. Synth. Catal.*, 2004, **346**, 1087–1089.

58. Sanda, F., Sanada, H., Shibasaki, Y., Endo, T. Star polymer synthesis from ε-caprolactone utilizing polyol/protonic acid initiator. *Macromolecules*, 2001, **35**, 680–683.

59. Zeng, F., Lee, H., Chidiac, M., Allen, C. Synthesis and characterization of six-arm star poly(δ-valerolactone)-block-methoxy poly(ethylene glycol) copolymers. *Biomacromolecules*, 2005, **6**, 2140–2149.

60. Persson, P. V., Casas, J., Iversen, T., Córdova, A. Direct organocatalytic chemoselective synthesis of a dendrimer-like star polyester. *Macromolecules*, 2006, **39**, 2819–2822.

61. Nederberg, F., Connor, E. New paradigms for organic catalysts: The first organocatalytic living polymerization. *Angew. Chem.*, 2001, **42**, 2712–2715.

62. Nyce, G. W., Glauser, T., Connor, E. F., Möck, A., Waymouth, R. M., Hedrick, J. L. In situ generation of carbenes: A general and versatile platform for organocatalytic living polymerization. *J. Am. Chem. Soc.*, 2003, **125**, 3046–3056.

63. Dove, A. P., Pratt, R. C., Lohmeijer, B. G. G., Waymouth, R. M., Hedrick, J. L. Thiourea-based bifunctional organocatalysis: Supramolecular recognition for living polymerization. *J. Am. Chem. Soc.*, 2005, **127**, 13798–13799.

64. Dove, A. P., Pratt, R. C., Lohmeijer, B. G. G., Culkin, D. A., Hagberg, E. C., Nyce, G. W., et al. N-Heterocyclic carbenes: Effective organic catalysts for living polymerization. *Polymer (Guildf)*, 2006, **47**, 4018–4025.

65. Lohmeijer, B. G. G., Pratt, R. C., Leibfarth, F., Logan, J. W., Long, D. A., Dove, A. P., et al. Guanidine and amidine organocatalysts for ring-opening polymerization of cyclic esters. *Macromolecules*, 2006, **39**, 8574–83.

66. Kamber, N. E., Jeong, W., Waymouth, R. M., Pratt, R. C., Lohmeijer, B. G. G., Hedrick JL. Organocatalytic ring-opening polymerization. *Chem. Rev.*, 2007, **107**, 5813–5840.

67. Brule, E., Guerineau, V., Vermaut, P., Prima, F., Balogh, J., Maron, L., et al. Polymerization of cyclic esters using N-heterocyclic carbene carboxylate catalysts. *Polym. Chem.*, 2013, **4**, 2414–2423.

68. Zhang, L., Nederberg, F., Pratt, R. C., Waymouth, R. M., Hedrick, J. L., Wade, C. G., et al. Phosphazene bases: A new category of organocatalysts for the living ring-opening polymerization of cyclic esters. *Macromolecules*, 2007, **40**, 4154–4158.

69. Goldys, A. M., Dixon, D. J. Organocatalytic ring-opening polymerization of cyclic esters mediated by highly active bifunctional iminophosphorane catalysts. *Macromolecules*, 2014, **47**, 1277–1284.

70. Bourissou, D., Martin-vaca, B., Dumitrescu, A., Graullier, M. Controlled cationic polymerization of lactide. *Macromolecules*, 2005, **38** 9993–9998.

71. Makiguchi, K., Kikuchi, S., Satoh, T., Kakuchi, T. Synthesis of block and end-functionalized polyesters by triflimide-catalyzed ring-opening polymerization of ε-caprolactone, 1,5-dioxepan-2-one, and rac-lactide. *J. Polym. Sci. Part A Polym. Chem.*, 2013, **51**, 2455–2463.

72. Bailey, W. J. Free radical ring-opening polymerization. *Polym. J.*, 1985, **17**, 85–95.

73. Sanda F, Endo T. Radical ring-opening polymerization. *J. Polym. Sci. Part A Polym. Chem.*, 2001, **39**, 265–276.

74. Agarwal, S. Chemistry, chances and limitations of the radical ring-opening polymerization of cyclic ketene acetals for the synthesis of degradable polyesters. *Polym. Chem.*, 2010, **1**, 953.

75. Plikk, P., Tyson, T. Mapping the characteristics of the radical ring-opening polymerization of a cyclic ketene acetal towards the creation of a functionalized polyester. *J. Polym. Sci. Part A Polym. Chem.*, 2009, **47**, 4587–4601.

76. Jin, S., Gonsalves, K. E. A study of the mechanism of the free-radical ring-opening polymerization of 2-methylene-1,3-dioxepane. *Macromolecules*, 1997, **30**, 3104–3106.

77. Undin, J., Plikk, P., Finne-wistrand, A., Albertsson, A. Synthesis of amorphous aliphatic polyester-ether homo- and copolymers by radical polymerization of ketene acetals. *J. Polym. Sci. Part A Polym. Chem.*, 2010, **48**, 4965–4973.

78. Delplace, V., Harrisson, S., Tardy, A., Gigmes, D., Guillaneuf, Y., Nicolas, J. Nitroxide-mediated radical ring-opening copolymerization: Chain-end investigation and block copolymer synthesis. *Macromol. Rapid Commun.*, 2014, **35**, 484–491.

79. Jiang, Z. Lipase-catalyzed synthesis of aliphatic polyesters via copolymerization of lactone, dialkyl diester, and diol. *Biomacromolecules*, 2008, **9**, 3246–3251.

80. Matsumoto, K., Taguchi, S. Biosynthetic polyesters consisting of 2-hydroxyalkanoic acids: Current challenges and unresolved questions. *Appl. Microbiol. Biotechnol.*, 2013, **97**, 8011–8021.

81. Yang, Y., Yu, Y., Zhang, Y., Liu, C., Shi, W., Li, Q. Lipase/esterase-catalyzed ring-opening polymerization: A green polyester synthesis technique. *Process Biochem.* 2011, **46**, 1900–1908.

82. Philip, S., Keshavarz, T., Roy, I. Polyhydroxyalkanoates: Biodegradable polymers with a range of applications. *J. Chem. Technol.*, 2007, **247**, 233–247.

83. Akaraonye, E., Keshavarz, T., Roy, I. Production of polyhydroxyalkanoates: The future green materials of choice. *J. Chem. Technol. Biotechnol.*, 2010, **85**, 732–743.

84. Orts, W. J., Nobes, G. A. R., Kawada, J., Nguyen, S., Yu, G., Raveneile, F. Poly(hydroxyalkanoates): Biorefinery polymers with a whole range of applications. The work of Robert H. Marchessault. *Can. J. Chem. Can. Chim.*, 2008, **86**, 628–640.

85. Wang, Q., Yang, P., Xian, M., Yang, Y., Liu, C., Xue, Y., et al. Biosynthesis of poly(3-hydroxypropionate-co-3-hydroxybutyrate) with fully controllable structures from glycerol. *Bioresour. Technol.*, 2013, **142**, 741–744.

86. Nakayama, A., Yamano, N., Kawasaki, N., Nakayama, Y. Synthesis and biodegradation of poly(2-pyrrolidone-co-ε-caprolactone)s. *Polym. Degrad. Stab.*, 2013, **98**, 1882–1888.

87. Ji, S., Bruchmann, B., Klok, H. Synthesis of side-chain functional polyesters via baylis–hillman polymerization. *Macromolecules*, 2011, **44**, 5218–5226.

88. Kissel, T., Li, Y., Unger, F. ABA-triblock copolymers from biodegradable polyester A-blocks and hydrophilic poly(ethylene oxide) B-blocks as a candidate for in situ forming hydrogel delivery systems for proteins. *Adv. Drug Deliv. Rev.*, 2002, **54**(1), 99–134.

89. Pan, Z., Ding, J. Poly(lactide-co-glycolide) porous scaffolds for tissue engineering and regenerative medicine. *Interface Focus*, 2012, **2**, 366–377.

90. Twibanire, J., Grindley, T. B. Polyester dendrimers. *Polymers (Basel)*, 2012, **4**, 794–879.

91. Jain, K., Kesharwani, P., Gupta, U., Jain, N. K. Dendrimer toxicity: Let's meet the challenge. *Int. J. Pharm.*, 2010, **394**, 122–142.

92. Vishwa Prasad, A., Yinghuai, Z. Syntheses of cyclic poly(lactones) by zwitterionic ring opening polymerization catalyzed by N-heterocyclic carbene. *J. Appl. Polym.*, 2013, **128**, 3411–3416.

93. Oledzka, E. Synthesis of genistein-containing star-shaped homo- and copolyesters by the ring-opening polymerization. *Polym. Bull.*, 2013, **70**, 2587–2597.

94. Rochette, J. M., Ashby, V. S. Photoresponsive polyesters for tailorable shape memory biomaterials. *Macromolecules*, 2013, **46**, 2134–2140.

95. He, C., Kim, S. W., Lee, D. S. In situ gelling stimuli-sensitive block copolymer hydrogels for drug delivery. *J. Control. Release*, 2008, **127**, 189–207.

96. Thomas, C. M. Stereocontrolled ring-opening polymerization of cyclic esters: Synthesis of new polyester microstructures. *Chem. Soc. Rev.*, 2010, **39**, 165–173.

97. Kayaman-Apohan, N., Akyürek, E. Synthesis and drug-release properties of biodegradable hydrogels having ε-cyclodextrin. *Polym. Bull.*, 2013, **70**, 2151–2167.

98. Rasal, R. M., Janorkar, A. V., Hirt, D. E. Poly(lactic acid) modifications. *Prog. Polym. Sci.*, 2010, **35**, 338–356.

99. Shah, A. A., Hasan, F., Hameed, A., Ahmed, S. Biological degradation of plastics: A comprehensive review. *Biotechnol. Adv.*, 2008, **26**, 246–265.

100. Lucas, N., Bienaime, C., Belloy, C., Queneudec, M., Silvestre, F., Nava-Saucedo, J.-E. Polymer biodegradation: Mechanisms and estimation techniques. *Chemosphere*, 2008, **73**, 429–442.

101. Leja, K., Lewandowicz, G. Polymer biodegradation and biodegradable polymers—a review. *Polish J. Environ. Stud.*, 2010, **19**, 255–266.

102. Ikada, Y., Tsuji, H. Biodegradable polyesters for medical and ecological applications. *Macromol. Rapid Commun.*, 2000, **21**, 117–132.

103. Kasirajan, S., Ngouajio, M. Polyethylene and biodegradable mulches for agricultural applications: A review. *Agron. Sustain. Dev.*, 2012, **32**, 501–529.

104. Singh, B., Sharma, N. Mechanistic implications of plastic degradation. *Polym. Degrad. Stab.*, 2008, **93**, 561–84.

105. Swift, G. Biodegradable polymers in the environment: Are they really biodegradable? In: (Gebelein, C., Carraher, Jr. C., ed.), *Biotechnology and Bioactive Polymers* SE—15, Springer US, 1994, pp. 161–168.

106. Hergenrother, R. W., Wabers, H. D., Cooper, S. L. The effect of chain extenders and stabilizers on the in-vivo stability of polyurethanes. *J. Appl. Biomater.*, 1992, **3**, 17–22.

107. Ratner, B. D., Gladhill, K. W., Horbett, T. A. Analysis of in vitro enzymatic and oxidative degradation of polyurethanes. *J. Biomed. Mater. Res.*, 1988, **22**, 509–527.

108. Siotto, M., Sezenna, E., Saponaro, S., Innocenti, F. D, Tosin, M., Bonomo, L., et al. Kinetics of monomer biodegradation in soil. *J. Environ. Manag.*, 2012, **93**, 31–37.

109. Nishida, H., Suzuki, S., Tokiwa, Y. Distribution of poly (p-propiolactone) aerobic degrading microorganisms in different environments. *J. Environ. Polym. Degrad.*, 1998, **6**, 43–58.

110. Eubeler, J. P., Bernhard, M., Knepper, T. P. Environmental biodegradation of synthetic polymers II. Biodegradation of different polymer groups. *TrAC Trends Anal. Chem.*, 2010, **29**, 84–100.

111. Eubeler, J. P., Zok, S., Bernhard, M., Knepper, T. P. Environmental biodegradation of synthetic polymers I. Test methodologies and procedures. *TrAC Trends Anal. Chem.*, 2009, **28**, 1057–1072.

112. Kijchavengkul, T., Auras, R. Compostability of polymers. *Polym. Int.*, 2008, **57**, 793–804.

113. Palmisano, A. C., Pettigrew, C. A. Biodegradability of plastics. *Bioscience*, 1992, **42**, 680.

114. Hiltunen, K., Sepp, J. V., It, M., Mika, H. The Biodegradation of Lactic Acid-Based Poly (ester-Urethanes). *J. Environ. Polym. Degrad.*, 1997, 5, 167–173.

115. Nakayama, A., Kawasaki, N., Arvanitoyannis, I., Aiba, S., Yamamoto, N. Synthesis and biodegradation of poly(γ-butyrolactone-co-l-lactide). *J. Environ. Polym. Degrad.*, 1996, **4**, 205–211.

116. Vieira, A. C., Guedes, R. M., Marques, A. T. Development of ligament tissue biodegradable devices: A review. *J. Biomech.*, 2009, **42**, 2421–2430.

117. Von Oepen, R., Michaeli, W. Injection moulding of biodegradable implants. *Clin Mater.*, 1992, **10**, 21–28.

118. Meng, D., Erol, M., Boccaccini, A. R. Processing technologies for 3D nanostructured tissue engineering scaffolds. *Adv. Eng. Mater.*, 2010, **12**, B467–B487.

119. Park, S., Kim, T., Kim, H., Yang, D., Park, T. Development of dual scale scaffolds via direct polymer melt deposition and electrospinning for applications in tissue regeneration. *Acta Biomater.*, 2008, **4**, 1198–1207.

120. Ma, P. X. Scaffolds for tissue fabrication. *Materialstoday*, 2004, **7**, 30–40.

121. Dasari, A., Quirós, J., Herrero, B., Boltes, K., García-Calvo, E., Rosal, R. Antifouling membranes prepared by electrospinning polylactic acid containing biocidal nanoparticles. *J. Memb. Sci.*, 2012, **405–406**, 134–140.

122. Chang, J.-J., Lee, Y.-H., Wu, M.-H., Yang, M.-C., Chien, C.-T. Preparation of electrospun alginate fibers with chitosan sheath. *Carbohydr. Polym.*, 2012, **87**, 2357–2361.

123. Baji, A., Mai, Y.-W., Wong, S.-C., Abtahi, M., Chen, P. Electrospinning of polymer nanofibers: Effects on oriented morphology, structures and tensile properties. *Compos. Sci. Technol.*, 2010, **70**, 703–718.

124. Ero-Phillips, O., Jenkins, M., Stamboulis, A. Tailoring crystallinity of electrospun PLLA fibres by control of electrospinning parameters. *Polymers (Basel)*, 2012, **4**, 1331–1348.

125. Kariduraganavar, M. Y., Kittur, A. A., Kamble, R. R. Polymer Synthesis and Processing. In: (Kumbar, S. G., Laurencin, C. T., Deng, M., eds.), *Natural and Synthetic Biomedical Polymers,* 1st ed., Elsevier; 2014, p. 19–26.

126. Park, S. A., Lee, S. H., Kim, W. Fabrication of hydrogel scaffolds using rapid prototyping for soft tissue engineering. *Macromol. Res.*, 2011, **19**, 694–698.

127. Ovsianikov, A., Deiwick, A., Van Vlierberghe, S., Dubruel, P., Möller, L., Dräger, G., et al. Laser fabrication of three-dimensional CAD scaffolds from photosensitive gelatin for applications in tissue engineering. *Biomacromolecules*, 2011, **12**, 851–858.

128. Billiet, T., Vandenhaute, M., Schelfhout, J., Van Vlierberghe, S., Dubruel, P. A. Review of trends and limitations in hydrogel-rapid prototyping for tissue engineering. *Biomaterials*, 2012, **33**, 6020–6041.

129. Nie, Z., Kumacheva, E. Patterning surfaces with functional polymers. *Nat. Mater.*, 2008, **7**, 277–290.

130. Suganuma, J., Alexander, H. Biological response of intramedullary bone to poly-L-lactic acid. *J. Appl. Biomater.*, 1993, **4**, 13–27.

131. Winet, H., Hollinger, J. O., Stevanovic, M. Incorporation of polylactide-polyglycolide in a cortical defect: Neoangiogenesis and blood supply in a bone chamber. *J. Orthop. Res.*, 1995, **13**, 679–689.

132. Khatiwala, V. K., Shekhar, N., Aggarwal, S., Mandal, U. K. Biodegradation of poly(ε-caprolactone) (PCL) film by alcaligenes faecalis. *J. Polym. Environ.*, 2008, **16**, 61–67.

133. Funabashi, M., Ninomiya, F., Kunioka, M. Biodegradation of polycaprolactone powders proposed as reference test materials for

international standard of biodegradation evaluation method. *J. Polym. Environ.*, 2006, **15**, 7–17.

134. Declercq, H. A., Desmet, T., Dubruel, P., Cornelissen, M. J. The role of scaffold architecture and composition on the bone formation by adipose-derived stem cells. *Tissue Eng. Part A,* 2014, **20**, 434–445.

135. Kim, G. H., Son, J. G. 3D polycarprolactone (PCL) scaffold with hierarchical structure fabricated by a piezoelectric transducer (PZT)-assisted bioplotter. *Appl. Phys. A*, 2008, **94**, 781–785.

136. Mariani, M., Rosatini, F., Vozzi, G., Previti, A., Ahluwalia, A. Characterization of tissue-engineered scaffolds microfabricated with PAM. *Tissue Eng.*, 2006, **12**, 547–557.

137. Weinand, C., Gupta, R., Weinberg, E. Human shaped thumb bone tissue engineered by hydrogel-β-tricalciumphosphate/poly-ε-caprolactone scaffolds and magnetically sorted stem cells. *Ann. Plast. Surg.,* 2007, **59**, 46–52.

138. Garkhal, K., Verma, S., Jonnalagadda, S., Kumar, N. Fast degradable poly(L-lactide-co-ε-caprolactone) microspheres for tissue engineering: Synthesis, characterization, and degradation behavior. *J. Polym. Sci. Part A Polym. Chem.*, 2007, **45**, 2755–2764.

139. Hoque, M. E., Hutmacher, D. W., Feng, W., Li, S., Huang, M.-H., Vert, M., et al. Fabrication using a rapid prototyping system and in vitro characterization of PEG-PCL-PLA scaffolds for tissue engineering. *J. Biomater. Sci. Polym. Ed.*, 2005, **16**, 1595–1610.

140. Pang, L., Hao, W., Jiang, M., Huang, J., Yan, Y., Hu, Y. Bony defect repair in rabbit using hybrid rapid prototyping polylactic co glycolic acid/β tricalciumphosphate collagen I/apatite scaffold and bone marrow mesenchymal stem cells. *Indian J. Orthop.*, 2013, **47**, 388–394.

141. Feng, Y., Guo, J. Biodegradable polydepsipeptides. *Int. J. Mol. Sci.*, 2009, **10**, 589–615.

142. John, G., Morita, M. Biodegradable cross-linked microspheres from poly-(ε-caprolactone-co-glycolic acid-co-L-serine) based polydepsipeptides. *Macromol. Rapid Commun.*, 1999, **20**, 265–268.

143. Chu, C.-C. Biodegradable polymeric biomaterials: An updated overview. In: (Bronzino, J. D., ed.), *The Biomedical Engineering Handbook: Second Edition*, 2000.

144. Williams, D. F., Zhong, S. P. Talking point. Are free radicals involved in biodegradation of implanted polymers? *Adv. Mater.*, 1991, **3**, 623–626.

145. Zhong, S. P., Doherty, P. J., Williams, D. F. A preliminary study on the free radical degradation of glycolic acid/lactic acid copolymer. *Plast. Rubber Compos. Process Appl.*, **21**, 89–97.

146. Badwey, J. A., Karnovsky, M. L. Active oxygen species and the functions of phagocytic leukocytes. *Annu. Rev. Biochem.*, 1980, **49**, 695–726.

147. Devereux, D. F., O'Connell, S. M., Liesch, J. B., Weinstein, M., Robertson, F. M. Induction of leukocyte activation by meshes surgically implanted in the peritoneal cavity. *Am. J. Surg.*, 1991, **162**, 243–246.

148. Engelhardt, S., Hoch, E., Borchers, K., Meyer, W., Krüger, H., Tovar, G. E. M., et al. Fabrication of 2D protein microstructures and 3D polymer-protein hybrid microstructures by two-photon polymerization. *Biofabrication*, 2011, **3**, 025003.

149. West, J. L. Protein-patterned hydrogels: Customized cell microenvironments. *Nat. Mater.*, 2011, **10**, 727–729.

150. Seppälä, J., Korhonen, H., Hakala, R., Malin, M. Photocrosslinkable polyesters and poly(ester anhydride)s for biomedical applications. *Macromol. Biosci.*, 2011, **11**, 1647–1652.

151. Saiz, L. M., Oyanguren, P. A., Galante, M. J., Zucch, I. A. Light responsive thin films of micelles of PS-b-PVP complexed with diazophenol chromophore. *Nanotechnology*, 2014, **25**, 65601.

152. Orofino, A. B., Arenas, G., Zucchi, I., Galante, M. J., Oyanguren, P. A. A simple strategy to generate light-responsive azobenzene-containing epoxy networks. *Polymer (Guildf)*, 2013, **54**, 6184–6190.

153. Alam, M. Z., Ohmachi, T., Ogata, T., Nonaka, T., Kurihara, S. Surface relief gratings on azo polymer films through reversible photoisomerization by the irradiation of a monochromatic light. *J. Appl. Polym. Sci.*, 2006, **102**, 3123–3126.

154. Baumann, L., De Courten, D., Martin, W., René, M. R., Lukas, J. S. Light-responsive caffeine transfer through porous polycarbonate. *Appl. Mater. Interfaces*, 2013, **5**, 5894–5897.

155. Cao, J., Huang, S., Chen, Y., Li, S., Li, X., Deng, D., et al. Near-infrared light-triggered micelles for fast controlled drug release in deep tissue. *Biomaterials*, 2013, **34**, 6272–6283.

156. Schumers, J.-M., Fustin, C.-A., Gohy, J.-F. Light-responsive block copolymers. *Macromol. Rapid Commun.*, 2010, **31**, 1588–1607.

157. Jochum, F. D., Theato, P. Temperature- and light-responsive smart polymer materials. *Chem. Soc. Rev.*, 2013, **42**, 7468–7483.

158. Hu, J., Meng, H., Li, G., Ibekwe, S. I. A review of stimuli-responsive polymers for smart textile applications. *Smart Mater. Struct.*, 2012, **21**, 53001.

159. Amass, W., Amass, A., Tighe, B. A review of biodegradable polymers: Uses, current developments in the synthesis and characterization

of biodegradable polyesters, blends of biodegradable polymers and recent advances in biodegradation studies. *Polym. Int.*, 1998, **47**, 89–144.

160. Ichikawa, Y., Mizukoshi, T. Bionolle (Polybutylenesuccinate). *Adv. Polym. Sci.*, 2012, **245**, 258–314.

161. Siracusa, V., Rocculi, P., Romani, S., Rosa, M. D. Biodegradable polymers for food packaging: A review. *Trends Food Sci. Technol.*, 2008, **19**, 634–643.

162. Kiersnowski, A., Dąbrowski, P., Budde, H., Kressler, J., Pigłowski, J. Synthesis and structure of poly(ε-caprolactone)/synthetic montmorillonite nano-intercalates. *Eur. Polym. J.*, 2004, **40**, 2591–2598.

163. Sinha Ray, S., Bousmina, M. Biodegradable polymers and their layered silicate nanocomposites: In greening the 21st century materials world. *Prog. Mater. Sci.*, 2005, **50**, 962–1079.

164. Cadebo, L., Feijoo, J. L., Villanueva, M. P., Lagarón, J. M., Giménez, E. Optimization of biodegradable nanocomposites based on aPLA/PCL blends for food packaging applications. *Macromol. Symp.*, 2006, **233**, 191–197.

165. Auras, B. R. A., Singh, S. P., Singh, J. J. Evaluation of oriented poly (lactide) polymers vs. Existing PET and oriented PS for fresh food service containers packaging technology and science. *Packag. Technol. Sci.*, 2005, **18**, 207–216.

166. Showa Denko Europe GmbH. Bionolle®: The pioneer in biodegradable 2010.

167. Holmes, P. A. Applications of PHB: A microbially produced biodegradable thermoplastic. *Phys. Technol.*, 1985, **16**, 32.

168. Chang, Y.-N., Mueller, R., Iannotti, E. Use of low MW polylactic acid and lactide to stimulate growth and yield of soybeans. *Plant Growth Regul.*, 1996, **19**, 223–232.

169. Rai, R., Tallawi, M., Grigore, A., Boccaccini, A. R. Synthesis, properties and biomedical applications of poly(glycerol sebacate) (PGS): A review. *Prog. Polym. Sci.*, 2012, **37**, 1051–1078.

170. She, M.-S., Lo, T.-Y., Hsueh, H.-Y., Ho, R.-M. Nanostructured thin films of degradable block copolymers and their applications. *NPG Asia Mater.*, 2013, **5**, e42.

Chapter 22

Synthetic–Natural Hybrid Polymers Based on Polyurethane: Structures, Properties, and Applications

Veronica Saucedo-Rivalcoba,[a,c] Ana Laura Martínez-Hernández,[b,c] and Carlos Velasco-Santos[b,c]

[a]*Subdivisión de Posgrado e Investigación, Instituto Tecnológico Superior de Tierra Blanca, Prolongación de Avenida Veracruz s/n esq. Héroes de Puebla, Colonia Pemex, 95180, Tierra Blanca, Veracruz, México*

[b]*Centro de Física Aplicada y Tecnología Avanzada, Universidad Nacional Autónoma de México, Campus Juriquilla, Boulevard Juriquilla 3001, 76230, Juriquilla, Querétaro, México*

[c]*División de Estudios de Posgrado e Investigación, Instituto Tecnológico de Querétaro, Av. Tecnológico s/n esq. Gral. Mariano Escobedo, Colonia Centro Histórico, 76000, Querétaro, Querétaro, México*

almh72@gmail.com

22.1 Introduction

Polyurethane (PU) represents an important class of thermoplastic and thermoset polymers whose mechanical, thermal, and chemical properties can be tailored with various polyols and isocyanates. Usually, both isocyanates and polyols are substances obtained

In memory of Dr. Adolfo M. Espíndola-González (1977–2014).

Handbook of Sustainable Polymers: Processing and Applications
Edited by Vijay Kumar Thakur and Manju Kumari Thakur

ISBN 978-981-4613-53-8 (Hardcover), 978-981-4613-54-5 (eBook)
www.panstanford.com

from petroleum, but, in recent years, natural sources such as polysaccharides, proteins, amino acids, lipids, fatty acids, and derivatives have attracted significant attention as raw materials for preparation of PUs. Thus, renewable resources could provide an interesting sustainable platform for substitution of petroleum-based polymers through the design of bio-based polymers that can compete or even surpass the existing materials on a cost-performance basis with high eco-friendly values [1]. PUs have a wide variety of properties and applications, which obey the step-growth polymerization of diisocyanates with diols, producing two separate segmented phases, i.e., hard and soft segments. The hard segments are derived from the diisocyanates and their chain extenders, while the soft segments consist of the polyol. It is possible to obtain very brittle and hard materials or soft, tacky, and viscous materials, depending on the properties and raw materials used in their synthesis [2].

Segmented PUs are generally composed of polyether and polyester soft segments and urethane-based hard segments along the polymer backbone giving rise to a microphase-separated morphology that is caused by the poor compatibility between the segments. The unique properties of these polymers are directly related to their two-phase microstructure, in which the hard domains act as reinforcing filler and as a thermally reversible cross-link. The soft segment is typically the block of the PU that is used to shift the behavior of conventional PUs to biodegradable ones. These changes in the PU macromolecular structure can also bring modifications in the microphase morphology. In the PUs at the soft segments, also there exist a versatility in hydrogen bonds formed, since, these are produced in the urethane linkages (inter-urethane hydrogen bonding), in the ester linkage (from a polyester moieties) or with the oxygen in the ethers (from the polyether moieties) [3]. Therefore, the chemical structure of the soft or hard segments can be modified, and biodegradable, biocompatible, or functional PU elastomer polymers can be obtained during the synthesis process, via incorporation of natural polymers [4], such as starch [5], chitosan [6], cereals (1,4:3,6-dyanhydrohexitols) [7], cellulose [8], protein from chicken feathers [9], and oil sources [10, 11]. Recent years have witnessed an increasing demand for natural products in industrial applications due to environmental and waste-disposal issues, as well as depletion of non-renewable resources. In non-food applications, the most widely applied renewable resources include plant oils, polysaccharides, and proteins [12].

Nowadays, the trend of research in this field is toward new reaction routes to establish the linking process among polyol and isocyanates obtained from biopolymers to synthesize hybrid PUs and, thus, reach to overcome thermal and mechanical properties of these materials. Also other important parameters such as functionality of diols, isocyanate type, and chain extenders are relevant in PU synthesis [13]. In these materials, the molecular structure varies from rigid cross-linked to linear or highly extensible elastomers; hence, their applications are versatile, such as coating or membranes in medical devices, materials in construction engineering, automotive parts, foams and adhesives.

This chapter is a compilation of the recent research on synthesis and characterization of natural-synthetic PU polymers from diverse natural resources, their properties and applications. Various methods to improve compatibility among biopolymer and PU and their effects on the mechanical and thermal properties are also discussed, as well as the use of PU membranes in environmental fields to remove industrial wastes.

22.1.1 Polyurethane Synthesis

PUs were first produced and investigated by Otto Bayer in 1937. PU is a polymer in which the repeating unit contains a urethane moiety. Urethanes are derivatives of carbamic acids ($RCOONH_2$). Variations in the R group and substitutions of the amide hydrogen produce multiple urethanes. Although all PUs contains repeating urethane groups, other moieties such as urea, ester, ether, and aromatic groups may be included. The urethane linkage results most readily from the reaction of an isocyanate (–N=C=O) with an alcohol (–OH) where the hydrogen atom of the hydroxyl group is transferred to the nitrogen atom of the isocyanate. The major advantage of PU is that the chain is not composed exclusively of carbon atoms but rather of heteroatoms, i.e., oxygen, carbon, and nitrogen [14], and the wide applicability of PU is due to the selection of monomeric materials from a huge list of macrodiols, diisocyanates, and chain extenders [15].

The PU synthesis process involves numerous reactions in which the isocyanates are highly reactive chemicals and create several chemically different products when combined with –OH and –NH functional substances. The desired products and side-products are formed in different amounts. The basic reactions of isocyanate with diverse reagents are shown in Fig. 22.1. The high reactivity of isocyanate groups toward nucleophilic reagents is mainly due to the pronounced positive character of the C atom in the cumulative double-bond sequence consisting of nitrogen, carbon, and oxygen, especially in aromatic systems. The electronegativity of the oxygen and nitrogen impart a large electrophilic character to the carbon in the isocyanate group. The common reactions of isocyanates can be divided into two main classes: (a) with compounds containing reactive hydrogen to give addition products; and (b) polymerization of isocyanates, i.e., self-addition. One of the main reactions is the isocyanate reaction with hydroxyl or polyol compounds to give urethanes, and with amines to give ureas [15]. The chain extender is usually a small molecule with either hydroxyl or amine end-groups. The isocyanate is a low-molecular-weight compound that can react with either the polyol or the chain extender. The combination of the chain extender and the isocyanate components is referred to as the hard segments of the polymer [16].

Polyol, or the so-called soft segment is an oligomeric macromonomer. Conventional polyols are usually polyethers (with repeating structure of –R–O–R′–) or polyesters (with repeating structure of –R–CO–O–R′–) with chain ends terminated by hydroxyl groups [16]. Variations in the number of substitutions and the spacing between and within branch chains produce PUs ranging from linear to branched and from flexible to rigid. Also, according to the molecular mass of the polyurethane prepolymers (isocyanates and polyols), various block copolymers can be produced. The terminal hydroxyl group allows alternating blocks, called segments, to be inserted into the PU chain. Modifying these segments give in varying degrees of tensile strength and elasticity; in addition, blocks also provide rigid crystalline phases, as function of the chain extender from the hard segments [14]. Donelly et al. proposed a schematic representation of the basic reaction to produce polyurethane copolymers, also the microstructure phase separation of both hard and soft segments [17].

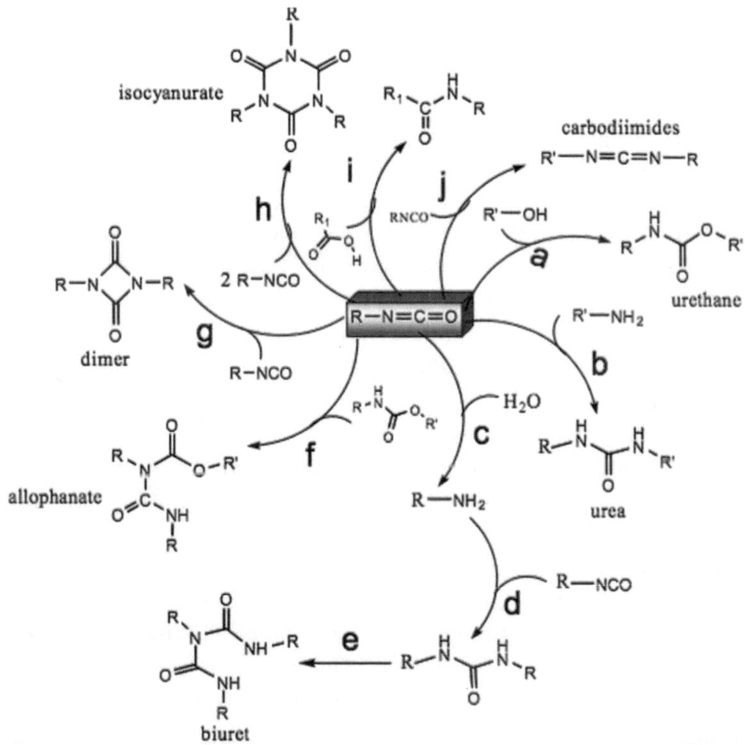

Figure 22.1 Basic reactions of isocyanate with different reactants: (a) hydroxyl compound to give urethanes, (b) amines to give ureas, (c) water to form an unstable compound, carbamic acid, which decomposes and produces an amine, (d) amines and excess isocyanate produces a urea linkage, (e) urea linkage and one molecule of isocyanate produces biuret, (f) urethane and asocyanate produce allophanates. Secondary reactions of isocyanate self-condensation produce: (g) uretidione ring (dimer-), (h) isocyanurate (trimer-), (j) carbodiimide (reprinted with permission of Chattopadhyay et al., 2007, Elsevier [15]).

Innovative technologies and competitive industrial products are reducing the dependence on petrochemicals for the production of polymers. Increasing concerns about the deteriorating environment caused by conventional polymers have directed worldwide research toward renewable resources. Therefore, bio-based materials are receiving wide attention and the polymer synthesis from these natural sources affords an alternative route.

22.2 Polyols, Isocyanates, and Chain Extenders Based on Natural Sources

22.2.1 Oils

Vegetable oils are one of the most important classes of bioresources for producing polymeric materials. This type of biomaterial can either be derived from nature or be synthesized in the laboratory via a variety of chemical approaches involving metallic components, polymers, ceramics, or composite materials. Bio-based polymers are defined as any polymer derived from living organisms such as plants, trees, or algae, while biopolymers refer to any polymer applicable for biomaterials, though bio-based polymers and biopolymers are used equivalently elsewhere. Due to their universal availability, inherent biodegradability and low price, vegetable oils have become an area of intensive interest for both academic and industrial research as platform chemicals for polymeric materials. Particular attention has been paid to investigating the suitability of vegetable-oil-based polymers (VOBPs) as future biomaterials. Compared to other bio-based biopolymers such as carbohydrates and proteins, VOBPs have several characteristics that make them more attractive as biomaterials: (1) although vegetable oils are not naturally present as polymers, they are precursors for monomer chains that can be used to synthesize various polymers including polyurethane, polyester, polyether, and polyolefine; (2) VOBP structures can be easily tuned by converting vegetable oil to different monomers; (3) vegetable oils are suitable for synthesis of hydrophobic polymers and complement other bioresources such as carbohydrates and proteins that are naturally hydrophilic; and (3) vegetable oils are suitable for producing monomers with structures similar to petroleum-based monomers, and so petroleum-based biopolymers could possibly be replaced by VOBPs with identical properties [18].

Vegetable oils are an excellent renewable source of raw materials for the manufacture of polyurethane components, such as polyols; hence, the chemical modification or the transformation of the double bonds of triglycerides of oils to hydroxyl moieties and their applications in polyurethanes has been the subject of many studies. The main technological advantages of these polyurethanes from vegetable oils are their high strength and stiffness, environmental resistance, and long life. Oil-based polyols are often oligomers with a wide distribution of molecular weights and a considerable degree of branching, which together affect the viscosity and processing properties of polyurethane foams. Precise characterization of the polyol composition and its properties are very important for understanding synthetic processes, as well as for quality control [19].

Montero et al. reported that the main constituents of plant oils are triacylglycerols (triglycerides), which are a product of esterification of glycerol with three fatty acids, and these compounds correspond to 95% of the total weight of triglycerides depending on the specific plant oil. Figure 22.2 shows the triglyceride structure and the main fatty acids used in polymer science. One research aim is to study chemical reactions to overcome the low reactivity of natural triglycerides, which usually contain double bonds, by introducing easily polymerizable functional groups [10]. The biggest challenge for the production of PUs using polyols from natural oils is the variation of the unsaturated content among natural oils. If the double-bond content in natural oils is not properly controlled, the nature of the polyol will be changed, affecting the performance of the final polyurethane product [19].

A

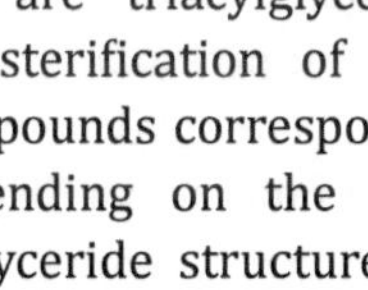

	14:0	14:1	16:0	16:1	17:0	17:1	18:0	18:1	18:2	18:3	20:0	20:1
Canola	0.1	0.0	4.1	0.3	0.1	0.0	1.8	60.9	21.0	8.8	0.7	1.0
Corn	0.1	0.0	10.9	0.2	0.1	0.0	2.0	25.4	59.6	1.2	0.4	0.0
Linseed	0.0	0.0	5.5	0.0	0.0	0.0	3.5	19.1	15.3	56.6	0.0	0.0
Olive	0.0	0.0	13.7	1.2	0.0	0.0	2.5	71.1	10.0	0.6	0.9	0.0
Palm	1.0	0.0	44.4	0.2	0.1	0.0	4.1	39.3	10.0	0.4	0.3	0.0
Soybean	0.1	0.0	11.0	0.1	0.0	0.0	4.0	23.4	53.2	7.8	0.3	0.0
Sunflower	0.0	0.0	6.1	0.0	0.0	0.0	3.9	42.6	46.4	1.0	0.0	0.0
High oleic	0.0	0.0	6.4	0.1	0.0	0.0	3.1	82.6	2.3	3.7	0.2	0.4

B

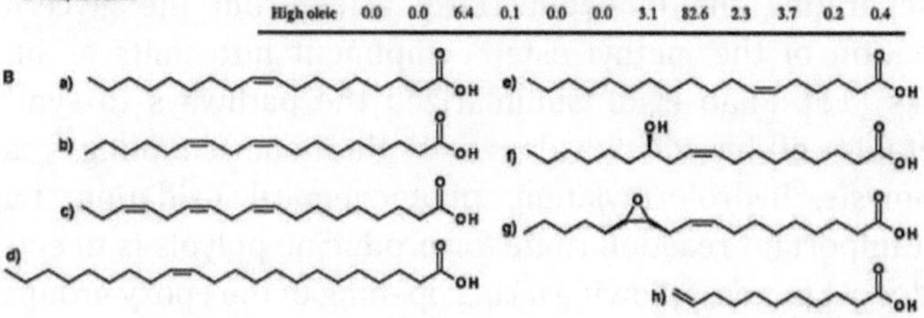

Figure 22.2 (A) General triglyceride structure and fatty acid percentage composition of common plant oil. (B) Fatty acids commonly used in polymer chemistry: (a) oleic acid, (b) linoleic acid, (c) linolenic acid, (d) erucic acid, (e) petroselinic acid, (f) ricinoleic acid, (g) vernolic acid, (h) 10-undecenoic acid (reprinted with permission of Montero de Espinosa et al., 2011, Elsevier [10]).

Triglyceride compounds of oils are highly functionalized molecules, and, therefore, have been used in the synthesis of cross-linked polymers via two main approaches. The first one takes advantage of the naturally occurring functional groups present in triglycerides, such as internal double bonds, alcohols, or epoxides, which can be polymerized using different methods. The second strategy depends on the chemical modifications prior to polymerization. This approach solves the drawback of the low reactivity of natural triglycerides, which usually contain double bonds, by introducing easily polymerizable functional groups, and thus widens the synthetic possibilities. The hydrolysis of triglycerides provides glycerol and a mixture of fatty acids. Glycerol is a widely used building block in polymer science, in which there has recently been a focus on the use of fatty acids as precursors of monomers, attributable to their renewability and the properties that they can provide to the final polymers. Metathesis and ethenolysis of fatty acids can be used not only for monomer synthesis, but also as polymerization techniques [10]. Following are some chemical reactions proposed to increase functionality of fatty acids and oils as renewable resources: Modification of castor oil via epoxidation reactions as monomer, prior to synthesis of polyurethane films [20], production of poly(ether urethane) networks from a polyether-polyol obtained by ionic-coordinative polymerization of epoxidized methyl oleate [1], transesterification of castor oil to synthesize an interpenetrating polymer network (IPN) [21], polyol derived from castor oil modified by alkoxylation with ethylene and propylene oxide in basic condition [22], or separating the individual fatty acids from the glycerine by formation of the methyl ester component into units as building blocks [11]. Miao et al. summarized the pathways to synthesize vegetable oil-based polyols as a thiol-ene coupling reaction, ozonolysis, hydroformylation, photochemical oxidation, but the most important reaction route for producing polyols is to epoxidize the double bonds, following a ring opening of the epoxy groups [18].

Castor oil is a clear (almost colorless), or pale yellow-colored viscous oil, where 90% of the content corresponds to ricinoleic acid. It presents aliphatic carboxylic acid with an 18-carbon chain, having a double bond between carbon 9 and 10 and a hydroxyl group on the carbon 12, which is exclusive to castor oil. This renewable resource also has high water resistance and, when it is dehydrated, is converted into fast-drying oil. All these properties combined with biodegradability, due to the fact that it comes from a natural molecule, make this material able to be applied in chemical industries, paints, varnishes, and adhesives, as well as wrapping and packaging for the food industry, cosmetics, and medicine, and is the raw material for the production of sebacic acid, which is the basic ingredient for nylon production. Taking this biomolecule as a building block, Ferreira et al. (2007) modified the hydroxyl groups from castor oil with isophorone diisocyanates in order to obtain the prepolymers with a terminal isocyanate group, and a degradable urethane (NCO) linkage, stable even for 7 days, was obtained. Also, this material had hydrophobic properties and was thermally stable above 139°C. This hybrid polymer was proposed as a bioadhesive in the medicine field [23]. Xu et al. studied the thermal and mechanical properties of a derivate from castor oil (polyricinoleate diol with dangling chains) at different concentrations (40, 50, 60, and 70% p/p) to obtain thermoplastic polyurethane. Morphology of this material tended to be "spherulitic-like" superstructures due to a possible nucleation and crystallization of the hard segments, and the properties of segmented polyurethanes depend not only on the structure of the soft segments (molecular weight and dangling chains of diol), but also on the soft-segment length and concentration. By differential scanning calorimetry analysis, when linking polyricinoleate diol at different concentrations, the T_g (glass transition temperature) moves from −77 to −61°C as a result of partial mixing of the hard segment into the soft segment; this behave could also be influenced by the formation of a bicontinuous morphology of hard and soft segments and the mobility among them. Changes in microphase morphology in segmented polyurethanes are more perceptible by dynamical mechanical analysis, where the rubbery plateau modulus of the segmented polyurethanes increases with the increase in the hard-segment content. While, in the increases in the soft segments, the modulus above the soft-segment T_g (−76°C) systematically decreases due to a diminished in the amount of the continuous hard-segment phase. On the other hand, in the strain-stress behavior, materials with lower content of hard-segment have low modulus and behavior of elastic/rubbery, while materials with the high hard-segment content are stiffer and displayed less recovery following deformation [24].

Corcuera et al. studied the microstructure and thermal properties considering hard/soft-segment microdomain by varying the nature and concentration of castor-oil-based soft segments and also the chain extender in the hard-segment domain. They reported, that the increase of hard-segments is related by an increase in the hard domain order (crystallinity and stiffness). Also, the elastic nature of soft segment provide enough physical cross-link sites to impart properties ranging from elastomeric to rigid behavior by increasing hard segment domain. In addition, if a biopolymer is used as a bio-based chain extender (corn, sugar), the tendency is a slightly lower crystallinity, compared with polyurethanes obtained from a petrochemical source. However, PUs synthesized with castor oil and chain-extender-based natural sources show a thermal stability comparable to that obtained from petroleum, due to that castor oil presents a maximum rate of decomposition at about 400°C, which is higher than conventional polyols [22]. Recently, interest in the protection of the environment has required testing of any form of recycling or re-usage of polymers, as well as studies about the biodegradability of synthetic polymers. Although polyurethanes were considered non-biodegradable for a long time, they can be degraded by hydrolysis, oxidation, and enzymatic and environmental attack, and polyurethane-based biomaterials are the answer to this issue by incorporating labile and hydrolyzable linkages through the soft segments, which are considered to be the starting point of degradation. The biodegradation of vegetable oil-based polyurethane is supported by the presence of ester groups in the structure of the plant oil that are susceptible to hydrolysis by enzymes secreted by microorganisms. Once the oil-based polyurethane moieties are exposed to the action of microorganisms, as well as in the biomolecules used in polyurethane elastomers cross-linked reactions, the mechanical properties are lower than those of the original polymer (previously to the microorganism attack); which indicates that the long-chain molecules have been broken down into shorter ones, and therefore hybrid bio-based polyurethanes are a new opportunity to overcome environmental issues [25, 26].

Not only can polyurethane-based polyol natural sources be used in medical and chemical industries, but recently the remarkable mechanical, thermal, and morphological properties of bio-based polyurethane have made their applications widespread.

In recent years, the synthesis of polyurethane membranes for the sorption and diffusion processes of organic molecules has become relevant. Castor-oil-based polyurethane as soft-segment moieties offers acceptable properties to act as a transport membrane, which was used by Ajithkumar et al. to synthesize interpenetrating polymer networks based on polyurethane to study the sorption and diffusion process of chlorobenzene by varying the cross-link density (NCO/OH ratio). The authors reported that the sorption/diffusion phenomena are explained by the NCO/OH ratio and the composition of the network, while increasing the cross-link density causes the chains in the structure to become more rigid and dense, and thus the ability of the polymer chain to accommodate solvent molecules decreases. At a lower cross-link density, the chains will have maximum flexibility due to there being fewer cross-link points, and, thus, the sorption capacity will be maximal. Thus, for a given NCO/OH ratio the diffusion coefficient increases and the sorption coefficient decrease [27]. Manjula and Siddaramaiah studied the sorption and diffusion of aromatic molecules through castor-oil-based polyurethane membranes by varying the type of chain extender (citric acid and maleic acid). This material shows significant transport characteristics as a result of phase separation behavior, which is a consequence of the type, chemical structure, molecular weight, length, and content of the compounds polyol, isocyanate, and chain extender. As well as the NCO/OH ratio, it is important to take into account the diffusion process through polyurethane membranes, the solvent type (solubility parameters), and the chemical nature of polymer, which seem to exert a significance influence on the molecular transport characteristics, i.e., when the difference in the solubility parameters of solvents and polymer membranes is low, there is an increase in sorption and permeation coefficients [28].

Castor oil has been extensively used as a polyol precursor in polyurethane synthesis because of the reactive hydroxyl functional groups (OH) contained in its molecule, but polyurethanes derived from hydroxylated epoxides of linseed, soybean, and rapeseed oils have also been reported. Soybean oil is another remarkable triglyceride rich in unsaturated fatty acids that are susceptible to chemical modification for desired polymeric materials. In comparison with castor oil, soybean oil also requires strategies to functionalize them. Lubguban et al. proposed the acylglycerol synthesis by thermal polymerization (bodying) of soybean oil (BSBO), followed by transesterification with glycerol, to manufacture modified polyol-based polyurethane, obtaining properties comparable to those petroleum-based polyurethane. Advantages of this functionalized polyol include the following: (a) simple non-hazardous processing to form polyol; (b) yields of polyol approaching 100%; and (c) feedstock costs that are considerably less than those of the propylene oxide and butylene oxide that are used to prepare petroleum-based polyols. In addition, this method is suggested to yield a soy-based polyol with high molecular weight, reduced saturated fatty acid moieties, increased primary alcohol moieties, and tunable hydroxyl equivalent weight. On the other hand enzymatic reactions, epoxidation (oxirane opening reactions), and ozonolysis show a series of drawbacks, including high cost and relatively low operational productivity and production of secondary hydroxyl moieties that react more slowly than primary moieties, and energy-intensive separations and disposal issues related with the generation of the monofunctional products of low-molecular-weight. Thus, the two-step process starting with the heat polymerization of BSBO results attractive method. This kind of reaction promotes the cross-linking of acylglycerols through a Diels-Alder reaction. During heat polymerization, the viscosity increases as a result of the participation of conjugated dienes in the Diels-Alder cross-linking; this reaction can be observed in Fig. 22.3. Subsequently, transesterification with glycerol molecules is used to impart functionality to hydroxy groups (Fig. 22.4). The reactions shown in Figs. 22.3 and 22.4 are possible when glycerol reacts with BSBO at 225°C. They include the following: (1) further polymerization of unreacted soybean oil (SBO) triglycerides to modified SBO; (2) formation of SBO monoglycerides and diglycerides; and (3) transesterification of SBO in the presence of glycerol. Polyurethane made of this modified soybean-based polyol presents thermal conductivity, density, and compressive strength similar to polymers synthesized with a petroleum resource [29].

Lu et al. reported the use of modified methoxylated soybean oil polyols that offer high OH functionality producing a new kind of thermoplastic PU with mechanical properties that are comparable to PUs from petroleum-based polyols, with the added advantage of being an environmentally friendly bio-based material, able to be used in decorative and protective coatings. Thermal stability due to cross-linking density and mechanical properties vary in the range from elastomeric polymers to tough and hard plastics, again because of the grafting and cross-linking behavior of PU [30].

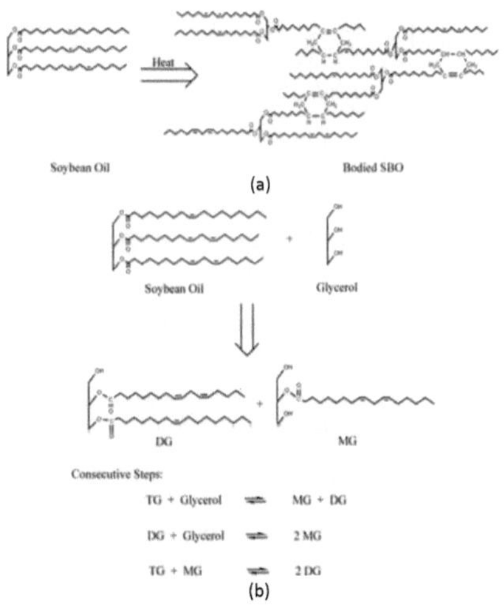

Figure 22.3 Reaction schemes for the hydroxylation process of modified soybean oil (BSBO): (a) unreacted soybean oil (SBO) is polymerized to BSBO. (b) SBO and glycerol react to form acylgycerols of SBO [monoglycerides (MG), diglycerides (DG), and triglycerides (TG)] (reprinted with permission of Lubguban et al., 2009, John Wiley and Sons [29]).

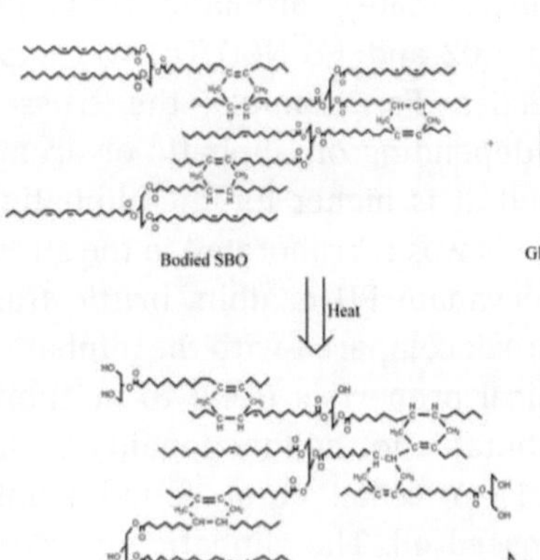

Figure 22.4 Major reaction in the hydroxylation process (BSBO and glycerol react to form GLYC-BSBO (acylglycerol)) (reprinted with permission of Lubguban et al., 2009, John Wiley and Sons [29]).

Table 22.1 Content of unsaturated fatty acids in rubber seed oil (RSO)

Fatty acid		%	Property of RSO	Value
Saturated				
C16:0	Palmitic acid	17.51	Color (Lovinbond)	22R, 23.2Y
C18:0	Stearic acid	4.82	Specific gravity (30°C)	0.916
	Total	22.33	Acid value (mg of KOH/g)	43.62
Unsaturated			Free fatty acid	21.4
C18:1	Oleic acid	25.33	Saponification value (mg of KOH/g)	202.91
C18:2	Linoleic acid	37.50	Iodine value (g of I_2/100g)	136.21
C18:3	Linolenic acid	14.21		
	Total	77.04		
	Others	0.63		

Source: Adapted with permission of Bakare et al., 2008, John Wiley and Sons [31].

An interesting alternative to classical oils used in polymer science: rubber seed oil, was studied by Bakare et al. This oil contents almost 77% unsaturated fatty acids (Table 22.1). These authors synthesized rubber seed oil bio-based PU using the alcoholysis method to obtain the monoglyceride showed in Fig. 22.5. Aliphatic (hexamethylene diisocyanate), aromatic (toluene diisocyanate) and molar ratios of 1.02 and 1.5 NCO/OH were applied, inducing variation in properties. For example, the cross density of PU differs in function depending on aliphatic or aromatic isocyanate molecules were used, it is higher for the aliphatic compounds at 1.5 NCO/OH ratio. This was corroborated in the stress-strain curves where aromatic isocyanate PU exhibits brittle fracture and very low elongation at break compared with the aliphatic isocyanate PU, where the mechanical properties trend to be more flexible. This behave is also attributable to the functionality of the hydroxylated oils (Fig. 22.6) [31]. Yücedag et al. found similar mechanical properties using linseed oil. The aliphatic or aromatic isocyanate structure causes the flexible or stiffness behavior, respectively, and the thermal stability is a consequence of the cross-linking density, i.e., the higher the cross-linking density, the higher glass transition temperature (T_g) in the PU. In this case, Yücedag et al.

reported that not only thermal and mechanical properties were enhanced, but also that the degradability and hydrophilicity can be modified. These characteristics are related to the use of an oil-based hydroxyl-containing material due to the effect on ester bonds in the polymer chain. Breaking of ester bonds results in the formation of hydroxyl groups on the film's surface; hence, hydrophilicity of the product increases. The synthesized material was non-cytotoxic, which is useful in the medical field [2].

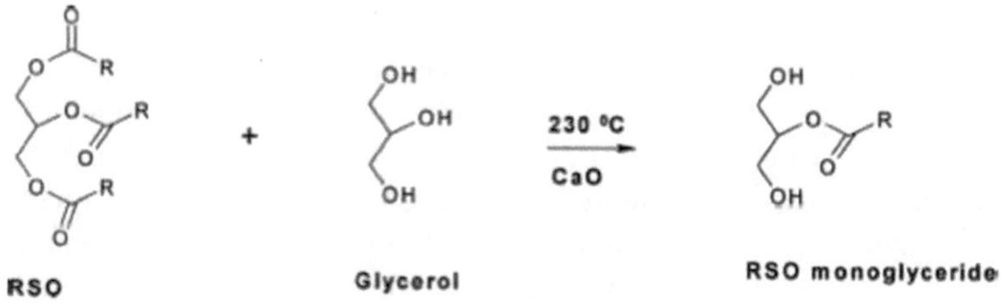

Figure 22.5 General synthesis of the rubber seed oil monoglyceride (where R represents the fatty acyl chain) (reprinted with permission of Bakare et al., 2008, John Wiley and Sons [31]).

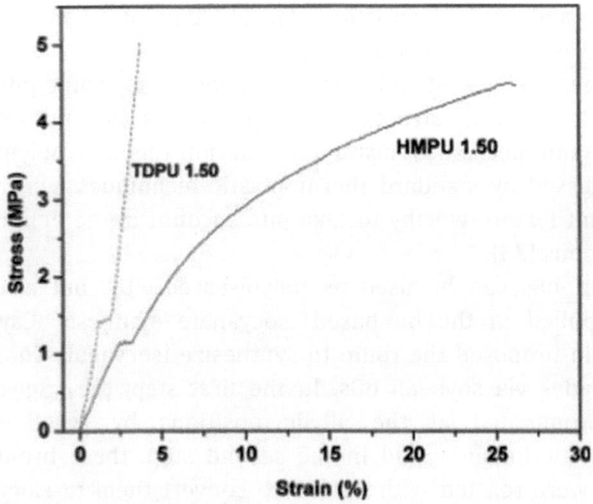

Figure 22.6 Typical stress-strain curve for the rubber-seed-oil-based PU at 25°C. TDPU 1.5 (toluene diisocyanates-polyurethane) and HMPU 1.5 (hexamethylene diisocyanates-polyurethane) (reprinted with permission of Bakare et al., 2008, John Wiley and Sons [31]).

The understanding of the mechanisms involved in the polymer synthesis with natural precursors is definitively a key factor for their appropriate exploitation. Taking into account this need, Ronda et al. explained recently different pathways to modify natural resources. These authors proposed three routes to modify vegetable oils to transform them into polymers: (1) direct polymerization (cationic, radical, or thermal polymerization); (2) functionalization and polymerization; and (3) monomer synthesized, chemical modification and polymerization [32].

The increasing requirement for renewable, sustainable and potentially biodegradable polymers has spurred significant efforts in developing new materials based on oleo-chemicals moieties. Recently, Petrovic et al. published their work about the structure and properties of segmented PUs based on modified oleic acid, diphenylmethane diisocyanate and butane diol [33]. This polymer has soft segments constituted by polyesters from hydroformylated oleic acid, which possibly impart certain biodegradability. This is a thermoplastic PU elastomer, with segmented block copolymers, whose properties are related with the composition and morphology, this last distinguished by the size and shape of the structural features and their interaction. Morphology of segmented PUs depends on their structure and sample preparation conditions; thus, these kinds of PUs are considered as non-equilibrium systems and their structure and properties depend on their thermal and mechanical history. In addition this new polymer can be processed by standard thermoplastic techniques, which is an important feature worthy to take into account in the definition of applications [33].

Plant oils can be used as polyol-based PUs, but also have been applied in the bio-based isocyanate synthesis. Cayli and Küsefoğlu proposed the route to synthesize isocyanate-containing triglycerides via soybean oils. In the first step, the triglycerides were brominated at the allylic positions by reaction with N-bromosuccinamide, and in the second step, these brominated species were reacted with AgNCO to convert them to isocyanate-containing triglycerides. At the end of the reaction, approximately 60–70% of the bromine was replaced by NCO groups, and the double bonds of the triglycerides were not consumed. When the amount of AgNCO was increased, the yield also increased. The reaction was carried out on soybean oil triglycerides, as shown in Fig. 22.7. Once obtained, the soybean oil isocyanate (SONCO) was mixed with castor-oil-based polyol, as well as with glycerine and triethylene tetramine for comparison. Synthesized bio-based PUs under the conditions mentioned above, present low mechanical strengths and high elongations because of the monomers only have aliphatic and soft segments, in addition they showed a high swelling ratio. Thermal stability of aliphatic urethanes is similar to those synthesized in the study; near to 250°C, but when both substituents on the urethane are aryl groups, the thermal stability drops to around 120°C [34].

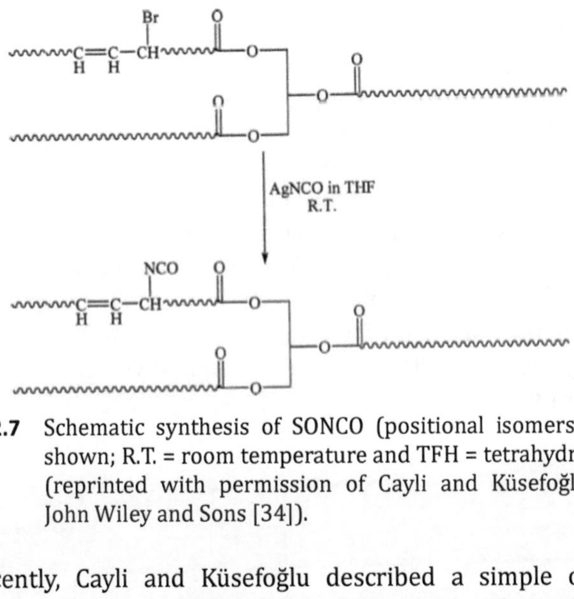

Figure 22.7 Schematic synthesis of SONCO (positional isomers are not shown; R.T. = room temperature and TFH = tetrahydrofurane) (reprinted with permission of Cayli and Küsefoğlu, 2008, John Wiley and Sons [34]).

Recently, Cayli and Küsefoğlu described a simple one-step route to synthesis a soybean oil based isocyanate. The soybean oil triglyceride double bonds reacted with iodine isocyanate to give soybean oil iodo isocyanate (ISONCO), which is similar to typical pseudo-halogens. The isocyanate group attaches the alkyl or aryl substituted olefinic bonds. Addition reactions of iodine isocyanate to alkenes occur in a stereospecific *trans* manner and it was suggested that an iodonium ion is involved as intermediate (Fig. 22.8). To obtain the PU, iodine isocyanate was mixed with the polyol (castor oil, glycerol, or 1,5-pentanediol). This system has the disadvantage that aliphatic isocyanates are less reactive than aromatic. Thus, synthesis of PUs requires longer reaction times of around 24 h for completion. Due to the nature of the monomers used for polyurethane/polyurea synthesis, good mechanical and thermal properties were not expected. The authors proposed some reasons to explain this behavior: (a) structure of the plant oil triglycerides, i.e., the long alkyl chains of the triglycerides may act as internal lubricants and thus cause a decrease in the mechanical strength of the material; (b) position of the isocyanate group in the middle of the fatty acid moiety, which would lower the reactivity of the isocyanate group; and (c) random distribution of the fatty acids in the triglyceride might vary, according to the soybean oil composition, i.e., some oil plants molecules are rich in saturated fatty acids, thus reducing the connectivity in the sample and acting as a lubricant. The swelling ratio was improved with the castor-oil-based monomer, the one with the highest values, as well as thermal stability was superior compared with the soybean oil isocyanate monomer (SONCO), reaching a degradation temperature of 206°C (Fig. 22.9). Iodine isocyanate derivative of plant oil triglycerides promises to be suitable for a fast reaction injection molding process [35]. In turn, More et al. (2013) transformed fatty acid derivatives into diisocyanates via a non-phosgene green method. These bio-based diisocyanates react with fatty-acid-based diols to obtain partially and fully bio-based PU, resulting in a semi-crystalline/amorphous behavior [36].

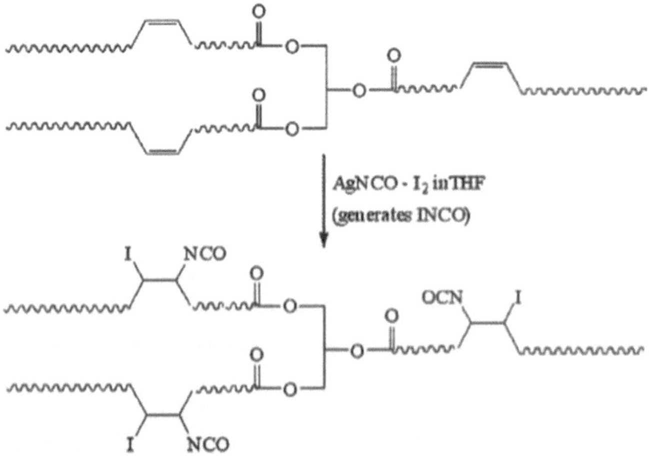

Figure 22.8 Schematic reaction for addition of iodine isocyanate to plant oil triglycerides (reprinted with permission of Cayli and Küsefoğlu, 2010, John Wiley and Sons [35]).

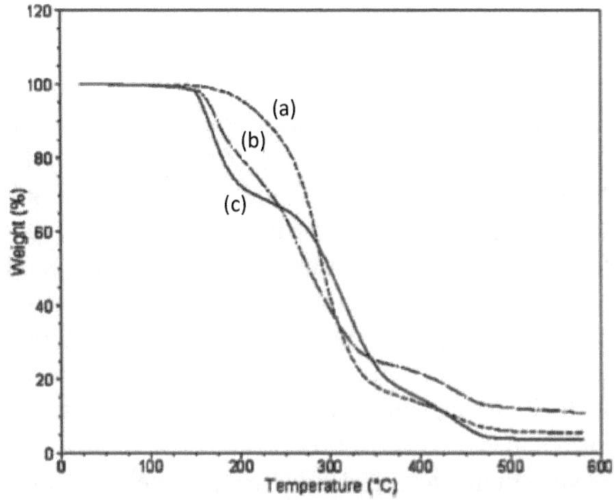

Figure 22.9 Thermogravimetric (TGA) curves of PU of ISONCO with (a) castor oil, (b) glycerol, and (c) 1,5-pentanediol (reprinted with permission of Cayli and Küsefoğlu, 2010, John Wiley and Sons [35]).

22.2.2 Polysaccharides

Polysaccharides are also a diverse group of natural polymers available in quantities commensurate with their envisaged use in the preparation of PUs. At first sight, these hydroxyl-containing biopolymers would seem to be eminently suitable; however, their crystalline nature, high polarity, lability to heat, acidity or alkalinity, their tendency to contain a high proportion of water and their ability to reduce end-groups to react with isocyanates tends to limit the use of native polysaccharides to reactive fillers. Nevertheless, unmodified polysaccharides have some advantages in PU foams such as increased high-temperature stability or increased hardness, for applications such as artificial beds, and biodegradability may be attained by the incorporation of starch. Consequently, for a wide range of polymer end-uses, modification of polysaccharides is required to obtain suitable polyols for incorporation into the PU. For the modification of polysaccharides, several approaches are possible so that reduction in polarity of the molecules occurs, hence decreasing their melting point and viscosity, and increasing miscibility and reactivity with relatively non-polar isocyanates. Reactions at the anomeric positions of the polysaccharide can also lead to an increase in the heat and alkali stability arising from the removal of the aldehydic or potential aldehydic groups via the formation of various glycosides. Some generalized routes can be described as follows: (a) depolymerization of the polysaccharide to monomers or oligomers followed by modification of the anomeric site, and a further optional stage of modification of the remaining hydroxyl groups may take place; (b) simultaneous depolymerization and anomeric-site modification (further reactions at the other hydroxyl may be required); (c) derivatization of the polysaccharide followed by depolymerization; and (d) conversion of the polysaccharide (with loss of the carbohydrate structure) into molecules such as diols that remain suitable for incorporation into PUs [17].

Cellulosic material derived from trees, wood wastes, agricultural residues and straw, offers another high tonnage source of carbohydrates. These materials contain cellulose, hemicellulose and lignin. The cellulose yields glucose while hemicellulose yields pentoses, especially xylose, when subjected to controlled acid hydrolysis in either a batch or continuous process. Sivakumar et al. blended cellulose acetate and PU to produce a protein separation ultrafiltration membrane that showed better protein flux compared to pure cellulose acetate membranes, this performance is attributed to porous formation in blended polymer membrane [37]. This group studied not only protein separation, but also metal [Cu(II), Ni(II), Zn(II), and Cd(II) at 1000 ppm] ion rejection, by adding an additive, polyvinylpyrrolidone (PVP), which was used to open the network structure. Increased additive concentration increased the pore density and pore size of the membrane, leading to higher flux. These novel membranes showed better results for the separation of proteins and metal ions by optimizing the hydrophilic nature and permeability/permselectivity balance [38]. Zavastin et al. also studied acetate cellulose–polyurethane (CA-PU) blends, and reported that the blend membrane surface is characterized by the existence of microcavities randomly distributed on the membrane surface. The pores had an unshaped geometry with dimensions varying from 0.5 to 1.1 μm (Fig. 22.10). Thermal stability of CA-PU blend membranes can be seen in Table 22.2, in which the PU and cellulose acetate material behaviors are compared to CA-PU performance. The results show entirely dissimilar thermal behavior. CA-PU blend membranes possess lower thermal stability, with their degradation process divided in three steps: (a) T_{onset}, temperature at which thermal decomposition starts; (b) T_{peak}, temperature at which maximal degradation is registered; and (c) T_{endset}, temperature at which the investigated process ends. Therefore, the blend membrane has characteristics that are different from those of the initial pure components from which it was obtained. According to these results, the blend membrane is stable up to 57°C. Besides these properties, this membrane is able to remove azo dye complexed with iron (III) ions, indicating that it is suitable for wastewater treatment in the textile industry [8]. Other researchers have focused on the use of PU as a support for copolymer coating films, taking advantage of the mechanical properties of this material. Copolymerized PU can be applied in water treatment and medical areas, where the coating formula provides an antimicrobial and biocidal function through the N-halamine moiety and acts on *Staphylococcus aureus* and *Escherichia coli* [39].

Table 22.2 Thermogravimetric characteristics of cellulose acetate–polyurethane blend membranes, cellulose acetate, and polyurethane

Material	TDS[1]	T_{onset} (°C)	T_{peak} (°C)	T_{endset} (°C)	Weight (%)	Residue	DTA[2] characteristic
Blend membrane	I	57	114	135	61.15		Endothermic
	II	261	320	335	21.85	5.63	Exothermic
	III	394	406	500	11.37		Exothermic
Cellulose acetate	I	51	69	89	4.62	9.54	Endothermic
	II	323	361	383	85.84		Exothermic
PU	I	51	65	88	2.05		Endothermic
	II	261	273	326	13.40	17.28	Exothermic
	III	326	357	372	37.49		Exothermic
	IV	372	398	423	29.78		Exothermic

Source: Adapted with permission of Zavastin et al., 2010, Elsevier [8].
[1]TDS: Thermal degradation step.
[2]DTA: Differential thermal analysis.

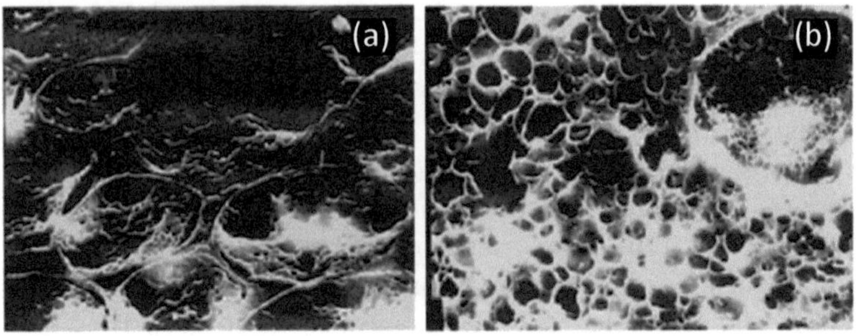

Figure 22.10 SEM micrographs of prepared cellulose acetate–polyurethane blend membranes recorded with magnification × 3000; (a) membrane top surface and (b) lower side (reprinted with permission of Zavastin et al., 2010, Elsevier [8]).

Blends between biopolymers and synthetic polymers are of particular significance because they can combine biocompatibility with good processability and mechanical resistance, and can be used as biomaterials to improve or modify physicochemical properties without altering the structure and function of individual polymers. Cellulose derivatives such as hydroxypropylmethylcellulose (HPMC), hydroxypropylcellulose (HPC) and methylcellulose (MC) are widely used excipients in pharmaceutical formulations, with applications in film coatings and as binders for solid dosage forms of pills or tablets. These components are cellulose ethers produced by partial or total etherification of the three hydroxyl groups present on the anhydroglucose repeating unit of the cellulose chain. Intra- and intermolecular interactions can occur between either unsubstituted hydroxyl or ether-oxygen, and the hydroxyl groups introduced by the incoming groups. Taking advantage of these possible reaction mechanisms PUs can be tailor-made by appropriate selection of variable constituents such as soft and hard segments and by using an appropriate amount of cross-linker. An interesting blend composed by a new polyester urethane with HPC was synthesized by Raschip et al. The goals of this work were to study the compatibility of this blend and to evaluate how the quantity of polysaccharide affected morphology and thermal properties of final PU. In addition the cytotoxicity and hemocompatibility of this material were studied. The results showed the existence of a biphasic system composed of a dispersed phase in a continuous matrix. The films containing mostly HPC or mostly PU were homogeneous and transparent, while, for the intermediate composition range, morphology with a fine dispersion in a continuous matrix was characteristic (Fig. 22.11). The blends obtained were not cytotoxic and exhibited good surface properties as well as hemocompatibility. Regardless of HPC being the most widely used cellulose derivative, its thermal properties are not easy to understand due to the ability of the component to form a complex morphological liquid crystalline order when blended into a PU matrix. The blend presented multiple transitions, some of which are overlapped (e.g., β-relaxation in HPC and glass transition of the soft segments in PU), causing a non-easy interpretation. The increase in the degree of compatibility is due to the creation of interactions between –OH groups from HPC and polar groups of the PU [40]. Another factor to improve biocompatibility in PU, besides the introduction of biomolecule moieties, is the surface modification. It can be done through energy such as plasma treatments, which change surface polarity, hydrophilicity, and adhesiveness. According to Macocinschi et al.,

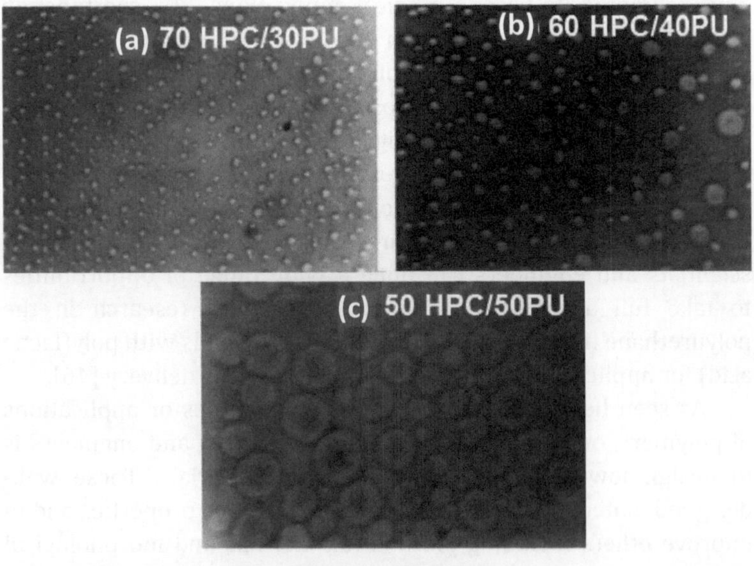

Figure 22.11 Optical microscopy images films: (a) 70 wt% of HPC and 30 wt% of PU; (b) 60 wt% of HPC and 40 wt% of PU; (c) 50 wt% of HPC and 50 wt% of PU (reprinted with permission of Raschip et al., 2009, John Wiley and Sons [40]).

changing the surface properties of poly(ether-urethanes) and HPC copolymer with a high-energy treatment (High frequency cold plasma) increases its compatibility with blood [41]. Nowadays, the use of polysaccharides; such as cellulose, has been successful to improve structure and mechanical properties [42] and shape-memory behavior [43]. Thus, polysaccharides are the center of attention in different research groups, mainly focused in the study of diverse pathways to introduce this type of biopolymer into polymer matrix, such as polyurethane composites [42].

Lactic acid or 2-hydroxy propionic acid is the building block of poly(lactic acid) (PLA), an important biodegradable polymer that belongs to the family of aliphatic polyesters derived from α-hydroxy acids. The chemical synthesis of PLA involves the processing and polymerization of lactic acid monomer through two possible pathways: (a) direct condensation (using solvents under high vacuum), or (b) formation of lactide, the cyclic dimer intermediate (none solvent is required). PLA has several advantages due to its thermoplastic processability, biocompatibility and biodegradability, which in addition to its mechanical properties have promoted its uses in medical applications. PLA can function as an outstanding matrix with degradation behavior, biological performance and suitable morphology, which could be useful for specific applications, such as prosthetic devices, implants, vascular grafts, catheters, sutures and ligament repair materials. Besides, this biocompatible polymer provides surfaces that promote the regeneration and reconstruction of human organs. Therefore multidisciplinary researches involving cell biologist, materials scientists and engineers can open a wide range of opportunities to take full advantage of this polymer. Also, research in the polyurethane field is focused on finding new blends with poly(lactic acid) for applications in medicine [44, 45] or drug delivery [46].

As seen before, in order to modify properties or applications of polymers, one main goal of materials scientist and engineers is to design low-cost and stable synthetic materials. These well-designed materials are intended to mimic specific properties and to improve others including mechanical, thermal, and morphological properties, as well as biodegradability, biocompatibility, and shape recovery. Wang et al. studied the synthesis of polylactide polyurethane materials by changing the molecular weight and determining their relationship with thermal and mechanical properties. The chemical system is showed in Fig. 22.12. These authors found that two thermal phenomena were present. One was the constraint received by the poly(L-lactide) segments from the hard segments, which leads to lower T_m and T_g of the poly(L-lactide) domains. The other was the phase separation, the degree of which determines the interactions between the soft and hard segments and may cause either an increase or a decrease in the T_g and T_m of the poly(L-lactide) phase, depending on the hard-segment content. Therefore, it is possible to adjust the T_g of the poly(L-lactide) phase by changing the poly(L-lactide) polyurethane composition. All this also has an effect on the mechanical properties, observed by the increase in tensile strength and Young's modulus with increasing molecular weight of the poly(L-lactide) diol or the hard-segment content; however, the elongation at break has the opposite trend [47]. Previously, Melnig et al. observed similar phenomena when working with lactate, an acid lactic derivate, to develop thermoplastic elastomers. The presence of lactate segments into the polyurethane caused an amorphous microphase organization. The surface ratio of (C=O)ester/(C=O)urethane peaks was probed by Attenuated Total Reflectance (ATR) and the values, suggested that the microphase organization of lactate-urethanes is more amorphous than polyurethane. Inside lactate-urethanes, the crystalline phases due to urethane groups are separated from the amorphous phases, including ester groups, and form rigid structures connected by urethane-urethane bonds. This behavior was also observed in thermal analysis, mainly in two of four common transitions in polyurethane (T_δ, T_γ, T_β, T_α) associated with a relaxation process and the glass and melting transition in soft and hard segments. T_γ and T_β are very much affected by the lactate segments, and both are related to the presence of an ester group and urethane-urethane-ester moiety. T_γ is −2.8 and −49.8°C for polyurethane and lactate-urethane respectively, whereas for the same segments T_β is 25.1 and 10.2°C. ATR-FTIR and thermal analyses indicate a strong hydrogen bonding between urethane N-H and C=O groups from polyurethane and lactate. This linkage provides compatibility between these two polymers and makes them more suitable for soft-tissue replacement and for drug delivery systems [48].

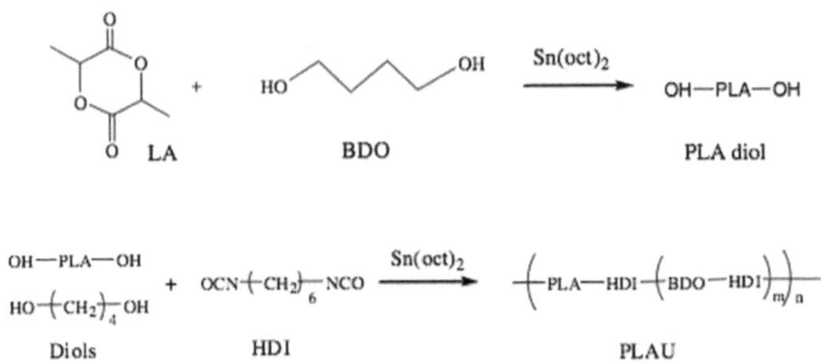

Figure 22.12 Schematic diagram of synthesis of poly(L-lactide) polyurethane; LA [L-lactide], BDO [1,4-butanediol], PLA diol [poly(L-lactide)], HDI [hexamethylene diisocyanate], PLAU [poly(L-lactide) polyurethane] (reprinted with permission of Wang et al., 2006, Elsevier [47]).

Other approach to modify polymer properties is the synthesis of cross-linking networks, starting on reactions to produce prepolymer structures. Subsequently, these can be cross-linked in situ to form both rigid and elastomeric polymers. Adhikari et al. applied this technique to synthesize biodegradable polyurethane: they developed two-part injectable prepolymer systems consisting of lactic and glycolic acid-based star-polyester polyols, pentaerythrol and ethyl lysine diisocyanates. This formulation allowed them to vary properties for a suitable application (i.e., orthopedic). The chemical structure of the polyol influenced the mechanical properties and DL-lactic-acid-based polyols yielded polymers with high compressive strength. The polymer networks formed were amorphous and exhibited glass transitions in the range 80–90°C. Also it was found that hydrophobicity/hydrophilicity plays a key role in copolymerization [49].

The major advantage in preparation of PUs is that they can be synthesized from various types of polymers that contain different mechanical and physical properties. In particular, poly(ester-urethane) elastomers are interesting because they can be designed synthetically to have hydroxyl functional groups at their chain terminals. These react readily with diisocyanate groups to form urethane linkages. In the same sense, chemical modification of the involved monomers allows incorporation of some functionality into a PU matrix, thus improving a specific property. An example of chemical modification of monomers was developed with chitosan. This polysaccharide is product of the alkaline deacetylation of chitin, is a biodegradable, biocompatible and inexpensive natural polymer, comprising of β-(1,4)-2-amino-2-deoxy-D-glucopyranose units. However, its limited solubility in water and common organic solvents has inhibited its study and utilization. The carboxymethylation reaction of chitosan with monochloroacetic acid and alkali solutions results in a water-soluble chitin derivative, carboxymethylchitosan (CMC). This modified chitosan was used by Kadnaim et al., who reported that PU chains interpenetrated into the carboxymethylchitosan network with 1,6-hexamethylene diisocyanate as cross-linker (Fig. 22.13). The resulting product is a hydrogel, which presented microphase separation, as is normal in this kind of hybrid natural-based polyurethane. The microphase separation is attributed to the hydrophobic characteristics of PU dispersed into the hydrophilic CMC continuous phase, all of which is related to the ratio of the phases. The presence of hydroxyl, amine, and carboxylic acid functional groups causes physical cross-linking throughout the material. PU incorporation into CMC resulted in an enhancement of tensile properties due to the presence of polyester soft segments in PU structure, as well as increases the percentage of elongation, indicating the enhancement of toughness properties of the materials [6]. On the other hand, Barikani et al. used chitin without any modification; thus, they took advantage of the trifunctionality of this molecule to enhance the thermal and mechanical properties. The sample with a high content of chitin implied that its cross-linked density was high in the molecular structure. An increase in chitin content results in a decrease in both melting temperature and melting enthalpy of the soft segment, which is probably because cross-linking increases the miscibility of the soft and hard segments. This miscibility modification could be result of the reducing in the crystallization of the soft segment, i.e., the more chitin is used; the less crystallization enthalpy is needed. At the same time, it was found that if chitin content is higher, the storage modulus in the material is also higher, and this behavior is attributable to chitin's composition of N-acetyl-D-glucosamine monomers (Fig. 22.14). The linked glucosamine rings on chitin had better miscibility with linked double rings in the hard segment and the NH group in the hard segments. In addition, chitin does not melt but degrades at elevated temperature and this can play a role in the enhancement of thermal properties [4]. One of the advantages of using chitosan and its derivatives is their biocompatibility, biodegradability, antibacterial properties, and non-toxicity, as well as the presence of functional bond-linking to interact with other polymers. These characteristics make this biopolymer versatile for development of polymeric materials with the objective of improving thermal, mechanical, and adsorbent properties, and to possess compatibility with organs in the medical field, for food packaging, or to act as smart materials [50].

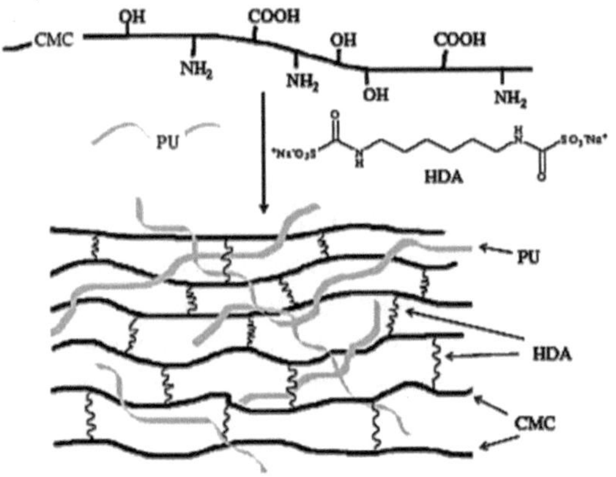

Figure 22.13 Preparation of CMC hydrogels modified with poly(ester-urethane) (reprinted with permission of Kadnaim et al., 2008, Elsevier [6]).

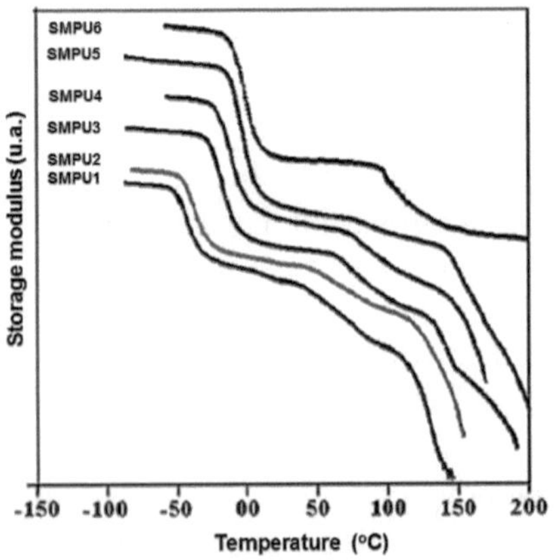

Figure 22.14 Storage modulus vs. temperature curves of PU samples. SMPU1, SMPU2, and SMPU3 (chitin content = 0 mol), SMPU4 (chitin content = 1 mol), SMPU5 (chitin content = 2 mol), and SMPU6 (chitin content = 3 mol) (reprinted with permission of Barikani et al., 2008, Elsevier [4]).

Awareness of ecological problems has led to a paradigm shift in the use of biodegradable materials, especially from renewable agriculture feedstock and marine food-processing industry wastes. Consequently, natural polymers and their derivatives have attracted considerable attention. Nowadays, polymer synthesis from renewable and biomass resources are essential from the viewpoint of being carbon neutral. Thus, various naturally occurring compounds have been frequently adopted as starting blocks for biodegradable and/or biomedical polymeric materials. For example, several kinds of polyurethanes have been synthesized from saccharides and uronic lactones to produce polyamides, but also bio-based polyurethanes from D-glucose-derived diols and from aldose derivative have been developed, as can be seen in Fig. 22.15. Polyurethanes containing dilactone or lactone rings can be hydrolyzed in a neutral phosphate buffer solution to the corresponding saccharide derivatives, diamines, and carbon dioxide at 27°C. The ring opening of lactone groups under these conditions is thought to increase the hydrophilicity of polyurethane chains, which may also enhance the hydrolysis of the urethane bonds in the main chains. In addition, other polyurethanes bear not only lactone rings but also hydroxyl groups in the repeating units, whose have been prepared through the polyaddition of diols derived from L-gulonic acid with diisocyanates, as it is observed in Fig. 22.16. Hydroxyl groups are capable of undergo hydrolysis reactions, as well as lactone rings [51]. In order to form urethane

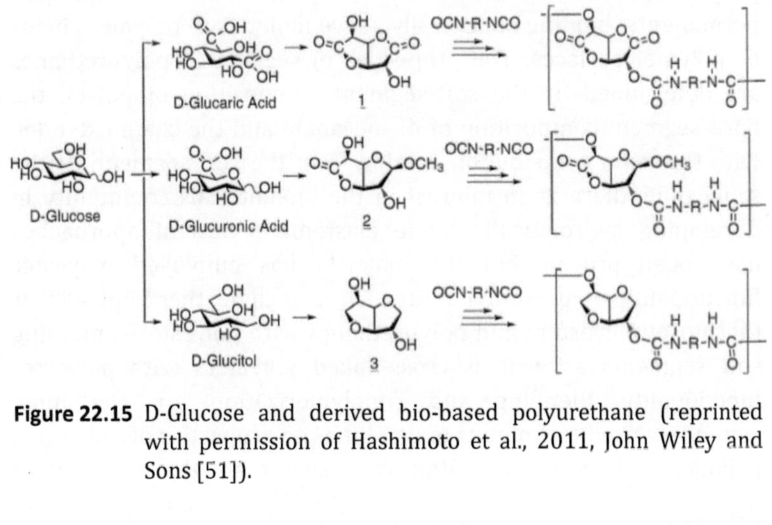

Figure 22.15 D-Glucose and derived bio-based polyurethane (reprinted with permission of Hashimoto et al., 2011, John Wiley and Sons [51]).

moieties, it is needed that saccharides present a specific characteristic, such as chemical cross-linking, which is a feasible method to adapt the structure of natural polymers, e.g., introducing hydroxyl or amino groups at the same time such as glycosylamines and glucosamines [52] or introducing multifunctional hydroxyl groups such as L-arabinitol [53]. Thus, it can be concluded that segmented polyurethanes synthesized from various polyols, diisocyanates, and chain extenders can be structurally manipulated to achieve a wide range of properties.

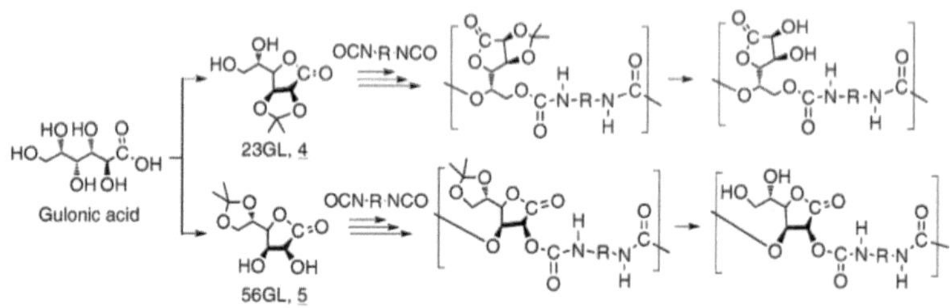

Figure 22.16 Gulonic acid bio-based polyurethane (reprinted with permission of Hashimoto et al., 2011, John Wiley and Sons [51]).

22.2.3 Amine and Amino-Acid Groups

Polyurethane-based biomaterials are used in the fabrication of medical devices and their biocompatibility and material strength properties have to be characterized; hence, studies are focused on permanently binding biologically active moieties to polymer chains or polymer surfaces. The properties of segmented polyurethanes are determined by the soft-segment composition of polyol, the hard-segment composition of diisocyanate and the chain extender, with the main contribution coming from the soft segment. At the same time, there is an interest in the biomaterials community in developing hydrolytically labile elastomers. Several approaches have been proven, but the majority has employed polyester functionality to present lability. These include thermoplastic or thermoset polyesters and polyurethanes with polyester-containing soft segments as well as cross-linked polymers with polyester functionality. Blending and copolymerization are the most commonly used techniques to combine the properties of individual polymers. However, blending is easier than the preparation of copolymers to obtain the advantageous properties of the constituent polymers.

Several researches using lysine diisocyanates have been developed in order to obtain more biodegradable and non-toxic polyurethanes. Lysine-derived polyisocyanates, such as lysine methyl ester diisocyanate (LDI) and lysine triisocyanate (LTI), are synthesized by phosgenation of amine-terminated lysine esters. These materials present the advantage of low vapor pressure relative to aliphatic polyisocyanates (e.g., HDI). Considering the inhalation toxicity of many isocyanates, the low vapor pressure of LDI and LTI offers significant handling and processing advantages. In addition, the lysine-based polyurethanes, upon degradation, generate diamines that are considered naturally occurring substances in the body and therefore they are less harmful to use. Urethane bonds, formed by reaction of hydroxyl functional molecules with lysine-derived molecules, can be degraded in vivo and in vitro by microorganisms producing lysine and acid-terminated urethane fragments. This polyurethane based in amino acid precursors exhibits mechanical properties comparable to commercially available poly(methyl methacrylate) bone cements and due to its natural origin biodegrades at a controlled rate to non-cytotoxic decomposition products [54]. Del Rio et al. compared two diisocyanates by varying their aromatic and aliphatic structure (a lysine derived was used as the aliphatic compound) and using methyl oleate as polyol. Their results showed that T_g values increased as the functionality of the polyol increased and were higher for the aromatic (MDI isocyanate) than for the aliphatic (LDI isocyanate) non-segmented polyurethanes. Moreover, phase segregation induces an amorphous morphology, since the use of non-symmetric methyl ester diisocyanate inhibits hard-segment crystallinity. All polyurethanes are thermally stable over 300°C, and their mechanical properties increase with increasing cross-linking density and hard-segment concentration, as well as the functionality of aromatic or aliphatic structure of diisocyanate [1]. In other study, ethyl lysine diisocyanate (ELDI) was used as a monomer for synthesis of cross-linked polyurethane, enhancing degradability properties due to the bio-based diisocyanate moieties [49].

The use of naturally derived polymer is limited due to insufficient mechanical strength and low compatibility compared to synthetic polymers, which represent a challenge to overcome for the use of these biomaterials. Proteins offer a promising alternative and provide an opportunity for commercial reproducibility. Sharifpoor et al. developed a functional tissue regenerating material by manipulating polar hydrophobic ionic polyurethane with a protein-derived and degradable lysine-based polycarbonate divinyl oligomer (DVO) and 2-hydroxyethyl methacrylate cross-linked moiety to obtain a porous scaffold that had optimal physical properties and was simultaneously cell compatible. The content of lysine-based cross-linker was also studied. Thus, these authors reported an increased elastic modulus (0.5–21 MPa) and decreased elongation-at-yield (45–5%); these were achieved with increasing cross-linker concentration. Scaffolds prepared with higher concentrations of cross-linked moieties possessed higher cross-linked density, greater restriction of polymer chain movement, and ultimately increased polymer stiffness. All this is related to the number of hydrogen bonds present between and within the hard and soft segments [55].

Most of the time, amine and amino-acid derivatives are used as chain extenders, such as L-tyrosine, which offers the potential for use in different biomaterial devices because of its biodegradability, and whose thermal and mechanical properties are controlled by the polyol molecules [56]. Polyurethane shows poor cell–material interactions, and surface modifications can overcome this property in order to get better cytocompatible material, in addition to the introduction of natural molecules like peptides. Gao et al. synthesized a cytocompatible polymer by grafting polymerization of hydrophilic poly(methacrylic acid) onto a PU membrane surface by UV irradiation and the further immobilization of a combination of bifunctional peptides (arginine-glycine-aspartic). This last was achieved using a water-soluble carbodiimide as condensing agent that accelerates the reaction between the COOH groups of poly(methacrylic acid) and NH_2 groups of amino acids [57]. Guang and Wagner reported a triblock alanine–alanine–lysine chain extender used to form a copolymer with polycaprolactone-*b*-poly(ethylene glycol)-*b*-polycaprolactone as the soft segment and 1,4-butanediisocyanate as the hard segment. This PU copolymer displayed a high molecular weight, low glass-transition temperature (≤54°C), and tensile strength of 15–28 MPa. In order to enhance cell adhesion, the PU surface was modified with radio frequency glow discharge followed by coupling of arginine–glycine–asparagine–serine, showing an increase in cell adhesion from 140% to >200% [58].

In the other hand, the layer-by-layer assembly of charged species can be used for surface modification according to specific requirements. In this technique an adjustment of introductory molecules is required before the assembly. This technique was applied by Zhu and Sun to synthesize a polyelectrolyte charged polyurethane membrane, where the modified amino groups come from collagen type I. They found that by varying the order in the layer-by-layer assembly, the cell adhesion could be improved. If a negative charge was at the surface, the endothelial cell adhesion phenomena did not take place due to repulsive interactions, but when the surface charge is reversed by one layer of collagen, cell adhesion occurred [59].

In spite of knowing the outstanding properties of amine, imide, and amide groups from natural-source-based polyurethanes, numerous researchers continue focused on introducing non-natural molecules to synthesize polyurethanes in order to enhance a specific property such as thermal and mechanical properties, or compatibility and degradability. Among this kind of polyurethane moieties, the following have been reported: (a) tertiary amine (2-diethylaminoethylamine) for a low-toxicity gene transfer vector [60]; (b) hydrophilic vinylpyrrolidone-dimethylacrylamide, which is thermodynamically compatible with polyurethane, able to interact with aromatic amino acids and serves to adsorb proteins [61]; (c) *N,N*,-dimethylformamide; which contains a large number of amine-terminated functional groups that allow synthesis of hyperbranched polyamides, as shown in Fig. 22.17 [62, 13]; (d) 5-maleimidoisophtalic diisocyanate mixed with dibenzyl diisocyanate, these increase the cross-linking density and improves mechanical properties [63].

In the future, development and synthesis of naturally based polyurethane will be largely driven by three factors: the need to derive more carbon for chemical process from renewable substances instead of oil reserves, to develop cleaner chemical processes, and to avoid disturbing the ecosystems, an important benefit of renewable feedstocks compared with petroleum.

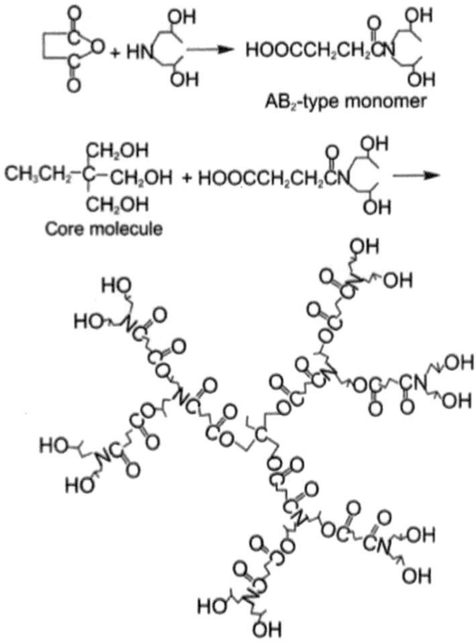

Figure 22.17 Synthetic process of hyperbranched polyesteramides (core molecule, trimethylolpropane) (reprinted with permission of Lin et al., 2011, John Wiley and Sons [13]).

22.3 Polyurethane Membranes

A membrane is an interphase between two adjacent phases acting as a selective barrier, regulating the transport of substances between the two compartments. The main advantages of membrane technology compared with other unit operations are related to this unique separation principle. Separations with membranes do not require additives, and they can be performed isothermally at low temperature and—compared to other thermal separation processes—with low energy consumption. Also, upscaling and downscaling of membrane processes, as well as their integration into other separation or reaction processes, is easy. In industrially established applications, some of the state-of-art synthetic membranes have a better overall performance than their biological counterpart. However, relatively few of the many possible separation principles and processes have yet been fully explored. Consequently, a strong motivation for improving established membrane materials and processes is driving the current research. The membrane process conditions must be engineered very carefully, but the performance limits are clearly determined by the membrane itself. This will be briefly explained by giving an overview of the main membrane processes and separation mechanisms. Current polymeric membranes are made with biopolymers (mainly cellulose derivatives) or with synthetic engineering polymers that were originally developed for different purposes. The development of synthetic membranes has always been inspired by the fact that the selective transport through biological membranes enables highly specialized macromolecular and supramolecular assemblies based on and involved in molecular recognition. Passive transport through membranes occurs as a consequence of a driving force, i.e., a difference in chemical potential by a gradient across the membrane in, for example, concentration or pressure, or by an electrical field. The barrier structure of membranes can be classified according to their porous character: non-porous, microporous with pore diameter (d_p) around 2 nm, mesoporous (d_p = 2–50 nm), and macroporous (d_p = 50–500 nm). Active development is also concerned with the combination of non-porous or porous membranes with additional separation mechanisms, and the most important ones are electrochemical potentials and affinity interactions [64].

Membrane technology has been applied to a number of separation processes, each using a membrane but differing in the details of the method. Membranes suited for technical applications may be classified by the following characteristics [64, 65]:

(1) *Membrane materials.* Organic polymers, inorganic materials (oxides, ceramics, metals), mixed matrix, or composite materials.
(2) *Membrane cross section.* Isotropic (symmetric), integrally anisotropic (asymmetric), bi- or multilayer, thin-layer, or mixed matrix composite.
(3) *Preparation method.* Phase separation (phase inversion) of polymer, sol-gel process, interface reaction, stretching, extrusion, track-etching, and micro-fabrication.
(4) *Membrane shape.* Flat sheet, hollow fiber, and hollow capsule.

Alternatively, membranes may also be divided into artificial and natural membranes based on their origin. An artificial or synthetic membrane is a synthetically created membrane intended for separation purposes in the laboratory or industry. Synthetic membranes have been successfully used for small- and large-scale industrial processes, in artificial organs, and in medical devices. They can be produced from organic materials, such as polymers and liquids, or from inorganic materials. Most of the commercially utilized synthetic membranes in the separation industry and in artificial organs are made of polymeric materials. Polymeric membranes take the form of polymeric interphases that can selectively transfer certain chemical species over others. The membranes have advanced or novel functions in various membrane separation processes for liquid and gaseous mixtures (gas separation, reverse osmosis, pervaporation, nanofiltration, ultrafiltration, and microfiltration) and in other important applications such as biomaterials, catalysis (including fuel-cell systems) or lab-on-a chip technology [65].

In medical applications, membrane technology is of major importance, in particular in a number of lifesaving treatment methods. Membranes are used in drug delivery, artificial organs, tissue regeneration, diagnostic devices, as coatings for medical devices, bioseparation, etc. The largest part of the medical market involves membranes in drug delivery, hemodialysis, artificial organs (oxygenators, pancreas, etc.), and tissue engineering. In all cases, biocompatible and, in some applications, biodegradable materials are required for the membrane fabrications. In this context some important considerations must be emphasized:

(a) Biomaterials can be defined as substances in therapeutic or diagnostic systems that are in contact with biological fluids. Therefore, biomaterial-based membranes that are in contact with biological fluids should prevent any type of infection and any immune response, or other biological response that could affect the properties of the fluid and, consequently, the patient. For this reason, it is important to know both the host and material response for a certain biomaterial. The host response is usually related to inflammation, fibrosis, coagulation, and hemolysis. The material response focuses on fracture, wear, corrosion, dissolution, swelling, and leaching.

(b) Biocompatibility has been defined as the ability of a material to perform its desired function with respect to a medical therapy, without eliciting any undesirable local or systemic effects in the recipient or beneficiary of that therapy, but generating the most appropriate beneficial cellular or tissue response in that specific situation, and optimizing the clinically relevant performance of that therapy. Additionally, cytotoxicity, genotoxicity, mutagenicity, carcinogenicity, and immunogenicity are considered as key components which constitute biocompatibility. Due to their inherent features, not all biomaterials have the same degree of biocompatibility. Often, surface properties have to be modified in order to enhance their interactions with the host or biological fluid and to suppress the immune response. Biomaterial surfaces can be modified either physically by methods such as plasma etching, corona discharge, and UV radiation, or by covalent attachment. For the latter, chemical grafting, photo-grafting, plasma polymerization, grafting with ionizing radiation, monolayer formation by self-assembly, or biological modification are some strategies that can be used to control the host response and to increase biocompatibility of the membrane surface.

(c) Degradability (erosion) can be a desirable property for materials that are used as implants and/or in tissue regeneration. Degradation is directly linked to drug release, and if a polymer degrades very fast, an elevated drug concentration is released to the patient, which can be disadvantageous or even fatal. Dissolution of polymers and ceramics is a more frequent phenomenon. In the case of polymers, the dissolution varies depending on the nature of the polymer (hydrophilic/hydrophobic). Hydrophobic polymers, for example, dissolve preferentially in the amorphous regions, which results in an increased surface area, integrity loss, and release of small particles. Bioresorbable polymers are designed to degrade within the body and be absorbed naturally when its function has been accomplished. These degradation characteristics differ from polymer to polymer, and can vary from swelling to dissolution by hydrolysis, for example, when exposed to bodily fluids. Bioresorbable materials degrade in products that are normal metabolites of the body. Some examples of degradable polymers are polylactide, polyglycolide, polycaprolactone, and polyhyaluronic acid esters, but also natural polymers such as collagen and chitosan [66].

The degradability of poly(ester-urethane) has been assessed and considered an advantageous property, especially for biomedical applications. At the same time, polymers derived from renewable resources, also known as green polymers, are gaining increasing attention due to the escalating environmental problems with polymer waste management. Poly(ester-urethane) is degradable by several mechanisms of which hydrolytic chain scission is one of the most important. Esther groups are more hydrolytically susceptible than urethane linkages, suggesting that biodegradation of poly(ester-urethane)s proceeds first via the hydrolysis of the ester linkage and second via the assimilation of polyester fragments by microorganisms. The urethane segments are believed to not be bioassimilated at all. Hence, the rate of biodegradation can be tuned by modifying the length and structure of the ester segments [67]. Nowadays, advances in membrane technology are even more competitive than traditional, high energetic consumption, environmentally undesirable, and costly membrane-based separation and purification processes. Membrane technology has become an essential part of human life because of its growing industrial applications in hi-tech areas such as biotechnology, nanotechnology, and membrane-based energy devices in addition to diverse uses based in separation and purification process [68].

Contamination of air, soil, and surface water with organic and inorganic (e.g., metals) pollutants is recognized as a problem of growing importance in recent years. However, many traditional cleaning techniques (chemical precipitation, ion exchange, biological remediation, electrochemical treatments) are highly costly processes. Hence, membranes have gained recognition as a remarkable technology at low energy consumption and with easy management. In spite of their advantages, membranes have not been fully exploited, e.g., according to Smitha et al., in the separation process of organic-organic mixtures by pervaporation techniques, the use of polyurethane membranes is still limited and few authors are studying this kind of material [69]. An important trend in this field is the research of polyurethane membranes for water treatment because of the ease in forming a porous barrier that is mechanically resistant. Various authors have developed systems based on polyurethane membranes and have applied them successfully in organic and metal substances separation processes. In these researches, the authors have synthesized a variety of polyurethane types by changing the ratio and type of the basic monomers; polyol and isocyanate [9, 70–82].

As a consequence of the outstanding properties of polyurethane as a porous material and therefore for synthesis of membranes, research continues focused not only on environmental fields but also in diverse areas such as medical devices or adhesives. Therefore, there is a potential area to develop polyurethane-separation materials as membranes. The use of naturally based polyurethane monomers during synthesis reactions (polyol, isocyanates, and chain extenders), are also key factors to explore in future to control morphological, mechanical, and thermal desirable properties. In addition the modification of polyurethane with potential natural absorbent biopolymers is an emergent research line that has demonstrated successful results [9, 81, 82].

22.3.1 A New Tendency in the Environmental Industry: Polyurethane Hybrid Membranes

One of the very promising and rapidly developing novel industrial applications of polymers is in the field of polymeric membrane separation for the production of pure chemicals and clean effluents. Currently, the most extensively employed purification procedure is adsorption of organic pollutants on active carbon although it is neither very effective nor economical. Industrial wastewater purification operations are a combination of expensive and time-consuming physical, chemical, and/or biological degradation procedures. However, membrane separation is a type of separation process that may replace the majority of conventional separation systems. The possibility of separations with various synthetic polymeric membranes is highly appealing because no conventional single-step purification treatment offers the same potential and versatility as those of synthetic membranes. The membranes used in such operations function as barriers separating two phases and as sorbing media. These membranes are permeable and often selective toward a particular species present in one of the phases. The separation is governed by both the chemical and the physical natures of the membrane material. Separation occurs because of the differences in shape, size, chemical properties, or charge of the species to be separated and is driven by a pressure or a concentration difference. In the case of polyurethane membranes, these are copolymers that due to the immiscibility of both soft and hard segments can be considered as a heterogeneous matrix. A suggested mechanism by which organic compounds are removed from solution by the polyurethane membranes can be divided into three separate phenomena: (a) the transfer of the species from bulk solution into the solution-membrane interface, (b) adsorption/desorption from the membrane surface, and (c) transport into the bulk of the polymer [80]. Taking into account this mechanism, the morphological characteristics and suitable adsorption sites are one of the main subjects of interest in the developing of novel polyurethane membranes. With this last purpose, Saucedo-Rivalcoba et al. studied the synthesis of a hybrid polyurethane membrane made of chicken-feather keratin, commercial toluene diisocyanate (TDI) and poly(propilenglycol) [81]. Keratin is a fibrillar protein composed of 18 amino acids that are the responsible for the protein's functionality and could be considered as reactive sites including their OH, C=O, NH, S–S, and S–H moieties. These chemical groups are susceptible to be modified through functional reactions to increase their natural affinity toward metals, organic compounds, and other materials. Polyurethane offers the advantage that it can be synthesized from a wide variety of polyols, isocyanates, and chain extenders, whether synthetic or naturally based monomers. Thus, the incorporation of keratin, which is able to attract metal or organic compounds, onto a polyurethane matrix could improve the functional adsorption in a separation membrane. Keratin was grafted in two diverse ways: first it was dissolved in a high urea solution and, second, this solution was dialyzed in order to eliminate the urea salt. Both the salt (urea-keratin) and dialyzed solutions were introduced onto a polyurethane matrix. The results revealed that the cell and porous opening are influenced by the quantity and chemical modification of denatured protein. When urea is present in the solution, the morphology of the membranes is stable and flexible, but when the dialyzed solution is added, the structure tends to open, thus breaking and becoming more rigid (Fig. 22.18). This was corroborated with the differential scanning calorimetry (DSC), showed in Figs. 22.19 and 22.20. The DSC curves display a decrease in thermal transitions because of the presence of protein and urea salt. This behavior demonstrates not only that urea linkages exist in the polyurethane matrix, but also that there are other interactions of amino and carboxylic bonds from amino acids of proteins with the synthetic polymer, in addition it is demonstrated that urea is a key moiety in the linking process. Besides, the synthetic polyurethane can be joined to a natural protein due to there is a structural similarity between peptide and urethane functional groups, and these reactive sites are able to interact with inorganic (metals) or organic molecules. This is so useful, since the use of these membranes, such as removing the toxic residues or pollutants from water, is widened [81, 82].

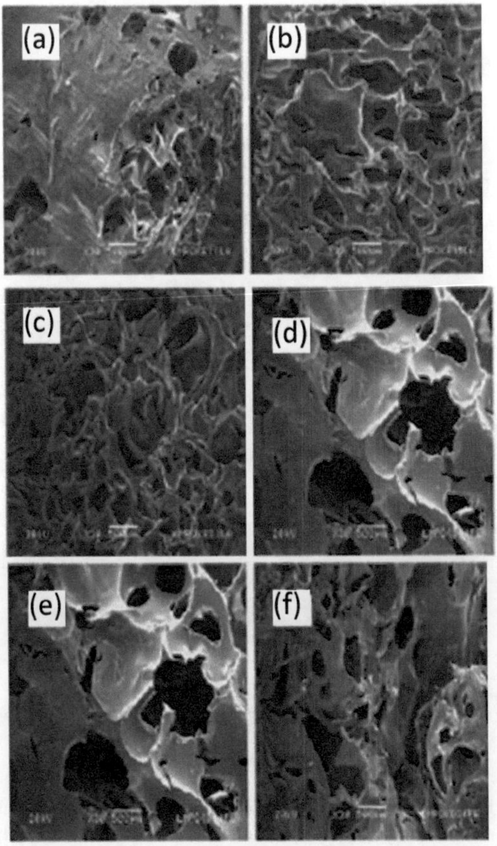

Figure 22.18 SEM images of polyurethane-keratin-salt membranes at 30x resolution, (a) PUcoKES11, (b) PUcoKES15, (c) PUcoKES21, and polyurethane-dialyzed keratin membranes at 30x resolution, (d) PUcoKED11, (e) PUcoKED15, and (f) PUcoKED21 (the numbers 11, 15, and 21 correspond to the grafting percentage of keratin solution onto the polyurethane matrix) (reprinted with permission of Saucedo-Rivalcoba et al., 2011, Springer [9]).

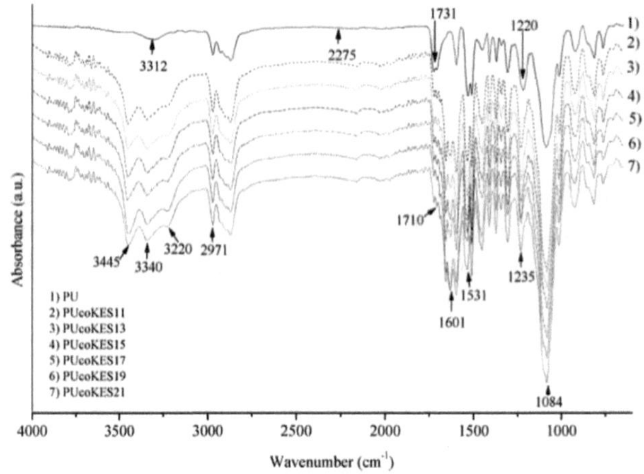

Figure 22.19 DSC Curves of polyurethane-keratin salt membranes at different grafting concentrations (reprinted with permission of Saucedo-Rivalcoba et al., 2011, Springer [9]).

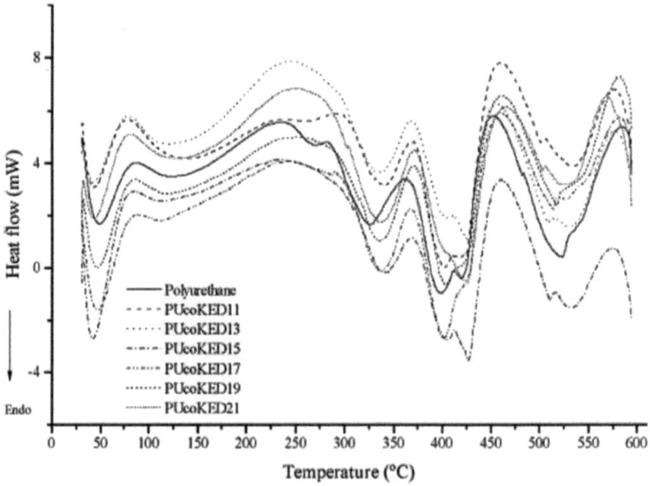

Figure 22.20 DSC Curves of polyurethane-dialyzed keratin membranes at different grafting concentrations (reprinted with permission of Saucedo-Rivalcoba et al., 2011, Springer [9]).

Once the membranes were characterized to guarantee a structural integrity and well performance on fluid and pressure conditions [9]. Saucedo-Rivalcoba et al. directed their research to the use of hybrid keratin-polyurethane membranes to remove hexavalent chromium via functional groups from urethane and amino acids of protein; thus, chemical structural changes of keratin and the membranes' behavior on metal removal have been studied and constitutes an important base in this research line. In order to study the effect of the structure on the removal procedure, keratin was grafted onto a polyurethane matrix in four different forms, which was done because protein is known to adopt different chemical structures according to the dissolving medium, e.g., from secondary to primary conformation, or anionic, cationic and zwiterion species of the keratin (as function of the pH). Besides the urea and dialyzed solutions that were mentioned before, keratin was also modified in acid and alkaline solutions to change the ionic state of the amino acids. The denaturation process of keratin caused changes in the membranes' morphology and pore size depending on the grafted quantity of the added solution. Synthesized membranes were in the mesoporous range and presented better adsorption when the porous size was 5 nm, reaching the maximum rate of metal removal (38%) at 15% of biofiber alkaline solution. The main functional groups of keratin linked to chromium are NH, CO, C-S, and broken S-S bonds. Incorporation of different kinds of keratin solution also caused changes in viscoelastic properties, obtaining elastomeric membranes with a diverse degree of stiffness observed via dynamical mechanical testing (Fig. 22.21). Here the effect is clearly observed when the solutions of keratin and biofibers are incorporated after alkaline treatments, these membranes showed lower values in the storage modulus (E'), which is interpreted as lower rigid behavior than just polyurethane membrane (Figs. 22.21a,d). However, when keratin is integrated into membranes after acidic treatments (Figs. 22.21b,c), membranes trend toward stiffness, showing higher values in E' than those observed for alkaline materials (Figs. 22.21a,d). All membranes were compared against pure polyurethane (PU). Figure 22.22 shows the damping factor, where tan δ peak is observed at higher temperatures for all membranes with respect to PU, revealing a higher thermal stability than PU, except for keratin salt membranes. In Figs. 22.21 and 22.22, it is observable that in PU, E' decreases gradually as temperature increases, whereas the value of tan δ at 50°C increases rapidly. This decrement in E' can be attributed to the beginning of a thermal transition, associated with hard segments phase. The changes presented between 80 and 110°C are due to a high presence of hydrogen-bonded aromatic urethane groups in the poly(ether-urethane) phase and hard segment domains which act as macroscopic cross-links. As the temperature is raised above 100°C, hydrogen bonds starts to dissociate and the hard segments suffer an order-disorder transition. If the temperature continues to increasing, the rate of cross-polymerization in the hard segments rises and the high degree of localized covalent cross-linking prevents the structural disruption. Therefore, the maximum value of tan δ around 125°C is attributed to the high cross-linking and elastomeric behavior. The presence of keratin in the several forms in which was added, definitively affects the cross-linking between the hard segments of PU, and the variety in the performance observed by dynamical mechanical analysis is the evidence of this structural modification.

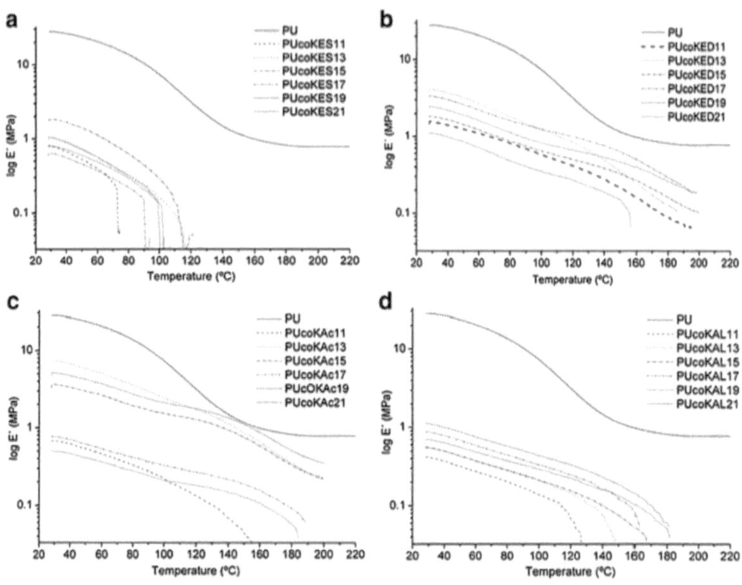

Figure 22.21 Storage modulus of membranes synthesized with (a) polyurethane-keratin salt, (b) polyurethane-dialyzed keratin, (c) polyurethane–acid keratin biofibers, and (d) polyurethane-alkaline keratin biofibers. Numbers 11, 13, 15, 17, 19, and 21 indicate grafted percentage of keratin solution added (reprinted with permission of Saucedo-Rivalcoba et al., 2011, Springer [81]).

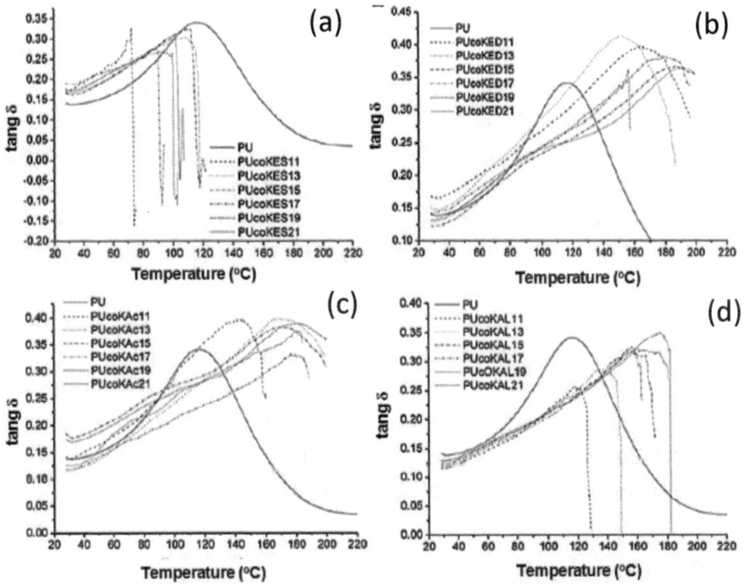

Figure 22.22 Tan δ of membranes synthesized with: (a) polyurethane-keratin salt, (b) polyurethane-dyalized keratin, (c) polyurethane–acid keratin biofibers, and (d) polyurethane–alkaline keratin biofibers. Numbers 11, 13, 15, 17, 19, and 21 grafted percentage of keratin solution added (reprinted with permission of Saucedo-Rivalcoba et al., 2011, Springer [81]).

On the other hand, variation of pH in keratin solutions causes changes in amino-acid behavior related to the electronic charge response. Thus, if keratin solution has a pH below keratin's isoelectric point (reported at pH of 4), a cationic response can be expected, whereas higher values of pH produce an anionic performance. Therefore keratin has the capability of interact with different kind of species pollutants, including not only chromium but also other metals that could be removed. Besides the pH of pollutant solutions also plays an important role as was demonstrated ecently by Manrique-Juárez et al. In this study, the authors reported that pH has a strong influence on the morphology, mechanical properties and Cr(VI) removal efficiency, since if pH in keratin solution is 2.5, the protein separates from water solution, and therefore the polymerization reaction is affected producing a more closed cell in the membrane. The Cr(VI) removal performance was also investigated at different pH values of chromium solutions. The results show that when pH of the chromium solution is 1.5, the removal percentage is high, reaching up to a 58%. Thus, these authors found that pH is the variable that most influences the removal efficiency, since when pH is reduced; the membrane is charged positively due to protons presence. At the same time, the chemical species of chromium in water are negative ions for all pH values: $HCrO_4^-$, CrO_4^{2-}, and $Cr_2O_7^{2-}$. Consequently, the attraction between the positive charges of PU-keratin membrane surface and the anionic species of chromium is strong, producing a more efficient removal process [82].

Based on these results, keratin extracted from feathers is a natural biosorbent that can be incorporated onto synthetic polymers to develop novel membranes and may have wide applications in the environmental field [81, 82].

22.4 Conclusions

Bio-based materials are receiving wide attention, in consequence innovative technologies and competitive industrial products are reducing the dependence on petrochemicals for the production of polymers. Increasing concerns about the environmental degradation caused by conventional polymers have directed worldwide research toward renewable resources. Natural polymers are one of the readily available alternatives for the synthesis path of polyurethanes. The functional groups present in this kind of polymer can be activated for condensation polymerizations, and polyurethane is produced by this route. The incorporation of moieties from natural polymers into the synthetic polymer chain allows tailoring of the properties of polyurethane products for widespread application.

In spite of polyurethane is a hazardous polymer, it can be modified through the basic chemistry of polyurethanes, which can modify a wide variety of soft and hard segments, morphological features, thermic and mechanical properties of structures, just by changing several conditions, such as the ratio NCO/OH, the aliphatic or aromatic isocyanate, the molecular weight, and the ester or ether form of the polyol, but especially the nature of the monomer, whether synthetic or natural. Among the natural options than can be used for synthesis are oil, polysaccharides, and amino acids, obtaining natural hybrid bio-based polyurethanes. These present outstanding characteristics that allow application in medical devices as they are biocompatible with the human body, but also present biodegradability because of natural origin of moieties of urethane, and soft and hard segments. Some advantages of incorporate natural polymers include low cost and high structural versatility, increase of cross-linking like hyperbranched or interpenetrating networks, variable performance as elastomeric or rigid depending on what is needed, thermal stability, enhance the mechanical properties of a determinate kind of polyurethanes. Additionally the biocompatibility and biodegradability leave behind the hazardous issues. The biggest challenge for the production of PU using natural sources is the study of chemical routes to incorporate the biopolymer. It is worthy of mention that the synthetic potential of nature is very high with this renewable feedstock and, consequently, only a few minor modification reactions have to be performed in order to obtain suitable monomers for many different applications. The strategy will depend on the chemical modifications needed in the biopolymer prior to hybrid polymerization. A new focus in polyurethane material is the synthesis of separation membranes, taking advantage of natural polymers as actives sites to adsorb toxic residues from liquid and air fluids, as well as improve mechanical and thermal properties of polyurethane, which allow producing a hybrid bio-based material contributing to the actual environmental worldwide research.

References

1. Del Rio, E., Ligadas, E., Ronda, J. C., Galià, M., Cádiz, V. (2010). Biobased Polyurethanes from polyether polyols obtained by ionic-coordinative polymerization of epoxized methyl oleate, *J. Polym. Sci. Part A Polym. Chem.*, **48**, 5009–5017.

2. Yücedag, F., Atalay-Oral, C., Erkal, S., Sirkecioglu, A., Karasartova, D., Sahin, F., Tantekin-Ersolmaz, S. B., Güner, F. S. (2010). Antibacterial oil-based polyurethane films for wound dressing applications, *J. Appl. Polym. Sci.*, **115**, 1347–1357.

3. Ayres, E., Oréfice, R. L., Yoshida, M. I. (2007). Phase morphology of hydrolysable polyurethane derived from aqueous dispersion, *Eur. Polym. J.*, **43**, 3510–3521.

4. Barikani, M., Zia, K. M., Bhatti, I. A., Zuber, M., Bhatti, H. N. (2008). Molecular engineering and properties of chitin based shape memory polyurethanes, *Carbohydr. Polym.*, **74**, 621–626.
5. Barikani, M., Mohammadi, M. (2007). Synthesis and characterization of starch-modified polyurethane, *Carbohydr. Polym.*, **68**, 773–780.
6. Kadnaim, A., Janvikul, W., Wichai, U., Rutnakornpituk, M. (2008). Synthesis and properties of carboxymethylchitosan hydrogels modified with poly(ester-urethane), *Carbohydr. Polym.*, **74**, 257–267.
7. Fenouillot, F., Rousseau, A., Colomines, G., Saint-Loup, R., Pascault, J. P. (2010). Polymers from renewable 1,4:3,6-dianhydrohexitols (isosorbide, isomannide and isoidide): A review, *Prog. Polym. Sci.*, **35**, 578–622.
8. Zavastin, D., Cretescu, I., Bezdadea, M., Bourceanu, M., Drăgan, M., Lisa, G., Mangalagiu, I., Vasić, V., Savić, J. (2010). Preparation, characterization and applicability of cellulose acetate-polyurethane blend membrane in separation techniques, *Colloids Surf. A: Physicochem. Eng. Aspects*, **370**, 120–128.
9. Saucedo-Rivalcoba, V., Martinez-Hernández, A. L., Martínez-Barrera, G., Velasco-Santos, C., Castaño, V. M. (2011a). (Chicken feathers keratin)/polyurethane membranes, *Appl. Phys. A Mater. Sci. Proc.*, **104**, 219–228.
10. Montero de Espinosa, L., Meier, M. A. R. (2011). Plant oils: The perfect renewable resource for polymer science?, *Eur. Polym. J.*, **47**, 837–852.
11. Sonnenschein, M. F., Ginzburg, V. V., Schiller, K. S., Wendt, B. L. (2013). Designing, polymerization, and properties of high performance thermoplastic polyurethane elastomers from seed-oil derived soft segments, *Polym*er, **18**, 1350–1360.
12. Raquez, M., Deléglise, Lacrampe, M. F., Krawczak, P. (2010). Thermosetting (bio)materials derived from renewable resources: A critical review, *Prog. Polym. Sci.*, **35**, 487–509.
13. Lin, D., Liu, B., Xu, L., Budzinski, K. L., Shou, C. (2011). Synthesis of hyperbranched polyester-amides and their applications as crosslinkers for polyurethane curing systems, *J. Appl. Polym. Sci.*, **121**, 957–963.
14. Howard, G. T. (2002). Biodegradation of polyurethane: a review, *Int. Biodeterior. Biodegrad.*, **49**, 245–252.
15. Chattopadhyay, D. K., Raju, K. V. S. N. (2007). Structural engineering of polyurethane coatings for high performance applications, *Prog. Polym. Sci.*, **32**, 352–418.

16. Santerre, J. P., Woodhouse, K., Laroche, G., Labow, R. S. (2005). Understanding the biodegradation of polyurethanes: From classical implants to tissue engineering materials, *Biomaterials*, **26**, 7457–7470.

17. Donnelly, M. J., Stanford, J. L., Still, R. H. (1991). The conversion of polysaccharides into polyurethanes: A review, *Carbohydr. Polym.*, **14**, 221–240.

18. Miao, S., Wang, P., Su, Zh., Zhang, S. (2013). Vegetable-oil-based polymers as future polymeric biomaterials, *Acta Biomater.*, **10**, 1692–1704.

19. Sharma, V., Kundu, P. P. (2008). Condensation polymers from natural oils, *Prog. Polym. Sci.*, **33**, 1199–1215.

20. Li, G., He, Y., Liu, X., Su, J., Qu, G. (2009). Preparation and blood compatibility of epoxidized castor oil-based polyurethane/chitosan biomedical materials, *Adv. Mater. Res.*, **79–82**, 219–222.

21. Athawale, V., Kolekar, S. (1998). Interpenetrating polymer networks based on polyol modified castor oil polyurethane and polymethyl methacrylate, *Eur. Polym. J.*, **34**, 1447–1451.

22. Corcuera, M. A., Rueda, L., Fernandez d'Arlas, B., Arbelaiz, A., Marieta, C., Mondragon, I., Eceiza, A. (2010). Microstructure and properties of polyurethane derived from castor oil, *Polym. Degrad. Stab.*, **95**, 2175–2184.

23. Ferreira, P., Pereira, R., Coelho, J. F. L., Silva, A. F. M., Gil, M. H. (2007). Modification of the biopolymer castor oil with free isocyanate groups to be applied as bioadhesive, *Int. J. Biol. Macromol.*, **40**, 144–152.

24. Xu, Y., Petrovic, Z., Das, S., Wilkes, G. L. (2008). Morphology and properties of thermoplastic polyurethanes with dangling chain in ricinoleate-based soft segments, *Polymer*, **49**, 4248–4258.

25. Oprea, S. (2010). Dependence of fungal biodegradation of PEG/castor oil-based polyurethane elastomers on the hard-segment structure, *Polym. Degrad. Stab.*, **95**, 2396–2404.

26. Aranguren, M. I., González, J. F., Mosiewicki, M. A. (2013). Biodegradation of vegetable oil based polyurethane and wood flour composites, *Polym. Test.*, **31**, 7–15.

27. Ajithkumar, S., Patel, N. K., Kansara, S. S. (2000). Sorption and diffusion of organic solvents through interpenetrating polymer networks (IPNs) based on polyurethane and unsaturated polyester, *Eur. Polym. J.*, **36**, 2387–2393.

28. Manjula, K. S., Siddaramaiah. (2010). Sorption and diffusion of substituted aromatic organic penetrants into carboxylic acid-based chain extended polyurethane membranes, *J. Chem. Eng. Data*, **55**, 2271–2278.

29. Lubguban, A. A., Tu, Y. Ch., Lozada, Z. R., Hsieh, F. H., Suppes, G. J. (2009). Functionalization via glycerol transesterification of polymerized soybean oil, *J. Appl. Polym. Sci.*, **112**, 19–27.

30. Lu, Y., Xia, Y., Larock, R. C. (2011). Surfactant-free core-shell hybrid latexes from soybean oil-based waterborne polyurethanes and poly(styrene-butyl acrylate), *Prog. Org. Coat.*, **71**, 336–342.

31. Bakare, I. O., Pavithran, C., Okieimen, F. E., Pillai, C. K. S. (2008). Synthesis and characterization of rubber-seed-oil-based polyurethanes, *J. Appl. Polym. Sci.*, **109**, 3292–3301.

32. Ronda, J. C., Lligadas, G., Galià, M., Càdiz, V. (2013). A renewable approach to thermosetting resins, *Polymer*, **73**, 381–395.

33. Petrović, Z. S., Hong, D. P., Javni, I., Erina, N., Zhang, F., Ilavský, J. (2013). Phase structure in segmented polyurethanes having fatty acid-based soft segments, *Polymer*, **45**, 372–380.

34. Cayli, G., Küsefoğlu, S. (2008). Biobased polyisocyanates from plant oils triglycerides: Synthesis, polymerization, and characterization, *J. Appl. Polym. Sci.*, **109**, 2948–2955.

35. Cayli, G., Küsefoğlu, S. (2010). A simple one-step synthesis and polymerization of plant oil triglyceride iodo isocyanates, *J. Appl. Polym. Sci.*, **116**, 2433–2440.

36. More, A. S., Lebarbè, T., Maisonneuve, L., Gadenne, G., Alfos, C., Cramail, H. (2013). Novel fatty acid di-isocyanates towards the synthesis of thermoplastic polyurethanes, *Eur. Polym. J.*, **49**, 823–833.

37. Sivakumar, M., Malaisamy, R., Sajitha, C. J., Mohan, D., Mohan, V., Rangarajan, R. (1999). Ultrafiltration application of cellulose acetate-polyurethane blend membranes, *Eur. Polym. J.*, **35**, 1647–1651.

38. Sivakumar, M., Malaisamy, R., Sajitha, C. J., Mohan, D., Mohan, V., Rangarajan, R. (2000). Preparation and performance of cellulose acetate-polyurethane blend membranes and their applications-II, *J. Membr. Sci.*, **169**, 215–228.

39. Barnes, K., Liang, J., Worley, S. D., Lee, L., Broughton, R. M., Huang, T. S. (2007). Modification of silica gel, cellulose, and polyurethane with a sterically hindered N-halamine moiety to produce antimicrobial activity, *J. Appl. Polym. Sci.*, **105**, 2306–2313.

40. Raschip, I. E., Vasile, C., Macocinschi, D. (2009). Compatibility and biocompatibility study of new HPC/PU blends, *Polym. Int.*, **58**, 4–16.

41. Macocinschi, D., Filip, D., Butnaru, M., Dimitriu, C. D. (2009). Surface characterization of biopolyurethanes based on cellulose derivatives, *J. Mater. Sci. Mater. Med.*, **20**, 775–783.

42. Ling, S., Huang, J., Chang, P. R., Wey, S., Xu, Y., Zhang, Q. (2013). Structure and mechanical properties of new biomass-based nanocomposite: Castor oil-based polyurethane reinforced with acetylated cellulose nanocrystal, *Carbohydr. Polym.*, **95**, 91–99.

43. Saralegi, A., Gonzalez, M. L., Valea, A., Eceiza, A., Corcuera, M. A. (2014). The role of cellulose nanocrystals in the improvement of the shape-memory properties of castor oil-based segmented thermoplastic polyurethanes, *Compos. Sci. Technol.*, **92**, 27–33.

44. Imnes, Bedò, D., Domjàn, A., Shön, P., Vancso, G. J., Pukànszky, B. (2013). Structure, properties and interfacial interactions in poly(lactic acid)/polyurethane blends prepared by reactive processing, *Eur. Polym. J.*, **49**(10), 3104–3113.

45. Ates, B., Koytepe, S., Karaaslan, M. G., Balcioglu, S., Gulgen, S. (2014). Biodegradable non-aromatic adhesive polyurethanes based on disaccharides for medical applications, *Int. J. Adhes. Adhes.*, **49**, 90–96.

46. Cheng, J. Y., Hou, T. Y., Shih, M. F., Talsma, H., Hennink, W. E. (2013). Polyurethane-based drug delivery systems, *Int. J. Pharm.*, **450**, 145–162.

47. Wang, W., Ping, P., Chen, X., Jing, X. (2006). Polylactide-based polyurethane and its shape-memory behavior, *Eur. Polym. J.*, **42**, 1240–1249.

48. Melnig, V., Apostu, M. O., Tura, V., Ciobanu, C. (2005). Optimization of polyurethane membranes morphology and structure studies, *J. Membr. Sci.*, **267**, 58–67.

49. Adhikari, R., Gunatillake, P. A., Griffiths, I., Tatai, L., Wickramaratna, M., Houshyar, Sh., Moore, T., Mayadunne, R. T. M., Field, J., McGee, M., Carbone, T. (2008). Biodegradable injectable polyurethanes: Synthesis and evaluation for orthopedic applications, *Biomaterials*, **29**, 3762–3770.

50. Chen, S. H., Tsao, Ch. T., Chang, Ch. H., Wu, Y. M., Liu, Zh. W., Lin, Ch. P., Wang, Ch. K., Hsieh, K. H. (2012). Synthesis and characterization of thermal-responsive chitin-based polyurethane copolymer as a smart material, *Carbohydr. Polym.*, **88**, 1483–1487.

51. Hashimoto, K., Hashimoto, N., Kamaya, T., Yoshioka, J., Okawa, H. (2011). Synthesis and properties of bio-based polyurethanes bearing hydroxyl groups derived from aditols, *J. Polym. Sci., Part A: Polym. Chem.*, **49**, 976–985.

52. Bachmann, F., Ruppenstein, M., Thiem, J. (2001). Synthesis of aminosaccharide-derived polymers with urea, urethane, and amide linkages, *J. Polym. Sci., Part A: Polym. Chem.*, **39**, 2332–2341.

53. Ferris, C., de Paz, M. V., Galbis, J. A. (2011). L-Arabinitol-based functional polyurethanes, *J. Polym. Sci. Part A Polym. Chem.*, **49**, 1147–1154.

54. Guelcher, S. A., Srinivasan, A., Dumas, J. E., Didier, J. E., McBride, S., Hollinger, J. O. (2008). Synthesis, mechanical properties, biocompatibility, and biodegradation of polyurethane networks from lysine polyisocyanates, *Biomaterials*, **29**, 1762–1775.

55. Sharifpoor, S., Labow, R. S., Santerre, J. P. (2009). Synthesis and characterization of degradable polar hydrophobic ionic polyurethane scaffolds for vascular tissue engineering applications, *Biomacromolecules*, **10**, 2729–2739.

56. Sarkar, D., Yang, J. C., Klettlinger, N., Lopina, S. T. (2007). Blends of L-tyrosine based polyurethanes for biomaterial applications, *Express Polym. Lett.*, **11**, 724–733.

57. Gao, C., Guan, J., Zhu, Y., Shen, J. (2003). Polyurethane for promotion of cytocompatibility to human endothelial cells, *Macromol. Biosci.*, **3**, 157–162.

58. Guang, J., Wagner, W. (2005). Synthesis, characterization and cytocompatibility of polyurethaneurea elastomers with designed elastase sensitivity, *Biomacromolecules*, **6**, 2833–2842.

59. Zhu, Y., Sun, Y. (2004). The influence of polyelectrolyte charges of polyurethane membrane surface on the growth of human endothelial cells, *Colloids Surf. B Biointerfaces*, **36**, 49–55.

60. Tseng, S. J., Tang, Sh. Ch. (2007). Synthesis and characterization of the novel transfection reagent poly(amino ester glycol urethane), *Biomacromolecules*, **8**, 50–58.

61. Abraham, G. A., de Queiroz, A. A. A., San Román, J. (2001). Hydrophylic hybrid IPNs of segmented polyurethanes and copolymers of vinylpyrrolidone for applications in medicine, *Biomaterials*, **22**, 1971–1985.

62. Nasar, A. S., Jikei, M., Kakimoto, M.aki. (2003). Synthesis and properties of polyurethane elastomers crosslinked with amine-terminated AB_2-type hyperbranched polyamides, *Eur. Polym. J.*, **39**, 1201–1208.

63. Gaina, C., Gaina, V., Ciobanu, C. (2009). Thermal and mechanical characterization of maleimide-functionalized copoly(urethane-urea)s, *J. Appl. Polym. Sci.*, **113**, 3245–3254.

64. Ulbricht, M. (2006). Advanced functional polymer membranes, *Polymer*, **47**, 2217–2262.

65. Zhao, Ch., Nie, Sh., Tang, M. Sun, Sh. (2011). Polymeric pH-sensitive membranes—A review, *Prog. Polym. Sci.*, **36**, 1499–1520.

66. Stamatialis, D. F., Papenburg, B. J., Gironés, M., Saiful, S., Bettahalli, S. N. M., Schmitmeier, S., Wessling, M. (2008). Review. Medical applications of membranes: Drug delivery, artificial organs and tissue engineering, *J. Membr. Sci.*, **308**, 1–34.

67. Edlund, U., Albertsson, A. C. (2003). Polyester based on diacid monomers, *Adv. Drug Deliv. Rev.*, **55**, 585–609.

68. Nagarale, R. K., Gohil, G. S., Shahi, V. K. (2006). Recent developments on ion-exchange membranes and electro-membrane processes, *Adv. Colloid Interface Sci.*, **119**, 97–130.

69. Smitha, B., Suhanya, D., Sridhar, S., Ramakrishna, M. (2004). Separation of organic-organic mixtures by pervaporation—a review, *J. Membr. Sci.*, **241**, 1–21.

70. Rzeszutek, K., Chow, A. (1998). Extraction of phenol using polyurethane membrane, *Talanta*, **46**, 507–519.

71. Moawed, E. A. (2006). Preparation of novel ion exchange polyurethane foam and its application for separation and determination of palladium in environmental samples, *Anal. Chim. Acta*, **580**, 263–270.

72. Ghosh, U. K., Pradhan, N. C., Adhikari, B. (2006). Separation of water and o-chlorophenol by pervaporation using HTPB-based polyurethaneurea membranes and application of modified Maxwell-Stefan equation, *J. Membr. Sci.*, **272**, 93–102.

73. Wolinska-Grabczyk, A. (2006). Effect of the hard segment domains on the permeation and separation ability of the polyurethane-based membranes, *J. Membr. Sci.*, **282**, 225–236.

74. Das, S., Banthia, A. K., Adhikari, B. (2006). Removal of chlorinated volatile organic contaminants from water by pervaporation using a novel polyurethane urea-poly/(methyl metacrilate) interpenetrating network membrane, *Chem. Eng. Sci.*, **61**, 6454–6467.

75. Weibel, D. E., Vilani, C., Habert, A. C., Achete, C. A. (2007). Surface modification of polyurethane membranes using acrylic acid vapour plasma and its effects on the pervaporation process, *J. Membr. Sci.*, **293**, 124–132.

76. Huang, Sh. L., Chang, P. H., Tsai, M. H., Chang, H. C. (2007). Properties and pervaporation performances of crosslinked HTPB-based polyurethane membranes, *Sep. Purif. Technol.*, **56**, 63–70.

77. Das, S., Banthia, A. K., Adhikari, B. (2008). Porous polyurethane urea membranes for pervaporation separation of phenol and clorophenol from water, *Chem. Eng. J.*, **138**, 215–223.

78. Kim, H. J., Pant, H. R., Choi, N. J., Kim, C. S. (2013). Composite electrospun fly ash/polyurethane fibers for absorption of volatile organic compound from air, *Chem. Eng. J.*, **230**, 244–250.

79. Wu, H., Lu, X., Li, X., Li, Y., Zhao, C., Jiang, Zh. (2014). Enhancing structural stability and pervaporation performance of composite membranes by coating gelatin onto hydrophilically modified support layer, *Chin. J. Chem. Eng.*, **22**, 19–27.

80. Rzeszutek, K., Chow, A. (2001). Extraction of metal-dye ion-association complexes by thin ether-type polyurethane membranes, *J. Membr. Sci.*, **181**, 265–277.

81. Saucedo-Rivalcoba, V., Martínez-Hernández, A. L., Martínez-Barrera, G., Velasco-Santos, C., Rivera-Armenta, J. L. Castaño, V. M. (2011b). Removal of hexavalent chromium from water by polyurethane-keratin hybrid membranes, *Water, Air, Soil Pollut.*, **218**, 557–571.

82. Manrique-Juarez, M. D., Martinez-Hernandez, A. L., Olea-Mejía, O. F., Flores-Estrada, J., Rivera-Armenta, J. L., Velasco-Santos, C. (2013). Polyurethane-keratin membranes: Structural changes by isocyanate and pH, and the repercussion on Cr(VI) removal, *Int. J. Polym. Sci.*, **2013**, article ID 892547.

Chapter 23

Biodegradable Poly(Lactic Acid) and Its Composites

Tao Yu and Yan Li

School of Aerospace Engineering and Applied Mechanics, Tongji University, No. 1239, Siping Road, Shanghai, 200092, China

liyan@tongji.edu.cn

23.1 Introduction

Poly(lactic acid) (PLA) as a sustainable alternative to petrochemical-derived products belongs to the family of aliphatic polyesters commonly derived from renewable resources, such as corn starch (in the United States), tapioca roots, chips or starch (mostly in Asia), or sugarcane (in the rest of the world). The basic materials will be transferred by means of a bacterial fermentation process in lactic acid, the basic chemical needed for PLA production. Compared to many commodity polymers (e.g., PP, PE, PVC, PS), PLA has good properties such as high stiffness, clarity, gloss, and UV stability. PLA is also a low-impact, greenhouse gas polymer because the CO_2 generated during PLA biodegradation is balanced by an equal amount taken from the atmosphere during the growth of plant feedstocks. In contrast, petrochemical-based

Handbook of Sustainable Polymers: Processing and Applications
Edited by Vijay Kumar Thakur and Manju Kumari Thakur

ISBN 978-981-4613-53-8 (Hardcover), 978-981-4613-54-5 (eBook)
www.panstanford.com

polymers contribute to volatile organic compound (VOC) emissions and CO_2 generation when incinerated. Furthermore, PLA is easily processed on standard plastic equipment to yield molded parts, film, or fibers. Therefore, it has been used as package materials and other products.

However, for PLA, the disadvantages such as the inherent brittleness, poor heat resistance, and low melt strength limit the applications. Modifications of PLA and PLA-based composites are common ways to solve these problems.

Three parts are covered in the chapter. First, the synthesis and properties of PLA are described. The modification and process of PLA are also discussed. Then, the composites with PLA as matrix and natural fiber or nanoparticles as reinforcement are reported in the second part. The processing and the properties of the composites are also given. The interface between PLA and the reinforcement and the surface treatment methods are discussed. Finally, the application and the development of PLA and PLA-based composites in the future are proposed.

23.2 Biodegradable PLA

23.2.1 Synthesis of PLA

Poly(lactic acid) was discovered by Carothers at DuPont in 1932. Only a low-molecular-weight PLA could be produced by heating lactic acid under vacuum while removing the condensed water [1].

Polymerization of lactic acid to high-molecular-weight PLA can be achieved in three ways: (1) Direct condensation, which involves solvents under high vacuum. (2) Ring-opening polymerization (ROP) of the cyclic dimer, lactide, which is made by depolymerization of the polycondensed lactic acid. (3) Utilizing the terminal groups of the prepolymer in linking process where a linking agent is employed. PLAs manufactured by different polymerization routes are showed in Fig. 23.1.

23.2.1.1 Direct condensation

The first route of producing PLA is generated using a polycondensation reaction of the hydroxyl and carboxylic acid group starting from lactic acid (shown in Fig. 23.2). Water produced during the polymerization has to be removed during the polymerization process, and soon an equilibrium between polymerization and depolymerization reaction is reached. Furthermore, the polymerization requires long reaction times in combination with high temperatures.

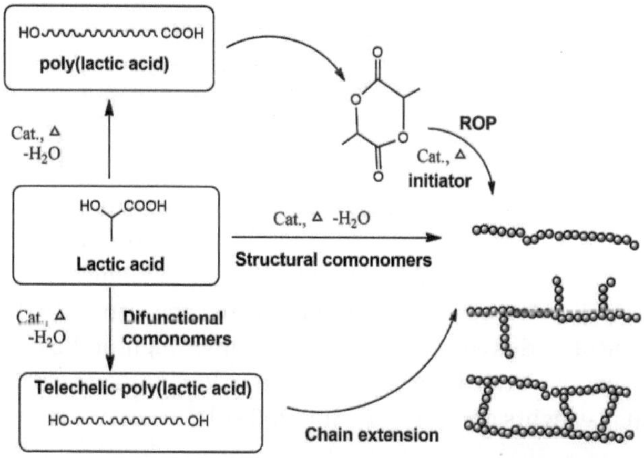

Figure 23.1 Manufacturing routes for lactic acid–based polymers [2].

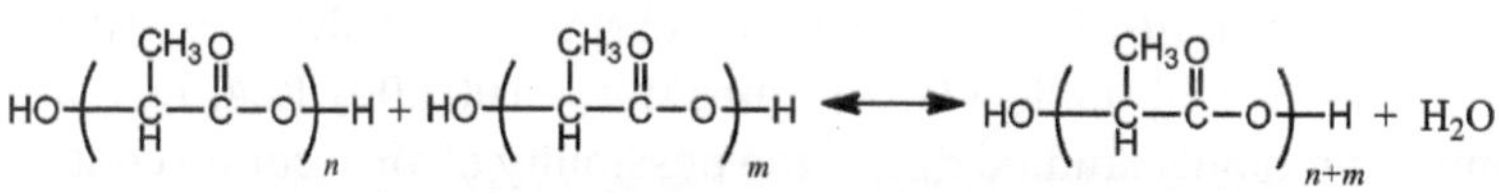

Figure 23.2 Synthesis PLA by direct polycondensation.

The disadvantages of the direct condensation are that the obtained PLA shows low molecular weight and mechanical properties, which will limit its applications. The low-molecular-weight precondensates may be of interest for the preparation of biodegradable glues or lacquers, because the –OH and –COOH end groups allow cross-linking with suitable inorganic or organic multivalent additives [3]. Some progress has been achieved to solve this problem. The molar mass of PLA is improved by sequential melt/solid polycondensation. A polycondensate with a molecular weight of 20,000 Da is first prepared by ordinary melt-polycondensation, crystallized by heat-treatment around 105°C, and heated at 140 or 150°C for 10–30 h for further polycondensation. A high-quality polymer of PLLA can be obtained in high yield in a relatively short reaction time and its molecular weight exceeds 500,000 Da, which is comparable with that of the PLLA obtained by the lactide method but has never been attained by the simple melt-polycondensation [4]. Lactic acid is polycondensated directly into high molar mass polymer by manipulating the equilibrium between lactic acid, water, and PLA in an organic solvent. PLA synthesized by one-step polycondensation of lactic acid can obtain high molecular weight if appropriate azeotropic solvents are employed. The azeotropic solution helps to decrease the distillation pressures and facilitates PLA separation from the solvent by application of molecular sieves. The variety and content of catalysts, solvent volume percentages, and the reaction time on the preparation of PLA have been studied. The results identify by using improved experimental equipment, the proper complex catalyst, and solvent volume ratio, in order to obtain a molecular weight of PLA of 6.6×10^4 [5]. Mitsui Toatsu Chemicals recently commercialized a process wherein lactic acid and catalyst are azeotropically dehydrated in a refluxing, high-boiling, aprotic solvent under reduced pressures to obtain PLA with weight-average molecular weights greater than 300,000 [6].

23.2.1.2 Ring-opening polymerization

The second route of producing PLA is to ring-open polymerize (ROP) lactide into high molecular weight (Mw > 100,000) PLA. ROP are most commonly studied due to the possibility of an accurate control of the chemistry, and the properties of the resulting polymers in a more controllable manner.

The ROP route includes polycondensation of lactic acid followed by a depolymerization into the dehydrated cyclic dimer, lactide (shown in Fig. 23.3). The depolymerization is conventionally done by increasing the polycondensation temperature and lowering the pressure and distilling off the produced lactide. Due to the two stereoforms of lactic acid, the corresponding optically active lactide can be found in two different versions. In addition, lactide can be formed from one D- and one L-lactic acid molecule yielding D,L-lactide(meso-lactide) [7].

$$HO\text{-}(CH(CH_3)COO)_o\text{-}H \longleftrightarrow HO\text{-}(CH(CH_3)COO)_{o-2}\text{-}H+$$

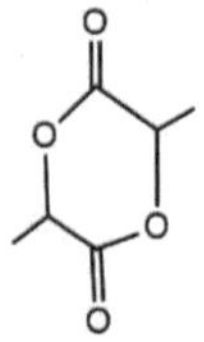

Figure 23.3 Depolymerization into the dehydrated cyclic dimer.

Many catalyst systems have been evaluated for the polymerization of lactide including complexes of aluminum, zinc, tin, and lanthanides. Even strong bases such as metal alkoxides have been used with some success. The catalyst currently used industrially is stannous octoate, which has solubility in molten lactide, high catalytic activity, and low rate of racemization of the polymer. Conversions of over 90% and less than 1% racemization can be obtained while providing polymer with high molecular weight. ROP of lactide can be carried out in melt, bulk, or in solution and by cationic, anionic, and coordination-insertion mechanisms depending on the catalyst. The choice of initiator system, co-initiator as chain control agent, catalyst concentration, monomer-to-initiator ratio, and reaction temperature and time significantly affect the properties of the polymer. These properties, such as the molecular weight, degree of crystallinity, and residual monomer content, in turn affect the physical-mechanical properties and range of temperature for use of the polylactide and its copolymers. The role of the racemization and the extent of transesterification in the polymerization and copolymerization process are also decisive for the enantiomeric purity and chain microstructure of the resulting polymer [8].

ROP of PLA is carried out most commonly by a stannous octoate catalyst, but for laboratory demonstrations, tin (II) chloride is often employed. Stannous alkoxide, a reaction product between stannous octoate and alcohol, is proposed as the substance initiating the polymerization through coordinative insertion of lactide. Alcohol could affect the polymerization through reactions leading to initiator formation, chain transfer, and transesterification. Carboxylic acids affect the polymerization through a deactivation reaction. Experiments have shown that alcohol increases PLA production rate while carboxylic acid decreases it. The higher the alcohol concentration, the lower is the polymer molecular weight. However, the final molecular weight of PLA is not sensitive to the carboxylic acid concentration [9].

Cargill Dow LLC has developed a patented, low-cost continuous process for the production of lactic acid–based polymers. The process combines the substantial environmental and economic benefits of synthesizing both lactide and PLA in the melt rather than in solution and, for the first time, provides a commercially viable biodegradable commodity polymer made from renewable resources. The process starts with a continuous condensation reaction of aqueous lactic acid to produce low-molecular-weight PLA pre-polymer. Next, the pre-polymer is converted into a mixture of lactide stereoisomers using tin catalysis to enhance the rate and selectivity of the intramolecular cyclization reaction. The molten lactide mixture is then purified by vacuum distillation. Finally, PLA high molecular-weight polymer is produced using a tin-catalyzed, ring-opening lactide polymerization in the melt, completely eliminating the use of costly and environmentally unfriendly solvents. This process is currently in operation at an 8.106 lb per year market development facility in Minnesota [9].

Lactide has been used as receive and after recrystallization it from solutions in solvents like chloroform, ethyl acetate, acetone, etc. It has been reported that recrystallized L-lactide gives higher intrinsic viscosities than those obtained for the once-recrystallized monomer [10].

Jacobsen et al. reported a new catalytic system and a reactive extrusion polymerization process, which can be used to produce PLA continuously in larger quantities and at lower costs than before. This extrusion polymerization process has been developed and tested with laboratory scale machines and the possibilities to extend this polymerization process to lactide-based block copolymers have been investigated [11].

23.2.1.3 Chain extension

The direct condensation polymerization is the least expensive route, but it is difficult in a solvent-free system to obtain high molecular weights. Therefore, the chain extenders or esterification-promoting adjuvants are used to produce the high-molecular-weight PLA, although adding cost and complexity. The self-condensation of lactic acid results in a low-molecular-weight product with an equimolar concentration of hydroxyl and carboxyl end-groups. To increase the molecular weight, chain extenders must be added, and these will preferentially react with either the hydroxyl or carboxyl group, which leads to different kinetic reaction rates of coupling. The condensed PLA can be modified to produce either all hydroxyl or all carboxyl groups. Hydroxyl-terminated PLA can be synthesized by the condensation of lactic acid in the presence of a small amount of multifunctional hydroxyl compounds such as 2-butenelactate 1,4-diol, glycerol, or 1,4-butanediol, which leads to preferential hydroxyl end-groups, or by the post condensation reaction of a monofunctional epoxide such as butyl glycidyl ether with the carboxylic acid to convert it to a hydroxyl group. This same concept can be used to synthesize an all-carboxyl-terminated PLA by the condensation reaction in the presence of multifunctional carboxylic acids such as maleic, succinic, adipic, or itaconicacid, leading to a carboxyl functional polymer. The PLA can also be post-reacted with acid anhydrides such as maleic or succinic to convert the hydroxyl to acarboxylic end-group. Chain extenders are usually bifunctional low-molecular-weight chemicals that can increase the molecular weight of polymers in a fast reaction without a separate purification step. Typical chain extenders for polyesters, which contain –OH and –COOH groups, are diisocycanates, diepoxides, bisoxazolines, dianhydrides, and bisketeneacetals.

A hydroxyl-terminated prepolymer was synthesized by Hiltunen et al. with 1,4-butanediol; isocyanate was then used as a chain extender to increase the molecular weight. According to their studies, the polyaddition reaction was influenced by the structure and functionality of the monomers, the characteristics of the hydroxyl group, polymerization temperature, polymerization time and –OH/–NCO ratio [12].

Gu et al. reported that a high-molecular-weight biodegradable lactic acid polymer was successfully synthesized in two steps. First, the lactic acid monomer was oligomerized to low-molecular-weight hydroxyl-terminated prepolymer; the molecular weight was then increased by chain extension using 1,6-hexamethylene diisocyanate as the chain extender. The results showed that the obtained polymer had an M_n of 27,500 $g \cdot mol^{-1}$ and an M_w of 116, 900 $g \cdot mol^{-1}$ after 40 min of chain extension at 180°C [13].

Ren et al. studied the effect of different diisocyanates as chain-extenders on chain extending of lactic acid oligomers. The results showed that the diisocyanate could be used as chain extenders to get high-molecular-weight chain-linked LA polymers. 1,6-hexamethylene diisocyanate (HDI) as chain-extender was better than methylenediphenyl diisocyanate (MDI) and isophorone diisocyanate (IPDI) [14].

Furthermore, carboxyl and hydroxyl reactive chain extenders were used together in lactic acid polymerization by Tuominen et al. The addition of highly reactive chain extenders during the final step of polycondensation led to polymers with higher molecular weight and lower polymer acid value (AV) at shorter reaction time and lower polymerization temperature. Bisoxazolines (BOX) was used as an end-group modifier and chain coupling agent of the lactic acid prepolymers before chain linking with diisocyanate. The effects of BOX on the carboxyl end-group content of different prepolymers and the progress of the diisocyanate reactions were investigated [15, 16].

23.2.2 Modification of PLA

PLA has several advantages: it is biocompatible and biodegradable, and can be readily broken down thermally by hydrolysis. It is available from renewable agricultural resources. This latter circumstance also helps to improve the farm economy. There is a reduction in carbon dioxide emissions in comparison with conventional petroleum-based commodity plastics because it helps by fixation of significant quantities of carbon dioxide. Although PLA is an eco-friendly bioplastic with excellent biocompatibility, processibility, and less energy dependence, it has drawbacks as well, such as poor toughness, slow degradation rate, hydrophobicity and lack of reactive side, etc., which limit its use in certain applications. To improve the quality and reduce the cost of production, lactic acid polymerizes with other monomers or PLA blend with other polymers. The successful implementation of PLA in consumer and biomedical applications relies not only on mechanical properties being better than or comparable to conventional plastics, but also on controlled surface properties (e.g., hydrophilicity, roughness, and reactive functionalities) [1]. PLA has been bulk modified mainly to improve toughness and degradation rate.

The surface modification of PLA has been attempted to control hydrophilicity, roughness, and to introduce reactive groups. The roughness improvement is a crucial necessity for many consumer applications, while the improvements in hydrophilicity and introduction of reactive groups are beneficial for biomedical applications. Poly(ε-caprolactone) (PCL) is one of the typical aliphatic polyesters, and it is fully biodegradable, biocompatible, and nontoxic to living organisms. Unfortunately, the miscibility of PLA and PCL is very poor due to their widely different physiochemical properties. Preparing a PCL-*b*-PLA block copolymer can give a copolymer with the good properties of PLA and PCL, especially improving the brittleness of PLA. Cohn et al. reported the synthesis of PCL/PLA biodegradable poly(ester-urethane). The synthesis consisted of ring-opening polymerization of L-lactide, initiated by hydroxyl end groups in the PCL chain, followed by chain extension of the triblocks using hexamethylene diisocyanate (HDI) as chain extender. The multiblock copolymer produced displayed enhanced mechanical properties with tensile strength of around 32 MPa, a Young's modulus around 30 MPa and elongation at break of above 600% [17]. Yu et al. prepared biodegradable copolymers of poly(lactic acid)-block-poly(ε-caprolactone) (PLA-*b*-PCL) by two steps, as shown in Fig. 23.4. In the first step, lactic acid monomer

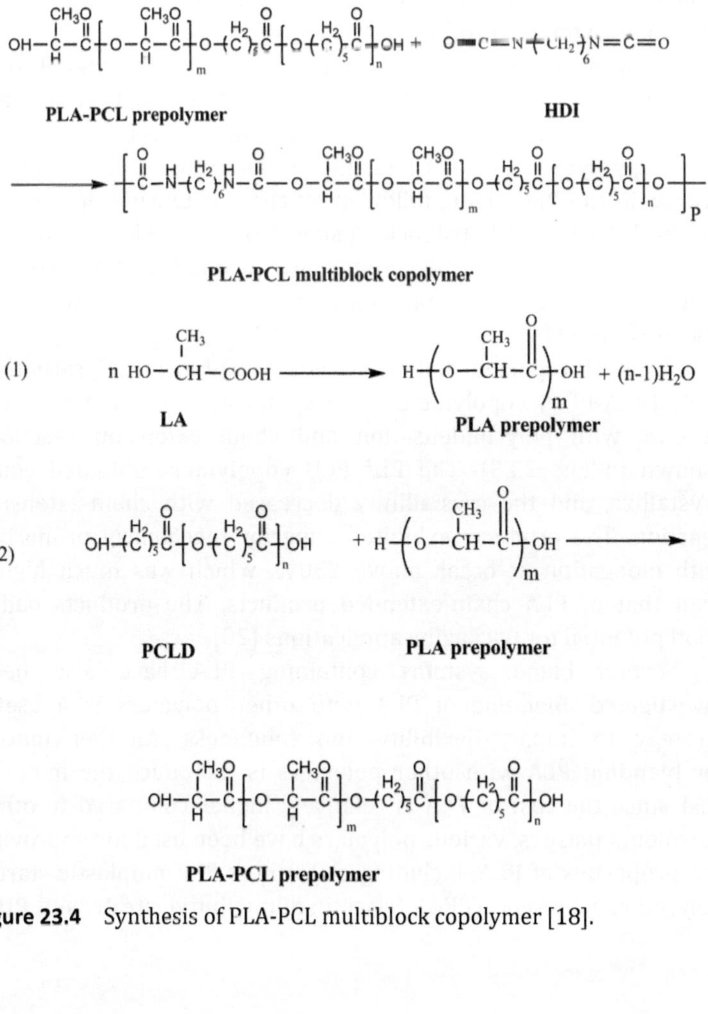

Figure 23.4 Synthesis of PLA-PCL multiblock copolymer [18].

was oligomerized to low-molecular-weight prepolymer and copolymerized with the (ε-caprolactone) diol to prepolymer, and then the molecular weight was raised by joining prepolymer chains together using 1,6-hexamethylene diisocyanate (HDI) as the chain extender. The results of 1H-NMR and GPC indicated that PLA-*b*-PCL prepolymer with number average molecular weights (M_n) of 4000–6000 were obtained. When PCL-diols were 10 wt%, copolymer was better for chain extension reaction to obtain the polymer with high molecular weight. After chain extension, the weight average molecular weight could reach 250,000 g/mol, as determined by GPC, when the molar ratio of –NCO to –OH was 3:1. The product exhibited superior mechanical properties with elongation at break above 297% that was much higher than that of PLA chain extended products [18].

Cohn et al. also used poly(ethylene oxide) (PEO) as flexible segment, and synthesized PEO- and PLA-based multiblock thermoplastic elastomers. The synthesis consisted of ring-opening polymerization of L-lactide, initiated by the hydroxyl end groups in the PEO chain, followed by chain extension of the PLA-block-PEO-block-PLA triblocks using HDI as a chain extender. The tensile strength of the product was around 30 MPa, Young's modulus was as low as 14 MPa and elongation at break is well above 1000% [19].

Yu et al. reported that poly(lactic acid)-block-(polycarbonate diol) (PLA-PCD) copolymers were synthesized using a two-step process with polycondensation and chain extension reactions (shown in Fig. 23.5). The PLA-PCD copolymers obtained could crystallize, and the crystallinity decreased with chain-extension reaction. The products exhibited superior mechanical properties with elongation at break above 230%, which was much higher than that of PLA chain-extended products. The products had a good potential for packaging applications [20].

Several blend systems containing PLA have also been investigated. Blending of PLA with other polymers is a useful strategy to impart flexibility and toughness. Another motive for blending PLA with other polymers is to reduce the material cost since the cost of PLA is relatively higher compared to other petroleum plastics. Various polymers have been used for improving the properties of PLA, including elastomers, thermoplastic starch, poly(ethylene glycol) (PEG), triacetin and tributyl citrate, and PHA.

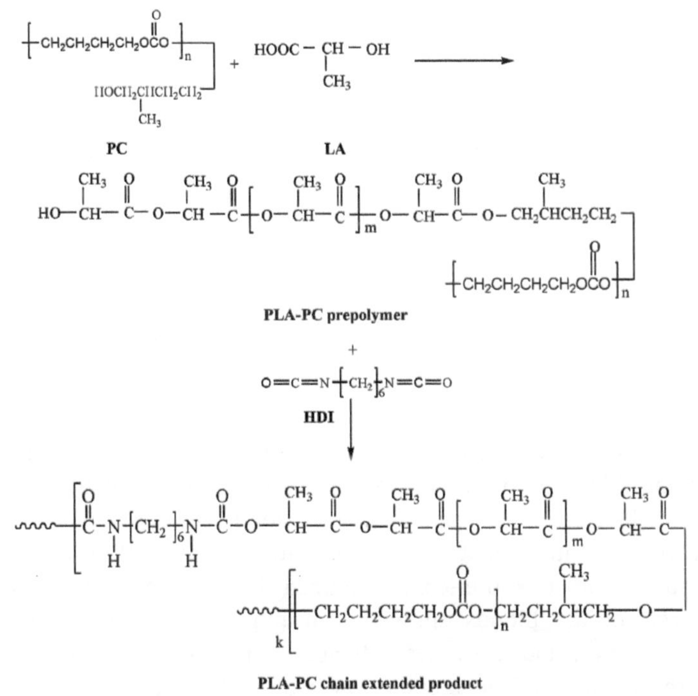

Figure 23.5 Synthesis and structure of PLA-PCD multiblock copolymer [20].

Ren et al. synthesized series of biodegradable polyesters poly(butylene adipate) (PBA), poly(butylene succinate) (PBS), and poly(butylene adipate-co-butylene terephthalate) (PBAT) by melt polycondensation. And the blends of PLA and biodegradable polyester were prepared using a twin screw extruder. PBAT, PBS, or PBA could be homogenously dispersed in PLA matrix at a low content (5–20 wt%), yielding the blends with much higher elongation at break than homo-PLA. Addition of PBAT into PLA may improve the toughness of PLA. The elongation at break increased at the higher polyester contents with the maximum value (>600%) occurring at a PBAT content of 15%. DSC analysis showed that the isothermal and nonisothermal crystallizabilities of PLA component were promoted in the presence of a small amount of PBAT [21]. Gu et al. reported that PLA and PBAT were melt blended using a twin-screw extruder. The phase morphology of PLA/PBAT blends was observed by scanning electronic microscopy (SEM). The linear and non-linear shear rheological behaviors of PLA/PBAT melts were investigated by an advanced rheology expended system (ARES). The results showed that PLA/PBAT blend was a kind of immiscible, two-phase system where PBAT dispersed evenly in PLA matrix. Linear viscoelastic limits of PLA/PBAT melts were smaller than that of neat PLA melt [22].

Most of the blends have been found to be immiscible with a phase-separated morphology. Ren et al. prepared binary and ternary blends of thermoplastic starch (TPS), PLA, and PBAT by using a one-step extrusion process. The concentration of TPS in both binary and ternary blends was fixed at 50 wt%, with the rest being PLA and PBAT. A compatibilizer with anhydride functional groups was used to improve the interfacial affinity between TPS and the synthetic polyesters. The addition of a small amount of compatibilizer greatly increased the mechanical properties of the blends. Mechanical properties of the blends exhibited a dramatic improvement in elongation at break with increasing PBAT content. Compared to the non-compatibilized blends, the morphology analysis of the blends showed that most of the TPS particles were melting and were well dispersed in the polyester matrix for the compatibilized blends. The water absorption data of the non-compatibilized blends increased more significantly than the compatibilized blends when PBAT content was increased [23]. Zhang et al. reported that PLA and PBAT were melt-blended in the presence of glycidyl methacrylate (GMA) by twin-screw extrusion. The results showed that with 2 or 5 wt% GMA, the tensile toughness of the PLA/PBAT blend was greatly increased without severe loss in tensile strength. The impact strength of the blend was also significantly improved at 1 wt% of GMA addition but ultimately trended to be saturated with increasing GMA. SEM micrographs revealed that better miscibility and more shear yielding mechanism were involved in the toughening of the blend. DSC results indicated that the blend was still a two-phase system in the presence of reaction agent and the addition of GMA was found to enhance the interfacial adhesion between PLA and PBAT [24].

Furthermore for the formation of a stereocomplex which blends of poly(L-lactide) and poly (D-lactide) could be found the interesting phenomena. The PLA stereocomplex was found to possess a racemic crystalline structure, where PDLA and PLLA chains were packed side by side with a D monomer unit to L monomer unit in the ratio of 1:1. One of the most interesting findings in PLLA/PDLA stereocomplexation studies was that the melting temperature of the stereocomplex was 230°C, approximately 50°C above that of the corresponding homopolymer. This enantiomeric polymer blend also exhibited substantially higher tensile properties and better thermal stability and hydrolysis resistance than the polylactides from which the blends were prepared [25].

23.2.3 Processing of PLA

Historically, the applications of PLA have been mainly limited to the biomedical areas due to its bioabsorbable characteristics. Over the past decade, the use of PLA in packaging and other consumer products has been considerably increased. The methods of manufacturing PLA have been used includes extrusion, injection molding, hot press molding, spinning, blow molding, foam molding, and electrospinning.

23.2.3.1 Extrusion

Extrusion is a common way for melting thermoplastics. Screw extruders are typically used in the polymer industry, and they can be part of the forming machine systems for injection molding, blow molding, melt spinning, and other polymer processes. They consist of an electrically heated metal barrel, a hopper for feeding the resin, a motor for rotating a screw, and a die where the polymer melt exists. So, the combination of thermal energy generated by a heater and frictional heat due to friction between the plastic and the screw and barrel provide sufficient heat to melt the pellets.

A typical screw consists of three sections: (1) feed section -acts as an auger which receives the polymer pellets and conveys the polymer into the screw; (2) transition section (also known as compression or melting sections)—flight depth decreases gradually, which compresses the pellets to enhance the friction and contact with the barrel. In order to segregate the molten polymer pool from the pellet unmelted pellets, various barrier flight designs have been adopted; (3) metering section—characterized by a constant and shallow flight depth, which acts as a pump to meter accurately the required quantity of molten polymer. The L/D ratio, which is the ratio of flight length of the screw to its outer diameter, determines the shear and residence time of the melt. Screws with large l/d ratio provide greater shear heating, better mixing, and longer melt residence time in the extruder. Another important screw parameter is the compression ratio, which is the ratio of the flight depth in the feed section to the flight depth in the metering section. The greater the compression ratio a screw possesses, the greater the shear heating it provides. Recommended extrusion conditions for PLA pellets include general purpose screws with L/D ratios from 24:1 to 30:1 and compression ratio of 2.5:1 to 3:1, melt temperature of 200 to 220°C, and also smooth barrels [26–28]. The combined thermal energy from the heater and frictional heat due to friction between the plastic and the screw and barrel, provide sufficient heat to raise the PLA polymer above its melting point (170–180°C) by the time it reaches the end of the barrel. To ensure that all the crystalline phases are melted and to achieve an optimal melt viscosity for processing, the heater is usually set at 200–210°C. Moreover, in the extruder, PLA can undergo thermal degradation, leading to the formation of lactide monomers and other byproducts. Therefore, the use of excessively high temperatures during melt processing should be avoided.

23.2.3.2 Injection molding

Injection molding is the most widely used converting process for thermoplastics, especially for those that are complex in shape and require high dimensional precision. Injection molding involves melting a thermoplastic by extrusion, injecting the polymer melt into a mold, cooling the part, and finally ejecting the part. Most thermoplastics including poly(ethylene terephthalate) (PET), poly(styrene) (PS), poly(propylene) (PP), high-density poly(ethylene) (HDPE), low-density poly(ethylene) (LDPE), nylon, and poly(vinyl chloride) PVC, etc., can be injection molded so long as they can flow and fill the mold cavity easily.

The injection molding machine is similar to the extruder and the main difference between them is in their screw operation. For extruder, the screw rotates continuously providing output of continuous and long product (pipe, rod, sheet), but the screw of an injection molding machine, which is called reciprocating screw, does not only rotate but also moves forward and backward according to the steps of the molding cycle. The mold is equipped with a cooling system providing controlled cooling and solidification of the material. Injection molding may be used to manufacture a wide variety of parts such as bottle caps, food trays, containers, and preforms for blow molding [29].

All injection molding machines have an extruder for plasticizing the polymer melt. Most injection molding machines for PLA are based on the reciprocating screw extruder. The beginning of mold close is usually taken as the start of an injection molding cycle. Immediately after the molds clamp up, the nozzle opens and the screw moves forward, injecting the polymer melt into the mold cavity. To compensate for the material shrinkage during cooling in the mold, the screw is maintained in the forward position by a holding pressure. At the end of the holding phase, the nozzle is shut off and the screw begins to recover, while the part continues to be cooled in the mold. During the recovery phase, the screw rotates and conveys the polymer forward along the screw. At the same time, the screw is allowed to slide backward within the barrel against a controlled back pressure exerted on the screw by a hydraulic cylinder. To ensure that the part is dimensionally stable enough to withstand the opening stroke the molds, sufficient cooling time must be given. In the molding cycle, heat removal takes place predominantly during the fill, hold, and cool phases, although mold opening phase also contributes to partial cooling since one side of the part (core-contacting side) is still being cooled prior to ejection. Cycle time is an important process parameter, which is often minimized to maximize the production throughput. To reduce the cycle time, it is quite common to transfer the partially cooled injection molded article to a post-mold cooling device, to provide an extended cooling of the part outside the molds, either by direct contact on a chilled surface and/or by forced air [27].

Injection mold-grade PLA can be injection molded on most conventional equipment, but there could be some torque limitations if the screw design has a high compression ratio. Compression ratios of 2.5 to 3 should be adequate and the recommended melting temperature is 200 to 205°C. Since PLA has a lower glass transition temperature (about 58°C) than PS or PET, it might take a little longer time to set up in the mold [30].

23.2.3.3 Cast film

The cast film process involves extruding a molten polymer through a slit die and drawing it around two or more highly polished high-speed rolls, typically chrome-plated and water-cooled. In less than one revolution, the chill roll solidifies the product as it draws it down to the correct thickness. Cast film is usually used in packaging, food wrap, substrate for coating, protective film, agricultural film for weed control, general-purpose polyethylene film as a protective barrier to prevent scratching of parts during shipment, and many other applications [31].

Due to rapid cooling by the chilled rolls, cast films typically have a low degree of crystallinity and transparent appearance. Besides providing good optical properties, cast film extrusion has the advantages of high production rate, good control of film thickness and uniformity, and little or no additive is required for processing [26].

Similar to PP, PET, and PS films, the physical properties of PLA films can be enhanced through orientation. Uniaxial orientation of PLA is achieved with conventional machine direction orientation (MDO) rolls. Since PLA tends to neck in (neck in happens by contacting the melt film with the first point of the die; the hot film shrinks on its way down so its width from the die to the chill roll is reduced. At the same time, beading or thickening of the edges occurs) during drawing, nipped rolls are usually required. It is possible to improve both the thermal resistance and impact resistance of PLA films or sheets by drawing, orientation, and crystallization to the same level of strength and stiffness as oriented polypropylene (OPP) or PET, while maintaining its high transparency. An oriented film is obtained by stretching it to 2 to 10 times its original length at 60–80°C, and further annealing it at temperatures between the stretching temperature and melting point. An oriented film may be processed for either dry lamination, printing, and heat seal or other applications, including various types of packaging [32].

The orientation in PLLA films depends on the draw rate, temperature, and ratio. High strain rate, low temperature, and high stretch ratio favor strain-induced crystallization during orientation. PLA has excellent optical properties and high modulus. However, it has low elongation, tear, and burst strengths. To overcome these shortcomings, PLA is often coextruded with other polymers to form multilayer structures to enhance its properties [27].

23.2.3.4 Stretch blow molding

With the increase in the recent environmental awareness, there is an improved interest from the food industry to replace the existing non-biodegradable thermoplastics with PLA for certain beverage products. PLA bottles are predominantly used for beverage containers, which are not sensitive to oxygen. While barrier properties of PLA bottles may be improved by various technologies, their implementation is currently limited due to higher production costs.

Injection stretch blow molding (ISBM) is an extension of injection blow molding with two modifications: (a) the preform is significantly shorter than the bottle and (b) a stretch rod is used to stretch the preform in the axial direction. This process became known in the blow molding industry with the introduction of plastic or PET soft drink bottles.

While all blow molding processes involve blowing air to stretch the parison or preform in some fashion, ISBM is designed to achieve and retain biaxial orientation to significantly improve gas barrier properties, impact strength, transparency, surface gloss, and stiffness. The process produces biaxial-orientated PLA bottle with much improved physical and barrier properties compared to injection molded amorphous PLA. The molecular orientation induced during the ISBM process decreases the effect of aging by stabilizing the polymer free volume [26].

ISBM-grade PLA resins are accessible and they are generally used for bottles for different foods like fresh dairy liquids, fruit juices, sport drinks, edible oils, and so on. ISBM-grade PLA resin is typically run at lower processing temperatures than bottle-grade PET and the blow molding conditions are preform temperature at 80 to 100°C, stretch rod speed 0.8 to 1.2 m/s, and blow mold temperature at 100 to 120°C. In fact, the heating of the preforms is critical in getting a container with good clarity and material distribution. Normal preform temperatures for running on a two-step process have been between 80 and 100°C. This temperature may be lower or higher depending on the preform design, bottle design, and reheating equipment that is being used [33].

23.2.3.5 Thermoforming

Thermoforming is a generic term encompassing many techniques for producing useful plastic articles from flat sheets. This process deals with the pressing or drawing of pliable plastic into final shape by vacuum or air pressure. Thermoforming is commonly used for forming products that do not have complicated features. Food packaging is the largest application for thermoformed containers, trays, cups, and tubs. Typical thermoforming steps are clamping, heating, shaping, cooling, and trimming [34].

In this process, PLA sheet is heated to soften the polymer, forced either pneumatic and/or mechanically against the mold, allowed to cool, removed from the mold, and then trimmed. The radiant heater of the thermoforming line for PLA must be adjusted to very low temperatures. Preheating is not absolutely necessary; however, it has the general advantage that the sheet is homogeneously preheated.

PLA sheet is quite brittle at room temperature and requires some special handling and storage considerations. There is a greater risk of cracking and breaking during transportation compared to oriented poly(styrene) (OPS) or PET. Typically, aluminum molds are used for thermoforming PLA containers. Molds, trim tools and ovens designed for thermoforming of PET, high-impact polystyrene (HIPS) and OPS can be used for forming PLA containers. However, molds for thermoforming of PP may not be used interchangeably for PLA, since PP shrinks more considerably than PLA during cooling. Neither the sheet nor the finished product can be stored at temperatures above 40°C or greater than 50% relative humidity. These conditions minimize moisture uptake and consequently sheet blocking, and resistance to unwinding. Exposure to high temperatures or humidity, even for a short period, can cause the material to deform and eventually break down. Sheet and formed products must be transported in cooled trucks and stored in a climate-controlled warehouse. The toughness of PLA increases with orientation, and therefore, thermoformed articles are less brittle than PLA sheet, particularly in the regions that have been highly stretched during the forming operations rather than flanges and lips. So, flange or lip areas that receive less orientation tend to be more brittle than the rest of the thermoformed part [26, 35, 36].

23.2.3.6 Foaming

PLA foams are first utilized in tissue engineering and medical implant applications, where the high cost of the materials is justified.

Their processing should be relatively simple and inexpensive if they want to be suitable for industrial foaming applications. However, thermal, oxidative, and hydrolytic degradations may occur during its processing-leading to the cleavage of polymer chains, and hence to a decrease in molecular weight. All these degradation processes result in a deterioration of the rheological properties, which should be avoided for processing purpose, especially when high levels of extensional viscosity and elasticity are required for extrusion foaming. Another shortcoming of PLA is its very low melt viscosity, which may also limit its blow molding and foaming processabilities. To overcome such shortcomings, many strategies to control the melt rheology of PLA and to increase the melt viscosity have motivated considerable research efforts [37].

Foaming of PLA is generally carried out by dissolving a blowing agent in the PLA matrix. The solubility of the blowing agent is then reduced rapidly by producing thermodynamic instability in the structure (e.g., temperature increase or pressure decrease), to induce nucleation of the bubbles. To stabilize the bubbles, the foam cells are vitrified when the temperature is reduced below the T_g of the polymer [38, 39].

Various foaming strategies have been adopted to reduce PLA density and improve foam mechanical properties. Loose-fill packaging materials provide cushioning, protection, and stabilization of packaged goods during shipping. Over the past decade, the use of expanded PS foams for loose-fill packaging has declined due to the replacement with the environmentally more benign starch-based expanded foams. To overcome the hydrophilic nature of starch, these biobased foams are often blended with petroleum polymers. PLA has been used for blending with starch during foaming. Recently, supercritical CO_2 has attracted considerable research attention as an environment-friendly solvent for many processes because it is inexpensive, nonflammable and can be easily removed from the products. A liquid becomes a supercritical fluid when both pressure and temperature are above the critical points. In this state, the substance possesses a combined gas-like viscosity and liquid-like density, making it an excellent solvent for various applications. For CO_2, its solubility and diffusivity in many polymers tend to increase considerably when it exists in the critical state, thereby facilitating the plasticization of many polymers and enabling the forming process to be conducted at lower temperatures [40]. The low-temperature supercritical CO_2 processes are expected to find their place in the manufacturing of structural foams for which mechanical properties are critical, since thermal and hydrolytic degradation encountered in the typical thermal processes can be avoided [27].

23.2.3.7 Spinning

PLA fibers are environment-friendly material and the demand for PLA fibers as a substitute for the present synthetic fibers has been growing greatly in recent years. Furthermore, high water vapor transmission rate of PLA makes it a good candidate for fabricating fibers used in garments (e.g., shirts, dresses, underwear, shoes, etc.) to improve their "breathability." While PLA fibers are not as wettable as cotton, they exhibit much greater water vapor transmission than aromatic polyesters or nylon fibers.

The manufacturing of PLA fiber is carried out by melt spinning, dry spinning, wet spinning, and dry-jet-wet spinning processes. Like other polyesters, such as polyethylene terephthalate (PET), PLA fibers can be manufactured by melt spinning owing to the thermoplastic nature of PLA. Melt spinning usually has advantages over wet spinning, i.e., it is a solvent-free process and provides a more economical and eco-friendly route. Production rate is normally higher than that in solution spinning. Sometimes, however, melt spinning is not possible, either because the polymer degrades while melting or the melt is thermally unstable. Commercially PLA fibers are generally produced using the melt spinning technique. PLA fibers are typically melt-spun at approximately 185–240°C through spinnerets with L/D ratios of 2–10. The processing temperature is comparable to polyolefins. The melting temperature used also depends on the optical purity of the polymer used. Lower processing temperatures can be used for lower optical purity polymer (i.e., greater d-isomer content), which can help reduce the thermal and hydrolytic degradation. Similar to the injection molding process, fiber-grade PLA needs to be dried to less than 0.005% (w/w) moisture before melting in the extruder to minimize molecular weight drop. In dry spinning, solvents are removed by thermal evaporation while in wet spinning the coagulation of the polymer is carried out in another fluid that is compatible with the spinning solvent but is not itself a solvent for the polymer. The dry spinning process involves dissolving the polymer in a solvent (typically chloroform, toluene or a mixture of the two solvents) and extruding the polymer solution in air or inert gas. Evaporation of the solvent causes the extruded filaments to solidify. The dry spinning technique is quite effective in preserving the molecular weight of the polymer due to the lower processing temperature used. Commercially PLA fibers are generally produced using the melt spinning technique. Moreover, ultra-fine PLA fibers can also be produced by electrospinning [33, 37].

23.3 PLA-Based Composite

PLA has been used as package materials and other products However, the physical properties of PLA such as brittleness limit the application of PLA. A way to improve the mechanical and thermal properties of PLA is the addition of fibers or nanofiller materials.

23.3.1 Natural Fiber-Reinforced Composites

Natural fiber (NF) is widely used as reinforcement to replace the conventional inorganic fibers in polymer matrix composites due to their advantages of being low weight, recyclable, biodegradable, low cost and high specific strength and modulus, etc. Natural fibers have different origins such as ramie, jute, hemp, sisal, and flax. These fibers are mainly made of cellulose, hemicelluloses, lignin and pectins, with a small quantity of extractives. The exact percentage of each chemical composition depends on the nature of the plant, location in which it is grown, age of the plant and extraction methods, etc. During the last decade, many research activities have been focusing on the natural fiber–reinforced thermoplastic composites, especially polypropylene (PP) matrix composites which are gaining increasing interest. As a result, natural fiber/PP composites have acceptable properties and have found wide use in the automobile industry [41]. Many studies have also been made to use PLA as matrix for natural fiber–reinforced composites.

The plants, which produce natural fibers, are classified as primary and secondary depending on their utilization. Primary plants are those grown for their fiber content while secondary plants are plants in which the fibers are produced as a by-product. Jute, hemp, kenaf, and sisal are examples of primary plants. Pineapple, oil palm, and coir are examples of secondary plants.

Flax is one of the oldest fiber crops in the world, which is grown in the temperate regions. Flax, *Linum usitatissimum*, belongs to the bast fibers. It is most frequently used in the higher value-added textile industry. Flax fibers have specific tensile properties greater than those of E-glass fibers. Nowadays, it is widely used in the fiber reinforced composites field. Bodros et al. reported that the mechanical properties of flax/PLA composites were quite promising compared to those of flax/polyporpylene (PP) composites for common commercial automotive use, which made them good candidates for plastics, automobile and packaging industries [42].

Ramie belongs to the family Urticaceae (Boehmeria), which includes about 100 species. The density of ramie fibers is much less than that of synthetics fibers; however, the specific strength and specific modulus of natural fibers are superior to E-glass fibers. Yu et al. fabricated ramie/PLA composites by hot pressing. The tensile strength of ramie/PLA composites was higher than that of jute/PLA composites. Since the strength of ramie was higher than that of jute. Elongation at break of the neat PLA or PLA-based composites decreased slightly by addition of the fibers due to their limitation of the mobility of the polymer matrix. Also the storage moduli and vicat softening temperature of PLA were improved slightly by addition of the fibers [43]. Yu et al. also studied the effect of PBAT on the mechanical properties and thermal properties of the composites. The results showed that the tensile and flexural strength decreased with the increase of PBAT content, and the addition of a small amount of PBAT increased the impact toughness (shown in Fig. 23.6) [44]. Chen et al. used press molding process to prepare the ramie fiber/PLLA composites, with short ramie fiber and ramie plain woven fabric as the reinforcement. The effect of ramie fiber, flame retardant and plasticizer on sound absorption property of ramie fiber reinforced PLA composites was studied. The results suggested that the special structures of ramie were the main reason for the better sound absorption property of the composite [45].

Jute is produced from plants of the genus Corchorus, which includes about 100 species. It is one of the cheapest natural fibers and is currently the bast fiber with the highest production volume. The fibers are extracted from the ribbon of the stem. Among all natural fibers, jute fibers are easily available in fabric and fiber forms with good mechanical and thermal properties. Plackett et al. determined the mechanical properties of the jute/PLA composites with the results showing that the tensile strength of composites was significantly higher than that of PLA. However, according to their results, the elongation at break of the composites was still very low which was about 2% [46]. Yu et al. reported that the storage moduli of the jute fiber/PLA composites increased with respect to the plain PLA [44].

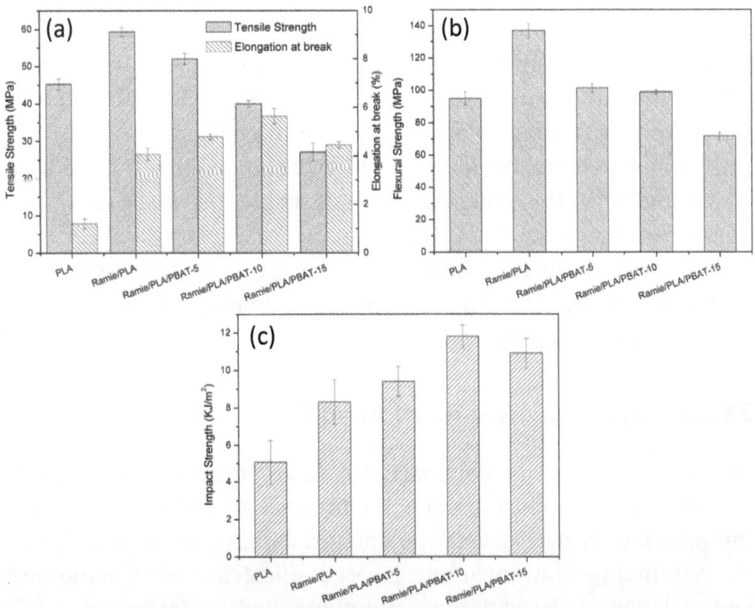

Figure 23.6 Mechanical properties of neat PLA, ramie/PLA without and with different PBAT content: (a) Tensile properties, (b) flexural properties, and (c) impact strength [45].

Kenaf is a warm season annual fiber crop closely related to cotton and jute, which belongs to the genus Hibiscus and there are about 300 species. Historically, kenaf has been used as a cordage crop to produce twine, rope, and sackcloth. Nowadays, there are various new applications for kenaf including paper products, building materials, absorbents and animal feeds [47]. Kenaf exhibits low density, non-abrasiveness during processing, high specific mechanical properties, and biodegradability. It is also used as nonwoven mats in the automotive industries, textiles.

Serizawa et al. reported high-performance biomass-based plastics that consist of PLA and kenaf fiber. Adding kenaf fiber to PLA greatly increased its heat resistance (distortion temperature under load) and modulus and also enhanced its crystallization [48].

Hemp is also bast fiber crop, which belongs to the Cannabis family. It is known to provide high mechanical strength and Young's modulus, and consists of cellulose (55–72%), hemicelluloses (8–19%), lignin (2–5%), wax (<1%) and minerals (4%). It is estimated that in recent years China has become the world leader in hemp fiber production, which accounts for nearly one third of total production. However, the application of hemp fiber is still mainly limited to the textile industry [49]. Masirek et al. prepared the composites of poly(L-lactide) (PLLA) with hemp fibers (Cannabis sativa) by batch mixing. Mechanical tests showed that the modulus of elasticity of the composites remarkably increased with the hemp content, reaching 5.2 GPa in the case of crystallized PLA reinforced with 20 wt% hemp, whereas the elongation and stress at break decreased with an increasing amount of fiber for all examined systems [50].

23.3.2 Nanoparticle-Reinforced Composites

Inorganic or organic nanoparticles have been incorporated to enhance the mechanical, barrier and thermal properties of PLA. Over the past few years, various nanomaterials have been investigated for reinforcing PLA, including layered silicates, carbon nanotube, hydroxyapaite, layered titanate, aluminum hydroxide, etc.

Among the nanomaterials investigated, the incorporation of organically modified layered silicate (OMLS) into polymers has been of particular interest because of their demonstrated significant enhancement of the physical properties of the polymers, including barrier, flammability resistance, thermal and environmental stability, solvent uptake, and rate of biodegradability of biodegradable polymers. Layered silicates generally have layer thickness at the order of 1 nm and very high aspect ratios (e.g., 10–1000). A few weight percent of OMLSs that are properly dispersed throughout the matrix thus create a much higher surface area for polymer filler interactions than conventional glass or mineral fibers do on their reinforced composites. For these reasons, polymer nanocomposites are lighter in weight than conventional composites, and make them competitive with other materials for specific applications. Thus, the incorporation of inorganic particulate fillers has been shown to be an effective way for the improvement of mechanical properties, although not always the toughness of polymeric materials. Sinha et al. reviewed the preparation, characterization, materials properties, crystallization behavior, melt rheology, and foam processing of pure polylactide (PLA) and PLA/layered silicate nanocomposites. They concluded this new family of composite materials frequently exhibited remarkable improvements in its material properties when compared with those of virgin PLA. Improved properties may include a high storage modulus in both solid and melt states, increased tensile and flexural properties, decreased gas permeability, increased heat distortion temperature, and increased rate of biodegradability of pure PLA [51].

Carbon nanotubes (CNTs) have been the subject of much attention because of their outstanding performance, including excellent mechanical, electrical, and thermal properties. The most promising area of nanocomposite research involves the reinforcement of polymers using CNTs as reinforcing filler. Li et al. introduced functionalized multiwalled carbon nanotubes (f-MWCNTs) into PLA to investigate the effect of such filler on the crystallization behavior of PLA. They concluded that the addition of f-MWCNTs accelerated the crystallization of PLA dramatically and induced formation of homogeneous and very small spherulites. The results of polarized optical microscopy showed that the average spherulite diameter was about 200 μm, but for nanocomposites, it was very difficult to differentiate the spherulites one by one [52].

Nakayama reported that TiO_2 nanoparticles were prepared and the surface of TiO_2 was modified using propionic acid and n-hexylamine, with the modified TiO_2 uniformly dispersed into PLA matrix without aggregation. They studied the photo degradation of PLA-TiO_2 nanocomposite under UV light and concluded photo degradability of nanocomposites could be efficiently promoted [53].

Kim et al. studied the effect of bacterial cellulose on the transparency of PLA/bacterial nanocomposites, since bacterial cellulose had shown good potential as reinforcement or preparing optically transparent materials due to its structure, which consists of ribbon-shaped fibrils with diameters in the range from 10 to 50 nm. They found that light transmission of the PLA/bacterial cellulose nanocomposite was quite high due to the size effect of the nanofibrillar bacterial cellulose. Additionally, the tensile strength and Young's modulus of the PLA/bacterial cellulose nanocomposite were increased by 203% and 146%, respectively, compared with those of the PLA [54].

23.3.3 Surface Treatment Method

23.3.3.1 For natural fiber–reinforced composites

The inherent polar and hydrophilic nature of lignocellulosic fibers and the nonpolar characteristics of most thermoplastics results in compounding difficulties leading to non-uniform dispersion of fibers within the matrix, which impairs the efficiency of the composite. This is a major disadvantage of natural fiber–reinforced composites.

Natural fibers are amenable to chemical modification due to the presence of hydroxyl groups. The hydroxyl groups may be involved in the hydrogen bonding within cellulose molecules thereby activating these groups or can introduce new moieties that form effective interlocks within the system. Many researches focused on using different physical and chemical methods, including silation, acylation, alkali treatment, etherification, isocynate treatment, and coupling agent treatments to improve interfacial bonding properties of natural fiber–reinforced PLA composites.

The popular chemical modification is silanation. The silanes are hydrophilic compounds with different groups appended to silicon such that one end will interact with hydrophilic compounds and the other end can react with the silane-coupling agent acting as a bridge between them. The uptake of silane is very much dependent on a number of factors including hydrolysis time, organofunctionality of silane, temperature and pH value. The different types of chemical treatment and its influence on properties of the composites have been described in detail by Yu et al. Ramie was treated by silane (3-aminopropyltriethoxy silane and c-glycidoxypropyltrimethoxy silane) and alkali (shown in Fig. 23.7), which improved the tensile strength, flexural strength, and impact strength. The Fourier transform infrared spectroscopy (FT-IR) showed that siloxane groups were grafted on ramie fibers. And alkali treatment improved the fiber surface adhesion characteristics by removing natural and artificial impurities, thereby producing a rough surface [55].

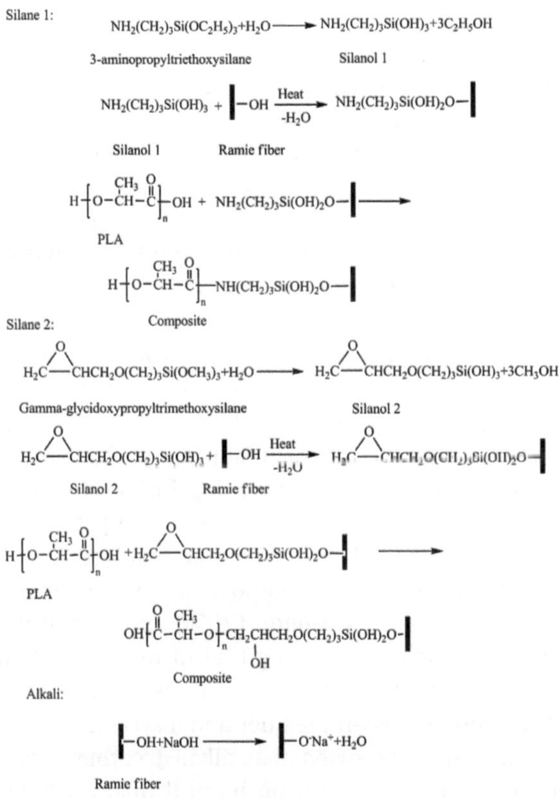

Figure 23.7 Reaction of surface treatment [55].

Goriparthi et al. studied the effect of fiber surface treatments on mechanical and abrasive wear performance of jute/PLA composites by hot pressing of solvent impregnated prepregs, and surface of the jute fiber was modified by silane, alkali, permanganate, and peroxide treatments. The results showed that surface treatments led to enhancements of tensile and flexural properties and a reduction in Izod impact strength. Composites treated with trimethoxymethyl silane resulted in the maximum improvement of tensile strength and moduli by 35% and 38%, respectively. Also results showed that treated composites had higher storage modulus and lower tangent delta, when compared to the rest of the composites [56].

Maleated coupling agents are also widely used to strengthen composites containing fillers and fiber reinforcements. Interactions between the anhydride groups of maleated coupling agents and the hydroxyl groups of natural fibers can overcome incompatibility problem to increase the tensile and flexural strengths of natural fiber–reinforced thermoplastic composites.

Garcia et al. fabricated PLA composites reinforced with kenaf fiber using extrusion and injection molding. They also added maleated-PLA in the composites as compatibilizer. It was found that tensile strength of the composites reinforced with 30 wt% fiber decreased about 10% compared to unreinforced PLA samples, which was inconsistent with other works. This could be due to the difference in reinforcement type and processing method [57]. Nyambo et al. reported that PLA-*g*-MA was used as a potential method for improving interfacial adhesion between agricultural residues and PLA, with the goal of enhancing mechanical properties. Compatibilization was achieved by using PLA-*g*-MA prepared via reactive extrusion. Green renewable and compatibilized wheat straw/PLA were extruded and injection molded. Addition of 3 and 5 phr PLA-*g*-MA to the composites resulted in significant improvements in tensile strength (20%) and flexural strength (14%) of the composites, matching that of the neat polymer. The observed improvement in strength was attributed to the good interfacial adhesion between the fiber and matrix [58].

Many researchers reported that alkali treatment can remove the lignin and hemicelluloses from natural fibers, which affected the tensile characteristics of the fibers, also decreased the moisture absorption, increased tensile strength, and the wettability of fibers. Ma et al. reported the biodegradable composites based on jute and PLA were prepared by a film-stacking process. The effect of jute fiber content, processing temperature, and alkali treatment on the structure and mechanical properties of jute/PLA composites were investigated. The optimum tensile properties of jute/PLA composites were obtained at 15 wt% fiber content and a processing temperature of 210°C. Both the maximum flexural modulus and strength of composites were obtained at 220°C and 15 wt% fiber contents. Meanwhile, the tensile moduli of single jute fiber and jute/PLA composites were increased by 29% and 76%, respectively. The modified equation of the rule of mixture was more suitable to predict tensile modulus of jute/PLA composites than the commonly used Halpin—Tsai model and the relative error was less than 8% [59].

Table 23.1 Mechanical properties of natural fiber–reinforced PLA composites [47]

Reinforcement	Processing	Fiber content (%)	Tensile strength (MPa)	Young's modulus (GPa)	Flexural strength (MPa)	Impact strength (KJ/m^2)
Flax/PLA (RO)	Compression molding	30	53	8.3	—	—
Flax/PLA (RO)	Injection molding	30	54.15	6.31	—	11.13
Flax/PLA (RO)	Injection molding	26	53.1	7.3	—	—
Flax/PLA (RO)	Film-stacking	25	81.3	8.85	—	—
Kenaf/PLA (MD)	Compression molding	40	82	7.6	126	14
Kenaf/PLA (fabric)	Compression molding	43	67	5.3	—	—
Kenaf/PLA (UD)	Compression molding	35	131	15	160	—
Kenaf/PLA (RO)	Compression molding	30	32	4.5	—	—
Kenaf/PLA (MD)	Compression molding	40	53	7.1	—	9.9
Jute/PLA(UD)	Compression molding	50	152	5.3	174	—
Jute/PLA (MD)	Film-stacking technique	40	93.5	—	—	14.3
Jute/PLA(RO)	Compression molding	30	48	—	101	8.3
Ramie/PLA (RO)	Compression molding	30	52.5	—	170	10
Ramie/PLA (RO)	Compression molding	30	52	—	105	9.3
Hemp/PLA (RO)	Injection molded	30	75	7.9	—	—
Hemp/PLA(MD)	Compression molding	35	85	12.7	—	7.5
Hemp/PLA(MD)	Compression molding	40	44.63	7.39	90	—
Sisal/PLA(RO)	Injection molded	30	—	—	97	3.3

Aydin et al. reported the influence of alkali treatment on the mechanical, thermal, and morphological properties of eco-composites made by short flax fiber reinforced PLA. Scanning electron microscopy (SEM) revealed that the packed structure of the fibrils was deformed by the removal non-cellulosic materials. The mechanical tests indicated that the modulus of the untreated flax fiber/PLA composites was higher than that of PLA; on the other hand, the modulus of alkali treated flax/PLA was lower than that of PLA. Thermal properties of the PLA in the treated flax fiber reinforced composites were also affected. T_g values of treated flax fiber reinforced composites were lowered by nearly 10°C for 10% NaOH treatment and 15°C for 30% NaOH treatment. A bimodal melting behavior was observed for treated fiber composites different than both of neat PLA and untreated fiber composites [60].

Improved fiber matrix adhesion and thereby improved mechanical properties can be accomplished by engineering a superior processing condition in preparing the bio-composites, by altering the polymer architecture of the matrix or by the surface treatment on the fiber [47]. Table 23.1 showed mechanical properties of natural fiber–reinforced PLA composites.

23.3.3.2 For nanoparticle-reinforced composites

Nanoparticles obtained from available natural resources generally need to be treated because the physical mixture of a polymer and layered particles for instance may not form a nanocomposite. In this case, a separation into discrete phases takes place. The poor physical interaction between the organic and the inorganic components leads to poor mechanical and thermal properties. In contrast, strong interactions between the polymer and the layered silicate nanocomposites would lead to the organic and inorganic phases being dispersed at the nanometer level. Organic modification of these nanofiller and addition of compatibilizer have been reported as successful methods.

Pluta et al. investigated PLA/MMT nanocomposites by studying the dispersions of modified and unmodified MMT prepared by melt blending. XRD showed that the good affinity between the organo-modified clay and the PLA was sufficient to form an intercalated structure in the nanocomposite. TGA showed that the PLA-based nanocomposites exhibited improvement in their thermal stability in air [61]. Ray et al. also reported that MMT modified with trimethyl octadecylammonium cation was used as an organically modified layered silicate for the nanocomposite preparation. All the nanocomposites exhibited superior improvement in practical materials properties such as storage modulus, flexural modulus, flexural strength, heat distortion temperature, and gas barrier property, compared to that of neat PLA. The biodegradability of neat PLA and a representative nanocomposite was also studied under compost, and the rate of biodegradation of neat PLA significantly increased after the addition of nanomaterials [62].

Chow et al. studied the effect of maleic anhydride-grafted ethylene propylene rubber (EPM-g-MA) on the thermal properties of organo-montmorillonite nanocomposites/PLA. They concluded that the addition of OMMT (organo-montmorillonite) and EPM-g-MA did not much influence the T_g and T_m (melting temperature) of PLA nanocomposites. The degree of crystallinity of PLA increased slightly in the presence of OMMT. It had been supposed that OMMT could act as a nucleating agent to increase the crystallinity of PLA. In contrast, the addition of EPM-g-MA may restrict the crystallization process and crystal formation of PLA, which subsequently reduced the degree of crystallinity of PLA/OMMT nanocomposites. Finally, they claimed that the thermal stability of PLA/OMMT was greatly enhanced by the addition of EPM-g-MA [63]. Ren et al. prepared a series of OMMT/PLA and PLA/PCL/OMMT nanocomposites using a simple melt extrusion. After twin-screw direct extrusion, OMMT interlayers spacing was enlarged, and with the addition of OMMT, the average spacing between layers decreased but was still larger than that of pristine OMMT. It indicated that the intercalated structure was formed in nanocomposites. And OMMT can help to increase the thickness of stacked internal structure but decrease the crystallinity of PLA. The addition of PCL had little effect on OMMT interlayers spacing, in addition, it made layers arrangement more regular and increased the degree of crystallinity of PLA (shown in Table 23.2). The addition of OMMT and PCL could both affect the thermal properties of the nanocomposites. Furthermore, annealing treatment was favorable to the better thermal properties [64].

Table 23.2 Thermal parameters of several PLA/PCL/OMMT nanocomposites of different PCL contents [64]

Characteristic parameters	100/0/5	100/1/5	100/3/5	100/5/5
Glass transition temperature (T_g) (°C)	50.8	50.8	50.2	50.1
Recrystallization temperature (T_c) (°C)	101.3	100.4	96.7	96.6
Crystalline melting temperature (T_m) (°C)	146.9	147.5	147.8	148.4
χ_c (%) ($H_m/H_{100\%}$)	31.9	34.9	34.6	35.6

Note: $H_{100\%}$ = 93 J/g.

Pillai et al. described a comparison between two chemical functionalization strategies for the amine functionalization of multiwalled CNTs. The modified CNTs with optimum amine content were used to prepare PLA/CNT nanocomposites through solution casting method. The polymer nanocomposite thus prepared showed improved thermal properties when compared to the neat PLA [65].

23.3.4 Processing of the Composites

23.3.4.1 Natural fiber–reinforced composite

Generally, natural fiber–reinforced plastic composites are manufactured by using traditional techniques designed for synthetic fiber reinforced polymer composites and thermoplastics. The processing techniques includes compounding, mixing, extrusion, injection molding, compression molding, which have been well developed and accumulated experience has proofed their success for producing composites with controllable quality. Innovative technologies and processing solutions should be intensively investigated to get the high performances engineering composites, which are related to their expanded application areas.

Blending of PLA with natural fibers such as kenaf is another possibility by which PLA can be converted to end products using slightly modified standard industrial machinery for thermoplastics such as thermoforming, injection molding, blow molding extrusion and importantly, film extrusion. High value films and rigid thermoformed containers are the most promising bulk applications.

Oksman et al. studied flax/PLA composite manufactured by a twin-screw extruder. Results have shown 50% increase of strength for flax/PLA composites compared to flax/PP composites. Micrographs of the composites showed that the adhesion between flax fibers and the matrix was quite poor which needed to be improved [66]. Bax et al. also manufactured PLA reinforced with cordenka rayon fibers and flax fibers by using hot-pressing followed by pelletizing and injection molding [67]. Tawakkal et al. studied kenaf deriver cellulose (KDC)-filled PLA composites, which were prepared via melt blending and compression modeling. The SEM analysis showed a poor adhesion between PLA matrix and kenaf fibers, due to the absence of a compatibilizer in the formulation [68]. Lee et al. demonstrated that long kenaf fibers derived from the bast part of the plant may be used to successfully reinforce PLA; these novel and useful bio-composites were made using a combination of carding and punching processes followed by hot press compression molding. It was demonstrated that such materials can be successfully utilized to manufacture automotive headliners [69].

23.3.4.2 Nanoparticle-reinforced composites

The manufacturing methods for PLA nanocomposites include intercalation of polymer from solution, polymer melt intercalation and intercalation of a suitable monomer and subsequent in situ polymerization.

The first route is the intercalation of polymer from solution. The layered host is exfoliated in a solvent, in which the polymer is soluble (water, toluene, etc). The polymer is adsorbed onto the single-layer surfaces and after evaporation of the solvent or a precipitation procedure, the single layers are restacked, trapping the polymer and the hydrated/solvated ionic species [70].

Chang et al. incorporated hexadecylamine, dodecyltrimethyl ammonium bromide, and quaternary ammonium salt-modified MMT clays in PLA films at 0–10% (w/w) MMT levels, based on wet casting method, using *N,N*-dimethylacetamide as a solvent. They observed that there was an optimal loading of organoclay (~3–

4%, w/w) for achieving the greatest improvement in mechanical properties. At higher MMT contents, the nanoparticles tended to agglomerate, leading to mechanical weakening. In contrast, within the 0–10% MMT loading levels, oxygen transmission rates decreased with increasing clay content, which was attributed to the increased tortuous diffusion path of oxygen molecules through the impermeable crystalline platelets [71]. Ogata et al. used a similar film casting methodology to evaluate the impact of organophilic MMT (0–10%, w/w) on PLA, except that chloroform was used as a solvent. These authors reported that only microcomposites were formed and intercalated structures were not achieved. It was unknown whether the formation of microcomposites was due to the incomplete exfoliation of MMT in the chloroform solution or caused by the agglomeration of MMT during the film drying step. Nevertheless, these studies did highlight the importance of solvent selection, which could affect the exfoliation of the layered silicates [72].

Another route is polymer melt intercalation. This method, an environmentally benign one, uses all types of polymers as well as being compatible with practicing polymer industrial processes such as injection molding, being the most popular procedure to prepare nanocomposites for industrial applications. In this method, polymers, and layered hosts are annealed above the softening point of the polymer [73].

Ray et al. studied various organically modified layered silicates, which was attributed to the elevated shear mixing of PLA in the presence of the silicate, and the presence of hydroxyl groups in the modified salt, both of which were capable of causing hydrolysis or transesterification reactions in PLA when exposed to elevated processing temperature. In their study, the nanocomposites were first prepared by dry-mixing of PLA resin with organoclay and extruding in twin-screw extruder (screw speed = 100 rpm, feed rate = 120 g/min) to form nanocomposite strands. The strand were pelletized and dried under vacuum at 60°C for 48 h, pressed into 0.7–2 mm sheet with 1.5 MPa at 190°C for 3 min, and then quenched and annealed at 110°C for 1.5 h before testing [74].

In the in situ polymerization method, the layered silicate is swollen within the liquid monomer or a monomer solution so the polymer formation can occur between the layered silicates. Polymerization can be initiated by either heat or radiation, by the diffusion of a suitable initiator, or by an organic initiator or catalyst fixed through cation exchange inside the interlayer before the swelling step. Subsequent polymerization takes place after combining the silicate layers and monomer, thus allowing the formation of polymer chains between the intercalated sheets [73].

Paul reported that both intercalated and exfoliated poly (L,L-lactide) (P(L,L-LA)/organo-modified montmorillonite nanocomposites were synthesized by in situ ring-opening polymerization of L,L-lactide, in bulk, directly in the presence of the nanofiller. Intercalation of polyester chains was found to appear even for natural unmodified montmorillonite, while exfoliation occurred when the aluminosilicate layers were modified by ammonium cations bearing primary hydroxyl groups. Clay delamination was effectively triggered by the grafting reaction of the growing PLA chains onto the hydroxyl groups [75].

23.4 Applications of PLA and PLA-Based Composites

The bio-based polymer and composites open new windows for becoming independent from petrochemical-based polymers and also free of environmental and health concerns.

In past years, PLA has been found practical medical applications as dissolvable structures, as matrices for drug delivery, and bone fracture internal fixation devices in surgery. Other applications include its role as agricultural plant growth promoter, in textiles, and non-woven applications such as fiberfill, crop covers, geotextiles, wipes, diapers, binder fibers, etc. However, the total volume of PLA production remains small compared to other polymers. Moreover, in previous years, the most negative point of PLA was its price in comparison with petrochemical-based polymers. Today, by using other sources of dextrose, optimizing lactic acid production processes and its costs, substituting electricity energy by wind and solar energy for PLA production, optimizing PLA production processes, and increasing PLA demands, reduction of its price can be attained. The present PLA price is much lower than in previous years.

Also, the brittleness of PLA may also limit its applications where toughness and impact resistance are critical. However, with the help of technology and providing safe PLA composites, many of its weakness compared to petrochemical-based polymer will be solved. PLA has the potential uses, for example as upholstery, disposable garments, awnings, feminine hygiene products, and nappies.

The commercial importance of polymers has been driving intense applications in the form of composites in various fields. Performance during use is a key factor of any composite material, which decides the real fate of products during services. Research and development in these areas may open up new opportunities for PLA for use as high performance biodegradable materials. It is predicted that PLA and PLA-based composites will be used in wide areas such as aerospace, automotive, marine, infrastructure, military, etc.

Acknowledgments

This work is supported by Major State Basic Research Development Program of China (973 Program) (No. 2010CB631105), Natural Science Foundation of China (No. 51103108 and No. 11172212).

References

1. Rasal, R. M., Janorkar, A. V., and Hirt, E. D. (2010). Poly(lactic acid) modifications, *Prog. Polym. Sci.*, **35**, 338–356.
2. Auras, R., Lim, L. T., Selke, S. E. M., and Tsuji, H. (2010). Industrial production of high molecular weight. In: *Poly(Lactic Acid): Synthesis, Structures, Properties, Processing and Applications* (Sődergărd, A., and Stolt, M., eds.), John Wiley & Sons, Inc, Chapter 3, pp. 27–41.
3. Hartmann, M. H. (1998). High-molecular-weight polylactic acid polymers. In: *Biopolymers from Renewable Resources* (Kaplan, D. L., ed.), Springer, pp. 367–411.
4. Moon, S. I., Lee, C. W., Taniguchi, I., Miyanoto, M., and Kimura, Y. (2001). Melt/solid polycondensation of L-lactic acid: An alternative route to poly(L-lactic acid) with high molecular weight. *Polymer*, **42**, 5059–5062.
5. Li, L., Tang, S. C., Wang, Q. H., Pan, Y. K., and Wang, T. L. (2006). Preparation of poly(lactic acid) by direct polycondensation in azeotropic solution. *J. East. China Univ. Sci. Technol.*, **32**, 672–675.
6. Garlotta, D. (2001). A literature review of poly(lactic acid). *J. Polym. Environ.*, **9**, 63–84.

7. Sodergard, A., and Stolt, M. (2002). Properties of lactic acid based polymers and their correlation with composition. *Prog. Polym. Sci.*, **27**, 1123–1163.

8. Mehta, R., Kumar, V., Bhunia, H., and Upadhyay, S. N. (2005). Synthesis of poly(lactic acid): A review. *J. Macromol. Sci. Part C Polym. Rev.*, **45**, 325–349.

9. Nampoothiri, K. M., Nair, N. R., and John, R. P. (2010). An overview of the recent developments in polylactide (PLA) research. *Bioresour. Technol.*, **101**, 8493–8501.

10. Kricheldorf, H. R., Berl, M., and Scharnagi, N. (1998). Polylactones: 9 Polymerization mechanism of metal alkoxide-initiated polymerizations of lacide and various lactones. *Macromolecules*, **21**, 286–293.

11. Jacobsen, S., Fritz, H. G., Degee, P., Dutois, P., and Jerome, R. (2000). New developments onthc ring opening polymerization of polylactide. *Ind. Crops Prod.*, **11**, 265–275.

12. Hiltunen, K., Seppl, J. V., and Hrkönen, M. (1997). Lactic acid based poly(ester-urethane)s: The effects of different polymerization conditions on the polymer structure and properties. *J. Appl. Polym. Sci.*, **64**, 865–873.

13. Gu, S., Yang, M., Yu, T., Ren, T., and Ren, J. (2008). Synthesis and characterization of biodegradable lactic acid-based polymers by chain extension. *Polym. Int.*, **57**, 982–986.

14. Ren, J., Yu, T., Yang, M., Gu, S., and Ren, T. (2008). Effect of different chain-extenders on chain extending of lactic acid oligomers. *Polym. Preprints*, **49**, 466–467.

15. Kylma, J., Tuominen, J., Helminen, A., and Seppala, J. (2001). Chain extending of lactic acid oligomers. Effect of 2,2′-bis(2-oxazoline) on 1,6-hexamethylene diisocyanate linking reaction. *Polymer*, **42**, 3333–3343.

16. Tuominen, J., Kylma, J., and Seppala, J. (2002). Chain extending of lactic acid oligomers. 2. Increase of molecular weight with 1,6-hexamethylene diisocyanate and 2,2′-bis(2-oxazoline). *Polymer*, **43**, 3–10.

17. Cohn, D., and Hotovely-Salomon, A. (2005). Designing biodegradable multiblock PCL/PLA thermoplastic elastomers. *Biomaterials*, **26**, 2297–2305.

18. Yu, T., Ren, J., Gu, S., and Yang, M. (2010). Preparation and characterization of biodegradable poly(lactic acid)-block-poly(ε-caprolactone) multiblock copolymer. *Polym. Adv. Technol.*, **21**, 183–188.

19. Cohn, D., and Hotovely-Salomon, A. (2005). Biodegradable multiblock PEO/PLA thermoplastic elastomers: Molecular design and properties. *Polymer*, **46**, 2068–2075.
20. Yu, T., Ren, J., Gu, S., and Yang, M. (2009). Synthesis and characterization of poly(lacticacid) and aliphatic polycarbonate copolymers. *Polym. Int.*, **58**, 1058–1064.
21. Zhao, P., Liu, W., Wu, Q., and Ren, J. (2010). Preparation, mechanical, and thermal properties of biodegradable polyesters/poly(lactic acid) blends. *J. Nanomater.*, **2010**, 287082-1–8.
22. Gu, S., Zhang, K., Ren, J., and Zhan, H. (2008). Melt rheology of polylactide/poly(butylene adipate-co-terephthalate) blends. *Carbohydr. Polym.*, **74**, 79–85.
23. Ren, J., Fu, H., Ren, T., and Yuan, W. (2009). Preparation, characterization and properties of binary and ternary blends with thermoplastic starch, poly(lactic acid) and poly(butylene adipate-co-terephthalate). *Carbohydr. Polym.*, **77**, 576–582.
24. Zhang, N., Wang, Q., Ren, J., and Wang, L. (2009). Preparation and properties of biodegradable poly(lactic acid)/poly(butylene adipate-co-terephthalate) blend with glycidyl methacrylate as reactive processing agent. *J. Mater. Sci.*, **44**, 250–256.
25. Brochu, S., Prud'homme, R. E., Barakat, I., and Jerome, R. (1995). Stereocomplexation and morphology of polylactides. *Macromolecules*, **28**, 5230–5239.
26. Jamshidian, M., Tehrany, E. A., Imran, M., Jacquot, M., and Desobry, S. (2010). Poly-lactic acid: Production, applications, nanocomposites, and release studies. *Compr. Rev. Food Sci. Food Saf.*, **9**, 552–571.
27. Lim, L. T., Auras, R., and Rubino, M. (2008). Processing technologies for poly(lactic acid). *Prog. Polym. Sci.*, **33**, 820–852.
28. Natureworks. (2005). Natureworks polymer technical data sheets—PLA polymer 7032D. Minnetonka, MN: Natureworks LLC.
29. Rosato, D. V., Rosato, D. V., and Rosato, M. G. (2000). *Injection Molding Handbook*. 3rd ed. Boston: Kluwer Academic Publishers 1488 p.
30. NatureWorks. 2006b. PLA injection molding guide for 3051D. Minnetonka, Minn.: NatureWorks LLC.
31. Giles, F. H., Wagner, J. R., and Mount, E. M. (2005). *Extrusion, the Definitive Processing Guide and Handbook*. 1st ed. New York: William Andrew Publishing. 547 p.
32. Kawashima, N., Ogawa, S., Obuchi, S., Matsuo, M., and Yagi, T. (2002). Poly lactic acid LACEA. In: *Biopolymers Polyesters III Applications*

and Commercial Products (Doi, Y., Steinbuchel, A., eds.), Weinheim: Wiley–VCH Verlag GmbH. pp. 251–274.

33. NatureWorks. 2005a. PLA ISBM bottle guide. Minnetonka, Minn.: NatureWorks LLC.
34. Throne, J. L. (1996). *Technology of Thermoforming*. 1st ed. New York: Hanser Publishers. p. 922.
35. NatureWorks. 2005c. Processing guide for thermoforming articles. Minnetonka, Minn.: NatureWorks LLC.
36. Patey, W. (2010). Thermoforming PLA: how to do it right. *Plastics Technol.*, **56**, 30–31.
37. Ren, J. (2010). *Biodegradable Poly(Lactic Acid): Synthesis, Modification, Processing and Applications*. Springer-Verlag Berlin Heidelberg, p. 186.
38. Di, Y., Iannace, S., Di, M. E., and Nicolais, L. (2005). Reactively modified poly(lactic acid): properties and foam processing. *Macromol. Mater. Eng.*, **290**, 1083–1090.
39. Mikos, A. G., Thorsen, A. J., Czerwonka, L. A., Bao, Y., Langer, R., Winslow, D. N., and Vacanti, J. P. (1994). Preparation and characterization of poly(L-lactic acid) foams. *Polymer*, **35**, 1068–1077.
40. Nalawade, S., Picchioni, F., and Janssen, L. P. B. M. (2006). Supercritical carbon dioxide as a green solvent for processing polymer melts: Processing aspects and applications. *Prog. Polym. Sci.*, **31**, 19–43.
41. Yu, T., and Li, Y. (2014). Influence of poly(butylenes adipate-co-terephthalate) on the properties of the biodegradable composites based on ramie/poly(lactic acid). *Composites Part A.*, **58**, 24–29.
42. Bodros, E., Pillin, I., Montrelay, N., and Baley, C. (2007). Could biopolymers reinforced by randomly scattered flax fibre be used in structural applications? *Compos. Sci. Technol.*, **67**, 462–470.
43. Yu, T., Li, Y., and Ren, J. (2009). Preparation and properties of short natural fiber reinforced poly(lactic acid)(PLA) composites. *Trans. Nonferrous Metals Soc. China*, **19**, S651–S655.
44. Chen, D. K., Li, J., and Ren, J. (2010). Study on sound absorption property of ramie fiber reinforced poly(L-lactic acid) composites: Morphology and properties. *Composites Part A.*, **41**, 1012–1018.
45. Yu, T., and Li, Y. (2014). Influence of poly(butylenes adipate-co-terephthalate) on the properties of the biodegradable composites based on ramie/poly(lactic acid). *Composites Part A.*, **58**, 24–29.
46. Plackett, D., Andersen, T. L., Pedersen, W. B., and Nielsen, L. (2003). Biodegradable composites based on L-polylactide and jute fibres. *Compos. Sci. Technol.*, **63**, 1287–1296.

47. Hassan, E., You, W., He, J., and Yu, M. (2012). Plant fibers reinforced poly(lactic acid) (PLA) as a green composites: Review. *Int. J. Eng. Sci. Technol.*, **4**, 4429–4439.
48. Serizawa, S., Inoue, K., and Iji, M. (2006). Kenaf-fiber-reinforced poly(lactic acid) used for electronic products. *J. Appl. Polym. Sci.*, **100**, 618–624.
49. Faruk, O., Bledzki, A. K., Fink, H., and Sain, M. (2012). Biocomposites reinforced with natural fibers: 2000–2010. *Prog. Polym. Sci.*, **37**, 1552–1596.
50. Masirek, R., Kulinski, Z., Chionna, D., Piorkowska, E., and Pracella, M. (2007). Composites of poly(L-lactide) with hemp fibers: Morphology and thermal and mechanical properties. *J. Appl. Polym. Sci.*, **105**, 255–268.
51. Sinha, R. S., and Okamoto, M. (2003). Polymer/layered silicate nanocomposites: A review from preparation to processing. *Prog. Polym. Sci.*, **28**, 1539–1641.
52. Li, Y., Wang, Y., Liu, L., Han, L., Xiang, F., and Zhou, Z. (2009). Crystallization improvement of poly(L-lactide) induced by functionalized multiwalled carbon nanotubes. *J. Polym. Sci. Part B Polym. Phys.*, **47**, 326–339.
53. Nakayama, N., and Hayashi, T. (2007). Preparation and characterization of poly(L-lactic acid)/TiO_2 nanoparticle nanocomposite films with high transparency and efficient photodegradability. *Polym. Degrad. Stab.*, **92**, 1255–1264.
54. Kim, Y., Jung, R., Kim, H. S., and Jin, H. J. (2009). Transparent nanocomposites prepared by incorporating microbial nanofibrils into poly(L-lactic acid). *Curr. Appl. Phys.*, **9**, S69–S71.
55. Yu, T., Ren, J., Li, S., Yuan, H., and Li, Y. (2010). Effect of fiber surface-treatments on the properties of poly(lactic acid)/ramie composites. *Composites Part A.*, **41**, 499–505.
56. Goriparthi, B. K., Suman, K. N. S., and Rao, N. M. (2012). Effect of fiber surface treatments on mechanical and abrasive wear performance of polylactide/jute composites. *Composites Part A*, **43**, 1800–1808.
57. Garcia, M., Garmendia, I., and Garcia, J. (2008). Influence of natural fiber type in eco-composites. *J. Appl. Polym. Sci.*, **107**, 2994–3004.
58. Nyambo, C., Mohanty, A. K., and Misra, M. (2011). Effect of Maleated Compatibilizer on performance of PLA/wheat straw-based green composites. *Macromol. Mater. Eng.*, **296**, 710–718.
59. Ma, H., and Joo, C. W. (2011). Structure and mechanical properties of jute—polylactic acid biodegradable composites. *J. Compos. Mater.*, **51**, 1451–1460.

60. Aydin, M., Tozlu, T., Kemaloglu, S., Aytac, A., and Ozkoc, G. (2011). Effects of alkali treatment on the properties of short flax fiber–poly(lactic acid) eco-composites. *J. Polym. Environ.*, **19**, 11–17.

61. Pluta, M., Galeski, A., Alexandre, M., Paul, M. A., and Dubois, P. (2002). Polylactide/montmorillonite nanocomposites and microcomposites prepared by melt blending: Structure and some physical properties. *J. Appl. Polym. Sci.*, **86**, 1497–1506.

62. Ray, S. S., Yamada, K., Okamoto, M., and Ueda, K. (2002). New polylactide-layered silicate nanocomposites: 2—concurrent improvements of material properties, biodegradability and melt rheology. *Polymer*, **44**, 857–866.

63. Chow, W. S., and Lok, S. K. (2009). Thermal properties of poly (lactic acid)/organomontmorillonite nanocomposites. *J. Therm. Anal. Calorim.*, **95**, 627–632.

64. Ren, J., Yu, T., Li, H., Ren, T., and Yang, S. (2008). Studies on morphologies and thermal properties of poly(lactic acid)/polycaprolactone/organic-modified montmorillonite nanocomposites. *Polym. Compos.*, **29**, 1145–1151.

65. Pillai, S. K., Ramontja, J., and Ray, S. S. (2011). Amine functionalization of carbon nanotubes for the preparation of CNT based polylactide composites-a comparative study. *Nanostruct. Mater. Nanotechnol. V Ceram. Eng. Sci. Proc.*, **32**, 43–52.

66. Oksman, K., Skrifvars, M., and Selin, J. F. (2003). Natural fibres as reinforcement in polylactic acid (PLA) composites. *Compos. Sci. Technol.*, **63**, 1317–1324.

67. Bax, B., and Mussig, J. (2008). Impact and tensile properties of PLA/ Cordenka and PLA/flax composites. *Compos. Sci. Technol.*, **68**, 1601–1607.

68. Tawakkal, I. S. M. A., Talib, R. A., Abdan, K., and Ling, C. N. (2012). Mechanical and physical properties of kenaf-derived cellulose (KDC)-filled polylactic acid (PLA) composites. *Bioresources*, **7**, 1643–1655.

69. Lee, B., Kim, H., Lee, S., Kim, H., and Dorgan, J. R. (2009). Bio-composites of kenaf fibers in polylactide: Role of improved interfacial adhesion in the carding process. *Compos. Sci. Technol.*, **69**, 2573–2579.

70. Balakrishnan, H., Hassan, A., Imran, M., and Wahit, M. U. (2012). Toughening of polylactic acid nanocomposites: A short review. *Polym. Plastics Technol. Eng.*, **51**, 175–192.

71. Chang, J. H., An, Y. U., and Sur, G. S. (2003). Poly(lactic acid) nanocomposites with various organoclays. I. Thermomechanical properties,

morphology, and gas permeability. *J. Polym. Sci. Part B Polym. Phys.*, **41**, 94–103.

72. Ogata, N., Jimenez, G., Kawai, H., and Ogihara, T. (1997). Structure and thermal/mechanical properties of poly(L-lactide)-clay blend. *J. Polym. Sci. Part B Polym. Phys.*, **35**, 389–396.
73. Ray, S. S., and Bousmina, M. (2005). Biodegradable polymers and their layered silicate nanocomposites: In greening the 21st century materials world. *Prog. Mater. Sci.*, **50**, 962–1079.
74. Ray, S. S., and Okamoto, M. (2003). Biodegradable polylactide and its nanocomposites: Opening a new dimension for plastics and composites. *Macromol. Rapid Commun.*, **24**, 815–840.
75. Paul, M., Delcourt, C., Alexandre, M., Degee, P., Monteverde, F., Rulmont, A., and Dubois, P. (2005). (Plasticized) Polylactide/(organo-)clay nanocomposites by in situ intercalative polymerization. *Macromol. Chem. Phys.*, **206**, 484–498.

Chapter 24

The Main Characteristics, Properties, Improvements, and Market Data of Polyhydroxyalkanoates

Elodie Bugnicourt,[a] Patrizia Cinelli,[b] Andrea Lazzeri,[b] and Vera Alvarez[c]

[a]*Innovació i Recerca Industrial i Sostenible (IRIS), Parc Mediterrani de la Tecnologia, Avda. Carl Friedrich Gauss n° 11, 08860 Castelldefels, Barcelona, Spain*
[b]*UdR Consortium INSTM, Department of Civil and Industrial Engineering, University of Pisa, Largo Lucio Lazzarino 1, 56126 Pisa, Italy*
[c]*INTEMA, Composite Materials Group (CoMP), Engineering Faculty, National University of Mar del Plata, Juan B. Justo 4302 (B7608FDQ), Mar del Plata, Argentina*

ebugnicourt@iris.cat

24.1 Introduction

Because of waste accumulation at the end of the life cycle of traditional polymer products, the development of environment-friendly, degradable, polymeric materials has attracted extensive interest in the past decades.

Packaging is the biggest polymer processing industry with the food sector being its principal customer. The materials used for this application often have short service time but end up mostly in landfills and stay there for over 100 years [1–5]. Cheap and biodegradable polymeric products are increasingly used in this

Handbook of Sustainable Polymers: Processing and Applications
Edited by Vijay Kumar Thakur and Manju Kumari Thakur

ISBN 978-981-4613-53-8 (Hardcover), 978-981-4613-54-5 (eBook)
www.panstanford.com

market to reduce the weight and volume of these products at end of life and keep with related environmental legislations [6].

Biodegradable polymers are a specific type of polymer that break down after they have been used in order to result in natural byproducts, like gases (CO_2, N_2), water, biomass, and inorganic salts [7, 8]. These polymers can be found both naturally and synthetically made and mainly consist of amide, ester, and ether functional groups. Their properties and breakdown mechanism are largely determined by their chemical structure and characteristics. These polymers are often synthesized by different kinds of reactions, such as condensation reactions and ring opening polymerization. There are a large number of examples not only of biodegradable polymers but also of their applications.

As it was previously said, biodegradable polymers tend to consist of ester, amide, or ether bonds. In general, biodegradable polymers can be separated into two main groups based on their structure and synthesis: Agro-polymers, or those derived from biomass [7] and biopolyesters, which are those derived from microorganisms or synthetically made from either naturally or synthetic monomers (Fig 24.1).

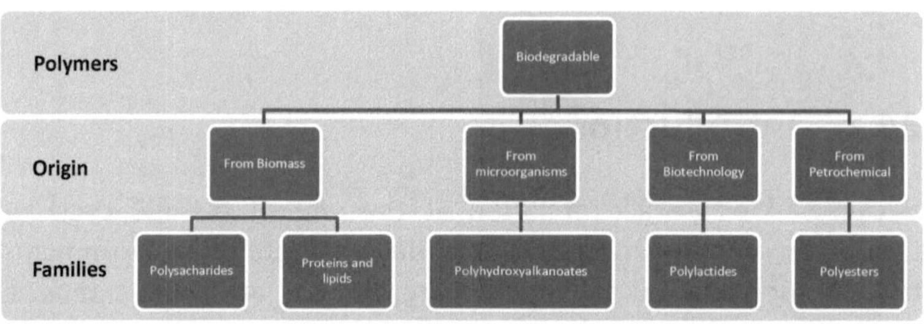

Figure 24.1 Biodegradable polymers organization based on structure and occurrence.

Agro-polymers include polysaccharides, like starches found in potatoes, corn, or maize, and proteins, like animal-based whey or plant-derived gluten [7]. Polysaccharides consist of the glycosidic bond, which takes a hemiacetal of a saccharide and binds it to alcohol via loss of water. Proteins are made from amino acids, which contain several known functional groups [9]. These amino acids come together again through condensation reactions producing peptide bonds, which consist of amide functional groups [9].

Examples of biopolyesters include polyhydroxybutyrate (PHB) and polylactic acid (PLA) [7].

Among all biodegradable polymers, unlike number of biopolymers limited to industrial compositing, polyhydroxyalkanoate is known to have the potential to degrade in every environment [10]. This was demonstrated by certificates of various PHA producers for industrial composting and also home composting, as well as in soil, fresh water, and marine seawater conditions. Furthermore, the complete biodegradation profile is often faster than for other biopolymers [11].

This chapter reviews the different classes of PHAs, their main properties, processing aspects, commercially available ones, and limitations and related improvements being researched related with the production nanocomposites to improve the final properties, as well as selected applications.

24.2 Polyhydroxybutyrate

Owing to their previously discussed benefits, interest in biodegradable polymers produced from renewable resources has increased significantly in recent years. Polyhydroxyalkanoate polymers are naturally produced by bacteria in general cultivated on agricultural raw materials. They can be processed to make a variety of useful products, where their biodegradability and naturalness are quite beneficial in particular for application in single-use packaging and agriculture. Poly(3-hydroxybutyrate) (Fig. 24.2) is a homopolymer of 3-hydroxybutyrate and is the most widespread and best characterized member of the polyhydroxyalkanoate family.

Poly-3-hydroxybutyrate is a linear polyester of D-(-)-3-hydroxybutyric acid which was first discovered in bacteria by Lemoigne in 1925. It is accumulated in intracellular granules by a wide variety of Gram-positive and Gram-negative organisms under conditions of a nutrient limitation other than the carbon source [12]. The molecular weight of PHB differs depending on the organism, conditions of growth, and method of extraction and can vary from about 50,000 to well over a million. The polymer possesses the important properties of thermo plasticity and biodegradability in compost and different environments comprising marine water, and, in consequence, has attracted considerable commercial interest.

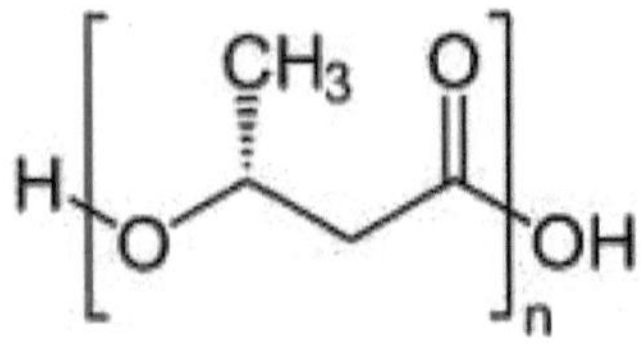

Figure 24.2 Chemical structure of polyhydroxybutyrate.

PHB is produced by microorganisms (such as *Ralstonia eutrophus* or *Bacillus megaterium*) in response to conditions of physiological stress and can be produced either by pure culture than mixed culture of bacteria [13].

Biologically produced, PHB is a semicrystalline isotactic stereo regular polymer with 100% R configuration that allows a high level of degradability [14].

24.2.1 General Properties of Polyhydroxybutyrate

PHAs polymers are thermoplastic and they differ in their properties depending on their chemical composition (homo-or copolyester, contained hydroxy fatty acids). Some grades of additivated PHB are similar in their material properties to polypropylene (PP), and offer good resistance to moisture and aroma barrier properties. Polyhydroxybutyric acid synthesized from pure PHB is relatively brittle (i.e., elongation at break typically below 15%) and stiff (i.e., E modulus above 1 GPa). PHB polymer chains are regular and isotactic, they don't contain residues of catalyst and don't have chain branching, and for these reasons form perfect crystals. The PHB glass transition temperature is near to °C, and consequently at room temperature polymers chains are free to move for acquiring a higher crystal strcuture. This process can last for many days, and the higher density of crystalline phase versus the amorphous phase induces cracks in between the crystals. These modification in structure with ageing time are responsible for the slow transition from ductile freshly moulded parts to very brittle parts after storing. For example, mechanical properties of specimen prepared with PHB Biomer LoT13 without any additive were reported (original data from the authors). PHB is a fragile material due to re-crystallization with ageing at room temperature as will be extensively reported in the following sections. Thus, mechanical properties change with time and, as reported in the example, samples stored at room temperature for 60 days have lower values for elongation at break than samples stored for 30 days. For these reasons, efforts in compounding PHB are mainly focused on the search of plasticizers and nucleating agents capable of reducing the crystallization process and improving flexibility and elongation in the final product.

Some general characteristics of PHB are summarized in the Table 24.1.

Table 24.1 General characteristics advantage and drawbacks) of virgin PHB

Advantages	Drawbacks
Insoluble in water	Poor resistance to acids and bases
Resistant to hydrolytic degradation	Poor mechanical properties in terms of elongation at break, brittle
Resistant to UV exposure	Non transparency due to high crystallinity
Soluble in chlorinated hydrocarbons	
Biocompatible	
Nontoxic	

24.2.2 Thermal and Mechanical Properties of Polyhydroxyalkanoates

Isostatic PHB displays a number of properties comparable to petroleum-based polymers (e.g., polypropylene), such as high melting temperature (175°C) and relatively high tensile strength (30–35 MPa). However, pure PHB has had only limited use mainly because of its intrinsic brittleness (presenting low strain at break) and the narrow processing window of this plastic. Indeed, the elongation at break is very different between PHB (5%) and PP (400%). PHB/V (polyhydroxybutyrate/valerate) is also brittle, its elongation at break is less than 15%, its modulus and fracture stress are 1.2 GPa and 25 MPa [15]. There are many reasons for the brittleness of PHB and PHB/V: the secondary crystallization of the amorphous phase takes place during storage at room temperature; the glass transition temperature (T_g) of PHB is close to room temperature; PHB has a low nucleation density, therefore large spherulites exhibit inter-spherulitic cracks. On addition of plasticizers, the molecular motion is enhanced, and the glass transition is lowered. To achieve high elongation at break and a higher flexibility for modified/formulated PHB, the glass transition temperature must reach a lower value than the testing temperature. The elongation and impact strength depend on T_g as well as the morphology. In the blends, the nucleation rate and spherulite size depend on the cooling rate and nucleation density, i.e., fast cooling after melting increases the crystallization rate. That forms fine spherulites and suppresses crystallinity. This is required for the achievement of the necessary mechanical properties.

Moreover, PHB thermally decomposes at temperatures just above its melting point. A short exposure of PHB to temperature near 180°C could induce a severe degradation accompanied by the production of the degraded products of olefinic and carboxylic acid compounds, e.g., crotonic acid and various oligomers: through the random chain scission reaction that involves a cis-elimination reaction of β-CH and a six-member ring transition [16–18].

The very low resistance to thermal degradation seems to be the most serious problem related to the processing of PHB.

During processing, the degradation of the chains may be reduced by the addition of a lubricant that prevents the degradation of the chains in processing, so that the material can be processed at 170–180°C, because PHB is sensitive to high processing temperatures. This leads to a decrease in the molecular weight, as well as a reduction in the melt viscosity. The crystallization temperature shifts to lower values, and crystallization takes longer. One possibility to improve this drawback is to incorporate nanofillers, such as nanoclays.

Other approaches to improve the properties of PHB is the production of derivatives based on PHB via the biosynthesis of copolyesters containing PHB units with other 3-hydroxyalkanoates units [19], such as poly (3-hydroxybutyrate-co-hydroxyvalerate) (PHBV) [20] or poly (3-hydroxybutyrate-co-3-hydroxyhexanoate) [21], with different molar ratios of hydroxycarboxylic acids. This approach has been investigated extensively [22] because it can effectively improve mechanical properties [23] and lower the melting point, avoiding degradation during processing. However, blends with other biodegradable polymers and composites based on PHB that exclude any synthetic component are also very convenient as industrial materials due to their easier processability and lower cost [24, 25].

As reported above, the addition of plasticizers is considered a relatively simple route to modify the thermal and mechanical properties of polymers. Blending polymers with plasticizers may modify the physical properties of polymers and a decrease in processing temperature can be achieved. Thus, PHB is commonly blended with plasticizers and nucleation agents that lead to a lower glass temperature and lower crystallinity due to the formation of numerous, small, and imperfect crystallites.

24.2.3 Improvements of PHB: Blending of PHB with Other Polymers or Plasticizers

Blending PHAs and in particular PHB with other polymers, or with plasticizers, may offer opportunities to improve processability by lowering the processing temperature and reducing the brittleness of PHA-based plastics. So far, many blends containing PHB/PHAs have been studied and also many types of plasticizers have been proposed [26, 27].

The literature reports the use as plasticizers from materials that are cheap and readily available on the market and generally also of natural origin, as summarised in Table 24.2.

Table 24.2 Some plasticizers used for PHAs [28, 29]

Oxypropylated glycerin	Glycerol	Dibutyl phthalate
Acetyl tributyl citrate	Salicylic ester	Triethyl citrate
Glycerol triacetate	Acetylsalicylic acid ester	Dioctyl phthalate
4-Nonylphenol	Soybean oil	Dioctyl sebacate
4,40-Dihydroxydiphenyl-methane	Epoxidized soybean oil	Acetyl tributyl citrate
bis(2-Ethylhexanoate)	di-2-Ethylhexylphthalate	Tri(ethylene glycol)

A good strategy to modify properties of PHB/PHAs in the solid state is blending the polyester with a second polymeric component as summarised in Table 24.3 [30]. This route has been broadly explored, testing as the other blend component a vast range of thermoplastic polymers [31–36]. Besides changing degree of crystallinity and width of the processing window, blending also offers the opportunity of lowering production costs that are at present the major problem for large-scale applications of bacterial polymers based plastics.

Table 24.3 Some examples of blending of PHB with other polymers and plasticizers

Blending with	Result	Ref
Polyethylene glycol (solvent casting)	–Increase in the elongation at break by up to about four times versus the original elongation at break of PHB and a reduction in the tensile strength.	[19, 37]
	– The plasticizing effect of PEG that weakened the intermolecular forces between the adjacent polymer chains.	
	–Reduction in the melting temperatures of the system.	
PEG/wood (by extrusion)	–PEG addition makes processing easier having a lubricant effect on melted PHB/wood formulations.	[10, 38]
	–PEG was not efficiently blended with PHB and with time leached to the surface and reduced its possible plasticizing effect.	
	–PEG1500 is not completely compatible with PHB, as confirmed by separation observed in the samples with ageing.	
PEG/Cellulose Nanowhiskers (CNW)	– The adhesion of PEG to the natural fibers can explain the lubricating effect of PEG in the preparation of composites with PHB and natural fibers.	[39]
Acetyl tributyl citrate	–Weak positive influence on thermal properties of PHB in the melt.	[40]
Glycerol, tributyrin, triacetin, acetyltriethylcitrate, acetyltributylcitrate		[15]

24.3 Comparison with Other Biodegradable and Traditional Polymers

The main properties of some important biodegradable polymers, including PHBV, are summarized in Table 24.4.

Table 24.4 Comparison of the properties of the main biodegradable polyesters

Property/Material	PHB	PLA	PCL	PEA
Physical Properties				
Density (g/ml)	1.25	1.25	1.10	1.05
Thermal Properties				
Melting temperature (°C)	165	151	62	114
Glass transition temperature (°C)	-5	58	-60	-32
Degree of crystallinity (%)	40–60	0–2	60–70	30–36
Mechanical Properties				
Young's modulus (MPa)	1000–1200	2000–2100	180–240	250–270
Tensile strength (MPa)	15–20	16–103	13–15	18
Elongation at break (%)	14–16	8–10	Higher than 600	400–460
Barrier Properties				
WVTR at 25 °C (g/m^2·day)	20	170	180	700
Surface Properties				
Surface tension (mN/m)	60	50	53	62
Polar component	32	14	12	24
Biodegradability				
Biodegradation 60 days (%)	100	100	100	100

Source: Adapted from [41].

One of the difficulties with the pure PHB and its copolymers (for example, PHBV) is it high melting point, similar to that of PLA, with the additional problem that the degradation temperature is quite near, but that problem could be solved by using plasticizers (as was previously explained) or by incorporating nanoparticles (as explained in the following section).

Regarding the mechanical properties, the stiffness of this material is high (the highest with the exception of PLA), but the elongation at break is quite low, which can be also solved by means of the incorporation of plasticizers.

It is really interesting that the water vapor transmission rate is the lowest of the commonly used biodegradable polymers; this characteristic is truly relevant, for example, in packaging applications.

24.4 PHB-Based Nanocomposites

Because of waste accumulation at the end of the life cycle of traditional polymer products, the development of environment-friendly, degradable polymeric materials has attracted extensive interest [6]. PHB is one of these types of polymers.

In order to obtain a competitive product, the PHB performance can be greatly enhanced with the addition of nanometer-size inorganic fillers. This kind of materials are called nanocomposites and have an interesting characteristic: The mechanical properties [42], the barrier properties [43], the thermal properties [44], and some others such as the flammability [45], water adsorption [46], and creep resistance [47] can be greatly enhanced with the addition of a small amount of filler (usually less than 10 wt%) [6, 42–48].

24.4.1 PHAs/Clay Nanocomposites

Among all the nanofillers that have been studied within nanocomposites' combinations with PHBs, clay is the most widely reported in the literature. This chapter will therefore focus mainly on the former as well as briefly afterward on nanocellulose.

One kind of nanometer-size reinforcement is the montmorillonite, which is a layered silicate whose interlayer ions can be exchanged by organ-ions in order to produce an increment in the interlayer spacing (d_{001}) and to improve the polymer/clay compatibility. These improvements allow the dispersion of clay platelets to be easier [49].

The selection of the best clay for a specific matrix is crucial to obtain nanocomposites with optimized properties. A series of the most commonly used montmorillonites are summarized in Table 24.5.

Table 24.5 Characteristics of several clays used as nanofillers

Clay	Organic modifier	Modifier concentration (meq/100 g clay)	Specific gravity (g/cm³)	d_{001} (Å)
Montmorillonite (CNa^+)	None	—	2.86	14.0
Cloisite 25A (C25A)	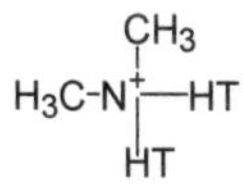	95	1.87	18.6
Cloisite 30B (C30B)	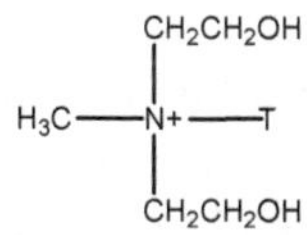	90	1.98	18.8
Cloisite 10A (C10A)	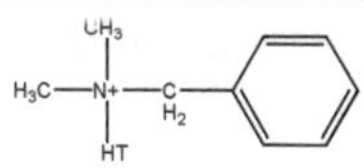	125	1.90	20.0
Cloisite 93A (C93A)	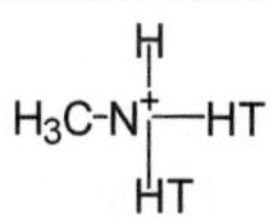	90	1.88	23.6
Cloisite 20A (C20A)	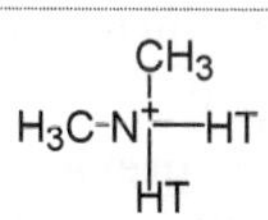	95	1.77	24.2
Cloisite 15A (C15A)	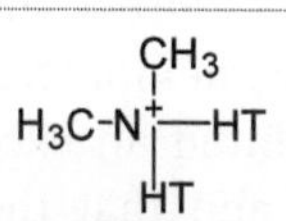	125	1.66	31.5

Note: HT stands for hydrogenated tallow (~65% C18; ~30% C16; ~5% C14).

In that field, several authors have been working on PHB/clay nanocomposites, studying different aspects.

Botana et al. [50] have prepared polymer nanocomposites, based on a bacterial biodegradable thermoplastic polyester, PHB and two commercial montmorillonites (MMT), unmodified and modified by melt-blending technique at 165°C. PHB/Na and PHB/C30B were characterized by differential scanning calorimetry (DSC), polarized optical microscopy (POM), X-ray diffraction (XRD), transmission electron microscopy (TEM), mechanical properties, and burning behavior. Intercalation/exfoliation observed by TEM and XRD was more pronounced for PHB30B than PHB/Na, indicating the better compatibility between the modified clay and the PHB matrix. An increase in the crystallization temperature and a decrease in spherulites size were observed for PHB/C30B nanocomposites. The intercalation/exfoliation, observed by TEM and XRD patterns were correlated with the higher moduli of the nanocomposites. They have also observed that the burning behavior of PHB/C30B was influenced by the aggregation of the clay mineral particles.

D'Amico et al. [51] have studied the isothermal and non-isothermal crystallization kinetics of nanocomposites of PHB with original and chemically modified commercial clays prepared by solution casting. They have demonstrated that, independently of the type of clay, the incorporation of nanoparticles did not significantly affect the crystallinity and melting temperatures of the PHB. They have also shown that CNa^+ and C93A have only a slight effect on the crystallization kinetics parameters of pure PHB matrix. Nevertheless, a retardant effect was found when C15A was added which was in turn attributed to the high dispersion of this clay in the polymer. All the results clearly suggested that the type of clay and consequently the dispersion inside PHB matrix (which is related with polymer/clay compatibility) influence the crystallization kinetics of the biodegradable polymer, PHB.

Rocha e Silva et al. [52] have prepared layered PHB nanostructured materials by adding different amounts of commercial clays on PHB matrix. They have demonstrated that the obtained nanostructured materials exhibited an excellent degree of dispersion (intercalation/exfoliation) and that the clay was able to promote changes in the structure of neat polymer by reducing its crystallinity degree.

Bordes et al. [53] have studied the effect of ammonium surfactants, commonly used as clay modifiers, on the thermal- and thermo-mechanical degradation of polyhydroxyalkanoates (PHAs) by using two PHBV grades and performing thermogravimetric and molecular weight changes after processing analysis. The obtained data revealed that all surfactants enhance the PHBV degradation. The results also highlighted the preponderant effect of the initial Mw rather than the HV content on the thermal stability. On the other hand, the thermo-mechanical study confirmed the role of surfactants and their different behavior towards PHBV degradation. They have concluded that all surfactants improve the PHA degradation because their decomposition products could act as catalytic agents.

Erceg et al. [54] have studied the non-isothermal degradation behavior of Poly(3-hydroxybutyrate)/organically modified montmorillonite nanocomposites (PHB/C25A) prepared by the solution-intercalation method. Kinetic analysis of the non-isothermal degradation of pure matrix and nanocomposites was performed using isoconversional (Flynn–Wall–Ozawa, Kissinger–Akahira–Sunose and Friedman) methods and the invariant kinetic parameters method. All results suggested that the addition of Cloisite 25A reduces the rate of the non-isothermal degradation of PHB.

Achilias et al. [55] have prepared, by the melt mixing method, and characterized PHB/organically modified clay nanocomposites. The thermal degradation kinetics of matrix and nanocomposites was investigated using thermogravimetric analysis and by applying isoconversional methods. They have found that intercalated structures are formed and the thermal stability of the material was improved by the addition of the nanoclay. The kinetic rate constant decrease as a function of clay content up to 3 wt%, while for higher fillers that parameter increased reaching a value greater than that of pure PHB for the 10 wt% nanocomposites.

Puglia et al. have prepared nanocomposites based on PHB, and unmodified and chemically modified montmorillonites through a solution route. The materials were initially characterized and disintegrability in composting conditions was tested after several durations. XRD results suggest a better dispersion of modified clay based nanocomposites, due to the higher compatibility between the matrix and the clay, which was also related with disintegrability behavior of PHB nanocomposites. Visual observation, chemical, thermal and morphological investigations have proved that the disintegration in composting conditions was faster for PHB/unmodified clay nanocomposites implying the possibility to modulate the disintegrability capacity of PHB by the selection of the filler. In all the previous studies, it was established that the polymer and surfactant degradations are interdependent.

24.4.2 PHAs/Nanocellulose Nanocomposites

Plant fibers are mainly composed of cellulose, hemicellulose, and lignin. Cellulose, which awards the mechanical properties of the complete natural fiber, is ordered in micro-fibrils enclosed by the other two main components: hemicellulose and lignin [57]. Cellulose micro-fibrils can be found as intertwined micro-fibrils in the cell wall (2–20 μm diameter and 100–40,000 nm long depending on its source). In addition to these microfibrils, there exist nanofibers (composed by cellulose) with diameters of 5–50 nm and lengths of several millimeters confirmed by nanocrystalline domains and amorphous regions [58]. A controlled acid hydrolysis can separate both regions driving to crystalline domains with an elastic modulus of 150 GPa, which is higher than that of the S-glass (85 GPa) and Aramid fibers (65 GPa) [59]. Because of the good mechanical properties, nanocellulose has generated a great deal of interest as a source of nanometer-sized reinforcement. In the past years, these fibers also attracted much attention due to environmental concerns especially as the reinforcement of biodegradable polymers to produce fully biodegradable nanocomposites with enhanced mechanical properties [60–63].

Zhijiang et al. [64] have prepared bionanocomposite by impregnation of bacterial cellulose into PHB. They have studied the morphology of the obtained nanocomposites by field emission scanning electron microscopy, finding that bacterial cellulose nanofibrils were well dispersed in PHB. They have also observed changes in the crystallinity of PHB caused by the addition of the bacterial cellulose, which in turns suggested that the nanoparticles probably help the PHB molecules to crystallize. In addition, the prepared material displayed enhanced mechanical properties suggesting that bacterial cellulose acted as reinforcement into the PHB matrix.

24.5 Market Aspects of Polyhydroxyalkanoates

24.5.1 Industrial Production of PHAs: History and Current Landscape

Among all PHA polymers existing, PHB, its copolymers, and PHBV are today the most widely commercially available. The first commercialization of PHB is found back in the 1950s by W.R. Grace Co., USA, but the trial soon failed [27]. Subsequently in the 1970s, Imperial Chemical Industries (ICI), United Kingdom, developed

PHB obtained via fermentation under trade name Biopol. It was distributed in the United States by Monsanto and then Metabolix, USA, which formed later a joint venture with Archer Daniels Midlands, called Telles, commercializing now PHA products under the brand Mirel. More recently Metabolix announced termination of Telles joint venture and signed a distribution and supply agreements with the Chinese Tianjin GreenBio Materials Co [65].

In parallel, Chemie Linz, Austria, (now Borealis), together with Petrochemia Danubia (PCD), started pilot scale production of PHB in the 1980s. The business was later acquired by Biomer, Germany, in the 1980s.

Procter & Gamble (P&G), USA, also started production of PHA (polyhydroxybutyrate-co-hexanoate, PHBH) in the 1980s under proprietary technology. They later formed a joint venture with Kaneka, Japan. Meredian, USA, acquired the P&G patents and recently its announced intention to open in 2024 the world's largest PHA Production Facility with projected over 30,000 tons of PHA per year at the Bainbridge facility [66].

PHB Industrial Usina da Pedra-Acucar e alcool, Serrana Sao Paulo, is running a 10,000 T per year (tpa) plant since 2006 and sells under the reference Biocycle.

In Japan, Mitsubishi Gas Chemical (MGC) has made an in-depth development study of the production of PHB from methanol fermentation (trade name BioGreen).

The Chinese Tianan [67] with an annual capacity of 2000 metric tons describes itself the world's leading PHBV producer.

As can be seen from the previous discussion, whereas the PHA business is not new, it is becoming increasingly dynamic in the recent years. According to a very recent report, China and Japan are on the way to dominating the PHA market [65].

24.5.2 Feedstocks Used in PHA Production

The production of PHA generally uses pure microbial cultures grown on renewable but refined and possibly food competing feedstocks under sterile conditions. Although in the literature, a wide range of culture media are found depending on the microorganism and PHA to be obtained [68] (carbohydrates: glucose, fructose, sucrose; alcohols: methanol, glycerol; alkanes: hexane to dodecane; organic acids: butyrate upwards), in the industry, mainly sugar of different sources and vegetable oil are reported as a raw material for the fermentation.

Researchers in industry have also reported working on methods with which transgenic crops will be developed that express PHA synthesis routes from bacteria to produce PHA as energy storage in their tissues [16].

Different research efforts also deal with the valorization of wastes as non-food competing sources for cultivating a range of bacteria with the purpose of obtaining PHA biopolymers with improved sustainability. For example, ongoing research deals with the growth of cyanobacteria in olive mill wastewaters to produce PHAs including their possible genetic modification to increase the yield in PHA [10]. Another group of researchers at Micromidas is working to develop methods of producing PHA from municipal wastewater.

Different other wastes can also be considered [27], such as carbohydrates: molasses, starch and whey [69], cellulose hydrolysates (e.g., paper industry waste); alcohols: wastes from biodiesel production, methanol plus glycerol, methanol; fats and oils: lipids from plant and animal wastes; organic acids: lactic acid from the dairy industry.

Except sources indicating that Biomatera Inc, Canada, is specialized in the manufacture of PHA by fermentation of agricultural residues [70], almost aforementioned efforts were mainly at the academic level. Nevertheless, they could have a very important impact if transferred to the industry. Indeed the feedstock used such as sugar accounts for roughly 40% of the price of derived PHA [70].

The sugar market is dominated by Brazil and Asian countries and its price has been quite volatile in the last decade. As can be appreciated from Nasdaq markets reports [71], a ca. fivefold increase was observed from 2004 until 2011 and now, in 2014, the price is about 2.5 times its value in 2007. All this has a great impact on PHA price since as much as 4–5 kg of sugar is needed to produce 1 kg of PHA [72].

More than 300 types microorganism accumulating PHAs have been identified, e.g., *E. coli* fed with a range of oils (lipids, saccharides, etc.) as feedstock produces different compositions of P(3HB-co-HHx); *R. eutropha* fed with a combination of glucose and propionate produces P(3HB-co-HV). Nevertheless, it is not easy to find out the exact strain, of generally heterotrophic bacteria, used by each company in the industrial production of PHA.

The fermentation results directly in polymer with no need for further polymerization as opposed to, for example, PLA. Nevertheless, the polyester needs to be extracted from the bacteria (often using chloroform or enzymes) and purified, and this is again a high contributor to the cost of PHA. Tianan also reports a patented commercial production of PHBV using a water-based extraction technology [67].

24.5.3 PHA Production Capacity and Economic Aspects

In spite of announced production increase, PHA still represents today very few percent (1.6% in 2013) vs. all bioplastics. According to the European Bioplastics Association, in 2012 production capacities amounted to approximately 1.4 million ton (defining bioplastics as partly or fully bio-based plastics, biodegradable, or both). Market data of European Bioplastics forecasts production capacities will multiply by 2017—to more than 6 million ton. While PHA still has a minor production whereas drop in solutions such as bio-based PE and PET have taken a major place [73].

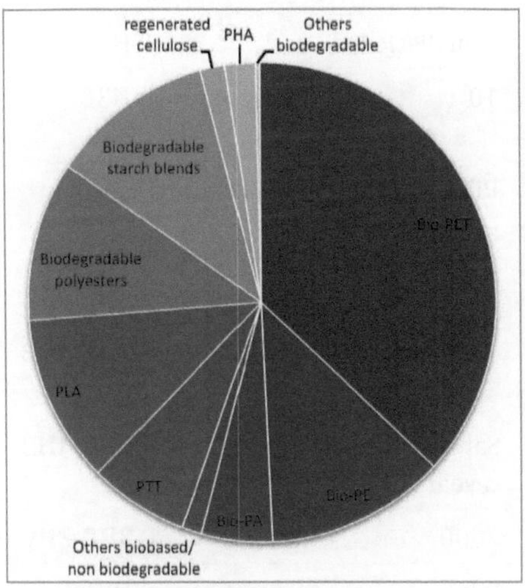

Biobased/non biodegradable: 62.4%

Bio-PET	37%
Bio-PE	12.3%
Bio-PA	4.9%
Others biobased/non biodegradable	1.4%
PTT	6.8%

Biodegradable: 37.6%

PLA	11.4%
Biodegradable polyesters	10.8%
Biodegradable starch blends	11.3%
regenerated cellulose	1.7%
PHA	2.1%
Others biodegradable	0.3%

Figure 24.3 Bioplastics production capacity by type in 2013. Adapted from [73].

According to the Table 24.6 below, PHA capacity was around 80,000 tpa in 2009. PHAs are forecast to achieve 41% growth over the 2010–2020 period according to a study from Pira [74]. Nevertheless, European capacities are quite limited with mainly two small producers Bio-On and Biomer (e.g., 500 tpa for Biomer), and as previously discussed, the greatest plans for growth are in Japan and China as well as in the United States to some extent.

The main drivers of this growth are consumer demand for environment-friendly products and the volatile price of petroleum, different policies enforced in different countries that support the increase use renewable sourced raw materials (e.g., Lead Market Initiative for Bio-based Products) or directives for sound environment-friendly end of life (e.g., European Directive on waste (2008/98/CE) and packaging and packaging waste (94/62/CE and 2004/12/CE)).

Table 24.6 Production capacity for different global PHA manufacturers in 2009

Producer	Country	Production	Extraction process	PHA produced
Metabolix (Telles)	USA	Ton for year	Solvent	P-3HB-4HB
MHG Meredian	USA	Pilot	Solvent	PHBHx
BioMatera	Canada	Non disclosed	n.d.	P-H3B, PHBV
Mitsubishi Gas Chemical	Japan	10	Solvent	P-H3B
Kaneka	Japan	Pilot	Solvent	PHBHx
Biomer	Germany	Small scale	Solvent	P-3HB
PHB Industrial	Brazil	50	Solvent	P-3HB PHB
Tianjin Green Bioscience/DSM	China	10000 in 2009	Solvent	P-3HB-4HB
Shenzhen Ecomann Technology	China	Samples for research	Solvent	P-3HB-4HB
Tianan Biologic	China	2000	Water	P-3HB-3HV
Thianzu	China	Pilot	Sovent	PHBHx

Source: Adapted from [74].

Table 24.7 reflects the SOWT analysis (strength, weaknesses, opportunities, and threats) for PHAs vs. other bioplastics on the market.

Table 24.7 SWOT analysis for PHA vs. other bio-based plastics

Strengths	**Weaknesses**
Technical	
• High modulus and strength • Processability using number of converting technologies • Good barrier properties	• PHB crystallinity and kinetic of crystallization • PHB brittleness • Molecular weight may vary depending on culture conditions • Difficulty to process films
Environmental	
• Biodegradable in the environment and marine water	• NA
Marketing	
• Suitability for a huge range of applications	• High cost
Opportunities	**Threats**
Technical	
• Innovative formulations based on reactive plasticizers and eventually nanoadditives, use of blends can enhance PHAs properties	• NA
Environmental	
• Use of PHA based items to replace non bio-based/degradable counterparts	• Sometimes need to use of higher amount of material than conventional plastic to achieve same properties
Marketing	
• Large market for packaging where collection is difficult and composting is the favorite option • Plastic for use for short-life applications on ships.	• Reach a competitive cost vs. alternatives • Insufficient production capacity

Source: Adapted by the authors from the Plan for Use and Dissemination of the foreground of the OliPHA project [10].

Although the price of PHAs has considerably decreased compared to five years ago, it is in general higher than other bio-based polyesters due to aforementioned high raw material costs and high processing costs. PHAs prices are extremely variable depending on the origin, source, type, etc. They are reported in the literature to be between €4/kg and €20/kg with Chinese PHBV produced at large scale being the low range and European PHB produce at "pilot" scale being the high range [27], As previously discussed, the use of waste as feedstock or of genetically engineered of bacteria should all lead to reductions in production cost.

24.5.4 PHA Applications and Commercial Products

In spite of its higher cost, due to its good mechanical and barrier properties, its biodegradability and recyclability, PHA has a good potential among all bioplastics and could substitute several of its non bio-based counterparts. Depending on the composition of the copolymers, grafting co-monomers or tailoring of the formulations (using plasticizers and others additives), PHAs' properties are very versatile. As shown in Fig. 24.4, the PHB design space is claimed to cover from elastomers to rigid thermoplastics [70].

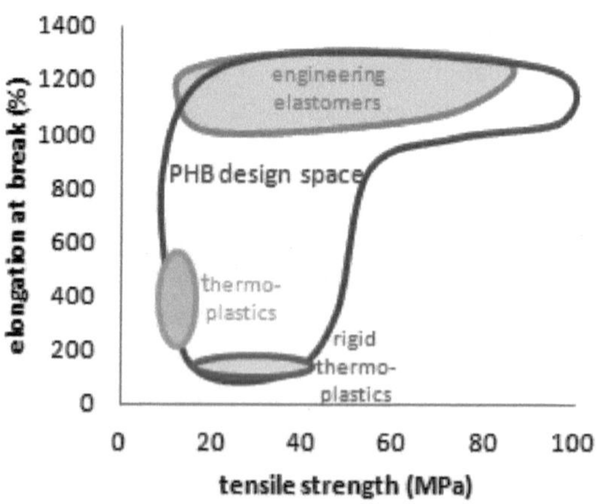

Figure 24.4 Mechanical properties of PHB (regarding the grades of PHAs from Metabolix). Adapted from [70].

PHB is claimed to be suitable to replace PVC, PE-HD, PE-LD and PP [27], as shown in Table 24.8.

Table 24.8 Comparison of PHB properties with PP and PE of different densities

Polymer	Tensile strength (MPa)	Elongation at break (%)	Modulus (MPa)
Biomer® P226	24–27	6–9	1140–1900
PP	22	12–20	600–1200
Biomer® P209	15–20	8–15	900–1200
PE-LD	15–20	600	150–450
PE-HD	25–32	600–900	700–1200

Source: Adapted from [75].

PHB is used in applications ranging from surgical sutures, tissue engineering [76], and agricultural foils to packaging for the storage of food products [77]. Like PVC and PET, it exhibits good barrier properties [78] and can be used in the packaging industry as a biodegradable plastic for contributing to solving environmental pollution problems [79]. Due to these interesting properties, PHB is expected to be a good candidate to substitute [80] PP and PE, but also PET in certain applications. In medicine, PHB is compatible with the blood and tissues of mammals. The monomer of PHB is a normal metabolic in the human blood. As the body reabsorbs PHB, it could be used as a surgical implant, in surgery, as seam threads for the healing of wounds and blood vessels. In pharmacology, PHB can be used as microcapsules in therapy or as materials for cell and tablet packaging. It can also be used for packaging applications for deep drawing articles in the food industry, for example, bottles, laminated foils, fishnets, flowerpots, sanitary goods, fast foods, disposable cups, agricultural foils, and fibers in textiles.

24.6 Conclusions

PHA is a very promising polymer for a wide range of applications that can be accumulated from multiple microbial sources cultivated on number of media. In comparison with other bio-based polymers such as PLA, it has better barrier and mechanical properties. To overcome its brittleness as well as improve its processability, PHB is often formulated with additives, nanofillers, and blends leading. Research is also ongoing to obtain more flexible grades of PHAs and to control its crystallization to obtain transparent films. Nevertheless, limitations still persist, such as the elevated cost of commercial PHA (€7–10/kg), limited market availability and the use of refined/food competing feedstock. To decrease PHA cost, trials were done using waste feedstock and/or genetically modifying the bacteria to improve PHA yield and increase the industrial sustainability and commercialization of PHAs [81]. The use of nanoscale and natural fillers with high availability and low cost allows the production of bio-composite for packaging and agriculture.

The versatile properties and processability (injection molding, extrusion, thermoforming and sheet extrusion, film blowing, foaming, blow molding, fiber spinning, paper coating) of PHB makes it suitable for widening wide range of potential uses. Packaging and agricultural film are the most important markets for PHAs, and their use is predicted to widen into buildings, textiles, transportation, electric and electronic applications, and household applications, among others.

Acknowledgments

The authors acknowledge the financial support from the European Community's Seventh Framework Programme (FP7/2007–2013) under NMP grant agreement n° 280604.

References

1. Biron, M., The plastics industry: Economic overview. In *Thermoplastics and Thermoplastic Composites* (Elsevier, ed.), 2nd ed., William Andrew Publishing, Chapter 2, pp. 31–131 (2013).
2. Dikgang, J., Leiman, A., Visser, M., Analysis of the plastic-bag levy in South Africa. Resources, conservation and recycling, **0**, 59–65 (2012), DOI: http://dx.doi.org/10.1016/j.resconrec.2012.06.009.
3. Meneses, J., Corrales, C. M., Valencia, M., Síntesis y cartacterización de in polímeros biodegradable a partir del almidón de yuca. Revista EIA, 57–67 (2007).
4. Mutha, N. H., Patel, M. K., Premnath, V., Plastics materials flow analysis for india. *Conservation and Recycling*, **3**, 222–244 (2006).

5. Mulder, K. F., Sustainable consumption and production of plastics? *Technological Forecasting and Social Change*, **1–2**, 105–124 (1998), DOI: http://dx.doi.org/10.1016/S0040-1625(97)00129-7.
6. Lepoittevin, B., Devalckenaere, M., Pantoustier, N., Alexandre, M., Kubies, D., Calberg, C., Jérôme, R., Dubois, P., Poly(ε-caprolactone)/clay nanocomposites prepared by melt intercalation: Mechanical, thermal and rheological properties. *Polymer*, **14**, 4017–4023 (2002), DOI: http://dx.doi.org/10.1016/S0032-3861(02)00229-X.
7. Luc Avérous, E. P., *Environmental Silicate Nano-Biocomposites*, Springer, London (2012).
8. Bastioli, C., *Handbook of Biodegradable Polymers*, Rapra Technology, Shawbury, Shrewsbury, Shropshire, U.K. (2005).
9. Cox, D., Nelson, L. M. M., *Lehninger Principles of Biochemistry*, W. H. Freeman, New York (2008).
10. www.olipha.eu. Accessed: 10/07/2014.
11. Ebnesajjad, S., *Handbook of Biopolymers and Biodegradable Plastics: Properties, processing and Applications*, William Andrew, New York (2013).
12. Dawes, E. A., Senior, P. J., The role and regulation of energy reserve polymers in micro-organisms. *Advances in Microbial Physiology*, **10**, 135–266 (1973).
13. Laycock, B., Halley, P., Pratt, S., Werker, A., Lant, P., The chemomechanical properties of microbial polyhydroxyalkanoates. *Progress in Polymer Science*, **3–4**, 536–583 (2013), DOI: http://dx.doi.org/10.1016/j.progpolymsci.2012.06.003.
14. Noda, I., Marchessault, R. H., Terada, M., *Polymer Data Handbook*. Oxford University Press, London (1999).
15. El-Hadi, A., Schnabel, R., Straube, E., Müller, G., Henning, S., Correlation between degree of crystallinity, morphology, glass temperature, mechanical properties and biodegradation of poly(3-hydroxyalkanoate) PHAs and their blends. *Polymer Testing*, **6**, 665–674 (2002), DOI: http://dx.doi.org/10.1016/S0142-9418(01)00142-8.
16. Doi, Y., Mukai, K., Kasuya, K., Yamada, K., Biodegradation of biosynthetic and chemosynthetic polyhydroxyalkanoates. In *3rd International Scientific Workshop, Biodegradable Plastics and Polymers, Amsterdam, The Netherlands*, vol. 12. (1994).
17. Grassie, N., Murray, E. J., Holmes, P. A., The thermal degradation of poly(-(d)-β-hydroxybutyric acid): Part 1. Identification and quantitative analysis of products. *Polymer Degradation and Stability*, **1**, 47–61 (1984), DOI: http://dx.doi.org/10.1016/0141-3910(84) 90016-8.

18. Grassie, N., Murray, E. J., Holmes, P. A., The thermal degradation of poly(-(d)-β-hydroxybutyric acid): Part 2. Changes in molecular weight. *Polymer Degradation and Stability*, **2**, 95–103 (1984), DOI: http://dx.doi.org/10.1016/0141-3910(84)90075-2.

19. Zhang, L., Deng, X., Zhao, S., Huang, Z., Biodegradable polymer blends of poly(3-hydroxybutyrate) and hydroxyethyl cellulose acetate. *Polymer*, **24**, 6001–6007 (1997), DOI: http://dx.doi.org/10.1016/S0032-3861(97)00158-4.

20. Ahmed, T., Marçal, H., Lawless, M., Wanandy, N. S., Chiu, A., Foster, L. J. R., Polyhydroxybutyrate and its copolymer with polyhydroxyvalerate as biomaterials: Influence on progression of stem cell cycle. *Biomacromolecules*, **10**, 2707–2715 (2010), DOI: 10.1021/bm1007579.

21. Qiu, Y.-Z., Han, J., Guo, J.-J., Chen, G.-Q., Production of poly(3-hydroxybutyrate-co-3-hydroxyhexanoate) from gluconate and glucose by recombinant *Aeromonas hydrophila* and *Pseudomonas putida*. *Biotechnology Letters*, **18**, 1381–1386 (2005), DOI: 10.1007/s10529-005-3685-6.

22. Caballero, K. P., Karel, S. F., Register, R. A., Biosynthesis and characterization of hydroxybutyrate-hydroxycaproate copolymers. *International Journal of Biological Macromolecules*, **2**, 86–92 (1995), DOI: http://dx.doi.org/10.1016/0141-8130(95)93522-Y.

23. Zhao, K., Deng, Y., Chun Chen, J., Chen, G.-Q., Polyhydroxyalkanoate (PHA) scaffolds with good mechanical properties and biocompatibility. *Biomaterials*, **6**, 1041–1045 (2003), DOI: http://dx.doi.org/10.1016/S0142-9612(02)00426-X.

24. Avella, M. M. E., Raimo, M., Review: Properties of blends and composites based on poly(3-hydroxy)butyrate (PHB) and poly(3-hydroxybutyrate-hydroxyvalerate) (phbv) copolymers. *Journal of Material Science*, **3**, 523–545 (2000), DOI: DOI 10.1023/A:1004740522751.

25. Scaffaro, R., Dintcheva, N. T., Marino, R., La Mantia, F. P., Processing and properties of biopolymer/polyhydroxyalkanoates blends. *Journal of Polymers and the Environment*, **2**, 267–272 (2012), DOI: 10.1007/s10924-011-0385-2.

26. Innocentini-Mei, L. H., Bartoli, J. R., Baltieri, R. C., Mechanical and thermal properties of poly(3-hydroxybutyrate) blends with starch and starch derivatives. *Macromolecular Symposia*, **1**, 77–88 (2003), DOI: 10.1002/masy.200350708.

27. Shen, L., Haufe, J., Patel, M., Product overview and market projection of emerging bio-based plastics. In *PRO-BIP 2009. Utrecht, The Netherlands,*

(2009). http://news.bio-based.eu/media/news-images/20091108-02/Product_overview_and_market_projection_of_emerging_bio-based_plastics,_PRO-BIP_2009.pdf.

28. Hong, S.-G., Hsu, H.-W., Ye, M.-T., Thermal properties and applications of low molecular weight polyhydroxybutyrate. *Journal of Thermal Analysis and Calorimetry*, **2**, 1243–1250 (2013), DOI: 10.1007/s10973-012-2503-3.

29. Parra, D. F., Fusaro, J., Gaboardi, F., Rosa, D. S., Influence of poly (ethylene glycol) on the thermal, mechanical, morphological, physical-chemical and biodegradation properties of poly (3-hydroxybutyrate). *Polymer Degradation and Stability*, **9**, 1954–1959 (2006), DOI: http://dx.doi.org/10.1016/j.polymdegradstab.2006.02.008.

30. Yu, L., Dean, K., Li, L., Polymer blends and composites from renewable resources. *Progress in Polymer Science*, **6**, 576–602 (2006), DOI: http://dx.doi.org/10.1016/j.progpolymsci.2006.03.002.

31. Scandola, M., Ceccorulli, G., Pizzoli, M., Miscibility of bacterial poly(3-hydroxybutyrate) with cellulose esters. *Macromolecules*, **24**, 6441–6446 (1992), DOI: 10.1021/ma00050a009.

32. Scandola, M., Focarete, M. L., Adamus, G., Sikorska, W., Baranowska, I., Świerczek, Sabina, Gnatowski, M., Kowalczuk, M., Jedliński, Z., Polymer blends of natural poly(3-hydroxybutyrate-co-3-hydroxyvalerate) and a synthetic atactic poly(3-hydroxybutyrate). Characterization and biodegradation studies. *Macromolecules*, **9**, 2568–2574 (1997), DOI: 10.1021/ma961431y.

33. Choi, H. J., Kim, J., Jhon, M. S., Viscoelastic characterization of biodegradable poly(3-hydroxybutyrate-co-3-hydroxyvalerate). *Polymer*, **14**, 4135–4138 (1999), DOI: http://dx.doi.org/10.1016/S0032-3861(98)00627-2.

34. Avella, M., Martuscelli, E., Raimo, M., The fractionated crystallization phenomenon in poly(3-hydroxybutyrate)/poly(ethylene oxide) blends. *Polymer*, **15**, 3234–3240 (1993), DOI: http://dx.doi.org/10.1016/0032-3861(93)90396-R.

35. Gassner, F., Owen, A. J., Physical properties of poly(beta-hydroxybutyrate)-poly(epsilon-caprolactone) blends. *Polymer*, **10**, 2233–2236 (1994), DOI: http://dx.doi.org/10.1016/0032-3861(94)90258-5.

36. Abe, H., Doi, Y., Kumagai, Y., Synthesis and characterization of poly [(r,s)-3-hydroxybutyrate-b-6-hydroxyhexanoate] as a compatibilizer for a biodegradable blend of poly[(r)-3-hydroxybutyrate] and poly(6-hydroxyhexanoate). *Macromolecules*, **21**, 6012–6017 (1994), DOI: 10.1021/ma00099a012.

37. Rodrigues, J. A. F. R., Parra, D. F., Lugão, A. B., Crystallization on films of PHB/PEG blends. *Journal of Thermal Analysis and Calorimetry*, **2**, 379–381 (2005), DOI: 10.1007/s10973-005-0069-z.
38. Anguillesi, I., Cinelli, P. A. L., Composite based on polyhydroxybutyrate, lignin and wood fibres. In *Proceedings of 2nd Congress on Biodegradable Polymer Packaging BioPolPack. Milano, Italy*, vol. 49 (2012).
39. S. De, O. Patrício, P., Pereira, F. V., dos Santos, M. C., de Souza, P. P., Roa, J. P. B., Orefice, R. L., Increasing the elongation at break of polyhydroxybutyrate biopolymer: Effect of cellulose nanowhiskers on mechanical and thermal properties. *Journal of Applied Polymer Science*, **5**, 3613–3621 (2013), DOI: 10.1002/app.37811.
40. Erceg, M., Kovačić, T., Klarić, I., Thermal degradation of poly(3-hydroxybutyrate) plasticized with acetyl tributyl citrate. *Polymer Degradation and Stability*, **2**, 313–318 (2005), DOI: http://dx.doi.org/10.1016/j.polymdegradstab.2005.04.048.
41. www.matweb.com. Accessed 24/04/2015.
42. Ludueña, L. N., Alvarez, V. A., Vazquez, A., Processing and microstructure of PCL/clay nanocomposites. *Materials Science and Engineering: A*, **460**, 121–129 (2007), DOI: http://dx.doi.org/10.1016/j.msea.2007.01.104.
43. Messersmith, P. B., Giannelis, E. P., Synthesis and barrier properties of poly(ε-caprolactone)-layered silicate nanocomposites. *Journal of Polymer Science Part A: Polymer Chemistry*, **7**, 1047–1057 (1995), DOI: 10.1002/pola.1995.080330707.
44. Kojima, Y., Usuki, A., Kawasumi, M., Okada, A., Fukushima, Y., Kurauchi, T., Kamigaito, O., Mechanical properties of nylon 6-clay hybrid. *Journal of Materials Research*, **05**, 1185–1189 (1993), DOI: doi:10.1557/JMR.1993.1185.
45. Gilman, J. J., C., Morgan, A., Harris, R., Manias, E., Giannelis, E., Wuthenow, M., Hilton, D., Phillips, S., Flamability properties of polymer-layered silicate nanocomposites: Polypropylene and polystyrene nanocomposites. *Chemistry of Matererials*, **12**, 1866–1873 (2000).
46. Gorrasi, G., Tortora, M., Vittoria, V., Pollet, E., Lepoittevin, B., Alexandre, M., Dubois, P., Vapor barrier properties of polycaprolactone montmorillonite nanocomposites: Effect of clay dispersion. *Polymer*, **8**, 2271–2279 (2003), DOI: http://dx.doi.org/10.1016/S0032-3861(03)00108-3.
47. Zhang, Z., Yang, J.-L., Friedrich, K., Creep resistant polymeric nanocomposites. *Polymer*, **10**, 3481–3485 (2004), DOI: http://dx.doi.org/10.1016/j.polymer.2004.03.004.

48. Kunioka, M., Ninomiya, F., Funabashi, M., Novel evaluation method of biodegradabilities for oil-based polycaprolactone by naturally occurring radiocarbon-14 concentration using accelerator mass spectrometry based on iso 14855-2 in controlled compost. *Polymer Degradation and Stability*, **7**, 1279–1288 (2007), DOI: http://dx.doi.org/10.1016/j.polymdegradstab.2007.03.028.

49. Maiti, P., Influence of miscibility on viscoelasticity, structure and intercalation of oligo-poly(ε-caprolactone)/layered silicate nanocomposites. *Langmuir*, **19**, 5502–5510 (2003).

50. Botana, A., Mollo, M., Eisenberg, P., Torres Sanchez, R. M., Effect of modified montmorillonite on biodegradable PHB nanocomposites. *Applied Clay Science*, **3–4**, 263–270 (2010), DOI: http://dx.doi.org/10.1016/j.clay.2009.11.001.

51. D'Amico, D. A., Manfredi, L. B., Cyras, V. P., Crystallization behavior of poly(3-hydroxybutyrate) nanocomposites based on modified clays: Effect of organic modifiers. *Thermochimica Acta*, **544**, 47–53 (2012), DOI: http://dx.doi.org/10.1016/j.tca.2012.06.012.

52. Bruno Rocha e Silva, M., Tavares, M. I. B., Oliveira da Silva, E., Neto, R. P. C., Dynamic and structural evaluation of poly(3-hydroxybutyrate) layered nanocomposites. *Polymer Testing*, **1**, 165–174 (2013), DOI: http://dx.doi.org/10.1016/j.polymertesting.2012.09.006.

53. Bordes, P., Hablot, E., Pollet, E., Avérous, L., Effect of clay organomodifiers on degradation of polyhydroxyalkanoates. *Polymer Degradation and Stability*, **5**, 789–796 (2009), DOI: http://dx.doi.org/10.1016/j.polym degradstab.2009.01.027.

54. Erceg, M., Kovačić, T., Perinović, S., Kinetic analysis of the non-isothermal degradation of poly(3-hydroxybutyrate) nanocomposites. *Thermochimica Acta*, **476**(1–2), 44–50 (2008), DOI: http://dx.doi.org/10.1016/j.tca.2008.07.009.

55. Achilias, D. S., Panayotidou, E., Zuburtikudis, I., Thermal degradation kinetics and isoconversional analysis of biodegradable poly(3-hydroxybutyrate)/organomodified montmorillonite nanocomposites. *Thermochimica Acta*, **514(1–2)**, 58–66 (2011), DOI: http://dx.doi.org/10.1016/j.tca.2010.12.003.

56. Puglia, D., Fortunati, E., D'Amico, D. A., Manfredi, L. B., Cyras, V. P., Kenny, J. M., Influence of organically modified clays on the properties and disintegrability in compost of solution cast poly(3-hydroxybutyrate) films. *Polymer Degradation and Stability*, **99**, 127–135 (2014), DOI: http://dx.doi.org/10.1016/j.polymdegradstab.2013.11.013.

57. Placet, V., Characterization of the thermo-mechanical behaviour of hemp fibres intended for the manufacturing of high performance

composites. *Composites Part A: Applied Science and Manufacturing*, **8**, 1111–1118 (2009), DOI: http://dx.doi.org/10.1016/j.compositesa.2009.04.031.

58. Itoh, T., Brown, R. M., Jr., The assembly of cellulose microfibrils in valonia macrophysa kütz. *Planta*, **4**, 372–381 (1984), DOI: 10.1007/bf00393419.
59. Baillie, C., *Green Composites: Polymer Composites and the Environment*. Woodhead Publishing Limited, England (2004).
60. Iwatake, A., Nogi, M., Yano, H., Cellulose nanofiber-reinforced polylactic acid. *Composites Science and Technology*, **9**, 2103–2106 (2008), DOI: http://dx.doi.org/10.1016/j.compscitech.2008.03.006.
61. Morin, A., Dufresne, A., Nanocomposites of chitin whiskers from riftia tubes and poly(caprolactone). *Macromolecules*, **6**, 2190–2199 (2002), DOI: 10.1021/ma011493a.
62. Petersson, L., Kvien, I., Oksman, K., Structure and thermal properties of poly(lactic acid)/cellulose whiskers nanocomposite materials. *Composites Science and Technology*, **11–12**, 2535–2544 (2007), DOI: http://dx.doi.org/10.1016/j.compscitech.2006.12.012.
63. Cho, M.-J., Park, B.-D., Tensile and thermal properties of nanocellulose-reinforced poly(vinyl alcohol) nanocomposites. *Journal of Industrial and Engineering Chemistry*, **1**, 36–40 (2011), DOI: http://dx.doi.org/10.1016/j.jiec.2010.10.006.
64. Zhijiang, C., Guang, Y., Kim, J., Biocompatible nanocomposites prepared by impregnating bacterial cellulose nanofibrils into poly(3-hydroxybutyrate). *Current Applied Physics*, **2**, 247–249 (2011), DOI: http://dx.doi.org/10.1016/j.cap.2010.07.016.
65. Global and chinese polyhydroxyalkanoates (PHA) industry report 2014, QYresearch (February 2014).
66. http://www.meredianpha.com/news/. Accessed: 10/07/2014.
67. http://www.tianan-enmat.com/#. Accessed: 10/07/14.
68. Braunegg, G., Bona, R., Koller, M., Wallner, E., Graz University of Technology, Austria: Sustainable polymeric materials from renewable resources and agro-industrial waste. In *Expert Group Meeting on Environmentally Degradable Plastics and Sustainable Development. Institute of Biotechnology, Trieste, Italy*, (2002).
69. Koller, M., Hesse, P., Bona, R., Kutschera, C., Atlić, A., Braunegg, G., Potential of Various Archae-and Eubacterial Strains as Industrial Polyhydroxyalkanoate Producers from Whey. *Macromolecular Bioscience*, **7**(2), 218–226 (2007).

70. Kosior, E., Braganca, R., Fowler, P., Lightweight compostable packaging: Literature review, WRAP project, June 2006.

71. http://www.nasdaq.com/markets/sugar.aspx?timeframe=10y. Accessed: 10/07/2014.

72. Endres, H.-J., Siebert-Raths, A., *Engineering Biopolymers, Markets, Manufacturing, Properties and Applications*, Hanser (2011).

73. Data from European Bioplastics. Source European Bioplastics, Institute for Bioplastics and Biocomposites, nova-Institute (2014). http://en.european-bioplastics.org/download/labelling/Global%20Prod%20Market%20Segment_Total_2011_en.jpg. Accessed: 02/05/2014.

74. http://www.icis.com/resources/news/2011/02/15/9433445/pha-shows-great-promise-in-packaging-applcation/. Accessed: 10/07/2014.

75. www.biomer.de. Accessed 24/02/2015.

76. Misra, S. K., Ohashi, F., Valappil, S. P., Knowles, J. C., Roy, I., Silva, S. R. P., Salih, V., Boccaccini, A. R., Characterization of carbon nanotube (MWCNT) containing P(3HB)/bioactive glass composites for tissue engineering applications. *Acta Biomaterialia*, **3**, 735–742 (2010), DOI: http://dx.doi.org/10.1016/j.actbio.2009.09.023.

77. Bucci, D. Z., Tavares, L. B. B., Sell, I., PHB packaging for the storage of food products. *Polymer Testing*, **5**, 564–571 (2005), DOI: http://dx.doi.org/10.1016/j.polymertesting.2005.02.008.

78. Miguel, O., Fernandez-Berridi, M. J., Iruin, J. J., Survey on transport properties of liquids, vapors, and gases in biodegradable poly(3-hydroxybutyrate) (PHB). *Journal of Applied Polymer Science*, **9**, 1849–1859 (1997), DOI: 10.1002/(sici)1097-4628(19970531) 64:9 <1849::aid-app22>3.0.co;2-r.

79. Chen, G. Q., Polyhydroxyalkanoates (PHA) and their applications advances in biochemical engineering/biotechnology. In *Biofunctionalization of Polymers and their Applications* (Advances in Biochemical Engineering Biotechnology) Gibson Stephen Nyanhongo, Walter Steiner, Georg Gubitz, Springer, August, 2011, 304 pages.

80. http://www.bc.bangor.ac.uk/_includes/docs/pdf/lightweight%20compostable%20packaging.pdf. Accessed: 02/05/2014.

81. Bhatt, R., Patel, K., Trivedi, U., Biodegradation of poly(3-hydroxyalkanoates). In *A Handbook of Applied Biopolymer Technology: Synthesis, Degradation and Applications* (Sharma, S. K., Mudhoo, A., eds.), The Royal Society of Chemistry, Chapter 10, pp. 311–331 (2011), RSC publishing.

Chapter 25

Polymers and Polymer Composites for Electrolyte Application

Aleksandra Kiyanitsa,[a] Sergey Gaidukov,[a] and Gerda Gaidukova[b]

[a]*Institute of Polymer Materials, Riga Technical University, Paula Valdena street 3/7, Riga, LV 1048, Latvia*
[b]*Latvan State Institute of Wood Chemistry, Dzerbenes street 27, Riga, LV 1006, Latvia*

sergejs.gaidukovs@rtu.lv

25.1 The Basic Theory of Polymer Electrolytes

Nowadays a large amount of synthetic polymers is used as packaging material, structural materials, as electric insulators [1–5]. The demand for synthetic polymers is rapidly increasing for portable electronic consumer devices. The most popular portable products are notebook computers, cellular and cordless phones, and camcorders.

It should be noted that all portable devices depend on the battery, which is their power source. In view of battery costs and environmental considerations, the batteries used to power these portable devices should preferably be rechargeable. This means, that there is the need of engineering solid polymer electrolyte films

Handbook of Sustainable Polymers: Processing and Applications
Edited by Vijay Kumar Thakur and Manju Kumari Thakur

ISBN 978-981-4613-53-8 (Hardcover), 978-981-4613-54-5 (eBook)
www.panstanford.com

with acceptable conductivity, mechanical property, electrochemical stability, and ease of handling [5].

That is why in the past 20 years, conductors having been designed that can be used as electron or ion. Moreover, a polymer material can be combined with appropriate salts, thus obtaining a composite whose ionic conductivity can be put to use as an electrolyte [6, 7].

25.1.1 Definitions

A wide variety of definitions are often used in the literature and data sheets to specify the properties of various battery systems. These definitions are described as follows.

25.1.1.1 Polymer

The word polymer comes from the Greek words "poly," which means many, and "meros," which means a part or a unit. Polymers are high-molecular-weight chemical compounds whose molecules consist of a large number of bonded (up to 105 or more) element units. Polymer density is 3–10 times smaller than that of metals [8].

25.1.1.2 Composite

If two or more types of different materials are mixed up and treated in defined conditions (varying with temperature, pressure, and other chemical and physical processes), a composite material with a clear interfacial boundary will be obtained. If a major part of the produced composite consists of polymer, then it is called a polymeric composite. A polymeric composite material is one of the most developed areas of modern science and technology. In addition to composite materials, modern science and technology use nano-sized materials. Such composites are called nanocomposites, whose main attraction is related to very high operation properties, such as flexibility, elasticity, recycling, hardness, resistance to abrasion, and optical and electrical transmission [9].

Polymeric nanocomposites (PNCs) are a new class of composite materials. The main attraction of PNCs is their very high operation properties, such as flexibility, elasticity, recycling, hardness, resistance to abrasion, and optical and electrical transmission. The preparation of PNCs with finely controlled structure is still one of the prospective ways of the modification of the properties of polymeric composites [9].

25.1.1.3 Cell

The cell is the basic electrochemical unit used to generate electrical energy from stored chemical energy or to store electrical energy in the form of chemical energy. A cell consists of two electrodes (anode, or the negative electrode, and cathode, or the positive electrode) in a container filled with an electrolyte. The anode is the reducing electrode, gives up electrons to the external circuit, and is oxidized during the electrochemical reaction. The main desired properties of the anode are efficiency as a reducing agent, high coulombic output (the electrical energy delivered by a battery when in use), good conductivity, stability, ease of fabrication, and low cost. Typically, the anode material can use hydrogen, but practically, metals such as zinc, tin, and lithium, and also other materials such as graphite and silicon, can be used. The cathode is the oxidizing electrode, which accepts electrons from the external circuit and is reduced during the electrochemical reaction. The main requirements for the cathode are efficient oxidizing properties, stability in contact with the electrolyte, and suitable working voltage. It is possible to use oxygen, but practically metallic oxides, less often halogens and the oxyhalides, sulfur and its oxides, can also be used as the cathode material [10–12].

25.1.1.4 Electrolyte

The electrolyte is the medium that provides the essential ionic transfer between the cathode and the anode of a cell, which is why the electrolyte is called the ionic conductor. Electrolyte systems can be divided into several different categories. Electrolytes are mainly in liquid or solid states. It is necessary for the electrolyte to have ionic conductivity but not be electronically conductive, stability to electrode material, little change in properties with change in temperature, safety in handling, and low weight and cost [12, 13].

Table 25.1 shows the typical ranges of specific conductivities for various electrolyte systems used in batteries [12].

Table 25.1 Conductivity ranges of various electrolytes at ambient temperature

Electrolyte system	Specific conductivity, Ω^{-1} cm^{-1}
Aqueous electrolytes	1×10^{-1} to 5×10^{-1}
Molten salt	$\sim 10^{-1}$
Inorganic electrolytes	2×10^{-2} to 2×10^{-1}
Organic electrolytes	10^{-3} to 10^{-2}
Polymer electrolytes	10^{-7} to 10^{-3}
Inorganic solid electrolytes	10^{-8} to 10^{-5}

One of the important properties of a polymer electrolyte leading to its development activity is the ionic conductivity. Temperature dependence on the conductivity of amorphous polymer electrolytes generally follows the Vogel–Tammann–Fulcher (VTF) equation [14]:

$$\sigma = AT^{1/2}\exp\left[\frac{E}{T-T_o}\right], \tag{25.1}$$

where T_o is the glass transition temperature of the polymer electrolyte measured by differential scanning calorimetry (DSC), T the temperature of measurement, A the pre-exponential factor, and E the activation energy, which can be evaluated either from the configurational entropy theory or from the free-volume theory and hence relates to the segmental motion of polymer chains.

25.1.1.5 Battery

The battery is a device that can store energy and provide it when required. It consists of single and basically of two or more cells connected in an appropriate series or parallel arrangement, to obtain the necessary voltage and capacity for a certain load.

The term "battery" refers to a device that converts the chemical energy contained in its active materials directly into the electric energy by means of an electrochemical oxidation–reduction (redox) reaction. There are two possible battery types: primary and secondary. Primary batteries are non-rechargeable, whereas secondary batteries are rechargeable. Examples of primary batteries material are zinc–carbon–zinc–manganese dioxide ($ZnMnO_2$), zinc–alkaline-MnO_2 (alkaline batteries), zinc–air, mercuric-oxide, and lithium batteries. Examples of secondary batteries are sealed lead-acid (SLA), NiCd, NiMH, Li-ion, Li-ion–polymer, Li-metal, zinc–alkaline-MnO_2, and zinc–air batteries [12, 13, 15].

The processes that occur in that time are called electrochemical reduction–oxidation (red-ox) reactions. Red-ox reactions in the battery cell produce electrical and chemical energy during charging and discharging. These reactions occur in the two electrodes and the electrolyte through an electrical circuit external to the battery. Current flow is held by electrons inside the electrodes (cathode and anode) and by ions inside the electrolyte. Current flows through the charger (by putting energy into the battery device) or load (by utilization energy for maintaining electrical device).

Battery charging and discharging process is schematically illustrated in Fig. 25.1. It shows that during the charging of the battery cell, the oxidation (loss of electrons) occurs at the positive electrode and reduction (gain of electrons) at the negative electrode. During the discharging process, the current flow is reversed and the cathode plays the role of the negative electrode and anode the role of the positive electrode [12, 15].

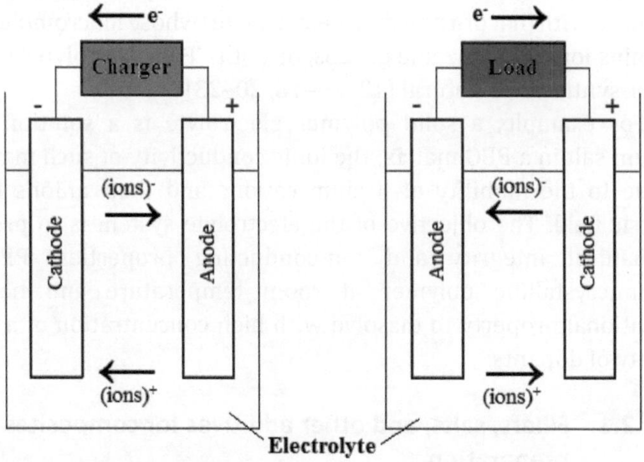

Figure 25.1 Schematic illustration of the battery charging and discharging processes [12, 15].

25.2 Polymers and Polymer Composites as Material for Electrolyte Application

25.2.1 Polymer Electrolyte

The field of polymer electrolytes has gone through three stages: "dry solid systems," "polymer gels," and "polymer composites." The "dry systems" use the polymer host as the solid solvent and do not include any organic liquids. The "polymer gels" contain organic liquids as plasticizers, which with a lithium salt remain encapsulated in a polymer matrix, whereas the "polymer composites" include high-surface-area inorganic solids in proportion with a "dry solid polymer" or "polymer gel" system [10, 12, 13, 16–19].

25.2.2 Solid Polymer Electrolyte

A solid polymer electrolyte may be defined as electrically conducting solution of a salt in a polymer. Solid polymer electrolytes are popular and practical applications in portable telecommunication devices, computers, and hybrid electric vehicles and for the possibility of using them in compact, lightweight, high-energy-density rechargeable lithium batteries.

The definition of a solid polymer electrolyte should not be confused with that of a polymeric electrolyte (whose macromolecules contains ionic or ionizable groups, or both). Polyelectrolytes can be either synthetic or natural [12, 16–18, 20–23].

For example, a solid polymer electrolyte is a solution of a lithium salt in a PEO matrix; the ionic conductivity of such material is due to the mobility of lithium cations and their anions in an electric field. The objective of the electrolyte system is to provide mechanical integrity and ion-conducting properties. PEO is a semicrystalline polymer at room temperature and has an exceptional property to dissolve with high concentration of a wide variety of dopants.

25.2.2.1 Fillers, salts, and other additives for composites preparation

An understanding of the interactions of the various species in the polymer electrolytes has made it possible to choose appropriate polymer hosts, complexing salts, and salt concentrations so that the ionic conductivities are optimized. Evidence of salt dissolution into a polymer host is mainly based on spectroscopic methods such as IR and Raman spectroscopies [24].

The common way to classify additives is by their nature and particle size. Additives can be classified as inorganic, organic, metallic, and carbon. Based on the scale of the additives' particle size, they can be described as macro-, meso-, micro- and nano-sized. The main interest is in nano-sized additives, in which at least one of the particle dimensions is equal to 100 nm (or 1×10^{-7} m) or smaller. At the same time, particle may have no dimensions (0D) in the nano-range, or one or two or three dimensions in the nano-range.

Studies indicate the formation of an intercalated morphology in the nanocomposites due to favorable interactions between the polymer matrix and the clay. The morphology of the nanocomposite is intricately linked to the amount of silicate in the system. With clay content ≥15 wt%, mechanical properties are further improved but the formation of an apparent superlattice structure correlates with a loss in the electrical properties of the nanocomposite [25].

Many polymer matrices are mostly studied as a host for ions, i.e., lithium ions from different LiX salts, because of its flexible macromolecular structure and good ability of the ion dissociation. It should be noted that lithium salt exhibits great charge delocalization favorable to ionic dissociation in solvating polymers such as PEO. Besides, LiX presents several significant advantages in the PEO–LiX composite, such as good mechanical stability and "plasticizing" effect (decreases the crystallinity of the host polymer and facilitates the transportation capability of ions) [26].

However, these polymer–LiX composites, which are mainly used as polymer electrolyte for lithium rechargeable batteries, have still some unsolved problems, i.e., bad interface property and a conductivity that remains at practical levels only at medium-high temperature, i.e., 60–70°C. These disadvantages may be solved by addition to the system inorganic fillers, such as SiO_2, MMT, and carbon nanotube (CNT), with high surface area. The second aim of nano-filler addition is to reduce the crystallizing ability of the polymer without reducing the mechanical properties of the PEO composite [27, 28].

The MMT structure has a net negative charge that is compensated with sodium cations. In the presence of water, the compensating Na^+ cations on the clay layer surfaces can be easily exchanged for other cations when available in excess amount in solution [29, 30]. CNT with its exceptional electrical properties has become one of the most potential materials to be used as conductive filler in polymer composites. The CNT network–polymer composite exhibits electron tunneling conductivity, which shows that the CNT network offers a promising application as a filler material for creating superconductive composite [27, 31–36]. Among the many potential applications of CNT, its use to strengthen polymers has been paid considerable attention due to the exceptional stiffness, excellent strength, and the low density of the CNT. Dispersion of CNTs within the polymer matrix, effective filler–matrix interfacial interactions, and alignment/orientation of polymer chains/CNT contribute to the composite fibers' superior properties. Composites incorporating CNTs have received a great deal of attention in both academia and industry for their potential replacement of carbon fibers in polymer-based reinforced materials. CNTs have been heralded as a game changer for producing next-generation high-performance materials. Additionally, due to the unique combination of properties, use of CNTs in polymer composites not only improves strength and modulus but also can result in enhancements in chemical resistance, thermal conductivity, electrical conductivity, and dimensional stability. Polymer–SiO_2 nanocomposites have been extensively used in various fields such as aerospace industries, automobiles, marine, and defense industries. Their main advantages are good corrosion resistance, lightweight, dielectric characteristics, and better damping characteristics than metals. The nano-sized filler particles such as SiO_2 in polymer nanocomposite increase the mechanical strength and enhance the ionic conductivity [27, 37–39].

In addition, Lewis acid–base interactions between the polar surface groups of the inorganic oxide filler and the electrolyte ionic species yield a better extent of salt dissociation through formation of ion–inorganic oxide complex. Due to the heavy mass of the ion–inorganic oxide complex, the conductivity in this composite is not likely to be contributed directly from the movement of these complexes. Instead, the major conducting path still originates from the local diffusion within the amorphous polymer matrix and the ion transport is realized through sequential replacement of ion to adjacent vacancy. Due to their large surface area, inorganic fillers prevent local PEO chain reorganization with the result of locking in at ambient temperature a higher degree of disorder, which in turn favors fast ionic transport. At the same time, obtained results from [27] demonstrate that smaller filler particles show insignificant influence in the improvement of the conductivity. This phenomenon is probably due to the immobilization of the long polymer chains. Addition of fine (nano-sized) fillers into the PEO–LiX composite film may cause the filler grains to get closer to each other so that the blocking effect imposed by the more abundant filler grains could make the long polymer chains more immobilized, leading to the decrease in conductivity. However, the mechanism by which these inorganic particles influence the properties of the native materials is still not satisfactorily understood.

Most of the recent papers deal with PEO and other polymers incorporated with inert fillers [30, 40–53], which also have poor mechanical properties. These films have to be hardened by chemical or physical curing, which leads to high processing costs. Recent studies reveal that only composite polymer electrolytes incorporated with inorganic filler can offer mechanically strong and safe batteries with improved electrolyte-electrode compatibility. One of the functions of the fillers is also the reduction of crystallinity of the host matrix.

For example, after addition of the filler, the ionic conductivity of the electrolyte reached 2.15 mS/cm at room temperature, a more than twofold increase over the electrolyte without filler [37]. The results of electrical conductivity comparison between PEO and various nanocomposites are listed in Table 25.2, contributed by Kiyanitsa et al. [27].

Polymer nanocomposite materials as a new insulation material also have been given attention because the properties of the original material can be drastically improved by adding a few percent of nano-sized filler. Polymer nanocomposites for HV insulation have been extensively being studied in recent years because of their unique properties and ability to enhance electrical performance. The analysis result showed that DC conductivity exhibits significant reduction with the addition of any types of nanofiller in right amount. However, the right percentage of nanofiller added is not same for different type of materials. The conductivity analysis shows that addition of nanofiller into the polymer improved its insulation properties [54].

Table 25.2 Electric conductivity (at 24°C) of the pure PEO and PEO-based polymer composite films

	PEO/SiO_2 (3%)		PEO/MMT (3%)		PEO/CNT <0.1%>	
	Conductivity, Sm/m	Increase compare to pure PEO, %	Conductivity, Sm/m	Increase compare to pure PEO, %	Conductivity, Sm/m	Increase compare to pure PEO, %
Pure PEO	2.57×10^{-10}	—	2.57×10^{-10}	—	2.57×10^{-10}	—
0% TEOS	1.67×10^{-09}	648	1.83×10^{-09}	710	2.55×10^{-08}	9910
10% TEOS	1.71×10^{-09}	666	1.42×10^{-09}	553	1.11×10^{-09}	431

25.2.3 Requirements for Polymer Electrolyte

Wide specific properties for use in batteries, high specific energy and specific power, safe operation and offer electrochemical stability of an electrolyte, flexibility in packaging, and low cost of manufacturing should be achieved. However, the polymer electrolytes should exhibit ionic conductivities with a minimum value approximately of 10^{-3} to 10^{-2} $S \cdot cm^{-1}$ at room temperature.

During charge and discharge, heat is generated in the battery, which increases the surface area of an existing passive layer at the electrode surface. The heat can also melt and degrade the polymer electrolyte within the battery and cause internal short circuits. Increase in heat due to environmental factors during storage increases the self-discharge reactions in the battery and shorten life. This means that the polymer should exhibit good cycle lives and good thermal and mechanical strengths in order to resist external and internal factors (i.e., temperature and pressure fluctuations) during battery operation.

It should be mentioned that lightweight and non-combustible polymer materials, during electrolyte manufacturing, may assume predetermined requirements of size and shape, thus offering a wide range of designs [14, 15, 19, 55–57].

25.3 Applications of Polymer Electrolytes

Polymer electrolytes are extensively used in a variety of industries due to their various advantageous properties. At the same time, polymer electrolytes mechanically behave like solids, but their internal structures, and therefore the conductivity, behave almost as in the liquid state [16].

One of the advantages of polymer electrolytes is their ability to produce high electrical conductivity (in the range of 10^{-3} to 10^{-2} $S \cdot cm^{-1}$). Moreover, they have the ability to form thin films with such properties as flexibility, light weight, elasticity, and high energy density, which make them suitable for many electrochemical devices [16, 17].

Polymer electrolytes have interesting properties for practical use in batteries, smart windows, and electrochemical solar cells, fuel cells, and water desalination [58]. The electrolyte system lets the batteries to be produced in countless structures and has been used in devices as well as cellular telephones and laptop computers [18].

Polymer electrolyte membrane (PEM) fuel cells, which convert the chemical energy stored in hydrogen fuel directly and efficiently to electrical energy with water as the only byproduct, have the potential to reduce our energy use, pollutant emissions, and dependence on fossil fuels. Some industrial companies such as Toyota, Honda, Hyundai, Daimler, and General Motors are planning to manufacture their fuel-cell vehicles by 2015. PEM can be applied in heat-power co-generation for household/residential use and uninterruptable power supply (UPS) [23, 59].

Developing micro-scale technologies have allowed the production of advanced small-scale devices, commonly referred to as microelectromechanical (MEM) systems, which find applications in many areas, such as optics and sensors [23, 60]. The need to power these devices is a driving force for the development of small-sized portable power systems. The development of novel thin polymer film electrolytes by UV cross-linking self-assembly of the electrolyte enabled them to be applied conformally onto a complex substrate, which gives an opportunity to produce 3D-microbatteries [61].

It should be noted that polymer electrolyte technology is currently a widely discussed topic in the fields of engineering, military, and satellites. It is necessary for soldiers to carry more electronic equipment, but at the same time reduce the mass of the accompanying battery. At the same time, the space industry is one of the most funded industries in manufacture [62].

An example of an organization exploring deep space the U.S. National Aeronautic and Space Agency (NASA). In April 2014, NASA announced that they would start program to develop something that can store energy better than today's battery technology. This next-generation energy battery should "store significant amounts of energy per unit mass, ... must also function across a wide temperature range, maintain capacity after frequent charge/discharge cycles, and maintain safe operation in close proximity to humans and mission critical equipment" [63, 64].

In this electronic age, it is mandatory to use solid polymer electrolytes for different applications in science and technology. Polymer electrolytes can be shaped in the form of thin film, thereby reducing the internal resistance leading to application as gas sensing material. Few reports appeared on proton-conducting polymer films and their application to gas sensors [65].

Classic methods of drug delivery have embraced a number of routes into the site of pharmacological action. Skin patches containing active drug that is allowed to diffuse passively across the external skin barrier into the bloodstream now enjoy wide application, but a problem is that the rate of egress is often slow. For certain ionic drugs, including local anesthetics and, more recently, peptides and gene-based, biotechnological engineered pharmaceuticals, it is possible to substantially enhance transdermal transport by iontophoresis. The technique of iontophoresis facilitates the passage of ionic drugs through the skin using an electric current. Ionically conducting polymers (polymer electrolytes) are potential candidates as hosts for drugs to be delivered iontophoretically [66].

References

1. Lupinski, J. H., Moore, R. S. (1990). Polymeric materials for electronics packaging and interconnection, *Anal. Chem.*, **62(4)**, 274A–274A.
2. Unterhofer, K., Preu, H., Walter, J., Lorenz, G., Mack, W., Petzold, M. (2013). Thermo-mechanical material characterization of organic polymer films in advanced packages using nanoindentation. *Microelectronics Packaging Conference (EMPC), 2013 European*, pp. 1–6.
3. Charlas, M., Nassiet, V., Habas, J. P., Vieillard, S., Meuret, R. (2009). Definition of new polymer packaging solutions for airborne power electronic system applications. *IEEE International Conference on Industrial Technology, 2009 ICIT 2009*, pp. 1–6.
4. Zhuo, L., Yi, G., Kyoung-Sik, M., Tannenbaum, A., Wong, C. P. (2012). Nano filler dispersion in polymer composites for electronic packaging. *Electronic Components and Technology Conference (ECTC), 2012 IEEE 62nd*, pp. 1752–1757.
5. Ramesh, S., Arof, A. K. (2009). A study incorporating nano-sized silica into PVC-blend-based polymer electrolytes for lithium batteries, *J. Mater. Sci.*, **44**(23), 6404–6407.
6. Nookala, M., Scanlon, L. G., Marsh, R. A. (1997). Preparation and characterization of a hybrid solid polymer electrolyte consisting of poly(ethyleneoxide) and poly(acrylonitrile) for polymer-battery application. *Energy Conversion Engineering Conference, 1997 IECEC-97, Proceedings of the 32nd Intersociety*, vol. 1, pp. 13–18.
7. Nakajima, Y., Aihara, Y., Kato, S., Murata, K. (1996). Development of film-like lithium battery using solid polymer electrolyte. *IEEE International Symposium on Personal, Indoor and Mobile Radio Communications, 1996 PIMRC'96, Seventh*, vol. 3, pp. 1059–1063.
8. Kalniņš, M., Neimanis, Ē., Kaļķis, V. (1981). *Lielmolekulārie Savienojumi* (Zvaigzne, Rīga).
9. Friedrich, K., Fakirov, S., Zhang, Z. (2005). *Polymer Composites: From Nano- to Macro-Scale,* Springer.
10. Barbir, F. (2013). *PEM Fuel Cells,* Elsevier Inc.
11. Young, W.-S., Kuan, W.-F., Epps, T. H. (2014). Block copolymer electrolytes for rechargeable lithium batteries, *J. Polym. Sci. Part B Polym. Phys.*, **52**(1), 1–16.
12. Linden, D., Reddy, T. B. (2002). *Handbook of Batteries,* McGraw-Hill.
13. Pop, V. (2008). *Battery Management Systems: Accurate State-of-Charge Indication for Battery-Powered Applications*, Springer.

14. Jacobs, P. W. M., Lorimer, J. W., Russer, A., Wasiucionek, M. (1989). Impedance studies on the system $LiClO_4$—MEEP, *J. Power Sources*, **26**(3–4), 483–489.

15. Bergveld, H., Kruijt, W., Notten, P. L. (2002). Basic information on batteries. *Battery Management Systems*, vol. 1, Springer Netherlands; pp. 31–53.

16. Agrawal, R. C., Pandey, G. P. (2008). Solid polymer electrolytes: Materials designing and all-solid-state battery applications: An overview, *J. Phys. D: Appl. Phys.*, **41**(22), 223001.

17. Hallinan, D. T., Balsara, N. P. (2013). Polymer electrolytes, *Ann. Rev. Mater. Res.*, **43**(1), 503–525.

18. Lepage, D., Kuss, C., Schougaard, S. B. (2013). Conducting polymers in lithium batteries. *Conducting Polymers: Synthesis, Properties and Applications*, Nova Science Publishers, Inc.; pp. 291–314.

19. Yuan, X. Z., Wang, H. (2008). PEM fuel cell fundamentals. *PEM Fuel Cell Electrocatalysts and Catalyst Layers: Fundamentals and Applications*, Springer London; pp. 1–87.

20. Bouchet, R., Phan, T. N. T., Beaudoin, E., Devaux, D., Davidson, P., Bertin, D., et al. (2014). Charge transport in nanostructured PS-PEO-PS triblock copolymer electrolytes, *Macromolecules*, **47**(8), 2659–2665.

21. Gurevitch, I., Buonsanti, R., Teran, A. A., Gludovatz, B., Ritchie, R. O., Cabana, J., et al. (2013). Nanocomposites of titanium dioxide and polystyrene-poly(ethylene oxide) block copolymer as solid-state electrolytes for lithium metal batteries, *J. Electrochem. Soc.*, **160**(9), A1611–A1617.

22. Young, W. S., Kuan, W. F., Epps Iii, T. H. (2014). Block copolymer electrolytes for rechargeable lithium batteries, *J. Polym. Sci. Part B*, **52**(1), 1–16.

23. Wang, Y., Chen, K. S., Mishler, J., Cho, S. C., Adroher, X. C. (2011). A review of polymer electrolyte membrane fuel cells: Technology, applications, and needs on fundamental research, *Appl. Energy*, **88**(4), 981–1007.

24. Jacobs, P. W. M., Lorimer, J. W., Russer, A., Wasiucionek, M. (1989). Impedance studies on the system $LiClO_4$-MEEP, *J. Power Sources*, **26**(3–4), 483–489.

25. Kurian, M., Galvin, M. E., Trapa, P. E., Sadoway, D. R., Mayes, A. M. (2005). Single-ion conducting polymer-silicate nanocomposite electrolytes for lithium battery applications, *Electrochim. Acta*, **50**(10), 2125–2134.

26. Appetecchi, G. B., Hassoun, J., Scrosati, B., Croce, F., Cassel, F., Salomon, M. (2003). Hot-pressed, solvent-free, nanocomposite, PEO-based

electrolyte membranes: II. All solid-state Li/$LiFePO_4$ polymer batteries, *J. Power Sources*, **124**(1), 246–253.

27. Kiyanitsa, A., Gaidukov, S., Zukulis, E. (2014). *Preparation and Properties of Polyethylene Oxide Composites*. CYSENI 2014, May 29–30, Kaunas, Lithuania.

28. Appetecchi, G. B., Croce, F., Hassoun, J., Scrosati, B., Salomon, M., Cassel, F. (2003). Hot-pressed, dry, composite, PEO-based electrolyte membranes: I. Ionic conductivity characterization, *J. Power Sources*, **114**(1), 105–112.

29. Choudhary, S., Sengwa, R. J. (2014). Intercalated clay structures and amorphous behavior of solution cast and melt pressed poly(ethylene oxide)-clay nanocomposites, *J. Appl. Polym. Sci.*, **131**(4), DOI: 10.1002/app.39898.

30. Zhang, Y., Zhao, Y., Gosselink, D., Chen, P. (2014). Synthesis of poly-(ethylene-oxide)/nanoclay solid polymer electrolyte for all solid-state lithium/sulfur battery, *Ionics*, DOI: 10.1007/s11581-014-1176-2.

31. Ibrahim, S., Johan, M. R. (2011). Conductivity, thermal and neural network model nanocomposite solid polymer electrolyte system (PEO-LiPF 6-EC-CNT), *Int. J. Electrochem. Sci.*, **6**(11), 5565–5587.

32. Ibrahim, S., Mohd Yasin, S. M., Nee, N. M., Ahmad, R., Johan, M. R. (2012). Conductivity and dielectric behaviour of PEO-based solid nanocomposite polymer electrolytes, *Solid State Commun.*, **152**(5), 426–434.

33. Johan, M. R., Ibrahim, S. (2012). Optimization of neural network for ionic conductivity of nanocomposite solid polymer electrolyte system (PEO-LiPF6-EC-CNT), *Commun. Nonlinear Sci. Numer. Simul.*, **17**(1), 329–340.

34. Shaheer Akhtar, M., Park, J. G., Lee, H. C., Lee, S. K., Yang, O. B. (2010). Carbon nanotubes-polyethylene oxide composite electrolyte for solid-state dye-sensitized solar cells, *Electrochim. Acta*, **55**(7), 2418–2423.

35. Uddin, N. M., Capaldi, F. M., Farouk, B. (2011). Molecular dynamics simulations of the interactions and dispersion of carbon nanotubes in polyethylene oxide/water systems, *Polymer*, **52**(2), 288–296.

36. Zhao, F., Wang, M., Dong, S. (2004). Conductive property of multiwall carbon nanotubes-PEO-salt nanocomposite film, *Electrochem. Solid State Lett.*, **7**(11), E48–E50.

37. Ketabi, S., Lian, K. (2013). Effect of SiO_2 on conductivity and structural properties of PEO-$EMIHSO_4$ polymer electrolyte and enabled solid electrochemical capacitors, *Electrochim. Acta*, **103**, 174–178.

38. Ramesh, S., Ramesh, K., Arof, A. K. (2013). Fumed silica-doped poly(vinyl chloride)-poly(ethylene oxide) (PVC/PEO)-based polymer electrolyte for lithium ion battery, *Int. J. Electrochem. Sci.*, **8**(6), 8348–8355.

39. Yap, Y. L., You, A. H., Teo, L. L., Hanapei, H. (2013). Inorganic filler sizes effect on ionic conductivity in polyethylene oxide (PEO) composite polymer electrolyte, *Int. J. Electrochem. Sci.*, **8**(2), 2154–2163.

40. Janna, K. M. (2012). Polyelectrolytes for batteries: Current state of understanding. *Polymers for Energy Storage and Delivery: Polyelectrolytes for Batteries and Fuel Cells*, vol. 1096, American Chemical Society; pp. 1–17.

41. Moon Jeong, P., Jaewan, H., Seong Yeon, K. (2012). Role of Nanostructures in Polymer Electrolytes for Energy Storage and Delivery. *Polymers for Energy Storage and Delivery: Polyelectrolytes for Batteries and Fuel Cells*, vol. 1096, American Chemical Society; pp. 129–146.

42. Somisetti, V. S. (2014). Thermal stability and ionic conductivity of high-temperature proton conducting ionic liquid? Polymer Composite Electrolyte Membranes for Fuel Cell Applications. *Polymer Composites for Energy Harvesting, Conversion, and Storage*, vol. 1161, American Chemical Society; pp. 111–126.

43. Cai, F., Zuo, X., Wang, L., Zhai, W., Liu, X. M., Yang, H., et al. (2013). Study and fabrication of lithium ion batteries using tris (methoxypolyethyleneglycol) aluminate esters as plasticizer for polymer electrolyte, *Gongneng Cailiao*, **44**(19), 2844–2848.

44. Chandra, A. (2013). PEO-PVP blended Na^+ ion conducting solid polymeric membranes, *Chin. J. Polym. Sci. (Engl. Ed.)*, **31**(11), 1538–1545.

45. Das, A., Thakur, A. K., Kumar, K. (2013). Exploring low temperature Li^+ ion conducting plastic battery electrolyte, *Ionics*, **19**(12), 1811–1823.

46. Hanson, B., Pryamitsyn, V., Ganesan, V. (2013). Mechanisms Underlying Ionic Mobilities in Nanocomposite Polymer Electrolytes, *ACS Macro Lett.*, **2**(11), 1001–1005.

47. Kim, M., Lee, L., Jung, Y., Kim, S. (2013). Study on ion conductivity and crystallinity of composite polymer electrolytes based on poly(ethylene oxide)/poly(acrylonitrile) containing nano-sized Al_2O_3 Fillers, *J. Nanosci. Nanotechnol.*, **13**(12), 7865–7869.

48. Brandell, D., Kasemägi, H., Tamm, T., Aabloo, A. (2014). Molecular dynamics modeling the Li-Polystyrene TFSI/PEO blend, *Solid State Ionics*, **262**, 769–773.

49. Chaurasia, S. K., Singh, R. K., Chandra, S. (2014). Ionic liquid assisted modification in ionic conductivity, phase transition temperature and

crystallization kinetics behaviour of polymer poly(ethylene oxide), *Solid State Ionics*, **262**, 790–794.

50. Money, B. K., Hariharan, K., Swenson, J. (2014). Relation between structural and conductivity relaxation in PEO and PEO based electrolytes, *Solid State Ionics*, **262**, 785–789.

51. Kwon, S. J., Kim, D. G., Shim, J., Lee, J. H., Baik, J. H., Lee, J. C. (2014). Preparation of organic/inorganic hybrid semi-interpenetrating network polymer electrolytes based on poly(ethylene oxide-co-ethylene carbonate) for all-solid-state lithium batteries at elevated temperatures, *Polymer*, **55**(12), 2799–2808.

52. Safanama, D., Damiano, D., Rao, R. P., Adams, S. (2014). Lithium conducting solid electrolyte $Li_1 + {}_xAl_xGe_2 - X(PO_4)_3$ membrane for aqueous lithium air battery, *Solid State Ionics*, **262**, 211–215.

53. Terasawa, N., Hayashi, Y., Koga, T., Higashi, N., Asaka, K. (2014). High-performance polymer actuators based on poly(ethylene oxide) and single-walled carbon nanotube-ionic liquid-based gels, *Sens. Actuators B Chem.*, **202**, 382–387.

54. Jamail, N. A. M., Piah, M. A. M., Muhamad, N. A., Zainir, R. A., Kasri, N. F. B., Kamarudin, Q. E. (2013). DC conductivity of polymer nanocomposites for different types and amount of nanofiller, *Int. J. Electr. Eng. Informatics*, **5**(2), 217–225.

55. Dias, F. B., Plomp, L., Veldhuis, J. B. J. (2000). Trends in polymer electrolytes for secondary lithium batteries, *J. Power Sources*, **88**(2), 169–191.

56. Shaplov, A. S., Lozinskaya, E. I., Vygodskii, Y. S. (2011). Polymer ionic liquids: Synthesis, design and application in electrochemistry as ion conducting materials. *Electrochemical Properties and Applications of Ionic Liquids*, Nova Science Publishers, Inc.; pp. 203–298.

57. Singh, M., Singh, V. K., Surana, K., Bhattacharya, B., Singh, P. K., Rhee, H.-W. (2013). New polymer electrolyte for electrochemical application, *J. Ind. Eng. Chem.*, **19**(3), 819–822.

58. Wiencierz, M., Stolwijk, N. A. (2012). Systematics of ionic transport and pair formation in amorphous PEO–NaI polymer electrolytes, *Solid State Ionics*, **212**, 88–99.

59. Tarascon, J. M., Armand, M. (2001). Issues and challenges facing rechargeable lithium batteries, *Nature*, **414**(6861), 359–367.

60. Súait, M. S., Ahmad, A., Badri, K. H., Mohamed, N. S., Rahman, M. Y. A., Ricardo, C. L. A., et al. (2014). The potential of polyurethane bio-based solid polymer electrolyte for photoelectrochemical cell application, *Int. J. Hydrogen Energy*, **39**(6), 3005–3017.

61. Tan, S., Walus, S., Gustafsson, T., Brandell, D. (2011). 3-D microbattery electrolyte by self-assembly of oligomers, *Solid State Ionics*, **198**(1), 26–31.

62. Craig, S. C., Evelyne, S. (2007). Evaluation of lithium polymer technology for small satellite applications. *21st Annual AIAA/USU Conference on Small Satellites*, vol. SSC07-X-9.

63. Pearson, C., Schrantz, K., Neubauer, J. (2010). Lithium-ion space battery technology development and infusion. *Aerospace Conference, 2010 IEEE*, pp. 1–12.

64. David, E. S., Chris, R. (2014). NASA Looks to Go Beyond Batteries for Space Exploration. vol. 2014 http://www.nasa.gov/press/2014/april/nasa-looks-to-go-beyond-batteries-for-space-exploration/-.U6BDS_mSySo: NASA.

65. Reddy, V. M. (2011). Studies on thin film polymer electrolytes: Application to sensors. http://hdl.handle.net/10603/3485. Jawaharlal Nehru Technological University.

66. Latham, R. J., Linford, R. G., Schlindwein, W. S. (2003). Pharmaceutical and medical applications of polymer electrolytes, *Ionics*, **9**(1–2), 41–46.

Index

Acetobacter xylinum 495–496, 512
acetyl groups 19, 26, 309, 351–353, 355, 357, 359, 361, 371, 379, 662, 666–667, 680
acid
  acetic acid 219, 234, 314, 539–540, 561, 661–662, 666, 669, 677, 679, 684–685, 702–703, 705, 721–722
  acrylic acid 45, 54, 107–108, 114–115, 117, 248, 406, 571, 592, 601, 627–628, 631, 744–745, 748–749, 834
  adipic acid 194–195, 786–787
  amino acids 9, 27–29, 110, 158, 165, 204, 206, 601–602, 605, 608, 834, 842, 845, 848, 900
  dicarboxylic acid 194, 755–757
  fatty acids 460, 804, 809–810, 816, 820
  glycolic acid 754–755, 780
  lactic acid 192–193, 398, 400, 404–405, 434–435, 437, 826, 857–858, 860, 862, 864–866, 868, 870, 872, 874
  polyacrylic acid 45, 54, 107–108
addition polymerization 6, 9
alginate 13–15, 22–23, 85, 223, 239, 306, 500–504, 543–544, 546, 551, 554, 660
alkaline pectinases 155–156, 159, 163, 219
amino groups 204, 227, 513, 515, 660, 662, 670–671, 680, 695, 699, 704, 727, 832
  free 661, 666, 668, 688
amorphous phases 419, 433, 827, 902, 904
amorphous PLLA 420–421, 426, 429
amylopectin 14–16, 152, 189
antimicrobial activity 228–231, 235, 241, 513, 518, 524, 562, 704, 707
antimicrobial properties 123, 232, 706, 709, 743, 745
atom transfer radical polymerization (ATRP) 101, 603
ATRP, *see* atom transfer radical polymerization

bacterial cellulose (BC) 110–111, 122, 223, 484, 495–497, 499, 501–503, 505, 507–511, 513–523, 881–882, 912
barrier properties 246, 433, 787, 873, 907–908, 918
BC, *see* bacterial cellulose
BC-based composites 497, 499–501, 503, 505, 507, 509, 511, 513, 515, 517, 521, 524
BC fibrils 502, 511, 516–517, 522
BC film 507–509, 511–512, 516–517
BC membrane 504, 519
BC pellicle 496, 499, 507, 510, 516, 522
BC–alginate 500, 502–504
BC–aloe vera 510–513

BC–chitosan 515–519
BC–gelatin 497, 505–510
BC–PVA composites 522
bio-based polymers 754, 804, 808, 891, 919
bio-based polyurethanes 812, 831–832
bioavailability 44, 64
biocatalysts 142, 154
biodegradable composites 884
biodegradable PLA 858–859, 861, 863, 865, 867, 869, 871, 873, 875
biodegradable plastics 182, 186, 200, 919
biodegradable polyesters 193, 753–754, 756, 758, 760, 762, 764, 766, 768, 770, 772, 774, 776, 778, 788–789
biodegradable polymers 181–182, 184–186, 188, 190, 192, 194, 196, 198, 200–208, 210, 222–223, 767, 769–770, 786, 900–901
biopolymers 63–64, 81–84, 182–183, 197–198, 202–203, 208–209, 241–245, 247–248, 305–306, 513, 660, 805, 808, 849, 901
biosynthesis 7, 10, 32, 122, 158, 309–310, 498–503, 505–506, 508, 510–511, 514–518, 520–521, 904
biosynthesis approach 499, 502, 505, 507, 509, 517
birefringence 341–344, 346–351, 353–354, 356–357, 359–360, 363–364, 366–368, 371, 373, 377
  in-plane 344, 350–351, 356, 359, 370
  negative 348–352, 373, 379
  orientation 342, 344–349, 351–355, 359, 366–367, 372–378
  out-of-plane 341–342, 344–346, 350, 355–356, 359, 361, 363–368, 370–371, 378–379
  positive 349, 351–352, 355–357, 359, 361, 363, 376, 379
blood vessel 109–112, 458, 497, 919
bone tissue 55, 497, 505–506, 509, 683
BTCA, *see* butane tetra carboxylic acids
bulk polymerization method 632–633
butane tetra carboxylic acids (BTCA) 233, 236, 245, 700, 704
butyryl groups 350–352, 354

CAB, *see* cellulose acetate butyrate
cadmium 308, 320–325, 328–334
cadmium ions 321–322, 326–329
calcium pectinate 456–457
CAP, *see* cellulose acetate propionate
carbon nanotubes (CNTs) 551, 880–881, 935–936
carboxymethyl cellulose 107
cationic dialdehyde cellulose (CDAC) 105–106
CDAC, *see* cationic dialdehyde cellulose
cellulases 145, 151, 154–155, 160–161, 163–164, 168–169
cellulolytic microflora 704–705
cellulose
  hydroxyethyl 53, 266–267, 280
  hydroxypropylmethyl 53–54

microcrystalline 25–26, 122–123
modified 100, 104, 189
cellulose acetate 113–114, 233, 236, 266, 272–273, 278, 341, 823
cellulose acetate butyrate (CAB) 342, 351–352, 354, 356, 378
cellulose acetate propionate (CAP) 342, 351–354, 356, 371–373, 376–378
cellulose crystallites 484
cellulose esters 25, 341–342, 344, 346, 348, 350–352, 354–356, 358, 360, 362, 364, 366–368, 370, 372–376, 378–379
cellulose ethers 25, 53, 266, 269, 272, 277–278, 824
cellulose fibers 25–26, 123, 189, 216, 285, 293, 295, 305
cellulose graft copolymerization 100–101, 103
cellulose modification 100, 126
cellulose nanocrystals 475–476, 480–481
cellulose nanowhiskers (CNWs) 109, 123, 126, 906
cellulose triacetate 342
cellulose whiskers 125
cellulosic fabrics 101, 220, 222, 227, 247
cellulosic fibers 216, 218, 225, 686, 691
cellulosic substrates 218–222, 231, 237, 247
cellulosic textiles 169, 223, 225–226, 252
charge density 105, 230, 236–239, 243, 247–248, 321–322, 487, 573, 822
chiral separations 587–588, 594–595
chitin 26–27, 51, 189–190, 223–224, 227, 230, 533, 549, 551, 560, 659–666, 668, 680, 829
chitosan 51, 226–234, 242–243, 513–519, 533–539, 541–554, 560–568, 570–574, 659–673, 675–693, 695–701, 703–708, 710–715, 723–724, 726–729
amino groups of 539, 682, 692, 701, 718
depolymerization of 536, 670
depolymerized 671–672, 688, 705
high-molecular-weight 561, 670, 677–678, 687–690, 695–697, 705
hydroxyl groups of 680, 701
low-molecular-weight 535–537, 670–671, 673, 687–688, 690, 693, 696–697, 706–708, 719, 726–727, 729
polycationic nature of 562, 704, 706, 715
viscosity of 673, 675
chitosan-based materials 551
chitosan cotton fabric 245, 247
chitosan derivatives 51, 227, 539–541, 548, 705, 721
chitosan fibers 546, 668, 684–686
chitosan films 513, 519, 566, 679, 681, 689, 691
chitosan hydrogel 565–566, 678, 713
chitosan membrane 566–567, 679
chitosan microparticles 542, 569–570
chitosan nanoparticles 230, 543–544, 546, 550–551, 568
chitosan polymers 533–534, 536, 538, 540, 542, 544, 546, 548, 550, 552, 554, 671

chitosan pretreatment 690–692, 697, 708, 727
chitosan salts 228, 677, 682–683
chitosan-treated cotton fabric 687, 689, 691, 699, 701–702, 707, 727–728
CNTs, *see* carbon nanotubes
CNWs, *see* cellulose nanowhiskers
composite scaffolds 519
condensation polymerization 6–9, 474, 848
conducting polymers 483–484, 601, 940
copper complexes 296–300
corn cob hemicellulose 314–317, 320–326, 328–330, 333–334
  characterization of 315, 317
cotton fabric
  desizing of 149, 151
  treated 688, 695–696, 699, 701, 722–723
cotton fabric processing 149, 157–159
cotton fabrics
  desizing of 150–151
  organic 142
  scouring of 152–153, 155, 157
crystallinity index 298, 516
cyclodextrins 226, 236, 540, 548, 568

density functional theory (DFT) 621, 626–628, 630–631
depolymerization 195, 246, 290, 297, 536, 552, 665, 670, 672, 822, 858, 860
DFT, *see* density functional theory
diisocyanates 804–805, 820, 831–833, 863–864
DNA 34–41, 47–48, 201, 544, 572, 706, 724
DNA replication 38–40, 724
DNA synthesis 37–38
drug delivery systems 42, 46, 54, 108–109, 319, 435, 550, 552, 587, 644–645, 683, 827

electrical conductivity 281–284, 291, 484–485, 490, 710, 745, 936
enantiomers 402, 418, 423, 588–589, 593–594, 601–602, 605, 643
enzymatic scouring 154
epithelial cells 564–565
ester groups 271, 342, 350–355, 374, 378–379, 758, 812, 827

fabrics
  chitosan-treated 688, 690, 692–693, 695–696, 699
  dyed 696–698, 705, 728
fertilizers, slow-release 114, 117
fiber–reinforced composites 388, 390–391
  natural 877, 882
flocculants, cellulose-based 104, 107
food industry 52, 55, 64, 66, 85, 318–319, 663, 679, 811, 873, 919
fuel cells 939
functional monomers 589–592, 601, 603, 607, 616–617, 623–624, 626–627, 632, 637, 642, 644

galactomannans 17–18, 61–62, 64–65, 70, 72, 76–77, 81, 307, 309, 311, 318
gelatin 52, 191, 223–224, 461–462, 505–510, 542, 545, 660, 778
glass transition temperature 195, 371, 418–420, 428–429, 773, 811, 888, 904, 907, 932
glycerol 51, 190, 208, 809–810, 814, 819, 821, 862, 905–906, 913–914
graft copolymerization 100–103, 227, 265, 680
green chemistry 313–314
green composites 210, 386, 389–391
guar gum 13–15, 17, 53, 60–61, 66, 69–70, 73, 76–77, 81, 461–462

HDT, *see* heat distortion temperature
heat distortion temperature (HDT) 427–429, 887
hemicellulose extraction 312–313
hemicelluloses 156, 300, 305–330, 332, 334, 496, 822, 877, 880, 884, 911–912
  isolation of 312–313
hydrogel/silver nanocomposite 745, 748
hydrogels 48, 55, 461, 500, 563, 565–566, 645, 678–679, 743, 745–746, 748–749, 829
  bare 746–747, 749
hydrogen peroxide 146, 158, 162, 664, 670, 690, 726, 782
hydroxypropylmethylcellulose 456, 461–462, 824

impregnation technique 499, 515, 517–518
imprinted polymers 598, 601–602, 608, 616, 620, 624, 628, 632, 636, 640, 644
injection molding 195, 434, 436, 869–871, 884–885, 888–890, 920
interpenetrating polymer network (IPN) 810, 813
IPN, *see* interpenetrating polymer network
isocyanates 120–121, 803, 805–809, 811, 813, 815, 817, 819, 821, 823, 825, 827, 829, 833, 841–842
isotactic polypropylene 125

KDC, *see* kenaf deriver cellulose
kenaf deriver cellulose (KDC) 889

L-lactic acid 402, 435, 770, 778, 781
L-lactide 401–402, 405, 411, 413–415, 434, 827–828, 860, 868, 880, 891
lactic acid monomer 398, 826, 863, 865
lactic acid oligomers 400, 863
lactide 193, 398–403, 406–407, 413, 418, 434, 438, 780–781, 784, 826, 858, 860–862

crude 399–400
lactones 413, 757–759, 762, 765
lignocellulosic biomass 305–306

mannose 11, 19, 59, 61, 306–311
membrane technology 836–838, 840
microbial cellulose 108–109, 122
MIP
see molecularly imprinted polymer
synthesis of 593, 625, 627, 632–635, 637–640
MISPE, *see* molecularly imprinted solid phase extraction
molecular dynamics 404, 620–621, 625–626, 885
molecular imprinting 588–591, 594, 616–619, 622–623, 630, 635, 640
molecularly imprinted polymer (MIP) 588–589, 591, 593–597, 600–602, 604, 606, 608, 615–616, 618–620, 622, 624–647
molecularly imprinted solid phase extraction (MISPE) 625, 633, 640–642

Na-alginate 239–241, 500
nano chitosan 660, 662, 664, 666, 668, 670, 672, 674, 676, 678, 680, 712–714, 716–720, 722–726, 728
synthesis of 677, 711–712, 728
nano chitosan dispersion 714, 718–719, 728
nanocomposite films 123, 125, 416, 485, 519
nanocomposite hydrogels 746
nanocomposites 112, 125–126, 234–236, 415–417, 485, 488, 568–569, 749, 881, 886–887, 890, 908, 910–911, 930, 935–937
nanocrystalline cellulose 119–121, 123, 125
nanoparticles, metal 743–745
natural cellulose 209, 473–474
natural fiber–reinforced PLA composites 882, 885–886
natural fibers 119, 202–203, 385–386, 388, 391, 703, 858, 877–878, 882, 884–886, 888, 906, 912
natural polymers 3, 5, 7, 9–10, 27, 48, 55, 109, 113, 188, 306, 308, 831–832, 839, 848–849
natural polysaccharides 16, 71, 239
nematic interaction 350, 367–368, 372–374, 376–377
non-starch polysaccharides (NSP) 306, 334
NSP, *see* non-starch polysaccharides

optical anisotropy 342, 344, 346, 350–351, 353, 355
optical anisotropy of cellulose esters 341–342, 344, 346, 348, 350, 352, 354, 356, 358, 360, 362, 364, 366, 368, 370
optical films 341–343, 348, 350, 377–378

PBPs, *see* petroleum-based plastics
pectin 13–14, 23, 49–50, 85, 146, 149, 152, 156, 189–190, 217, 224, 306, 325–329, 453–458, 460–465
  amidated 454–457, 461
pectin in drug delivery 454–455, 457
pectin in transdermal drug delivery 458–459, 461, 463
pectinase scouring 163, 165
pectinases 145, 149, 154–155, 157, 162–164
PEM, *see* polymer electrolyte membrane
petroleum-based plastics (PBPs) 182, 202
PHAs, *see* polyhydroxyalkanoates
PHB, *see* polyhydroxybutyrate
PHB-based nanocomposites 908–909, 911
PLA
  see polylactic acid
  degradation of 416, 424–425, 434
  properties of 406, 418–419, 858, 866
PLA-based composite 877–879, 881, 883, 885, 887, 889, 891
PLA fibers 413, 437–438, 876–877
PLA films 404, 424, 431, 434, 436–437, 872, 889
PLA matrix 126, 415–417, 867–868, 875, 881, 889
PLA nanocomposites 415, 887, 889
plant proteins 206, 210
plastic materials 182, 187, 189, 753, 777
plasticizers 190–191, 207, 210, 366, 373, 432–433, 436, 462–463, 878, 903–906, 908, 934
plastics 181–184, 186–187, 191, 195, 200, 202–203, 205–210, 422, 753–754, 765, 767, 785–786, 869–870, 873, 878
PLGA 435, 543–544, 780, 783–784
PLLA, *see* poly(L-lactic acid)
PLLA fibers 435
PLLA films 430–431, 872
plumbum 308, 320–325, 327–328, 330, 332–334
plumbum ions 321–322, 324–329
polarizability anisotropy 345, 351, 353, 359, 361, 371
poly(L-lactic acid) (PLLA) 402, 413–415, 418–421, 423–424, 426–432, 434–435, 778, 859, 880
polyacrylamide 45–46, 107
polycaprolactone 188, 193, 834, 839
polycondensation 405, 434, 755, 765, 859, 863, 866
  direct 397, 404–405, 859
polycondensation reactions 398, 405, 757, 858
polyelectrolytes 66, 457, 676, 778, 835, 934
polyesters 186, 223–224, 231–232, 754–755, 757, 759–761, 763–765, 767–768, 772–773, 775, 777–778, 784–785, 804, 808, 900
  degradable 755, 766, 788
  processing of 772–773, 775, 777
polyethers 804, 806, 808

polyethylene 3, 6, 118, 184, 189, 200, 414, 436, 754, 787
polyethylene glycol 42, 65, 433, 476, 566, 701, 906
polyhydroxyalkanoates (PHAs) 186, 192–193, 203, 223–224, 765–766, 769, 786–788, 866, 899–901, 903, 905, 910, 912–920
polyhydroxybutyrate (PHB) 192, 204, 223–224, 769, 780, 901–913, 918–920
polylactic acid (PLA) 126, 192–193, 372–373, 397–398, 401–407, 416–420, 422–427, 429–430, 432–438, 787–788, 826, 857–882, 884, 886–892, 907–908
polymer biodegradation 194, 201, 771–772
polymer chains 7, 71, 80, 83, 148, 185, 344, 364, 367, 372, 417, 425–426, 454, 813, 817
polymer composites 207, 386–388, 929–930, 932, 934–938, 940
  fiber–reinforced 388, 390, 392
polymer degradation 185, 435, 767–768, 771–772
polymer electrolyte membrane (PEM) 939
polymer electrolytes 929, 931–935, 938–940
  basic theory of 929, 931, 933
  composite 937
  solid 934, 940
polymer formation 1, 6–7, 598, 890
polymer gels 934
polymer granules 774–778
polymer nanocomposites 119, 880, 888, 936
polymeric hydrogels 113, 744, 749
polymeric materials 108, 195, 197, 342–343, 345, 373, 423, 425, 687, 771, 808, 830, 838, 881, 899
  optical properties of 342–343, 345
polymeric membranes 837–838
polymerization methods 404, 615, 632, 634, 637, 640, 647
polymers
  basics of 2–3, 5, 7
  charged 47, 684
  condensation 9, 12, 69, 195
  degradable 185–186, 209, 435, 754, 772, 784, 788, 839
  petrochemical-based 891–892
polymersome 565, 571
polynucleotides 9, 30–36, 38
polyolefins 3, 197, 766, 876
polyols 52, 194, 803–806, 808–811, 813–815, 817, 819, 821, 823, 825, 827–829, 831–833, 835, 841–842, 848
polypeptide chains 28–29, 199
polypeptides 27–29, 505
polyphosphates 47, 245
polypropylene 3, 6, 118, 210, 419, 754, 877, 902–903
polypyrrole 483
polysaccharide adsorption 71
polysaccharide chains 18, 20, 25, 305
polysaccharide classification 12–13, 15
polysaccharides 9–18, 20–21, 23–25, 69–71, 82–85, 188–189, 305–306, 308–311, 318–319, 496, 660–661, 670, 804–805, 821–822, 826
  neutral 15, 17–19

polystyrene 3, 184, 344, 397, 419, 426
polyurethane 803–804, 806, 808, 810, 812–814, 816, 818, 820, 822–824, 826–828, 830–836, 838, 840–842, 844, 848–849
polyurethane membranes 813, 836–837, 839–843, 845, 847
polyurethanes, segmented 811, 832
polyvinyl acetate 45, 194, 520
polyvinyl alcohol 45, 151, 194, 218, 519–520, 522
polyvinyl pyrrolidone 43–44
porogen 616, 623–627, 630–633, 635, 637–638
proteins 9, 27–29, 50–51, 82–84, 158, 190–191, 204, 206–207, 223–224, 804–805, 808, 834–835, 842–843, 845, 900
  functional 28–30
  soy 191, 202
pseudoplastic behavior 17–20, 49, 53

ring-opening polymerization (ROP) 193, 398, 757–761, 763–765, 858, 860, 891
RNA 34–36, 38, 40–41
RNA synthesis 40–41
ROP
  see ring-opening polymerization
  metal-insertion 759–760

SAM, *see* super absorbent material
silver 597, 724, 743–744, 747–749
silver/hydrogel composite 744–745
  synthesis of 744–745
silver ions 567, 745, 747
silver nanocomposites 744–745
silver nanoparticle-incorporated hydrogels 743–744, 746, 748
silver nanoparticles 724, 744–749
sponges, composite 502, 504
starch 11, 13, 15–16, 49, 55–56, 69–71, 150–151, 166–167, 186–189, 198, 200, 207–208, 218, 223, 875
stratum corneum 458–460, 463–465, 558–560, 563–566, 570, 573
super absorbent material (SAM) 118, 597
superabsorbent polymer 113, 115, 117
synthetic membranes 836–838, 841
synthetic polymers 2–4, 7, 42, 45, 83, 101, 108, 186–188, 193–194, 199, 435, 587, 843, 848, 929
synthetic water-soluble polymers 42–43, 45, 47

tara gum 18, 59–70, 72–78, 80–82, 84–85
TCP, *see* tricresyl phosphate
TCP molecules 367–368
tensile strength 200–201, 388–390, 426–428, 430, 432–433, 506, 508–509, 512, 514–515, 520–523, 865–866, 868, 878–879, 882–885, 906–907

textile processing 143, 145, 659–660
thermoplastic polyurethanes 811
thermoplastics 9, 520, 803, 832, 870, 888, 902
TPP
  see triphenylphosphine
  concentration of 718, 728
transdermal drug delivery 453–454, 456, 458–465, 558, 560, 562–574
tricresyl phosphate (TCP) 350, 362, 367–368, 373–376, 783
triglycerides 191, 542, 808–810, 813–815, 818, 820
triphenylphosphine (TPP) 407, 409–410, 415, 438, 677, 679, 711, 713–719
tunicate cellulose 473–474, 476–478, 480, 482, 484, 486, 488, 490, 888

urethanes 3–4, 805–807, 819, 827, 843, 845, 849

vegetable-oil-based polymers (VOBPs) 808
vegetable oils 65, 787, 808, 818, 914
vinyl monomers 101–102
VOBPs, *see* vegetable-oil-based polymers

WAC, *see* water adsorption capacity
water adsorption capacity (WAC) 497, 501, 503, 506, 509, 511–512, 515, 517–518
water-soluble polymers 42–43, 46
water vapor permeability (WVP) 251, 433
water vapor transmission (WVTR) 501, 503, 506, 509, 511, 513, 907
WVP, *see* water vapor permeability
WVTR, *see* water vapor transmission

xanthan gum solutions 49

Young's modulus 120, 416, 430, 503, 512, 514, 517, 521–522, 827, 865–866, 880, 882, 885, 907